DICTIONARY OF BUILDING AND CIVIL ENGINEERING

DICTIONNAIRE DU BÂTIMENT ET DU GÉNIE CIVIL

ABOUT THE AUTHOR

Don Montague is a Chartered Engineer (MICE, MIMechE) with experience in many fields of engineering, building and construction. He read Engineering Science at Oxford (MA) and then did pioneering work on machine design using the early DEUCE digital computer in the 1960s. Subsequently he worked in the public service in timber research (AJWSc), became deeply involved in environmental conservation and began a love affair with the practice of building.

In 1971 he joined Ove Arup & Partners and spent his first five years with them co-ordinating multi-disciplinary building design teams. After a period running his own structural component firm he rejoined Arups and became responsible for their engineering specifications, design guides and feedback notes. He has published articles on a wide range of topics, from environmental conservation to structural components, and integrated design to maintenance of building services. He retired as Technical Director of Arups in 1991.

A life-long interest in the French language and life led him and his wife to move to France, where he is now as busy as ever working for himself with local French building professionals.

L'AUTEUR

Don Montague est un ingénieur (MICE, MIMechE) dont l'expérience couvre de nombreux domaines du génie, du bâtiment et de la construction. Après avoir fait ses études d'ingénieur à Oxford (MA), il entreprit des travaux innovateurs sur la conception des machines en faisant appel à l'ordinateur numérique DEUCE dans les années 1960. Par la suite, il travailla dans le secteur public dans la recherche sur le bois (AJWSc). Il devint alors un fervent défenseur de la protection de l'environnement et se passionna pour le bâtiment.

En 1971, il entra chez Ove Arup & Partners où il fut pendant cinq ans responsable de la coordination des équipes multidisciplinaires de conception du bâtiment. Ayant quitté Arup pour diriger sa propre société de pièces de génie civil, il entra à nouveau dans cette société en tant que responsable des cahiers des charges, des guides de conception et de l'information en retour. Il publia des articles sur de très nombreux sujets allant de la protection de l'environnement aux pièces de génie civil en passant par la conception intégrée et les services de maintenance des bâtiments. Lorsqu'il prit sa retraite en 1991, il était Directeur technique d'Arup.

S'étant toujours intéressé à la langue et au mode de vie français, il vit désormais en France avec son épouse. Il est plus occupé que jamais et travaille en indépendant avec des professionnels français du bâtiment.

DICTIONARY OF BUILDING AND CIVIL ENGINEERING DICTIONNAIRE DU BÂTIMENT ET DU GÉNIE CIVIL

Don Montague

Taylor & Francis
Taylor & Francis Group

LONDON AND NEW YORK

Published by Taylor & Francis
2 Park Square, Milton Park, Abingdon, Oxon, OX14 4RN
270 Madison Ave, New York NY 10016

Transferred to Digital Printing 2009

First edition 1996

Typeset in Times 11/14pt by Columns Design & Production Services Ltd, Reading

ISBN 0 419 19910 1

A catalogue record for this book is available from the British Library

Library of Congress Catalog Card Number: 95-71096

Publisher's Note
The publisher has gone to great lengths to ensure the quality of this reprint but points out that some imperfections in the original may be apparent.

CONTENTS

ACKNOWLEDGEMENTS

No dictionary of this kind is the work of one person, and I am delighted to acknowledge the help and encouragement I have had, particularly from former colleagues at Ove Arup & Partners, Consulting Engineers. Chris Barber, Mike Bussell, Paul Craddock, Steve Dyson, Steven Groák, Martyn Harrold, Mike Holmes, David Lush, Alain Marcetteau, Gordon Puzey, Brian Simpson, Bob Venning and others have all contributed, provided lists of words or translations from their work in the UK and France, or offered good advice or further contacts.

In France I have had the pleasure of working with Christian Barreau, Yves de Bazin, Maître Gilles Cathus, Albert Cologni, Jacques Deltreil, Laurent Dupuy, Francis Framarin, Guy Lung, Maître Jacques Marie, Roger Mirabel, Laurence and René Orazzio, Roger Roccon, Maurice Teulet and André Vincent, their colleagues and staff, who have taught me a great deal of French architectural, building, legal and property terminology, in the office and on site.

The Translation Service – Terminology – of the European Commission has been very helpful, allowing me to collate terms from its multilingual glossaries *Roads* (1978) and *Construction* (in draft 1995), the latter being particularly helpful in the field of health and safety at work. Cooperation with Madame Fiamozzi of this Service has been enjoyable and fruitful.

I am grateful to REHVA (the Federation of European Heating and Airconditioning Associations) for allowing me to make use of technical terms from its *International Dictionary of Heating, Ventilating and Air Conditioning* (see Appendix B).

Manufacturers' catalogues have provided examples in their own languages, in particular Legrand SA, whose information on electrical equipment has been invaluable. Single language dictionaries and encyclopaedias have also proved very useful in the difficult task of defining meanings and selecting and discarding terms in each language. The more important works that I have referred to are listed in Appendix B.

Finally, without the encouragement and support of my editors, Caroline Mallinder and Fiona Weston, and my wife Maureen, I would not have retained my sanity or ever have completed this dictionary. I offer my grateful thanks to them and everyone else who has helped, in particular at Chapman & Hall, and absolve them all for any mistranslations – these are all my own work!

If you, kind readers, spot any of these errors, or can think of words that should have been included but were not, please write to me and let me know, c/o Caroline Mallinder at E & FN Spon.

Don Montague
Aquitaine, 1996

REMERCIEMENTS

Aucun dictionnaire de ce type n'est le travail d'une seule personne et je tiens à exprimer ma gratitude pour l'aide et les encouragements que j'ai reçus, en particulier de la part d'anciens collègues à Ove Arup & Partners, Consulting Engineers. Chris Barber, Mike Bussell, Paul Craddock, Steve Dyson, Steven Groák, Martyn Harrold, Mike Holmes, David Lush, Alain Marcetteau, Gordon Puzey, Brian Simpson, Bob Venning, et d'autres encore – tous ont contribué et fourni des listes de mots ou de traductions tirées de leurs travaux au Royaume-Uni et en France, ou m'ont donné de bons conseils et des contacts supplémentaires.

En France j'avais le plaisir de travailler avec Christian Barreau, Yves de Bazin, Maître Gilles Cathus, Albert Cologni, Jacques Deltreil, Laurent Dupuy, Francis Framarin, Guy Lung, Maître Jacques Marie, Roger Mirabel, Laurence et René Orazzio, Roger Roccon, Maurice Teulet et André Vincent, aussi bien qu'avec leurs collègues et leur personnel. Ils m'ont tous enseigné de nombreux termes français dans les domaines de l'architecture, de la construction, du droit et de l'immoblier, autant au bureau que sur le terrain.

Le Service de Traduction – Terminologie – de la Commission Européenne a été très obligeant, me permettant de collationner des termes dans ses glossaires multilingues *Roads* (1978) et *Construction* (à l'état d'ébauche, 1995), ce dernier ouvrage étant en particulier très utile dans le domaine de la santé et sécurité au travail. Ma coopération avec Madame Fiamozzi (de ce service) a été agréable et fructueuse.

Je suis reconnaissant à REHVA (the Federation of European Heating and Airconditioning Associations) qui m'a autorisé à utiliser des termes techniques tirés de son *International Dictionary of Heating, Ventilating and Air Conditioning* (voir l'Appendice B).

Les catalogues de fabricants ont fourni beaucoup d'exemples dans leur propre langue, en particulier Legrand SA, dont les renseignements sur l'équipement électrique ont été inestimables. Dictionnaires unilingues et encyclopédies ont aussi été très utiles dans la tâche difficile de définir les sens des mots, sélectionner et écarter certains termes dans chaque langue. Les ouvrages les plus importants auxquels je me suis référé sont listés dans l'Appendice B.

Enfin, sans les encouragements de mes éditrices, Caroline Mallinder et Fiona Weston, et de ma femme Maureen, je n'aurais jamais conservé ma sérénité et je n'aurais jamais fini ce dictionnaire. Je les remercie sincèrement ainsi que tous ceux qui m'ont aidé, en particulier chez Chapman & Hall et je les décharge de toute responsabilité pour les fautes de traduction; si vous en trouvez, ce sont les miennes!

Si vous, chers lecteurs, remarquez des erreurs ou pensez à des mots qui auraient dû être inclus, veuillez m'écrire me le faire savoir, aux bons soins de Caroline Mallinder chez E & FN Spon.

Don Montague
Aquitaine, 1996

INTRODUCTION

This bilingual dictionary is aimed at both professionals and private individuals, English and French speaking, who are involved in property, building and civil engineering construction in each other's countries.

Deciding what to include and what to leave out has often been difficult. My first rule has been to try to put myself in the place of someone involved in a practical way with property dealing, development, architectural design, building, civil engineering and construction generally, and to ask myself what words I would meet in communicating with professionals and tradespeople in the office and on site, what questions I need to ask, and what I would want to be able to express. Having worked in these fields in the UK and France for many years, I have come to realize that simple lists of nouns in the two languages are of limited use, and that many technical terms are simply not given in standard bilingual dictionaries. My aim in this book is to fill some of these gaps.

To start with, the built environment exists within the natural environment, and so I have included some basic words for geographical and topological features. Such terms may be used in describing land and property for sale, and also in the investigations needed before buying land for building or development.

Before one can start to build, surveys, planning and official approvals, architectural and engineering design, budgeting, raising finance and drawing up of specifications and contract documents may all be necessary, so I have included some of the more commonly used technical terms in these fields.

Any construction must be based on foundations, so I have started in the ground with geological and soil mechanics terms, and worked upwards through drainage, structures, mains supplies, services installations and cladding, to roofing and finishes terminology, covering the materials, tools and techniques involved, plus terms for parts of different kinds of buildings and civil engineering structures.

Modern buildings may be engineered to a high degree by consultants or other specialists. I have included some vocabulary from the fields of acoustics, air-conditioning, communications, computers, controls, electrical and fire engineering, geotechnics, public health engineering, timber structures, welding and so on. Repair and restoration are increasingly important, so I have included many terms for existing features of buildings and their rehabilitation.

During the construction process weather may affect progress in different ways, so I have included some meteorological terms. Quality control and testing are parts of the modern construction process, and I have included some relevant vocabulary. Similarly, purchasing, transport, approvals of work, contract administration and payments, and health and safety at work are parts of the process of building, and have their specialist terminology.

In building, things may go wrong, delivery and progress problems may occur, differences in interpretation of specifications may lead to discord, and ultimately the parties involved may reach for their lawyers and even arrive in court. I have included some terms which could come in handy if things do turn out badly.

To round off, I have tried to cover some of the things you may want to say to architects, builders and others in the office and on site, in terms of how things should be done and questions you may need to ask: 'make this higher, move that light switch, this colour is too dark, do you have this in stock, if I pay cash will you give me a discount, when will the tiles arrive on site?' and so on. The words you need are not always found in conventional technical dictionaries, and I hope that including some of them will make life easier.

Technical vocabulary develops all the time. Not only are new terms created, but new variations on old terms arise when people with different educational or technical backgrounds try to describe the same thing. Which is the definitive translation? In many cases it is simply not possible to decide, and I have given several different versions that readers may meet in practice.

To cover all these fields is clearly a tall order, and different specialists will always be able to add particular, perhaps less used terms from their own fields. There is the problem of where to stop: which apparently everyday words to include with the particular meanings given to them by building professionals and which to leave out. I hope you will find the balance and coverage of this dictionary both useful and refreshing.

Please read the short section that follows on how to use the dictionary; it will help you to get the best out of this book. Two appendices are also provided: Appendix A covers 'False Friends', words whose translation is not always what one might think, and Appendix B contains useful general references.

INTRODUCTION

Ce dictionnaire bilingue est destiné aux professionnels et aux particuliers anglais et français qui travaillent dans l'immobilier, la construction et les travaux publics dans leurs pays réciproques.

Il m'a souvent été difficile de décider quelle information inclure et laquelle omettre. Ma première règle a été de me mettre à la place de quelqu'un travaillant d'une manière pratique dans les transactions et la promotion immobilière, les plans architecturaux, le bâtiment, les travaux publics et la construction en général, et de me demander quels mots je serais à même de rencontrer dans mes rapports avec les professionnels et les gens du métier dans les bureaux ou sur les chantiers, quelles questions j'aurais à poser et ce que je voudrais être en mesure d'exprimer. Travaillant dans ces domaines depuis plusieurs années, au Royaume-Uni et en France, je me suis rendu compte que de simples listes de mots sont de peu d'utilité et que de nombreux termes techniques ne sont pas donnés dans les dictionnaires bilingues de base. Le but de ce livre est de combler certaines de ces lacunes.

Tout d'abord, le milieu bâti existe au sein du milieu naturel; j'ai donc inclus quelques termes géographiques et topologiques de base. De tels termes peuvent être utilisés pour décrire un terrain ou une propriété à vendre et aussi au cours des recherches nécessaires avant l'achat d'un terrain à bâtir ou à développer.

Avant de commencer la construction, expertises, planification, approbations, plans architecturaux et techniques, budgétisation, obtention de financement, établissement de spécifications et contrats peuvent être nécessaires; j'ai donc inclus quelques-uns des termes techniques les plus usités dans ces domaines.

Tout bâtiment est construit sur des fondations; j'ai donc commencé au niveau du sol avec des termes de géologié et de la mécanique des sols, et continué avec le drainage, les charpentes, les réseaux de distribution et leur raccordement, et les revêtements, en terminant avec les toitures et la terminologie de finition. J'ai aussi mentionné les matériaux, les outils et les techniques utilisés ainsi que les termes pour différents types de construction d'ouvrages d'art.

Les bâtiments modernes sont conçus à un très haut niveau de technicité par les consultants ou autres spécialistes. J'ai inclus du vocabulaire des domaines de l'acoustique, de la climatisation, des communications, de l'informatique, des contrôles, de l'électrotechnique, de l'ingénierie du feu, de la géotechnique, de l'hydraulique urbaine, des charpentes, du soudage, et ainsi de suite.

La réparation et la restauration étant de plus en plus importantes, j'ai inclus de nombreux termes se référant aux caractéristiques subsistantes des bâtiments et à leur réhabilitation.

Pendant la construction, le temps peut affecter la progression des travaux de différentes manières. J'ai donc inclus quelques termes météorologiques. Le contrôle de qualité et les essais font partie du processus de construction moderne, et j'ai inclus du vocabulaire approprié. De même, achat, transport, réceptions, administration des contrats et paiements, et santé et sécurité au travail font partie du processus de construction et ont leur terminologie spécialisée.

Dans la construction, les choses peuvent mal se passer, des problèmes de livraison et de progression peuvent survenir, des différences d'interprétation de spécifications peuvent amener à des désaccords, et les parties intéressées peuvent en arriver à recourir à leurs avocats et même aux tribunaux. J'ai inclus quelques termes qui pourraient être utiles si les choses tournaient mal.

Pour conclure, j'ai essayé de mentionner certaines choses que vous voudriez peut-être dire aux architectes, entrepreneurs et autres au bureau d'études ou sur le chantier en ce qui concerne la manière dont les choses devraient être faites et des questions qui pourraient être utiles. 'Faites ça plus haut, déplacez cet interrupteur, cette couleur est trop foncée, avez-vous ça en stock, si je vous paie en liquide me ferez-vous une remise, quand vont les tuiles arriver?' et ainsi de suite. Les mots dont vous aurez besoin ne se trouvent pas toujours dans les dictionnaires techniques conventionnels, et j'espère que les inclure dans cet ouvrage vous rendra la vie plus facile.

Le vocabulaire technique évolue constamment. Non seulement de nouveaux termes voient le jour, mais de nouvelles variations d'anciens termes résultent quand des personnes de différentes formations essaient de décrire la même chose. Quelle est la traduction qui fait autorité? Dans beaucoup de cas, il est impossible de décider, et j'ai donné plusieurs versions que le lecteur peut rencontrer.

C'est beaucoup demander que de embrasser tous ces domaines, et différents spécialistes pourront toujours ajouter d'autres termes spécifiques et moins usités puisés dans leur domaine d'expertise. Où s'arrêter? Quels mots d'usage apparemment courant inclure avec leurs sens spécifiques attribués par les professionnels du bâtiment, et lesquels omettre, pose un problème. J'espère que le lecteur trouvera l'équilibre et l'ampleur de ce dictionnaire à la fois utile et stimulant.

Afin de profiter au mieux de cet ouvrage, veuillez lire la courte section sur l'utilisation de ce dictionnaire, qui suit. Vous trouverez aussi deux appendices, un sur les faux amis, mots dont la traduction n'est pas toujours ce que l'on pourrait croire (Appendice A) et l'autre (Appendice B) contenant d'utiles références générales.

USING THIS DICTIONARY

Key words

Key words are given in bold type and translations appear in normal type. Where the same key word may be used as an adjective, noun or verb, the translations are given in that order. For compound expressions such as **clean air** and **air blower**, the noun **air** is the key word, and they appear in the following order:

> **air, clean**
> **air blower**

with the comma in the key compound indicating that the word(s) after the comma modify the first word. Similarly, in French, with **arrêt** *m* as the key word we have:

> **arrêt** *m*, **zone** *f* **d'**
> **arrêt** *m* **d'autobus**

English does not have hard-and-fast rules about whether two-part compound terms are separated by spaces or by hyphens or run together, and there are sometimes similar problems in French. If you cannot find such a compound with a space, try further on where it may be given as one word or hyphenated. For simplicity we have treated hyphenated words as single words when deciding where they should appear alphabetically. For example:

> **fireproof**   contre-limon *m*
> **fire-stop**   contremaître *m*
> **firework**   contre-mur *m*

More complicated compound phrases such as **direct expansion refrigeration** may sometimes be found under more than one key noun, to save time, but will generally not be found under a key adjective. In English **compressed air** and **reinforced concrete** are among the exceptions.

Alternatives and additional meanings

Alternative translations are separated by a slash (solidus) or a comma:

> **abutment**   appui *m*, butée *f*
> **air brick**   brique *f* alvéolée/à alvéoles/perforée/à air

There are two French words for **abutment**, and **air brick** can be **brique alvéolée**, **brique à alveoles**, etc.

Where words are included in brackets in both English and French, these correspond to additional meanings:

> **water tank, (hot)**   ballon *m* d'eau (chaude)

Water tank is **ballon d'eau** and **hot water tank** is **ballon d'eau chaude**.

> **vibrator, (external/internal)**   vibrateur *m* (externe/interne)

We can have a vibrator, an external vibrator or an internal vibrator in either language.

If there is only a bracketed word in one language, this generally means that the bracketed word can be or is sometimes omitted:

> **water supply**   adduction *f* (d'eau)

In French **adduction d'eau** is sometimes abbreviated to **adduction**.

Masculine and Feminine

A word does not, of course, have a biological gender, but nouns in French do have gender, described conventionally as 'masculine' and 'feminine', and any adjectives that qualify them generally have correspondingly different endings. In some cases, a French noun may have entirely different meanings depending only on its gender. For example, the feminine noun **grave** means aggregate for road construction, while the masculine noun **grave** means an accent on a letter or a low register in music.

Unfortunately it is impossible to give absolute rules for determining whether a noun is 'masculine' or 'feminine'. For example, a tractor is **un tracteur**, while a drill is **une perceuse**, a trench-digging machine is **une trancheuse**, while a labourer who digs trenches is **un trancheur**. Thus the gender of each French noun has to be given throughout this dictionary. Where professions are concerned the masculine or common forms are given; the feminine versions are formed in accordance with the rules given in standard French dictionaries and grammars. Similarly, for space reasons, only the masculine singular forms of adjectives are given, as standard works give the rules for forming the feminine and plural versions.

Grammatical information

There are many comprehensive grammatical guides to French and English, and this dictionary does not try to do their job. Thus irregular, reflexive, transitive and intransitive verbs, adverbial and adjectival phrases are not identified as such. For advice on these topics the standard works listed in Appendix B are invaluable.

The basic, necessary grammatical information in this dictionary is given in abbreviated form, in italics, after the word it refers to:

adj	=	adjective
adv	=	adverb
f	=	feminine noun in French
f,pl	=	feminine plural noun in French

m	=	masculine noun in French
m/f	=	noun which is the same in the masculine and feminine in French
m,pl	=	masculine plural noun in French
vb	=	verb

Explanations and context notes

These are given in square brackets. For example:

> **spread** [sewage, fertilizer]   épandre *vb*
> **spread out** [e.g. gravel, objects]   étaler *vb*
> **spread to** [e.g. of fire]   atteindre *vb*

Where everyday words are used with a particular technical sense in a limited area of building or civil engineering, the area or discipline may also be given in square brackets. For example:

> **strike** [geological]   direction *f*

Some commonly used initials are given in round brackets after the phrase they abbreviate.

Warning

A symbol is used, after the word it refers to, as additional guidance:

> † = an abbreviation or an informal or slang expression

UTILISATION DU DICTIONNAIRE

Mots-clefs

Les mots-clefs sont écrits en caractères gras et leur traduction en caractères normaux. Quand le même mot-clef peut être utilisé comme adjectif, nom ou verbe, les traductions sont données dans cet ordre. Pour les expressions composées telles que **clean air** (air pur) et **air blower** (compresseur), le nom **air** est le mot-clef, et les expressions sont données dans l'ordre suivant:

> **air, clean**
> **air blower**

La virgule au milieu du mot composé indique que le/les mot(s) après le virgule modifie(nt) le premier mot. Pareillement, en français, avec un mot-clef comme **arrêt**, les composées sont données dans l'ordre:

> **arrêt** *m*, **zone** *f* **d'**
> **arrêt** *m* **d'autobus**

L'anglais n'a pas de règles absolues qui dictent que les mots composés doivent être séparés par un espac ou un trait d'union ou écrits en un seul mot, et il y a parfois des problèmes semblables en français. Si vous ne trouvez pas un mot composé avec un espace, cherchez plus loin dans la liste où il pourra être écrit en un seul mot ou avec un trait d'union.

Pour raison de simplicité, nous avons traité les mots à trait d'union comme un seul mot en ce qui concerne l'endroit où ils apparaissent en ordre alphabétique. Par exemple:

> **fireproof**   contre-limon *m*
> **fire-stop**   contremaître *m*
> **firework**   contre-mur *m*

Des expressions composées plus compliquées, telles que **direct expansion refrigeration** (système de réfrigération à détente directe) se trouveront parfois sous plus d'un mot-clef, pour gagner du temps, mais ne seront pas généralement listées sous un adjectif-clef. En anglais, **compressed air** (air comprimé) et **reinforced concrete** (béton armé) sont parmi les exceptions.

Choix et sens supplémentaires

Des traductions différentes sont séparées par un trait oblique ou une virgule, par exemple:

> **abutment**   appui *m*, butée *f*
> **air brick**   brique *f* alvéolée/à alvéoles/perforée/à air

Il y a deux mots en français pour **abutment**, et **air brick** peut être traduit par **brique alvéolée**, **brique à alveoles**, etc.

Quand des mots sont entre parenthèses en anglais et en français, ceci correspond à des sens supplémentaires:

> **water tank, (hot)**   ballon *m* d'eau (chaude)

Water tank signifie **ballon d'eau** et **hot water tank** signifie **ballon d'eau chaude**.

> **vibrator, (external/internal)**   vibrateur *m* (externe/interne)

On peut avoir un vibrateur, un vibrateur externe ou un vibrateur interne dans les deux langues.

Si un seul mot est entre parenthèses dans une langue, cela signifie généralement que le mot entre parenthèses peut être omis:

> **water supply**   adduction *f* (d'eau)

En français, le terme **adduction d'eau** est souvent abrégé en **adduction**.

Renseignements grammaticaux

Il existe de nombreux guides complets de grammaire française et anglaise, et ce dictionnaire ne prétend pas faire leur travail. Par conséquent, les verbes irréguliers, réfléchis, transitifs et intransitifs, et les locutions adverbiales et adjectivales ne sont pas identifiés en tant que tels. Pour plus de renseignements à leur sujet, les ouvrages de base listés dans l'Appendice B sont inestimables.

Le renseignement grammatical de base dans ce dictionnaire est donné en abrégé et en italique après le mot auquel il correspond:

adj	=	adjectif
adv	=	adverbe
f	=	nom féminin en français
f,pl	=	nom féminin pluriel en français
m	=	nom masculin en français
m/f	=	nom qui est le même au masculin et au féminin en français
m,pl	=	nom masculin pluriel en français
vb	=	verbe

Explications et notes de contexte

Celles-ci sont données entre crochets. Par exemple:

> **spread** [sewage, fertilizer]   épandre *vb*
> **spread out** [e.g. gravel, objects]   étaler *vb*
> **spread to** [e.g. of fire]   atteindre *vb*

Quand des mots d'usage courant sont utilisés avec un sens technique précis dans un domaine particulier de la construction ou des travaux publics, ce domaine ou cette discipline sont donnés entre crochets. Par exemple:

> **strike** [geological]   direction *f*

Certaines initiales d'usage courant sont écrites entre parenthèses après le syntagme qu'elles abrègent.

Avis

Un symbole est utilisé après le mot auquel il se réfère, comme information supplémentaire.

† = une expression abrégée ou une expression familière ou argotique

PART ONE

English/French
Anglais/Français

A

abacus [slab on top of a column]   abaque *m*, tailloir *m*
abandoned [e.g. redundant building]   abandonné *adj*,
  délaissé *adj*
abattoir   abattoir *m*
ability   pouvoir *m*
ablation   ablation *f*
abnormal   anormal *adj*
abode   domicile *m*
above   dessus *adv*
above ground   non enterré *adj*
above sea level   au-dessus du niveau de la mer
abrade   éroder *vb*
abrasion   abrasion *f*, attrition *f*, usure *f* par frottement
abrasive *adj*   abrasif *adj*
abrasive   abrasif *m*
abscissa   abscisse *f*
absolute zero   zéro *m* absolu
absorb   absorber *vb*
absorbent *adj*   absorbant *adj*, hydrophile *adj*
absorbent   absorbant *m*
absorption   absorption *f*
absorption bed [for purifying septic tank effluent]
  plateau *m* absorbant/tellurien
absorptivity, acoustic   absorptivité *f* acoustique
abstract [summary]   relevé *m*
abutment   appui *m*, butée *f*
abutment [of a dam]   culée *f* poids
abutment [of bridge]   culée *f*
abutment hinge   articulation *f* aux naissances/de culée
abutment piece [of wood, under a prop, etc]   semelle *f*
abutment pier   arc *m* boutant
acanthus [plant or architectural decoration]   acanthe *m*
accelerating   accélérateur *adj*
acceleration   accélération *f*
accelerator [for concrete]   accélérateur *m* de prise *f*
accelerator [in a heating or ventilation system]
  accélérateur *m*
accept   agréer *vb*
acceptance   acceptation *f*, agrément *m*, réception *f*
acceptance certificate   certificat *m* de réception
acceptance of works (and final account)   réception *f* des
  travaux (RDT)
acceptance of works, final   réception *f* définitive
acceptance of works, provisional   réception *f* provisoire
acceptance test   essai *m* de recette
acceptance tests on samples   recette *f*
access [general]   accès *m*
access [to land]   abord *m*
access, cleaning   accès *m* de nettoyage
access, limited   accès *m* réglementé
access, sloping   accès *m* en pente
access balcony   coursive *f*
access control   contrôle *m* d'accès

access control/management   gestion *f* d'accès
access control system (ACS)   système *m* de contrôle
  d'accès (SCA)
access cover/hatch   trappe *f* de visite
access cover or eye [plumbing]   bouchon *m* de
  dégorgement, tampon *m* hermétique
access gallery or tunnel   galerie *f* d'accès
access hatch   trappe *f* d'accès
access panel   panneau *m* d'accès
access platform, aerial   élévateur *m* à nacelle
access platform, lorry mounted   élévateur *m* (monté)
  sur camion
access platform, telescopic   élévateur *m* télescopique
access point [for motorway entry or exit]   zone *f* d'accès
access (point), limited   accès *m* spécial d'entrée
access scaffolding with angled support   éventail *m*
access/service road   voie *f* de desserte
access/service road, minor   voie *f* secondaire de
  desserte
access walkway   chemin *m* de service
access/work platform, articulated aerial   nacelle *f*
  élévatrice articulée
access/work platform, telescopic boom   nacelle *f* à
  bras/à flèche télescopique
access to, provide [e.g. one room to another]
  commander *vb*
accessibility   accessibilité *f*
accessible   accessible *adj*
accessible [easily]   abordable *adj*
accessible, easily   d'accès facile
accession [to ownership]   accession *f*
accessory *adj*   accessoire *adj*
accessory   accessoire *m*
accessory, fixing   accessoire *m* de pose
accessories   appareillage *m*
accessories, range of   gamme *f* d'accessoires
accident [mishap]   accident *m*
accident [loss under insurance policy]   sinistre *m*
accident prevention   prévention *f* des accidents
accommodation   logement *m*
accordance with, in   conformément à, suivant
according to   suivant *adj*
according to (the) regulations   conformément aux
  règlements
according to contract   suivant contrat
according to stock-list   suivant inventaire
account [at a bank]   compte *m*
account [invoice]   facture *f*, note *f*
account [of facts, etc]   exposé *m*
account [report on something]   notice *f*
accounts payable   dette *f*
account, establishment of the amount of final
  règlement *m*

account, fee   note *f* d'honoraires
account, final   décompte *m* final
account, pro rata   compte *m* de prorata
accountancy   comptabilité *f*
accountant   comptable *m*
accountant, chartered [near French equivalent]   expert *m* comptable
accounting   comptabilité *f*
accounting, cost   comptabilité *f* analytique
accounting principles   principes *m,pl* comptables
accretion   accroissement *m*
accumulation [of electrical energy]   emmagasinement *m*
accumulator [electrical]   accumulateur *m*
accumulator, heat   accumulateur *m* de chaleur
accuracy   exactitude *f*, justesse *f*, précision *f*
accuracy, setting   précision *f* d'affichage
acetylene   acétylène *m*
acetylene, dissolved   acétylène *m* dissous
acetylene generator   générateur *m* d'acétylène
acetylene welding   soudage *m* autogène/oxyacétylénique
achievement   réalisation *f*
acid *adj*   acide *adj*
acid   acide *m*
acid resistant   résistant aux acides
acknowledgement of receipt [of letter, parcel]   accusé *m* de réception
acoustic   acoustique *adj*
acoustic baffle   baffle *m* acoustique
acoustic bridge   pont *m* phonique
acoustic coupler [computer]   coupleur *m* acoustique
acoustic performance   effet *m* acoustique
acoustic power   intensité *f* acoustique
acoustician   acousticien *m*
acoustics   acoustique *f*
acquire   acquérir *vb*
acquisition   acquisition *f*
acquisition [of land for public purposes]   emprise *f*
acre [approx]   arpent *m*
across   en travers
Acrow type prop   chandelle *f* à crémaillère, étrésillon *m* à vérin
acrylic   acrylique *adj*
act(ion)   action *f*
action [legal]   procès *m*
action, biochemical   effet *m* biochimique
action, oxidation   action *f* oxydante
activate [release, set in motion]   déclencher *vb*
activated carbon   charbon *m* actif
activated charcoal filter   filtre *m* en charbon actif
activated sludge   boues *f,pl* activées
activated sludge process   procédé *m* des boues *f,pl* activées
active earth pressure   poussée *f* des terres
activities of the *maître d'œuvre*   maîtrise *f* d'œuvre
activity   activité *f*
actual   réel *adj*
actuator   actionneur *m*
adapt   adapter *vb*, ajuster *vb*
adaptation   adaptation *f*, aménagement *m*

adapted (for)   adapté *adj* (à)
adapter [person]   adaptateur *m*
adaptor [connecting piece]   raccord *m*
adaptor [plumbing]   adapteur *m*
adaptor plug [electrical]   prise *f* multiple
adaptor plug [plumbing]   bouchon *m* de raccord
add   ajouter *vb*
add a filler   fillériser *vb*
add water to [e.g. mixing concrete, plaster, etc]   délayer *vb*, gâcher *vb*
add water [to stiffened mortar]   rafraîchir *vb*
adding (up)   addition *f*
adding to   addition *f* à
addition [of sth]   adjonction *f*
additional   additionnel *adj*
additional [reserve] **tank**   réservoir *m* supplémentaire
additive   additif *m*, adjuvant *m*
additive, air-entraining   adjuvant *m* entraîneur d'air
additive for producing foamed or aerated concrete   moussant *m*
adhere   adhérer *vb*
adherent   adhérent *adj*
adhesion   adhérence *f*, adhésion *f*, collage *m*
adhesion [of layers, e.g. paint, rendering, road metal]   accrochage *m*
adhesive *adj*   adhérent *adj*, adhésif *adj*
adhesive   adhésif *m*, colle *f*
adhesive (material)   matière *f*/substance *f* adhésive
adhesiveness   adhésivité *f*
adiabatic   adiabatique
adiabatic change of state   changement *m* adiabatique d'état
adiabatic compression   compression *f* adiabatique
adiabatic lapse rate   chute *f* de température adiabatique
adit [dam]   galerie *f* d'accès
adit [geology]   galerie *f* d'exploration
adjacent (to)   adjacent/contigu *adj* (à)
adjoin   jouxter *vb*
adjoining   attenant *adj*, contigu *adj*
adjournment   ajournement *m*
adjudication   adjudication *f*
adjust   ajuster *vb*
adjust [a material, e.g. a piece of stone or wood]   dresser *vb*
adjust [a machine, a valve, etc]   régler *vb*
adjust [a machine]   agencer *vb*
adjust [an instrument]   rectifier *vb*
adjust [reduce size to make sth fit]   délarder *vb*
adjustable   réglable *adj*
adjustable [in direction, angle]   orientable *adj*
adjuster   régleur *m*
adjustment   adaptation *f*, réglage *m*, régulation *f*
adjustment [of an instrument]   rectification *f*
adjustment [of size, position, shape]   dressage *m*
adjustment [reduction of size to make sth fit]   délardement *m*
adjustment [running in]   mise *f* au point
adjustment, manual   réglage *m* manuel
adjustment, rough   réglage *m* approximatif

adjustment range   plage *f* de réglage
administer   gérer *vb*
administration   gestion *f*
administrative court   tribunal *m* administratif
administrative department   service *m*
administrative director   directeur *m* administratif
administrative head of a *département* in France   préfet *m*
administrative steps   démarches *f,pl* administratives
administrative system   régime *m*
admission   admission *f*
admixture   produit *m* d'addition
admixture (concrete)   adjuvant *m* (du béton)
adornment   ornement *m*
adsorb   adsorber *vb*
adsorbate   substance *f* adsorbée
adsorption   adsorption *f*
advance [loan]   avance *f*
advance notice   préavis *m*
advance payment   paiement *m* d'avance
advertisement   annonce *f*
advertisement hoarding   panneau *m* d'affichage, panneau *m* publicitaire
advertising campaign   promotion *f*
advice   conseil *m*
advice of delivery [of a letter, parcel]   avis *m* de réception (AR)
advise (someone to do sth)   conseiller *vb* (à qn de faire qch)
adviser, legal [similar to solicitor]   conseiller *m* juridique
adviser, technical   conseiller *m* technique
adviser, tax   conseil *m* fiscal
advocate [lawyer]   avocat *m*
advocate   préconiser *vb*
adze   doloire *f*, herminette *f*
adzed timber   bois *m* refait
aeolian   éolien *adj*
aerate   aérer *vb*
aerated   aéré *adj*
aerial [overhead]   aérien *adj*
aerial   antenne *f*
aerobic   aérobie *adj*
aerobic filter [in septic tank or sewage system]   filtre *m* épurateur
aerodynamic   aérodynamique *adj*
aerogenerator   aérogénérateur *m*, éolienne *f*
aerosol   aérosol *m*
aerosol particles   particules *f,pl* d'aérosols
aesthetic harmonization [in architecture, furniture, etc]   design *m*
aesthetically, treated   ouvragé *adj*
affix light wooden mouldings to decorate or cover joints   habiller *vb*
affix metal grill/mesh to a surface to reinforce its rendering   habiller *vb*
affluent *adj* [stream, etc]   affluent *adj*
after-cooler   refroidisseur *m* aval
after-cooling   refroidissement *m* aval
after-heat   chaleur *f* résiduelle
after-heater   réchauffeur *m* aval

after-treatment   traitement *m* ultérieur
age   vieillir *vb*
ageing   vieillissement *m*
ageing [to make sth look old]   vieillissage *m*
agency   agence *f*
agency commission   commission *f* d'agence
agency fees   frais *m,pl* d'agence
agenda   ordre *m* du jour
agent   agent *m*
agent, air-entraining   adjuvant *m* entraîneur d'air
agent, anti-corrosion   moyen *m* de protection contre la rouille, inhibiteur *m* de corrosion
agent, anti-foaming   agent *m* antimousse
agent, cleaning   détergent *m*
agent, curing [e.g. for concrete]   produit *m* de cure
agent, drying [siccative]   siccatif *m*
agent, estate   agent *m* immobilier
agent, fixing   agent *m* de fixation
agent, foaming   produit *m* moussant
agent, real estate   agent *m* immobilier
agent, retarding [for concrete]   retardateur *m* de prise
agent, rust protective   moyen *m* de protection contre la rouille
agglomerate   aggloméré *m*
aggregate   granulats *m,pl*
aggregate [for concrete]   agrégat *m*
aggregate [for macadam or tar macadam]   macadam *m*
aggregate [for road construction]   grave *f*
aggregate, angular particles of   granulats *m,pl* anguleux
aggregate, broken   grave *f* concassée, granulats *m,pl* concassés
aggregate, coarse   gros agrégats *m,pl*, gros granulats *m,pl*
aggregate, crushed   grave *f* concassée, agrégats *m,pl* concassés
aggregate, cubic particles of   granulats *m,pl* cubiques
aggregate, extremely fine grain [flour-like]   farine *f*
aggregate, fine   granulats *m,pl* fins
aggregate, lightweight   agrégats *m,pl* légers
aggregate, mineral   granulats *m,pl* minéraux
aggregate, pre-mixed   agrégats *m,pl* enrobés
aggregate, river   agrégats *m,pl* roulés
aggregate, well-graded   granulats *m,pl* bien gradués, grave *f* bien graduée
aggregate feed unit, cold   prédoseur *m* des enrobés
aggressive   agressif *adj*
aggressive environment   milieu *m* agressif
aggressiveness or aggressivity   agressivité *f*
agitation   agitation *f*
agitator   agitateur *m*, mélangeur *m*
agree [to do sth]   s'engager *vb*
agreement [about sth]   accord *m*
agreement [to do sth]   engagement *m*
agreement [generally written]   contrat *m*, convention *f*
agreement, architect–client   contrat *m* d'architecte
agreement, preliminary sale [to sell]   promesse *f* de vente
agreement, preliminary sale [to buy and sell]   compromis *m* de vente
agreement, selling [between owner and agent]   mandat *m* de vente

agricultural   agricole *adj*
agriculturalist   agronome *m*
agriculture   agronomie *f*
aim [e.g. of lamp]   but *m*
air   air *m*
air, ambient   air *m* ambiant
air, aspirated   air *m* d'aspiration
air, clean   air *m* pur
air, clean [filtered]   air *m* filtré
air, combustion   air *m* comburant, air *m* de combustion
air, compressed, *see* compressed air
air, cooling   air *m* rafraîchissant/de refroidissement
air, discharge or exhaust   air *m* expulsé/de refoulement
air, dust-laden   air *m* chargé de poussière
air, entrained   air *m* entraîné
air, entrapped   air *m* inclus
air, excess   air *m* en excès
air, exhaled   air *m* exhalé
air, exhaust   air *m* expulsé/rejeté
air, extract   air *m* évacué/extrait
air, forced   air *m* pulsé
air, fresh   air *m* frais/extérieur
air, impermeable to   imperméable à l'air
air, indoor   air *m* intérieur
air, infiltrated   air *m* parasite
air, make up   air *m* additionnel/d'appoint/de
   compensation/supplémentaire
air, outdoor/outside   air *m* extérieur
air, outlet   air *m* de soufflage
air, primary   air *m* neuf/primaire
air, recirculated   air *m* de circulation/de
   reprise/recirculé/recyclé
air, return   air *m* repris/de reprise
air, secondary   air *m* secondaire
air, standard   air *m* normal
air, vitiated   air *m* vicié
air blower   compresseur *m*, ventilateur *m*
air brick   brique *f* alvéolée/à alvéoles/perforée/à air
air brick [foundation ventilator]   bouche *f* d'aération
air bubble   bulle *f* d'air
air bubble in glass   bouillon *m*
air chamber [container]   réservoir *m* d'air
air change coefficient   coefficient *m* brassage
air change rate   taux *m* de renouvellement d'air/de l'air
air change rate [scavenging]   balayage *m*
air changes per hour   renouvellements *m,pl* d'air à
   l'heure
air circulation   circulation *f* d'air
air circulation, natural   circulation *f* naturelle
air cleaner   filtre *m* à air, épurateur *m* d'air
air cleaning/purification   épuration *f* de l'air
air cock or valve [vent]   purgeur *m*/robinet *m* d'air,
   robinet *m* purgeur
air compressor   compresseur *m* d'air
air compressor set   groupe *m* électro-compresseur
air condition   état *m* d'air/de l'air
air conditioner, *see* air-conditioner
air conditioning, *see* air-conditioning
air container   réservoir *m* d'air

air cooler   refroidisseur *m* à air/d'air
air cooling installation or system   installation *f*
   de refroidissement à air
air current or flow   courant *m*/écoulement *m* d'air
air curtain (warm)   rideau *m* d'air (chaud)
air cushion or mattress   couche *f*/coussin *m*/matelas *m*
   d'air
air density   densité *f* de l'air
air distribution   distribution *f* d'air/de l'air
air distribution box   boîtier *m* de diffusion d'air
air distribution envelope   profil *m* des vitesses d'air
air distribution system   système *m* de diffusion d'air
air dried   séché *adj* à l'air
air drying apparatus   appareil *m* de séchage de l'air
air drying equipment, installation or plant   installation
   f de séchage de l'air
air entraining agent [for concrete]   entraîneur *m* d'air,
   (adjuvant *m*)
air entrainment   entraînement *m* d'air
air exfiltration or leakage   fuite *f* d'air
air extract system   système *m* d'extraction d'air
air extraction   extraction *f* d'air
air filter, oil bath   filtre *m* à air à bain d'huile
air filter cell   filtre *m* multicellulaire aspirateur
air filter unit/section [in an air handling unit]   caisson *m*
   de filtre à l'air
air flow [quantity/rate]   débit *m* d'air/d'écoulement d'air
air flow, direction of   sens *m* de l'écoulement de l'air
air flow, recirculated [quantity/rate]   débit *m* d'air de
   reprise
air flow, stratified   écoulement *m* d'air stratifié
air flow, supply [quantity/rate]   débit *m* d'air soufflé
air flow path   voie *f* d'écoulement d'air
air flow pattern   schéma *m* d'écoulement d'air
air flow resistance   résistance *f* aéraulique/au passage de
   l'air
air gap or space   espace *m* d'air
air gap [in a cavity wall]   couche *f* d'air
air guide   guide *m* d'air
air handling   traitement *m* d'air
air handling plant [air-conditioning]   centrale *f* de
   climatisation
air handling unit (AHU)   unité *f* de traitement d'air
air heater, gas-fired   réchauffeur *m* d'air à gaz
air heating, gas-fired   chauffage *m* d'air à gaz
air hole, valve or vent   ventouse *f*
air hole [in casting]   bulle *f*
air humidifier   humidificateur *m*
air humidifying installation or plant   installation *f*
   d'humidification
air inlet   entrée *f* d'air
air inlet slot   fente *f* de prise d'air
air inlet to a flue to reduce excessive draught   coupe-
   tirage *m*
air intake   prise *f* d'air
air intake, fresh   prise *f* d'air neuf
air link [transport]   liaison *f* aérienne
air lock   écluse *f* d'air
air motion, movement or pattern   mouvement *m* d'air

6

air noise   bruit *m* d'air
air nozzle   bouche *f* d'air
air operated   pneumatique *adj*
air outlet   échappement *m*/sortie *f* d'air
air passage   voie *f* d'air
air pocket   poche *f* d'air
air pollution   pollution *f* d'air/de l'air
air pollution concentration, level or rate   taux *m* de pollution de l'air
air preheating   préchauffage *m* d'air
air pressostat   pressostat *m* de manque d'air
air purging or vent(ing)   purge *f* d'air
air purifier   épurateur *m* d'air
air quality   qualité *f* de l'air
air quality, indoor   qualité *f* d'air intérieur
air quantity measurement   mesure *f* de la quantité d'air
air quantity or volume   quantité *f* d'air
air rate, outside   taux *m* d'air frais
air receiver   réservoir *m* à air
air recirculating plant or system   système *m* de circulation d'air
air relief cock   robinet *m* décompresseur
air resistance   résistance *f* de l'air
air scrubber   épurateur *m* d'air
air seal   barrière *f* à l'air
air separator   séparateur *m* d'air
air separator, centrifugal   séparateur *m* d'air centrifuge
air shut-off valve   vanne *f* d'arrêt d'air
air stratification   stratification *f* de l'air
air stream   courant *m* d'air, veine *f* d'air
air supply   apport *m* d'air, soufflage *m*
air supply fixtures   équipement *m* de diffusion d'air
air temperature   température *f* ambiante
air throw   portée *f* d'air
air tool   outil *m* pneumatique
air treatment   traitement *m* de l'air
air vent   bouche *f* de ventilation, orifice *m* d'aération
air vent [tap or cock]   évent *m*
air vessel   réservoir *m* d'air
air washer   laveur *m* d'air
air well   courette *f*
airborne   aérien *adj*
air-bubble in glass   crachat *m*
air-condition   climatiser *vb*
air-conditioned   climatisé *adj*
air-conditioned chamber   chambre *f* climatisée
air-conditioner   climatiseur *m*, conditionneur *m* d'air, installation *f* de conditionnement d'air
air-conditioner, packaged   climatiseur *m* autonome, conditionneur *m* d'air autonome/monobloc
air-conditioner, roof-mounted/top   conditionneur *m* d'air en toiture
air-conditioner, room   aérateur-climatiseur *m*, climatiseur *m* individuel
air-conditioner, through-the-wall   conditionneur *m* d'air de paroi
air-conditioner, unitary   conditionneur *m* d'air indépendant
air-conditioner, window   conditionneur *m* d'air de fenêtre

air-conditioning   climatisation *f*, conditionnement *m* d'air
air-conditioning, room   conditionnement *m* d'air de pièce
air-conditioning engineering   génie *f* climatique
air-conditioning equipment, installation or plant   centrale *f* de climatisation, installation *f* de climatisation (d'air)/de conditionnement d'air
air-conditioning plant, modular   centrale *f* modulaire de climatisation
air-conditioning plant, single-duct   installation *f* de climatisation à un conduit
air-conditioning process   procédé *m* de conditionnement d'air
air-conditioning system   système *m* de climatisation (d'air)/de conditionnement d'air
air-conditioning system, all-air   système *m* de conditionnement d'air tout air
air-conditioning system, dual duct   système *m* de climatisation à double conduit
air-conditioning system, packaged terminal   conditionneur *m* d'air autonome
air-conditioning system, self-contained   système *m* monobloc de conditionnement d'air
air-conditioning system, variable air volume (VAV)   système *m* de climatisation à débit d'air variable
air-conditioning unit   appareil *m* de conditionnement d'air
air-conditioning unit, room   aérateur-climatiseur *m*
Air-conditioning and Refrigeration Institute (ARI)   Institut *m* International du Froid
air-cooled   refroidi par air
air-cooled water chiller   groupe *m* d'eau glacée refroidi par l'air
aircraft hangar   hangar *m* d'avions
aircraft parking area   aire *f* de stationnement
air-dried   desséché à l'air
airfield, *see also* airport   terrain *m* d'aviation
air-gap [between metal surfaces, e.g. rotor and stator]   entrefer *m*
air-gap [between two sensibly parallel faces]   lame *f* d'air
air-lagged   à chemise d'air
airless (paint) spraying   peinture *f* au pistolet sans air
airlock [e.g. of pressurized caisson, lobby, etc]   sas *m* à air
airlock [in a pipe]   bouchon *m* d'air
air-passage   conduite *f* à air
airport   aéroport *m*
airport freight terminal   aérogare *f* (de) fret
airport passenger terminal   aérogare *f* (de) passagers
airport runway   piste *f*
airtight   étanche *adj*/imperméable *adj* à l'air, hermétique *adj*
airtightness   étanchéité *f* (à l'air)
airway   voie *f* d'air
alarm   alarme *f*, avertisseur *m*
alarm, break-glass fire   boîtier *m* brise-vitre
alarm, false   alarme *f* intempestive
alarm, fire, *see* fire alarm
alarm bell   sonnerie *f* d'alarme, sonnette *f* d'alarme

alarm device   dispositif *m* d'alarme
alarm mechanism   mécanisme *m* d'alarme
alarm panel   tableau *m* des alarmes
alarm relay   relai(s) *m* de contrôle
alarm system   système *m* d'alarme
alcohol   alcool *m*
alcove   alcôve *f*
alder   au(l)ne *m*
algae   algues *f,pl*
algicide   algicide *m*
algorithm, control   algorithme *m* de contrôle
alidade [surveying]   alidade *f*
alight   allumé *adj*
alight, be   brûler *vb*
alight, catch   s'allumer *vb*
align   aligner *vb*
align [a machine]   dresser *vb*
alignment [the line]   alignement *m*
alignment [i.e. lining up]   mise *f* en ligne
alignment [of a road, etc]   tracé *m*
alignment [the action of, of a machine, etc]   dressage *m*
alignment, boundary   alignement *m*
alignment, straight   alignement *m* droit
alkali   alcali *m*
alkali resistant   résistant aux alcali(n)s
alkaline   alcalin *adj*
alkalinity   alcalinité *f*
all trades   tout corps *m* d'état/de métier
alley(way)   passage *m*, ruelle *f*
all-in [e.g. of fee, payment]   forfaitaire *adj*
allocate   attribuer *vb*
allocate [share, etc]   allouer *vb*
allocation   adjudication *f*
allow [an expense, etc]   allouer *vb*
allow [sth]   admettre *vb*
allow someone to do sth   admettre *vb* qn à faire qch
allowable   admissible *adj*, permissible *adj*
allowable, maximum   maximum *m* admissible/permissible
allowance [play]   jeu *m*
allowance [reduction in price]   remise *f*
allowance, machining   tolérance *f* d'usinage
allowed [allowable]   admis *adj*
alloy   alliage *m*
alloy, ferrous   alliage *m* ferreux
alloy, light   alliage *m* léger
alloy, wrought   alliage *m* forgé
alloy   allier *vb*
all-purpose [e.g. spanner]   universel *adj*
alluvial   alluvial *adj*, alluvionnaire *adj*
alluvium   alluvions *f,pl*, limon *m*
alter   modifier *vb*, remanier *vb*
alterability   altérabilité *f*
alteration   altération *f*, modification *f*, remaniement *m*, transformation *f*
altered   altéré *adj*, modifié *adj*, remanié *adj*, transformé *adj*
alternat(iv)e   variant *adj*
alternative   variante *f*

alternative product   produit *m* de substitution
alternator   alternateur *m*
altimeter   altimètre *m*
altitude   altitude *f*
altitude above a reference height   cote *f* de niveau
altitude recorder   altitraceur *m*
alumina   alumine *f*
aluminate   aluminate *m*
aluminium   aluminium *m*, alu[†] *m*
aluminium faced insulating paper   papier *m* d'aluminium
aluminium foil   feuille *f* d'aluminium
ambient   ambiant *adj*
amend   modifier *vb*
amend [an account]   rectifier *vb*
amended   modifié *adj*
amended area [for rent calculation]   surface *f* corrigée
amendment   modification *f*
amendment [to a contract]   avenant *m*
amenities   aménagements *m,pl*
ammeter   ampèremètre *m*
ammeter, clamp type   pince *f* de courant
ammeter, recording   ampèremètre *m* enregistreur
ammonia   ammoniac *m*
ammonia solution   ammoniaque *f*
amortization [of a loan]   amortissement *m*
amortize   amortir *vb*
amount   quantité *f*
amount [content, rate of application]   dose *f*
amount [money]   montant *m*
amount [total sum]   chiffre *m*
amount, approximate   quantité *f* approximative
amount still owing   résidu *m* de compte
ampere   ampère
amphitheatre   amphithéâtre *m*
amplification   amplification *f*
amplifier   amplificateur *m*
amplify   amplifier *vb*
amplitude   amplitude *f*
anaerobic   anaérobie *adj*
analogical   analogique *adj*
analogue/analogous *adj*   analogue *adj*
analogue   analogue *m*
analogue indication   indication *f* analogique
analogue transmission   transmission *f* analogique
analogy   analogie *f*
analyse   analyser *vb*
analyse [by calculation]   calculer *vb*
analyse [tenders, an account]   dépouiller *vb*
analyser   analyseur *m*
analyser, acoustic   analyseur *m* acoustique
analyser, chemical flue gas   analyseur *m* de gaz de fumée
analyser, infra-red   analyseur *m* infrarouge
analysis   analyse *f*
analysis [e.g. of accounts, tenders]   dépouillement *m*
analysis, check   analyse *f* de contrôle
analysis, chemical water   analyse *f* chimique de l'eau
analysis, cost-benefit   analyse *f* coûts-avantages
analysis, dimensional   analyse *f* dimensionnelle

analysis, elastic calcul *m* élastique
analysis, fuel gas analyse *f* des gaz combustibles
analysis, grain size analyse *f* granulométrique
analysis, melt analyse *f* de coulée
analysis, network, and PERT programming planning *m* pert
analysis, random [on finished material] analyse *f* de produit
analysis, sieve analyse *f* granulométrique
analysis, slip-circle analyse *f* de stabilité en rupture circulaire
analysis, stability étude *f* de stabilité
analysis, stress détermination *f* des efforts, vérification *f* des contraintes
analysis, thermal analyse *f* thermique
analysis, waste gas analyse *f* des gaz brûlés
analysis of costs analyse *f* des coûts
analysis of grain or particle size analyse *f* granulométrique
analysis of networks analyse *f* des réseaux
analysis of structures calcul *m* de structures
analytical balance/scales balance *f* de précision
anchor [fixing] ancre *f*
anchor (in place) ancrer *vb*
anchor block massif *m* d'ancrage
anchor block [pipe bend] butée *f* de coude
anchor bolt boulon *m* d'ancrage
anchor(age) ancrage *m*
anchorage, dead ancrage *m* mort
anchorage bond [in concrete, masonry, etc] scellement *m*
anchorage length longueur *f* d'ancrage
anchorage point point *m* d'ancrage
anchorage seating recul *m* à l'ancrage
anchored in [masonry, etc] scellé *adj*
anchoring cone cône *m* d'ancrage
anchoring rod tirant *m* d'ancrage
ancient [placed after the noun in French] ancien *adj* [après le substantif]
andiron chenet *m*
anemometer anémomètre *m*
anemometer, deflecting vane anémomètre *m* à palette
anemometer, hot wire anémomètre *m* à fil chaud
anemometer, recording anémomètre *m* enregistreur
anemometer, revolving vane anémomètre *m* à moulinet
aneroid barometer baromètre *m* anéroïde
angle *adj* [i.e. at meeting of two planes] cornier *adj*
angle [iron bar, etc] coude *m*
angle, acute angle *m* aigu
angle, azimuth angle *m* azimut
angle, bevel angle *m* de chanfrein
angle, clearance [of a cutting tool] angle *m* de dépouille
angle, contact [of a driving belt] angle *m* d'enroulement
angle, cut-off angle *m* de coupe
angle, cutting [of a cutting tool] angle *m* d'attaque
angle, discharge angle *m* de diffusion
angle, friction angle *m* de frottement
angle, inclination angle *m* d'inclinaison
angle, inlet angle *m* d'entrée

angle, internal friction angle *m* de frottement interne
angle, phase angle *m* de phase
angle, re-entrant angle *m* rentrant
angle, right angle *m* droit
angle, sharpness [of a cutting tool] angle *m* d'affûtage
angle, slope angle *m* de talus
angle, wall friction angle *m* de friction sur paroi
angle between the two planes of a curb or mansard roof brisis *m*
angle between two rendered surfaces, re-entrant cueillie *f*
angle of contact [of a driving belt] angle *m* d'enroulement
angle of dip [geology] pendage *m*
angle of discharge angle *m* de sortie
angle of friction angle *m* de frottement
angle of incidence angle *m* d'incidence
angle of inclination angle *m* d'inclinaison
angle of repose [of a slope] angle *m* d'éboulement/de talus naturel
angle of shear angle *m* de cisaillement
angle of slope angle *m* de talus
angle of view angle *m* de champ
angle (section), bulb cornière *f* à boudin
angle (section), equal-leg cornière *f* à ailes égales
angle (section), flange cornière *f* de membrure
angle (section), round-edged cornière *f* à angles arrondis
angle (section), square-edged cornière *f* à angles vifs
angle (section), unequal-leg cornière *f* à ailes inégales
angled cut in stone arch springer or abutment abattage *m* de retombée
angled cut(ting) of slates or tiles [to fit next to ridges or valleys] tranchis *m*
angledozer angledozer *m*, bouteur *m* biais, tracteur *m* à lame orientable
angle-iron reinforcement of corner of a wall cornette *f*
angular angulaire *adj*
angular break [e.g. in a kerb] décrochement *m* anguleux
anhydrous anhydre *adj*
anisotropic anisotrope *adj*
ankle boot demi-botte *f*
anklet boot brodequin *m*
anneal détremper *vb*, recuire *vb*, stabiliser *vb*
annealed détrempé *adj*, recuit *adj*
annealing stabilisation *f*
annex [extension] annexe *f*
annual [e.g. accounts, charges] annuel *adj*
annual closure fermeture *f* annuelle
annual premium [insurance] cotisation *f*
annually annuellement *adv*
annuity annuité *m*, rente *f*
annulment [of an insurance policy] résiliation *f*
annulus anneau *m*
anode anode *f*
anodized anodisé *adj*
answering machine, telephone répondeur *m*
anthracite anthracite *m*
anti-capillary layer [road construction] sous-couche *f* anti-capillaire

anti-clockwise   à gauche, en sens inverse des aiguilles d'une montre
anti-corrosion   antirouille *adj*
anti-corrosive   anticorrosif *adj*
anti-frost   antigel *adj*
anti-glare   antiéblouissant *adj*
anti-rust *adj*   antirouille *adj*
anti-rust   antirouille *f*
anti-rust agent   moyen *m* de protection contre la rouille
anti-scale   antitartre *adj*
anti-skid   antidérapant *adj*
anti-splash tap nozzle   brise-jet *m*
anti-static device   dispositif *m* antistatique
anti-vibration   antivibratile *adj*
anti-vibration device   dispositif *m* antivibratoire
anticline   anticlinal *m*
anti freeze solution   antigel *m*, solution *f* incongelable
anvil   enclume *f*
apartment   appartement *m*, duplex *m*
apartment [one room]   studio *m*
apartment, duplex   duplex *m*
apartment building, high rise   tour *f* d'habitation
aperture   orifice *m*, ouverture *f*
apparatus, *see also* appliance, device, equipment, machine, plant
apparatus [collective]   équipage *m*
apparatus [individual]   appareil *m*
apparatus, breathing   appareil *m* respiratoire
apparatus, direct shear   appareil *m* de cisaillement direct
apparatus, lifting   appareil *m* élévateur
apparatus, recording   appareil *m* enregistreur
apparatus, ring shear   appareil *m* de cisaillement circulaire
appendix [to a document]   annexe *f*
apple tree   pommier *m*
appliance, *see also* apparatus, device, equipment
appliance, domestic   appareil *m* ménager
appliance, heating   appareil *m* de chauffage
appliance, sanitary   appareil *m* sanitaire
appliance, towed   engin *m* tracté
appliance connection [water]   branchement *m* d'appareil
application   application *f*, cas *m* d'utilisation
application, area or field of   domaine *m* d'application/d'emploi
application for   demande *f* de
application for planning permission   demande *f* de permis de construire
application of a load   application *f* d'une force
application point   point *m* d'application
applied   appliqué *adj*, rapporté *adj*
apply [a coat of paint]   passer *vb* une couche
apply a thin coat of paint prior to filling a surface   déguiser *vb*
apply preliminary rendering   renformir *vb*
apply stone pitching, riprap, etc   enrocher *vb*
appointment [e.g. for a meeting]   rendez-vous *m*
appointment [nomination, selection]   désignation *f*, nomination *f*
apportion [share, etc]   allouer *vb*

appraisal [e.g. of a project]   évaluation *f*
appraisal [independent, of work, materials, etc]   expertise *f*
appraisal [of value]   estimation *f*
appraisal, financial   bilan *m* financier
appraise [work, materials, etc, independently]   expertiser *vb*
appraiser   expert *m*
appreciation [gain in sale value]   plus-value *f*
appreciation [of value of property, etc]   amélioration *f*
apprentice   apprenti *m*
approach   approche *f*
approach [to land]   abord *m*
approachable   abordable *adj*
approval   agrément *m*
approval [e.g. of drawings, plans]   approbation *f*
approval [of work done]   réception *f*
approval, official [in accordance with standards, etc]   homologation *f*
approval, official [of drawings, plans, etc]   approbation *f*
approval, official [of products, etc]   agrément *m*
approval, sale on   vente *f* à l'essai
approval certificate   certificat *m* d'homologation
approve   agréer *vb*, approuver *vb*
approve [e.g. quality of workmanship]   sanctionner *vb*
approved   approuvé *adj*
approved control authority or body   organisme *m* agréé de contrôle
approximate   approximatif *adj*
approximately   environ *adv*
appurtenance   appartenance *f*
apron [airfield]   aire *f* de stationnement, aire *f* d'envol
apse [of a cathedral, church]   abside *f*
aquaplaning   aquaplanage *m*
aquastat   aquastat *m*
aqueduct   aqueduc *m*, canal *m* d'adduction
aqueous   aqueux *adj*
aquifer   aquifère *m*
arbitration   arbitrage *m*
arbitrator   arbitre *m*
arbour [grindstone, etc]   axe *m*
arc [curve, in welding, etc]   arc *m*
arc, electric   arc *m* électrique
arc length [in welding]   longueur *f* de l'arc
arc tube [lighting]   brûleur *m*
arcade   arcades *f,pl*
arcade, decorative, on the face of a wall   arcature *f*
arcade, decorative, separated from backing wall   arcature *f* en claire voie
arcade with shops on each side   passage *m*
arch   arc *m*, arche *f*
arch [of a tunnel]   cintre *m*
arch [of a vault]   arceau *m*
arch [vault]   voûte *f*
arch, asymmetrical   arc *m* rampant
arch, balanced   arc *m* équilibré
arch, basket   arc *m* anse de panier
arch, built-in [rotation restrained at springs]   arc *m* encastré

arch, convex   arc *m* en doucine
arch, depressed   arc *m* surbaissé
arch, distance between springs of   ouverture *f* d'un arc
arch, discharging   arc *m* de décharge
arch, flat   arc *m* droit, plate-bande *f*
arch, flat, with rounded corners   arc *m* déprimé
arch, gothic or pointed   ogive *f*
arch, hinged   arc *m* articulé
arch, horseshoe   arc *m* outrepassé
arch, inflected   arc *m* infléchi
arch, jack   voût(a)in *m*
arch, keel   arc *m* en accolade
arch, lance-head or lanciolate   arc *m* lanciolé
arch, lancet   arc *m* en tiers-point/en lancette
arch, multi-lobed   arc *m* polylobé
arch, ogee   arc *m* en accolade, arc *m* gothique prolongé
arch, ogive   arc *m* en ogive
arch, pointed   ogive *f*, arc *m* en ogive
arch, relieving   arc *m* de décharge
arch, relieving [of a bridge]   voûte *f* de décharge
arch, reversed [with key beneath springs]   arc *m* renversé
arch, romanesque   arc *m* roman
arch, round   arc *m* en plein cintre
arch, segmental   arc *m* bombé
arch, semi-circular   arc *m* roman/plein cintre
arch, semi-circular, with centre above springs   arc *m* plein cintre surhaussé
arch, semi-circular, with centre above imposts   arc *m* exhaussé
arch, shouldered   arc *m* épaulé
arch, small   voût(a)in *m*
arch, stilted   arc *m* surhaussé
arch, tented   arc *m* infléchi
arch, three-centred   arc *m* anse de panier
arch, three-pin   arc *m* à trois rotules
arch, tied   arc *m* à tirant
arch, transverse   arc *m* doubleau
arch, trefoil   arc *m* trilobé
arch, triangular   arc *m* angulaire, arc *m* brisé
arch, truncated triangular   arc *m* angulaire tronqué
arch, trussed   arc *m* à/en treillis
arch, tudor   arc *m* en carène, arc *m* gothique surbaissé, arc *m* tudor
arch, two-hinged   arc *m* à deux articulations
arch, voussoirs and keystone of an   bandage *m*
arch built into wall above a lintel to distribute loads   arc *m* de décharge
arch segment bearing horizontally on wall   crossette *f*
arch stone   claveau *m*, voussoir *m*
arch [form into an arch]   cintrer *vb*
arch, assemble the voussoirs and keystone of an   bander *vb*
arched   cintré *adj*, en arc, voûté *adj*
arched brickwork between steel beams   voûtain *m*
arching   effet *m* de voûte
arching [of a ceiling, etc]   voussure *f*
architect   architecte *m*
architect, consultant/consulting   architecte-conseil *m*
architect, landscape   paysagiste *m/f*

architect in charge of the works   maître *m* d'œuvre
architect of higher qualifications and experience   [France]   architecte *m* agréé
architect with basic qualifications [France]   architecte *m* diplômé
architect with diploma from the *École Spéciale d'Architecture*   architecte DESA (Diplômé de l'École Spéciale d'Architecture)
architect with state architectural school diploma   [France]   architecte DPLG (Diplômé par le gouvernement)
architect's diploma [in France]   diplôme *m* d'architecte
architect's instruction (AI) to contractor   ordre *m* de service
architects' council [in France]   conseil *m* des architectes
architectural   architectonique *adj*, architectural *adj*
architecturally treated   ouvragé *adj*
architecture   architecture *f*
architrave   architrave *f*, chambranle *m*, épistyle *m*
architrave, unmoulded   champ *m* saillant
archives   archives *f,pl*
archway   arcade *f*
archway consisting of multi-lobed arches   arcade *f* composée
area   aire *f*, région *f*, surface *f*, zone *f*
area [amount]   superficie *f*
area [of a town, e.g. commercial]   quartier *m*
area, aircraft parking   aire *f* de stationnement
area, bearing   surface *f* portante/d'appui
area, borrow   zone *f* d'emprunt
area, built-up   agglomération *f*, zone *f* bâtie
area, clean working   zone *f* de travail blanche
area, conservation [UK, approximate French equivalent]   site *m* classé
area, contact   surface *f* de contact
area, covered [roofed]   terrain *m* couvert
area, cross-section(al)   superficie *f* de section, aire *f* de la section, aire *f*/section *f* transversale
area, danger(ous)   zone *f* dangereuse
area, delivery   zone *f* de livraison
area, dining   coin *m* repas
area, drainage   aire *f* de drainage, zone *f* de drainage
area, duct   surface *f* de conduit
area, face   surface *f* frontale
area, floor   surface *f*, surface *f* couverte/de plancher/des étages
area, glazed   surface *f* vitrée
area, grate   surface *f* de grille
area, habitable   surface *f* habitable
area, handling   zone *f* de manutention
area, high/low pressure   région *f* de haute/basse pression
area, highway maintenance   centre *m* d'entretien routier
area, housing priority   zone *f* à urbaniser en priorité
area, kitchen–dining   coin *m* à manger
area, level   plate-forme *f*
area, light   clair *m*
area, loading   aire *f*/zone *f* de chargement
area, low pressure   région *f* de basse pression
area, machine working   emprise *f* au sol d'un engin

area, motorway maintenance   centre *m* d'entretien autoroutier

area, open   espace *m* découvert/ouvert

area, open [uncovered]   terrain *m* nu

area, parking   aire *f* de parking, aire *f*/parc *m*/zone *f* de stationnement, parking *m*

area, pedestrian   piazza *f*

area, prohibited   zone *f* interdite

area, planned development   zone *f* d'aménagement concerté (ZAC)

area, planned development, where the public authority has compulsory purchase powers   zone *f* d'aménagement différé (ZAD)

area, radiant   surface *f* de rayonnement

area, reception   aire *f* de réception

area, recreation   aire *f* récréative

area, residential   quartier *m* résidentiel

area, rest [roadside]   aire *f* de repos

area, road maintenance   centre *m* d'entretien routier

area, service [motorway]   aire *f* de service(s)

area, storage   aire *f* de stockage, zone *f* de stockage et d'entreposage

area, surface   superficie *f*, surface *f*

area, traffic   surface *f* de circulation

area, unloading   aire *f* de déchargement

area, urban   agglomération *f*

area, waiting, for pedestrians/cyclists   refuge *m* pour piétons/cyclistes

area, working [e.g. of a crane]   rayon *m* d'action

area below street level giving light to basement windows   cour *f* anglaise, saut-de-loup *m*

area of application   domaine *m* d'application/d'emploi

area of compression/tension reinforcement   section *f* d'acier comprimé/tendu

area of ground occupied, required or requisitioned   emprise *f* au sol

area of work site   emprise *f* du chantier

area with no building development   zone *f* non-aedificandi

areas, common [of buildings]   parties *f,pl* communes

areas, floor   surface *f* des sols

argillaceous   argilacé *adj*, argileux *adj*

argon   argon *m*

arid   aride *adj*

arithmetic   arithmétique *adj*

armature [of DC electric motor]   armature *f*

armature [of electric motor, dynamo]   induit *m*

armchair   fauteuil *m*

armoured [door, etc]   blindé *adj*

armouring [dam]   blindage *m* extérieur

arrange [lay out]   agencer *vb*, aménager *vb*

arrange [put in order]   arranger *vb*, ranger *vb*

arrange [a date, a meeting]   arranger *vb*, fixer *vb*, organiser *vb*

arrangement   agencement *m*, aménagement *m*, combinaison *f*

arrangement [decision]   arrangement *m*, décision *f*

arrangement, general [of a building]   ordonnance *f*

arrangements   dispositions *f,pl*, mesures *f,pl*

array   rangée *f*

array of figures/numbers   rangée *f* de chiffres

arrestor, emergency or safety [of lift or hoist]   parachute *m*

arris [of a beam, etc]   arête *f*

arris fillet [roof]   coyau *m*

arrow   flèche *f*

arrow, direction of the   sens *m* de la flèche

artery [traffic]   artère *f*

artesian   artésien *adj*

articulated   articulé *adj*

articulation [between two rigid members]   genouillère *f*

asbestos   amiante *m*, asbeste *m*

asbestos cord   cordon *m* d'amiante

asbestos removal   désamiantage *m*

asbestos sheeting, bitumen   feutre *m* bituminé chargé d'amiante

as-built drawings   dessins/plans *m,pl* de recollement

as-built face   parement *m* brut

as-built record of work subsequently covered up   attachement *m*

as-built records   dossier *m* des ouvrages exécutés (DOE)

ascending   montant *adj*

ascent   montée *f*

ash [timber, tree]   frêne *m*

ash, fly   cendres *f,pl* volantes

ash bin/box/pan   cendrier *m*

ash box door   porte *f* de cendrier

ash content   teneur *f* en cendres

ash pit   cendrier *m*, fosse *f* à cendres

ash removal   évacuation *f* des cendres

ash transport   transport *m* des cendres

ashes   cendres *f,pl*

ashlar   pierre *f* de taille

Asiatic closet (WC)   siège *m* à la turque

aspect [e.g. of a house]   orientation *f*

asphalt   asphalte *m*, bitume *m*

asphalt, ground or powdered natural rock   poudre *f* d'asphalte naturelle

asphalt, guss   béton *m* bitumineux coulé

asphalt, hot-rolled, with a tar binder   béton au bitume-goudron *m*

asphalt, mastic   asphalte *m* coulé, mastic *m* d'asphalte

asphalt, natural   asphalte *m* naturel

asphalt, sand   enrobé *m* fin, sable *m* enrobé

asphalt drum mix(ing) plant   tambour *m* sécheur enrobeur

asphalt finisher or paver   finisseur *m* à asphalte

asphalt plant   centrale *f* d'enrobage, poste *m* d'enrobage

asphalt plant, batch type   poste *m* d'enrobage à mélange discontinu

asphalt recycler   recycleur *m* d'asphalte

asphalt recycling plant   poste *m* de recyclage de chaussée bitumineuse

asphalt shingle   bardeau *m* d'asphalte

asphalt   bitumer *vb*

asphalter   asphalteur *m*

asphalting [the process]   bitumage *m*

as-rolled condition [steel]   état *m* brut de laminage

assemble   assembler *vb*, construire *vb*, monter *vb*

assemble a tongue-and-groove joint   affourcher *vb*

assemble two ladders in an inverted V   éventailler *vb*

assembling or assembly [the action]   assemblage *m*, jointement *m*, montage *m*

assembly [the result]   assemblage *m*

assembly [the whole]   ensemble *m*

assembly, bolted   assemblage *m* par boulons

assembly, guide vane   ensemble *m* d'aubes directrices

assembly, laminated   lamellation *f*

assembly, link-rod   assemblage *m* à biellette

assembly, rigid   assemblage *m* rigide

assembly, riveted   assemblage *m* rivé/à rivets, rivure *f*

assembly, shop   assemblage *m* à l'atelier, montage *m* en atelier

assembly, site   montage *m* sur chantier

assembly instructions   notice *f* d'assemblage/de montage

assembly line   chaîne *f* de fabrication

assembly sequence   ordre *m* de montage

assembly shop   atelier *m* de montage

assessing/assessment [numerically]   chiffrage *m*

assessment   bilan *m*, appréciation *f*

assessment [e.g. of a project]   évaluation *f*

assessment [e.g. of data]   appréciation *f*

asset, depreciable   immobilisation *f* amortissable

asset, fixed   immobilisation *f*

asset, intangible   immobilisation *f* incorporelle

asset, tangible   immobilisation *f* corporelle

assets   masse *f* active

assets [of a company]   actif *m*

assets [property]   biens *m,pl*

assets, capitalization of   immobilisation *f*

assets, circulating   actif *m* circulant

assets, fixed   actif *m* immobilisé

assets, liquid   liquidités *f,pl*

assets, net   actif *m* net

assets, net current   fonds *m* de roulement net

assignee   cessionnaire *f*

assignment [ceding]   cession *f*

assignment [task]   tâche *f*

assignment [of architect, engineer, etc]   mission *f*

assignment [of shares, patents, rights, etc]   transmission *f*

assignor   cédant *m*

assistance in the tender analysis   assistance *f* au dépouillement des offres

assistance with commissioning   assistance *f* aux réceptions

assistance with implementation   assistance *f* à l'exploitation

association   syndicat *m*

association of consulting engineers [France] [near equivalent of UK ACE]   Chambre *f* des Ingénieurs Conseils de France (CICF)

association of designing and consulting engineers [France] [generally those owned by contractors]   Chambre *f* Syndicale des Sociétés d'Études et de Conseils Technique (SYNTEC)

association, trade   association *f* professionnelle

assumption   hypothèse *f*

assurance, life, *see* insurance, life

astragal [of a column, facade, etc]   astragale *m*

asymmetric(al)   asymétrique *adj*

asymmetry   asymétrie *f*

asynchronous [e.g. motor]   asynchrone *adj*

at high level [on a drawing]   en plafond

at the edge of   en bordure de

at the same time   ensemble

at the top of [a building, a ladder, stairs]   en haut de

at the top of [a hill, mountain]   au sommet de

at work [e.g. forces]   en jeu

atmosphere   atmosphère *f*

atmosphere, standard   atmosphère *f* normale

atmospheric   atmosphérique *adj*

atomize [liquid]   atomiser *vb*, pulvériser *vb*

atomizer   atomiseur *m*, pulvérisateur *m*

atomizer [fuel jet in oil-fired boiler]   gicleur *m*

atrium   atrium *m*

attach   attacher *vb*

attachment [accessory]   accessoire *m*

attachment [for electric drill, etc]   adaptation *f*

attack [by corrosion, etc.]   attaque *f*

attempt   tentative *f*, effort *m*

attempt to do sth   tâcher *vb* de faire qch

attenuate   affaiblir *vb*, amortir *vb*

attenuation   affaiblissement *m*, atténuation *f*

attenuation [e.g. of oscillations]   amortissement *m*

attenuation, acoustic or sound   affaiblissement *m* acoustique/phonique

attenuation, crosstalk   affaiblissement *m* diaphonique

attenuator [sound]   silencieux *m*

Atterberg flow curve   courbe *f* de liquidité Atterberg

attic   combles *m,pl*, grenier *m*

attic [small, above and behind an entablature]   faux-attique *m*

attic [in a mansard roof]   mansarde *f*

attic [top storey, generally set back]   attique *m*

attic storey   étage *m* d'attique

attract [of magnets]   solliciter *vb*

attribute   caractéristique *f*

attrition   attrition *f*

auction, by   par voie d'adjudication *f*

auction, sale by   vente *f* aux enchères

auction room   salle *f* des ventes

auction sale   adjudication *f*, vente *f* aux enchères

audibility   audibilité *f*

audible   audible *adj*

audit   audit *m*, vérification *f*

audit [accounts]   reviser/réviser *vb*

audit, energy   audit *m* énergétique

auditor   contrôleur *m*

auditor, statutory   Commissaire *m* aux Comptes

auditorium   salle *f*

auger   tarière *f*

auger, carpenter's   roanne/rouanne *f*

auger, cutting tip of an   mouche *f*

auger, scotch screw   tarière *f* américaine

auger, screw   tarière *f* à vis
austenite   austénite *f*
authenticate   authentifier *vb*
authority   autorité *f*
authority [to do sth]   mandat *m*, pouvoir *m*
authorization (for or to do)   autorisation *f* (de)
authorization of payment   ordonnance *f* de paiement
authorize (someone to)   autoriser *vb* (qn à)
authorized control body   organisme *m* agréé de contrôle
auto-checked   autosurveillé *adj*
automatic   automatique *adj*
automatically   automatiquement *adv*
automatically closing switch [electrical]   conjoncteur *m*
auto-transformer   autotransformateur *m*
availability   disponibilité *f*
available   disponible *adj*
avalanche   avalanche *f*
average *adj*   moyen *adj*
average   moyenne *f*
average [numerical]   moyenne *f* numérique
award [e.g. a sum as damages]   allouer *vb*
award(ing) [of a contract]   adjudication *f*
awarding authority [for a construction contract]   maître *m* d'ouvrage
away from   hors de

awning   banne *f*, store *m*, vélum *m*
awning mounted on hoops   store *m* capote
axe   cognée *f*, hache *f*
axe [bricklayer's hammer]   marteau *m* de maçon
axe [small]   hachereau *m*
axe, felling   hache *f* de bûcheron
axial   axial *adj*, centré *adj*
axing off [of a lump, stone, etc]   piochement *m*
axis   axe *m*
axis, centroidal or gravity   ligne *f* des centres de gravité
axis, longitudinal   axe *m* longitudinal
axis, major/minor   grand/petit axe *m*
axis, neutral   axe *m* neutre
axis, reference   axe *m* de référence
axis, vertical   axe *m* vertical
axis of rotation   axe *m* de rotation
axis of symmetry   axe *m* de symétrie
axis of symmetry [architectural]   axe *m* de balancement
axle   arbre *m*, axe *m*, essieu *m*
axle, front/rear   essieu *m* avant/arrière
axonometric *adj*   axonométrique *adj*
axonometric (drawing)   dessin *m* axonométrique
axonometric perspective   perspective *f* axonométrique
azimuth   azimut *m*

bachelor's degree   licence *f*
back *adj*   arrière *adj*
back   dos *m*
back [e.g. of a building]   arrière *m*
back [of a tenon saw]   dosseret *m*
back [of a vehicle]   faire *vb* marche arrière
back, put or set   retarder *vb*
back to back   adossé *adj*
back to front   sens devant derrière
back to solid material [when dressing stone, plaster, etc]   à vif
back- or top-iron of a bench plane   contre-fer *m*
backacter   pelle *f* (équipée) en rétro
back-draught [in a furnace]   contre-tirage *m*, contrevent *m*
backfill   remblai *m*
backfill   remblayer *vb*
backfill, layers of   couches *f,pl* de remblai
backfilling [action of]   remblaiement *m*
backfilling [of underground quarries, mines]   bourrage *m*
backfilling, trench   remplissage *m* de tranchée
backfill(ing) material   matériau *m* de remplissage *m*
backfiller   remblayeuse *f*
background   arrière-plan *m*, fond *m*

background noise level   intensité *f* de bruit de fond
backhoe   pelle *f* (équipée) en rétro
backing (on to sth)   adossé *adj* (à qch)
backing (bar) [welding]   bande-support *f*
back-plate of a chimney hearth   contre-cœur *m*
back-up system   réseau *m* de réserve
backwards [back first]   à rebours
backwards [in time or space]   en arrière
backwards [of flow of gas, liquids]   à contre-courant
backwards movement   recul *m*
backyard   arrière-cour *f*
bacteria bed [in sewage works]   lit *m* bactérien
bacterial   bactérien *adj*
bacterial decay   altération *f* bactérienne
bactericidal   bactéricide *adj*
bactericide   bactéricide *m*
bacterium   bactérie *f*
badly   mal *adv*
baffle   chicane *f*
baffle, acoustic   baffle *m* acoustique
baffle(r) or baffle plate   déflecteur *m*
bag   sac *m*
bail (bond)   caution *f*
bailiff   huissier *m*

balance   balance *f*, équilibre *m*
balance [in an account]   solde *m*, résidu *m* de compte
balance, debit   solde *m* débiteur
balance, energy   bilan *m* énergétique
balance, spring-loaded   peson *m*
balance, temperature   équilibre *m* des températures
balance, thermal   équilibre *m* thermique
balance condition   condition *f* d'équilibre
balance lever [of bascule bridge]   arrière-bras *m*
balance point   point *m* d'équilibre
balance sheet   bilan *m*
balance tank   réservoir *m* d'équilibre
balance weight   contrepoids *m*
balance   équilibrer *vb*
balance [make symmetrical]   balancer *vb*
balanced   équilibré *adj*
balanced [as in balanced cantilever]   en porte-à-faux
balanced [of a wooden panel]   contrebalancé *adj*
balancing   compensateur *adj*
balancing   équilibrage *m*, stabilisation *f*
balancing [of forces, etc]   compensation *f*
balcony   balcon *m*
balcony [of a theatre]   galerie *f*
balcony, inset   loggia *m*
balcony, small, on columns   méniane *f*
balcony [guard] rail   garde-corps *m*, garde-fou *m*
ball mill   broyeur *m* à boulets
ballast [load, e.g. for pile test]   lest *m*
ballast [of road, railway track, or lamp]   ballast *m*
ballast [stone]   blocaille *f*
ballast bed [of railway track]   coffre *m*
ballast   ballaster *vb*
ballast [a roadway]   caillouter *vb*
ballcock   flotteur *m*, robinet *m* (à) flotteur
baluster   balustre *m*, barreau *m*
balustrade   appui *m*, balustrade *f*
balustrade, elbow height   accotoir *m*, accoudoir *m*
balustrade, handrail of   glisse-main *m*, main *f* courante
balustrade of a stair   rampe *f* (d'escalier)
band [of material]   bande *f*
bandsaw   scie *f* à ruban, scie *f* sans fin
bandsaw blade   lame *f* de scie à ruban
bandwidth   largeur *f* de bande
banister(s)   appui *m* d'escalier, balustres *m,pl*, rampe *f* d'escalier
banister post   balustre *m*, barreau *m*
banister rail   glisse-main *m*, main *f* courante
banister support bracket on side of a stair string   piton *m*
bank *adj*   bancaire
bank [financial]   banque *f*
bank [of a river, etc]   rive *f*, bord *m*
bank [of earth]   banquette *f*, levée *f* de terre
bank [regular and with even slope]   glacis *m*
bank [sloping ground]   talus *m*
bank [steep, of a river]   berge *f*
bank charges   agios *m*
bank deposit   dépôt *m* bancaire
bank rate   taux *m* officiel d'escompte

bank [road at a bend]   déverser *vb*, relever *vb*
bank (up)   remblayer *vb*
banked [of a road]   déversé *adj*
banking *adj*   bancaire *adj*
banking [of a road]   dévers *m*
banking [of ground]   retroussis *m*
bankrupt   failli *adj*
bankruptcy   faillite *f*
bankruptcy, file one's petition in   déposer *vb* son bilan
bar [door reinforcement]   bâcle *f*
bar [door]   verrou *m*
bar [metal, wood]   barre *f*
bar [small cross-section iron/steel]   tringle *f*
bar [unit of pressure]   bar *m*
bar, claw   pied-de-biche *m*
bar, deformed or ribbed [reinforcement]   acier *m* crénelé/barre *f* crénelée/haute adhérence
bar, deformed round reinforcement   rond *m* à béton crénelé
bar, glazing, *see* glazing bar
bar, merchant   fer *m* marchand
bar, notched   barreau *m* entaillé
bar, pinch   pied-de-biche *m*, pince *f*
bar, plain (steel)   acier *m* lisse
bar, pry   pied-de-biche *m*
bar, reinforcement, *see also* reinforcement bar(s)   barre *f* de renforcement
bar, ribbed (steel)   acier *m* nervuré
bar, round (steel)   rond *m* (en acier)
bar, short   barrette *f*
bar, small   barreau *m*
bar, square iron   fer *m* carré
bar, starter, *see* starter bar(s)
bar, test   barreau *m* d'essai
bar, wrecking   pied-de-biche *m*
bar bender [worker], *see* steelfixer
bar bending schedule   nomenclature *f* d'acier/des aciers
bar cropper   cisailleuse *f* (à aciers)
bar mill   broyeur *m* à barres
bars [window]   grille *f*
bar [a door]   barrer *vb*
bar-bending   cintrage *m* des aciers
barbed wire   fil *m* de fer barbelé, ronce *f*
barbican [of a fortress]   barbacane *f*
bare [e.g. strip a wire]   dénuder *vb*
bargain   affaire *f*
barium   barium *m*
bark   écorce *f*
barn   grange *m*
barn [open-sided]   hangar *m*
barograph   baromètre *m* enregistreur
barometer   baromètre *m*
barometer, aneroid   baromètre *m* anéroïde
barometer, recording   baromètre *m* enregistreur
barracks   caserne *f*
barrage   barrage *m*
barrel   fût *m*
barrel [for tar]   gonne *f*
barren [ground, etc]   aride *adj*

barrier   barrage *m*
barrier [e.g. level crossing]   barrière *f*
barrier, cavity   coupe-vapeur *m* dans un mur creux
barrier, crash or safety [highways]   glissière *f* de
   sécurité
barrier, double-sided crash or safety [highways]
   glissière *f* double de sécurité
barrier, flame   pare-flamme *m*
barrier, single-sided crash or safety [highways]
   glissière *f* simple de sécurité
barrier, steel beam crash or safety [highways]   glissière
   f métallique de sécurité
barrier, vapour   barrière *f* de vapeur, coupe-vapeur *m*,
   écran *m*/protection *f* pare-vapeur, pare-vapeur *m*
barrier, waterproof   écran *m* étanche
barrier, wind   barrière *f* au vent, coupe-vent *m*
barrier cream   crème *f* barrière/de protection
barrier paper, wind   pare-vent *m*
barrier rail [level crossing]   lisse *f*
barring [of a road]   barrage *m*
barrow for carrying wet concrete   brouette *f* à bascule
barytes   barytine *f*
BAS (building automation system)   automatisation *f* des
   systèmes des bâtiments
basal   de base
basalt   basalte *m*
base   base *f*
base [foundation]   fondation *f*
base [lowest part]   fond *m*
base [of a dam]   assise *f*
base [of an instrument, switch, chimney, etc]   embase *f*
base [structural]   embasement *m*, socle *m*
base, bulb/lamp   culot *m*
base, column or stanchion   base *f* de colonne, pied *m* de
   poteau
base, concrete   radier *m*/socle *m* en béton
base, embankment   pied *m* de remblai
base, insulating   socle *m* isolant
base, screw [for a lamp]   culot *m* à vis
base, spread, of a column   patte *f* d'éléphant
base, stone [of foundations]   couche *f* d'enrochements
base course   soubassement *m*
base course [of a road]   assise *f* de chaussée, couche *f* de
   base
base exchange process   méthode *f* à échange d'ions
base lending rate   taux *m* de base bancaire
base plate [of a door lock]   palastre/palâtre *m*
base unit [e.g. in a kitchen]   élément *m* bas
base (on)   fonder *vb* (sur)
base (on) [e.g. theory]   baser *vb* (sur)
basement   cave *f*, soubassement *m*, sous-sol *m*
basement course   soubassement *m*
basement parking   parking *m* en sous-sol
basement storey   étage *m* en sous-sol
baseplate   plaque *f* d'assise/de base, semelle *f*
baseplate [for scaffolding, a structural frame or wall]
   patin *m*
baseplate [of a machine]   embase *f*
baseplate [railway track]   selle *f* de rail

basic concept or idea   idée *f* de départ
basin   bassin *m*
basin [built in]   vasque *f*
basin [geographical]   cuvette *f*
basin, catchment   aire *f* de captage
basin, hand or wash   cuvette *f*, lavabo *m*, vasque *f*
basin, sedimentation   bassin *m* de décantation,
   décanteur *m*
basis   base *f*, fondement *m*
basis of design   base *f* de calcul
basket, siphon   panier *m* siphon
basket filter [in a siphon]   panier *m*
bastide [fortified town in SW France]   bastide *f*
batch [of concrete]   gachée *f*
batch [of materials, tools, etc]   lot *m*
batch [concrete]   doser *vb*
batcher, concrete   grande bétonnière *f*
batcher, weight [concrete]   doseur *m*
batching equipment   dispositif *m* de dosage *m*
batching of constituents [of concrete]   dosage *m* de
   constituants
batching plant [concrete]   centrale *f* à béton, usine *f* à
   béton/de bétonnage
bath   bain *m*
bath(tub)   baignoire *f*
bath, pickling   bain *m* de décapage
bath, shower, *see also* shower   baignoire-douche *f*
bath, whirlpool   baignoire *f* à brassage
bath drain   bouche *f* d'évacuation du bain
bath geyser [instantaneous water heater]   chauffe-bain *m*
bath water heater [instantaneous]   chauffe-bain *m*
bathroom   salle *f* de bain(s), les sanitaires *f,pl*
bathroom cabinet or cupboard   armoire *f* de toilette
bathroom suite or plumbing   sanitaires *f,pl*
batten   tasseau *m*
batten [for stage lights]   herse *f*
batten [slating or tiling]   liteau *m*, latte *f*
batten, chamfered, to support ends of cross joists
   lambourde *f* chanlattée
batten, chamfered, under metallic roofing   chanlatte *f*
batten, cover(ing)   couvre-joint *m*
batten, eaves [double thickness, of a tiled roof]
   chanlatte *f*
batten, slating/tiling   liteau *m*
batten fixed to principal rafter [to prevent purlins
   slipping]   chantignole *f*
batten to support wood strip/parquet flooring
   lambourde *f*
battens [of a roof]   lattis *m*
batter [excavations]   talus *m*
batter [of a stepped embankment]   recoupement *m*
batter [opposite of overhang]   fruit *m*
batter [an embankment, etc]   taluter *vb*
battery   pile *f*
battery [accumulator]   accumulateur *m*
battery [of a vehicle]   batterie *f*
battery [lighting]   rampe *f*
battery, dry   pile *f* sèche
battery, electric   batterie *f* électrique

battery, lithium pile *f* au lithium
battery, re-fill or replacement batterie *f*/pile *f* de rechange
battery charger chargeur *m*
battery solution électrolyte *m*
battlement créneau *m*
batwing [light fitting] luminaire *m* extensif
ba(u)lk of fir/spruce timber sapine *f*
bay [architectural] baie *f*
bay [of a bridge or building] travée *f*
bay [of a building] hall *m*
bay [shoreline] golfe *m*
bay, loading aire *f*/zone *f* de chargement
bayonet fitting or socket culot *m* à baïonnette, douille *f* à baïonnette
be out of plumb porter *vb* à faux, surplomber *vb*
be suitable for convenir *vb* à
beach plage *f*
beaconing balisage *m*
bead [concave profile] congé *m*
bead, angle [wood] baguette *f* d'angle
bead, cover(ing) couvre-joint *m*
bead, glazing parclause/parclose *f*
bead, square [wood] tringle *f*
beading [quarter or half round] baguette *f*
beam solive *f*, poutre *f*
beam [heavy timber, at least 60–80mm thick] madrier *m*
beam [of a highway safety barrier] élément *m* de glissement
beam [of light] faisceau *m*, rayon *m*
beam [small] poutrelle *f*
beam, angle poutre *f* cornière/équerre
beam, brace of a décharge *m* d'une poutre
beam, built-up poutre *f* composée
beam, ceiling doubleau *m*
beam, collar [timber roof] entrait *m*, entrait *m* retroussé, faux entrait *m*
beam, composite poutre *f* composée
beam, continuous poutre *f* continue
beam, corbel avant-solier *m*
beam, cross entretoise *f*
beam, cross [linking piles] chapeau *m*, longrine *f*
beam, edge poutre *f* de bordure/de rive
beam, elastically supported poutre *f* sur appuis élastiques
beam, foundation/ground longrine *f*
beam, head sommier *m*
beam, joist trimmer chevêtre *m*
beam, laminated poutre *f* laminée
beam, landing poutre *f* palière
beam, lifting palonnier *m*
beam, lintel, spanning between two posts or columns linçoir *m*, linceau *m*
beam, lintel [single or composite] poitrail *m*
beam, main poutre *f* maîtresse/principale
beam, parallel-flange poutrelle *f* à ailes parallèles
beam, precast or prefabricated poutre *f* préfabriquée
beam, ridge faîte *m*, madrier *m* de faîtage, poutre *f* de faîte

beam, ring ceinture *f*, poutre *f* ceinture
beam, ring [on top of or within a wall] chaînage *m*
beam, rolled steel poutrelle *f* laminée (en acier)
beam, roof poutre *f* de toiture
beam, roof [horizontal] entrait *m*
beam, simply supported poutre *f* sur appuis libres/simples
beam, spandrel poutre *f* de bordure/de rive
beam, tie- [timber roof] entrait *m*
beam, tie- [timber roof; sometimes double] moise *f*
beam, trussed poutre *f* à treillis
beam, underpinning [temporary support] tête *f* de chevalement
beam, upstand poutre *f* renversée
beam, wide flange poutrelle *f* à larges ailes
beam, windbracing poutre *f* de contreventement
beams, exposed poutraison *f*, poutres *f,pl* apparentes
beams, framework of poutrage *m*
beam between two columns or posts forming a lintel over an opening linçoir *m*, linceau *m*
beam depth or height hauteur *f* de poutre
beam grillage or system poutraison *f*
beam on grade longrine *f*
beam scales balance *f* à fléau
beam soffit dessous *m* de poutre
beam used as a ram in demolition work bélier *m*
beam with fixed ends poutre *f* encastrée
bear supporter *vb*
bear the costs of something faire *vb* les frais de qch
bearer [structural] support *m*
bearer [of a letter, message, etc] porteur *m*
bearer [of a card, document, etc] titulaire *m*
bearing portant *adj*
bearing [metal, of an arch bridge] balancier *m*
bearing [of a beam or column] repos *m*
bearing [rotating machine] palier *m*
bearing [structural] support *m*
bearing [support] appui *m*
bearing [swivel] tourillon *m*
bearing, ball coussinet *m*/palier *m*/roulement *m* à billes
bearing, built-in appui *m* encastré
bearing, central appui *m* médian
bearing, conical rouleau *m* conique
bearing, elastic appui *m* élastique
bearing, expansion appui *m* mobile
bearing, fixed appareil *m* d'appui fixe, appui *m* fixe
bearing, floating appui *m* flottant
bearing, free appareil *m* d'appui multidirectionnel
bearing, guided appareil *m* d'appui unidirectionnel
bearing, intermediate appui *m* intermédiaire
bearing, laminated (neoprene) elastomeric appareil *m* d'appui néoprène fretté
bearing, laminated rubber appareil *m* d'appui en élastomère fretté
bearing, needle roulement *m* à aiguilles
bearing, outboard palier *m* en porte-à-faux
bearing, pedestal palier *m* support auxiliaire
bearing, plain palier *m* lisse
bearing, point of point *m* d'appui

bearing, pot [rubber]   appareil *m* d'appui à pot
bearing, rocker   appui *m* à rotule
bearing, roller [machinery]   palier *m* à roulement, roulement *m* à rouleaux
bearing, roller [e.g. bridge support]   appui *m* à rouleau(x), appui *m* cylindrique
bearing, sleeve   coussinet *m*
bearing, sliding   appui *m* à glissement, appui *m* glissant
bearing, split bush   coussinet *m* à deux pièces
bearing, thrust   coussinet *m* palier, palier *m* de butée
bearing area or surface   surface *f* d'appui, surface *f* portante
bearing bush(ing)   coussinet *m*
bearing cap   chapeau *m* de palier
bearing capacity   capacité *f* portante
bearing capacity [e.g. of a pile]   force *f* portante
bearing capacity [per unit area]   portance *f*
bearing capacity index/value   indice *m* portant
bearing face, horizontal [course at top of wall]   arase *f*
bearing mechanism [e.g. for a bridge]   appareil *m* d'appui
bearing pedestal   chevalet *m*
bearing plate   plaque *f* d'appui
bearing support   support *m* de soutien
bearing surface or area   surface *f* d'appui, surface *f* portante
bearing surface [of a joist or beam]   tablette *f*
beat [earth, concrete, etc]   pilonner *vb*
beat [earth, for rammed earth construction]   piser *vb*
Beaufort scale   échelle *f* de Beaufort
bed [under footings, paving stones, etc]   formation *f*
bed [geological]   banc *m*
bed [of a road or railway]   infrastructure *f*
bed [of mortar, river, etc]   lit *m*
bed, bacteria [in sewage works]   lit *m* bactérien
bed, ballast [of railway track]   coffre *m*
bed, clinker   lit *m* de mâchefer
bed, river   lit *m* de fleuve/rivière
bed base [divan, etc]   sommier *m*
bed [e.g. a beam into a wall]   sceller *vb*
bed [foundations]   asseoir *vb*
beds, bunk   lits *m,pl* superposés
bed down [e.g. a foundation]   prendre *vb* son assiette
bed down [earth, ground, etc]   se tasser *vb*
bedded rubble stone [needing only light tooling]   moellon *m* grisant
bed(ding) [for paving stones, tiling, etc]   forme *f*
bedding [of a pipe, etc]   enrobage *m*
bedding, concrete   enrobage *m* bétonné
bedding, stone   hérisson *m*
bedding and jointing in a single operation   joint *m* fait en montant
bedding concrete   béton *m* de propreté
bedding of parquet backing strips   auget *m*
bedding [of sand, under paving, etc]   couchis *m*
bedding plane   plan *m* de stratification
bedding with putty [of glazing]   contre-mastiquage *m*
bedplate [of a machine]   embase *f*
bedplate, motor   socle *m* de moteur
bedrock   assise *f* rocheuse, roche *f* saine, soubassement *m*, substratum *m* rocheux

bedroom   chambre *f*
beech [timber, tree]   hêtre *m*
beetle, wood-boring   vrillette *f*
beetle head [pile driver]   marteau *m* dame, mouton *m* à déclic, singe *m*
begin [civil works]   amorcer *vb*
beginning [e.g. of road or tunnel construction]   amorce *f*
behaviour   comportement *m*
behaviour in fire   tenue *f* au feu
bel (10 decibels)   bel *m*
belfry   beffroi *m*, clocher *m*
Belgian national railways   Société *f* Nationale des Chemins de Fer Belges (SNCB)
bell [large]   cloche *f*
bell [small, generally electric]   sonnerie *f*
bell, alarm   sonnerie *f*/sonnette *f* d'alarme
bell, door   carillon *m* de porte
bell, electric   sonnerie *f* électrique
bellmouth entry or intake   trompe *f* d'entrée
bellows   soufflet *m*
bellows, expansion   compensateur *m*/soufflet *m* de dilatation, dispositif *m* de compensation *f* de dilatation
bell-push   bouton *m* de carillon (de porte)/de sonnerie/de sonnette
below   dessous *adv*
belt [drive belt]   courroie *f*
belt [of hills]   ceinture *f*
belt [of land]   bande *f*
belt, conveyor, *see* conveyor belt, belt conveyor
belt, fan   courroie *f* de ventilateur
belt, running   courroie *f* courante
belt, V-/vee-   courroie *f* trapézoïdale
belt conveyor   transporteur *m* à bande/courroie/tapis
belt (conveyor), feed   tapis *m* alimenteur
belt conveyor/elevator, mobile/portable   élévateur-transporteur *m*, sauterelle *f*
belt feeder   alimentateur *m* à courroie
belt loader   chargeuse *f* à bande
belt weigher   tapis *m* peseur
belvedere   belvédère *f*
bench [in cut or fill]   banquette *f*, berme *f*
bench [work]   établi *m*
bench, collapsible [e.g. Workmate]   établi *m* étau
bench, small joiner's   bidet *m*
bench, woodworker's   établi *m* de menuisier
bench mark   point *m* fixe, repère *m* de niveau
bench stop   griffe *f* d'établi
bench top   plateau *m* d'établi
benching, concrete [of a sewer]   cunette *f* en béton
bend   courbe *f*, courbure *f*
bend [of a road]   contour *m*, coude *m*, courbe *f*, tournant *m*, virage *m*
bend [of a watercourse]   coude *m*, détour *m*, méandre *m*
bend [of an arch]   voussure *f*
bend [of a bar, pipe- or ductwork, etc]   coude *m*
bend [deviation of a straight beam]   jarret *m*
bend, double [of a road]   virage *m* en S
bend, expansion   arc *m* compensateur, coude *m* de dilatation

bend, pipe   coude *m* de tube
bend, return   coude *m* en U
bend, S-   coude *m* de renvoi
bend, smooth   coude *m* lisse
bend, socket and spigot (>90°)   coude *m* à emboîtement et cordon
bend, socket and spigot (90°)   coude *m* au quart à emboîtement et cordon
bend, U-   courbe *f* de retour
bend, U- [pipework]   coude *m*/tube *m* en U
bend, welded   coude *m* soudé
bend [fold]   plier *vb*
bend [make curved, a beam, pipe, road, etc]   courber *vb*
bend [physically, a bar, pipe, etc]   cintrer *vb*
bend [of a beam, under load]   fléchir *vb*
bend [take on a bent shape]   se couder *vb*
bend sth at right angles   couder *vb*
bend sth back   recourber *vb*, replier *vb*
bend down [yield]   s'affaisser *vb*
bend sth down [by loading]   affaisser *vb*
bendable   cintrable *adj*
bender, bar [tool]   cintreuse *f* (à aciers), clef *f* à griffe
bender, pipe [tool]   cintreuse *f* de tube
bending [of pipes, etc]   cintrage *m*
bending [under axial load]   flexion *f*
bending [under transverse load]   fléchissement *m*
bending, bi-axial   flexion *f* déviée
bending, compound   flexion *f* composée
bending, simple   flexion *f* simple
bending machine   machine *f* à cintrer
bending moment   moment *m* de flexion, moment *m* fléchissant
bending radius   rayon *m* de courbure
bending schedule [reinforcement]   nomenclature *f* d'acier/des aciers, table *f* à couder
bending tool [reinforcement]   griffe *f* à couder
beneath   sous *adv*
bent   coudé *adj*, courbé *adj*
bent [frame]   palée *f*
bentonite   bentonite *f*
bentonitic   bentonitique *adj*
bent-up bar [reinforcement]   barre *f* relevée
bent(wood)   bois *m* courbe
Berlin wall [strutting system for deep excavations]   paroi *f* berlinoise
berm   banquette *f*, berme *f*
best [tender]   mieux-disant *adj*
berth [port]   poste *m* d'accostage/d'amarrage
berthing quay   quai *m* d'amarrage
betterment   amélioration *f*
between inside faces [of measurements]   dans œuvre (D/O)
bevel *adj*   en biais *adj*
bevel   biseau *m*, chanfrein *m*
bevel, sliding/square   angle *m* oblique, chanterelle *f*, fausse équerre *f*, sauterelle *f*
bevel   biseauter *vb*, chanfreiner *vb*
bevelled   biseauté *adj*, chanfreiné *adj*
bi-axial   biaxial *adj*

bi-coloured   bicolore *adj*
bicycle   cycle *m*
bid [e.g. at auction]   enchère *f*
bid [tender]   offre *f*, soumission *f*
bid package   lot *m*, tranche *f* de travaux
bid (for)   soumissionner *vb*
bid, submit a   faire *vb* une soumission, soumissioner *vb*
bidder, successful   adjudicataire *m*, bénéficiaire *m* de l'adjudication, entreprise *f* adjudicataire, titulaire *m* du marché
bidder, successful [at auction]   adjudicataire *m*
bidet   bidet *m*
bight [in a shore]   anse *f*
bill [invoice]   facture *f*
bill of exchange   lettre *f* de change
bill of quantities (BOQ or BQ)   détail/document *m* quantitatif estimatif (DQE), devis *m* quantitatif, tableau *m* des quantités des matériaux
bill of sale   acte *f* de vente
bill   facturer *vb*
billet [of timber or metal]   bille *f*
billing   facturation *f*
bimetallic   bimétallique *adj*
bimetallic strip   bilame *f*
bin   bac *m*
bin [bunker]   trémie *f*
bin, waste   poubelle *f*
binary   binaire *adj*
bind oneself [to do something]   s'engager *vb*
bind [of machines, etc]   se coincer *vb*
bind [to prevent splitting or swelling]   fretter *vb*
binder [concrete reinforcement]   étrier *m*
binder [fine sand or crushed stone]   agrégation *f*
binder [in concrete or road materials]   liant *m*
binder [road material]   enrobé *m*
binder [timber roof]   entrait *m*
binder, air-setting [aerobic]   liant *m* aérien
binder, bitumen   binder *m*, enrobé *m* bitumineux, liant *m* hydrocarboné
binder, dense bitumen   enrobé *m* dense au bitume
binder, hydraulic   liant *m* hydraulique
binder, hydrocarbon based   liant *m* hydrocarboné
binder, tar-based   binder *m*
binder, tar-bitumen   enrobé *m* au bitume-goudron
binding [around something]   frettage *m*
binding [e.g. of road material]   agrégation *f*
binding [ring or spiral, to restrain splitting or swelling]   frette *f*
binding material [road or groundworks]   agglomérant *m*
binding piece [timber structure]   moise *f*
biochemical action   effet *m* biochimique
biological oxygen demand (BOD)   demande *f* biochimique en oxygène (DBO)
birch [timber or tree]   bouleau *m*
bird guard [grille]   grille *f* anti-volatile
bit [for ground/rock drilling]   taillant *m*, trépan *m*
bit [of a pneumatic hammer]   fleuret *m*
bit, auger-type   taillant *m* en tarière

bit, centre   mèche *f* anglaise
bit, diamond   couronne *f* à diamants
bit, drill   mèche *f*
bit, drilling   couronne *f*
bit, flat [for a screwdriver]   pointe *f* plate
bit, hard metal   couronne *f*/trépan *m* au métal dur
bit, plain [for a screwdriver]   pointe *f* plate
bit, rock drill   trépan *m*
bit, roller   trépan *m* à molette
bit, spoon   mèche *f* à cuiller, tarière *f* cuiller
bit, tungsten (tipped)   taillant *m* au tungstène
bit, twist drill   foret *m* (hélicoïdal)
bitumen   bitume *m*
bitumen, asphaltic [bitumen + fine clay or sand]   bitume *m* asphaltique
bitumen, cut-back   bitume *m* fluidifié
bitumen binder   binder *m*
bitumen emulsion   émulsion *m* de bitume
bitumen(-based) emulsion [tack coat]   émulsion *m* d'accrochage
bitumen paper   papier *m* bitumé
bitumen storage system   centrale *f* à bitume
bitumen tank with spraying equipment   citerne *f* à bitume avec rampe de répandage, citerne *f* thermique répandeuse de liants bitumineux
bitumen-tar asphaltic concrete   bitume-goudron *m* (BG)
bituminous   bitumineux *adj*
black   noir *adj*
black body   corps *m* noir
blacksmith's (hand) hammer   marteau *m* de forgeron
blade   pale *f*
blade [knife, saw]   lame *f*
blade [of a fan, turbine]   ailette *f*
blade [of a hand plane]   fer *m*
blade [of a turbine]   aube *f*
blade [of electrical switch]   couteau *m*
blade, bandsaw   lame *f* de scie à ruban
blade, circular saw   lame *f* de scie circulaire
blade, fan or impeller   ailette *f* de ventilateur
blade, saw   lame *f* de scie
blade, venetian blind   lame *f* de persienne
blade guard [circular saw]   chapeau *m* de scie, coiffe *f*, protège-lame *m*
blading   aubage *m*
blank [without openings, as in blank wall]   borgne *adj*
blanket, impermeable [groundworks]   tapis *m* imperméable
blanket [with earth]   recouvrir *vb* de terre
blast   faire *vb* sauter à l'explosif
blast [rock]   faire *vb* sauter
blaster   artificier *m*, boutefeu *m*
blastfurnace   haut-fourneau *m*
blastfurnace slag, crushed   laitier *m* de haut-fourneau concassé
blastfurnace slag, expanded   laitier *m* expansé, mousse *f* de laitier
blasting [ground working]   abattage *m* à l'explosif
blasting, bog   déplacement *m* des marais à l'aide d'explosifs

blasting, liquid oxygen   tir *m* à l'air/l'oxygène liquide
blasting cap   amorce *f*
blasting equipment   engin *m* de mise à feu
blasting shot   coup *m* de mine
bleach   eau *f* de Javel
bleed   soutirer *vb*
bleed(ing) [e.g. of brakes]   purge *f*
bleeding [water to surface of concrete, or one colour through another]   exsudation *f*
bleed-through   ressuage *m*
blind [of a blocked opening]   aveugle *adj*
blind [shade]   store *m*
blind, roller   store *m* à rouleaux
blind, shop   banne *f*
blind, venetian   store *m* vénitien
blind [in road construction]   caler *vb*
blind [with sand]   ensabler *vb*
blinding (layer)   couche *f* de propreté
blinding concrete   béton *m* de propreté
blinking [light]   clignotant *adj*
blister [on casting]   bulle *f*
blister [on paint]   boursouflure *f*
blistering [of paint]   cloquage *m*
block   bloc *m*
block [fixed to a principal rafter to prevent a purlin slipping]   échantignole *f*
block [piece added to raise something]   hausse *f*
block [pulley]   poulie *f*
block, anchor   massif *m* d'ancrage
block, anchor [pipe bend]   butée *f* de coude
block, breeze or precast concrete   parpaing *m*
block, building, *see* building block
block, chimney or flue   boisseau *m*
block, foundation [e.g. for a pole or post]   bloc *m* de fondation
block, hollow   parpaing *m* creux, bloc *m* perforé
block, hollow clay   bloc *m* creux (de terre cuite), brique *f* creuse/tubulaire
block, hollow (concrete)   aggloméré *m* (de béton), bloc *m* creux (de béton de ciment), corps *m* creux
block, office   immeuble *m* de bureaux
block, pulley   palan *m*
block, residential   immeuble *m* d'habitation
block, stone   moellon *m*
block, terminal   bloc *m* de jonction
blocks, hollow clay   hourdis *m*
block and tackle   moufle *f*
block of buildings   ensemble *m* de bâtiments
block of flats   immeuble *m* d'appartements
block of flats, high-rise   tour *f* d'habitation
block of stone   moellon *m*
block paving, concrete   pavé *m* de béton
block   bloquer *vb*, obstruer *vb*
block [a hole]   boucher *vb*
block [a road]   barrer *vb*
blockage   obstruction *f*
blockboard   latté *m*, panneau *m* latté/à âme lattée
blocked [a duct, a pipe, an aperture]   obturé *adj*, obstrué *adj*

blocked [a road]   barré *adj*, obstrué *adj*
blocked [openings, holes]   bouché *adj*
blocked up, get or become   s'obstruer *vb*
blocking   engorgement *m*
blocking [of a duct, a pipe, an aperture]   obturation *f*
blockwork, hollow clay   hourdis *m* (creux)
blondin   téléphérique *f*, transporteur *m* aérien à câble
blow [knock]   coup *m*
blow   souffler *vb*
blow [a fuse, a circuit]   griller *vb*
blow a fuse   fondre *vb* un fusible
blow off, out or through   purger *vb*
blow out/down [water from a boiler]   évacuer *vb*
blow up [inflate]   gonfler *vb*
blow up [with explosives]   exploser *vb*, faire *vb* sauter
blow-down   purge *f* de déconcentration
blow-down pit   fosse *f* de vidange
blow-down tank   réservoir *m* de décompression
blower   compresseur *m*, soufflante *f*, soufflerie *f*,
  ventilateur *m*
blower, gas   soufflerie *f* à gaz
blowhole [casting defect]   soufflure *f*
blowholes on the formed surface of concrete   bullage *m*
blowlamp   chalumeau *m*, lampe *f* à braser
blow-out indicator [electrical]   voyant *m* de fusion
blow-out port   orifice *m* d'échappement
blowpipe   chalumeau *m*
blowtorch   torche *f* à souder
blow-valve [on a boiler]   reniflard *m*
blue   bleu *adj*
blue, cobalt   bleu *m* de cobalt
blue, Prussian   bleu *m* de Prusse
blueprint   bleu† *m*, photocalque *m* bleu
bluestain [timber defect]   bleuissement *m*
board, *see also* plank, timber, wood   bord *m*
board [10–13mm thick, for door panels, etc]   feuillet *m*
board [30–60mm thick]   ais *m*
board [for horizontal cladding]   clin *m*
board [for notices, etc]   tableau *m*
board [plank less than 54mm thick]   planche *f*
board, barge   bordure *f* du pignon
board, cellular (hollow) core   panneau *m* à âme creuse
board, cladding [timber]   bardeau *m*
board, drawing   planche *f* à dessin(er), table *f* à
  dessin
board, eaves [roof]   coyau *m*
board, fascia   volige *f*
board, fibre-cement   plaque *f* de fibres-ciment
board, flat-sawn   bord *m* débité sur dosse
board, floor   parquet *m*, plancher *m*
board, guard   garde-pieds *m*, planche *f* de garde, latte *f*
  de garde, garde-gravois *m*
board, gypsum or plaster-   panneau *m*/plaque *f* de plâtre,
  Placoplâtre *m* [trade name]
board, insulating or insulation   panneau *m* isolant
board, ironing   table *f* à repasser
board, lower/upper edge [excavations]   règle *f* de niveau
  inférieur/supérieur
board, luffer   abat-son *m*

board, notice   tableau *m* d'annonces
board, particle   aggloméré *m*, panneau *m* de particules
board, perforated ceiling   panneau *m* perforé pour
  plafond
board, profile [in excavations]   chevalet *m* pour tirer au
  cordeau
board, quarter-sawn [normally 130–150×40–60mm]
  merrain *m*
board, ridge   panne *f* faîtière
board, ribbon [timber frame structure]   sommier *m*
board, roof sheathing   bardeau *m*, volige *f*
board, sarking   volige *f*
board, scaffolding   planche *f* d'échafaudage
board, scaffolding guard   latte *f*/planche *f* de garde
board, square-edged   planche *f* avivée/équarrie/à arêtes
  vives
board, terminal   plaque *f* à/de bornes
board, trimmed   planche *f* avivée/équarrie/à arêtes
  vives
board, untrimmed   planche *f* en grume/non équarrie
board, waney-edged   planche *f* en grume/non équarrie
boards, scaffolding   platelage *m*
board of directors   conseil *m* d'administration
board platform   plate-forme *f* de madriers, platelage *m*
boarding   planchéiage *m*
boatswain's/bos'n's chair   sellette *f*
bodge up†   bricoler *vb*
body [of a lamp, etc]   corps *m*
body [of a structure]   massif *m*
body [of a centrifugal pump]   coquille *f*
body, pump   corps *m* de pompe
body harness [for working at height]   harnais *m* de
  sécurité
body of the dam   corps *m* du barrage
body of water   masse *f* d'eau
bogie [of a crane on rails]   châssis *m* de translation
boil   bouillir *vb*
boiler   chaudière *f*
boiler, alternative fuel   chaudière *f* mixte
boiler, auxiliary   chaudière *f* auxiliaire
boiler, central   chaufferie *f* centrale
boiler, coke   chaudière *f* à coke
boiler, combination   chaudière *f* mixte
boiler, convertible   chaudière *f* convertible
boiler, cross-tube   chaudière *f* à bouilleurs croisés
boiler, fire tube   chaudière *f* à tube de
  fumée/multitubulaire
boiler, forced circulation   chaudière *f* à circulation d'eau
  forcée
boiler, gas-fired   chaudière *f* à gaz
boiler, high capacity   chaudière *f* à grande capacité
boiler, high pressure   chaudière *f* (à) haute pression
boiler, hot water (heating)   chaudière *f* à eau chaude
boiler, inclined tube   chaudière *f* à tubes inclinés
boiler, low-pressure   chaudière *f* (à) basse pression
boiler, low-pressure steam   chaudière *f* à vapeur à basse
  pression
boiler, multi-pass   chaudière *f* à plusieurs parcours
boiler, multi-tubular   chaudière *f* multitubulaire

boiler, oil-fired   chaudière *f* à fuel
boiler, packaged   chaudière *f* équipée
boiler, pressure-fired   chaudière *f* à foyer surpressé
boiler, Scotch   chaudière *f* à tubes de fumée
boiler, standby   chaudière *f* de réserve
boiler, steam [steam generator]   générateur *m* de vapeur
boiler, steam (high/low-pressure)   chaudière *f* à vapeur
(à haute/basse pression)
boiler, vertical tube   chaudière *f* à tubes-foyers verticaux
boiler, waste heat   chaudière *f* à/de récupération
boiler, water tube   chaudière *f* à tubes d'eau
boilers   chaudronnerie *f*
boiler casing   enveloppe *f* de chaudière
boiler feed   alimentation *f* de chaudière
boiler feedwater heater   réchauffeur *m* d'eau
d'alimentation de chaudière
boiler fittings   raccorderie *f* de chaudière
boiler heating surface   surface *f* de chauffe d'une
chaudière
boiler house   bâtiment *m* des chaudières, chaufferie *f*
boiler inspection   surveillance *f* des chaudières
boiler mounting   socle *m* de chaudière
boiler output or rating   puissance *f* de chaudière
boiler plant   chaufferie *f*, installation *f* de chaufferie
boiler plant, central   centrale *f* de chauffage
boiler room   chaufferie *f*
boiler scale   incrustation *f*/tartre *f* de chaudière, dépôt *m*
d'une chaudière
boiler shell   mur *m* de chaudière
boiler water temperature control   contrôle *m* de
température de chaudière
boiler water treatment   traitement *m* des eaux de
chaudière
boilermaker or boilermaking plant   chaudronnerie *f*
boiling   bouillant *adj*
boiling   ébullition *f*
boiling point   point *m* d'ébullition
bole [of a tree]   fût *m*
bollard   bitte *f* d'amarrage, bollard *m*
bolometer   bolomètre *m*
bolster [chisel]   ciseau *m* de briqueteur
bolster [on top of posts or props]   chapeau *m*
bolt   boulon *m*
bolt [cupboard, door, gate]   verrou *m*
bolt [of a lock or door catch]   pêne *m*
bolt, anchor, holding down or rock   boulon *m* d'ancrage
bolt, bevelled-end [of a lock, latch]   pêne *m* biseauté
bolt, black   boulon *m* brut
bolt, casement   crémone *f*
bolt, coach   tire-fond *m*
bolt, Cremona   crémone *f*
bolt, dead   dormant *m*, pêne *m* dormant
bolt, erection   boulon *m* de montage
bolt, espagnolette   espagnolette *f*
bolt, expansion/rag   boulon *m* de scellement
bolt, eye-   boulon *m* d'œil, vis *f* à œillet
bolt, fixing/securing   boulon *m* de fixation
bolt, hexagon head   boulon *m* à tête hexagonale
bolt, high-strength   boulon *m* à haute résistance

bolt, holding down   boulon *m* d'ancrage/de fixation
bolt, rag   boulon *m* barbelé/de scellement
bolt, rag [used for lifting a piece of stone]   louve *f*
bolt, retaining   boulon *m* de retenue
bolt, rock   boulon *m* d'ancrage
bolt, round-head   boulon *m* à tête ronde
bolt, sash   birloir *m*, targette *f*
bolt, screw   boulon *m* fileté/à écrou
bolt, self-locking [operating without a key]   pêne *m* à
demi-tour
bolt, shutter   espagnolette *f*
bolt, sliding   verrou *m* glissant
bolt, small flat [for gate, etc]   targette *f*
bolt, spring   verrou *m* à ressort
bolt, stud   goujon *m*, prisonnier *m*, tige *f* filetée
bolt, threaded shank of a   tige *f* filetée
bolt cropper or cutter   coupe-boulons *m*
bolt head   tête *f* de boulon
bolt hole circle   cercle *m* de perçage
bolt staple   verterelle/vertevelle *f*
bolt [door, window]   mettre *vb* les verrous, verrouiller *vb*
bolt [with bolts]   boulonner *vb*
bolt down   boulonner *vb*
bolt-cutters   cisaille *f* américaine/coupe-boulons
bolted   boulonné *adj*
bolting   boulonnage *m*
Boltzmann constant   constante *f* de Boltzmann
bond [adhesion]   adhérence *f*
bond [brickwork or masonry]   appareil *m*
bond [debenture]   obligation *f*
bond [in a contract]   cautionnement *m*
bond [link]   liaison *f*, lien *m*
bond, anchorage [in concrete, masonry, etc]   scellement
m
bond, bid/tender   caution *f* de soumission
bond, chimney   appareil *m* de cheminée
bond, common   appareil *m* à assises réglées/en
panneresses
bond, completion   cautionnement *m*, garantie *f* de bonne
fin
bond, English   appareil *m* alterné simple, appareil *m*
anglais
bond, English cross   appareil *m* croisé
bond, Flemish   appareil *m* flamand, losange *m* flamand
bond, header or heading   appareil *m* en boutisses
bond, masonry   appareil *m* de construction
bond, performance   cautionnement *m*, garantie *f* de
bonne fin
bond, quoin [at corners of stonework]   appareil *m* en
besace
bond, stretcher or stretching   appareil *m* à assises
réglées/en panneresses
bond strength   force *f* de liaison
bonderize   bondériser *vb*
bonding [appearance, of brickwork or masonry]
appareillage *m*
bonding [link, connection]   liaison *f*
bonding [together, of masonry, tiling]   jointement *m*
bone dry   absolument sec *adj*

bonnet [chimney]   capot *m*
book-keeper   comptable *m*
book-keeping   tenue *f* des livres
booklet   brochure *f*
boom [crane]   flèche *f*
boom, lattice   flèche *f* en treillis
boomheader   abatteuse *f*
boost   allure *f* poussée
booster, pressure   surpresseur *m*
booster, voltage   survolteur *m*
boot, ankle   demi-botte *f*
boot, anklet   brodequin *m*
boot, half   demi-botte *f*
boot, safety   botte *f* de sécurité, chaussure *f* (avec coquille) de sécurité
boot scraper [metal]   décrottoir *m*
border   bord *m*
border [of a map, plan]   cadre *m*
border [boundary, of an area]   limite *f*
border   border *vb*
border on   côtoyer *vb*
bordered by   bordé de
bordering [e.g. on a road]   riverain *adj*
borderline case   cas *m* limite
bore [ground investigation]   forage *m*, sondage *m*, trou *m* de forage/sondage/sonde
bore [tunnel, pile]   creuser *vb*, forer *vb*
bore [hole]   forer *vb*
bore a hole through   percer *vb*
bore (with a gimlet)   vriller *vb*
bore-dust or -hole [of wood-boring insects]   vermoulure *f*
borehole [bored, drilled]   forage *m*, sondage *m*, trou *m* de forage/sondage/sonde
borehole, cased or lined   forage *m* tubé
borehole cover   tampon *m* de forage
borehole investigation   sondage *m* du sol
borehole sounding   sondage *m* par puits
borer, bolt-hole [for timber]   boulonnière *f*
boring [a hole]   perçage *m*
boring [a tunnel, pile, hole]   forage *m*
boring [a well or shaft]   fonçage *m*
boring [into the ground]   puits *m* de sondage
boring [with a drill]   percement *m*
boring (out)   alésage *m*
boring, trial   forage *m* d'essai, sondage *m* par puits/du sol
boring, tube sample   forage *m* au tube carottier
boring, wash-out   forage *m* par délavage
boring bit   couronne *f* de forage
boring machine   foreuse *f*
boring machine, tunnel   taupe *f*, tunnelier *m*
boring through [a material]   perçage *m*
boring tool for peg holes in timber structures   laceret *m*
borings, make   sonder *vb*
borrow   emprunter *vb*
borrow material   matériau *m* d'emprunt
borrow (of material)   emprunt *m* (de matériau)
borrow pit   puits *m* d'emprunt
borrow zone   zone *f* d'emprunt

borrower   emprunteur *m*
boss [e.g. of a fan, impeller]   moyeu *m*
bottle   bouteille *f*
bottle trap   siphon *m* à culotte démontable
bottom [base, foundation]   base *f*, socle *m*
bottom [lowest part]   bas *m*
bottom [under side]   dessous *m*
bottom [valley, swimming pool, well, etc]   fond *m*
bottom, trench   fond *m* de fouille
boulder   rocher *m*
boundary   ligne *f* de démarcation
boundary [of a garden, plot, site, etc]   limite *f*
boundary, catchment   ligne *f* de faîte/de partage des eaux
boundary, plot or site   limite *f* du terrain
boundary alignment   alignement *m*
boundary between, set up a   établir *vb* une séparation entre
boundary conditions   conditions *f,pl* aux limites
boundary layer   couche *f* limite
boundary layer thickness   épaisseur *f* de couche limite
boundary marker   témoin *m*, borne *f*
boundary markers, install   borner *vb*
boundary stone   pierre *f* de bornage
bounds   cadre *m*
Bourdon gauge or tube   tube *m* de Bourdon
bow [of timber]   voilement *m* longitudinal de face
bowl   cuvette *f*
box   boîte *f*, caisson *m*
box, air distribution   boîtier *m* de diffusion d'air
box, cardboard   carton *m*
box, conduit   boîte *f* de jonction
box, connecting   boîte *f* de jonction
box, cut-out [electrical]   coffret *m* de coupe-circuit
box, distribution   boîte *f* de branchement
box, distribution [electrical]   coffret *m* de distribution
box, junction   boîte *f* de dérivation/jonction
box, mitre   boîte *f* à coupe(s), boîte *f* à onglets/d'onglet
box, mixing   boîte *f* de mélange
box, mixing [in an air handling unit]   caisson *m* de mélange
box out   réservation *f*
boxroom   débarras *m*
Boyle's law   loi *f* de Mariotte
brace   jambe *f* de force
brace [diagonal]   lien *m*, lierne *f*
brace [of brace and bit]   fût *m*
brace [on a boarded shutter or door]   écharpe *f*
brace [prop]   étai *m* stabilisateur
brace [roof]   contre-fiche *f*
brace [steel roof]   bielle *f*
brace, angle or corner   aisselier *m*, entretoise *f*, étrésillon *m*, lien *m* d'angle
brace, corner   moise *f*
brace, cross   croisillon *m*
brace, diagonal   entretoise *f*, étrésillon *m*, lien *m* d'angle, lierne *f*, moise *f* en écharpe
brace, ratchet   vilebrequin *m* à cliquet
brace-and-bit   vilebrequin *m*

brace of a beam　décharge *m* d'une poutre
brace　ancrer *vb*, entretoiser *vb*, étayer *vb*
brace [a wall]　amarrer *vb*
brace [diagonally]　étrésillonner *vb*
brace [with guys]　hauban(n)er *vb*
bracing　entretoisement *m*, étaiement *m*
bracing [with guys]　hauban(n)age *m*
bracing, diagonal　étrésillonnage *m*
bracing, diagonal [the action]　étrésillonnement *m*
bracing, wind　contreventement *m*
bracket　console *f*, support *m*, tasseau *m*
bracket [of prices]　fourchette *f*
bracket [range within a series or table]　plage *f*
bracket [supporting a shelf]　gousset *m*
bracket, cable　support *m* de câble
bracket, gutter　crochet *m*/patte *f* de gouttière
bracket, handrail [built into a wall]　écuyer *m*
bracket, offset [for safety barriers]　dispositif *m* d'écartement
bracket, right angle　équerre *f*
bracket, roofing　console *f* de couvreur
bracket, stirrup　support *m* à étrier
bracket, triangulated　potence *f*
bracket, wall　console *f* (murale)
bracket on side of a string to support banisters　piton *m*
brackish　saumâtre *adj*
brad　clou *m* sans tête
bradawl　poinçon *m*
braid　tresse *f*
brake　frein *m*
brake, emergency　frein *m* de secours
brake, hand　freins *m,pl* à main
brake, vacuum　frein *m* à vide
brake block or shoe　sabot *m* de frein
brake drum　tambour *m* de frein
brake lever　levier *m* de frein
brake lining　garniture *f* de frein
brake pad [disc brake]　plaquette *f*
brake power　puissance *f* au frein
braking　freinage *m*
braking, regenerative　freinage *m* par récupération
braking equipment to absorb kinetic energy
　　équipement *m* à frein absorbeur d'énergie cinétique
Bramah lock　serrure *f* batteuse
Bramah safety lock　serrure *f* de sûreté à pompe
bramble　ronce *f*
branch [of a company or organization]　filiale *f*
branch [of a river, a tree, a subject]　branche *f*
branch [pipe]　branchement *m*
branch [current]　dériver *vb*
branch, outlet　branchement *m* de sortie
branch circuit　branchement *m*
branch line [main or pipe]　conduite *f* secondaire
branch piece　raccord *m* de branchement
branching　ramification *f*, branchement *m*
branching [e.g. of ductwork]　embranchement *m*
branching system [no loops]　réseau *m* maillé
brand [make]　marque *f*
brass　laiton *m*

braze　braser *vb*, souder *vb*
brazed　brasé *adj*
brazing　brasage *m*, brasure *f*, soudage *m*, soudure *f*
breach [in a wall]　brèche *f*, percée *f*
breach of contract　rupture *f* de contrat
breach(ing) [of a dam, etc]　rupture *f*
breadth　largeur *f*, travers *m*
break [fracture]　cassure *f*, fracture *f*
break [in a wall]　brèche *f*
break [interruption]　interruption *f*
break [of a circuit, a line]　coupure *f*, rupture *f*
break [of an electrical switch]　déclenchement *m*
break, full-load [electrical]　coupure *f* en charge
break　casser *vb*, briser *vb*
break [a circuit]　couper *vb*, interrompre *vb*
break [a contract, etc]　rompre *vb*
break [a lock]　fausser *vb*
break [a pipe, circuit, contact, etc]　ouvrir *vb*
break [fall apart, come to pieces]　se briser *vb*, se casser *vb*, se rompre *vb*
break a joint　briser *vb* un joint
break down [a log to square edged timber]　aviver *vb*
break ground　entamer *vb*
break in continuity of an arch or vault profile　jarret *m*
break off [meeting, negotiations, etc]　rompre *vb*
break up [ground or road]　défoncer *vb*
breakage　rupture *f*
breakage [of current]　coupure *f*, interruption *f*
breakage of stone edge or surface　épaufrure *f*
breakdown　panne *f*
breakdown of a system　écroulement *m* d'un système
breakdown of communication(s)　panne *f* de communication
breakdown of costs　analyse *f* des coûts
breakdown of insulation　défaillance *f* de l'isolement
breakdown of wood attacked by fungi　échauffure *f*
breakdown [rescue] **vehicle**　dépanneuse *f*
breaker [electrical]　interrupteur *m*
breaker, circuit, *see* circuit-breaker
breaker, contact　disjoncteur *m*, rupteur *m*
breaker, rock　brise-roc(he) *m*, dérocheuse *f*
breaker-contactor　discontacteur *m*
break-even point or threshold　seuil *m* de rentabilité
break-glass fire alarm　boîtier *m* brise-vitre
breaking　bris *m*, fracture *f*, rupture *f*
breaking unit [for aggregates]　groupe *m* de broyage
breaking up　désintégration *f*, dislocation *f*
breakwater　brise-lames *m*, môle *m*
breakwater wall [bridge pier protection]　arc *m* de radier
breast [wall under window ledge]　allège *f*
breast stone [under a window or door sill]　parpaing *m* d'appui
breastwork [of a fortress]　parapet *m*
breathe (in)　aspirer *vb*
breather pipe　reniflard *m*
breathing　respiration *f*
breathing apparatus　appareil *m* respiratoire
breathing apparatus, fresh air hose　appareil *m* à adduction d'air, appareil *m* isolant à prise d'air libre

breathing mask   masque *m* respiratoire
breccia   brèche *f*
breccia marble   marbre *m* brèche
breech fitting [pipe junction]   culotte *f*
breech fitting with cleaning eye   culotte *f* à regard
breeze [light wind]   brise *f*
brewery   brasserie *f*
bribe   pot *m* de vin[+]
brick   maçonner *vb*, murer *vb*
brick [normally hollow in France]   brique *f*
brick 220×110×30mm [used for fireplaces in France]   chantignole *f*, échantignole *f*
brick 220×60×60mm [sometimes used for chimneys in France]   clozot *m*
brick, air   brique *f* à air
brick, air/perforated   brique *f* alvéolée/à alvéoles/perforée
brick, air-dried   brique *f* crue
brick, burnt   brique *f* cuite
brick, calcium silicate   brique *f* silico-calcaire
brick, cavity/hollow   brique *f* creuse
brick, chimney   boisseau *m*
brick, compressed, made with clay and ggbfs   brique *f* de laitier
brick, facing   brique *f* de parement
brick, glazed   brique *f* vernissée
brick, half- [special, 220×55×60mm]   mulot *m*
brick, hollow   brique *f* tubulaire
brick, hollow [used to form ceiling between steel joists]   bardeau *m*
brick, hollow clay   bloc *m* creux (de terre cuite), brique *f* tubulaire
brick, hollow clay [30–50mm thick]   brique *f* plâtrière
brick, paving   brique *f* de pavage
brick, perforated   brique *f* alvéolée/à alvéoles/ajourée/perforée
brick, refractory   brique *f* réfractaire
brick, reinforced   brique *f* armée
brick, solid   brique *f* pleine, brique *f* non ajourée
brick, standard   brique *f* pleine calibrée
brick, toothing   harpe *f*
brick, ventilating/ventilation   brique *f* perforée
brick, vitrified   brique *f* vitrifiée
bricks and mortar   bâtisse *f*
bricks, course of diagonal   assise *f* en épi
brick axe   grelet *m*, hachette *f*
brick facing [of a wall]   faux parement *m*
brick masonry   briquetage *m*
bricklayer   briqueteur *m*, maçon *m*
bricklayer's float   bouclier *m*
bricklayer's hammer   grelet *m*, hachette *f*, marteau *m* de maçon
bricklayer's labourer   manœuvre *m*
bricklayer's line   cordeau *m*
bricklayer's pencil   crayon *m* de maçon *m*
bricklayer's trowel   briqueteuse *f*, truelle *f* à mortier
bricklaying   maçonnage *m*
brickwork   briquetage *m*, maçonnerie *f*, ouvrage *m* en briques

brickwork, hollow clay   hourdis *m* (creux)
brickworks   briqueterie *f*
bridge   pont *m*
bridge [small, over a culvert]   ponceau *m*
bridge, acoustic   pont *m* phonique
bridge, arch   pont *m* en arc(s)
bridge, arch(ed)   pont *m* voûté
bridge, bascule or lifting   pont *m* basculant
bridge, bowstring truss   pont *m* à tablier inférieur
bridge, cable-stayed   pont *m* à haubans
bridge, cantilever   pont *m* cantilever
bridge, cold or thermal   pont *m* thermique
bridge, concrete slab deck   pont *m* dalle
bridge, double bascule   pont *m* ouvrant
bridge, double-deck   pont *m* à deux étages
bridge, flat beam   pont-poutres *m*
bridge, girder   pont *m* à poutres
bridge, launched   pont *m* poussé
bridge, lift   pont *m* levant
bridge, lifting [bascule]   pont *m* basculant
bridge, movable   pont *m* mobile
bridge, portal (frame)   pont *m* à béquille
bridge, rail   pont *m* rail, pont-rail *m*
bridge, road   pont *m* route, pont-route *m*
bridge, skew   pont *m* biais
bridge, steel   pont *m* métallique
bridge, steel lattice   pont *m* en treillis métallique
bridge, suspension   pont *m* suspendu
bridge, swing   pont *m* tournant
bridge, temporary [e.g. over an excavation]   pont *m* de service
bridge, thermal   pont *m* thermique
bridge, toll   pont *m* à péage
bridge, transporter   pont *m* transbordeur
bridge, trussed arch   pont *m* en arcs en treillis
bridge building   pontage *m*
bridge deck   aire *f* d'un pont, tablier *m*
bridge decking or surfacing   revêtement *m* de pont
bridging [in a structural frame]   entretoisement *m*
brief, technical   programme *m* technique
bright [colour, light]   vif *adj*, lumineux *adj*
bright [day, weather]   clair *adj*
brightness   clarté *f*, luminance *f*, luminosité *f*
brine   saumure *f*
briny   saumâtre *adj*
brine tank   bac *m* à saumure
Brinell hardness number   chiffre *m* de dureté Brinell
Brinell hardness test   essai *m* de dureté Brinell
bringing into line   alignement *f*
bringing to bear [of forces]   mise *f* en jeu
bristle [of a paintbrush]   poil *m*, soie *f*
British Standard (BS)   norme *f* anglaise
brittle   fragile *adj*
brittleness   fragilité *f*
broad   large *adj*
broaden [grow broader]   s'élargir *vb*
broaden [make broader]   élargir *vb*
brochure   brochure *f*
broken down [machinery, vehicle]   en panne

broken particles of aggregate   granulats *m,pl* concassés
broken stones   pierraille *f*
broken-stone [road material]   matériau *m* concassé, pierre *f* à macadam
bromine   brome *m*
bronze   airain *m*, bronze *m*
brook   ruisseau *m*
brooklet   ruisselet *m*
broom   balai *m*
broomstick   manche *m* à balai
brown   brun *adj*
bruise [sth]   bosseler *vb*
brush   brosse *f*
brush [paint]   pinceau *m*
brush, chimney-sweep's   hérisson *m*
brush, cutting-in   pinceau *m* à rechampir
brush, dry [paperhanger's]   brosse *f* à étaler
brush, flat paint   pinceau *m* plat
brush, flue   écouvillon *m*, hérisson *m*
brush, lining   traînard *m*
brush, paperhanger's   balai *m* à encoller
brush, paste   brosse *f* à encoller
brush, radiator paint   pinceau *m* pour radiateurs
brush, round paint   pinceau *m* rond
brush, scrubbing   brosse *f* dure
brush, wire   brosse *f* métallique
brush clamp [electrical]   sabot *m* de balai
brush cleaning container   pincelier *m*
brush coat   couche *f* protectrice
brush for smoothing inside of joints in concrete pipe   écouvillon *m*
brush [a surface, finish]   brosser *vb*
brushwood   abattis *m*, brousailles *f,pl*
brushwood raft [road foundation on peat]   tapis *m* de branchage
bucket   seau *m*
bucket [of a dredger, excavator, etc]   godet *m*
bucket [of dredger]   benne *f*
bucket, bottom opening   benne *f* à fond ouvrant
bucket, dragline   benne *f* racleuse
bucket, excavator   godet *m* de pelle
bucket digger   excavateur *m* à godets
bucket loader   chargeuse *f* à godets
bucket-wheel reclaimer   engin *m* de reprise à roue-pelle
buckle   se déformer *vb*, gauchir *vb*, flamber *vb*
buckle [a wheel]   voiler *vb*
buckle [metal]   boucler *vb*, déjeter *vb*, gauchir *vb*, fausser *vb*
buckled   faussé *adj*, tordu *adj*, voilé *adj*
buckled [metal]   déjeté *adj*
buckling   déformation *f*, flambage *m*, flambement *m*, gauchissement *m*
buckling, local   voilement *m*
buckling, flexural   flambement *m* avec flexion
buckling, lateral   déversement *m*
buckling, safety against local   sécurité *f* au voilement
buckling, torsional   flambement *m* avec torsion
buckling length   longueur *f* de flambement
bucksaw   scie *f* à bûches

budget   budget *m*
budget, total   budget *m* global
budget estimate or figure   montant *m* du devis
buffer   tampon *m*
buffer [computer]   mémoire *f* tampon
buffer(s) [railway]   butoir *m*, heurtoir *m*
buffer tank   réservoir *m* tampon
build   bâtir *vb*, édifier *vb*, construire *vb*
build in   encastrer *vb* (dans)
build in cob   torcher *vb*
build in or let in [end of a joist, into a main beam]   enclaver *vb*
build up [a collection, deposit, etc]   former *vb*
build up [of magnetic field]   s'amorcer *vb*
build up by welding   recharger *vb* par soudure
buildable   constructible *adj*
builder   constructeur *m*, entrepreneur *m*
builder, jobbing   maçon *m*
builder's hoist/lift   élévateur *m*/monte-charge *m* de chantier
builder's labourer   manœuvre *m*, ouvrier *m* du bâtiment
builder's line   cordeau *m*
builder's line used for marking out circles   simbleau *m*
builder's merchant   commerçant *m* en matériaux de construction
builder's price book   série *f* des prix du bâtiment
builder's rubbish/rubble   décombres *m*
builder's rubbish/rubble removal   élimination *f* des décombres
builder's rubbish/rubble storage   stockage *m* de décombres
building [e.g. a house]   bâtisse *f*, immeuble *m*
building [a structure, and the process]   bâtiment *m*, construction *f*
building [a tunnel, i.e tunnelling]   percement *m*
building [imposing, large, important]   édifice *f*
building, commercial   bâtiment *m* commercial
building, factory   fabrique *f*, usine *f*
building, farm   construction *f* agricole, ferme *f*
building, high rise [with special security provisions]   immeuble *m* de grande hauteur
building, historic [UK]   monument *m* historique [France]
building, house   construction *f* d'habitation
building, industrial   bâtiment *m* industriel, construction *f* industrielle
building, link   bâtiment *m* de liaison/de jonction
building, listed [UK]   monument *m* classé/historique [France]
building, multi-storey   bâtiment *m* à étages multiples/à plusieurs étages
building, residential   bâtiment *m* résidentiel
building, school   bâtiment *m* scolaire
building, set back part of a   arrière-corps *m*
building, sick   bâtiment *m* malsain
building, single-storey   bâtiment *m* sans étage
building, steel frame   bâtiment *m* à ossature métallique
building, stone   bâtiment *m* en pierre
building, tower   bâtiment *m* très élevé

building and civil engineering works   bâtiment *m* et travaux *m,pl* publics (BTP)

building automation system (BAS)   automatisation *f* des systèmes des bâtiments

building block   bloc *m*

building block, burnt clay [brick]   brique *f*

building block, concrete   parpaing *m*

building block, plaster [for partitions, etc]   carreau *m* de plâtre

building by-law   règle *f* de construction

building code   code *m* du bâtiment, législation *f* de la construction, règlement *m* de la construction

building component   composant *m* d'ouvrage

building connection   branchement *m* d'immeuble

building contractor   entrepreneur *m* de bâtiment/de construction

building defect   faute *f*/vice *m* de construction

building design quality control office   bureau *m* de contrôle

building element/plane [wall, floor, roof, etc]   paroi *f*

building energy management system   système *m* de gestion de l'énergie des bâtiments

building envelope   enveloppe *f* du bâtiment

building industry   industrie *f* du bâtiment

building inspector employed by client   contrôleur *m* technique

building inspector, local authority   inspecteur *m* municipal des travaux

building land   terrain *m* constructible

building lease   bail *m* à construction

building line   alignement *m*

building management system (BMS)   (système *m* de) gestion *f* technique centralisée (GTC), gestion *f* technique de bâtiment

building material   matériau *m* de construction

building paper   papier *m* de construction

building permit   permis *m* de construire/de bâtir

building plan   plan *m* de construction

building plot   lot *m*, terrain *m* à bâtir

building regulation   règle *f* de construction

building regulations   code *m* du bâtiment

building regulations [in France]   Code *m* de la Construction

building regulations, high rise   Codes *m,pl* pour Structures Élevées [in France]

building rubbish/rubble   décombres

building rubbish/rubble removal   élimination *f* des décombres

building rubbish/rubble storage   stockage *m* de décombres

building site   chantier *m* (de construction), lieu *m* de construction

building structure and envelope   gros ouvrage/œuvre *m*

building trade [masonry, carpentry, plumbing, etc]   corps *m* d'état

building under construction   tas *m*

building up [magnetic field]   amorçage *m*

building-in   encastrement *m*

build-up or deposit of sediment   formation *f* de sédiments

built   bâti *adj*, construit *adj*

built-in   encastré *adj*

built-in [e.g. chimney into a wall]   engagé *adj*, incorporé *adj*

bulb [horticultural]   bulbe *m*, oignon *m*

bulb [lamp]   ampoule *f*, lampe *f*

bulb, bayonet cap   ampoule *f* (a culot) à baïonnette

bulb, filament   lampe *f* à incandescence

bulb, frosted   ampoule *f* dépolie

bulb, screw cap   ampoule *f* (à culot) à vis

bulb, silvered crown lamp   calotte *f* argentée

bulb base or cap   culot *m*

bulb holder   porte-ampoule *m*

bulb socket   douille *f*

bulge   bosse *f*, bombement *m*

bulk *adj*   en vrac *adj*

bulk [size]   grosseur *f*, grandeur *f*

bulk [volume]   masse *f*, volume *m*

bulk   foisonner *vb*

bulk density   densité *f* en vrac, masse *f* volumique, poids *m* spécifique

bulk materials store   accumulateur *m* de matières

bulkhead [for ground support]   rideau *m* de soutènement

bulkhead, watertight   cloison *f* étanche

bulkhead light   hublot *m*

bulking [of soil]   décompactage *m*

bulking material [non-binding] **in concrete**   charge *f*

bulky [of a building]   massif *adj*

bull's eye glass   verre *m* en cul de bouteille

bulldozer   bouldozeur *m*, bouteur *m*, bull(dozer) *m*, tracteur *m* à lame droite

bump   bosse *f*

bundle (of) [e.g. reinforcement bars]   paquet *m* (de)

bundle [e.g. of pipes, wires]   faisceau *m*

bung   bouchon *m*

bunk beds   lits *m,pl* superposés

bunker   accumulateur *m* de matières, silo *m*, soute *f*, trémie *f*

bunker, coal   soute *f* à charbon

bunker, fuel   soute *f* à combustible

buoyancy   flottabilité *f*, poussée *f* d'Archimède

buried   enterré *adj*

burn   brûlure *f*

burn [stream, in northern UK]   ruisseau *m*

burn (up, down, away, out)   brûler *vb*

burn out [electric motor, a portable tool, a lamp bulb]   claquer *vb*, griller *vb*

burner [gas, oil]   brûleur *m*

burner [on a cooking stove]   feu *m*

burner, atomizing   brûleur *m* à évaporation/à atomisation

burner, Bunsen   bec *m* Bunsen

burner, cup   coupelle *f*

burner, gas   bec *m* de gaz, brûleur *m*

burner, gas jet   brûleur *m* à jet de gaz

burner, high pressure   brûleur *m* à haute pression

burner, multi-jet   brûleur *m* à plusieurs becs

burner, nozzle mixing   brûleur *m* à jets croisés

burner, parallel flow   brûleur *m* à courants parallèles
burner, pot   brûleur *m* à coupelle/à pot
burner, pressure jet   brûleur *m* à injection
burner, vaporizing   brûleur *m* à vaporisation
burner array   rangée *f* des brûleurs
burner mounting   support *m* du brûleur
burner nozzle or tip   tête *f* de brûleur
burner pot   pot *m* de brûleur
burner sight glass   voyant *m* du brûleur
burning off [of paint]   brûlage *m*
burning out [electrical]   claquage *m*
burr [on a casting]   bavure *f*
burst [boiler]   éclater *vb*, sauter *vb*
burst [pipe, tyre]   crever *vb*, éclater *vb*
burst [sth]   faire *vb* éclater
burst its banks [watercourse]   rompre *vb* ses berges
bursting [of a boiler, pipe, tyre, etc]   éclatement *m*
bursting [of a dam, etc]   rupture *f*
bursting disc   disque *m* de sécurité, pastille *f*
bury   enfouir *vb*, enterrer *vb*
burying depth [e.g. of services]   profondeur *f* d'enterrement
bus [urban transport]   autobus *m*, bus *m*
bus [coach]   autocar *m*, car *m*
bus layby [bus stop]   zone *f* d'arrêt *m*
bus shelter   aubette *f*
bus station   gare *f* routière
bus stop   arrêt *m* d'autobus
bus stopping bay   zone *f* d'arrêt
busbar   bande *f* commune, barre *f* collectrice
busbar, distribution   barre *f* omnibus de distribution
bus(bar), electrical   bus *m* (électrique)
busbar, main   barre *f* omnibus principale
bush [bearing, collar]   bague *f*, douille *f*
bush, anti-friction [of a bearing]   foyer *m* d'antifriction
bush, bearing   coussinet *m*
bush hammer   boucharde *f*
bush hammer [stone]   boucharder *vb*, étonner *vb* à la boucharde
bush hammered   bouchardé *adj*
bush hammering   bouchardage *m*, taille *f* bouchardée
bushing, anti-friction   foyer *m* d'antifriction
bushing, reducing   manchon *m* réducteur
business [firm]   affaire *f*, commerce *m*, entreprise *f*
business [volume of]   chiffre *m* d'affaires
business associate   collègue *m/f*
business concern   exploitation *f* commerciale
business day   jour *m* ouvrable

business expenses   frais *m,pl* généraux
business machine   machine *f* de bureau
business manager   directeur *m* commercial
business trip   voyage *m* d'affaires
businessman/woman   homme *m*/femme *f* d'affaires
busy   occupé *adj*
butane   butane *m*
butt [end]   bout *m*, about *m*
butt [containing liquid, e.g. rainwater]   tonneau *m*
butt [place end-to-end]   ab(o)uter *vb*
butt [two lengths of sth together]   rabouter *vb*
butt hinge   charnière *f*
butt strap   éclisse *f*
butt joint   assembler *vb* bout à bout
butt together [two planks, pipes]   aboucher *vb*
butted [of wallpaper edges]   posé bord à bord
butted together [boards or planks, side-to-side]   à plat joint
button, call   touche *f* d'appel
button, control   bouton *m* de réglage
button, push or press   bouton *m* de poussoir, poussoir *m*
button, reset   bouton *m* de réarmement
button, starting   bouton *m* de commande
button, stop   arrêt *m*
button head [of a screw]   tête *f* ronde
buttress   butée *f*, contrefort *m*, dosseret *m*, éperon *m*
buttress, flying   arc-boutant *m*
buttress   arc-bouter *vb*, mettre *vb* des contreforts
butt-welded   soudé en bout
buy   acheter *vb* acquérir *vb*
buy for cash   acheter *vb* comptant
buy on credit   payer *vb* à terme
buyer   acheteur *m*, acquéreur *m*
buzzer   avertisseur *m*, ronfleur *m*
by (mutual) agreement   de gré à gré
by auction or tender   par voie d'adjudication
by hand   manuellement *adv*
by-law   arrêté *m* municipal
by-law, planning   règle *f* d'urbanisme
bypass *adj*   de dérivation
bypass [pipework, or valve controlling it]   dérivation *f*
bypass [relief road]   rocade *f*
bypass, town   contournement *m* d'une ville
bypass (pipe)   by-pass *m*, conduite *f*/tube *m* de dérivation
bypass(ing)   contournement *m*
bypass [regulations, a place]   contourner *vb*
by-product   sous-produit *m*
by-road   chemin *m* vicinal

cabin [building]   cabane *f*
cabin [of machine, equipment]   cabine *f*
cabin [small hut]   cabanon *m*
cabin, control   cabine *f* de commande
cabin, crane driver's   cabine *f* du grutier
cabin, driver's   cabine *f* du conducteur

cabin cableway/ropeway   télécabine *f*
cabinet [box]   coffret *m*
cabinet [cupboard]   placard *m*
cabinet, bedside   chevet *m*
cabinet, drying [for clothes]   sèche-linge *m*
cabinet, electrical   armoire *f* électrique

cabinet, fireproof   armoire *f* ignifuge
cabinet, low-voltage   armoire *f* basse tension
cabinet maker   ébéniste *m*
cabinet making   ébénisterie *f*
cable   câble *m*
cable [wire]   fil *m*
cable, armoured   câble *m* armé
cable, buried   câble *m* enterré
cable, bypass   câble *m* de dérivation
cable, coaxial   câble *m* coaxial/concentrique
cable, control   câble *m* de commande
cable, earth(ing)   câble *m* de masse/de terre/de mise à la terre
cable, electrical   câble *m* (rigide)
cable, extension   prolongateur *m*, cordon *m* (de) prolongateur
cable, four-core   câble *m* en quatre âmes
cable, haulage [cableway]   câble *m* tracteur
cable, looped   câble *m* bouclé
cable, main   câble *m* principal
cable, main supply   conducteur *m* principal
cable, main (suspension) [bridge]   câble *m* porteur principal
cable, mains   câble *m* de distribution
cable, MICC   câble *m* du type pyrocâble
cable, multi-conductor   câble *m* multiple
cable, multi-core   câble *m* multifilaire
cable, overhead   câble *m* aérien
cable, pilot   câble *m* pilote
cable, power   câble *m* de transmission
cable, screened   câble *m* blindé
cable, steel   câble *m* d'acier
cable, stranded   câble *m* toronné
cable, supply   câble *m* d'amenée de courant/du secteur
cable, suspension   câble *m* porteur/de suspension
cable, tension   câble *m* tendeur
cable, trailing [lift/elevator]   câble *m* pendentif
cable, welding   câble *m* de soudage
cable box   coffret *m* électrique
cable bracket   support *m* de câble
cable channel   conduite *f* de câbles
cable clamp   serre-câbles *m*
cable clip   attache *f*, collier *m* de fixation de câble
cable connecting clamp   pince *f* de raccordement
cable connection box   boîte *f* de jonction
cable connector   connecteur *m* (de câble)
cable cover   couvre-câble *m*
cable cutter   coupe-câble *m*
cable detector   localisateur *m* de câbles
cable distribution box   boîte *f* de raccordement
cable distribution plug   sous-répartiteur *m*
cable distribution system   système *m* de distribution de câbles
cable drawing-in box   boîte *f* de tirage
cable drum   bobine *f* de câble, touret *m*
cable duct   caniveau *m*, conduit *m*
cable duct or its contents   canalisation *f*
cable duct or protection sleeve   gaine *f* de câbles
cable duct system   système *m* de canalisation pour câbles

cable ducting   goulotte *f*
cable end or tail   tête *f* de câble
cable entry   entrée *f* de câble
cable failure   rupture *f* de câble
cable fitting   fixation *f* de câble
cable gland   presse-étoupe *m* de câble
cable hanger   porte-câble *m*
cable joint   raccord *m* de câble, jointure *f* de câbles
cable layout or plan   plan *m* de câblage
cable marking system   système *m* de repérage pour filerie
cable outlet   sortie *f* de câbles
cable pit or manhole   puits *m* à câbles
cable protection sleeve   gaine *f* de câbles
cable railway   funiculaire *m*
cable screening   blindage *m* de câble
cable socket   borne *f* de câble
cable stay [bridge]   hauban *m*
cable strand   toron *m* de câble
cable tape   câble-ruban *m*
cable television   télévision *f* par câble
cable terminal   borne *f*
cable tie   collier *m*
cable transporter, overhead   transporteur *m* aérien
cable tray   chemin *m* de câbles
cable trench   tranchée *f* à câbles
cable tunnel   galerie *f* des câbles
cableway   blondin *m*, téléphérique *f*, transporteur *m* aérien à câble
cableway, cabin   télécabine *f*
cabling   câblage *m*
cabling, voice and data   câblage *m* voix données image
cadastral survey   levé *m* cadastral
cadastral register   cadastre *m*
cadmium   cadmium *m*
cadmium plating   cadmiage *m*
cage [reinforcement]   cadre *m*
cage, lift   cabine *f* d'ascenseur
cage guide [lift/elevator]   guide *m* de la cabine
cage lift   élévateur *m* (monté) sur camion
cage slide [lift/elevator]   coulisseau *m* de cabine
caisson   caisson *m*
caisson, compressed air or pneumatic   caisson *m* à air comprimé
caisson, sheet pile   batardeau *m*
calcareous   calcaire *adj*
calcify   calcifier *vb*, pétrifier *vb*
calcite   calcite *m*
calcium   calcium *m*
calcium carbide   carbure *m* de calcium
calcium carbonate, natural   pierre *f* à chaux
calcium fluoride   fluorine *m*
calculate   calculer *vb*
calculate [amounts, numbers, etc]   chiffrer *vb*
calculating, calculation [the process]   chiffrage *m*
calculating machine   machine *f* à calculer
calculation   calcul *m*
calculation, approximate   calcul *m* approché
calculation, current consumption   calcul *m* de la consommation de courant

calculation, pro rata   compte *m* de prorata
calculation, probability   calcul *m* de probabilités
calculation(s) [record of]   note *f* de calcul
calculation(s), approximate   calcul *m* approché
calculation(s), supporting   justification *f*
calculation method   méthode *f* de calcul
calculation of costs   calcul *m* des coûts
calculation of current consumption   calcul *m* de la consommation de courant
calculation of heat losses   calcul *m* des déperditions calorifiques, calcul *m* des pertes de chaleur
calculation of probabilities   calcul *m* de probabilités
calculation of vertical loads on a structure   descente *f* des charges
calculator   machine *f* à calculer
calculator (pocket)   calculatrice *f* (de poche)
calendar   calendrier *m*
calibrate [an instrument, etc]   calibrer *vb*, étalonner *vb*
calibration   calibrage *m*, étalonnage *m*, étalonnement *m*
California Bearing Ratio (CBR)   indice *m* CBR, indice *m* portant californien (CBR)
call   appel *m*
call sign   indicatif *m*
call wire   fil *m* d'appel
callipers, external/outside   compas *f* d'épaisseur
callipers, inside/internal   compas *m* d'alésage
callipers, sliding   équerre *f* à coulisse
callipers, vernier   pied *m* à coulisse
calorie   calorie *f*
calorific   calorifique *adj*
calorific value   pouvoir *m* calorifique, valeur *f* calorifique
calorific value, gross   pouvoir *m* calorifique supérieur (PCS)
calorific value, net   pouvoir *m* calorifique inférieur (PCI)
calorifier, domestic   appareil *m* de chauffage ménager
calorifier, storage   ballon *m* à chauffage mixte
calorifier (coil)   serpentin *m* rechauffeur
calorimeter   calorimètre *m*
cam   came *f*, ergot *m*
camber [of a beam]   contre-flèche *f*, courbure *f*
camber [of a road]   profil *m* circulaire
camber [of an arch, a piece of wood]   cambrure *f*
camber(ing)   bombement *m*
cambium   cambium *m*
came [of a leaded light]   baguette *f*, résille *f*
camera   appareil *m*, caméra *f*
camera, photogrammetric   appareil *m* de photogrammétrie
camera, stereometric   appareil *m* de stéréométrie
camera, thermal imaging   caméra *f* thermique de surveillance
campaign [marketing, etc]   campagne *f*
campanile   campanile *m*
camp(ing) site   camping *m*, terrain *m* de camping
canal   canal *m*
canal, diversion   canal *m* de dérivation
canal length between two locks   bief *m*
cancel [a contract of insurance]   résilier *vb*
cancel [e.g a contract]   annuler *vb*

cancellation [e.g. of a contract]   annulation *f*
cancellation [of an insurance policy]   résiliation *f*
canopy   marquise *f*
canopy [over a porch]   auvent *m*
canteen   local *m* de restauration, réfectoire *m*
canteen, site   cantine *f*
cantilever   cantilever *m*, console *f*, encorbellement *m*
cantilever arm   bras *m* en porte-à-faux
cantilevered   en bascule, en encorbellement, en porte-à-faux
cantonal   cantonal *adj*
canyon   canyon *m*
cap   chapeau *m*, couvercle *m*
cap [bulb/lamp]   culot *m*
cap [e.g. vehicle radiator]   bouchon *m*
cap(ping) [of a bridge pier]   chapiteau *m*
cap, oil filler   bouchon *m* de remplissage d'huile
cap, screw(ed)   bouchon *m* à vis
cap [a pile]   coiffer *vb*
cap [chimney, wall, etc]   couronner *vb*
capacitance   capacitance *f*
capacitor [electrical]   condensateur *m*
capacity   capacité *f*, pouvoir *m*, puissance *f*
capacity [volumetric]   contenance *f*
capacity, absorbing or absorptive   pouvoir *m* absorbant
capacity, bearing, *see* bearing capacity
capacity, breaking [electrical]   pouvoir *m* de coupure
capacity, carrying   capacité *f* portante, portance *f*
capacity, covering [of a paint]   pouvoir *m* couvrant
capacity, daily   capacité *f* journalière
capacity, dust absorption   capacité *f* d'absorption de la poussière
capacity, dust holding/retaining   capacité *f* de rétention des poussières
capacity, effective storage (water)   eau *f* utilisable
capacity, fan   capacité *f*/débit *m* de ventilateur
capacity, filter   capacité *f* d'un filtre
capacity, filtering   capacité *f* de filtrage
capacity, heat   capacité *f* calorifique
capacity, heat storage   capacité *f* de stockage thermique
capacity, holding   capacité *f* de stockage
capacity, installed   capacité *f* installée
capacity, lifting [maximum load]   charge *f* limite
capacity, lifting [of a crane]   puissance *f* de levage
capacity, load [e.g. of a lorry]   charge *f* utile
capacity, load-carrying   capacité *f* portante, portance *f*
capacity, load-carrying [of road material]   résistance *f* mécanique
capacity, official   capacité *f* officielle
capacity, overload   capacité *f* de surcharge
capacity, production   capacité *f* de production
capacity, reservoir   volume *m* de retenue
capacity, self-purifying   pouvoir *m* auto-épurateur
capacity, standard   capacité *f* normale
capacity, storage   capacité *f* de stockage
capacity, suction   débit *m* d'aspiration
capacity, switching   pouvoir *m* de coupure
capacity, thermal   capacité *f* thermique
capacity, total   capacité *f* totale

capacity, useful puissance *f* utile
capacity, well capacité *f* de puits
capillarity capillarité *f*
capillary *adj* capillaire *adj*
capillary capillaire *m*
capillary action attraction *f* capillaire, capillarité *f*
capillary rise ascension *f* capillaire
capital capital *m*, fonds *m*
capital [of a column] chapiteau *m*
capital [share] capital *m*
capital, share capital *m* social
capital, working fonds *m* de roulement
capital gain (on disposal) plus-value *f* (de cession)
capital loss (on disposal) moins-value *f* (de cession)
capital outlay mise *f* de fonds
capital, lay out or put up some faire *vb* une mise de fonds
capitalization of assets immobilisation *f*
capped [of a pile] coiffé *adj*
capping [end of a conductor with insulation] capotage *m*
capping [of a chimney, pier, dam, etc] couronnement *m*
capping [of row of piles] chapeau *m*
capping, chimney, with side openings and closed top lanterne *f*
capping, profiled profilé *m* de couverture
capping, ridge [lead, of a slate roof] brifier *m*
capping, ridge [tile or metal] enfaîtement *m*, faîtage *m*
capstone [of a wall] crête *f*
capture the light capter *vb* la lumière
carbide carbure *m*
carbon carbone *m*
carbon, activated charbon *m* actif
carbon content teneur *f* en carbone
carbon dioxide anhydride *m* carbonique, gaz *m* carbonique
carbon monoxide oxyde *m* de carbone
carbonate carbonate *m*
carbonation carbonatation *f*
carbonization carbonisation *f*
carborundum carborundum *m*
carcass (carcase) [of built-in furniture] caisson *m*
carcass (carcase) [of a house] cage *f*
carcass (carcase) [of electric motor] carcasse *f*
card index box fichier *m*
card punch [computer] perforateur *m* de cartes
card reader [computer] lecteur *m* de cartes
cardboard (box) carton *m*
caretaker gardien *m*
caretaker's house maison *f* de garde/gardien
cargo fret *m*
carillon carillon *m*
Carnot cycle cycle *m* de Carnot
carpark parking *m*, parc *m* de stationnement *m*
carpark, multi-storey parking *m* en élévation
carpark, staff parking *m* personnel
carpark, underground parking *m* souterrain
carpark, visitors' parking *m* visiteurs
carpark, workmen's parking *m* ouvriers
carpenter charpentier *m*

carpenter, formwork coffreur-boiseur *m*, coffreur *m* (pour béton armé)
carpenter's auger roanne/rouanne *f*
carpenter's beam compass troussequin *m*
carpenter's clamp sergent *m*, valet *m*
carpenter's hammer marteau *m* de charpentier
carpenter's line used for marking out circles simbleau *m*
carpenter's marking gauge trusquin *m*
carpenter's rule réglet *m*
carpet tapis *m*
carpet, fitted moquette *f*
carpet, flexible [road foundations] tapis *m* bitumineux
carpet, needle-punch tapis *m* aiguilleté
carpet, tufted tapis *m* tufté
carpet recouvrir *vb* d'une moquette
carport abri *m* voiture
Carrara marble carrare *m*
carriage [bringing] amenage *m*
carriage [of goods] transport *m*
carriage costs frais *m,pl* de transport
carriage entrance porte *f* charretière/cochère
carriage gate porte *f* charretière/cochère
carriage paid franc de port, port payé
carriageway chaussée *f*
carriageway, concrete chaussée *f* en béton *m* (de ciment)
carriageway edge rive *f* de chaussée *f*
carriageway surfacing revêtement *m* de chaussée
carry [a load, etc] supporter *vb*
carry [goods, etc] transporter *vb*, camionner *vb*
carry away enlever *vb*
carry out [e.g. works] effectuer *vb*, exécuter *vb*
carry out a survey [topological] lever *vb* un plan
carry out tests or trials faire *vb* des essais
carrying portant *adj*
carrying out réalisation *f*
carrying up montage *m*
cart chariot *m*
cart [goods] camionner *vb*
cartouche [architectural] cartouche *m*
cartridge [for blasting] cartouche *m*
cartridge fuse fusible *m* à cartouche, fusible-cartouche *m*
cartridge fuse, switched coupe-circuit *m* à cartouche
cascade cascade *f*
case, *see also* casing boîte *f*, boîtier *m*, caisse *f*, caisson *m*
case [jacket, of a boiler, etc] chemise *f*
case [instance] cas *m*
case [of a shaft, well] cadre *m*
case [a borehole] tuber *vb*
case [a borehole, shaft, well] cuveler *vb*
case [a structure] blinder *vb*
case [equipment] chemiser *vb*, envelopper *vb*
case, borderline cas *m* limite
case, extreme cas *m* exceptionnel
case history exposé *m* d'un cas concret
case study étude *f* de cas
cased chemisé *adj*
case-hardening cémentation *f*
casement châssis *m*

casement (cremona) bolt mechanism boîte *f* de crémone

casement bolt crémone *f*

casement bolt closure fermeture *f* à crémone

casement stay entrebailleur *m*

cash, buy for acheter *vb* comptant

cash, pay payer *vb* (au) comptant

cash, spot argent *m* comptant

cash a cheque toucher *vb* un chèque

cash desk caisse *f*

cash flow cash-flow *m*, flux *m* de trésorerie

cash payment paiement *m* comptant, paiement *m* en liquide

cashier caissier *m*

casing caisson *m*, capot *m*, carter *m*, manchon *m*

casing [formwork] coffrage *m*

casing [jacket, of a boiler, etc] chemise *f*, enveloppe *f*

casing [jacketing, action and result] chemisage *m*

casing [of a borehole] tubage *m*

casing [of a door or window] dormant *m*, chambranle *m*

casing [of a motor] coquille *f*

casing [of equipment] habillage *m*

casing, flat, for surface electric wiring baguette *f* électrique

casing, pump carter *m* de pompe

casing, sheet metal carter *m* en tôle

casing of a well blindage *m* de puits

cask fût *m*

cast *adj* coulé *adj*

cast [of concrete, in a mould or form] banché *adj*

cast mouler *vb*

cast [concrete] couler *vb*

cast in place/in situ coulé *adj* en/sur place

cast in place/in situ couler *vb* sur place

cast iron *adj* en fonte

cast iron fer *m* coulé/de fonte, fonte *f*

cast iron, ductile fonte *f* ductile

cast iron, glazed fonte *f* émaillée

cast iron, grey fonte *f* grise

cast iron, malleable fonte *f* malléable

cast iron fitting [coupling/connection] raccord *m* en fonte

cast iron sleeve or socket manchon *m* en fonte

casting [concrete, metal] pièce *f* coulée

casting [piece of cast metal] pièce *f* fondue

casting [the action] coulage *m*, coulée *f*, fonte *f*

castle château *m*

cat hole chatière *f*

cat's-eye plot *m* (rétro-)réfléchissant/réflectorisé

catalogue catalogue *m*, nomenclature *f*

catalyst catalyseur *m*

catch fermeture *f*

catch [door, shutter, cupboard, etc] loqueteau *m*

catch [for a shutter or other non-rebated closure] arrêt *m*

catch [of a latch] bec *m*

catch [of a pivoted/lifting type latch] mentonnet *m*

catch, magnetic loqueteau *m* magnétique

catchment captage *m*

catchment, spring captage *m* de source

catchment area bassin *m* d'apport *m*, bassin *m* versant

catchment basin aire *f* de captage

catchment boundary ligne *f* de faîte/de partage des eaux

catchpit puisard *m*

category catégorie *f*, classe *f*

catenary *adj* caténaire *adj*

catenary caténaire *f*

cathedral cathédrale *f*

cathode cathode *f*

cathodic protection protection *f* cathodique

cation cation *m*

catwalk passerelle *f*

caulk calfater *vb*, mater *vb*

caulking calfatage *m*, calfeutrage *m*, calfeutrement *m*, matage *m*

caulking, lead calfatage *m* de plomb

caulking compound ciment *m* à mater

caulking tool ciseau *m* à mater

cause to deflect affaisser *vb*

causeway levée *f*

cave [natural] grotte *f*

cave in s'affaisser *vb*

cave(rn) caverne *f*

cavernous caverneux *adj*

caving in éboulement *m*

cavitation cavitation *f*

cavity cavité *f*, évidement *m*, fouille *f*

cavity [in masonry, stone] alvéole *f*

cavity barrier coupe-vapeur *m* dans un mur creux

cedar cèdre *m*

ceiling plafond *m*

ceiling, boarded plafond *m* en lambris

ceiling, coffered plafond *m* à/en caissons

ceiling, false or suspended faux plafond *f*

ceiling, floating [isolated from structure above] plafond *m* flottant

ceiling, heated plafond *m* chauffant

ceiling, high-vaulted toit *m* cathédrale/chapelle

ceiling, isolated plafond *m* flottant

ceiling, panelled plafond *m* lambrissé

ceiling, perforated plafond *m* perforé

ceiling, suspended faux plafond *f*, plafond *m*/plancher *m* suspendu

ceiling, ventilated plafond *m* diffuseur d'air

ceiling, put in a plafonner *vb*

ceiling, reach a plafonner *vb*

ceiling board, perforated panneau *m* perforé pour plafond

ceiling coil batterie *f* de plafond

ceiling fixture plafonnier *m*

ceiling height hauteur *f* sous plafond

ceiling infill between joists or girders entrevous *m*

ceiling lamp plafonnier *m*

ceiling lampholder douille *f* à plafond

ceiling light plafonnier *m*

ceiling paperhanger porte-lé *m*

ceiling rose rosace *f*

ceiling socket fitting douille *f* à plafond

ceiling structure plancher *m* haut

ceiling system, integrated   système *m* de plafond intégré
ceiling ventilator   bouche *f* de plafond
ceiling void   vide *m* sous plafond
cell   cellule *f*
cell [of a battery]   élément *m*
cell, creep   cellule *f* de fluage
cell, load   cellule *f* de contrainte
cell, photo-electric   cellule *f* photo-électrique
cell, solar, *see* solar cell
cellar   cave *f*
cellar, small   caveau *m*
cellular   cellulaire *adj*
cement   ciment *m*
cement, adhesive tile fixing   ciment *m* colle
cement, aluminous   ciment *m* alumineux/fondu
cement, artificial Portland   ciment *m* Portland artificiel (CPA)
cement, asbestos   asbeste-ciment *m*, fibrociment *m*
cement, asphalt   béton *m* asphaltique
cement, blastfurnace (ggbfs) with higher lime sulphate content   ciment *m* de haut fourneau (CHF)
cement, blastfurnace (ggbfs) with lower lime sulphate content   ciment *m* de fer (CF)
cement, bulk   ciment *m* en vrac
cement, expansive   ciment *m* expansif
cement, grouting   mortier *m* joint
cement, high alumina (HAC)   ciment *m* alumineux/fondu
cement, high strength   ciment *m* à haute résistance
cement, hydraulic   ciment *m* hydraulique
cement, lime slag [fat or hydraulic lime + ggbfs]   ciment *m* de laitier à la chaux (CLX)
cement, low heat   ciment *m* à faible chaleur d'hydratation
cement, magnesium [refractory magnesite + magnesium chloride]   ciment *m* magnésien
cement, metallic [magnesium cement with zinc replacing magnesium]   ciment *m* métallique
cement, metallurgic [artificial cement + bfs]   ciment *m* métallurgique
cement, mixed metallurgic [equal parts OPC and ggbfs + lime sulphate]   ciment *m* métallurgique mixte (CMM)
cement, natural   ciment *m* naturel (CN)
cement, ordinary Portland (OPC)   ciment *m* Portland ordinaire
cement, Portland pozzolana   ciment *m* à la pouzzolane
cement, quick-setting   ciment *m* à prise rapide, ciment *m* prompt
cement, rapid-hardening   ciment *m* à durcissement rapide/à haute résistance initiale
cement, rapid-hardening Portland   ciment *m* Portland à haute résistance initiale (HRI)
cement, refractory   ciment *m* réfractaire
cement, Roman   ciment *m* romain
cement, secondary constituent of   ajout *m*
cement, slag [OPC + ggbfs]   ciment *m* de laitier
cement, slow-setting   ciment *m* à prise lente
cement, stone [for artificial stone facings]   ciment *m* pierre

cement, sulphate resisting   ciment *m* résistant aux sulfates
cement, super [OPC mixed to improve long term strength]   superciment *m*
cement, supersulphated   ciment *m* sursulfaté
cement, waterproof or water repellent   ciment *m* hydrofuge
cement, white   ciment *m* blanc
cement with 20% min OPC + 80% max basic slag   ciment *m* Portland de laitier
cement with 70% min OPC + 30% max basic slag   ciment *m* Portland de fer
cement   cimenter *vb*
cement content [of concrete]   dosage *m* en ciment
cement finisher   cimentier *m*, cimentier-ravaleur-applicateur *m*
cement grout   coulis *m* de ciment
cement grout applied to cured concrete before pouring fresh   couche *f* de reprise
cement grout applied to protect reinforcing steel   barbotine *f*
cement grouting   injection *f* de ciment
cement gun   canon *m* à ciment
cement silo   silo *m* à ciment
cement works   cimenterie *f*
cement-rubber latex   ciment *m* latex
cementation   cimentation *f*, injection *f* de ciment
cemetery   cimetière *m*
central [position]   central *adj*
central [e.g. of a component]   centré *adj*
central heating   chauffage *m* central
central heating, full   chauffage *m* central général
central heating fittings   raccorderie *f* de chauffage central
central heating plant   installation *f* de chauffage central
centre   centre *m*
centre   centrer *vb*
centre [in lathe tailstock]   pointe *f* vive
centre, dead   point *m* mort
centre, live [for wood turning]   mèche *f* à bois à trois pointes
centre, recreation   base *f* de loisirs
centre of gravity or mass   centre *m* de gravité
centre of rotation   centre *m* de rotation
centre-line   axe *m* (central), axe *m* d'implantation
centre-line of road   axe *m* de chaussée
centrifugal   centrifuge *adj*
centrifuge   centrifugeur *m*
centrifuge   centrifuger *vb*
centroid   centre *m* de gravité
ceramic   céramique *adj*
ceramic tile with a length twice the width   carrelet *m*
ceramic(s)   céramique *f*
certificate   attestation *f*, certificat *m*
certificate, acceptance   certificat *m* de réception
certificate, approval [compliance with standard(s)]   certificat *m* d'homologation
certificate, test   certificat *m* d'épreuve/d'essai
certificate, workshop   certificat *m* d'usine

cesspit fosse f d'aisance
cesspool fosse f d'aisance, puisard m, puits m perdu
chain chaîne f
chain, driving chaîne f de commande
chain [measure] chaîner vb
chain man [surveying] porte-mire m, porte-chaîne m
chain tongs serre-tubes m
chainage [surveying] arpentage m, chaînage m
chainsaw scie f à chaîne(tte), scie f articulée,
   tronçonneuse f
chair chaise f
chair [railway track] selle f de rail
chair, bos'n's sellette f port
chair, desk fauteuil m de bureau
chair, office siège m de bureau
chairlift télésiège m
chairman [e.g. of a meeting] président m
chalet chalet m
chalk craie f
chalk a line battre vb la ligne, tringler vb
chalk-line cordeau m
chamber chambre f
chamber, anechoic chambre f sourde
chamber, collecting chambre f collectrice, cuve f
chamber, drying étuve f de séchage
chamber, environmental chambre f climatique
chamber, inspection chambre f/pièce f/puits m de visite
chamber, inspection [small] regard m de visite
chamber, mixing chambre f/plenum m de mélange
chamber, reverberation chambre f réverbérante
chamfer biseau m, chanfrein m
chamfer biseauter vb, chanfreiner vb
chamfered biseauté adj, chanfreiné adj
chancel [of a cathedral, church] chœur m
chandelier [ornamented] lustre f
change changement m, modification f
change [disturbance] remaniement m
change [money] monnaie f
change modifier vb, varier vb
change in changement m de
change in plane [of a facade or architectural feature]
   décrochement m
change of direction renversement m
change of gradient [point of] point m d'inflexion
change of scale changement m d'échelle
change of polarity inversion f de (la) polarité
changeable variable adj
changed modifié adj
changeover, seasonal changement m (de température)
   été-hiver
changing remaniage m
channel canal m
channel [architectural] cannelure f
channel [duct] conduit m
channel [groove] rainure f
channel [metal section] poutre f (en) U
channel [pipe] buse f, conduite f
channel [port] chenal m
channel [river bed] lit m

channel [road gutter] caniveau m
channel [trench or ditch] rigole f
channel, cable conduite f de câbles
channel, collecting gouttière f collectrice
channel, discharge canalisation f de refoulement
channel, discharge [water] canal m de rejet
channel, drainage fossé m de drainage, rigole f
   d'évacuation
channel, drainage [road] caniveau m
channel, enclosed rigole f couverte
channel, feeder canal m d'alimentation
channel, fixing rail m de scellement
channel, flushing canal m de chasse
channel, harbour/navigable passe f
channel, hollow kerb bordure f creuse
channel, intake [water] canal m d'amenée
channel, intercepting [road construction] fossé m de
   crête
channel, irrigation fossé m d'irrigation
channel, navigable canal m navigable
channel, open rigole f ouverte
channel, services caniveau m technique
channel, sliding glissière f
channel, slotted [road] caniveau m à fentes
channel, supply canal m d'alimentation
channel, valley [roofs] noue f
channel, water intake canal m d'amenée
channel insert [fixing channel] rail m de scellement
channel iron fer m (en) U, barre f à U
channel [cut a groove] évider vb
channelled joint [in building stone] anglet m
chapel chapelle f
characteristic adj caractéristique adj
characteristic caractéristique f
characteristic curve, performance courbe f de
   fonctionnement
characteristics, operating caractéristiques f,pl de
   fonctionnement
characteristics, pump caractéristiques f,pl d'une pompe
charcoal charbon m (de bois)
charcoal, filter(ing) charbon m à filtrer
charge charge f
charge [fee, etc] frais m,pl, prix m
charge, fixed tarif m forfaitaire
charge, mortgage privilège m d'hypothèque
charges [administrative] droits m,pl
charges, bank agios m
charges, deferred charges f,pl différées
charges, direct charges f,pl directes
charges, fixed charges f,pl fixes
charges, list of tarif m
charges, scale of tableau m des tarifs
charges attached to a property servitudes f,pl
charges fixed in advance charges f,pl constatées
   d'avance
charge charger vb
charge [a battery] mettre vb en charge
chargehand chef m d'équipe, maître m ouvrier
charging [of a furnace, etc] chargement m

charging rate [electrical]   taux *m* de charge
Charles' law   loi *f* de Gay-Lussac
chart   abaque *m*, tableau *m*, graphique *m*
chart, colour   carte *f* de coloris
chart, comfort   diagramme *f* de confort
chart, design   abaque *m*
chart, fault finding   tableau *m* de dépannage
chart, lubrication   tableau *m* de graissage
chart, organization   organigramme *m*
chart, plasticity   diagramme *f* de plasticité
chart, psychrometric   diagramme *f* psychrométrique/de Carrier
chart, shade [paint colours]   carte *f* de coloris
chase [groove]   saignée *f*
chase [on outside of jamb providing key for walling]   nervure *f*
chase, pipe [in masonry, concrete work]   fourreau *m*
chase out [for an inlay, or to form bas relief]   champlever *vb*
chassis   châssis *m*
château   château *m*
cheaply   à bon marché
check [of a door]   arrêt *m*
check, drying or seasoning [timber defect]   fente *f* de retrait
check, potability [for drinking water]   contrôle *m* de potabilité
check   contrôler *vb*, vérifier *vb*
check (off) [on or from a list]   pointer *vb*
checker [checks the measurements of a *métreur*]   vérificateur *m*
check(ing)   contrôle *m*, vérification *f*
check(ing) [on or from a list]   pointage *m*
checking agency [design and construction, in France]   bureau *m* de contrôle
checking certificates [for payment]   vérification *f* de factures
checking engineer [for structural soundness, construction and safety]   contrôleur *m* technique
checkpoint   poste *m* de contrôle
checkpoint, security   poste *m* de contrôle sécurité
cheek [of a bearing, mortice, plane, etc]   joue *f*
cheek [of a dormer window]   jouée *f*
cheese head [of a screw or bolt]   tête *f* cylindrique
chemical *adj*   chimique *adj*
chemical   produit *m* chimique
chemical engineer   ingénieur *m* chimiste
chemical name   nom *m* chimique
cheque   chèque *m*
cheque, bank   chèque *m* bancaire
cheque, blank   chèque *m* en blanc
cheque, stopping a   arrêt *m* de paiement d'un chèque
cheque book   carnet *m* de chèques, chéquier *m*
cheque counterfoil or stub   talon *m*
cheque without funds to meet it   chèque *m* sans provision
cheque, cash a   toucher *vb* un chèque
cheque, draw or write a   tirer *vb* un chèque
cheque, endorse a   endosser *vb* un chèque

cheque, payment by   paiement *m* par chèque
cheque, rubber[†]   chèque *m* de bois[†]
cherry picker   élévateur *m* (monté) sur camion, élévateur *m* télescopique, engin *m* élévateur à nacelle, nacelle *f* à bras/flèche télescopique
cherry [wood]   merisier *m*
chert   silex *m*
chest of drawers   commode *f*
chestnut [wood, and sweet chestnut tree]   châtaignier *m*, marronnier *m*
chestnut, horse   marronier *m* d'Inde
chiller   machine *f* à eau glacée, refroidisseur *m*
chiller, air-cooled (packaged)   refroidisseur *m* à condensation par air
chiller, air-cooled water   groupe *m* d'eau glacée refroidi par l'air
chiller, water   groupe *m* d'eau glacée
chilling   réfrigération *f*
chime   carillon *m*
chimes   sonnerie *f*
chimney   cheminée *f*
chimney [for loading, e.g. solid fuel hopper]   trémie *f*
chimney, house   cheminée *f* de maison
chimney, self-supporting   cheminée *f* auto-porteuse
chimney, waste gas   cheminée *f* de happement
chimney capping with closed top and side openings   lanterne *f*
chimney draught   tirage *m* de cheminée
chimney effect   effet *m* de cheminée/de tirage
chimney flue   conduit *m* de cheminée
chimney flue [pipe]   conduite *f*/tuyau *m* de cheminée
chimney pot   pot *m* de cheminée
chimney pot, rectangular   mitre *f*
chimney pot, round   mitron *m*
chimney stack above roof level   souche *f* (de cheminée)
chimney-sweep   ramoneur *m*
chimney-sweep's brush   hérisson *m*
chimney-sweeping   ramonage *m*
chink [in wall, plaster, etc]   lézarde *f*
chip [of wood, stone]   éclat *m*
chip [wood, metal]   copeau *m*
chipboard (wood)   (bois *m*) aggloméré, panneau *m* de copeaux/de particules
chipping [scaling a boiler]   piquage *m*
chippings [stone]   gravillons *m,pl*
chippings, coarse   gravillons *m,pl* gros
chippings, coated   gravillons *m,pl* enrobés
chippings, gravel or crushed rock   gravier *m* concassé, gravillons *m,pl* concassés
chippings, layer of   couche *f* de gravillons
chippings, loose   gravillons *m,pl*
chippings, pre-coated   gravillons *m,pl* pré-enrobés/pré-traités
chippings spreader   gravillonneuse *f*
chips (steel, iron)   limaille *f* (d'acier, de fer)
chisel   ciseau *m*
chisel [long, for cutting mouldings in stonework]   repoussoir *m*
chisel, bevel-edged   ciseau *m* biseauté

chisel, cold   burin *m*
chisel, firmer or wood   ciseau *m* ordinaire/à bois
chisel, framing   ciseau *m* à bords biseautés, ciseau *m* biseauté à brides
chisel, heavy [for cramping floorboards for nailing] fermoir *m*
chisel, hollow   gouge *f*
chisel, joiner's   ciseau *m* de menuisier
chisel, mason's narrow   ognette *f*
chisel, mason's toothed   gradine *f*
chisel, mortice   bec *m* d'âne, bédane *m* (à bois)
chisel, wood   ciseau *m* à bois
chisel   ciseler *vb*
chisel [work with a]   buriner *vb*
chiselled decorations, and art of making these ciselure *f*
chlorinate   chlorer *vb*
chlorination   chloration *f*
chlorine   chlore *m*
chlorite   chlorite *m*
chlorofluorocarbon (CFC)   chlorofluorocarbone *m* (CFC)
chock   cale *f*, coin *m*
chock (up)   caler *vb*
choir [of a cathedral, church]   chœur *m*
choke (coil)   bobine *f* de réaction/réactance
choke [an outlet]   obstruer *vb*
choke [drain, etc]   colmater *vb*
choke [electrical]   self *f*
choke (up) [clog]   encrasser *vb*
choke (up) [block]   boucher *vb*
choked (up)   encrassé *adj*
choking [clogging]   encrassement *m*
choking (up) [drains, etc]   colmatage *m*
chord   membrure *f*
chord, compression   membrure *f* comprimée
chord, lower   aile *f*/membrure *f*/semelle *f* inférieure
chord, tension   membrure *f* tendue
chord, upper   aile *f*/membrure *f*/semelle *f* supérieure
choose and lay out timber and mark and cut joints appareiller *vb*
chromaticity [lighting]   chromaticité *f*
chromaticity coordinates   coordonnées *f,pl* de chromaticité
chromium   chrome *m*
chromium plating   chromage *m*
chuck [of a drill]   mandrin *m*
chuck, hollow   mandrin *m* creux
chuck, quick-action, multi-head   mandrins *m,pl* à serrage rapide
chuck, two-jaw   mandrin *m* à deux mors
chuck key   clef *f* de mandrin
church   église *f*
church tower   clocher *m*
chute [of wood, for building materials or rubbish] coulisse *f*, coulotte *f*, goulotte *f*
chute, rubbish [domestic rubbish]   vide-ordures *m*
chute, rubbish [building rubbish]   goulotte *f*
chute, waste   vide-ordures *m*

CIE colour triangle   diagramme *m* x–y de CIE
cinders   cendre *f*
circle   cercle *m*, rond *m*
circuit   circuit *m*
circle template or stencil   trace-circles *m*
circuit, branch   branchement *m*
circuit, bypass   circuit *m* de dérivation
circuit, call   circuit *m* d'appel
circuit, closed   circuit *m* fermé
circuit, control   circuit *m* de contrôle
circuit, earth   circuit *m* de terre
circuit, extra-low voltage   circuit *m* à très basse tension (fonctionnelle), circuit *m* TBTF
circuit, inductive   circuit *m* inductif
circuit, high-voltage   circuit *m* à haute tension
circuit, lighting   circuit *m* d'éclairage
circuit, local   circuit *m* local
circuit, main   circuit *m* d'alimentation
circuit, mixing   circuit *m* de mélange
circuit, out of   hors circuit
circuit, parallel   circuit *m* en parallèle
circuit, primary   circuit *m* primaire
circuit, printed   circuit *m* imprimé
circuit, refrigerant   circuit *m* frigorifique
circuit, resistive   circuit *m* résistant
circuit, sealed [pipework]   circuit *m* étanche à vase fermé
circuit, secondary   circuit *m* secondaire
circuit, short, *see* short circuit
circuit, shunt   circuit *m* dérivé
circuit diagram, drawing, plan or layout   schéma *m* de câblage, plan *m* de circuit
circuit label   label *m* de circuit
circuit tester [electrical]   vérificateur *m* de circuit
circuit wiring   câblage *m* de circuit
circuit-breaker   brise-circuit *m*, contacteur-disjoncteur *m*, coupe-circuit *m*, disjoncteur *m*, interrupteur *m*
circuit-breaker, cartridge fuse   coupe-circuit *m* à cartouche
circuit-breaker, consumer outlet   disjoncteur *m* de branchement
circuit-breaker, earth leakage   disjoncteur *m* différentiel/de fuite de terre/de perte à la terre
circuit-breaker, fuse   coupe-circuit *m* à fusibles
circuit-breaker, high voltage   appareil *m* de coupure HT
circuit-breaker, main   disjoncteur *m* général/principal
circuit-breaker, miniature   disjoncteur *m* miniature
circuit-breaker, two-pole   disjoncteur *m* bipolaire
circular   circulaire *adj*
circulate   circuler *vb*
circulate [sth]   faire *vb* circuler
circulation   circulation *f*
circulation, direction of   sens *m* de circulation
circulation pattern   modèle *m* de circulation
circumference   circonférence *f*, contour *m*, tour *m*
circumference [e.g. surrounding a building]   pourtour *m*
circumferential   circonférentiel *adj*
cistern   bâche *f*, citerne *f*, cuve *f*
cistern, toilet/lavatory flush   chasse *f* d'eau, réservoir *m* de chasse (d'eau)

city   cité *m*, ville *f*

civil court [dealing with major matters]   tribunal *m* de grande instance

civil court [dealing with minor matters]   tribunal *m* d'instance

civil engineer   ingénieur *m* (en) génie civil, ingénieur *m* des travaux publics

civil engineer [from l'École des Ponts et Chaussées]   ingénieur *m* des ponts et chaussées

civil engineering   génie *m* civil

civil engineering firm   entreprise *f* de travaux publics

civil engineering structure/work, major   ouvrage *m* d'art

civil servant   fonctionnaire *m*

clad   revêtir *vb*

clad (with)   bardé *adj* (de)

cladding   revêtement *m*

cladding [with boards or panels]   bardage *m*

cladding, decorative   habillage *m*

cladding, roof   couverture *f*

cladding, weatherboard   bardage *m* à clin/en planches

cladding board [timber]   bardeau *m*

cladding of the underside of a roof   sous-toiture *f*

cladding unit   élément *m* de façade

claim [by a contractor, with quantities + costs]   mémoire *f* (de réclamation), réclamation *f*

claim, contractual   réclamation *f*

claim, make or put in a   soulever *vb* une réclamation

claim (for)   demande *f* (de)

claim for damages   demande *f* de dommages-intérêts

claim form for expenses   note *f* de frais

clamp   bride *f* de serrage, collier *m* de fixation, crampon *m*, griffe *f*, patte *f*, pièce *f* de serrage, pince *f*

clamp [to hold a rule used in dressing a rendering flat]   chevillette *f*

clamp, cable   serre-câbles *m*

clamp, cable connecting   pince *f* de raccordement

clamp, carpenter's   sergent *m*, valet *m*

clamp, earth(ing)   collier *m* de mise à la terre

clamp, gluing   sergent *m*, serre-joint *m* à coller

clamp, screw   serre-joint *m* à vis

clamp, spacing   collier *m* d'écartement

clamp, vertical [of a theodolite]   blocage *m* d'inclinaison

clamp support   suspente *f* à collier

clamp   serrer *vb*

clamp [a piece of machinery]   bloquer *vb*

clamped [wheel, machinery, etc]   bloqué *adj*

clamping [in position]   blocage *m*

clamping device   dispositif *m* de fixation

clamping ring   collier *m* de serrage

clamshell grab   benne *f* à demi-coquilles

clarity   clarté *f*

class   catégorie *f*, classe *f*

classification   classement *m*, classification *f*

classification, water   classification *f* des eaux

classify   classer *vb*

clastic   clastique *adj*

clause [in agreement, contract, etc]   article *m*, clause *f*

clause, penalty   clause *f* pénale

claw   griffe *f*

claw bar   pied-de-biche *m*

claw peen [of a hammer]   panne *f* fendue

clay   argile *f*, glaise *f*

clay, boulder or glacier   argile *f* à blocaux

clay, expanded burnt   argile *f* expansée

clay, fired   terre *f* cuite

clay, fissured   argile *f* fissurée

clay, glacier   argile *f* glacière/à blocaux

clay, layer of   lit *m* d'argile

clay, puddle   corroi *m* d'argile

clay, puddled   argile *f* compactée

clay, quick   argile *f* sensible

clay, sandy   argile *f* sableuse

clay, vitrified   grès *m* cérame

clay blanket   tapis *m* d'argile

clay block, hollow   brique *f* creuse

clay blocks, hollow   hourdis *m*

clay pit   glaisière *f*

clay pocket   poche *f* d'argile

clay shale   argile *f* schisteuse, schiste *m* argileux

clayey   argilacé *adj*, argileux *adj*, glaiseux *adj*

clean   net *adj*

clean room   chambre *f* propre, salle *f* blanche

clean   nettoyer *vb*

clean [a gas]   épurer *vb*

clean [unclog]   décrasser *vb*

clean down/off/up [e.g. brickwork, stonework, etc]   ragréer *vb*

clean off/down [used materials]   décrotter *vb*

clean up   assainir *vb*

clean (up) [depollute, e.g. a watercourse]   dépolluer *vb*

clean(li)ness   propreté *f*

cleaner   nettoyeur *m*

cleaner, air   épurateur *m* d'air, filtre *m* à air

cleaner, high pressure   nettoyeuse *f* à haute pression

cleaner's sink   évier *m* agent de ménage

cleaning   assainissement *m*, décrassage *m*, nettoyage *m*

cleaning, air   épuration *f* de l'air

cleaning, sewer   curage *m* des égouts

cleaning access   accès *m* de nettoyage

cleaning agent   détergent *m*

cleaning apparatus or machine   nettoyeuse *f*

cleaning down/off/up [e.g. brickwork, stonework, etc]   ragréage *m*, ragrément *m*

cleaning eye   bouchon *m* de dégorgement, tampon *m* hermétique

cleaning off mortar, etc, from used masonry materials   décrottage *m*

cleaning out [of a drain, etc]   curage *m*

cleaning out (of joints) [prior to pointing]   dégradation *f* (des joints)

cleanse   décrasser *vb*

cleanse [by purging through]   purger *vb*

cleanse [water, air, etc]   purifier *vb*

cleansing   décrassage *m*

cleansing [purification]   épuration *f*

cleansing [washing]   lavage *m*

clean(s)ing product   produit *m* d'entretien

clear   libre *adj*
clear [sky]   clair *adj*
clear, in [uncoded]   clair, en *adj*
clear [empty]   vider *vb*
clear [to free or open]   dégager *vb*
clear fell   faire *vb* un abattis
clear of rocks   dérocher *vb*
clearance   dégagement *m*, écartement *m*, espace *m* libre,
   intervalle *m*, tolérance *f*
clearance [overhead]   tirant *m* d'air
clearance [play]   jeu *m*
clearance, running   jeu *m* libre
clearance for placing a structural fixing
   dévêtissement *m*
clearance (height)   hauteur *f* échappée
clearing [in woodland]   clairière *f*
clearing [of a drain, etc]   curage *m*
clearing [of ground, etc]   nettoiement *m*
clearness [of water]   netteté *f*
cleat [fixed to a principal rafter to prevent a purlin slipping]
   chantignole *f*, échantignole *f*
cleat, flat fixing [one end is grouted in]   patte *f* à
   scellement
cleat, L-shaped [for holding two walls or timbers together]
   harpon *m*
cleat, L-shaped metal   équerre *f* d'assemblage
cleavage   clivage *m*
cleavage, plane of, of a stone bed   lit *m*
cleavage plane   plan *m* de clivage
cleavage strength   résistance *f* au clivage
cleave   fendre *vb*
cleave [e.g. slate]   cliver *vb*, refendre *vb*
cleft   fente *f*
clench [a nail]   abattre *vb*, river *vb*
clerestory   claire-voie *f*, coupole *f* vitrée
clerk of works   surveillant *m* de travaux
clerk or registrar of the court   greffier *m*
clevis   manille *f*
client   client/e *m/f*, maître *m* d'ouvrage, maître *m* de
   l'ouvrage
cliff (face)   falaise *f*
climate   climat *m*
climate, indoor   climat *m* intérieur
climate, temperate   climat *m* tempéré
climatic chamber   chambre *f* climatique
climatic zone [defined in French regulations]   zone *f*
   climatique
climatological   climatologique *adj*
climatology   climatologie *f*
climb   montée *f*
climb   monter *vb*
climb up a ladder   monter *vb* sur une échelle
clinch [a nail]   abattre *vb*, river *vb*
cling (to)   adhérer *vb* (à)
clinic   clinique *f*
clinker   scorie *f*
clinker [for use as aggregate]   mâchefer *m*
clinker bed   lit *m* de mâchefer
clinometer   clinomètre *m*, éclimètre *m*

clip   patte *f*, pince *f* de fixation
clip, cable   attache *f*, collier *m* de fixation de câble
clip, crocodile   pince *f* crocodile/multiprise
clip, fixing   attache *f*/pince *f* de fixation
clip, gutter   crochet *m*/patte *f* de gouttière
clip, Jubilee   collier *m* à vis
clip, L-shaped   équerre *f* d'assemblage
clip, pipe [collar]   collier *m* (d'attache), gâche *f*
clip, rail or sleeper   crapaud *m* de rail
clip, retaining [in inclined metal frame glazing]   agrafe *f*
clip, spring   collier *m* à ressort
clip used to fix down support of a *gouttière anglaise*
   pontet *m*
clip [fix]   attacher *vb*
clip [cut off]   découper *vb*
clip [pare edge]   rogner *vb*
clip (together)   agrafer *vb*
clip-on   enclipsable *adj*
cloakroom   cabinet *m* d'aisances/de toilette, vestiaire *m*
clock   horloge *f*
clock, control   horloge *f* de réglage
clock, seven-day   horloge *f* hebdomadaire
clock, time [time switch]   horloge *f* à contacts
clock in or out [employee]   pointer *vb*
clog (up) [filter, etc]   colmater *vb*
clogging   encrassement *m*, engorgement *m*
clogging (up) [filters, etc]   colmatage *m*
cloister   cloître *m*
close [enclosed space]   clos *m*, enclos *m*
close [of a cathedral, etc]   enceinte *f*
close [a door, a circuit]   fermer *vb*
close [a road]   barrer *vb*
close [door, window]   joindre *vb*
close the flood-gates or sluices   mettre *vb* les vannes
close up   serrer *vb*
close-grained or textured   à texture serrée/fermée
close-jointed [rock, etc]   à joint fermé
closed [of a road]   barré *adj*
closed or 'on' position [of a switch]   position *f* de
   fermeture
closed to heavy traffic/loads/lorries   interdit *adj* aux
   poids lourds
closer, door   ferme-porte *m*
closet   cabinet *m*, garde-robe *f*
closet, dry/earth   cabinet *m* sec
closet, squatting (WC)   siège *m* à la turque
closing [of a cavity]   obturation *f*
closing [of a road]   barrage *m*
closing (down, off, up)   fermeture *f*
closing date   date *f* limite
closing down of site   repliement *m*
closure, espagnolette   fermeture *f* à crémone
closure over the end of a seam in metallic roofing
   talon *m*
cloth   toile *f*
cloth, glass fibre   toile *f* de fibre de verre
cloth, rubberized   caoutchouc *m* toilé
clothes cupboard   armoire *f* (de chambre à coucher),
   garde-robe *f*

clothes dryer   sécheuse *f*

clothing, protective   vêtement *m* protecteur

clothoid [spiral transition curve]   clothoïde *f*

clouding [glass]   ternissement *m*

clout nail   clou *m* calot(t)in/à tête de diamant, pointe *f* à tête large

clue   indice *m*

cluster [e.g. of houses, piles, trees]   groupe *m*

cluster, pile   groupe *m* de pieux

clutch plate or disc   disque *m* d'embrayage

coagulate   coaguler *vb*

coagulation tank   bassin *m* de coagulation

coal   charbon *m*, houille *f*

coal, boiler or steam   charbon *m* à vapeur

coal, brown   lignite *m*

coal, medium volatile   charbon *m* flambant

coal, non-bituminous   charbon *m* maigre

coal, pulverized   charbon *m* pulvérisé

coal, smokeless   charbon *m* sans fumée

coal burning/firing   chauffe *f* au charbon

coal tar [used as protective or waterproof coating]   coaltar *m*

coarse   grossier *adj*

coarse-textured [sand, aggregate]   de grosse granularité *f*

coast   côte *f*

coat [layer, of paint, tar, etc]   couche *f*

coat, bituminous   enduit *m* bitumeux

coat, brush   couche *f* protectrice

coat, decorative finish(ing)   couche *f* d'habillage

coat, decorative top   couche *f* d'habillage

coat, final   couche *f* de teinte

coat, finish(ing)   couche *f* de finition

coat, glaze [paintwork]   lustre *f*

coat, primary   couche *f* de fond

coat, priming   première couche *f*

coat, priming [largely absorbed, on wood or masonry]   imbue *f*

coat, priming [oil-based, on a porous surface]   huilage *m*

coat, priming [painting]   couche *f* d'impression

coat, protective   couche *f* de protection

coat, protective [paint]   peinture *f* protectrice

coat, sealing [blinding]   béton *m* /couche *f* de propreté

coat, sealing [road construction]   couche *f*/enduit *m* de scellement

coat, tack [road construction]   émulsion *m* d'accrochage

coat, top   couche *f* de finition

coat hook   patère *f*

coat [apply one or more coats]   coucher *vb*

coat (with)   enduire *vb* (de), enrober *vb* (de)

coat with binder   enrober *vb* de liant

coated [road material]   traité *adj*

coated, powder   thermolaqué *adj*

coating   revêtement *m*, enduit *m*

coating [application of]   enduisage *m*

coating [e.g. of mortar, cement]   chape *f*

coating [e.g. road material]   enrobage *m*

coating [of a lamp]   poudrage *m*

coating, adhesive   enduit *m* colle

coating, bitumen   revêtement *m* de bitume

coating, decorative   habillage *m*

coat(ing), plaster skim   enduit *m* fin

coating, plastic   couche *f* en matière plastique

coating, protective   enduit *m* protecteur/de protection, couche *f*/revêtement *m* de protection

coating, protective [paint]   peinture *f* protectrice

coating, tar-based wood protective   galipot *m*

coating, thermoplastic   revêtement *m* thermoplastique

coating, waterproof   enduit *m* étanche

coating, zinc   galvanisation *f*, zingage *m*

coating plant   poste *m* d'enrobage

coating providing base for surface treatment   enduit *m* de surfaçage

coaxial   coaxial *adj*

cob   bousillage *m*, torchis *m*

cob, construct in   bousiller *vb*

cobble   gros caillou *m*, galet *m*

cock [tap]   prise *f* d'eau *f*, robinet *m*

cock, air   robinet *m* d'air

cock, air relief   robinet *m* décompresseur

cock, bleed(er)   robinet *m* de purge

cock, drain   robinet *m* de vidange

cock, junction   robinet *m* de branchement

cock, main   robinet *m* principal

cock, (main) gas   robinet *m* (principal) de gaz

cock, outlet   robinet *m* d'écoulement

cock, purging   robinet *m* purgeur

cock, regulating   robinet *m* de réglage

cock, sampling   robinet *m* d'échantillonnage

cock, sludge   robinet *m* de vidange de boue

cock, stop   robinet *m* d'arrêt

cock, test   robinet *m* d'essai/de contrôle

cock, through way   robinet *m* à passage direct

cock, waste   robinet *m* purgeur

code   code *m*

code of practice   norme *f*

code of practice for building works [in France]   document *m* technique unifié (DTU)

code word   mot-code *m*

coefficient   coefficient *m*

coefficient, differential   dérivée *f*

coefficient, discharge   coefficient *m* de décharge

coefficient, heat transfer   coefficient *m* de transfert thermique/de transmission de chaleur

coefficient, leakage [HVAC]   coefficient *m* d'exfiltration

coefficient, regression   coefficient *m* de régression

coefficient, roughness   coefficient *m* de rugosité

coefficient, run-off   coefficient *m* de ruissellement

coefficient, shading   coefficient *m* de protection du soleil

coefficient, shape   coefficient *m* de forme

coefficient, sound transmission   coefficient *m* de transmission du son

coefficient of elasticity   coefficient *m* d'élasticité

coefficient of expansion   coefficient *m* de dilatation

coefficient of friction   coefficient *m* de frottement

coefficient of heat loss of a habitation (watts/m^3/°C)   coefficient-G *m* d'un logement

coefficient of performance   facteur *m* de fonctionnement
coefficient of performance (COP)   coefficient *m* de performance
coefficient of permeability   coefficient *m* de perméabilité
coefficient of resistance   coefficient *m* de résistance
coefficient of thermal conductivity   coefficient *m* conductibilité
coefficient of thermal expansion   coefficient *m* de dilatation thermique
coefficient of uniformity   coefficient *m* d'uniformité
coffer [e.g. in coffered slab]   caisson *m*
coffer, pre-formed [between beams/joists]   corps *m* creux
cofferdam   batardeau *m*, caisson *m* cellulaire, coffre *m*
coherent   cohérent *adj*
cohesion   cohésion *f*
cohesionless [geotechnics]   pulvérulent *adj*
cohesive   cohérent *adj*
coil   bobine *f*
coil [filament]   filament *m* électrode
coil [of wire]   botte *f*
coil [pipe, tubing]   serpentin *m*
coil, calorifier   serpentin *m* rechauffeur
coil, ceiling   batterie *f* de plafond
coil, condenser   serpentin *m* de condenseur
coil, cooling   groupe *m* serpentin refroidisseur, serpentin *m* de refroidissement/de réfrigération
coil, direct expansion   serpentin *m* à détente directe
coil, evaporation   batterie *f* d'évaporation
coil, evaporator   serpentin *m* évaporatoire
coil, expansion   serpentin *m* de dilatation
coil, heating   serpentin *m* de chauffage
coil, pipe   serpentin *m*
coil, preheat   batterie *f* de préchauffage
coil, refrigeration/ing   serpentin *m* de réfrigération
coil, reheat   batterie *f* de réchauffage
coil, resistance   bobine *f*
coil, steam   serpentin *m* à vapeur
coil unit, direct expansion   batterie *f* à détente directe
coil (up)   enrouler *vb*
coke   coke *m*
coke burning/firing   chauffe *f* à coke
coke oven   four *m* à coke
cold   froid *adj*
cold   froid *m*
cold room   armoire *f* frigorifique, chambre *f* frigorifique/froide
cold store   entrepôt *m* frigorifique
cold snap or wave   coup *m* de froid
cold, get   refroidir *vb*
cold draw   étirer *vb* à froid
cold roll   laminer *vb* à froid
cold start   démarrer *vb* à froid
cold weld   souder *vb* à froid
collapse   écroulement *m*, effondrement *m*
collapse [floor, etc]   affaissement *m*
collapse [under load]   rupture *f*
collapse, progressive   effondrement *m* en cascade
collapse   s'affaisser *vb*, s'écrouler *vb*, s'effondrer *vb*
collapse, cause to   affaisser *vb*

collapsing   éboulement *m*
collar [on metallic downpipe]   bague *f*
collar [pipe clip]   collier *m*, gâche *f*
collar [pipe, tube]   bride *f*
collar, insulating   bague *f* isolante
collar, thrust   rondelle *f* de butée
collar beam [timber roof]   entrait *m*, entrait *m* retroussé, faux entrait *m*
collar brazed on to metal downpipe to prevent it slipping down   nez *m*
collateral security   nantissement *m*
colleague   collègue *m/f*
collect [debts, etc]   recouvrer *vb*
collect [documents, facts, information]   rassembler *vb*, recueillir *vb*
collect [evidence, proof]   rassembler *vb*
collect [surface water]   recueillir *vb*
collect [taxes, etc]   percevoir *vb*
collection [action of, of facts, etc]   rassemblement *m*
collection [action of, of rents]   encaissement *m*
collection [action of, of taxes]   perception *f*, recouvrement *m*
collection [set or group, e.g. of houses]   ensemble *m*
collection [of water, or current from overhead wire]   captage *m*
collection, rubbish   ramassage *m* d'ordures (ménagères)
collection of data   recueil *m* des données
collective   collectif *adj*
collector [drain, road, sewer]   collecteur *m*
collector [of taxes, fines, etc]   percepteur *m*
collector [of taxes]   receveur *m*
collector, flat plate   collecteur *m*/capteur *m* plan
collector, solar   capteur *m* solaire
colloid   colloïde *m*
colloidal   colloïdal *adj*
colonnade   colonnade *f*
colorimetric   colorimétrique *adj*
colour   coloration *f*, couleur *f*
colour [dye]   teint *m*
colour [of paint]   coloris *m*
colour card or chart   carte *f* de coloris
colour code, standard   code *m* de couleurs normalisées
colour rendering index (CRI)   indice *f* de rendu des couleurs (IRC)
colour wash   badigeon *m*
colour washing   badigeonnage *m*
colour wash   badigeonner *vb*
colour with bright tints   enluminer *vb*
coloured   coloré *adj*
colouring [dye]   colorant *m*
colouring matter   matière *f* colorante
column   colonne *f*, poteau *m*
column [straight, not ornamented]   pilier *m*
column, braced   poteau *m* tenu en tête
column, cast iron   colonne *f* en fonte
column, half [applied to face of a wall]   demi-ceint *m*
column, small   pédicule
column, spread base of a   patte *f* d'éléphant
column, steel   poteau *m* métallique

column head tête *f* de poteau
column head [in flat-slab construction] champignon *m*
column supporting a small balcony or veranda
 colonne *f* méniane
columnar en forme de colonne
comb [of an escalator, moving walkway] peigne *m*
comb or corrugate the surface of a rendering
 peigner *vb*
combination combinaison *f*
combined heat and power (CHP) cogénération *f*
combined heat and power (CHP) station centrale *f*
 combinée chaleur force/de chaleur/force
comb-plate [escalator] plaque-peigne *f*
combustible combustible *adj*
combustion combustion *f*
combustion, fluid bed combustion *f* par lit fluidisé
combustion, heat of chaleur *f* de combustion
combustion chamber chambre *f* de combustion
combustion chamber, forced draught foyer *m* soufflé/à
 tirage mécanique
combustion medium comburant *m*
come loose se desceller *vb*
comfort commodité *m*
comfort [as in air-conditioning] confort *m*
comfort, thermal confort *m* thermique
comfort chart diagramme *m* de confort
comfort index (thermal) indice *m* de confort thermique
comfort zone zone *f* de confort
comment observation *f*, remarque *f*
commercial commerçant *adj*, commercial *adj*
comminute concasser *vb*
commission mettre *vb* en service
commission [mandate to do something] mandat *m*
commission [percentage] commission *f*
commission, agency commission *f* d'agence
commissioning mise *f* en service
commissioning assistance assistance *f* aux réceptions
commissioning authority [for acceptance of works]
 commission *f* de réception
commissioning plan plan *m* de mise en route
commit [someone to something] engager *vb*
commit oneself [to do something] s'engager *vb*
commitment engagement *m*
common collectif *adj*, commun *adj*
common [as in common flue] unitaire *adj*
communal collectif *adj*, communal *adj*
communal forest forêt *f* domaniale
communal or shared facilities installations *f,pl*
 collectives
commune [lowest tier of local government in France]
 commune *f*
communicate [ideas, information, etc] communiquer *vb*
communicate [e.g. rooms with each other] communiquer *vb*
communication breakdown or failure panne *f* de
 communication
communication network réseau *m* de communication
communication system système *m* de communication
communications conductor [for controls, alarms, etc]
 conducteur *m* pilote

communication(s) equipment matériel *m* de
 communication
community collectivité *f*
community centre centre *m* communautaire
compact compact *adj*
compact compacter *vb*
compact [ground, to avoid later settlement] affermir *vb*
compacted [e.g. of fill] compacté *adj*
compactibility compactibilité *f*
compacting, final cylindrage *m* définitif
compacting equipment or plant engins *m,pl* de
 compactage
compacting machine engin *m* de compactage
compacting tyre pneu *m* compacteur
compaction compactage *m*
compaction [amount of] énergie *f* de compactage
compaction, degree of degré *m* de compactage
compaction, initial précompactage *m*
compaction, relative compacité *f*
compaction, soil compactage *m* de sol
compaction, vibration compactage *m* par vibration
compaction by rolling/vibration/watering compactage
 m par cylindrage/vibration/arrosage
compaction of concrete serrage *m* du béton
compaction test, Proctor essai *m* de compactage Proctor
compaction testing contrôle *m* du compactage
compactive effort [amount of compaction] énergie *f* de
 compactage
compactive effort [of a roller or vibratory compactor]
 effort *m* de compactage
compactness [of ground, road pavements, etc]
 compacité *f*
compactor compacteur *m*, engin *m* de compactage
compactor, embankment or slope compacteur *m* de
 berges
compactor, landfill compacteur *m* pour enfouissement
compactor, plate pilonneuse *f*
compactor, refuse compacteur *m* de déchets
compactor, rubber-tyred compacteur *m* à
 pneu(matique)s, rouleau *m* (compacteur) à pneu(s)
compactor, rubbish compacteur *m* d'ordures ménagères
compactor, vibrating plate compacteur *m* vibrant à
 plaque/à plaque vibrante, plaque *f* vibrante, vibro-
 compacteur *m*
compactor, vibrating roller compacteur *m* vibrant à
 rouleau(x)
company entreprise *f*, société *f*
company, joint public/private société *f* d'économie
 mixte (SEM)
company, limited liability société *f* à responsabilité
 limitée (SARL)
company, management compagnie *f*
company holding a concession [e.g. statutory undertaker]
 société *f* concessionnaire
compartment compartiment *m*
compartment [of a forest] parcelle *f*
compartmentation compartimentage *m*
compartmentation, fire cloisonnement *m* de sécurité
compass boussole *f*

compass [drawing]   compas *m*
compass [in the sense of bounds]   cadre *m*
compass, beam   compas *m* à verge
compasses, pair of   compas *m*
compatible   compatible *adj*
compensate [for loss]   indemniser *vb*
compensating   compensateur *adj*
compensating loop   boucle *f* de dilatation
compensation [for loss]   indemnisation *f*
compensation [mechanical, electrical]   compensation *f*
compensator   compensateur *m*
compensator, pressure   compensateur *m* de pression
competence   ressort *m*
competent person   personne *f* compétente
competition [e.g. architectural]   concours *m*
competition [commercial]   concurrence *f*
complain   réclamer *vb*
complaint   réclamation *f*
complement [= number of people]   effectif *m*
complete   terminer *vb*
completion [e.g. of a section of work]   achèvement *m*
completion date   date *f* d'achèvement
completion of curing of concrete   maturation *f*
completion of the works   achèvement *m* des travaux
completion on schedule/on time   achèvement *m* dans les délais voulus
component   élément *m*, pièce *f*
component [element]   constituant *m*
component [of voltage, force, etc]   composante *f*
component [part]   composant *m*
component, building   composant *m* d'ouvrage
composition [e.g. of sewage]   composition *f*
compost   terreau *m*
composting   compostage *m*
compound   composé *adj*
compound, caulking   ciment *m* à mater
compound, curing [e.g. for concrete]   produit *m* de cure
compound, fireproofing   ignifugeant *m*
compound, joint sealing   garniture *f* de joint
compound, jointing   mastic *m* pour joints
compound, sealing [joint]   garniture *f* de joint, produit *m* de garnissage
compound growth   croissance *f* composée
compress   comprimer *vb*
compressed   comprimé *adj*
compressed air   air *m* comprimé
compressed air caisson   caisson *m* à air comprimé
compressed air tank or vessel   réservoir *m* à/d'air comprimé
compressed air working   travaux *m,pl* dans l'air comprimé/en caisson/en milieu hyperbare
compressibility   compressibilité *f*
compressibility, modulus of   module *m* de compressibilité *f*
compressible   compressible *adj*
compression   compression *f*, pression *f*
compression, adiabatic   compression *f* adiabatique
compression, heat of   chaleur *f* de compression
compression, unconfined   compression *f* simple

compression diagram   courbe *f* de compression
compression flange   membrure *f* comprimée
compression fitting [pipe jointing]   raccord *m* montage bicône
compression ratio   rapport *m*/taux *m* de compression
compression refrigeration system   système *m* de réfrigération à compression
compression reinforcement, area of   section *f* d'acier comprimé
compression stroke   course *f* de compression
compressive force   effort *m*/force *f* de compression
compressive yield point   limite *f* d'écrasement, limite *f* élastique à la compression
compressor   compresseur *m*
compressor, air   compresseur *m* d'air
compressor, centrifugal turbo-   turbo-compresseur *m* radial
compressor, double/single-acting   compresseur *m* à double/simple effet
compressor, multi-stage   compresseur *m* multi-étage
compressor, multi-vane rotary   compresseur *m* rotatif multicellulaire
compressor, reciprocating   compresseur *m* alternatif/à piston
compressor, refrigeration   compresseur *m* frigorifique
compressor, return flow   compresseur *m* à flux inversé
compressor, screw   compresseur *m* à vis
compressor, scroll   rouleau *m* compresseur
compressor, sliding vane   compresseur *m* à ailettes
compressor, twin   compresseur *m* jumelé
compressor, two-stage   compresseur *m* à deux étages
compressor unit   moto-compresseur *m*
compromise   compromis *m*
compromise solution   solution *f* de compromis
compulsory   obligatoire *adj*
compulsory medical examination before hiring   examen *m* médical d'embauche/avant embauche
compulsory purchase   expropriation *f*
computation   calcul *m*
computer *adj*   informatique *adj*
computer   ordinateur *m*
computer design   étude *f* par ordinateur
computer equipment or hardware   matériel *m* informatique
computer input   entrée *f* d'ordinateur
computer keyboard key   touche *f*
computer output   sortie *f* d'ordinateur
computer print-out   édition *f* d'ordinateur
computer program   logiciel *m*
computer run   exécution *f* d'un programme informatique
computer science   informatique *f*
computer storage size   taille *f* mémoire d'un ordinateur
computer-aided   assisté par ordinateur
computer-aided design (CAD)   conception *f* assistée par ordinateur (CAO)
computer-aided design of formwork   coffrage *m* assisté par ordinateur (CAO)
computer-aided draughting (CAD)   dessin *m* assisté par ordinateur (DAO)

computer-assisted planned maintenance   maintenance *f* assistée par ordinateur (MAO)
computer-based system   système *m* piloté par ordinateur
computing   informatique *f*
concave   incurvé *adj*
conceal [a minor defect, a difference of level]   racheter *vb*
conceive [design]   concevoir *vb*
concentrated   concentré *adj*
concentration   concentration *f*
concentration, equilibrium   concentration *f* d'équilibre
concentric   concentrique *adj*
concept   concept *m*
concept design   étude *f* d'avant-projet
conception   conception *f*
concessionaire   société *f* concessionnaire
conclude an agreement   passer *vb* un accord
conclusion   terme *m*
concourse [e.g. of station]   hall *m*
concourse, passenger   hall *m* des passagers
concrete   béton *m*
concrete, added   béton *m* coulé en deuxième phase
concrete, aerated or air-entrained   béton *m* à entraînement d'air/à air occlus/à occlusion d'air, béton *m* avec entraîneur d'air/avec occlusion d'air, béton *m* aéré/léger
concrete, asphalt, with oven-dried gravel aggregate   béton *m* d'asphalte
concrete, asphaltic   béton *m* bitumineux
concrete, asphaltic, with a tar binder   béton *m* au bitume-goudron
concrete, as-struck   béton *m* brut (de décoffrage)
concrete, bedding   béton *m* de propreté
concrete, bitumen binder   béton *m* hydrocarboné
concrete, bitumen-tar asphaltic   bitume-goudron *m* (BG)
concrete, bituminous   béton *m* (asphaltique) bitumineux
concrete, blinding   béton *m* de propreté
concrete, brick aggregate   béton *m* de briquaillons
concrete, bush hammered   béton *m* bouchardé
concrete, cast   béton *m* moulé
concrete, cast in situ   béton *m* coulé sur place
concrete, cellular   béton *m* alvéolaire/cellulaire
concrete, centrifugally cast   béton *m* centrifugé
concrete, compaction of   serrage *m* du béton
concrete, completion of curing of   maturation *f*
concrete, cyclopean   béton *m* cyclopéen
concrete, demolition waste aggregate   béton *m* de gravats
concrete, diatomaceous (lightweight)   béton *m* de diatomée/de kieselguhr
concrete, dry-mixed   béton *m* malaxé à sec
concrete, easily workable   béton *m* plastique
concrete, electrically heated [poured in cold weather]   électro-béton *m*
concrete, exposed   béton *m* apparent
concrete, exposed aggregate   béton *m* à granulats apparents
concrete, exposed aggregate [obtained by hosing]   béton *m* lavé

concrete, facing   béton *m* de parement
concrete, fair-faced   béton *m* lisse de parement
concrete, fibre-reinforced   béton *m* fibré
concrete, fly-ash   béton *m* de cendres volantes
concrete, foamed   béton-mousse *m*, béton *m* mousse
concrete, foamed glass (aggregate)   béton *m* de verre multicellulaire/vésiculaire
concrete, formed   béton *m* banché
concrete, fresh   béton *m* jeune/frais
concrete, graded exposed aggregate   béton *m* de gravillons lavés
concrete, green   béton *m* jeune/frais
concrete, ground granulated blastfurnace slag (ggbfs)   béton *m* de mâchefer/de mousse de laitier
concrete, hard [metallic or carborundum aggregate]   béton *m* dur
concrete, hardened   béton *m* durci
concrete, heavy   béton *m* compact
concrete, heavy aggregate [e.g. for radiation shielding]   béton *m* lourd
concrete, high alumina cement (HAC)   béton *m* alumineux/fondu
concrete, hydraulic   béton *m* hydraulique
concrete, lean   béton *m* maigre
concrete, lightly sandblasted   béton *m* gommé
concrete, lightweight   béton *m* léger/de granulats légers
concrete, lightweight (diatomaceous)   béton *m* de diatomée/de kieselguhr
concrete, lightweight (expanded clay aggregate)   béton *m* d'argile expansé
concrete, lightweight (ggbfs)   béton *m* de (mousse de) laitier
concrete, lightweight (pumice aggregate)   béton *m* aux cendres volcaniques
concrete, lightweight (Siporex)   béton *m* Siporex
concrete, lightweight aggregate   béton *m* d'agrégats légers
concrete, mass   béton *m* de masse/non armé
concrete, measured constituents ready for mixing   charge *f*
concrete, metallic waste aggregate   béton *m* de riblons
concrete, no-fines   béton *m* caverneux/drainant/sans éléments fins/sans sable
concrete, percolation of water to the surface on setting   remontée *f* d'eau
concrete, plain   béton *m* non armé
concrete, pouring of [using formwork or shuttering]   banchage *m*
concrete, pozzolana aggregate   béton *m* volcanique
concrete, pozzolanic   béton *m* de pouzzolane
concrete, precast   béton *m* manufacturé/préfabriqué
concrete, prepack   béton *m* prépack/prépakt
concrete, prestressed   béton *m* précontraint
concrete, projecting irregularity in   balèvre *f*
concrete, pulverized fuel ash (PFA)   béton *m* de cendres volantes/de scorie
concrete, pumice aggregate   béton *m* volcanique/de ponce
concrete, pumped [generally with plastifier]   béton *m* pompé

concrete, rammed or tamped   béton *m* damé

concrete, ready-mix(ed)   béton *m* préparé/prêt à l'emploi

concrete, refractory   béton *m* réfractaire

concrete, reinforced, *see also* reinforced concrete   béton *m* armé

concrete, resin [i.e using resin binder]   béton *m* de résine

concrete, resin modified   béton *m* aux résines

concrete, rich   béton *m* gras

concrete, sand and gravel aggregate   béton *m* de graviers

concrete, sand only aggregate   béton *m* de sable

concrete, sandblasted   béton *m* sablé

concrete, sprayed   gunitage *m*

concrete, stabilized earth   béton *m* de terre stabilisé

concrete, tamped   béton *m* damé

concrete, tar binder   béton *m* hydrocarboné

concrete, ultra-lightweight (expanded perlite aggregate)   béton *m* de perlite

concrete, unreinforced   béton *m* non armé, gros béton *m*

concrete, untreated   béton *m* brut (de décoffrage)

concrete, vacuum dewatered [accelerated setting]   béton *m* au vide/essoré sous vide

concrete, vermiculite   béton *m* de vermiculite

concrete, vibrated   béton *m* vibré

concrete, visible   béton *m* apparent

concrete, weak   béton *m* maigre

concrete, wet   béton *m* jeune/frais

concrete, wood sawdust aggregate   béton *m* de sciure de bois

concrete, woodwool   béton *m* de bois, fibragglo *m*

concrete heated in situ by heated formwork   béton *m* chauffé

concrete made by mortar injection into aggregate in situ   béton *m* à remplissage

concrete made with pre-heated constituents   béton *m* chaud

concrete using aggregate containing stone (>30mm)   béton *m* de pierres

concrete accelerator   accélérateur *m* de prise *f*

concrete base course   socle *m* de béton

concrete batcher   grande bétonnière *f*

concrete batching and mixing plant   centrale *f* à béton, installation *f* de préparation du béton, installation *f* de dosage et de mélange du béton

concrete bedding   enrobage *m* bétonné

concrete benching [of a sewer]   cunette *f* en béton

concrete blinding   béton *m* de propreté

concrete block, precast   parpaing *m*

concrete block paving   pavé *m* de béton

concrete breaker   brise-béton *m*

concrete carriageway   chaussée *f* en béton *m* (de ciment)

concrete compaction   serrage *m* du béton

concrete cover [of reinforcement]   enrobage *m* des aciers

concrete diaphragm   diaphragme *m* en béton

concrete finish, washed fine gravel   mignonnette *f* lavée

concrete finisher [in a paving train]   finisseur *m* à béton

concrete footing, mass   semelle *f* en gros béton

concrete insert [fixing channel]   rail *m* de scellement

concrete layer, protective   couche *f* protectrice en béton

concrete masonry block   parpaing *m*

concrete mix   composition *f* du béton

concrete mixer   bétonnière *f*, malaxeur *m* de béton

concrete mixer, tilting   bétonnière *f* à cuve basculante

concrete mixing plant   centrale *f*/usine *f* à béton, usine *f* de bétonnage

concrete pavement   revêtement *m* bétonné

concrete pipe, prestressed   tuyau *m* en béton précontraint

concrete placing boom   flèche *f* de distribution de béton

concrete plank, precast [permanent shuttering]   prédalle *f*

concrete pouring, restart of   reprise *f* de bétonnage

concrete pump   pompe *f* à béton

concrete pump, mobile   pompe *f* automobile à béton

concrete recycling plant   poste *m* de recyclage de béton résiduel

concrete saw   scie *f* à béton

concrete section   section *f* brute

concrete shell   voile *m* en béton

concrete shield [nuclear power station]   bouclier *m* en béton

concrete skip or bucket   benne *f* à béton/de bétonnage

concrete topping, in situ   hourdis *m*

concrete wall [thin]   voile *m* (de béton)

concrete works   travaux *m,pl* de bétonnage

concrete   bétonner *vb*

concrete, pour [using formwork or shuttering]   bancher *vb*

concreter [worker]   bétonneur *m*

concreting   bétonnage *m*, travaux *m,pl* de bétonnage

concreting, hot/cold weather   bétonnage *m* par temps chaud/froid

concreting skip   benne *f* de bétonnage

concretion   concrétion *f*

condensate   condensat *m*, eau *f* condensée

condensation   buée *f*, condensation *f*, rosée *f*

condensation, heat of   chaleur *f* de condensation

condensation, interstitial   condensation *f* interstitielle

condensation, surface   condensation *f* en surface

condense   condenser *vb*

condenser   condensateur *m*, condenseur *m*

condenser, air-cooled   condenseur *m* refroidi par l'air

condenser, double bundle   condensateur *m* à double faisceau

condenser, double pipe   condenseur *m* à tubes concentriques/jumelés

condenser, evaporative   condenseur *m* évaporatif

condenser, plate-type   condensateur *m* à plaques

condenser, shell and coil   condenseur *m* à calandre et serpentin

condenser, sprayed coil   condenseur *m* à aspersion

condenser main   conduite *f* de condenseur

condenser output   rendement *m* de condenseur

condenser tube   tube *m* de condenseur

condensing unit   groupe *m* condenseur

condensing unit, air-cooled   appareil *m*/unité *f* de condensation refroidi(e) par air

condensing unit, sealed   groupe *m* compresseur-condenseur hermétique

condition   condition *f*

condition [state]   état *m*
condition, balance   condition *f* d'équilibre
condition, balance(d)   régime *m* d'équilibre
condition, precedent   condition *f* suspensive
condition, surface   état *m* de surface
condition of the air   état *m* d'air/de l'air
conditions, ambient   conditions *f,pl*
  ambiantes/d'ambiance
conditions, boundary   conditions *f,pl* aux limites
conditions, contract, *see* contract condition(s), etc
conditions, controlled   conditions *f,pl* contrôlées
conditions, design   données *f,pl* du projet
conditions, existing   état *m* des lieux
condition(s), failure   conditions *f,pl* à la rupture
conditions, local   conditions *f,pl* locales
conditions, operating   conditions *f,pl* de fonctionnement,
  mode *m* opératoire
conditions, reference   conditions *f,pl* standards/de
  référence
conditions, sales   conditions *f,pl* de vente
conditions, weather   conditions *f,pl* climatiques
condition   conditionner *vb*
conductance   conductance *f*, conductibilité *f*/conductivité
  f spécifique
conductance, thermal   conductance *f* thermique
conductibility   conductibilité *f*
conduction   conduction *f*
conduction, heat or thermal   conduction *f* thermique
conduction loss   perte *f* par conduction
conductivity   conductibilité *f*, conductivité *f*
conductivity, coefficient of thermal   coefficient *m*
  conductibilité
conductivity, heat or thermal   conductivité *f* thermique
conductivity, thermal   conductibilité *f*/conductivité *f*
  thermique
conductivity (k-value), thermal   coefficient *m* de
  transmission thermique, coefficient-k *m*
conductor [electrical]   conducteur *m*
conductor, communications [for controls, alarms, etc]
  conducteur *m* pilote
conductor, current-carrying   conducteur *m* actif
conductor, earth(ing)   conducteur *m* à la masse/de terre
conductor, live   conducteur *m* actif
conductor, neutral   conducteur *m*/fil *m* neutre
conductor, phase   conducteur *m* de phase
conductor, protective   conducteur *m* de protection
conductor, resistance of a   résistance *f* d'un conducteur
conduit   canal *m*, conduit *m*, conduite *f*, gaine *f*
conduit, connecting   conduite *f* de raccordement
conduit, electrical   canalisation *f* électrique
conduit, flat, for surface electric wiring   baguette *f*
  électrique
conduit box   boîte *f* de jonction
conduit coupling   couplage *m*
conduit section, prefabricated, for building into a wall
  wagon *m*
cone   cône *m*
cone, anchoring   cône *m* d'ancrage
cone, inlet [on fan impeller]   pavillon *m* d'aspiration

cone, truncated   cône *m* tronqué
cone of diffusion   cône *m* de diffusion
configuration   configuration *f*
confirm [approval, a report, etc]   sanctionner *vb*
confirm   confirmer *vb*
confirmation   confirmation *f*
confirmation [official]   homologation
conglomerate   conglomérat *m*, aggloméré *m*,
  poudingue *m*
conifer   conifère *m*
connect   connecter *vb*, coupler *vb*, relier *vb*
connect [a branch pipe or duct into a main]   piquer *vb*
connect [electrically]   accoupler *vb*, brancher *vb*
connect in parallel   coupler *vb* en parallèle
connect in series   coupler *vb* en série
connect (to, with)   raccorder *vb* (à)
connect up [sth electrically]   mettre *vb* en circuit
connect up [electrical wiring, etc]   brancher *vb*,
  monter *vb*
connect up [pipes]   brider *vb*
connect wrongly   mal coupler *vb*
connected   raccordé *adj*
connecting   accouplement *m*, raccordement *m*
connecting together   couplement *m*
connecting up [to main supplies]   branchement *m*
connecting up [sth electrically]   mise *f* en circuit
connecting up [wiring]   montage *m*
connection   joint *m*, liaison *f*, raccord *m*, raccordement *m*
connection [electrical]   connexion *f*
connection [node]   nœud *m*
connection [to main supplies]   branchement *m*
connection [structural]   assemblage *m*
connection, air   branchement *m* d'air
connection, air inlet   branchement *m* de prise d'air
connection, appliance [water]   branchement *m* d'appareil
connection, building or house   branchement *m*
  d'immeuble
connection, bypass   raccord *m* de dérivation
connection, connecting up   branchement *m*
connection, consumer   branchement *m* d'abonné/de
  l'abonné
connection, delta   couplage *m*/montage *m* en triangle
connection, electricity   branchement *m* d'électricité
connection, elbow   conduit *m* coudé
connection, flange(d)   joint *m* à brides, raccordement *m*
  par brides
connection, flexible   connexion *f*/raccord *m* flexible
connection, gas main   branchement *m* particulier de gaz
connection, hose   raccordement *m* par tuyau flexible
connection, house or individual   branchement *m*
  particulier
connection, mains   connexion *f* au réseau de distribution,
  raccordement *m* aux réseaux publics
connection, mains [electrical]   branchement *m*
  d'électricité
connection, parallel   connexion *f*/couplage *m* en
  parallèle, raccordement *m* parallèle
connection, pipe   jonction *f* de tuyaux
connection, rail   liaison *f* ferroviaire

connection, rainwater   branchement *m* pluvial
connection, rigid   connexion *f* rigide
connection, riveted   assemblage *m* rivé
connection, road   liaison *f* routière
connection, screwed   filetage *m*, raccord *m* vissé
connection, screwed pipe   raccord *m* vissé pour tuyau
connection, series   connexion *f*/couplage *m*/raccordement *m* en série
connection, side [between two ducts or pipes]   empattement *m*
connection, site   joint *m* de chantier
connection, socket [pipe]   jonction *f* par manchon
connection, star [electrical]   étoile *f*
connection, star-delta   montage *m* en étoile-triangle
connection, suction   aspiration *f*
connection, supplier   branchement *m* de distributeur
connection, supply [pipework]   raccord *m* d'alimentation
connection, threaded   filetage *m*, raccord *m* fileté
connection, welded   assemblage *m* soudé
connection box   boîte *f* de raccordement/de connexion
connection in parallel   couplage *m* en parallèle
connection in series   couplage *m* en série
connection member or piece   pièce *f* de jonction
connection to (the) public mains   raccordement *m* aux réseaux publics
connection with fishplates   éclissage *m*
connector   raccord *m*
connector [connecting piece]   connecteur *m*
connector [electrical]   connecteur *m*, serre-fil *m*
connector [of galvanized metal, for timber structures]   étrier *m*
connector, bolted [timber structure]   aiguille *f*
connector, cable   connecteur *m* (de câble)
connector, circuit   connecteur *m* de circuit
connector, flexible   flexible *m*
connector, ring [between two pieces of wood]   anneau *m*
connector, shear [composite construction]   goujon *m*
connector, threaded   raccord *m* fileté
consent   agrément *m*
consequences   bilan *m*
conserve [protect or safeguard]   sauvegarder *vb*
consistency [e.g. of concrete before setting]   consistance *f*
consolidate   consolider *vb*
consolidate [the footing of a wall]   rechausser *vb*, rempiéter *vb*
consolidated   compacté *adj*, consolidé *adj*
consolidation   consolidation *f*
consolidation [of ground, etc]   compacité *f*
consolidation, initial   consolidation *f* initiale/instantanée
consolidation meter   œdomètre *m*
consortium   consortium *m*, groupement *m* d'entreprises
constant *adj*   constant *adj*
constant [mathematical, physical]   constante *f*
constant, dielectric   constante *f* diélectrique
constituent   constituant *m*
constraints on the design or use of a property   servitudes *f,pl*
constriction   étranglement *m*
construct   bâtir *vb*, construire *vb*, édifier *vb*, élever *vb*

construct [to fit something in, contrive something]   pratiquer *vb*
construct in cob   bousiller *vb*
constructed   construit *adj*
construction   bâtiment *m*, construction *f*, mise *f* en œuvre
construction, composite   construction *f* mixte
construction, cut-and-cover   construction *f* en fouille (ouverte)/en tranchée (ouverte)
construction, fast track   construction *f* en régime accéléré
construction, frame   construction *f* en charpente
construction, panel   construction *f* à panneaux
construction, prefabricated   construction *f* en éléments préfabriqués
construction, principle of modular   principe *m* de la construction modulaire
construction, rammed earth   pisé *m*
construction, random rubble   blocage *m*
construction, reinforced concrete   construction *f* en béton armé
construction, steel   charpente *f*/construction *f* métallique
construction, temporary   construction *f* provisoire
construction, timber frame   construction *f* en bois
construction, tubular   construction *f* tubulaire
construction, welded   construction *f* soudée
construction coordinator   pilote *m*
construction costs index   indice du coût de la construction (ICC)
construction drawing   plan *m* d'exécution
construction engineer   ingénieur *m* des travaux publics
construction engineering   génie *m* civil
construction method   méthode *f* de construction
construction programme, *see* programme, construction
construction programme, master   planning *m* directeur de réalisation
construction regulations [in France]   Code *m* de la Construction
construction schedule   planning *m* général
construction site   chantier *m*, chantier *m* de construction
construction system   système *m* de construction
constructor   constructeur *m*
consultancy office, technical [generally contractor owned in France]   bureau *m* d'études techniques (BET)
consultant   consultant *m*, expert-conseil *m*
consultant, lead [in construction]   maître *m* d'œuvre
consultant/consulting architect   architecte-conseil *m*
consulting engineer [approximate French equivalent]   ingénieur *m* conseil, ingénieur-conseil *m*
consulting engineer's office   bureau *m* d'études
consumption, air   consommation *f* d'air
consumption, current or electric power   consommation *f* de courant
consumption, energy   consommation *f* d'énergie
consumption, excess   excédent *m* de consommation
consumption, (fuel) oil   consommation *f* de mazout
consumption, oil   consommation *f* d'huile
consumption, per capita   consommation *f* par tête

consumption, power   énergie *f* consommée, puissance *f* absorbée
consumption curve   courbe *f* de consommation
consumption measurement   mesure *f* de la consommation
consumption rate   taux *m* de consommation
contact   contact *m*
contact [electrical]   plot *m*
contact, direct [with a bare conductor]   contact *m* direct
contact, door   contact *m* de porte
contact, earth   contact *m* de terre
contact, electric   contact *m* électrique
contact, indirect [through faulty insulation]   contact *m* indirect
contact, normally closed   contact *m* de repos, contact *m* normalement fermé
contact, normally open   contact *m* de travail, contact *m* normalement ouvert
contact, opening   contact *m* ouvert
contact angle [of a driving belt]   angle *m* d'enroulement
contact area or surface   surface *f* de contact
contact breaker   disjoncteur *m*, interrupteur *m*, rupteur *m*
contact breaker, bimetallic   disjoncteur *m* à bilame
contact point   point *m* de contact
contact rating   pouvoir *m* de coupure
contact shoe   sabot *m* de prise de courant
contact [a person]   joindre *vb*
contactor   interrupteur *m*
contactor [electrical]   contacteur *m*
contactor, main   contacteur *m* principal
contain   contenir *vb*
container [box, case, etc]   récipient *m*
container [tank, etc]   réservoir *m*
container [transport]   conteneur *m*
container, refuse   poubelle *f*
contaminant   contaminant *m*
contaminate   contaminer *vb*
contamination   contamination *f*
contemporary   contemporain *adj*
content   contenu *m*
content [relative, percentage]   teneur *f* en
content, ash   teneur *f* en cendres
content, carbon   teneur *f* en carbone
content, cement [of concrete]   dosage *m* en ciment
content, dust [of air]   teneur *f* en poussière(s)
content, equilibrium moisture   teneur *f* en eau d'équilibre
content, heat   contenance *f* thermique, contenu *m* de chaleur, enthalpie *f*
content, moisture   teneur *f* en humidité
content, voids   teneur *f* en vides
content, water   teneur *f* en eau
content of a project   programme *m*
contents   contenance *f*, contenu *m*
contents, list or table of   sommaire *m*, table *f* de matières
contiguous   attenant *adj*, contigu *adj*
contiguous to, be   jouxter *vb*
continuity   continuité *f*

continuity tester [electrical]   détecteur *m* de tension, vérificateur *m* de continuité
continuous   continu *adj*
continuous [of a block or plot of land]   d'un seul tenant
continuous line [road marking]   ligne *f* continue/infranchissable
contort   contourner *vb*
contour   contour *m*, ligne *f*, profil *m*
contour [of an architectural element]   galbe *m*
contour interval   équidistance *f*
contour, velocity   profil *m* de vitesse
contour line   courbe *f* de niveau
contour map   carte *f* avec courbes de niveau
contouring [surveying]   dénivellation *f*
contra-rotation   contra-rotation *f*
contract   contrat *m*, engagement *m*, marché *m*
contract, according to   suivant contrat
contract, breach of   rupture *f* de contrat
contract, competitive   marché *m* par adjudication/sur appel d'offres
contract, construction   marché *m* de travaux
contract, draft   projet *m* d'acte
contract, items covered by the   éléments *m,pl* compris dans le contrat
contract, lump sum   marché *m* à forfait
contract, negotiated   marché *m* négocié/de gré à gré
contract, sale   contrat *m* de vente
contract, turnkey   contrat *m*/marché *m* clef en main
contract awarded to a contractor   contrat *m* passé à une entreprise
contract clause/condition   clause *f* de contrat
contract conditions   cahier *m* des charges, conditions *f,pl* du contrat
contract conditions, general   cahier *m* des clauses (administratives) générales (CCAG)
contract conditions, project   cahier *m* des charges
contract conditions, specific   cahier *m* des clauses administratives particulières (CCAP), cahier *m* des clauses particulières/spécifiques
contract design   solution *f* de base
contract documents   pièces *f,pl* contractuelles
contract manager   chef *m* de chantier
contract of sale   contrat *m* de vente
contract package   lot *m*
contract period   délai *m* d'exécution, durée *f* de contrat
contract value   valeur *f* de contrat
contract, enter into or sign a   passer *vb* un contrat
contracting   rétrécissement *m*
contracting with joint responsibility of all contractors   groupement *m* solidaire
contracting with individually responsible trade contractors   groupement *m* conjoint
contracting firm   entreprise *f*
contraction   contraction *f*, retrait *m*
contractor   entrepreneur *m*, entreprise *f*
contractor, appointed   titulaire *m* du marché
contractor, coordinating   entreprise *f* pilote
contractor, demolition   démolisseur *m*
contractor, direct   entreprise *f* contractante

contractor, general entreprise *f*/entrepreneur *m* général

contractor, main or prime entrepreneur *m* général/principal, entreprise *f* principale

contractor for minor or small works tâcheron *m*

contractor responsible for coordination of the whole works entreprise *f* pilote

contractor's name board or name plate plaque *f* d'entreprise

contractual contractuel *adj*

contraflexure, point of point *m* d'inflexion

contrast contraste *m*

contrast mettre *vb* en contraste

contrast with faire *vb* contraste avec

contributor intervenant *m*

control *adj* témoin *adj*

control commande *f*

control [adjustment, regulation] régulation *f*

control [check] contrôle *m*, vérification *f*

control [device] appareil *m* de contrôle, régulateur *m*

control [of a machine, etc] contrôle *m*

control [setting, tuning] réglage *m*

control [supervision] surveillance *f*

controls organes *m,pl* de commande

control, automatic régulation *f* automatique

control, automatic speed réglage *m* de vitesse automatique

control, capacity régulation *f* de puissance

control, cascade régulation *f* en cascade

control, changeover commande *f* de changement de température été-hiver

control, closed loop contrôle *m* en boucle fermée

control, constant value régulation *f* à valeur de consigne fixe

control, defrost contrôle *m* de dégivrage

control, differential pressure contrôle *m* de pression différentielle

control, direct digital programme *m*/régulation *f* numérique directe

control, feed commande *f* d'alimentation

control, fixed setting régulation *f* à pointe de consigne fixe

control, float flotteur *m* de commande

control, hand/manual commande *f*/régulation *f* manuelle, réglage *m* manuel

control, local government régie *f*

control, noise/sound contrôle *m* acoustique

control, oil pressure cut-out contrôle *m* par pressostat de sécurité d'huile

control, oil temperature cut-out contrôle *m* par thermostat de sécurité d'huile

control, on-off régulation *f* par tout ou rien

control, pneumatic contrôle *m* pneumatique

control, power régulation *f* de puissance

control, press-button commande *f* par bouton poussoir

control, remote commande *f*/contrôle *m*/régulation *f* à distance, télécontrôle *m*, télécommande *f*

control, room régulation *f* de local

control, sequence régulation *f* séquentielle

control, state régie *f*

control, step contrôle *m*/régulation *f* pas à pas

control, temperature réglage *m* de température

control, time clock commande *f* par horloge

control, traffic régulation *f* de la circulation

control, two-position régulateur *m* à deux positions

control, variable air volume régulation *f* de débit d'air variable

controls, electric(al) commandes *f,pl* électriques

controls, timed commandes *f,pl* temporisées

control action action *f* de contrôle

control algorithm algorithme *m* de contrôle

control block organe *m* de commande/de contrôle

control board or panel panneau *m*/tableau *m* de régulation, tableau *m* de commande/de contrôle

control cabin cabine *f* de commande

control cabinet or cupboard armoire *f* de contrôle/de commande

control console or desk [e.g. for stage lighting] pupitre *m* de commande/de contrôle

control cupboard armoire *f* de contrôle/de commande

control deviation écart *m* de réglage

control device or equipment équipement *m* de commande

control equipment [for testing and verification] appareillage *m* de contrôle

control gear dispositif *m* de commande

control installation installation *f* de contrôle

control lever levier *m* de commande/manœuvre

control mechanism mécanisme *m* de contrôle

control module module *m* de contrôle

control module, electronic module *m* électronique de contrôle

control panel tableau *m* de commande/contrôle, panneau/tableau *m* de régulation

control panel, central tableau *m* général de commande

control panel, remote tableau *m* de télécommande

control point point *m* de contrôle

control point, security poste *m* de contrôle sécurité

control range plage *f* de régulation

control relay relai(s) *m* de contrôle

control response réponse *f* de contrôle

control rod tringle *f* de commande

control sequence séquence *f* de contrôle

control strategy stratégie *f* de contrôle

control system système *m* de commande

control system, (automatic) système *m* de contrôle (automatique)

control system, energy management système *m* de gestion énergétique

control time switch minuterie *f* à commande/de contrôle

control unit organe *m* de commande/de contrôle

control valve robinet *m* modulateur, soupape *f* de régulation/de contrôle, vanne *f* de réglage

control valve, circuit vanne *f* de réglage de circuit

control valve, diaphragm soupape *f* à diaphragme

control valve, main vanne *f* principale de contrôle

control valve, pressure soupape *f* de surpression

control valve, solenoid électrovanne *f*

control valve, thermostatic　soupape *f* de contrôle thermostatique
control variable　variable *f* de contrôle
control　commander *vb*, contrôler *vb*, réguler *vb*
control [set, adjust, regulate, etc]　régler *vb*
control [supervise]　surveiller *vb*
controlled locally　manœuvré à pied d'œuvre
controller [device]　appareil *m* de contrôle, régulateur *m*
controller [person]　contrôleur *m*
controller, cascade　régulateur *m* à cascade
controller, electro-pneumatic　contrôleur *m* électro-pneumatique
controller, high pressure　pressostat *m* haute pression
controller, humidity　régulateur *m* d'humidité
controller, (liquid) level　contrôleur *m* de niveau (de liquide)
controller, low pressure　pressostat *m* basse pression
controller, pressure　régulateur *m* de pression
controller, programme　régulateur *m* à programme
controller, proportional action　contrôleur *m* proportionnel
controller, room temperature　contrôleur *m* de température ambiante
controller, slave or sub-master　régulateur *m* esclave
controller, step　contrôleur *m*/régulateur *m* pas à pas
controller, temperature　thermorégulateur *m*
controller, two-position　régulateur *m* à deux positions
controller, volume　contrôleur *m* de volume
convection　convection *f*
convection, forced　convection *f* forcée
convection, heat　convection *f* de chaleur
convection, natural　convection *f* naturelle
convector　convecteur *m*
convector, natural　convecteur *m* naturel
convector, skirting board　plinthe *f* à convecteur
convenience [for living, etc]　commodité *m*
convenience (WC)　commodités *m,pl*
convenience, public　toilettes *f,pl*
convenient　commode *adj*
conventional　classique *adj*
conversation　entretien *m*
conversion　aménagement *m*, conversion *f*, transformation *f*
conversion work of a minor nature　corvée *f*
convert (into)　transformer *vb* (en)
convert (into) [building, etc]　aménager *vb* (en)
converted (into)　aménagé *adj* (en)
convertible　convertible *adj*
convertible into　aménageable en
convertor　convertisseur *m*
convertor, d.c./a.c.　alternateur *m*, convertisseur *m* c.c. en c.a.
convertor, frequency and voltage　convertisseur *m* de fréquence et de tension
converter, rotary　groupe *m* convertisseur
convex　convexe *adj*
convexity [of a road surface]　bombement *m*
convey [goods, etc]　transporter *vb*, camionner *vb*
convey [property]　céder *vb*

conveyance　acte *f* authentique de vente, acte *f* finale de vente
conveyance [of goods]　transport *m*
conveyance [of property, etc]　transmission *f*
conveyor　convoyeur *m*, transporteur *m*
conveyor, apron　sauterelle *f*
conveyor, belt, *see* conveyor belt
conveyor, bucket　transporteur *m* à godets
conveyor, bucket [for sand, aggregates]　sauterelle *f*
conveyor, cement screw　vis *f* à ciment
conveyor, chain　chaîne *f* transporteuse
conveyor, collecting [pavement planing machine]　tapis *m* de récolte
conveyor, continuous (belt)　convoyeur *m* en continu
conveyor, loading [pavement planing machine]　tapis *m* élévateur
conveyor, mobile/portable belt　élévateur-transporteur *m*, sauterelle *f*
conveyor, pneumatic　transporteur *m* à air comprimé
conveyor, sand screw　vis *f* à sable
conveyor, screening　sauterelle *f* cribleuse
conveyor belt　bande *f*/courroie *f* transporteuse, convoyeur *m* à bande, transporteur *m* à bande/courroie/tapis
cooker　cuisinière *f*, poêle *m* de cuisine
cooker, gas　cuisinière *f* à gaz, gazinière *f*
cooker hood　hotte *f*
cooking stove　cuisinière *f*, poêle *m* de cuisine
cooking top unit　table *f* de cuisson
cool　frais *adj*, froid *adj*
cool　rafraîchir *vb*
cool (down)　refroidir *vb*
coolant　fluide *m* frigoporteur, réfrigérant *m*
cooled　réfrigéré *adj*, refroidi *adj*
cooler　appareil *m* réfrigérant, refroidisseur *m*
cooler [of liquids]　rafraîchisseur *m*
cooler, air　refroidisseur *m* à air/d'air
cooler, drinking water　refroidisseur *m* d'eau potable
cooler, evaporative　refroidisseur *m* par évaporation
cooler, tunnel　tunnel *m* de réfrigération
cooler battery　batterie *f* de refroidissement
cooling *adj*　refroidisseur *adj*
cooling　rafraîchissement *m*, refroidissement *m*
cooling, absorption　refroidissement *m* par absorption
cooling, air　refroidissement *m* à air
cooling, closed circuit　refroidissement *m* par circulation
cooling, comfort　rafraîchissement *m* d'ambiance/de confort
cooling, district　distribution *f* urbaine de froid
cooling, evaporative　refroidissement *m* par évaporation
cooling, expansion　refroidissement *m* par dilatation
cooling, forced-draught　refroidissement *m* à convection forcée
cooling, free　rafraîchissement *m* libre
cooling, gas　refroidissement *m* à gaz
cooling, liquid　refroidissement *m* par liquide
cooling, pressure　refroidissement *m* par pression d'air
cooling, radiant　refroidissement *m* par rayonnement
cooling, space　rafraîchissement *m* de locaux

cooling, spot   refroidissement *m* local
cooling, spray   rafraîchissement *m* par pulvérisation
cooling capacity   pouvoir *m* réfrigérant, puissance *f* de refroidissement
cooling chamber   chambre *f* de réfrigération
cooling curve   courbe *f* de refroidissement
cooling disc   disque *m* de refroidissement
cooling down time or period   période *f* de refroidissement
cooling efficiency ratio   coefficient *m* d'effet frigorifique
cooling element   élément *m* de refroidissement
cooling installation or plant   installation *f* de réfrigération
cooling jacket   chemise *f* de refroidissement
cooling jacket (water)   chemise *f* réfrigérante (d'eau)
cooling liquid   liquide *f* de refroidissement
cooling medium   agent *m* de refroidissement, réfrigérant *m*
cooling method   méthode *f* de refroidissement
cooling range   chute *f* de température entrée-sortie
cooling system, (direct)   système *m* de refroidissement (direct)
cooling system, radiant   système *m* de rafraîchissement par rayonnement
cooling tower   réfrigérant *m* à cheminée/atmosphérique, tour *f* de refroidissement/de réfrigération/réfrigérante
cooling tower, cross-flow   tour *f* à courants croisés
cooling tower, film   tour *f* de refroidissement par film
cooling tower, forced draught water   tour *f* de refroidissement à convection forcée/à tirage forcé
cooling tower, natural draught water   tour *f* de refroidissement à tirage naturel
cooling tower, spray   tour *f* de refroidissement à pulvérisation
cooling tower, water   tour *f* de refroidissement d'eau
cooling tower fogging   brouillard *m* de tour de refroidissement
cooling unit   bloc *m* refroidisseur, groupe *m* de réfrigération
cooling water circulating pump   pompe *f* de circulation d'eau de refroidissement
coolness   fraîcheur *f*
coordinate   coordonnée *f*
coordinating contractor   entreprise *f* pilote
coordination [of plans, works, etc]   coordination *f*
coordination, modular [dimensional, in design]   coordination *f* modulaire
coordination, modular [the action, or its result]   modulation *f*
coordination, safety   coordination *f* de la sécurité
coordinator [e.g. of construction team]   coordinateur *m*
coordinator of works on site   pilote *m*
coordinator 'project design'   coordinateur *m* 'projet de l'ouvrage'
coordinator 'project execution'   coordinateur *m* 'réalisation de l'ouvrage'
co-owner   copropriétaire *m*
co-ownership   copropriété *f*
coping   entablement *m*

coping [of a bridge]   chape *f*
coping [of a wall]   chaperon *m*, couronnement *m*, crête *f*
coping [of a well or swimming pool]   margelle *f*
coping, profiled   profilé *m* de couverture
copper *adj*   en cuivre
copper   cuivre *m*
copper plate   cuivrer *vb*
coppice/copse   bocage *m*, taillis *m*
co-proprietor   copropriétaire *m*
copy   copie *f*
copy [e.g. of document or drawing]   exemplaire *m*
copy, certified   ampliation *f*
copy   copier *vb*
copy [a drawing, etc]   reproduire *vb*
corbel   console *f*, corbeau *m*
corbel beam   avant-solier *m*
corbel supporting a mantlepiece   courge *f*
corbelling [out]   encorbellement *m*
cord   cordeau *m*, cordon *m*
cord, asbestos   cordon *m* d'amiante
cord, seven-strand [for blinds, or as sash cord]   septain *m*
cord marking face line of a wall under construction   ligne *f*
cordless [of portable electric tools]   sans fil
core   cœur *m*, noyau *m*
core [of a cable, wall, etc]   âme *f*
core, clay   noyau *m* d'argile
core, cut-off or impermeable [of a dam]   noyau *m* étanche/imperméable
core, reactor   cœur *m* du réacteur
core, rock   carotte *f* de roche
core, service   noyau *m* de service
core, solid   âme *f* massive
core barrel   tube *m* carottier
core sample   carotte *f*, échantillon *m* carotté
core sample extraction or removal   carottage *m*
coring bit   carottier *m*
cork   liège *m*
cork slab   plaque *f* de liège
corner *adj* [i.e. at a corner]   cornier *adj*
corner   coin *m*
corner [of a stone, mantlepiece]   carne *f*
corner, chipped [of a stone, etc]   écornure *f*
corner bead [wood] or moulding   baguette *f* d'angle
corner of two walls, chamfered   pan *m* coupé
corner piece [wood or stone]   écoinçon *m*
corner pier [to stiffen junction of two walls]   chaîne *f* d'angle
corner stone   pierre *f* de refend
corner tower   tour *f* d'angle
corners cut off, with   à pans coupés
cornice   acrotère *m*, corniche *f*
cornice, rectangular and without mouldings   corniche *f* en chanfrein
cornice consisting of successive arc(he)s   corniche *f* cintrée
cornice interrupted by other elements   corniche *f* coupée
correct   corriger *vb*

correct [drawing, error, price, etc]   rectifier *vb*
correction   correction *f*
correction [to a document, drawing]   rectificatif *m*
correction, temperature   correction *f* de température
correction factor   coefficient *m* de correction
corridor   corridor *m*, couloir *m*
corrode   corroder *vb*, rouiller *vb*
corroded   corrodé *adj*
corroded, become   se corroder *vb*, se rouiller *vb*
corrosion   corrosion *f*
corrosion, atmospheric   corrosion *f* atmosphérique/aux
   intempéries
corrosion, metallic   corrosion *f* des métaux
corrosion, stress   corrosion *f* sous contrainte/tension
corrosion inhibitor   inhibiteur *m* de corrosion
corrosion prevention   mesures *f,pl* anti-corrosives
corrosion protection   protection *f* contre la corrosion
corrosion protection layer   couche *f* antirouille
corrosion resistant   non-corrosif *adj*, résistant à la
   corrosion
corrosive   corrosif *adj*
corrugated   ondulé *adj*
corten (steel)   acier *m* corten
corundum   corindon *m*
cosmetic [e.g. changes]   de surface
cost   coût *m*, prix *m*
cost, basic   prix *m* de revient
cost, capital   dépenses *f,pl* d'investissement
cost, construction   coût *m* de la construction
cost, conversion   coût *m* de conversion
cost, direct   coût *m* direct
cost, energy   prix *m* de l'énergie
cost, erection   coût *m* de la construction
cost, estimated   coût *m* estimé/estimatif
cost, extra   coût *m* supplémentaire
cost, final   coût *m* final
cost, first   dépenses *f,pl* d'investissement
cost, indirect [e.g. manufacturing overhead]   coût *m*
   indirect
cost, installation   coût *m* de l'installation
cost, investment   dépenses *f,pl* d'investissement
cost, labour   charge *f* de main d'oeuvre
cost, life-cycle   coût *m* du cycle de vie
cost, marginal   coût *m* marginal
cost, operating   coût *m* d'exploitation, coût *m* de
   l'opération
cost, plant   coût *m* de centrale
cost, prime   coût *m* direct, prix *m* initial
cost, purchase   coût *m* d'achat
cost, replacement   coût *m* de remplacement
cost, reproduction   coût *m* de reproduction
cost, running   coût *m* d'exploitation/de l'opération
cost, weighted average   coût *m* moyen pondéré
costs   frais *m,pl*
costs, analysis or breakdown of   analyse *f* des coûts
costs, buying   frais *m,pl* d'acquisition
costs, calculation of   calcul *m* des coûts
costs, carriage   frais *m,pl* de transport
costs, fixed   frais *m,pl* fixes

costs, heating   frais *m,pl* de chauffage
costs, legal [buying/selling property]   frais *m,pl* de notaire
costs, maintenance   frais *m,pl* d'entretien/de maintenance
costs, operating/running   frais *m,pl* d'exploitation
costs, purchase   frais *m,pl* d'acquisition
costs, transport   frais *m,pl* de transport
costs and quantities, estimate of   métré *m* estimatif
cost accounting   comptabilité *f* analytique
cost allocation   imputation *f* des coûts
cost breakdown   analyse *f* des coûts
cost calculation   calcul *m* des coûts
cost control   contrôle *m* des frais, maîtrise *f* des coûts
cost estimate   devis *m* estimatif
cost excluding profit and overheads   débours(é) *m*
cost increase [e.g. during course of a contract]
   plus-value *f*
cost of works for (site) services   coûts *m,pl* de
   viabilisation
cost or price revision   revision/révision *f* des prix
cost-benefit analysis   analyse *f* coûts-avantages
cost-cutting   réduction *f* des coûts
cost-effective   rentable *adj*
cost-effectiveness   rentabilité *f*
costing   calcul *m* de frais
cost-plus-(fixed-)fee [type of contract]   prix *m* de revient
   plus honoraires
cost-plus-percentage [type of contract]   travaux *m,pl* en
   régie
cost-reimbursement [type of contract]   dépenses *f,pl*
   contrôlées
cost   coûter *vb*
co-tenant   colocataire *m/f*
cottage   fermette *f*
cottage, country   maison *f* de campagne, chalet *m*
cottage, thatched   chaumière *f*
cotter pin   clavette *f*
cottering   coinçage *m*
counsel [advice]   conseil *m*
counterbalance   contrepoids *m*
counterbalance   équilibrer *vb*
counterfall   contre-pente *f*
counterflow *adj*   à contre-courant *adj*
counterflow   contre-courant *m*
counter-nut   contre-écrou *m*
counterpart [copy]   ampliation *f*
countersink   fraise *f*
countersink   fraiser *vb*
countersink bit   fraise *f*
countersinking   fraisage *m*
countersinking for nut and bolt head   lumière *f*
countersunk head   tête *f* fraisée/noyée
counterweight   contrepoids *m*, lest *m*
counting   comptage *m*
counting, pulse   comptage *m* d'impulsions
counting dial(s)   minuterie *f*
country *adj*   campagnard *adj*
country [as opposed to town]   campagne *f*
country, back   arrière-pays *m*
country, out in the   en pleine campagne

country cottage   maison *f* de campagne, chalet *m*
country lodging to let   gîte *m* rural
country seat   gentilhommière *f*, maison *f* de maître
countryside   campagne *f*, paysage *m*
county planning, housing and highways department
   Direction *f* Départementale de l'Équipement (DDE) [in
   France]
county [third layer of local government above canton in
   France]   département *m*
couple [of forces]   couple *m*
couple   accoupler *vb*
coupler [plumbing], *see* coupling
coupler [reinforcement]   coupleur *m*, manchon *m*, raccord
   m
coupler [scaffolding, etc]   raccord *m*
coupler, acoustic [computer]   coupleur *m* acoustique
coupling   accouplement *m*, couplage *m*, liaison *f*, raccord
   m
coupling, close [electrical]   couplage *m* fermé
coupling, conduit   couplage *m*
coupling, direct   accouplement *m* direct
coupling, double male reducing/reduction   double
   manchon *m* mâle/mâle
coupling, flange   accouplement *m* à plateaux
coupling, flexible   manchon *m* (élastique)
coupling, reducing/reduction   accouplement *m*/raccord
   m de réduction
coupling together   couplement *m*
course [of a journey]   trajet *m*
course [of a river, lectures, etc]   cours *m*
course [of bricks or stone]   assise *f*
course [of tiles or slates]   rang *m*
course, band/string [in masonry]   bandeau *m*, cordon *m*
course, basement   soubassement *m*
course, binder   couche *f* de liaison
course, bituminous surfacing or wearing   couche *f* de
   roulement hydrocarbonée, revêtement *m* bitumineux/
   hydrocarboné, couche *f* de surface hydrocarbonée
course, brick   assise *f*/couche *f* de briques
course, brick-on-edge   assise *f* de champ
course, brick-on-end   assise *f* debout
course, damp-proof (DPC)   assise *f* imperméable (à
   l'humidité), barrière *f*/couche *f* d'étanchéité, couche *f*
   isolante, masque *m* étanche
course, eaves [of roof tiles or slates]   rang *m* de gouttière,
   battellement *m*
course, foundation [road construction]   couche *f* de
   fondation
course, diagonal brick   assise *f* en épi
course, heading   assise *f* de boutisses
course, levelling [of masonry or thin stone]   arase *f*
course, lintel [of stone]   plate-bande *f*
course, plinth [of a stone wall]   plinthe *f*
course, ridge [of roof tiles or slates]   rang *m* de faîtage
course, stretching   assise *f* de panneresses
course, surface [road construction]   couche *f* de surface
course depth/height [of bricks, stone]   hauteur *f* d'assise
court   tribunal *m*
court, administrative   tribunal *m* administratif

court, civil [dealing with major matters]   tribunal *m* de
   grande instance
court, civil [dealing with minor matters]   tribunal *m*
   d'instance
court, clerk or registrar of the   greffier *m*
court, commercial   tribunal *m* de commerce
court process server   huissier *m*
courtyard   cour *f*
courtyard, enclosed   cour *f* close
courtyard, small enclosed   courette *f*
courtyard, square, etc, in front of public building
   parvis *m*
courtyard or covered area of a courtyard   préau *m*
cove [in a shore]   anse *f*
cove between wall and ceiling   voussure *f*
covenant   convention *f*
covenant [to do something]   s'engager *vb*
cover   abri *m*, capot *m* (de protection), masque *m*
cover [e.g. manhole]   trapillon *m*
cover [e.g. of a lamp]   capuchon *m*
cover [lid]   couvercle *m*
cover [of a road gulley]   couronnement *m*
cover [of reinforcement]   enrobage *m*
cover [plate or slab]   tampon *m*
cover, access, entry or inspection   trappe *f* de visite
cover, borehole   tampon *m* de forage
cover, cast iron   trappe *f* en fonte, tampon-fonte *m*
cover, cast iron access/inspection   hermétique *f*
cover, flanged   couvercle *m* à bride
cover, heavy duty cast iron   trappe *f* en fonte série
   lourde
cover, inspection   couvercle *m* d'inspection, tampon
   m/trappe *f* de visite
cover, inspection, sealed   tampon *m* hermétique
cover, manhole   couvercle *m* de trou d'homme, plaque
   f/tampon *m* d'égout
cover, reinforcement   enrobage *m* des armatures
cover, removable   couvercle *m* démontable
cover, sealed inspection   tampon *m* hermétique
cover, sealed/watertight   tampon *m* avec étanchéité
cover bead or batten   couvre-joint *m*
cover fillet/moulding, flat   limande *f*
cover flap [on a socket outlet]   volet *m*
cover moulding [e.g. between wall and door frame]
   calfeutrement *m*
cover plate for unused hole in a sanitary appliance
   mascaron *m*
cover seating   siège *m* de trapillon
cover strip   bande *f* de recouvrement, calfeutrement *m*,
   couvre-joint *m*
cover strip [over joint or crack]   parclause *f*, parclose *f*
cover strip [wiring]   baguette *f*
cover   recouvrir *vb*
cover [furniture, a wall]   tapisser *vb*
cover [generally, and one colour by another]   couvrir *vb*
covered   couvert *adj*
covered [from the sky]   à ciel couvert
covered [surfaced, coated]   revêtu *adj*
covered (in) [shingles, weatherboarding]   bardé *adj* (de)

covering   couverture *f*
covering [e.g. of a manhole]   fermeture *f*
covering [interior wall, floor]   revêtement *m*
covering, felt   revêtement *m* de feutre
covering, floor   couverture *f* de sol, revêtement *m* du sol
covering, roof   couverture *f*
covering, temporary, to shelter repairs/works
    parapluie *m*
covering, wall   revêtement *m* mural, revêtement *m* de
    murs
covering, wall [wallpaper]   tapisserie *f*
covering capacity [of a paint]   pouvoir *m* couvrant
covering of a joint or crack to ensure surface continuity
    pontage *m*
cowl   capot *m*
cowl [anti-back draught, on flue or chimney]
    antirefouleur *m*
cowl [to improve draught of a chimney or flue]   aspirateur
    m statique
cowl [of a chimney]   capote *f*, capuchon *m*, chapeau *m* (de
    cheminée)
cowl [on top of a chimney]   abat-vent *m*
cowl, open topped [on chimney]   mitre *f*
cowl, weatherproof   chapeau *m* pare-pluie
cowshed   étable *f*, vacherie *f*
crab [portable winch]   singe *m*
crack   crevasse *f*, crique *f*, fente *f*, fissure *f*
crack [in metal]   crique *f*
crack [in plaster, etc]   lézarde *f*, fil *m*
crack [in stone]   poil *m*
crack [in timber]   gerce *f*
crack, drying/seasoning [timber defect]   fente *f* de retrait,
    gerce *f*
crack, hair(line)   gerçure *f*, fissure *f* capillaire/fine,
    microcrique *f*, microfissure *f*
crack, incipient   amorce *f* de crique/de fissure
crack, long and thin, in a piece of stone   coup *m* de
    sabre
crack, root   fissure *f* à la base
crack, shear   fissure *f* de cisaillement
crack, shrinkage   crique *f*/fente *f*/fissure *f* de retrait
crack, shrinkage [in welds]   tapure *f*
crack, stress   crique *f* de tension
crack, surface   fissure *f* superficielle
crack, surface [in paintwork]   craquelure *f*
crack, thermal stress   crique *f* due aux tensions
    thermiques
crack, transverse   fissure *f* transversale
crack in a ceiling or arch [construction defect]
    couleuvre *f*
crack initiating breakage of a piece of glass   langue *f*
crack length   longueur *f* de fente
crack, fill/seal/stop up a   boucher *vb*/obturer *vb* une
    fissure
crack   se fendre *vb*, se fissurer *vb*
crack [an arch]   étonner *vb*
crack [ground, etc]   fendre *vb*
crack [ground, timber]   gercer *vb*
crack [of metals]   criquer *vb*

cracked   fendu *adj*, fissuré *adj*
cracking   fissuration *f*
cracking [array of fine cracks]   craquelure *f*
cracking [of timber]   gercement *m*
cracking, surface   faïençage *m*
cracking, surface [of tiles, etc]   craquelage *m*
cracking, weld   fissuration *f* dans la soudure
cracking in a road surface   marronnage *m*
crackle/crackling [radio/PA/phone interference]   friture *f*,
    grésillement *m*
cradle [of a flying scaffolding]   plateau *m*
cradle, travelling   échafaudage *m* volant
cradle, window cleaners'   nacelle *f* de nettoyage
craftsman, *see also* labour, labourer, workman   artisan *m*
craftsman, qualified   compagnon *m*, ouvrier *m* qualifié
craftsman, skilled   ouvrier *m* qualifié
cramp   bride *f* de serrage, serre-joint *m*
cramp [holding stones together in masonry]   agrafe *f*
cramp [iron]   crampon *m*
cramp, G   bride *f* à capote
cramp iron [stone, timber]   happe *f*
cramp iron [timber]   clameau *m* (à deux points)
crane   grue *f*
crane, all-terrain (AT)   grue *f* routière tout-terrain
crane, crawler   grue *f* sur chenilles
crane, erection   grue *f* de montage
crane, fixed jib   grue *f* sapine
crane, floating   grue *f* flottante, ponton-grue *m*
crane, gantry   grue *f* à portique
crane, gantry [jib]   pont *m* portique
crane, grab(bing)   benne *f* preneuse
crane, handling   grue *f* de manutention
crane, lattice boom   grue *f* à flèche treillis
crane, loading [vehicle mounted]   grue *f* de chargement
crane, lorry/truck-mounted, *see also* crane, vehicle
    mounted   grue *f* auxiliaire hydraulique, grue *f*
    mobile/sur camion
crane, luffing jib   grue *f* à flèche relevable
crane, mobile   grue *f* mobile
crane, overhead   pont *m* roulant, portique *m* de
    manutention
crane, portal [jib]   pont *m* portique
crane, rail-mounted   grue *f* sur rails
crane, rotary tower   grue *f* à tour pivotante
crane, rough-terrain (RT)   grue *f* automotrice tout
    terrain
crane, self-propelled   grue *f* automobile/automotrice
crane, site   grue *f* de chantier
crane, telescopic (boom)/(jib)   grue *f* à flèche
    télescopique
crane, tower   grue *f* à tour
crane, tower (slewing)   grue *f* à mât pivotant, grue *f* à
    tour à flèche distributrice
crane, travelling (gantry)   pont *m* roulant, portique *m* de
    manutention
crane, trolley jib   grue *f* à flèche horizontale à chariot
crane, truck   camion-grue *m*
crane, truck-mounted   grue *f* mobile/sur
    camion/auxiliaire hydraulique

crane, vehicle mounted (telescopic boom)   grue *f* sur porteur (à flèche télescopique)
crane driver or operator   grutier *m*
crane driver's cabin   cabine *f* du grutier
crane runway   chemin *m* de la grue
crane track   voie *f* de grue
crane working area [radius]   aire *f* de balayage de la grue
crane working radius   portée *f* de grue
crane-excavator (crawler mounted)   pelle-grue *f* (sur chenilles)
crank or crank-handle   manivelle *f*
crank [a pipe, shaft, etc]   couder *vb*
crankcase   carter *m*
cranked   coudé *adj*
cranking [offset]   coudage *m*
crank-pin   bouton *m* de manivelle
crankshaft   arbre *m* manivelle
cranny   fente *f*
crate   cadre *m*, caisse *f*
crater [welding]   cratère *m*
crawl way [beaneath a floor]   vide *m* sanitaire
crawler dozer   bouteur *m* sur chenilles
crawler loader   chargeuse *f* à/sur chenilles
crawler tractor   chenillard *m*
creation   création *f*
credit   crédit *m*
credit, revolving   crédit *m* par acceptation renouvelable
credit, short term   crédit *m* à courte terme
credit facility   facilité *f* de crédit
credit note   avoir *m*
credit, buy on   payer *vb* à terme
creditor   créancier *m*
creek   crique *f*
creep(ing)   fluage *m*
Cremona bolt   crémone *f*
Cremona bolt guide   conduit *m* de crémone
creosote   créosote *f*
crest [of a dam]   couronnement *m*, crête *f*
crest [of a road, embankment]   sommet *m*, point *m* haut
crest, dam   crête *f* de barrage
crest, embankment   crête *f* de remblai
crest, embankment [dike]   crête *f* de la digue
crest, free [of a dam]   couronnement *m* libre
crest, spillway   crête *f* déversante
crest level [of a dam]   cote *f* de la crête
crest width   épaisseur *f* en crête
crevasse   crevasse *f*
crevice   crevasse *f*, fissure *f*
crevice [in wall, plaster, etc]   lézarde *f*
cribwall   mur *m* cribwall
crimp [an electrical connection]   sertir *vb*
crimper [electrical]   sertisseur *m*
crimping tool   pince *f* de sertissage
criteria, design   hypothèses *f,pl* de calcul
criterion   critère *m*
critical   critique *adj*
critical range   domaine *m* critique
criticism   critique *f*
crook [of timber]   voilement *m* longitudinal de rive

cross-   transversal *adj*
cross [plumbing fitting]   croix *f*
cross   traverser *vb*
cross [a cheque]   barrer *vb*
cross a bridge   passer *vb* un pont
cross a thread   fausser *vb* une vis
crossable [e.g. of a flush kerb]   franchissable *adj*
cross-bar   entretoise *f*, traverse *f*
cross-brace   diagonale *f*
cross-brace   entretoiser *vb*
cross-checking [surveying]   recoupement *m*
crossed [e.g. lines]   quadrillé *adj*
crossfall   pente *f* transversale
crossfall [profile]   profil *m* en travers en toit
crossfall [of a road]   dévers *m*
crossfall, straight [road]   profil *m* en travers à pente unique
cross-flow *adj*   à courants croisés
cross-flow   courants *m,pl* croisés
cross-grained [of wood]   rebours *adj*
cross-hatch   contre-hacher *vb*
cross-hatching   contre-hachure *f*
cross-head [Phillips/Pozidrive/Supadrive type screw]   tête *f* cruciforme
crossing   croisement *m*, passage *m*
crossing, diamond [railway lines]   croisement *m* oblique
crossing, half-barrier [railway]   passage *m* à demi-barrière
crossing, level [railway]   passage *m* à niveau
crossing, pedestrian or zebra [UK]   passage *m* clouté/pour piétons
crossing for cyclists   passage *m* pour cyclistes
crossover [railway]   bretelle *f*, croisement *m* de voie
cross-piece   croisillon *m*, croix *f*, traverse *f*
cross-piece [sometimes double]   moise *f*
crossroads   carrefour *m*
crossroads, signal controlled   carrefour *m* à feux
cross-ruling   quadrillage *m*
cross-section   coupe *f* transversale/en travers, profil *m* en travers, section *f* (droite)/(transversale)
cross-section, cambered [of a road]   profil *m* en forme de toit, profil *m* en travers en toit
cross-section, effective   section *f* utile
cross-slot head [for a screwdriver]   pointe *f* cruciforme
crosstalk [sound transmission between spaces]   transmission *f* phonique
crosstalk attenuation   affaiblissement *m* diaphonique
crosswall   mur *m* de refend
crosswall unit   élément *m* mural transversal
crosswise   travers, en
crow's beak [any projection resembling one]   bec *m* de corbin
crowbar   levier *m*, pince *f*
crown [of an arch]   sommet *m*
crown post   poinçon *m*
crowning [of a wall]   couronnement *m*
crucial   critique *adj*
cruciform   cruciforme *adj*
crude [unworked, without finishes]   brut *adj*

crumble away [ground, a structure]   s'écrouler *vb*
crumbling [e.g. building, wall]   caduc *adj*, ruineux *adj*
crumbling [of solid materials]   effritement *m*
crumbling away [of ground, a structure]   écroulement *m*
crush   broyer *vb*
crush [rock, etc]   concasser *vb*
crushed   écrasé *adj*
crushed [of stone or rock]   concassé *adj*
crusher   broyeur *m*
crusher [for asphalt, rock, etc]   concasseur *m*
crusher, impact   concasseur *m* à percussion
crusher, jaw   concasseur *m* à mâchoires
crushing   broyage *m*, écrasement *m*
crushproof   résistant à l'écrasement
cryogenic   cryogénique *adj*
crypt   caveau *m*, crypte *f*
cube   cube *m*
cul-de-sac   impasse *m*
cul-de-sac with closed loop   impasse *m* avec boucle
cultivation   culture *f*
cultivator [earthworks]   émietteuse *f*
culvert   canal *m*, conduite *f* enterrée, ponceau *m*
culvert [dam]   galerie *f*, passage *m* inférieur
culvert [underground]   conduit *m* souterrain
culvert, open   canal *m* à ciel ouvert
cumulative   cumulé *adj*
cup burner   coupelle *f*
cupboard   armoire *f*
cupboard, (built in)   placard *m*, armoire *f* murale
cupboard, clothes   armoire *f* (de chambre à coucher), garde-robe *f*
cupboard, drying   étuve *f* de séchage
cupboard, electrical   armoire *f* électrique
cupboard, fireproof   armoire *f* ignifuge
cupboard, fume   hotte *f* de laboratoire
cupboard, hanging   penderie *f*
cupboard, meter   coffret *m* de compteur
cupboard, staircase   caveau *m*
cupboard, wall   placard *m* suspendu
cupboard space   rangement *m*
cupola   coupole *f*
cupping [of timber]   voilement *m* transversal
curb [purlin between slopes of a mansard roof]   panne *f* de bris
cure [remedy]   remède *m*
cure (sth) [i.e. some defect]   remédier *vb* (à qch)
curing   cure *f*
curing [of concrete]   traitement *m* après prise
curing, completion of [of concrete]   maturation *f*
curing agent/compound [e.g. for concrete]   produit *m* de cure
curing period or time   durée *f* de prise
current [e.g. price]   actuel *adj*
current [of code of practice, etc]   en vigueur
current [present, running]   courant *adj*
current [of air, fluids, electricity, etc]   courant *m*
current [electrical]   ampérage *m*, intensité *f*
current [gas, liquid]   écoulement *m*
current [of water, direction of]   fil *m* de l'eau

current [of water]   courant *m* d'eau
current, against the [of water]   à contre-fil
current, air   courant *m*/écoulement *m* d'air
current, alternating (a.c.)   courant *m* alternatif (c.a.)
current, back   contre-courant *m*
current, convection   courant *m* de convection
current, direct (d.c.)   courant *m* continu (c.c.)/ direct (c.d.)
current, electric   courant *m* électrique, jus[+] *m*
current, induced   courant *m* induit
current, inrush [of a lamp]   courant *m* d'allumage
current, light   courant *m* faible
current, low-tension/low-voltage   courant *m* basse tension
current, no-load   courant *m* à vide
current, peak   courant *m* de pointe, courant *m* de crête
current, polyphase   courant *m* polyphasé
current, primary   courant *m* primaire
current, starting   courant *m* de démarrage
current, stray   courant *m* vagabond
current, surge   courant *m* d'appel
current, three-phase   courant *m* triphasé
currents, cross   courants *m,pl* croisés
currents, eddy   courants *m,pl* de Foucault
current consumption calculation   calcul *m* de la consommation de courant
current decrease [electrical]   abaissement *m* de courant
current meter [electrical]   ampèremètre *m*
current meter [paddle type]   moulinet *m*
current, be   avoir *vb* cours
currently   actuellement *adv*
curtailment (of a reinforcement bar)   arrêt *m* de barre
curtain   rideau *m*
curtain, impermeable [in groundworks]   rideau *m* d'étanchéité
curtain, safety [in a theatre]   rideau *m* de fer
curtain, stage   rideau *m* de scène
curtain, watertight   masque *f* étanche
curtain loop   embrasse *f* (de rideau)
curtain push-rod   lance-rideau *m*
curtain ring   anneau *m*/boucle *f* de rideau
curtain rod   barre *f* à rideau, tringle *f* (de rideau)
curvature   cintrage *m*, courbure *f*
curvature [out of required plane]   gauchis *m*
curvature, radius of   rayon *m* de courbure
curve [in an arch, plate, vault, etc]   cintre *m*
curve [of a beam, etc]   cambrure *f*
curve [of a road]   virage *m*
curve [of an arch]   voussure *f*
curve [on a graph, of a road, etc]   courbe *f*
curve, Atterberg flow   courbe *f* de liquidité Atterberg
curve, basket-handle [approximating to an ellipse]   anse *f* de panier
curve, catenary   arc *m* en chaînette, funiculaire *m*
curve, characteristic   courbe *f* caractéristique
curve, compound   courbe *f* composée
curve, consumption   courbe *f* de consommation
curve, counter   contre-courbe *f*
curve, deflection-load   courbe *f* efforts-déformations

curve, deformation-load   courbe *f* efforts-déformations
curve, drawing or French   pistolet *m*
curve, frequency   courbe *f* de fréquence
curve, grading   courbe *f* granulométrique
curve, hysteresis   courbe *f* d'hystérésis
curve, light distribution   courbe *f* d'intensité lumineuse
curve, load-deflection/deformation   courbe *f* efforts-déformations
curve, loading   courbe *f* de chargement
curve, moisture-density   courbe *f* de teneur en eau-densité
curve, mortality [lighting]   courbe *f* de mortalité
curve, operating   courbe *f* d'efficacité
curve, penetration resistance   courbe *f* de résistance à la pénétration
curve, performance characteristic   courbe *f* de fonctionnement
curve, pressure-void ratio   courbe *f* pression-indice des vides
curve, Proctor   courbe *f* Proctor
curve, rebound   courbe *f* de déchargement
curve, settlement-load   courbe *f* de tassement/de chargement/efforts-déformations
curve, settlement-time   courbe *f* de tassement en fonction du temps
curve, spacing   courbe *f* d'espacement
curve, spiral transition   clothoïde *f*
curve, transition, *see* transition curve
curve (sth)   cintrer *vb* (qch)
curve [a beam, pipe, etc]   courber *vb*
curved   courbe *adj*
curved steel rod for bending or bell-mouthing pipe   broche *f*
customer   client *m*
custom-made   fait sur commande *adj*
cut [excavation]   déblai *m*, excavation *f*, tranchée *f*
cut [wound]   coupure *f*
cut, bevel   coupe *f* fausse
cut, depth of   hauteur *f* de coupe
cut, mitre   coupe *f* biaise
cut, straight [at 90°]   coupe *f* d'équerre/droite
cut and fill   déblais *m,pl* et remblais, remblai *m* et déblai, réutilisation *f* des déblais en remblais
cut   couper *vb*
cut [a key]   refendre *vb*
cut [design on a material]   graver *vb*
cut [earthworks]   déblayer *vb*
cut [excavate or slice]   trancher *vb*
cut [stone in its bedding plane]   déliter *vb*
cut down [a tree]   abattre *vb*
cut down on [e.g. heating]   économiser *vb*
cut down the expenses/costs   rogner *vb* les dépenses/frais
cut down the top of [a wall, a piece of stonework]   déraser *vb*
cut into [incise]   entamer *vb*
cut into [stone, to heighten a sculpture or relief]   refouiller *vb*
cut into lengths, pieces or sections   tronçonner *vb*
cut off   découper *vb*

cut off [the heads of piles]   araser *vb*
cut off the gas/steam/water   couper *vb* le gaz/le vapeur/l'eau
cut out [design, pattern, etc]   découper *vb*
cut to a required profile or shape   chantourner *vb*
cut to length   couper *vb* de longueur
cut to size   coupé aux dimensions nécessaires *adj*
cut up [ground or road]   défoncer *vb*
cut up [logs, etc]   débiter *vb*
cut up [subdivide land, a site, etc]   découper *vb*
cut-and-cover [carrying out the works]   réalisation *f* d'ouvrages enterrés en tranchée ouverte
cut-and-cover [the works]   ouvrages *m,pl* enterrés en tranchée ouverte
cut(-off)   coupure *f*
cut-off [lighting]   défilement *m*
cut-off, core [of a dam]   noyau *m* imperméable
cut-off, sheet pile   parafouille *f* en palplanches
cut-off (of a reinforcement bar)   arrêt *m* de barre
cut-off point   point *m* de coupure
cut-out [electrical]   brise-circuit *m*
cut-out [for fitting or passage of services]   découpe *f*
cut-out, high discharge temperature   thermostat *m* de sécurité au refoulement
cut-out, high-/low-pressure   pressostat *m* haute/basse pression
cut-out, high-/low-pressure safety   pressostat *m* de sécurité haute/basse pression
cut-out, low water level   dispositif *m* de sécurité de manque d'eau
cut-out, overload   interrupteur *m* de surcharge
cut-out, pressure differential   pressostat *m* différentiel
cut-out, safety   disjoncteur *m* de sécurité
cut-out, thermal   interrupteur *m* thermique
cut-out device   dispositif *m* de déconnexion
cut-out in string(er) to receive the end of a stair tread   emmarchement *m*
cut-out point   point *m* de déclenchement
cut-out setting   position *f* de coupure
cutter   fraise *f*
cutter, bolt   coupe-boulons *m*
cutter, cable   coupe-câble *m*
cutter, knot-hole   fraise *f* à dénoder
cutter, pipe   coupe-tubes *m*, coupe-tuyaux *m*
cutter, universal   couteau *m* universel
cutters, wire   coupe-fil *m*
cutting *adj*   tranchant *adj*
cutting   coupage *m*, coupement *m*, fraisage *m*
cutting [excavation, not the process]   déblai *m*, tranchée *f*
cutting [of stone, brick, etc]   taille *f*
cutting [shearing]   cisaillage *m*
cutting, arc   coupage *m* à l'arc
cutting, flame   découpage *m* au chalumeau
cutting, gas   découpage *m* au chalumeau, oxycoupage *m*
cutting, oxyacetylene   découpage *m* autogène
cutting, oxygen   découpage *m* à l'oxygène, oxycoupage *m*
cutting, water-jet   découpe *f* au jet d'eau
cutting down [of trees]   abattage *m*

cutting into [stone, to heighten a sculpture or relief]
  refouillement *m*
cutting out [patterns, etc] découpage *m*
cutting up [of logs, etc] débit *m*
cutting up [of wood, etc] découpe *f*
cutting (work) [metal cutting] travaux *m,pl* de coupage
cutwater [of a bridge] avant-bec *m*, bec *m*, éperon *m*
cycle cycle *m*
cycle, Carnot/Rankine/Stirling/Otto cycle *m* de
  Carnot/Rankine/Stirling/Otto
cycle, cooling cycle *m* de transfert de chaleur
cycle, defrosting cycle *m* de dégivrage
cycle, heat transfer cycle *m* de transfert de chaleur
cycle, life cycle *m* de vie
cycle, loading/unloading cycle *m* de
  chargement/déchargement
cycle, operation cycle *m* opératoire
cycle, refrigeration cycle *m* frigorifique

cycle, reverse cycle *m* inversé
cycle, switching cycle *m* de commutation, fréquence *f* de
  mise en circuit
cycle, water circuit *m*/cycle *m* de l'eau
cycle track piste *f* cyclable
cycling/cyclic operation opération *f* cyclique/périodique
cyclist cycliste *m/f*
cyclists' crossing passage *m* pour cyclistes
cyclone cyclone *m*
cylinder cylindre *m*
cylinder [of a lock] barillet *m*
cylinder, direct réservoir *m* d'eau à chauffe directe
cylinder, gas bouteille *f* à gaz/de gaz, cylindre *m* à gaz
cylinder, hinged cylindre *m* articulée
cylinder, pumping cylindre *m* de pompage
cylinder, slewing [crane, etc] cylindre *m* d'orientation
cylinder displacement cylindrée *f*
cyma cimaise *f*, cymaise *f*, doucine *f*

dab [a surface, with a pad, etc] tamponner *vb*
dab on [paint, etc] taper *vb*
dado lambris *m* d'appui
dado [architecture] cymaise *f*
dado [of a pedestal] dé *m*
daily or per day journalier *adj*
dairy laiterie *f*
dam barrage *m*
dam, arch barrage *m* (en) voûte
dam, arched barrage *m* en arc(he)
dam, buttress barrage *m* à contreforts
dam, clay barrage *m* en argile
dam, concrete barrage *m* en béton
dam, control barrage *m* de retenue/de régulation
dam, earth(fill) barrage *m* en terre/en remblai
dam, gravity barrage *m* poids
dam, impounding retenue *f*
dam, masonry barrage *m* en maçonnerie
dam, multi-arch barrage *m* à voûtes multiples
dam, retaining barrage *m* de retenue
dam, rockfill barrage *m* en enrochement(s)
dam, sluice barrage *m* à vannes
dam, spillway barrage *m* déversoir
dam body corps *m* du barrage
dam crest crête *f* de barrage
dam lake lac *m* de barrage
damage dégâts *m,pl*, dommage *m*
damage, concealed/hidden/latent défaut *m*/vice *m*
  caché/latent/occulte
damage, property dégats *m,pl*/dommages *m,pl* matériels
damage, structural dommages *m,pl*
damage to wood surface resulting from an impact
  frotture *f*

damages, claim for demande *f* de dommages-intérêts
damages for delay pénalité *f* de retard
damages [legal] dommages-intérêts *m,pl*
damage abîmer *vb*, dégrader *vb*, détériorer *vb*,
  endommager *vb*
damaged dégradé *adj*
damming [of a valley] barrage *m*
damp *adj* humide *adj*, moite *adj*
damp humidité *f*
damp humecter *vb*, humidifier *vb*, mouiller *vb*
damp [deaden, noise] étouffer *vb*
dampen [moisten] humecter *vb*
dampening mouillage *m*
dampening medium agent *m* humectant
damper [of shocks, vibration] amortisseur *m*
damper [in ductwork, flues, etc] clapet *m* de réglage,
  registre *m*, volet *m*
damper, air clapet *m* de ventilation, volet *m* d'aération
damper, air inlet clapet *m* de prise d'air
damper, air regulating volet *m* de réglage d'air
damper, butterfly registre *m* papillon/tournant
damper, bypass volet *m* de dérivation
damper, control registre *m*/volet *m* de réglage
damper, control [in a fireplace chimney] trappe *f* de
  fumée
damper, draught control [in a flue pipe] registre *m* de
  tirage
damper, equalizing volet *m* d'équilibrage
damper, exhaust air clapet *m* d'étranglement d'air
damper, fire, *see* fire damper
damper, motorized registre *m* motorisé
damper, multi-leaf registre *m* à lames/à volets
  multiples/à persiennes

damper, non-return   volet *m* anti-retour
damper, opposed blade   registre *m* à volets opposés
damper, outside air   registre *m* extérieur
damper, parallel blade   registre *m* à volets parallèles
damper, return air   clapet *m* de réglage de retour
damper, single-leaf   registre *m* à papillon/à volet unique mobile
damper, sliding   registre *m* coulissant
damper, smoke outlet/vent   trappe *f* de désenfumage
damper, splitter   volet *m* de répartition
damper, volume control   registre *m* de réglage de débit
damper actuator or motor   moteur *m* de registre
damper plate [in ductwork]   clapet *m* à guillotine
dampness   humidité *f*
damp-proof   hydrofuge *adj*
damp-proof course (DPC)   assise *f* imperméable (à l'humidité), barrière *f*/couche *f* d'étanchéité, couche *f* isolante, masque *m* étanche
damp-proof membrane   chape *f* d'étanchéité
danger, contamination   risque *f* de contamination
dangerous [situation]   critique *adj*
dark [colour]   foncé *adj*, sombre *adj*
dark [room, etc]   obscur *adj*
darken a colour   foncer *vb*
darkness, hours of   heures *f,pl* d'obscurité
data   données *f,pl*
data, analogue   données *f,pl* analogiques
data, characteristic   données *f,pl* caractéristiques
data, critical   données *f,pl* critiques
data, field   données *f,pl* en place
data, operating   bases *f,pl* d'exploitation
data, survey   indications *f,pl* topographiques
data, weather   données *f,pl* météorologiques
data acquisition   saisie *f* de données
data bank   banque *f* de données
data collection   recueil *m* des données
data converter   convertisseur *m* de données
data display module   module *m* d'affichage de données
data link [to or between computers]   liaison *f* de transmission
data logger   chaîne *f* d'acquisition (de données)
data logging   concentration *f* de données
data processing   traitement *m* de(s) données
data recorder   enregistreur *m* de données
data sheet   fiche *f* technique, notice *f* de documentation
data storage [computer]   stockage *m* des données
database   base *f* de données
date   date *f*
date [of maturity, for payment, etc]   échéance *f*
date, completion   date *f* d'achèvement
date, due [for payment, etc]   échéance *f*
date, redemption [of a loan]   échéance *f*
date, target   échéance *f*
datum level   plan *m* de comparaison
datum line   ligne *f* de repère
datum point [for setting out, etc]   point *m* de repère
datum surface   surface *f* de référence
day, working   jour *m* ouvrable
day shift   poste *m* de jour

daylight   lumière *f* du jour
daywork   travail *m* à la journée, travaux *m,pl* en régie
dazzle   éblouir *vb*
dazzle [from headlights]   éblouissement *m*
dazzling   éblouissant *adj*
de-aeration   désaération *f*
dead corner [in a rectilinear space]   angle *m* mort
dead leg   branchement *m* fermé/en attente
deadlight   fausse fenêtre *f*, fenêtre *f* fixe
deadline   date *f* limite, délai *m* d'exécution
deal   affaire *f*, marché *m*
deal (timber)   bois *m* blanc
deal, do a   traiter *vb*
dealer, real-estate   marchand de biens *m*
death-watch beetle   vrillette *f*
debenture   obligation *f*
debit   débit *m*
debit balance   solde *m* débiteur
debit   débiter *vb*
debris   débris *m*, détritus *m*
debris [earth, etc]   éboulis *m*
debris [from buildings]   décombres *m,pl*
debt   créance *f*, dette *f*
debt redemption   amortissement *m* d'une dette
debtor   débiteur *m*
deburr [castings]   ébarber *vb*
decametre [10 metres]   décamètre *m*
decarbonization   décalaminage *m*
decarbonize   décalaminer *vb*
decay [of a material, etc]   décomposition *f*
decay [of a building, a structure]   ruine *f*
decay, bacterial   altération *f* bactérienne
decay rate [rate of decrease]   taux *m* de décroissance
decay rate [rate of deterioration]   vitesse *f* de détérioration
decay   se décomposer *vb*, se dégrader *vb*
decay [sth]   décomposer *vb*, dégrader *vb*
deceleration   décélération *f*
deceleration section or zone   zone *f* de décélération
decibel   décibel *m*
decide (on) [a date, a programme, etc]   régler *vb*
deciduous [bushes, trees]   caduc *adj*
deciduous forest   forêt *f* feuillue
deciduous tree   arbre *m* à feuillage caduc
decimal   décimal *adj*
decimal point [comma in French numbers]   virgule *f*
decisive [crucial]   critique *adj*
decisive [of an experiment, test]   concluant *adj*, décisif *adj*
deck [e.g. bridge, floor]   tablier *m*
decking (lightweight) [of a bridge]   platelage *m* (léger)
declaration that no other payments are involved [in property purchase in France]   déclaration *f* de sincérité
declination   déclinaison *f*
decline [e.g. in prices, etc]   baisse *f*
declining balance depreciation   amortissement *m* dégressif
decompose   se décomposer *vb*
decompose [sth]   décomposer *vb*

decomposition   décomposition *f*
decompression   décompression *f*
decontaminate   assainir *vb*, décontaminer *vb*
decorate   décorer *vb*
decorated   décoré *adj*
decoration   ornement *m*
decoration [of a room, house, etc]   décor *m*
decoration, shell-like   coquille *f*
decoration, thin fired clay   tuileau *m*
decorations [works]   travaux *m,pl* de décoration
decorative   décoratif *adj*
decorative element at ridge ends, raised   antéfixe *f*
decorator   décorateur *m*
decorator, interior   décorateur *m* d'intérieur, tapissier *m*
decrease   abaissement *m*, décroissance *f*, réduction *f*
decrease [amounts, numbers, etc]   diminution *f*
decrease [intensity, strength]   diminution *f*, décroissance *f*
decrease [price]   baisse *f*
decrease in speed   ralentissement *m*
decrease [purity of a colour by adding another tint]   casser *vb* un ton
decrease [pressure, price, water level, etc]   baisser *vb*
decreasing   décroissant *adj*
decree [administrative]   arrêté *m*, décret *m*
decrement   décroissance *f*, décroissement *m*
decrepit   vermoulu *adj*
decrepit [building]   caduc *adj*
deduct [from a price]   rabattre *vb*
deduct in advance   prélever *vb*
deduction [from salary, wages]   retenue *f*
deduction [from sum to be paid]   décompte *m*
deduction in advance   prélèvement *m*
deed   acte *f*
deed of sale   acte *f* authentique de vente, acte *f* finale de vente
de-energize [electrically]   mettre *vb* hors circuit/tension
de-energized   hors-tension *adj*
deep   profond *adj*
deepening [of cuts in stone]   refouillement *m*
deep-freezer   congélateur *m*
deepness   profondeur *f*
deface   dégrader *vb*
defect   défaut *m*, vice *m*
defect [in mortar, plaster, paintwork, etc]   bavure *f*
defect [in workmanship]   malfaçon *m*
defect, building   vice *m* de construction
defect, concealed/hidden/latent   défaut *m*/vice *m* caché/latent/occulte
defect, deliberately concealed   faute *f* dolosive
defect, intermittent   défaut *m* fugitif
defect, manufacturing   défaut *m* de fabrication
defect, roll(ing)   défaut *m* de laminage
defect, structural   défaut *m* de construction
defect, surface   défaut *m* de surface
defect, visible   défaut *m*/vice *m* apparent
defect due to bad workmanship   malfaçon *f*
defect in face of cut stone [hole, containing earth]   moie *f*/moye *f*
defective   défectueux *adj*

defect(iveness)   défectuosité *f*
defer   différer *vb*
definition   définition *f*
definitive   définitif *adj*
deflect   fléchir *vb*
deflect, cause to   affaisser *vb*
deflecting vane anemometer   anémomètre *m* à palette
deflection [beam, etc]   affaissement *m*, fléchissement *m*
deflection [bending]   cintrage *m*
deflection [deflected distance]   flèche *f*
deflection [displacement]   déplacement *m*
deflection [distortion]   déformation *f*
deflection [structural movement]   déflexion *f*
deflection, permanent   flèche *f* permanente
deflection indicator   fleximètre *m*
deflection under load   fléchissement *m*
deflection-load curve   courbe *f* efforts-déformations
deflector   chicane *f*, déflecteur *m*
deflector [metal, for rainwater on the slope of a roof]   becquet *m*
deforestation   déboisage *m*
deform   déformer *vb*
deformability   aptitude *f* à la déformation, capacité *f* de déformation, deformabilité *f*
deformation   déformation *f*
deformation, elastic   déformation *f* élastique
deformation, permanent   déformation *f* permanente
deformation, plastic   déformation *f* plastique
deformation properties of a soil   deformabilité *f* d'un sol
deformation work   énergie *f* de déformation
deformation-load curve   courbe *f* efforts-déformations
deformed   déformé *adj*, gauche *adj*
defrost   dégeler *vb*, dégivrer *vb*, déglacer *vb*
defrost control   contrôle *m* de dégivrage
defrosting   dégivrage *m*
defrosting device   dispositif *m* de dégivrage
defrost(ing) method   méthode *f* de dégivrage
defrosting process   procédé *m* de dégivrage
defrosting system   système *m* de dégivrage
de-gas   dégazer *vb*
degradability   altérabilité *f*
degradation   dégradation *f*
degrade   dégrader *vb*
degrease [prior to painting]   dégraisser *vb*
degreasing of sewage   écumage *m*
degree   degré *m*
degree day   degré-jour *m*
degree of [e.g. protection, saturation]   degré *m* de
degrees of freedom   degrés *m,pl* de liberté
dehumidification   déshumidification *f*
dehumidifier   déshumidificateur *m*
dehumidify   déshumidifier *vb*
dehydratant   déshydratant *m*
dehydrate   déshydrater *vb*
dehydration   déshydratation *f*
dehydrator   déshydrateur *m*
de-ice   déglacer *vb*
de-icing   fusion *f* de glace
deionize   déioniser *vb*

delay délai *m*, retard *m*
delay, time temporisation *f*
delay in starting [e.g. of work] retard *m* de commencement
delay penalty pénalité *f* de retard
delay retarder *vb*
delimitation délimitation *f*
deliver livrer *vb*
delivery livraison *f*
delivery [handing over, of a letter, parcel, etc] remise *f*
delivery area zone *f* de livraison
delivery network réseau *m* de distribution
delivery note bordereau *m* de livraison
delivery of a pipe refoulement *m*
delivery period [time for delivery] délai *m* de livraison
delivery point poste *m* de livraison
delivery road accès *m* livraison
delivery side [of a fan, etc] côté *m* de refoulement
delta [of a river] delta *m*
demand besoin *m*
demand (for) demande *f* (de)
demand, limiting of maximum [electrical supply] écrêtage *m* des pointes
demand, maximum/peak demande *f* de pointe
demand, refrigeration demande *f* de froid
demand exiger *vb*
demarcated délimité *adj*
demineralization déminéralisation *f*
demineralize déminéraliser *vb*
demineralized déminéralisé *adj*
demodulation démodulation *f*
demolish démolir *vb*, défaire *vb*
demolish upper part of a structure écimer *vb*
demolition abattage *m*, démolition *f*
demolition contractor démolisseur *m*
demolition permission/permit permis *m* de démolir
demolition rubbish gravats *m,pl*, gravois *m*
demolition waste mouchette *f*
demolition (works) travaux *m,pl* de démolition
demountable démontable *adj*
demulsify désémulsionner *vb*
dense compact *adj*, dense *adj*
densification [of wood] densification *f*
densimeter densimètre *m*
density densité *f*, masse *f* volumique, poids *m* spécifique
density, air densité *f* de l'air
density, dry densité *f* (apparente) sèche
density, flux densité *f* de flux
density, gas densité *f* d'un gaz de fumée
density, housing densité *f* des logements
density, particle densité *f* des particules
density, population densité *f* de la population
density, Proctor densité *f* Proctor
density, smoke densité *f* d'un gaz de fumée, opacité *f* de fumée
density, vapour, *see* vapour density
density, wet densité *f* apparente humide
dent bosseler *vb*
deodorant déodorant *m*, désodorisant *m*

department [area of activity] domaine *m*
department [ministry] ministère *m*
department, administrative service *m*
departure hall hall *m*
departure point point *m* de départ
dependability fiabilité *f*
depollute dépolluer *vb*
deposit [down payment] acompte *m*
deposit [e.g. of dust] dépôt *m*, précipitation *f*, précipité *m*
deposit [guarantee, security] caution *f*
deposit [on a purchase] arrhes *f,pl*
deposit [sediment] dépôt *m*, sédiment *m*
deposit [sludge] boue *f*
deposit, alluvial terrain *m* de transport
deposit, bank dépôt *m* bancaire
deposit, dust dépôt *m*/précipitation *f* de poussière(s)
deposit, guarantee dépôt *m* de garantie
deposit, sediment [the process] formation *f* de sédiments
deposit déposer *vb*
depot dépôt *m*, parc *m*
depot, lorry/truck gare *f* routière
depot, oil fuel parc *m* à mazout
depreciate [write off] amortir *vb*
depreciation dépréciation *f*, moins-value *f*
depreciation [amount written off] amortissement *m*
depreciation, declining balance amortissement *m* dégressif
depreciation, linear amortissement *m* linéaire
depreciation, lumen chute *f* de flux lumineux
depression dépression *f*
depression [in a road, pathway, etc] flache *f*
depressurization dépressurisation *f*
depth profondeur *f*
depth [e.g. of an arch] hauteur *f*
depth [e.g. of slab] épaisseur *f*
depth, effective [e.g. of beam] hauteur *f* utile
depth, excavation hauteur *f* de déblai
depth, laying profondeur *f* de pose
depth, mean hydraulic profondeur *f* hydraulique moyenne
depth, overall hauteur *f* totale
depth, throat épaisseur *f* de gorge
depth, web hauteur *f* de l'âme
depth of base of foundations below ground to avoid freezing garde *f* au gel
depth of cut hauteur *f* de coupe
depth of excavation hauteur *f* de déblai
depth of foundation profondeur *f* de la fondation
depth of laying profondeur *f* de pose
depth of penetration (profondeur *f* de) pénétration *f*
depth of section hauteur *f* du profil
depth of water hauteur *f* d'eau
depth of water [draught] tirant *m* d'eau
depth of web hauteur *f* de l'âme
derating [of electrical equipment] déclassement *m*
derelict abandonné *adj*
derivative dérivée *f*
derive [a mathematical function] dériver *vb*
derrick chèvre *f*

derrick, guy or pole   mât *m* inclinable/de levage, derrick *m*

derust   dérouiller *vb*

descale   décalciner *vb*, désincruster *vb*, détartrer *vb*

descale [a casting, etc]   décalaminer *vb*

descaler   détartreur *m*

descaling   désincrustation *f*, détartrage *m*

description   description *f*

description, detailed   description *f* en détail

description and estimate, detail   détail *m* estimatif

description of finished work   dossier *m* des ouvrages exécutés (DOE)

description of works   descriptif *m*, description *f* de travaux

description of work to be subsequently covered up   constat *m*

desiccant   désséchant *m*

desiccation   dessiccation *f*

desiccator   dessiccateur *m*

design [i.e. contractual]   contractuel *adj*

design   conception *f*, projet *m*

design [in the sense of study]   étude *f*

design [of machines, etc]   construction *f*

design [pattern]   dessin *m*

design, basic (project)   étude *f* de base (du projet)

design, basis of   base *f* de calcul

design, composite   construction *f* mixte

design, computer   étude *f* par ordinateur

design, computer-aided (CAD)   conception *f* assistée par ordinateur (CAO)

design, computer-aided formwork   coffrage *m* assisté par ordinateur (CAO)

design, concept   étude *f* d'avant-projet

design, contract   solution *f* de base

design, detailed   projet *m* détaillé

design, final   projet *m* d'exécution

design, finished   épure *f*

design, light gauge   construction *f* légère

design, lighting   conception *f* d'éclairage

design, preliminary   avant-projet *m* détaillé (APD), dessin *m* préliminaire

design, project   étude *f* de/du projet

design, reinforced concrete   étude *f* de béton armé

design, welded structural   projet *m* de construction soudée

design, working   étude *f*/projet *m* d'exécution

design agency or office   bureau *m* d'études

design agency or office [generally contractor owned in France]   bureau *m* d'études techniques (BET)

design and supervision of the works   étude *f* et supervision des travaux

design conditions   données *f,pl* du projet

design chart or diagram   abaque *m*

design criteria   caractéristiques *f,pl* du projet, hypothèses *f,pl* de calcul

design cut into a rendering [decoration]   sgraffite *f*

design flood   crue *f* nominale

design formula   formule *f* de calcul/de dessin

design load arrangement or case   combinaison *f* de cas de charges

design of formwork, computer-aided   coffrage *m* assistée par ordinateur (CAO)

design of the works   étude *f* des travaux

design performance   performance *f* prévue

design problem   problème *m* de dessin/de conception

design report, detail   dossier *m* technique

design report, scheme stage   avant-projet *m* sommaire (APS), dossier *m* de la solution d'ensemble préconisée

design sketch or study, preliminary   avant-projet *m* sommaire (APS)

design studies   études *f,pl* de conception

design study   étude *f* de/du projet

design study, final   avant-projet *m* détaillé (APD)

design table   abaque *m*

design work   études *f,pl* (de conception), travail *m* d'étude

design   concevoir *vb*, dessiner *vb*, projeter *vb*

design-build   conception-réalisation *f*

designation   désignation *f*

designer   maître *m* d'œuvre, projeteur *m*

designer [architect or engineer]   concepteur *m*

designer, interior   architecte *m* d'intérieur

designing [of machines, etc]   construction *f*

desk   secrétaire *m*

desludge   extraire *vb* les boues

desorption   désorption *f*

despatch   expédition *f*

despatcher   expéditeur *m*

destabilize   déstabiliser *vb*

destroy   détruire *vb*, ruiner *vb*

destruction   destruction *f*

detachable   détachable *adj*

detail   détail *m*

detail   détailler *vb*

detailed   détaillé *adj*

detect [leaks]   déceler *vb*

detecting element   élément *m* détecteur

detection system   système *m* de détection

detector   détecteur *m*

detector, cable   localisateur *m* de câbles

detector, fire   détecteur *m* d'incendie

detector, gas   capteur *m* à gaz

detector, halide leak   détecteur *m* de fuites d'haloïdes

detector, immersion   détecteur *m* immergé

detector, intruder   détecteur *m* d'intrus

detector, leak [generally for gas]   cherche-fuite(s) *m*

detector, leak   déceleur *m*/détecteur *m* de fuite(s)

detector, level   capteur *m* de niveau

detector, movement   détecteur *m* de mouvement

detector, occupancy   détecteur *m* de présence

detector, smoke   détecteur *m* de fumée

deter intruders   dissuader *vb* les intrus

detergent   détergent *m*

deteriorate   détériorer *vb*

deterioration   altération *f*, détérioration *f*, usure *f*

deterioration, powdery, of stone surface   effleurissement *m*

determinate, statically   isostatique *adj*

determination of level [surveying]   dénivellation *f*

determine déterminer *vb*, fixer *vb*
determined fixé *adj*
detonating fuse cordeau *m* détonant
detonator [blasting] détonateur *m*
detonator charge amorce *f*
develop développer *vb*, mettre *vb* en exploitation *f*
develop [the economy of an area] valoriser *vb*
developable [can be built upon] constructible *adj*
developer promoteur *m*
developer [client for building/construction work] maître *m* de l'ouvrage/d'ouvrage
developer, property promoteur *m* immobilier/de construction
development [building, construction] aménagement *m*, développement *m*
development [of land] mise *f* en valeur
development [of prices, a market] évolution *f*
development, back-land (building) construction *f* en fond de parcelle
development plan, area plan *m* de secteur
development plan [long term] schéma directeur *m*
development plan [regional, urban, etc, long term] schéma *m* directeur d'aménagement et d'urbanisme
development regulations [in France] Code *m* de l'Environnement
deviation déviation *f*, écart *m*
deviation, allowable écart *m* admissible
deviation, average écart *m* moyen
deviation, mean square écart *m* quadratique moyen
deviation, permissible écart *m* admissible, tolérance *f*
deviation, standard déviation *f* standard, écart *m* type
deviation or departure from planarity gauchissement *m*
device appareil *m*, dispositif *m*
device, alarm dispositif *m* d'alarme
device, changeover inverseur *m*
device, clamping dispositif *m* de fixation
device, cut-out dispositif *m* de déconnexion
devil float gratton *m*
dew rosée *f*
dew point point *m* de condensation, point *m* de rosée
dew point depression abaissement *m*/écart *m* du point de rosée
dew point rise élévation *f* du point de rosée
dewater assécher *vb*
dewatering assèchement *m*
dewatering [e.g. of concrete] essorage *m*
dewatering [of excavations] épuisement *m*
dewatering screen grille *f* essoreuse
dewatering unit groupe *m* d'essorage
diabase diabase *f*
diagonal diagonale *f*
diagram abaque *m*, diagramme *m*, épure *f*, schéma *m*, tableau *m*, tracé *m*
diagram [graph] courbe *f*
diagram, compression courbe *f* de compression
diagram, design abaque *m*
diagram, enthalpy diagramme *m* enthalpique
diagram, entropy diagramme *m* entropique
diagram, flow, *see* flow diagram

diagram, load-extension diagramme *m* effort-déformation
diagram, moment aire *f* des moments
diagram, polar diagramme *m* en coordonnée polaire
diagram, shear diagramme *m* de l'effort tranchant
diagram, vapour pressure diagramme *m* de pression de vapeur
diagram, wiring schéma *m* de câblage/de montage
dial cercle *m*
dial [of a gauge or meter] cadran *m*
dial gauge [measuring tool] comparateur *m* à cadran
dial(s), recording [e.g. of a watt–hour meter] minuterie *f*
dialling code indicatif *m* téléphonique
diameter diamètre *m*
diameter, clear diamètre *m* absolu
diameter, core [of a bolt] diamètre *m* à fond de filet
diameter, effective diamètre *m* efficace
diameter, exterior diamètre *m* extérieure
diameter, hole diamètre *m* du trou
diameter, internal/internal alésage *m*, diamètre *m* intérieure
diameter, nominal diamètre *m* nominale
diameter, pipe diamètre *m* d'un tuyau
diameter, pitch [of a gear wheel] diamètre *m* primitif
diametrical diamétral *adj*
diamond *adj* [motorway interchange] en forme de losange
diamond diamant *m*
diaphragm diaphragme *m*, membrane *f*
diaphragm [watertight] voile *m* d'étanchéité
diatomaceous earth or diatomite diatomées *f,pl*, kieselguhr *m*, terre *f* d'infusoires
diatomite diatomite *f*
dichroic dichroïque *adj*
die [for threading] filière *f*
die [of a pedestal] dé *m*
die, riveting bouterolle *f*
die-cast couler *vb* sous pression
dielectric diélectrique *adj*
dielectric constant constante *f* diélectrique
diesel [engine, or fuel] diesel *m*
diesel engine moteur *m* diesel
differ différer *vb*
difference différence *f*
difference [between numbers, temperatures, etc] écart *m*
difference, mean différence *f* moyenne
difference of pressure, etc différence *f* de pression, etc
differential différentiel *m*
differential coefficient dérivée *f*
differentiate différencier *vb*
diffraction diffraction *f*
diffuse diffuser *vb*
diffuser [light or ventilation] diffuseur *m*
diffuser, air diffuseur *m* d'air
diffuser, ceiling diffuseur *m* de plafond/plafonnier
diffuser, slot diffuseur *m* à fente
diffuser, supply air bouche *f* d'alimentation
diffuser neck col *m* de diffuseur
diffusing [lighting] diffusant *adj*

diffusion   diffusion *f*
diffusion cone   cône *m* de diffusion
diffusion effect   effet *m* de diffusion
diffusion-absorption system   système *m* à absorption-diffusion
diffusivity, thermal   diffusivité *f* thermique
dig   fouiller *vb*, excaver *vb*
dig [e.g. a tunnel or pile]   creuser *vb*
dig out   excaver *vb*, terrasser *vb*
dig up [opposite of bury]   déterrer *vb*
digester [sewage]   digesteur *m*
digestion (sludge)   digestion *f* (des boues)
digger   excavateur *m*, pelle *f*
digger, bucket   excavateur *m* à godets
digger, mechanical   excavateur *m*, pelle *f* mécanique, pelleteuse *f*
digger, skid-steer   pelle *f* à direction par ripage
digger-loader   tracto-pelle *f*
digging   fouille *f*
digit [number]   chiffre *m*
digital   digital *adj*, numérique *adj*
digital transmission   transmission *f* numérique
digitizer   digitaliseur *m*
dike or dyke [riverside]   digue *f* (de rivage), levée *f*, turcie *f*
dilapidate   dégrader *vb*
dilapidate [become degraded]   se dégrader *vb*
dilapidated   dégradé *adj*, délabré *adj*, ruineux *adj*
dilapidation   dégradation *f*, délabrement *m*
dilatation   dilatation *f*
dilutant   diluant *m*
dilute   diluer *vb*
diluting medium   substance *f* diluante
dilution   dilution *f*
dim, go or grow   pâlir *vb*
dimension   dimension *f*
dimension [indication of]   cote *f*
dimension, external   mesure *f* hors-œuvre
dimension, internal   dimension *f* intérieure, mesure *f* dans œuvre
dimension, nominal/outside   cote *f* nominale/extérieure
dimension, overall   dimension *f* hors tout
dimension arrow [on a drawing]   flèche *f* de cote
dimension line [on a drawing]   trait *m* de cote
dimension without finishes   cote *f* brute
dimension   coter *vb*
dimension in terms of a module   moduler *vb*
dimensioned   coté *adj*
dimensioning [of ducts, pipes, structural members, etc]   dimensionnement *m*
dimensions   taille *f*
dimensions [of a squared piece of timber]   équarrissage *m*
dimmer   réducteur *m* d'éclairage, variateur *m* de lumière
dimmer (switch)   gradateur *m*
dimming [light]   atténuation *f* de la lumière
dining area   coin *m* repas
dining hall   réfectoire *m*
dining room suite [also dining room]   salle *f* à manger
dip [in a road]   point *m* bas

dip-pipe   siphon *m* renversé
dipstick   jauge *f*
direct debit [on a bank account]   prélèvement *m*
direct debit [from a bank account]   prélever *vb*
direct expansion   expansion *f* directe
direct expansion coil unit   batterie *f* à/de détente directe
direct expansion refrigeration system   système *m* de réfrigération à détente directe
direct expansion system   système *m* à détente directe/d'expansion directe
direct   diriger *vb*
direction   direction *f*, sens *m*
direction [e.g. of axis of a building]   orientation *f*
direction [management]   conduite *f*
direction, change/changing of   renversement *m*
direction, in the wrong   à contre-sens
direction of air flow   sens *m* de l'écoulement de l'air
direction of circulation   sens *m* de circulation
direction of motion   sens *m* de déplacement
direction of rolling   sens *m* du laminage
direction of rotation   sens *m* de rotation
direction of the arrow   sens *m* de la flèche
direction of traffic   sens *m* de circulation
directional [e.g. lighting]   directionnel *adj*
directions for use   mode *m* d'emploi
director   directeur *m*/directrice *f*
director, administrative   directeur *m* administratif
director, project   directeur *m* de travaux
director, technical   directeur *m* technique
director of administration   directeur *m* administratif
directors, board of   conseil *m* d'administration
directory enquiries [telephone]   service *m* des renseignements
dirt   saleté *f*
dirt pocket   évacuation *f* de(s) boues
dirtiness   saleté *f*
disagreement (about)   différend *m* (à)
disassembling   démontage *m*
disc, *see* disk
discharge   évacuation *f*, refoulement *m*
discharge [a debt]   quitter *vb*
discharge [electrical]   décharge *f*
discharge [gas, liquid]   écoulement *m*
discharge [of liquid]   déversement *m*
discharge [output]   débit *m*
discharge, characteristic average   débit *m* caractéristique moyen
discharge angle   angle *m* de diffusion
discharge channel   canalisation *f* de refoulement
discharge channel [water]   canal *m* de rejet
discharge chimney [of fan coil unit]   cheminée *f* de refoulement
discharge coefficient   coefficient *m* de décharge
discharge drain pipe   conduit *m* d'évacuation
discharge measurement   mesure *f* de décharge/de débit
discharge rate   débit *m* d'évacuation, degré *m* de décharge
discharge side [of a fan, etc]   côté *m* de refoulement
discharge stroke   course *f* de refoulement

discharge tank   réservoir *m* de décharge
discharge vent   évent *m* de décharge
discharge   décharger *vb*, évacuer *vb*
discharging [electrical]   décharge *f*
discoloration   décoloration *f*
discolour   décolorer *vb*
disconnect   déclencher *vb*
disconnect [electrically]   débrancher *vb*, déconnecter *vb*
disconnect [mechanically]   débrayer *vb*
disconnecter [of an electrical breaker]   déclencheur *m*
disconnection   coupure *f*, déconnexion *f*
disconnection [electrical]   débranchement *m*
discontinuous   discontinu *adj*
discordance [acoustics]   discordance *f*
discount   escompte *m*, rabais *m*
discount [on a price]   remise *f*
discuss   débattre *vb*
discussed, to be   à débattre
disengage   déclencher *vb*
disengagement [of gears]   déclenchement *m*
dished [cover, etc]   en cuvette *adj*
dishing of a horizontal surface   flache *f*
dishwasher   lave-vaisselle *m*
disinfection   désinfection *f*
disintegrate   désagréger *vb*
disintegration   décomposition *f*, désintégration *f*
disintegration [of solid materials]   effritement *m*
disintegration shield [in a lamp]   bouclier *m* d'électrode
disk   disque *m*
disk, hard [computer]   disque *m* dur
disk, sanding/polishing [backing disk]   plateau *m* de polissage
disk memory [computer]   mémoire *f* à disques
disk unit [computer]   unité *f* de disques
disk(ette), floppy [computer]   disquette *f*
diskette unit [computer]   unité *f* de disquette
dislocation   dislocation *f*
dislodgement [of parts of a structure]   dislocation *f*
dismantle   démanteler *vb*, démonter *vb*
dismantling   démontage *m*, dislocation *f*
dismantling (works)   travaux *m,pl* de démantèlement
dispatch   expédier *vb*
dispensation, special   dérogation *f*
dispersant   dispersant *m*
disperse   disperser *vb*
dispersion   dispersion *f*
displace   déplacer *vb*
displace [e.g. brushes of electric motor]   décaler *vb*
displace backwards/forwards [in position, time]   décaler *vb*
displace sideways   riper *vb*
displacement   déplacement *m*
displacement backwards/forwards [in position, time]   décalage *m*
display [computer]   affichage *m*
display cabinet, refrigerated   vitrine *f* réfrigérante
display unit [shelf]   présentoir *m*
display unit, visual (VDU)   console *f* (de visualisation)
display [a notice, etc]   afficher *vb*

disposal, refuse   traitement *m* des ordures ménagères
disposal, waste   élimination *f* des déchets
dispose of [a lease]   céder *vb*
disposition   disposition *f*
dispute (about)   différend *m* (à)
disrepair   décrépitude *f*, délabrement *m*
disrepair, fall into   se dégrader *vb*
dissimilar   dissemblable *adj*
dissipation [e.g. of floodwater]   dissipation *f*
dissolve   dissoudre *vb*
dissolved   dissous *adj*
dissolving [of sth in a liquid]   dissolution *f*
dissonance [acoustics]   discordance *f*
distance   distance *f*
distance [apart]   écart *m*, intervalle *m*
distance [of journey, travel]   trajet *m*
distance, braking   distance *f* de freinage
distance, cumulative   distance *f* cumulée
distance, edge   distance *f* au bord
distance, sight or visibility   distance *f* de visibilité
distance, stopping   distance *f* d'arrêt
distance (between)   écartement *m* (de, entre)
distance between a builder's line and the wall surface   jour *m* de ligne
distance between battens for metal sheet roofing   travée *f*
distance between columns   entrecolonnement *m*
distance between inside of a flue and nearest wooden elements   écart *m* au feu
distance between springs of an arch   ouverture *f* d'un arc
distance between supports   distance *f* entre les supports, travée *f*
distance marker   borne *f*
distance piece   écarteur *m*
distance used as a basis for calculation of haulage costs   relais *m*
distant   éloigné *adj*
distant signal [railway]   signal *m* d'avertissement
distemper [paint]   badigeon *m*
distemper   badigeonner *vb*
distempering   badigeonnage *m*
distil   distiller *vb*
distillation   distillation *f*
distilled   distillé *adj*
distort   déformer *vb*
distort [a calculation or statistic]   fausser *vb*
distort [by twisting, buckling]   tordre *vb*
distort [warp, of or by itself]   se gauchir *vb*
distorted   déformé *adj*
distorted [report, etc]   faux *adj*
distorted [twisted]   tordu *adj*
distorted [wood]   déjeté *adj*
distortion   déformation *f*, distorsion *f*
distortion, permanent   déformation *f* permanente
distortion, shrinkage   déformation *f* de retrait
distribute   distribuer *vb*
distribute [e.g. document]   diffuser *vb*
distribute [e.g. loads]   répartir *vb*

distribute [supplies, etc]   aménager *vb*
distributed   réparti *adj*
distributed, uniformly   uniformément réparti
distributing frame [telephone connections]   répartiteur *m*
distribution   distribution *f*
distribution [e.g. of air]   diffusion *f*
distribution [e.g. of loads, stresses]   répartition *f*
distribution, air   distribution *f* d'air/de l'air
distribution, heat   distribution *f* de chaleur
distribution, overhead [of services]   distribution *f* en
   parapluie/par le haut
distribution, pressure   distribution *f* de pression
distribution, spectral energy   répartition *f* d'énergie
   spectrale
distribution, stress   distribution *f* des contraintes
distribution, temperature   distribution *f* de la
   température
distribution bar [reinforcement]   acier *m*/armature *f*/fer
   m de répartition
distribution block [electrical]   répartiteur *m*
distribution board [electrical]   distributeur *m*, répartiteur
   m, tableau *m* électrique
distribution board or panel, low voltage   tableau *m*
   général basse tension (TGBT)
distribution box   boîte *f* de branchement
distribution box, air   boîtier *m* de diffusion d'air
distribution box [electrical]   coffret *m* de distribution
distribution busbar   barre *f* omnibus de distribution
distribution from above   distribution *f* par le haut
distribution from below   distribution *f* inférieure
distribution network   réseau *m* de distribution
distribution panel   tableau *m* électrique/de distribution
distribution system   système *m* de distribution
distribution system, ring   réseau *m* circulaire
distribution system fed from both ends   réseau *m*
   alimenté en retour/maillé/smillé
distribution system fed from one end   réseau *m* ramifié
distributor   distributeur *m*
distributor [in a sewage treatment filter bed]   épandoir *m*
distributor, district [roads]   réseau *m* secondaire
distributor, primary [roads]   réseau *m* primaire
distributor, soap   distributeur *m* de savon
distributor (steam, etc)   distributeur *m* (de vapeur, etc)
distributor (road)   voie *f* de distribution
distributor road, district or local   route *f* collectrice
distributor road, primary   voie *f* de distribution
district [i.e. cantonal]   cantonal *adj*
district   arrondissement *m*, secteur *m*
district [of a town, e.g. commercial]   quartier *m*
district [second layer of local government in France]
   canton *m*
district, shopping   quartier *m* commerçant
district cooling   distribution *f* urbaine de froid
district heating   chauffage *m* de quartier/collectif/urbain,
   distribution *f* de chaleur à distance
district heating main   conduite *f* principale de chauffage
   à distance
district heating installation or plant   centrale
   f/installation *f* de chauffage urbain

disturb [a sample]   remanier *vb*
disturbance   dérangement *m*
disturbance [alteration, change]   remaniement *m*
disturbance [atmospheric, magnetic]   perturbation *f*
disturbed [of a soil sample]   remanié *adj*
disused   abandonné *adj*
ditch   fossé *m*, tranchée *f*
ditch, drainage   saignée *f*, fossé *m* collecteur/de drainage
ditch, enclosed   rigole *f* couverte
ditch, grassed   fossé *m* herbacé
ditch, intercepting [road construction]   fossé *m* de crête
ditch, open   fossé *m*, rigole *f* ouverte
ditcher [machine]   excavateur *m* à godets, trancheur *m*,
   trancheuse *f*
ditcher, tracked   trancheur *m* sur chenilles
ditcher, wheeled   trancheur *m* sur pneus
diurnal   diurne *adj*
diver   plongeur *m*, scaphandrier *m*
diver, helmet   scaphandrier *m*
diver's or diving suit   scaphandre *m*
diversion   déviation *f*
diversion [of a pipe, flue, etc]   dévoiement *m*
diversion [of a watercourse]   dérivation *f*
diversity   diversité *f*
diversity factor   coefficient *m* d'utilisation, facteur *m* de
   diversité/de simultanéité
divert [the course of]   dériver *vb*
divide   séparer *vb*
divide into sheets   feuilleter *vb*
divide up [into compartments, rooms, etc]   cloisonner *vb*
divided into four [of an opening, by mullions]   bigéminé
   adj
divided into parcels or plots   parcellaire *adj*
dividend   dividende *m*
dividers   compas *f* de mesure, compas *m* à pointes sèches
dividing [separation]   séparation *f*
diving (work)   travaux *m,pl* en plongée
divisible [e.g. a plot]   divisible *adj*
division [distribution, sharing]   distribution *f*, répartition *f*
division [mathematical]   division *f*
division [separation]   séparation *f*
division [separator]   barrière *f*, cloison *f*
division (into)   division *f* (en)
division of a site into building plots   lotissement *m*
do a repair (on)   faire *vb* des réparations (à)
do again   refaire *vb*, répéter *vb*
do odd jobs oneself   bricoler *vb*
do-it-yourself   bricolage *m*
do-it-yourself enthusiast   bricoleur *m*
dock [basin]   bassin *m*
dock [quay]   quai *m*
dock, dry   cale *f*/forme *f* sèche
dock, graving   forme *f* de radoub
dock, loading   quai *m* de chargement
dock, unloading   dock *m*
dock wall   mur *m* bajoyer
docking [of boats]   garage *m*
document   acte *f*, document *m*
document signed before a notary   acte *f* authentique

document signed under hand   acte *f* sous seing privé
document transporter   transporteur *m* des documents
documentation/documents   documentation *f*
documents [file]   dossier *m*
dog [cramp]   crampon *m*
dog [holding stones together in masonry]   agrafe *f*
dog [holding stones or timbers together]   happe *f*
dog, timber   clameau *m* (à deux points)
dolly [piling]   avant-pieu *m*
dome   dôme *m*
dome [architectural]   coupole *f*
dome of a wall niche [architectural]   coquille *f*
domed (raised, rounded)   bombé *adj*
domestic   domestique *adj*
domicile [legal]   domicile *m*
done up [renewed]   refait *adj*
door   porte *f*
door, access   porte *f* d'accès
door, access [for inspection]   porte *f* de visite
door, ash box   porte *f* de cendrier
door, batten   porte *f* en planches
door, cleaning   porte *f* de nettoyage
door, cleaning [flue or chimney]   porte *f* de ramonage
door, cleaning [in the side of a duct]   trappe *f* de nettoyage
door, double   double porte *f*, porte *f* à deux battants
door, electrically opened   portier *m* électrique
door, emergency or escape   porte *f* de secours
door, false or imitation   porte *f*, fausse
door, fire   porte *f* anti-incendie/(de) coupe feu
door, fire-resisting   porte *f* ignifugée
door, flush   porte *f* isoplane/plane
door, flush, flush with wall surface   porte *f* dérobée/perdue
door, folding   porte *f* pliante
door, folding, with two leaves   porte *f* à deux battants, porte *f* portefeuille
door, front   porte *f* d'entrée
door, glass   porte-glace *f*
door, glazed   porte *f* à claire-voie, porte *f* croisée/vitrée
door, hinged   porte *f* à charnière, porte *f* tournante
door, inspection   porte *f* de visite
door, landing   porte *f* palière
door, ledged and braced   porte *f* à barres et écharpes, porte *f* en planches
door, multi-folding   porte *f* accordéon
door, non-functional [blind]   fausse porte *f*
door, panic bolt   porte *f* anti-panique/coup de poing
door, reinforced   porte *f* blindée
door, revolving   porte-revolver *f*, tambour *m*
door, screen or storm   contre-porte *f*
door, sliding   porte *f* à glissière/à coulisse/coulissante
door, small, in a larger door   guichet *m*
door, stable   porte *f* coupée
door, swing   porte *f* battante/va-et-vient
door, up-and-over or swing-up   porte *f* basculante
door casing   chambranle *m*
door closer   ferme-porte *m*
door curtain   portière *f*

door fastening   verrouillage *m* de porte
door flush with [concealed in] **wall panelling**   porte *f* perdue
door folding away to one side of an opening   porte *f* pivotante
door frame   bâti *m*/châssis *m* de porte, huisserie *f*
door jamb [pier or post]   pied-droit *m*, piédroit *m*, poteau *m* d'huisserie/de baie/de porte
door jamb [splay]   ébrasement *m*, ébrasure *f*
door hardware   ferrure *f*
door panel painted darker than surround   anglaise *f*
door set [complete with frame, hinges, etc]   bloc-porte *m*
door sill   seuil *m*
door stud   jambage *m* de porte
door threshold (strip)   barre *f* de seuil
door unit [prefabricated]   bloc-porte *m*
door within a carriage gate   portillon *m*
doorbell   sonnette *f*, carillon *m* de porte
doorhandle   poignée *f* de porte
doorkeeper   portier *m*
doorknob   bouton *m* (de porte)
doormat   paillasson *m* (d'entrée)
doorpost   poteau *m* d'huisserie/de baie
doorstep   seuil *m*
doorstop   butoir *m* de porte/d'une porte
doorway   (ouverture *f* de) porte *f*
dormer, eyebrow   chatière *f*, lucarne *f* bombée/à tabatière
dormer, hip(ped)   lucarne *f* capucine/à croupe
dormer, internal   lucarne *f* en défoncé
dormer, shed   lucarne *f* rampante
dormer skylight   châssis *m* tabatière
dormer window, *see* window
dormer with backward sloping roof   chien-assis *m*, lucarne *f* à demoiselle/en chien assis, lucarne *f* retroussée
dormer with ridged roof   lucarne *f* à deux pentes
dormitory   dortoir *m*
dosage   dosage *m*
dose rate   débit *m* de dose
double   double *adj*
double   doubler *vb*
double-acting   à double effet
double-pole   bipolaire *adj*
Douglas fir [timber]   pin *m* d'Oregon
dovecote   colombier *m*, pigeonnier *m*
dovetail   adent *m*, queue *f* d'aronde
dowel   cheville *f*, goujon *m*
dowel [for timber structure]   fenton *m*
dowel [loose, e.g. to support a shelf]   taquet *m*
dowel, wooden   cheville *f* à bois
downdraught   courant *m* d'air descendant
downfall [e.g. of a building]   ruine *f*
downlighter   luminaire *m* intensif
downpipe   descente *f*, tube *m*/tuyau *m* de descente
downpipe, rain water   descente *f* eaux pluviales, gouttière *f*
downstream   en aval
dozer, crawler   bouteur *m* sur chenilles
dozer, wheeled   bouteur *m* à pneus
draft, first   jet *m* premier

draft, rough [document, plan]   brouillon *m*
draft [documents, etc]   établir *vb*
draft [e.g. an agreement]   minuter *vb*
drafting [a plan, a report]   dressage *m*
drafting [documents, etc]   établissement *m*
drag   force *f* frictive
drag, French   brette *f*, chemin *m* de fer, guillaume *m*
drag [e.g. a mould board]   traîner *vb*
drag [finish, with French/toothed drag]   bretteler *vb*
dragging   traînant *adj*
dragline   drag(ue)line *f*, drague *f* à câble
dragline [excavator]   pelle *f* à benne traînante/en dragueline
dragline, walking   dragline-marcheuse *f*, pelle *f* mécanique marcheuse
dragline bucket   benne *f* racleuse
dragline excavator   pelle *f* à câbles
drain [underground]   conduit *m* souterrain
drain   drain *m*
drain [bottom outlet]   bouche *f* d'égout (BE)
drain [of a sewer]   cunette *f*
drain [open]   rigole *f*
drain [outlet]   évacuation *f*
drain [pipe]   tuyau *m* d'écoulement
drain, bath   bouche *f* d'évacuation du bain
drain, buried   drain *m* souterrain
drain, catch   drain *m* d'interception
drain, collector   drain *m* d'interception, fossé *m* collecteur
drain, condensation [in joinery]   trou *m* de buée
drain, earthenware   conduit *m*/drain *m* en grès
drain, floor   puisard *m*
drain, French   drain *m* français/de pierraille, fossé *m* couvert
drain, main   canal *m* d'écoulement principal, collecteur *m*, égout *m* principal
drain, main collector   collecteur *m* principal, grand collecteur *m*
drain, open   fossé *m* collecteur/de drainage
drain, outfall   évacuateur *m*
drain, road   caniveau *m*
drain, sand   drain *m* de/en sable
drain, storm (water)   collecteur *m* (eaux) pluviales, conduite *f* des eaux pluviales, égout *m*/collecteur *m* pluvial
drain, street   caniveau *m*
drain, subsoil   drain *m* profond/de la fondation
drain, subsoil [road construction]   système *m* drainant
drain, surface   fossé *m*
drain, underground   conduit *m* souterrain
drain, weeping   drain *m* pleureur
drain beneath a dam, cut-off   parafouille *f*
drain channel [of a bridge]   barbacane *f*
drain down pipe   tuyau *m* de vidange
drain of loose stones   pierrée *f*
drain pan   bac *m* de récupération
drain pipe, collector, from several appliances   collecteur *m* d'appareils
drain pipe, discharge   conduit *m* d'évacuation

drain pipe, perforated   tuyau *m* de drainage perforé
drain pit [sewage]   bassin *m* de décantation
drain tank   puisard *m*
drain tap   robinet *m* purgeur/de vidange
drain trap   siphon *m*
drain tube   tube *m* de vidange
drain well   puits *m* drainant/filtrant
drain   assécher *vb*, drainer *vb*, vider *vb*
drain [a cylinder]   purger *vb*
drain [empty]   vidanger *vb*
drain [empty itself]   se vider *vb*
drain (off)   évacuer *vb*
drainage *adj*   drainant *adj*
drainage   assainissement *m*, drainage *m*
drainage [flow]   écoulement *m*
drainage [of ground]   drainage *m*
drainage [of water]   évacuation *f*
drainage [sewerage system]   égout *m*
drainage, domestic   assainissement *m* d'habitation
drainage, foul water   évacuation *f* des eaux vannes
drainage, house   assainissement *m* d'habitation
drainage, main   tout à l'égout *m*
drainage, mole   drainage *m* taupe
drainage, rain water   assainissement *m* pluvial, évacuation *f* des eaux pluviales
drainage, roof   évacuation *f* des eaux du toit
drainage, storm water   assainissement *m* pluvial, canalisation *f* des eaux de pluie
drainage, subsurface   drainage *m* souterrain
drainage, surface   drainage *m* de surface
drainage, town   assainissement *m* urbain
drainage, waste water   évacuation *f* des eaux usées
drainage area   aire *f* de drainage, zone *f* de drainage
drainage basin   aire *f*/bassin *m* de drainage
drainage basin (catchment area)   bassin *m* versant
drainage blanket [groundworks]   tapis *m* drainant
drainage channel   fossé *m* de drainage, rigole *f* d'évacuation
drainage channel [road]   caniveau *m*
drainage ditch, collector   fossé *m* collecteur
drainage fitting   matériel *m* de drainage
drainage gallery   galerie *f* de drainage
drainage inlet [road]   bouche *f* à grille
drainage layer   couche *f* drainante
drainage layer [groundworks]   tapis *m* de drainage
drainage network   réseau *m* d'évacuation/de drains
drainage path   chemin *m* de drainage
drainage pipe [in a dam]   colonne *f* de drainage
drainage pipe-layer, mole   pose-drains *m*
drainage pit [e.g. for draining oil from a vehicle]   fosse *f* de vidange
drainage plan   plan *m* d'évacuation des eaux
drainage system   ouvrage *m* d'assainissement, réseau *m* d'évacuation/de drains, système *m* de drainage
drainage system, combined/separate   système *m* d'assainissement unitaire/séparatif
drainage works   ouvrage *m* de drainage/d'assainissement
drain-cock, automatic   purgeur *m* automatique
drained   drainé *adj*

draining   purge *f*
draining [emptying]   vidange *f*
draining board [of a sink]   égouttoir *m*
draining culvert or sluice   pertuis *m* de vidange
draining device   dispositif *m* de vidange
draining off   évacuation *f*
draught   courant *m* d'air
draught [depth of water]   tirant *m* d'eau
draught [of a chimney, flue]   tirage *m*
draught, balanced   tirage *m* équilibré
draught, chimney or flue   tirage *m* de cheminée
draught, forced   tirage *m* mécanique
draught, induced   tirage *m* par aspiration
draught control damper [in a flue pipe]   registre *m* de tirage
draught excluder, *see also* draught strip   bourrelet *m*
draught indicator   indicateur *m* de tirage
draught stabilizer   régulateur *m* de tirage
draught strip   bande *f* de calfeutrement, bourrelet *m*, colmatage *m*
draughting, computer-aided (CAD)   dessin *m* assisté par ordinateur (DAO)
draughting machine   machine *f* à dessiner
draught-proofed   calfeutré *adj*
draught-proofing   calfeutrage *m*, calfeutrement *m*
draughtsman   dessinateur *m*
draw wire   tire-fils *m*
draw knife   couteau *m* à deux manches, plane *f*
draw   dessiner *vb*, tracer *vb*
draw a cheque   tirer *vb* un chèque
draw (a furnace)   décrasser *vb* (un fourneau)
draw a plan   lever *vb* un plan
draw layout of joints [stonework, formwork, tiling, etc]   calepiner *vb*
draw off   soutirer *vb*
draw out [stretch]   étirer *vb*
draw up   aligner *vb*
draw up [documents, etc]   établir *vb*
draw up [water, etc]   aspirer *vb*
draw water from   puiser *vb*
drawbar pull   force *f* de traction
drawbridge   pont *m* levant, pont-levis *m*
drawdown   abaissement *m*
drawer   tiroir *m*
drawer slide   coulisse *f* de tiroir
drawing   dessin *m*, plan *m*, tracé *m*
drawing [of metals]   étirage *m*
drawing, approved   plan *m* approuvé
drawing, as-built or as-constructed   dessin *m*/plan *m* de recollement
drawing, axonometric   dessin *m* axonométrique
drawing, circuit   plan *m* de circuit, schéma *m* de câblage
drawing, cold   étirage *m* à froid
drawing, construction   plan *m* d'exécution/de construction
drawing, detail   dessin *m* de détail, plan *m* de détails
drawing, diagrammatic   dessin *m* schématique
drawing, dimensioned   dessin *m*/plan *m* coté

drawing, ductwork   plan *m* des canalisations
drawing, elevation   dessin *m* en élévation
drawing, formwork   plan *m* de coffrage
drawing, indicative   plan *m* guide
drawing, isometric   dessin *m* isométrique
drawing, joint layout [stonework, formwork, tiling, etc]   calepin *m*
drawing, layout   épure *f*, plan *m*
drawing, manufacturing   dessin *m* industriel
drawing, outline   plan *m*
drawing, pencil   dessin *m* au crayon
drawing, perspective   dessin *m* en perspective
drawing, plan   dessin *m* des plans
drawing, preliminary   plan *m* provisoire
drawing, reinforcement   plan *m* de ferraillage
drawing, scale   dessin *m* à l'échelle
drawing, section   dessin *m* en coupe
drawing, schematic   schéma *m*
drawing, shop   dessin *m* d'atelier/d'exécution, plan *m* d'atelier
drawing, working   dessin *m*/plan *m* d'exécution, épure *f*
drawing, (work)shop   dessin *m*/plan *m* d'atelier, dessin *m* d'exécution, plan *m* de construction
drawings, as-built/as-constructed   dessins/plans *m,pl* de recollement
drawings, working   plans *m,pl* d'exécution des ouvrages (PEO)
drawing board   planche *f* à dessin(er), table *f* à dessin
drawing compass   compas *m*
drawing curve   pistolet *m*
drawing near   approche *f*
drawing office   bureau *m* de dessin
drawing office [design office]   bureau *m* d'études
drawing showing overall dimensions of a building   géométral *m*
drawing to scale   dessin *m* à l'échelle
drawing up [a plan, a report]   dressage *m*
drawing up [documents, etc]   établissement *m*
drawing up [e.g. manuals]   rédaction *f*
drawing-in box, cable   boîte *f* de tirage
drawn, cold   étiré à froid
drawn wire   fil *m* tréfilé
draw-off point   point *m* de tirage
draw-off point [water]   point *m* de puisage
draw-off valve   robinet *m* de puisage
drawshave   couteau *m* à deux manches, plane *f*
dredge   draguer *vb*
dredge(r)   drague *f*
dredger, bucket   drague *f* à godets
dredger, suction   drague *f* aspirante/suceuse
dredging   dragage *m*
drench   inonder *vb*
drench [masonry before rendering]   abreuver *vb*
dress circle [of a theatre]   premier balcon *m*
dress [a wall, stone, etc]   dégauchir *vb*
dress [joinery, timber]   blanchir *vb*
dress [masonry with a percussion tool]   piquer *vb*
dress [metal or timber]   corroyer *vb*
dress [metals]   apprêter *vb*

dress or edge timber to produce square edges dédosser *vb*

dress or trim timber with a drawing knife chapoter *vb*

dress stone to give a rustic (rough) finish rustiquer *vb*

dress stone with a *laye* layer *vb*

dressed with sharp edges by a percussion tool [of stone, etc] piqué *adj*

dressing [of a piece of stone, timber, etc] dressage *m*

dressing [of brick, etc] taille *f*

dressing [of sawn timber] corroyage *m*

dressing [of stone] dégauchissage *m*, piquage *m*, taille *f*, taillage *m*

dressing [squaring of stone at the quarry] prétaillage *m*

dressing, preparatory, for a moulding, etc taille *f* d'ébauche

dressing, rough taille *f* brute

dressing, rough [rustic], **of stone** rusticage *m*

dressing, smooth taille *f* lisse/plane

dressing, split-face taille *f* éclatée

dressing, stone taillage *m*

dressing, surface taille *f* de parement

dressing of a visible surface taille *f* de parement

dressing of joinery, final parage *m*

dressing of side of stone to form a joint taille *f* d'un joint

dressing of stone, final, before polishing préparation *f* au poli

dressing to bed a stone on another taille *f* d'un lit

dressing to give a 'rustic' finish taille *f* rustiquée

dressing with a toothed hammer taille *f* layée

dried up desséché *adj*

drift [from zero, of an instrument] décalage *m*

drift [of a fluid] courant *m*

drift [punch] mandrin *m*, poinçon *m*

drift [tool] broche *f*, chasse *f*, mandrin *m*

drifter, hydraulic perforateur *m* hydraulique

drift eliminator [cooling tower] séparateur *m* de gouttelettes [tour de refroidissement]

drill perceuse *f*

drill, electric perceuse *f* électrique

drill, hammer perceuse *f* à percussion

drill, hammer or pneumatic marteau-perforateur *m*

drill, masonry foret *m* de maçonnerie

drill, percussion perceuse *f* à percussion

drill, pinion hand chignole *f*

drill, pneumatic marteau *m* piqueur, marteau-perforateur *m*

drill, pneumatic [for breaking concrete] brise-béton *m*

drill, rock marteau-perforateur *m*, marteau *m* pneumatique

drill, twist foret *m* (hélicoïdal)

drill bit mèche *f*

drill core sample échantillon *m* carotté

drill hole [geotechnics] forage *m*, sondage *m*

drill rig engin *m* de forage

drill rig, hydraulic perforatrice *f* hydraulique

drill stand support *m* de perceuse

drill percer *vb*

drill [e.g. tunnel, pile, hole] forer *vb*

driller foreur *m*

drilling forage *m*, perçage *m*, percement *m*

drilling [a tunnel, pile, hole] forage *m*

drilling, percussion [in the ground] sondage *m* par percussion

drilling bit couronne *f*

drilling fluid fluide *m* de forage/de perforation

drilling mud boue *f* de forage

drilling machine or unit foreuse *f*

drilling spoil débris *m* de forage

drilling platform plate-forme *f* de forage

drilling rig installation *f* de forage

drilling rig of timber derrick *m*

drilling tool fraise *f*

drinkable potable *adj*

drinking fountain borne-fontaine *f*

drinking water supply distribution *f* d'eau potable

drip [drop] goutte *f*

drip [under a soffit, window ledge, etc] coupe-larme *m*, goutte *f* d'eau, larmier *m*

drip edge larmier *m*

drip moulding jet *m* d'eau

drip pipe pipette *f*

drip tray bac *m* de récupération, égouttoir *m*

drip égoutter *vb*

dripstone capucine *f*, goutte *f* d'eau, larmier *m*

dripstone, outer edge/fillet of a mouchette *f*

drive commande *f*, entraînement *m*

drive [lined with trees] allée *f*

drive, belt commande *f*/transmission *f* par courroie (trapézoïdale)

drive, bevel transmission *f* par pignons

drive, direct entraînement *m* direct

drive, fan transmission *f* de ventilateur

drive, friction transmission *f* à friction

drive, private voie *f* privée

drive, pump transmission *f* de pompe

drive, v-belt entraînement *m* à courroie trapézoïdal

drive chain [e.g. in an escalator] chaîne *f* d'entraînement

drive shaft arbre *m* de commande

drive actionner *vb*, entraîner *vb*

drive [a pile] battre *vb*, foncer *vb*

drive [a tunnel, etc] percer *vb*, excaver *vb*

drive [a vehicle] conduire *vb*

drive [rivets] poser *vb*

drive in enfoncer *vb*

drive in [a nail with a nail punch] chasser *vb*

drive in [nail, post, etc] ficher *vb*

drive in to refusal enfoncer *vb* à refus

driven in enfoncé *adj*

driver [device] actionneur *m*

driver [of a vehicle] chauffeur *m*, conducteur *m*

driver, pile, *see* pile driver

driver's cabin cabine *f* du conducteur

driving [a tunnel] percement *m*

driving [of a vehicle] conduite *f*

driving [piling] battage *m*

driving band [round a pile head] couronne *f*

driving chain chaîne *f* de commande

driving head [piling]   tête *f* de battage
driving in   enfoncement *m*
driving in [piles]   fonçage *m*
driving licence   permis *m* de conduire
driving resistance of the pile   résistance *f* du pieu au battage
driving shaft   arbre *m* de commande
driving shoe [piling]   sabot *m* de battage
drizzle   bruine *f*
drop [difference in level]   dénivellation *f*
drop [of water level, pressure, price, etc]   baisse *f*
drop [in the sense of fall]   chute *f*
drop in value   moins-value *f*
drop system of water distribution   distribution *f* en parapluie
drop [fall, an object]   tomber *vb*
drop [let fall]   laisser *vb* tomber
drop [temperature, price, etc]   baisser *vb*
droplet   gouttelette *f*
dropper [dropping conduit]   conduite *f* verticale
drought   sécheresse *f*
drum [for oil, etc]   bidon *m*
drum [for tar]   gonne *f*
drum [for winding up a hose, extension cable, etc]   enrouleur *m*
drum [of a column, under a cupola or dome, etc]   tambour *m*
drum, brake   tambour *m* de frein
drum, filter   tambour *m* filtrant
drum, mixing   tambour *m* mélangeur/de malaxage
drum mix(ing) plant, asphalt   tambour *m* sécheur enrobeur
dry   sec *adj*
dry [country, land]   aride *adj*
dry (place)   sec *m*
dry as a bone   sec comme de l'amadou
dry bulb   bulbe *m* sec
dry ice   neige *f* carbonique
dry rot   carie *f*/pourriture *f* sèche
dry rot fungus [*Merulius lacrymans*]   mérule *m*
dry [in a kiln, oven, etc]   étuver *vb*
dry (out)   sécher *vb*
dry up   assécher *vb*, dessécher *vb*, sécher *vb*
dry up [a spring, well, stream, etc]   tarir *vb*
dryer   déhydrateur *m*, dessiccateur *m*, sécheur *m*, installation *f* de séchage
dryer, clothes   sécheuse *f*
dryer, hand   sèche-mains *m*
dryer, tumble   sèche-linge *m*
drying *adj*   siccatif *adj*
drying   séchage *m*
drying, air   séchage *m* à l'air
drying agent [siccative]   siccatif *m*
drying batten [between pieces of green timber]   trésillon *m*
drying bed, sludge   lit *m* de séchage (des boues)
drying cabinet [for clothes]   sèche-linge *m*
drying crack or split [timber defect]   fente *f* de retrait, gerce *f*
drying kiln   étuve *f* de séchage

drying out   dessiccation *f*
drying unit   groupe *m* sécheur
dryness   siccité *f*
dryness [of weather]   sécheresse *f*
drystone facing [of an embankment]   perré *m*
drystone wall   mur *m* de pierres sèches
dual duct air-conditioning system   système *m* de climatisation à double conduit
dual duct installation   installation *f* à deux conduits
dual duct system   système *m* à deux conduits
duct   canal *m*, conduit *m*, gaine *f*
duct [for services]   caniveau *m*
duct [underground, for services]   galerie *f*
duct(work)   canalisation *f*
duct, air   conduit *m*/gaine *f* d'air
duct, air distribution   gaine *f* de distribution d'air
duct, air inlet   canalisation *f* de prise d'air
duct, auxiliary wiring [communications wiring, etc]   colonne *f* pilote
duct, branch   conduit *m* secondaire
duct, built-in   conduit *m* incorporé
duct, bypass   conduit *m* de dérivation
duct, cable   gaine *f* de câbles
duct, common   conduite *f* collective
duct, communications   gaine *f* informatique
duct, control cable   canalisation *f* des câbles de contrôle
duct, cooling   conduit *m* de refroidissement
duct, discharge   canal *m* de sortie
duct, exhaust air   canalisation *f* d'air évacué
duct, exhaust/extract   conduit *m*/gaine *f* d'évacuation/d'extraction
duct, flexible   conduit *m* d'air flexible, gaine *f* souple
duct, fresh air   canalisation *f* d'air frais
duct, high-velocity   gaine *f* haute vitesse
duct, inlet   gaine *f* d'entrée
duct, kitchen extract   conduit *m* de ventilation de la cuisine
duct, leakage   conduit *m* de fuite
duct, main   conduit *m* principal
duct, mixing   conduit *m* d'air mélangé
duct, pipe   canalisation *f* pour tuyaux
duct, recirculation   gaine *f* de recirculation
duct, return air   canalisation *f* d'air de retour, gaine *f* de reprise/de retour d'air
duct, riser   colonne *f* montante
duct, service(s)   alvéole *f* technique, gaine *f* de service, gaine *f* technique
duct, sheet metal   conduit *m* en tôle
duct, smoke extract or ventilation, for a whole building   shunt *m*
duct, supply   conduite *f* d'alimentation
duct, supply air   gaine *f* d'alimentation
duct, trunk   conduit *m* principal
duct, underfloor or underground   canal *m* souterrain
duct, ventilating/ventilation   canal *m* d'air, conduit *m*/gaine *f* d'aération/de ventilation
duct fixed against a wall   conduit *m* accolé à un mur
duct fixed to but standing free of a wall   conduit *m* adossé à un mur

duct leak(age)   fuite *f* de conduit
duct sealant   mastic *m* pour canalisations
duct size   dimension *f* de conduit
duct system   réseau *m* de conduits, système *m* de canalisation
ductile   ductile *adj*
ductility   ductilité *f*
ducting, cable   goulotte *f*
ductwork   canalisation *f*
ductwork drawing   plan *m* des canalisations
ductwork system   réseau *m* de distribution d'air
due [payment, etc]   exigible *adj*
due date [for payment, etc]   échéance *f*
dues   droits *m,pl*
dull [finish]   mat *adj*
dull   ternir *vb*
dullness of paint due to absorption by substrate   embu *m*
dumb waiter   monte-plats *m*
dump   décharge *f*
dumper [site]   moto-basculeur *m*
dumper, four-wheel drive   moto-basculeur *m* à quatre roues motrices
dumper or dump-truck   dumper *m*, tombereau *m* (automoteur)
dumper truck   camion *m* benne
dune   dune *f*
duplicate [a drawing, etc]   reproduire *vb*
durability   durabilité *f*, résistance *f* au vieillissement
durable   durable *adj*
duralumin(ium)   duralumin *m*
duration   durée *f*
dust   poudre *f*, poussière *f*
dust band or strip   bande *f* de poussière

dust collector   dépoussiéreur *m*, récupérateur *m* de poussière
dust collector, dry   dépoussiéreur *m* à sec
dust collector, multicyclone dry   dépoussiéreur *m* à multicyclones
dust collector, wet   dépoussiéreur *m* à voie humide
dust content [of air]   teneur *f* en poussière(s)
dust eliminator   capte-suie(s) *m*
dust extraction or removal   dépoussiérage *m*
dust extraction/removal installation or plant   installation *f* de dépoussiérage
dust extractor   aspirateur *m* de poussière(s), dépoussiéreur *m*
dust particle   grain *m* de poussière
dust particles   particules *f,pl* de poussières
dustbin   bac *m* à ordures, poubelle *f*
dust-tight   étanche à la poussière/aux poussières
duty [tax]   droit *m*
duty, letting   droit *m* de bail
duty, rated   puissance *f* nominale
duty, stamp   droit *m* de timbre
dwell in   habiter *vb*
dwelling   demeure *f*, habitation *f*, local *m* d'habitation, logement *m*, résidence *f*
dwelling place   domicile *m*
dwelling unit   unité *f* d'habitation
dye   colorant *m*, matière *f* colorante, teint *m*
dye(stuff)   teinture *f*
dyke, *see* dike
dynamic   dynamique *adj*
dynamic penetration test apparatus   appareil *m* de sondage à pénétration dynamique
dynamometer   dynamomètre *m*

ear [fixing lug]   oreille *f*
ear plugs   bouchons *m,pl*/boules *f,pl* d'oreilles
earth [of current]   tellurien *adj*
earth [electrical]   masse *f*, terre *f*
earth [ground]   terre *f*
earth, refractory   terre *f* réfractaire
earth, reinforced   terre *f* armée
earth bank   levée *f* de terre
earth (circuit)   circuit *m* de terre
earth closet   cabinet *m* sec
earth collector   collecteur *m* de terre
earth contact   contact *m* de terre
earth (continuity) tester   détecteur *m* de contact à la terre
earth lead, main   collecteur *m* de terre
earth leakage (current) loss   perte *f* de courant à la terre
earth leakage circuit-breaker   disjoncteur *m* différentiel/de perte à la terre/de fuite de terre
earth pressure   pression *f* des terres

earth pressure, active   poussée *f* des terres
earth pressure, passive   butée *f* des terres
earth protection   mise *f* à (la) terre, protection *f* de terre
earth (return)   retour *m* à la terre, retour *m* à la masse
earth road   piste *f*
earth tremor   léger tremblement *m* de terre, légère secousse *f* sismique
earth wire   fil *m* de terre/de masse
earth [electrically]   mettre *vb* à la masse, mettre *vb* à (la) terre
earth, act as an   faire *vb* masse
earthenware   faïence *f*, terre *f* cuite
earthing [electrical]   mise *f* à (la) terre
earthing, protective   mise *f* à (la) terre, protection *f* de terre
earth(ing) cable   câble *m* de masse/de terre/de mise à la terre
earth(ing) clamp   collier *m* de mise à la terre
earthing clip [welding]   prise *f* de masse
earth(ing) conductor   conducteur *m* à la masse/de terre

earth(ing) fault   défaut *m* de mise à la terre
earth(ing) lead   conducteur *m* à la masse/de terre
earth(ing) rod   piquet *m*/prise *f* de terre
earthing screw   vis *f* de mise à la terre
earthing system   prise *f* de terre
earthing terminal strip   bornier *m* de terre
earth-moving   mouvement *m* des terres/terrassement (général)
earth-moving equipment   engins *m,pl*/matériel *m* de terrassement
earth-moving records   mouvement *m* des terres/de terre générale
earthquake   séisme *m*, tremblement *m* de terre, secousse *f* sismique
earthquake resistant [foundation]   parasismique *adj*
earthwork   ouvrage *m* de terre, terrassement *m*
earthwork support   blindage *m*, boisage *m*
earthworks   (travaux *m,pl* de) terrassement *m*
ease [a curve]   adoucir *vb*
ease off a block and tackle   soulager *vb* un palan
easement to erect ladder(s) for repairs on another's land   échelage *m*
easements   servitudes *f,pl*
easily accessible   d'accès facile
east(ern)   est *adj*
east   est *m*
easy to run   fonctionnel *adj*
eat away   éroder *vb*
eaten away   corrodé *adj*
eaves [roof]   avant-toit *m*
eaves batten [double thickness, of a tiled roof]   chanlatte *f*
eaves board [roof]   coyau *m*
eaves course [of roof tiles or slates]   rang *m* de gouttière, battellement *m*
eaves gutter   chéneau *m*
eaves overhang   saillie *f* de rive
ebb(ing)   reflux *m*
ebullition   ébullition *f*
eccentric *adj*   excentré *adj*, excentrique *adj*
eccentricity   excentricité *f*
ecological   écologique
ecologist   écologiste *m/f*
ecology   écologie *f*
economic viability   rentabilité *f* économique
economic(al) [i.e. cheap or good value for money]   économique *adj*
economics   économie *f*, (science *f*) économique *f*
economics, energy   économie *f* d'énergie
economics, heating   économie *f* de chauffage
economist   économiste *m/f*
economize on [e.g. on heating]   économiser *vb*
economizer   économiseur *m*
economizer, feed water   économiseur *m* d'eau alimentaire
economy   économie *f*
economy, energy   économie *f* d'énergie
economy, fuel   économie *f* de combustible
economy, heat or thermal   économie *f* thermique
economy, heating   économie *f* de chauffage
eddy   remous *m*
eddy currents   courants *m,pl* de Foucault

edge   bord *m*, bordure *f*, champ *m*
edge [of a beam, etc]   arête *f*
edge [of a board, a door leaf]   chant *m*
edge [of a field or forest, or wallpaper]   lisière *f*
edge [of a roof, a piece of wood, etc]   rive *f*
edge [of a stone, mantlepiece]   carne *f*
edge, bevelled or chamfered   arête *f* chanfreinée, bord *m* chanfreiné
edge, carriageway   rive *f* de chaussée *f*
edge, chipped [of a stone, etc]   écornure *f*
edge, cutting [of a tool]   bord *m* de coupe, taillant *m*
edge, drip   larmier *m*
edge, embankment   épaulement *m* de remblai
edge, folded and rolled, of metal sheeting   ourlet *m*
edge, pavement   bordure *f* du trottoir
edge, road   limite *f* de chaussée
edge, sharp or square   arête *f* vive
edge, straight [of wood]   limande *f*
edge, with a raised   à rebord
edges, tongued-and-grooved   coupe *f* embrevée
edge distance   distance *f* au bord
edge marking   marquage *m* latéral
edge of a roof panel, side/upper   rive *f* latérale/de tête
edge shake [timber defect]   fente *f* de rive
edge to edge   bord à bord
edge tooling [round edges of the face of a stone to be dressed]   plumée *f*
edge trimmer [for lawns]   coupe-bordure *m*
edge trimmer [plumber's tool]   débordoir *m*
edge [a slab of wood]   déligner *vb*
edge, set on   posé sur *adj* champ
edgewise   sur champ
edging   bordure *f*
edging strip   bande *f* de chant
edging strip [on a door]   couvre-chant *m*
edict   ordonnance *f*
editing   rédaction *f*
efface   effacer *vb*
effect   effet *m*
effect, chilling/cooling   effet *m* de refroidissement
effect, chimney   effet *m* de tirage/de cheminée
effect, cooling   effet *m* refroidissant
effect, diffusion   effet *m* de diffusion
effect, externally applied   sollicitation *f*
effect, filtering   effet *m* de filtre
effect, notch   effet *m* d'entaille
effect, quenching [on metals]   effet *m* de trempe
effect, refrigerating   effet *m* de refroidissement
effect, scale   effet *m* d'échelle
effect, stack   effet *m* de cheminée
effect, wind   effet *m* du vent
effects, terrain   effets *m,pl* du terrain
effective [efficient]   efficace *adj*
effective [real]   effectif *adj*
effective [useful]   utile *adj*
effective cross-section   section *f* utile
effective depth [e.g. of beam]   hauteur *f* utile
effective height of water seal in a siphon   garde *f* d'eau
effectiveness   efficacité *f*
efficacy   efficacité *f*

efficacy, **luminous**   efficacité *f* lumineuse
efficiency   efficacité *f*, rendement *m*
efficiency, **boiler**   rendement *m* de chaudière
efficiency, **burner**   rendement *m* du brûleur
efficiency, **combustion**   rendement *m* de combustion
efficiency, **compression**   rendement *m* de compression
efficiency, **conversion**   rendement *m* de conversion
efficiency, **degree of**   degré *m* d'efficacité
efficiency, **filter (separation)**   rendement *m* d'un filtre
efficiency, **high**   haut rendement
efficiency, **maximum or peak**   rendement *m* maximal
efficiency, **mechanical**   rendement *m* mécanique
efficiency, **overall**   rendement *m* global
efficiency, **volumetric**   rendement *m* volumétrique
efficiency measurement   mesure *f* du rendement
efficiency ratio, energy   rendement *m* énergétique
efficient   efficace *adj*
efflorescence   efflorescence *f*
effluent [from a cesspit, septic tank, factory]   eaux *f,pl*
   vannes (EV), effluent *m*
effort   effort *m*
effort, **compactive** [amount of compaction]   énergie *f* de
   compactage
effort, **compactive** [of a roller or vibratory compactor]
   effort *m* de compactage
egg-shaped decorative feature   ove *f*
ejector   éjecteur *m*
ejector, **hydraulic/silt**   hydro-éjecteur *m*
elastic   élastique *adj*
elastic shortening   raccourcissement *m* élastique
elasticity   élasticité *f*
elasticity, **coefficient of**   coefficient *m* d'élasticité
elasticity, **modulus of**   coefficient *m*/module *m* d'élasticité
elastomer   élastomère *f*
elbow [pipe- or ductwork, etc]   coude *m*
elbow, **large-radius**   coude *m* grand rayon
elbow, **offset**   coude *m* de renvoi
elbow, **reducing**   coude *m* réducteur/de réduction
elbow, **square**   coude *m* à angle droit
elbow connection   conduit *m* coudé
electric, *see also* electrical   électrique *adj*
electric blasting cap [detonator]   détonateur *m* électrique
electric firing [blasting]   tir *m* électrique
electric power absorbed   puissance *f* électrique absorbée
electric power consumption   consommation *f* de courant
electric power generation   production *f* de puissance
   électrique
electric power station   centrale *f* électrique
electric power supply [amount]   puissance *f* électrique
electric shock, *see* shock, electric
electric shock, get an   être *vb* électrisé, recevoir *vb* le
   courant/une décharge électrique; prendre *vb* le jus[†]
electrical   électrique *adj*
electrical demand   énergie *f* électrique requise, puissance
   f électrique appelée
electrician   électricien *m*
electricity   électricité *f*
electricity, static   électricité *f* statique
electrode   électrode *f*
electrode, **coated**   électrode *f* enrobée

electrode, **composite**   électrode *f* composée
electrode, **ignition**   électrode *f* d'allumage
electrode, **water level**   électrode *f* de niveau
electrolysis   électrolyse *f*
electrolyte   électrolyte *m*
electrolytic   électrolytique *adj*
electro-magnet   électro-aimant *m*
electronic   électronique *adj*
electronics   électronique *f*
electro-osmosis   électro-osmose *f*
electrostatic precipitator   dépoussiéreur *m* électrostatique
element   élément *m*
element [e.g. in bill of quantities]   poste *m*
element [of an electrical heater]   résistance *f*
element, **building**   composant *m* d'ouvrage
element, **decorative, shaped like head of barley**   grain
   m d'orge *f*
element, **detecting**   élément *m* détecteur
element, **precast**   pièce *f* préfabriquée
elevated [e.g. of a railway]   surélevé *adj*
elevation [drawing]   élévation *f*, projection *f* verticale
elevation [height]   hauteur *f*
elevation [of a building]   façade *f*
elevation [of the sun]   élévation *f*
elevation, **end**   fin *f* d'élévation
elevation, **front**   élévation *f*/vue *f* de face
elevation, **side**   élévation *f*/vue *f* de côté, vue *f* latérale
elevator   élévateur *m*
elevator [lift]   ascenseur *m*
elevator, **belt**   sauterelle *f*
elevator, **bucket**   élévateur *m* à godets
elevator, **inclined**   ascenseur *m* à plan incliné
elevator, **mobile/portable belt**   élévateur-transporteur *m*,
   sauterelle *f*
elimination   élimination *f*
eliminator   éliminateur *m*
eliminator, **dust**   capte-suie(s) *m*
eliminator, **moisture**   éliminateur *m* de gouttelettes
ellipse   ellipse *f*
ellipse, **strain**   ellipse *f* des déformations
ellipse template or stencil   trace-ellipses *m*
elm [tree, wood]   orme *m*
elm grove or plantation   ormaie *f*
elongate   allonger *vb*
elongation   allongement *m*, élongation *f*
elongation, **fracture**   allongement *m* de rupture
elongation before reduction in cross-section
   allongement *m* avant striction
embank   remblayer *vb*
embankment   digue *f*, levée *f*, remblai *m*, talus *m*,
   terrassement *m*
embankment, on   en remblai
embankment base or foot   pied *m* de remblai *m*
embankment crest   crête *f* de remblai
embankment crest [dike]   crête *f* de la digue
embankment failure   glissement *m* de remblai
embankment slope   talus *m* en remblai
embed [hammer or push in]   enfoncer *vb*
embed [fix or grout]   sceller *vb*
embed (into)   encastrer *vb* (dans)

embedded   enfoncé *adj*, scellé *adj*
embedded in   noyé en
embedded (into)   encastré *adj* (dans)
embedding   encastrement *m*
embossed [generally of metal sheet]   embouti *adj*
embrittlement   apparition *f* de fragilité
emerge (on to)   déboucher *vb* (sur)
emergency, *see also* brake, emergency, etc   cas *m* d'urgence
emergency arrestor [of lift or hoist]   parachute *m*
emergency equipment   équipement *m* de secours
emergency evacuation   évacuation *f* d'urgence
emergency exit   issue *f*/sortie *f* de secours
emergency generating set   groupe *m* électrogène de secours
emergency measure   moyen *m* de secours
emergency services   police-secours *f*
emergency smoke vent   cheminée *f* d'appel
emergency spillway   évacuateur *m* de secours
emergency stop   arrêt *m* d'urgence
emery   émeri *m*
emery cloth   toile *f* émeri
emery paper   papier *m* d'émeri *m*
emery powder [for polishing metal]   émeri *m*, potée *f* d'émeri
emission   dégagement *m*, émission *f*
emission, heat   dégagement *m*/émission *f* de chaleur
emissivity   émissivité *f*
emit [heat, etc]   dégager *vb*
emittance   émittance *f*
emoluments   émoluments *m,pl*
empirical   empirique *adj*
employ [force, a piece of equipment, etc]   utiliser *vb*
employer   employeur *m*
employer [client for works]   maître *m* d'ouvrage
empty (of)   vide *adj* (de)
empty   se vider *vb*
empty [drain]   assécher *vb*
empty [sth]   vider *vb*
empty (out) [drain off]   vidanger *vb*
emptying   vidage *m*, vidange *f*
emulsifier   émulsifiant *m*
emulsion, bitumen   émulsion *m* de bitume
emulsion, bitumen(-based) [tack coat]   émulsion *m* d'accrochage
emulsion, tar   émulsion *m* de goudron
enamel   émail *m*
encasement   cachetage *m*
enclave [of land]   enclave *f*
enclose [hem in]   enserrer *vb*
enclose [surround] **(with)**   entourer *vb* (de), ceindre *vb* (de)
enclose [with a fence, wall, etc]   clôturer *vb*
enclose [with a letter, in a parcel]   joindre *vb*
enclosed [of space]   clos *adj*
enclosed [fenced, walled]   clôturé *adj*
enclosed [with a letter, etc]   inclus *adj*, joint *adj*
enclosed (by) [e.g. walls]   ceint *adj* (de)
enclosure   clos *m*
enclosure [enclosed space]   enclos *m*
enclosure [fence or wall]   clôture *f*, enceinte *f*
enclosure [ground enclosed]   enceinte *f*

enclosure [with letter]   pièce *f* jointe (PJ)
encumbering [in the way]   encombrant *adj*
end   bout *m*, fin *f*
end [conclusion]   terme *m*
end [of a cable, etc]   extrémité *f*
end [of a piece of timber or steel, a wall]   about *m*
end of party wall in stone, visible from the street   jambe *f* boutisse
end of travel [e.g. of a lift]   fin *f* de course
end of window ledge built into a pier   oreillon *m*
end restraint   encastrement *m*
end [work, a task]   achever *vb*, finir *vb*, terminer *vb*
end-to-end [butted]   bout à bout *adj*
endorse a cheque   endosser *vb* un chèque
endothermic   endothermique *adj*
energize [electrically]   mettre *vb* sous tension
energize [of generator]   s'amorcer *vb*
energy   énergie *f*
energy, conservation of   conservation *f* d'énergie/de l'énergie
energy, impact   énergie *f* de choc
energy, kinetic   énergie *f* cinétique, puissance *f* vive
energy, law of conservation of   loi *f* de la conservation d'énergie
energy, nuclear   énergie *f* nucléaire
energy, potential   énergie *f* potentielle
energy, radiant   énergie *f* de rayonnement
energy, recovered   énergie *f* récupérée
energy, solar   énergie *f* solaire
energy, sound   énergie *f* acoustique/sonore
energy, spectral   énergie *f* spectrale
energy, stored   énergie *f* accumulée/emmagasinée
energy, thermal [heat]   énergie *f* thermique
energy audit   audit *m* énergétique
energy balance   bilan *m* énergétique
energy conservation   conservation *f* d'énergie/de l'énergie
energy exchange   échange *m* d'énergie
energy input   apport *m* d'énergie
energy loss   perte *f* d'énergie
energy management control system   système *m* de gestion énergétique
energy recovered   énergie *f* récupérée
energy requirement   énergie *f* nécessaire
energy saving   économie *f*/maîtrise *f* d'énergie
energy source, alternative   source *f* alternative d'énergie
energy source(s), waste   source *f* d'énergie perdue
enforcement   mise *f* en vigueur
engage [drive, gear]   enclencher *vb*
engaged   occupé *adj*
engaged [in gear]   en prise
engaging [of drive, gear]   enclenchement *m*
engine   moteur *m*
engine [railway]   machine *f*
engine, diesel   moteur *m* diesel
engineer   ingénieur *m*
engineer, building   ingénieur *m* de bâtiment
engineer, chartered [approximate French equivalent]   ingénieur *m* agréé
engineer, checking [for structural soundness, construction and safety]   contrôleur *m* technique

engineer, chemical   ingénieur *m* chimiste
engineer, chief   ingénieur *m* en chef
engineer, civil   ingénieur *m* (en) génie civil
engineer, civil [from l'École des Ponts et Chaussées]
   ingénieur *m* des ponts et chaussées
engineer, civil or construction   ingénieur *m* des travaux
   publics
engineer, consulting [approximate French equivalent]
   ingénieur *m* conseil, ingénieur-conseil *m*
engineer, construction   ingénieur *m* des travaux publics
engineer, electrical   ingénieur *m* électricien
engineer, field or site   ingénieur *m* de chantier
engineer, heating   chauffagiste *m*
engineer, hydraulic   hydraulicien *m*
engineer, lighting   éclairagiste *m/f*
engineer, mining [from l'École des Mines]   ingénieur *m*
   des mines
engineer, public health or sanitary   ingénieur *m*
   sanitaire
engineer, qualified [title with legal protection in France]
   ingénieur *m* diplômé
engineer, resident   ingénieur *m* permanent
engineer, sanitary   ingénieur *m* sanitaire
engineer, section   conducteur *m* de travaux
engineer, structural   ingénieur *m* (en) structure/de
   structure
engineer-surveyor [generally for civil works]   ingénieur-
   géomètre *m*
engineering [in French construction, from concept and
   design through to supervision]   ingénierie *f*
engineering, air-conditioning   génie *f* climatique
engineering, civil or construction   génie *m* civil
engineering, construction   génie *m* civil
engineering, electrical   électrotechnique *f*
engineering, HVAC   génie *m* climatique
engineering, industrial   génie *m* industriel
engineering, maritime   génie *m* maritime
engineering, mechanical   construction *f* mécanique,
   génie *m* mécanique
engineering, public health   hydraulique *f* urbaine
engineering, refrigeration   technique *f*/technologie *f* du
   froid
engineering, sanitary   génie *m*/technique *f* sanitaire
engineering diploma [in France]   diplôme *m* d'ingénieur
engineering geology   géologie *f* de l'ingénieur
engineering services   services *m,pl* d'ingénierie
engineering structures   ouvrages *m,pl* d'art
engrave [a design on a material]   graver *vb*
engraving   gravure *f*
enhance the value of   valoriser *vb*
enjoyment, right of   jouissance *f*
enlarge   agrandir *vb*, élargir *vb*
enlarged   élargi *adj*
enlargement   élargissement *m*
enlargement [of a photograph, drawing]   agrandissement *m*
enquire into [causes, etc]   rechercher *vb*
enquire into sth   faire *vb* des recherches sur qch
enquiries, make   demander *vb* renseignements
enquiry, *see also* inquiry   demande *f* de renseignements
enrichment [decoration]   ornement *m*

entablature   entablement *m*
enter into a contract   passer *vb* un contrat
enterprise [company]   entreprise *f*
enthalpy   enthalpie *f*
enthalpy diagram   diagramme *m* enthalpique
entrain [air, not railway]   entraîner *vb*
entrainment of air   entraînement *m* d'air
entrance   entrée *f*, porte *f*
entrance, carriage   porte *f* charretière/cochère
entrance, side   entrée *f* latérale
entrance, tradesman's   porte *f* de service
entrance hall   hall *m* (d'entrée)
entropy   entropie *f*
entropy diagram   diagramme *m* entropique
entry   entrée *f*
entry, bellmouth   trompe *f* d'entrée
entry, cable   entrée *f* de câble
entry point [for traffic]   point *m* d'entrée
entry ramp   bretelle *f* d'accès
enumerate   détailler *vb*
envelope   enveloppe *f*
envelope [e.g. containing a letter]   pli *m*
environment   environnement *m*, milieu *m*
environment, aggressive   milieu *m* agressif
environment, human   milieu *m* humain
environment, natural   milieu *m* naturel
environment, physical   milieu *m* physique
environment and energy economy agency [France]
   Agence de l'Environnement et de la Maîtrise de
   l'Énergie (ADEME)
environmental chamber   chambre *f* climatique
epicentre   épicentre *m*
equal   égal *adj*
equal worker participation   participation *f* équilibrée des
   travailleurs
equalization [of forces, etc]   compensation *f*
equalize   égaliser *vb*
equalizer   égaliseur *m*
equalizer, flow   égalisateur *m*, répartiteur *m* d'écoulement
equalizing   compensateur *adj*
equalizing [of levels]   nivellement *m*
equation   équation *f*
equation, approximate   équation *f* approximative
equation, regression   équation *f* de régression
equation of state   équation *f* d'état
equilibrium   équilibre *m*
equilibrium, state of   état *m* d'équilibre
equilibrium, thermal   équilibre *m* thermique
equilibrium concentration   concentration *f* d'équilibre
equip   aménager *vb*, équiper *vb*
equipment, *see also* apparatus, machine, plant, unit
   équipement *m*, équipage *m*, appareillage *m*
equipment [apparatus, appliance]   appareil *m*
equipment [device, mechanism]   dispositif *m*
equipment [plant]   matériel *m*
equipment [radio, telephone, TV, etc]   poste *m*
equipment [tools, implements]   outillage *m*
equipment, building   matériel *m* de construction
equipment, communication(s)   matériel *m* de
   communication

equipment, compacting   engins *m,pl* de compactage
equipment, computer   matériel *m* informatique
equipment, de-icing   matériel *m* de dégel
equipment, domestic   appareils *m,pl* ménagers
equipment, earth-moving   engins *m,pl*/matériel *m* de terrassement
equipment, electrical   appareillage *m* électrique
equipment, emergency   équipement *m* de secours
equipment, excavating   matériel *m* de creusement
equipment, factory-assembled   matériel *m* assemblé en usine
equipment, filter(ing)   matériel *m* de filtrage
equipment, flame proof   matériel *m* anti-déflagrant
equipment, grouting   matériel *m* d'injection
equipment, handling   matériel *m* de manutention
equipment, land surveying   matériel *m* de topographie
equipment, lifting   appareil *m* élévateur, engin *m* de levage
equipment, outdoor   équipement *m* extérieur
equipment, measuring   appareil *m*/dispositif *m* de mesure
equipment, peripheral   appareil *m* périphérique
equipment, personal protective   équipement *m* de protection individuelle (EPI)
equipment, piece of   engin *m*
equipment, portable electrical   appareil *m*/machine *f*/matériel *m* électrique mobile/portatif
equipment, power   matériel *m* force
equipment, relay   matériel *m* de relayage
equipment, soldering   matériel *m* de brasage
equipment, surveying   matériel *m* de topographie
equipment, test(ing)   dispositif *m* d'essai, matériel *m* de contrôle
equipment, towed   engin *m* tracté
equipment, tunnelling   matériel *m* de perforation
equipment, X-ray   appareil *m* radiographique
equipotential   équipotentiel *adj*
equipped   aménagé *adj*, équipé *adj*
equivalent   *adj*   équivalent *adj*
equivalent   équivalent *m*
eraser, road marking   effaceuse *f* (de marquage routier)
erect [build, construct]   construire *vb*, édifier *vb*
erect [assemble, install]   assembler *vb*, monter *vb*
erecting   établissement *m*, montage *m*
erection   construction *f*, montage *m*
erection [of a monument, etc]   dressage *m*
erection [of a structural frame assembled flat]   levage *m*
erector   constructeur *m*, monteur *m*
erode   éroder *vb*
erode [ground, by flow of water]   affouiller *vb*
erosion   érosion *f*
erosion [of ground, by flow of water]   affouillement *m*
erratic   erratique *adj*
erratic performance   irrégularité *f*
erroneous   faux *adj*
error   erreur *f*
error, accidental   erreur *f* accidentelle
error, mean   erreur *f* moyenne
error, measuring   erreur *f* de mesure
error, random   erreur *f* aléatoire
error, reading   erreur *f* de lecture
error, source of   source *f* d'erreur

error band or range   plage *f* d'erreur
escalator   escalator *m*, escalier *m* mécanique/roulant/à marches mobiles
escape [emission of gas, heat, steam, etc]   dégagement *m*
escape [leak]   fuite *f*
escape route   voie *f* d'évacuation
escarpment   escarpement *m*
escutcheon   cache *f*, entrée *f* de serrure
espagnolette bolt, flat arm of [holding shutter closed]   panneton *m*
espagnolette closure   fermeture *f* à crémone
establish   établir *vb*
establishment [building, enterprise]   établissement *m*
establishment of the amount of the final account   règlement *m*
estate   domaine *m*
estate agency   agence *f* immobilière
estate agent   agent *m* immobilier
estate, housing   lotissement *m*
estate, small [enclosed]   closerie *f*
estimate   devis *m*, estimation *f*, évaluation f, prévision *f*
estimate [of cost]   devis *m* estimatif
estimate, approximate   estimation *f*
estimate, budget   montant *m* du devis
estimate, detailed   devis *m* détaillé
estimate, detailed quantitative   détail *m*/document *m* quantitatif estimatif (DQE)
estimate, preliminary   devis *m* estimatif
estimate, rough   approximation *f*
estimate and description, detail   détail *m* estimatif
estimate of quantities and costs   métré *m*, métré *m* estimatif, devis *m* descriptif
estimate of the cost of the works   devis *m* descriptif des travaux
estimate   estimer *vb*
estimated   estimatif *adj*
estimator   calculateur *m* de devis, technicien *m* en étude de prix
estuary   estuaire *m*
etching   gravure *f* à l'eau forte
etching [marking]   marquage *m*
European Technical Agreement (approval)   Agrément Technique Européen (ATE)
evacuate   évacuer *vb*
evacuation   évacuation *f*
evaluate   déterminer *vb*, évaluer *vb*
evaluation   évaluation *f*, interprétation *f*
evaluation of tests/test results   interprétation *f* des essais
evapo-transpiration   évapo-transpiration *f*
evaporate   évaporer *vb*
evaporation   évaporation *f*
evaporation, direct expansion   évaporation *f* à détente directe
evaporation, latent heat of   chaleur *f* latente d'évaporation
evaporation loss   perte *f* par évaporation
evaporation rate   taux *m* d'évaporation
evaporator   évaporateur *m*
evaporator, ceiling type   évaporateur *m* de plafond
evaporator, direct expansion   évaporateur *m* à détente directe

evaporator, finned   évaporateur *m* à ailettes
evaporator, plate-type   évaporateur *m* à plaques
evaporator, shell-and-coil   évaporateur *m* à calandre et serpentin
evaporator, shell-and-tube   évaporateur *m* à faisceau tubulaire
evaporator, showcase type   évaporateur *m* type vitrine
evaporator, vertical-type   évaporateur *m* à tubes verticaux
even [ground]   uni *adj*
even [of numbers]   pair *adj*
even [surface]   plan *adj*
even (with), be   affleurer *vb*
evening out [levelling]   nivellement *m*
evenness [result of levelling, making flush]   affleurement *m*
event   événement *m*
every x metres/at x metre centres   tous les x mètres
evict [tenants]   évacuer *vb*
eviction [of a tenant]   expulsion *f*, évacuation *f*
evil smelling   nauséabond *adj*
examination   examen *m*, vérification *f*
examination [of accounts]   revision/révision *f*
examination, radiographic   examen *m* radiographique
examination, visual   examen *m* visuel
examination, X-ray   examen *m* à rayons X, analyse *f* aux rayons X
examine   examiner *vb*
example   exemple *m*
excavate   excaver *vb*, fouiller *vb*, terrasser *vb*
excavate in steps   tailler *vb* en redans
excavation   excavation *f*, fouille *f*, terrassement *m*
excavation [the process]   réalisation *f* d'une fouille
excavation, deep, with vertical sides   cuve *f*
excavation, edge of an   berge *f*
excavation, foundation   fouille *f* de fondation
excavation, lined or timbered   fouille *f* blindée
excavation, open   fouille *f* ouverte
excavation, shallow   repiquage *m*
excavation, step-by-step during underpinning   fouille *f* en rigole
excavation, timbered   fouille *f* blindée
excavation, trench   fouille *f* en rigole
excavation, unlined   fouille *f* non blindée
excavation by removal of rectangular blocks   fouille *f* en abattage
excavation depth   hauteur *f* de déblai
excavation works   travaux *m,pl* de creusement/de terrassement
excavator   engin *m* de déblaiement, excavateur *m*, pelle *f*
excavator, bucket   pelle *f* à godets
excavator, bucket-wheel   excavateur *m* à roue-pelle
excavator, cable/dragline   pelle *f* à câbles
excavator, general purpose, *see also* excavator   excavatrice *f*
excavator, grab   pelle *f* en benne preneuse
excavator, hydraulic   pelle *f* hydraulique
excavator, mechanical   excavateur *m*, pelle *f* mécanique, pelleteuse *f*
excavator, skid-steer   pelle *f* à direction par ripage
excavator, trench [machine]   trancheuse *f*
excavator, walking   pelle-araignée *f*
excavator bucket   godet *m* de pelle

exceed   dépasser *vb*
excess *adj*   excédentaire *adj*
excess   surplus *m*
excess [on insurance claim]   franchise *f*
excess (of)   excès *m* (de)
excess (over)   excédent *m* (sur)
exchange   échange *m*
exchange control   contrôle *m* des changes
exchange rate [currency]   cours *m*/taux *m* de change
exchange   échanger *vb*
exchanger [heat]   échangeur *m*
excise stamp   timbre *m* fiscal
excitation [of a generator]   amorçage *m*, excitation *f*
excite [a generator]   amorcer *vb*, exciter *vb*
exclusivity [e.g. of selling agency]   exclusivité *f*
execution   exécution *f*
execution [e.g. of a project]   mise *f* à exécution
execution procedure   méthode *f* d'exécution
executive [staff]   cadre *m*
exemption   franchise *f*
exercise [a trade, skill or profession]   exercer *vb*, pratiquer *vb*
exercise [permission, powers, etc]   user *vb* de
exfiltration   exfiltration *f*
exfiltration of air   fuite *f* d'air
exhaust [by suction]   aspiration *f*
exhaust [escape]   échappement *m*
exhaust fan   aspirateur *m*, ventilateur *m* à aspiration
exhaust flap   clapet *m* d'échappement
exhaust manifold   collecteur *m* d'échappement
exhaust opening or outlet   orifice *m* de sortie/de soufflage
exhaust opening or vent   bouche *f* d'évacuation
exhaust pipe [vehicle]   tuyau *m* d'échappement
exhaust port   orifice *m* d'échappement
exhaust system   système *m* d'extraction d'air
exhaust [air, etc]   aspirer *vb*
exhaust [steam]   évacuer *vb*
exhaust [use up]   épuiser *vb*
exhausted   épuisé *adj*
exhaustion [of stocks]   épuisement *m*
exhibit   exposer *vb*
exhibition   exposition *f*
existing condition or state [of a building, site, etc]   état *m* des lieux
existing structures [when starting construction of new]   restes *m,pl* d'ouvrages
exit   issue *f*, sortie *f*
exit, emergency   issue *f*/sortie *f* de secours
exit point [for traffic]   point *m* de sortie
exit ramp or slip road   bretelle *f* de sortie
expand   s'agrandir *vb*, se dilater *vb*
expand [e.g. forest, town, by growing]   s'étendre *vb*
expand [inflate, swell]   gonfler *vb*
expand [increase in volume]   foisonner *vb*
expand [spread, spread out]   étendre *vb*
expand, cause to   dilater *vb*
expanded   expansé *adj*
expander [plumber's]   mandrin *m*
expansibility   expansibilité *f*
expansion   dilatation *f*, expansion *f*

expansion, coefficient of thermal   coefficient *m* de dilatation thermique

expansion, direct, *see* direct expansion

expansion, linear   allongement *m* longitudinal, dilatation *f*/expansion *f* linéaire

expansion, thermal   dilatation *f* thermique

expansion bellows   compensateur *m*/soufflet *m* de dilatation, dispositif *m* de compensation *f* de dilatation

expansion cooling   refroidissement *m* par dilatation

expansion joint   joint *m* glissant/de dilatation

expansion joint [highway/bridge]   joint *m* de chaussée

expansion joint cover [metallic roofing]   besace *f* de dilatation

expansion loop   lyre *f* de dilatation

expansion stroke   course *f* de détente

expansion tank or vessel   cumulus *m*, réservoir *m* de dilatation, vase *m* d'expansion/de dilatation

expansion tank or vessel, membrane   vase *m* de dilatation à membrane

expansion valve, thermostatic   détendeur *m* thermostatique

expenditure   dépense *f*, frais *m,pl*

expense   charge *f*, dépense *f*

expenses   frais *m,pl*

expenses, accrued   charges *f,pl* à payer

expenses, claim form for   note *f* de frais

expenses, legal [buying/selling property]   frais *m,pl* de notaire

expenses, operating   charges *f,pl*/coût *m*/frais *m,pl* d'exploitation

expenses, out-of-pocket   débours(é) *m*

expenses, travelling   frais *m,pl* de déplacement

experience   expérience *f*

experience, operating   expérience *f* en service

experience, practical   expérience *f* pratique, pratique *f*

experiment   expérience *f*

experiment [make an]   expérimenter *vb*

experimental   expérimental *adj*

expert, independent   expert *m*

expert opinion   expertise *f*

expire   arriver *vb* à terme

expired   périmé *adj*

expiry [of a time limit]   échéance *f*

explode   éclater *vb*, exploser *vb*

exploit   mettre *vb* en exploitation

exploitation   exploitation *f*

exploitation of the resource   mobilisation *f* de la ressource

explosion   explosion *f*

explosion [e.g. of a boiler]   éclatement *m*

explosive *adj*   explosif *adj*

explosive   explosif *m*

explosive store   dépôt *m* d'explosifs

explosive/explosible substance   substance *f* explosible

exponential *adj*   exponentiel *adj*

exponential   exponentielle *f*

exposed   apparent *adj*

exposed to   exposé à

exposition [of facts, etc]   exposition *f*

exposure [of a building, facade, etc]   exposition *f*

exposure, sheltered   exposition *f* abritée

exposure period or time   temps *m* d'exposition

exposure to chemical agents   exposition *f* aux agents chimiques

expropriate   exproprier *vb*

expropriation   expropriation *f*, emprise *f*

extend [e.g. of forest, town, by growing]   s'étendre *vb*

extend [lengthen itself]   s'allonger *vb*

extend [spread, spread out]   étendre *vb*

extend [make longer]   prolonger *vb*

extend the length of [piece of wood, ladder, etc]   enter *vb*

extension [building]   annexe *f*

extension [enlargement]   agrandissement *m*

extension [expansion]   dilatation *f*

extension [generally of time]   prolongation *f*

extension [of a building, road, etc]   prolongement *m*

extension [stretching or expansion]   extension *f*

extension, telephone   poste[†] *m*

extension cable or lead   prolongateur *m*, cordon *m* (de) prolongateur

extension of an insufficiently long pile   faux-pieu *m*

extension of time   délai *m*, prolongation *f* de temps

extension piece   rallonge *f*

extension pipe   tube *m* de rallonge

extensometer   extensomètre *m*

extent   étendue *f*, mesure *f*

exterior [outside]   en dehors

exterior   extérieur *m*

external   externe *adj*

external [overall, between external faces, of dimensions or measurements]   hors œuvre *adj* (H/O)

extinguish   éteindre *vb*

extinguisher   appareil *m* éteignoir, éteignoir *m*

extinguisher, fire   extincteur *m* (d'incendie)

extinguisher, portable   extincteur *m* à main

extra   additionnel *adj*

extract   extraire *vb*

extract [e.g. sheet piles]   arracher *vb*

extraction   extraction *f*

extraction [air, dust, liquid, etc]   évacuation *f*

extraction [uprooting, tearing out]   arrachage *m*

extraction, air   extraction *f* d'air

extraction, chip   évacuation *f* des copeaux

extraction, dust   dépoussiérage *m*

extraction of core samples   carottage *m*

extractor   extracteur *m*

extractor [dust]   aspirateur *m*

extractor, chip   aspirateur *m* de copeaux

extractor, dust   aspirateur *m* de poussière(s), dépoussiéreur *m*

extrados   extrados *m*, douelle *f* extérieure

extra-low voltage safety installation   installation *f* à très basse tension de sécurité/à TBTS

extrapolate (from)   extrapoler *vb* (de)

extremity   extrémité *f*

extremity [of a piece of timber or steel, a wall]   about *m*

extrusion   expulsion *f*, extrusion *f*, tréfilé *m*

eye [of an axe head, etc]   toyère *f*

eye protectors   lunettes *f,pl* de protection

eye-bolt   boulon *m* d'œil, vis *f* à œillet

eyelet   œillet *m*

eyepiece [e.g. of a surveying instrument]   viseur *m*

fabric [of a building]   gros ouvrage *m*
fabric, upholstery   tissu *m* d'ameublement
fabric reinforcement, welded   treillis *m* soudé
fabricate   fabriquer *vb*
fabrication   fabrication *f*
fabricator, steel   entreprise *f* de constructions métalliques
facade   façade *f*
facade, lightweight, installed between floors   façade *f* panneau
facade, lightweight, installed outside the floors   façade *f* rideau
facade, lightweight [e.g. metal, glass or light panel]   façade *f* légère
facade, projection in front of   avant-corps *m*
facade, set back [behind another]   façade *f* en retrait
facade, straight line   façade *f* en ligne droite
facade, ventilated [double skin with exterior air circulation]   façade *f* ventilée
facade or frontage, set back   façade *f* en retrait
facade with vapour permeable exterior skin   façade *f* respirante
face [in underground excavation]   front *m* d'attaque
face [of a building]   façade *f*
face [of a hammer]   table *f* aire *f*
face [of a hill, mountain, etc]   paroi *f*
face [of angular object, building, etc]   pan *m*
face [side]   face *f*
face, as-built   parement *m* brut
face, coursed [of a masonry wall]   parement *m* assisé
face, dressed [of a stone, stone wall, etc]   parement *m*
face, external/internal   paroi *f* extérieure/intérieure
face, long [of a building, a roof plane]   long-pan *m*
face, lower, of a paving block or stone   assiette *f*
face, rock   paroi *f* rocheuse
face, steep rock   falaise *f*
face, vertical, of trench excavation   front *m* d'abattage
face area   surface *f* frontale
face of a hammer   aire *f* d'un marteau
face shield   écran *m* facial
face shovel   pelle *f* (équipée) en butte
face [a wall]   paramenter *vb*
facilities   équipements *m,pl*
facilities, port/harbour   équipements *m,pl*/installations *f,pl* portuaires
facilities, sanitary   installations *f,pl* sanitaires
facilities, shopping   commerces *m,pl*
facilities, transport   facilités *f,pl* de transport
facing [e.g. facing the station]   face à
facing, south, etc   exposé *adj* sud, etc
facing   parement *m*
facing [exterior wall]   revêtement *m*
facing [interior]   chemise *f*
facing, brick [of a wall]   faux parement *m*

facing, drystone [of an embankment]   perré *m*
facing, reinforced concrete   masque *f* en béton armé
facing, rough finished [of a stone wall]   parement *m* débruti
facing, stone   placage *m* en pierre
facing, tooled finish rubble   parement *m* layé
facing, upstream/downstream [of a dam]   masque *f* amont/aval
facing, wall   parement *m*
facsimile [fax]   télécopie *f*
facsimile machine   télécopieur *m*
factor   coefficient *m*, facteur *m*, indice *m*
factor, conductance   facteur *m* de conductance
factor, correction   coefficient *m* de correction
factor, damping   facteur *m* d'amortissement
factor, daylight   facteur *m* de lumière du jour
factor, decay   indice *m* d'amortissement
factor, decisive   facteur *m* de décisif
factor, design   facteur *m* de sécurité/de calcul
factor, diversity   coefficient *m* d'utilisation, facteur *m* de simultanéité/de diversité
factor, flow   facteur *m* d'écoulement
factor, form   coefficient *m* de forme, facteur *m* de forme
factor, fouling   facteur *m* d'encrassement
factor, glare   indice *m* d'éblouissement
factor, load   facteur *m* de charge
factor, lumen maintenance   facteur *m* de dépréciation
factor, power   facteur *m* de puissance
factor, radiation   facteur *m* de rayonnement
factor, reflection [lighting]   facteur *m* de réflexion
factor, size   facteur *m* dimensionnel
factor, transmission [lighting]   facteur *m* de transmission
factor, utilization [lighting]   facteur *m* d'utilisation
factor of safety   facteur *m* de sécurité
factory   usine *f*
factory assembled or set   monté en usine
factory building   fabrique *f*
factor, reflection [lighting]   facteur *m* de réflexion
fade   pâlir *vb*
fading [colours]   ternissement *m*
Fahrenheit scale   échelle *f* de Fahrenheit
fail [beam]   fléchir *vb*
fail [brakes]   lâcher *vb*
fail [break down]   tomber *vb* en panne
fail [by breaking up or into pieces]   se rompre *vb*
fail [collapse]   s'affaisser *vb*, s'écrouler *vb*, s'effondrer
fail [negotiations]   ne pas aboutir *vb*
failure [breaking]   rupture *f*
failure [e.g. of insulation, lighting]   défaillance *f*
failure, brittle   rupture *f* fragile
failure, cable   rupture *f* de câble
failure, embankment   glissement *m* de remblai
failure, fatigue   rupture *f* d'endurance/par fatigue

failure, shear   rupture *f* par cisaillement
failure, strain at   déformation *f* à la rupture
failures, mean time between   temps *m* moyen entre
  pannes
failure condition(s)   conditions *f,pl* à la rupture
failure of communication   panne *f* de communication
failure zone   zone *f* de rupture
fall [drop]   chute *f*
fall [of earth, etc]   écroulement *m*
fall [of water level, pressure, price, etc]   baisse *f*
fall [slope]   pente *f*
fall, natural   pente *f* naturelle
fall from height   chute *f* de hauteur
fall in temperature   abaissement *m* de la température
fall of objects   chute *f* d'objets
fall on the same level   chute *f* de plain pied
fall (down) [earth, etc]   s'écrouler *vb*
fall due [e.g. payment]   arriver *vb* à terme
fall in [a structure]   s'écrouler *vb*
fall into ruin(s)   tomber *vb* en ruines
fall to pieces [a structure]   s'écrouler *vb*
falling   abaissement *m*
falling down/to pieces [e.g. of a building]   écroulement *m*
falling into ruins   ruineux *adj*
false [ceiling, etc]   faux *adj*
false alarm   alarme *f* intempestive
falsework   échafaudage *m*, étaiement *m*
falsework [curved, for arches, vaults, etc]   cintre *m*
fan   aérateur *m*, ventilateur *m*
fan [as in alluvial fan]   cône *m*
fan, airfoil   ventilateur *m* à aubes profilées
fan, alluvial   cône *m* de déjection
fan, axial (flow)   ventilateur *m* axial
fan, bifurcated   ventilateur *m* bifurqué/à fourche
fan, ceiling   ventilateur *m* de plafond
fan, centrifugal   ventilateur *m* centrifuge
fan, circulating   ventilateur *m* brasseur d'air
fan, cross-flow or tangential   ventilateur *m* tangentiel
fan, double entry or inlet   ventilateur *m* à deux/double
  ouïes
fan, exhaust   aspirateur *m*, ventilateur *m* à aspiration
fan, extract(or)   évacuateur *m*, ventilateur *m*
  extracteur/d'extraction (d'air)
fan, induced draught   ventilateur *m* de tirage par
  aspiration
fan, kitchen extract(or)   ventilateur *m* d'extraction de
  cuisine
fan, low pressure   ventilateur *m* basse pression
fan, mixed flow   ventilateur *m* hélico-centrifuge
fan, radial   ventilateur *m* radial
fan, single inlet   ventilateur *m* à une ouïe
fan, smoke extractor   ventilateur *m* de désenfumage
fan, suction   aspirateur *m*
fan, supply   ventilateur *m* de soufflage
fan, tangential   ventilateur *m* tangentiel
fan, ventilating [drawing in air]   ventilateur *m*
  d'aspiration d'air frais
fan, ventilating [blower]   soufflerie *f*
fan, wall   ventilateur *m* mural

fan belt   courroie *f* de ventilateur
fan blade   ailette *f* de ventilateur
fan capacity   capacité *f*/débit *m* de ventilateur
fan casing   enveloppe *f* de ventilateur
fan coil unit   (groupe *m*) ventilo-convecteur *m*
fan drive   transmission *f* de ventilateur
fan impeller   roue *f* de ventilateur
fan inlet   entrée *f* de ventilateur, ouïe *f* d'aspiration de
  ventilateur
fan laws   lois *f,pl* fondamentales des ventilateurs
fan outlet   sortie *f* de ventilateur
fan (performance) curve   courbe *f* caractéristique du
  ventilateur
fan power   puissance *f* de ventilateur
fan shroud   protection *f* du ventilateur
fan unit [in an air handling unit]   caisson *m* ventilateur
fanlight   imposte *f*, vasistas *m*
fanlight [bottom hung]   fenêtre *f* à soufflet
fanlight, fixed   dormant *m*
fanlight, semi-circular   imposte *f* en éventail
fanlight opener [remote]   ferme-imposte *m*
fanlight retaining strap [slotted]   coulisse *f* de vasistas
far [distant]   éloigné *adj*, lointain *adj*
farm   exploitation *f* agricole, ferme *f*
farm, tenanted [in France]   métairie *f*
farm track   chemin *m*/sentier *m* d'exploitation *f*
farmer   agriculteur *m*, fermier *m*
farmhouse   ferme *f*, maison *f* paysanne
farmhouse [South of France]   mas *m*
farmhouse [small]   fermette *f*
fascia board   volige *f*
fascia plate   plaque *f* de recouvrement
fascine mattress   fascine *f*, tapis *m* de branchage
fast [colour]   fixe *adj*
fasten   attacher *vb*
fasten (to)   fixer *vb* (à, sur)
fastened (to)   fixé *adj* (à, sur)
fastener   attache *f*, moyen *m* d'assemblage
fastening   fermeture *f*, fixation *f*, patte *f*
fat [of a plaster or cement rich in binder]   gras *adj*
fatigue [due to alternating stress]   fatigue *f*
fatigue, metal   fatigue *f* du métal
fatigue strength   résistance *f* aux efforts alternés
fault [blame]   faute *f*
fault [defect]   défaut *m*
fault [geological]   dislocation *f*, faille *f*
fault [mistake]   erreur *f*
fault, earth(ing)   défaut *m* de mise à la terre
fault, electrical   défaut *m* électrique
fault, insulation   défaut *m* d'isolement
fault, manufacturing   défaut *m* de fabrication
fault, operational   défaillance *f* opérationnelle
fault, thrust   faille *f* de chevauchement
fault, try to locate a   rechercher *vb* un dérangement
fault finding chart   tableau *m* de dépannage
fault line [geology]   ligne *f* de faille
fault location [electrical]   recherche *f* de dérangements
fault(iness)   défectuosité *f*
faulting [geology]   faille *f*

faulty défectueux *adj*
fax[†] télécopie *f*
fax machine télécopieur *m*
feasibility faisabilité *f*
feasible faisable *adj*
feature caractéristique *f*
fee account note *f* d'honoraires
fees [architect's, etc] émoluments *m,pl*, honoraires *m,pl*
fees, agency frais *m,pl* d'agence
fees based on time to complete an item of work vacation *f*
feed belt (conveyor) tapis *m* alimenteur
feed line conduite *f* d'alimentation
feed rate vitesse *f* d'alimentation
feed skip benne *f* d'alimentation, caisse *f* de chargement, chargeur *m*
feed tank bâche *f* alimentaire/d'alimentation, réservoir-nourrice *m*
feed water, boiler eau *f* d'alimentation de chaudière
feed water economizer économiseur *m* d'eau alimentaire
feed [with electric power, water, gas, etc] alimenter *vb*
feedback contre-réaction *f*
feedback system système *m* de contre-réaction
feeder alimentateur *m*
feeder [electrical] câble *m* d'alimentation
feeder, belt alimentateur *m* à courroie
feeder, cold alimentateur-doseur *m*
feeder, plate alimentateur *m* à plateau
feeder, reciprocating alimentateur *m* à mouvement alternatif/à chariot
feeder, vibrating alimentateur *m* à balourd
feeder, vibrating grizzly précribleur *m* vibrant
feeder mains câble *m* principal
feeding belt (conveyor) tapis *m* alimenteur
feeding hopper or skip benne *f* d'alimentation
feeding unit groupe *m* d'alimentation
fell [a tree] abattre *vb*
felling [of trees] abattage *m*
felling axe hache *f* de bûcheron
felt [roofing, carpet, etc] feutre *m*
felt, bitumen feutre *m* bitumé/bitumeux, toile *f* goudronnée
felt, roofing carton *m* feutre bitumé, feutre *m* à toit, toile *f* goudronnée
felt nail clou *m* à tête large
felt feutrer *vb*
felted feutré *adj*
felting couches *f,pl* de feutre
female [connection, thread, etc] femelle *adj*
fen marais *m*
fence clôture *f*
fence [of a power saw] butée *f*, guide *m* de refend
fence, angling or mitre guide *m* à onglets
fence, boarded palissade *f* de planches
fence, electric clôture *f* électrifiée
fence, openwork claire-voie *f*
fence, wooden palissade *f*
fence [land] clôturer *vb*
fence in [land, a site] palissader *vb*

fenced clôturé *adj*
fencing, wire grillage *m*
fender [on a bridge pier] défense *f*
fender, loose défense *f* mobile
fenestration fenestration *f*
ferritic ferritique *adj*
ferrocement ferrociment *m*
ferrule virole *f*
ferrule, electrical embout *m* de câblage
ferrule with insulated flange embout *m* à collerette isolante
ferry(boat) bac *m*
fettle [castings] ébarber *vb*
fettling reparage *m*
fibre fibre *f*
fibre, bottom fibre *f* inférieure
fibre, compression fibre *f* comprimée
fibre, extreme fibre *f* extrême
fibre, middle fibre *f* moyenne
fibre, neutral fibre *f* neutre
fibre, optical fibre *f* optique
fibre, tension fibre *f* tendue
fibre direction [in wood] fil *m* du bois
fibre optics fibre *f* optique
fibreboard panneau *m* de fibre (de bois)
fibreglass fibre *f* de verre
fibreglass, rigid plastique *m* armé de verre
fibreglass (insulation) laine *f* de verre
fibrous fibreux *adj*
fibrous plaster, *see* plaster, fibrous
field champ *m*
field [of action, of a science, etc] domaine *m*
field, electric champ *m* électrique
field, magnetic champ *m* magnétique
field, rotating (electrical) champ *m* tournant
field of application domaine *m* d'application/d'emploi
field of view champ *m* visuel
fields, in the open en pleins champs
figure [drawing] figure *f*
figure [number] chiffre *m*
figure out [amounts, numbers, etc] chiffrer *vb*
figuring [the process] chiffrage *m*
filament [of a lamp] filament *m*
filament, coiled coil filament *m* bispiral
filament bulb lampe *f* à incandescence
file [computer] fichier *m*
file [documents] dossier *m*
file [rasp] lime *f*
file, half-round demi-ronde *f*
file, rat-tail or round queue-de-rat *f*
file, roughing lime *f* à dégrossir, riflard *m*
file, three-cornered/triangular tiers-point *m*
file, wood lime *f* à bois
file [documents] classer *vb*
file [metal, wood, etc] limer *vb*
file one's petition in bankruptcy déposer *vb* son bilan
filings (steel, iron) limaille *f* (d'acier, de fer)
fill remblai *m*
fill, compacted remblai *m* compacté

fill, concrete   béton *m* de blocage, gros béton *m*
fill, hydraulic   remblai *m* hydraulique
fill, random   remblai *m* (en) tout venant
fill, random rubble   garni *m*
fill, stone   pierraille *f*
fill crest   crête *f* de remblai
fill material   matériau *m* de remplissage/de remblai
fill   remblayer *vb*
fill [a ditch, a well]   combler *vb*
fill [a road or bridge joint]   garnir *vb*
fill [cracks, etc, prior to painting or papering]   reboucher *vb*
fill [cracks, joints, etc]   calfeutrer *vb*
fill in [e.g. a form]   remplir *vb*
fill in [potholes in a road]   colmater *vb*
fill (up) [a crack, gap]   boucher *vb*
fill up a space [e.g. between ridge and sloping tiles]   fourrer *vb*
fill up [a tank]   faire *vb* le plein
fill up [a wall] with rubble   bloquer *vb*
fill very small holes [e.g. in plaster surface]   octer *vb*
fill (with)   remplir *vb* (de)
filled up with rubble [of a wall]   bloqué *adj*
filler   enduit *m*
filler [added to certain types of concrete, to plastics]   filler *m*
filler [in road material]   fines *f.pl*
filler [mastic, putty]   mastic *m*
filler, coarse   enduit *m* de rebouchage
filler, smooth   enduit *m* de lissage
filler, wood   pâte *f* à bois
filler applied in a single coat, painter's   enduit *m* non repassé
filler applied in several layers to produce smooth surface, painter's   enduit *m* repassé
filler for plaster, painter's [oil mastic + zinc white]   enduit *m* gras
filler for wood or old base, painter's   enduit *m* maigre
filler providing base for surface treatment   enduit *m* de surfaçage
filler content [of road material]   teneur *f* en fines
filler, add a   fillériser *vb*
fillet   baguette *f* d'angle, latte *f* de recouvrement
fillet [concave profile]   congé *m*
fillet [in a moulding]   listel *m*
fillet [of mortar, between roof slope and wall]   filet *m*
fillet [moulding separating two elements]   filet *m*
fillet [thin painted line]   filet *m*
fillet, arris [roof]   coyau *m*
fillet, glazing   parclause *f*, parclose *f*
fillet, half-round   angle *m* arrondi
fillet, tilting [roof]   coyau *m*
filling [i.e. adding a filler]   fillérisation *f*
filling [material]   bourrage *m*
filling [of cracks, etc, prior to painting or papering]   rebouchage *m*
filling between joists [on top of which floor is placed]   fausse-aire *f*
filling (in) [of a ditch, a well, etc]   comblement *m*
filling in [potholes in a road]   colmatage *m*

filling of holes, cracks or joints   masticage *m*
filling (up)   remplissage *m*
filling (up) [of a ditch, a well, etc]   comblement *m*
filling (with lead)   plombage *m*
film [deposited on and adhering to something]   feuil *m*
film [thin, deposited on something]   film *m*
film [thin layer or photographic]   pellicule *f*
film, protective   film *m* protecteur
filter   filtre *m*
filter, activated charcoal   filtre *m* en charbon actif
filter, aerobic [in septic tank or sewage system]   filtre *m* épurateur
filter, air   filtre *m* à air
filter, autoroll   filtre *m* automatique
filter, bacterial   filtre *m* bactérien
filter, bag   filtre *m* à manche
filter, basket [in a siphon]   panier *m*
filter, carbon   filtre *m* à charbon
filter, cartridge   filtre *m* à cartouche
filter, cellular   filtre *m* cellulaire
filter, charcoal   filtre *m* à charbon
filter, cloth   filtre *m* en tissu
filter, coarse   filtre *m* grossier, préfiltre *m*
filter, differential head of a   pression *f* filtrante différentielle
filter, disposable (element)   filtre *m* à éléments non régénérables
filter, dry fabric   filtre *m* en fibre sec
filter, duo   filtre *m* duplex
filter, dust   filtre *m* à poussières
filter, electrostatic   filtre *m* électrostatique
filter, exhaust air   filtre *m* d'air évacué
filter, fabric   filtre *m* sec
filter, fuel   filtre *m* à combustible
filter, grease   filtre *m* à graisse(s)
filter, high efficiency   filtre *m* à haute efficacité
filter, impact   filtre *m* à choc
filter, inverted   filtre *m* inversé
filter, lint [dryer or washing machine]   filtre *m* à charpie
filter, metal   filtre *m* métallique
filter, odour   filtre *m* désodorisant
filter, oil   filtre *m* à huile/d'huile
filter, oil bath air   filtre *m* à air à bain d'huile
filter, permutite   filtre *m* à permutite
filter, pipe   filtre *m* de tube
filter, pressure drop across a   pression *f* filtrante différentielle
filter, primary   filtre *m* primaire
filter, roll   filtre *m* à déroulement
filter, rotary   filtre *m* centrifuge/rotatif
filter, sand   filtre *m* à gravier
filter, sand-bed   clarificateur *m*
filter, secondary   filtre *m* secondaire
filter, sludge   filtre *m* à boues
filter, vacuum   filtre *m* à vide
filter, water   filtre *m* à eau
filter, wetted air   filtre *m* à air mouillé
filter, zeolite   filtre *m* à permutite
filter area   surface *f* du filtre/filtrante

filter bag   filtre *m* à manche, sac *m* filtrant
filter capacity   capacité *f* d'un filtre
filter charcoal   charbon *m* à filtrer
filter cloth   tissu *m* filtrant, toile *f* filtrante
filter drum   tambour *m* filtrant
filter element   élément *m* filtrant
filter element, disposable   élément *m* filtrant non régénérable
filter equipment   matériel *m* de filtrage
filter material   matière *f* de filtrage
filter media   médias-filtrants *m,pl*
filter medium   matière *f* filtrante
filter paper   papier *m* filtrant
filter plant   installation *f* de filtrage
filter replacement   remplacement *m* de filtre
filter (separation) efficiency   rendement *m* d'un filtre
filter tank   réservoir *m* à filtration/d'épuration
filter   filtrer *vb*
filter [gas, water]   épurer *vb*
filter [light]   tamiser *vb*
filter into   infiltrer *vb*
filter-bed   lit *m* de filtrage, lit *m* filtrant
filtering   filtrage *m*, filtration *f*
filtering [liquids]   épuration *f*
filtering [sieving]   tamisage *m*
filtering charcoal   charbon *m* à filtrer
filth   saleté *f*
filtration   filtrage *m*, filtration *f*
filtration, natural   filtration *f* naturelle
filtration, soil   filtration *f* par le sol
filtration, water   filtration *f* de l'eau
fin [cooling, radiating, etc]   ailette *f*
fin [on a casting]   bavure *f*
fin, cooling   ailette *f* de refroidissement
fin spacing   écartement *m*/pas *m* des ailettes
fin   nervurer *vb*
final [definitive]   définitif *adj*
final result [accounts, consequences]   bilan *m*
finance   finance *f*
financial   financier *adj*
financial appraisal   bilan *m* financier
financial contribution/input   apport *m*
financial system [national]   fiscalité *f*
financier   financier *m*
financing   financement *m*
financing scheme   plan *m* de financement
fine structure   microstructure *f*
fine-grained   à grain fin
fine-textured [sand, aggregate]   de granularité fine
fines [aggregate, etc]   fines *f,pl*
fines, crushed stone   sable *m* de concassage
fines content [of road material]   teneur *f* en fines
finger-plate [on a door]   plaque *f* de propreté
finial [on a roof]   épi *m* (de faîtage)
fining down [of a plank, etc]   affinage *m*, affinement *m*
finish   fini *m*
finish [surface treatment]   finition *f*
finish, anti-corrosion   protection *f* anti-rouille
finish, decorative   habillage *m*

finish, dragged   brettelure *f*
finish, float   fini *m* à la taloche
finish, floor   surface *f* du sol
finish, hammer [of glass]   martelé *adj*
finish, hammered   martelage *m*
finish, screeded   fini *m* au profileur
finish, spalled stone   parement *m* smillé .
finish, surface   couche *f* de surface, traitement *m* de surface
finish, trowelled   fini *m* à la lisseuse
finish feature, decorative   habillage *m*
finish   finir *vb*
finish [complete, e.g., a building, a task]   achever *vb*
finish [make an end]   terminer *vb*
finish off [a piece of work]   parfaire *vb*
finisher [road construction]   surfaceuse *f*
finisher, asphalt   finisseur *m* à asphalte
finisher, concrete, in a paving train   finisseur *m* à béton
finisher, road   finisseur *m* routier
finishing   finissage *m*, finition *f*
finishing [of a facade, stonework, etc]   ravalement *m*
finishing, surface   traitement *m* de surface
finishing accessory   accessoire *m* de finition
finishing coat, decorative   couche *f* d'habillage
finishing machine [road construction]   surfaceuse *f*
finishing operations or works   travaux *m,pl* de finissage
finishing with a thin sheet of material   placage *m*
finishing works [closure, services, partitions, finishes]   parachèvement *m*, second œuvre *m*
finite element method   méthode *f* des éléments *m,pl* finis
finned   à ailettes, nervuré *adj*
fir [wood, tree]   sapin *m*
fire   feu *m*
fire [e.g. burning building]   incendie *m*
fire, electric   radiateur *m* électrique
fire, seat of the   foyer *m* de l'incendie
fire alarm   alarme *f*/avertisseur *m*/signal *m* d'incendie
fire alarm, break-glass   boîtier *m* brise-vitre
fire barrier   coupe-feu *m*
fire behaviour   tenue *f* au feu
fire break   coupe-feu *m*, ligne *f* de feu, pare-feu *m*
fire brigade   pompiers *m,pl*, sapeurs-pompiers *m,pl*
fire brigade access   accès *m* pour les pompiers
fire compartmentation   cloisonnement *m* de sécurité
fire control panel   tableau *m* de contrôle incendie
fire damper   clapet *m*/volet *m* coupe-feu, dispositif *m* d'obturation coupe-feu
fire damper [smoke vent]   trappe *f* de désenfumage
fire detection system   détection *f* incendie
fire detection system, addressable   localisateur *m* d'incendie
fire detector   détecteur *m* d'incendie
fire division wall   mur *m* coupe-feu, cloison *m* coupe-feu/pare-feu
fire door [of a wood stove]   porte-foyer *f*
fire engine   voiture *f* de pompiers
fire engine (pump)   pompe *f* à incendie
fire escape (ladder)   échelle *f* à/d'incendie, échelle *f* de sauvetage

fire extinguisher　extincteur *m* (d'incendie)
fire extinguisher, hand operated/portable　extincteur *m* (d'incendie) à main
fire extinguishing equipment or installation　installation *f* d'extinction d'incendie
fire gable　pignon *m* coupe-feu
fire guard or screen　garde-feu *m*
fire hose　tuyau *m*/manche *m* d'incendie
fire hosereel or point　robinet *m* d'incendie armé (RIA)
fire hydrant [ground level]　bouche *f* d'incendie
fire hydrant [pillar type]　borne *f* d'incendie
fire hydrant, underground　bouche *f* d'incendie à clé sous trottoir
fire insurance　assurance *f* contre l'incendie, assurance *f* incendie
fire partition　cloison *f* coupe-feu/pare-feu
fire performance　réaction *f* au feu
fire point　robinet *m* d'incendie armé (RIA)
fire protection　protection *f* incendie/contre l'incendie/contre les incendies
fire protection strategy　stratégie *f* de protection *f* contre l'incendie
fire rake　pique-feu *m*
fire rated　coupe-feu *adj* (CF)
fire rating　degré *m* de résistance au feu
fire resistance　degré *m* de coupe-feu, résistance *f*/stabilité *f* au feu
fire resistance, required (period of)　durée *f* coupe-feu réglementaire
fire retardant compound　ignifugeant *m*
fire safety　sécurité *f* contre l'incendie
fire safety curtain　rideau *m* coupe-feu
fire screen　écran *m* de cheminée, garde-feu *m*
fire separation　isolement *m* coupe-feu
fire sprinkler installation　installation *f* d'extinction par sprinkle(u)rs
fire sprinkler system　système *m* d'extinction automatique
fire stable　stable au feu (SF)
fire station　caserne *f* des pompiers
fire stop　coupe-feu *m*
fire ventilation　désenfumage *m*
fire wall　mur *m* coupe-feu, cloison *m* coupe-feu/pare-feu
fire zone　zone *f* d'incendie
fire [a boiler]　échauffer *vb*, chauffer *vb*
fire, be on　brûler *vb*
fire, take/catch　s'allumer *vb*
fire to, set　incendier *vb*, mettre *vb* (le) feu à
fireback [ornamental cast iron]　plaque *f* (de cheminée)
firebox　chambre *f* de combustion, foyer *m*
fireclay　chamotte *f*
fire-dog　chenet *m*
fire-guard　pare-étincelles *m*
fire-lighter　allume-feu *m*
fireman's valve or hose connection　raccord *m* pompier/rapide
firemen's well, sump or water tank　puisard *m* incendie
fireplace　cheminée *f*, foyer *m*
fireplace, corner　cheminée *f* d'angle

fireplace, open　feu *m* ouvert
fireplace, open hearth　cheminée *f* à foyer ouvert
fireplace back　fond *m*
fireplace curtain [slides down to block off]　rideau *m*/tablier *m* (de cheminée)
fireplace facade immediately under mantlepiece　soubassement *m*
fireplace hearth　âtre *m*
fireplace surround [rectilinear and undecorated]　capucine *f*
fireproof　ignifuge *adj*, imbrûlable *adj*, incombustible *adj*
fireproof　ignifuger *vb*
fireproofing　ignifugeage *m*
fireproofing compound　ignifugeant *m*
fire-resistant　pare-flammes *adj* (PF), résistant au feu
firing [of a boiler]　chauffe *f*
firing [setting fire to]　mise *f* en feu
firing, electric [blasting]　tir *m* électrique
firing, hopper-fed　chauffe *f* avec alimentation par trémie
firing method [of a boiler]　méthode *f* de chargement
firing with a safety fuse [blasting]　tir *m* à la mèche
firm *adj*　ferme *adj*, stable *adj*
firm [company]　firme *f*, société *f*, entreprise *f*
firm, civil engineering　entreprise *f* de travaux publics
firm, contracting　entreprise *f*
firring　fourrure *f*
firring [at the edge of a timber roof]　coyau *m*
first　premier *adj*
first aid　premiers secours *m,pl*/soins *m,pl*, soins *m,pl* d'urgence
first aid point　poste *m* de secours
first aid room　infirmerie *f*
fiscal　fiscal *adj*
fish pass [dam]　passe *f* à poissons
fishbolt [railway line]　boulon *m* d'éclisse
fishplate　éclisse *f*
fishtail fixing lug, split end of a　queue-de-carpe *f*
fishtail tie　aronde *f*
fissure　crique *f*, criqûre *f*, fissure *f*
fissure [in stone]　fil *m*, poil *m*
fissure [an arch]　étonner *vb*
fissure [ground, etc]　fendre *vb*
fissured　fissuré *adj*
fissuring　fissuration *f*
fit for habitation　habitable *adj*
fit for its purpose　propre *adj* à sa destination
fit　adjustement *m*
fit, close　adjustement *m* précis
fit, loose　ajustage *m* à jeu/facile/lâche
fit, tight　adjustement *m* serré
fit [install, fix]　poser *vb*
fit [adapt, adjust]　adapter *vb*, ajuster *vb*
fit door or window hardware　ferrer *vb*
fit in [into a recess or housing]　enclaver *vb*
fit out or up [shop, etc]　aménager *vb*
fit sth to　ajuster *vb* qch à
fit up/together　agencer *vb*
fitted (for)　adapté *adj* (à)
fitted out (as)　agencé *adj* (en)

fitter   installateur *m*, monteur *m*
fitter, electrical   monteur *m* électricien
fitter, pipe   poseur *m* de tuyaux
fitting, *see also* fixture
fitting [adjustment]   ajustage *m*
fitting [assembling]   jointement *m*
fitting [light]   luminaire *m*
fitting [positioning, fixing]   posage *m*, pose *f*
fitting, bayonet   culot *m* à baïonnette, douille *f* à
   baïonnette
fitting, breech [pipe junction]   culotte *f*
fitting, breech, with cleaning eye   culotte *f* à regard
fitting, cable   fixation *f* de câble
fitting, compression [pipe jointing]   raccord *m* montage
   bicône
fitting, drainage   matériel *m* de drainage
fitting, four-way (pipe)   tuyau *m* en croix
fitting, iron   ferrure *f*
fitting, light, *see* fixture, or light fitting
fitting, pipe   appareillage *m* pour tube
fitting, pipe junction   culotte *f*
fitting, screw [bulb/lamp]   culot *m* à vis
fitting, solder [pipe jointing]   raccord *m* montage soudé
fitting, suspended light   luminaire *m* suspendu,
   suspension *f*
fitting, three-way pipe   tuyau *m* à trois voies
fittings [as in pipe fittings]   accessoires *m,pl*
fittings [equipment, etc]   appareillage *m*, installations *f,pl*
fittings [furnishings, e.g. office, house]   aménagements *m,pl*
fittings [taps]   robinetterie *f*
fittings, boiler   raccorderie *f* de chaudière
fittings, central heating   raccorderie *f* de chauffage
   central
fittings, shower   robinetterie *f* de douche
fitting door or window hardware   ferrage *m*
fitting out   aménagement *m*
fitting together   emboîtement *m*
fitting up   montage *m*
fix [a mast or derrick, by cables, etc]   amarrer *vb*
fix [attach, fasten]   attacher *vb*
fix [install]   poser *vb*
fix [sth, e.g. a nail, into sth]   ficher *vb*
fix [mend]   réparer *vb*
fix a price   établir *vb* un prix
fix a price or rate for   tarifer *vb*
fix in   sceller *vb*
fix (to)   fixer *vb* (à, sur)
fix together with a dog or cramp   agrafer *vb*
fix (up) [a date, a programme, etc]   régler *vb*
fixed   ferme *adj*, fixe *adj*
fixed in place   ancré *adj* en place
fixed into [masonry, etc]   scellé *adj*
fixing [accessory for]   fixation *f*, patte *f*
fixing [building in]   encastrement *m*
fixing [fitting, installing]   posage *m*, pose *f*
fixing [in concrete, masonry, etc, with cement, mortar, etc]
   scellement *m*
fixing [of a door, window, etc]   condamnation *f*
fixing, mirror   patte *f* à glace

fixing for metal sheet roofing   patte *f* d'agrafe
fixing from below of tiles against wind uplift
   pannetonnage *m*
fixings, second   finitions *f,pl*
fixture, *see also* fitting
fixture [light]   luminaire *m*
fixture, ceiling   plafonnier *m*
fixture, recessed [light fitting]   luminaire *m* encastré
fixture, suspended [light fitting]   luminaire *m* suspendu,
   suspension *f*
flag [with paving stones or slabs]   daller *vb*
flagstone   carreau *m* de pierre
flagstones   dallage *m*
flake [snow, vermiculite, etc]   flocon *m*
flake (off) [by itself]   s'écailler *vb*
flaking (off) [of paint or other surface]   écaillage *m*
flame   flamme *f*
flame, cutting   flamme *f* de coupe
flame, explosive   flamme *f* explosive
flame, naked   feu *m* nu, flamme *f* nue
flame, neutral   flamme *f* neutre
flame, oxidizing   flamme *f* oxydante
flame, pilot [of boiler, gas stove, etc]   veilleuse *f*
flame, reducing   flamme *f* réductrice
flame, welding   flamme *f* soudante
flame guard   pare-flamme *m*
flame cleaning   décapage *m* au chalumeau
flame cutting   découpage *m* au chalumeau, oxycoupage *m*
flame resistance   tenue *f* à la flamme
flame separation   décollement *m* de flamme
flame stability   stabilité *f* de la flamme
flame-resistant   pare-flammes *adj* (PF)
flammable substance   substance *f* inflammable
flange   collet *m*
flange [of girder]   membrure *f*, semelle *f*, table *f*
flange [of a wheel]   bourrelet *m*
flange [of angle iron, beam, girder]   aile *f*
flange [on pipe]   collerette *f*
flange [pipe, tube]   bride *f*
flange, blank   bride *f* non percée
flange, blind   bride *f* pleine
flange, compression   membrure *f* comprimée, table *f* de
   compression
flange, connecting/connection   bride *f* de raccordement
flange, coupling   bride *f* d'accouplement
flange, facing   bride *f* à emboîtement
flange, loose   bride *f* mobile
flange, lower   aile *f*/membrure *f*/semelle *f* inférieure
flange, male–female facing   bride *f* à emboîtement simple
flange, mating   contre-bride *f*
flange, pipe   bride *f* de tube
flange, projecting   aile *f* saillante
flange, screwed   bride *f* filetée
flange, slip   bride *f* folle
flange, support   bride *f* support
flange, tension   membrure *f* tendue
flange, tongue-and-groove facing   bride *f* à emboîtement
   double
flange, upper   aile *f*/membrure *f*/semelle *f* supérieure

flange connection joint *m* à brides/de bride, raccordement *m* par brides
flange coupling accouplement *m* à plateaux
flange jointing joint *m* à brides/de bride, assemblage *m* par brides
flange slope inclinaison *f* des ailes
flange thickness [of a beam] épaisseur *f* d'aile
flange width largeur *f* d'aile
flange brider *vb*
flash zinguer *vb*
flash [fit flashing] engraver *vb*
flash [of a signal light] clignoter *vb*
flashing [light] clignotant *adj*
flashing couvre-joint *m*, zinguerie *f*
flashing [lead] feuille *f* d'étanchéité/de recouvrement
flashing [metal, around a chimney or roof opening] alaise *f*
flashing [sheet metal] bavette *f*
flashing, lead ferblanterie *f* en plomb
flashing between wall and abutting roof solin *m*
flashing of mortar between a roof and wall ruellée *f*
flashing or slot to accept its edge engravure *f*
flashing (strip) bande *f* de solin
flat méplat *adj*, plat *adj*
flat [surface, etc] plan *adj*
flat [apartment] appartement *m*
flat [wide metal strip] large plat *m*
flat, show appartement *m* témoin
flat, small pied-à-terre *m*
flats, block of immeuble *m* d'appartements
flats, high rise block of tour *f* d'habitation
flat-laid [of a joist or beam] méplat *adj*
flatlet studio *m*
flat-slab flatslab *m*
flatten replanir *vb*
flattening replanissage *m*
flattening [of metal sheet by hammering] planage *m*
flatter [flat face hammer] marteau *m* à planer
flaw défectuosité *f*
fleet of lorries/trucks parc *m* de camions
flex, electrical câble *m* flexible
flexibility élasticité *f*, flexibilité *f*
flexibility [of metal] liant *m*
flexible flexible *adj*
flexing or flexion [under axial load] flexion *f*
flexion, lateral flambage *m*, flambement *m*
flicker papillotement *m*
flicker [of a light] clignoter *vb*
flier marche *f* carrée/droite
flies [of a theatre] cintres *m,pl*
flight [of stairs] foulée *f*, volée *f* (d'escalier)
flight of stairs escalier *m*
flight of steps [in front of a house] perron *m*
flint silex *m*
flint nodule rognon *m* de silex
flitch [large timber not yet resawn into boards] plateau *m*
float [in a cistern] flotteur *m*
float [tool] taloche *f*
float, bricklayer's bouclier *m*
float, devil gratton *m*

float, power hélicoptère *f*, lisseuse *f*, truelle *f* mécanique
float switch interrupteur *m* à flotteur
float flotter *vb*
float [plaster, etc] talocher *vb*
floating flottant *adj*
floating [of concrete or plaster surface] talochage *m*
flocculant floculent *adj*
flocculate floculer *vb*
flocculation floculation *f*
flood inondation *f*
flood [water] crue *f*
flood, design crue *f* nominale
flood bank digue *f*, levée *f*, turcie *f*
flood gate vanne *f* de décharge
flood peak attenuation écrêtement *m* des crues
flood plain bassin *m*/lit *m* de crue, lit *m* des hautes eaux, zone *f* d'inondation
flood protection protection *f* contre les crues
flood [land] inonder *vb*, submerger *vb*
flooded inondé *adj*
flooding, liable to inondable *adj*
flooding inondation *f*
floodlight projecteur *m*
floodlight illuminer *vb* au moyen de projecteurs
floodlighting illumination *f*
floodlit illuminé *adj*
floor aire *f*, plancher *m*
floor [of a valley] fond *m*
floor [solid] sol *m*
floor [storey] étage *m*
floor, boarded platelage *m* en bois
floor, concrete sol *m* en béton
floor, false faux plancher *m*
floor, finished sol *m* fini
floor, first premier étage *m*
floor, first [two storey building] étage *m*
floor, flat slab plancher *m* dalle
floor, ground rez-de-chaussée *m*
floor, hollow (block) slab plancher *m* à corps creux
floor, hollow block or coffer plancher *m* en hourdis creux
floor, mezzanine entresol *m*, mezzanine *f*
floor, mosaic dallage *m* en mosaïque
floor, on one de plain-pied
floor, polished or waxed parquet *m* ciré
floor, prefabricated [of beams and panels or slabs] plancher *m* préfabriqué
floor, projecting [visible between cladding panels] plancher *m* saillant
floor, raised faux plancher *m*
floor, solid plancher *m* massif
floor, solid, on formation terre-plein *m*
floor, tiled carrelage *m*
floor, top dernier étage *m*
floor, total structure supporting a gîtage *m*
floor, typical étage *m* courant
floor, unfinished sol *m* brut
floor, upper [two storey building] étage *m*
floor, vertical end of a tranche *f* d'un plancher

floor, wooden [boarded]   sol *m* en planches
floor area   surface *f*, surface *f* couverte/de plancher/des étages
floor areas   surface *f* des sols
floor covering   couverture *f* de sol, revêtement *m* du sol
floor finish   surface *f* du sol
floor grinder   meuleuse *f* de sols
floor height   hauteur *f* d'étage
floor joist   poutre *f* de plancher
floor joist, double   doubleau *m*
floor level, finished (FFL)   niveau *m* du sol fini
floor loading   charge *f* au sol/sur le plancher
floor of a bridge   aire *f* d'un pont
floor projecting between cladding panels   plancher *m* saillant
floor selector [lift/elevator]   sélecteur *m* d'étage
floor slab   dalle *f* de plancher
floor slab, hollow pot   dalle *f* en corps creux
floor space   surface *f* couverte/des étages
floor structure   plancher *m*
floor support   gîte *m*
floor tile   carreau *m*
floor tile [hexagonal, quarry]   tommette *f*
floor trap   siphon *m* de sol
floor [with boards]   planchéier *vb*
floorboard   planche *f*
floorboard or strip 60–80mm wide   frise *f*
floorboard, planed and grooved, 45–65mm wide   double frisette *f*
floorboard, planed and grooved, 65–80mm wide   frisette *f*
floorboards   parquet *m*
flooring [boarding]   planchéiage *m*
flooring [of a bridge]   platelage *m*
flooring [surfacing, covering]   revêtement *m* du sol
flooring [timber/wood]   parquet *m*
flooring, alternate joint strip   parquet *m* à l'anglaise à coupe de pierre
flooring, inlaid strip   parquet *m* (en) mosaïque
flooring, lightweight [of a bridge]   platelage *m* léger
flooring, parallel strip   parquet *m* à l'anglaise
flooring, plywood   parquet *m* contreplaqué/latté
flooring, random length strip   parquet *m* à l'anglaise à coupe perdue
flooring, rough, of rubble, broken bricks, etc   hourdage *m*
flooring, strip [parallel]   parquet *m* à lames/à l'anglaise
flooring, wood   parquet *m*
flooring, wood strip, in a fan-shaped pattern   parquet *m* en éventail
flooring strip 60–80 mm wide   lame *f*
flooring strip less than 1 m long   courson *m*
flotation   traitement *m* par flottation
flow   écoulement *m*, flux *m*
flow [of a river]   cours *m*
flow [of air, fluids, electricity]   courant *m*
flow [of open water]   affluence *f*
flow, air   écoulement *m* d'air
flow, annular   écoulement *m* annulaire

flow, back   refoulement *m*
flow, balanced   débit *m* équilibré
flow, cash   cash-flow *m*, flux *m* de trésorerie
flow, characteristic average   débit *m* caractéristique moyen
flow, daily   débit *m* journalier
flow, heat   écoulement *m* de chaleur
flow, laminar   écoulement *m*/flux *m* laminaire
flow, mass   flux *m* massique
flow, plastic   écoulement *m* plastique, fluage *m*
flow, recirculated air [quantity/rate]   débit *m* d'air de reprise
flow, return or reverse   courant *m* de retour
flow, reverse   écoulement *m* de retour, refoulement *m*
flow, spiral   écoulement *m* en spirales
flow, stratified air   écoulement *m* d'air stratifié
flow, streamline   écoulement *m* laminaire/non perturbé/selon la ligne de courant
flow, supply air   débit *m* d'air soufflé
flow, traffic   écoulement *m* du trafic
flow, transition   écoulement *m* de transition, régime *m* d'écoulement transitoire
flow, turbulent   écoulement *m* turbulent
flow, uniform   écoulement *m* uniforme
flow, viscous   écoulement *m* visqueux
flow, vortex   écoulement *m* en vortex
flow chart [of organization]   organigramme *m*
flow connection [on a boiler]   raccordement *m* de départ
flow diagram   schéma *m* de principe
flow diagram [computer]   ordinogramme *m*, organigramme *m*
flow diagram [pipework]   diagramme *m* de circulation
flow direction   direction *f* de l'écoulement
flow indicator   indicateur *m* de vitesse
flow limiting device   limiteur *m* de débit
flow line   ligne *f* d'écoulement
flow measurement   mesure *f* de débit
flow meter   débit-mètre *m*, débitmètre *m*
flow meter tapping   prise *f* de débit-mètre
flow pattern   configuration *f* d'écoulement
flow rate   débit *m*, régime *m*, vitesse *f* d'écoulement
flow rate [volumetric]   débit *m* de l'écoulement/volumique
flow rate [volumetric], discharge or outlet   débit *m* de volume de sortie
flow recorder   enregistreur *m* de débit
flow regulation   régulation *f* de(s) débit(s)
flow resistance,   résistance *f* à l'écoulement
flow resistance [frictional]   résistance *f* par friction
flow reversal   écoulement *m* de retour, renversement *m* de l'écoulement
flow variation   variation *f* de débit
flow [river, stream]   couler *vb*
flow [traffic, electricity, air, etc]   circuler *vb*
flow out [liquids]   s'écouler *vb*
flower bed   parterre *m*, plate-bande *f*
flowing *adj*   courant *adj*
flowing [of water]   affluence *f*
fluctuation   fluctuation *f*, variation *f*

fluctuation of load   fluctuation *f* de charge
fluctuations formula [costing, pricing]   formule *f* de révision des prix
flue   conduit *m*, fumisterie *f*
flue [between boiler and vertical flue]   carneau *m*
flue, balanced   circuit *m* de fumées équilibré
flue, boiler or smoke   conduit *m* de fumée
flue, built-in   conduit *m* incorporé
flue, chimney   conduit *m* de cheminée
flue, chimney [pipe]   conduite *f*/tuyau *m* de cheminée
flue, common [for several hearths]   conduit *m* unitaire
flue, exhaust   conduit *m* d'évacuation/d'extraction
flue, heating   carneau *m* de chauffage
flue, main   carneau *m* principal
flue, smoke   conduit *m* de fumée
flue adaptor pipe [joining different size pipes]   buse *f*
flue block   boisseau *m*
flue brick   boisseau *m*
flue cleaning   ramonage *m*
flue condenser   échangeur *m* à condensation
flue connection [metal ring built into masonry]   virole *f*
flue draught   tirage *m* de cheminée
flue fixed against a wall   conduit *m* accolé à un mur
flue fixed to but standing free of a wall   conduit *m* adossé à un mur
flue gas   gaz *m* de fumée
flue gas analyser   analyseur *m* de gaz de fumée
flue gas removal   évacuation *f* des fumées/gaz
flue liner   tubage *m* de cheminée
flue pipe   conduit *m* de fumée, tuyau *m* de cheminée
flue sweeping   ramonage *m*
fluid *adj*   fluide *adj*
fluid   fluide *m*
fluid, drilling   fluide *m* de forage/de perforation
fluid, heat transfer   fluide *m* caloporteur
fluid bed combustion   combustion *f* par lit fluidisé
fluidics   fluidique *f*
fluidity   fluidité *f*
flume   canal *m*
fluorine (gas)   fluor *m*
fluorocarbon   fluorocarbure *m*
flush [level]   affleur *adj*
flush [of water down a pipe or drain]   chasse *f*
flush, making   affleurage *m*
flush mounted   encastré *adj*
flush mounted in the ceiling   encastré *adj* dans le plafond
flush mounting   à encastrer
flush [rinse]   rincer *vb*
flush, make   affleurer *vb*
flush (with), be   affleurer *vb*
flushing [of a drain, etc]   curage *m*
flushing channel   canal *m* de chasse
flushing cistern   chasse *f* d'eau, réservoir *m* de chasse d'eau
flushing device   appareil *m* de chasse d'eau
flushing lever [WC cistern]   levier *m* de la chasse d'eau, manette *f* de chasse
flushing out   purge *f*

flushing tank   réservoir *m* de chasse d'eau
fluting [of a column]   cannelure *f*
flux   fondant *m*
flux [for brazing, etc]   flux *m*
flux, brazing   flux *m* de brasage
flux, luminous   flux *m* lumineux
flux, welding   flux *m*/poudre *f* à souder
flux density   densité *f* de flux
flyover   autopont *m*, croisement *m* à niveaux différents
flywheel   volant *m* mécanique/moteur
foam   mousse *f*
foam [on the surface of a fluid]   écume *f*
foam, phenolic   mousse *f* phénolique
foam, (rigid) plastic   plastique *f* expansée (rigide)
foam   écumer *vb*, mousser *vb*
focus [optical]   foyer *m*
focusing   mise *f* au point
focusing ring [e.g. of a surveying telescope]   bague *f* de mise au point
fog   brouillard *m*, brume *f*
fogging   nébulisation *f*
foil   feuille *f*
fold   pli *m*
fold in the ground   pli *m* de terrain
fold   plisser *vb*
fold (up, down, back)   plier *vb*
folded   plissé *adj*
folding *adj*   pliant *adj*
folding [of a door leaf or a shutter]   brisé *adj*
folding   pliage *m*
foot   pied *m*
foot, cubic/square   pied *m* cube/carré
footboard   garde-pieds *m*
footbridge   passerelle *f*
footing   embasement *m*, empattement *m*
footing [foundation]   semelle *f*
footing [of a wall]   empattement *m*
footing, continuous   semelle *f* continue
footing, individual or pad   semelle *f* isolée
footing, mass concrete   semelle *f* en gros béton
footing, pad   fondation *f*/semelle *f* isolée
footing, piled   semelle *f* sur pieu
footing, reinforced concrete   semelle *f* armé
footing, shallow   semelle *f* superficielle
footing, spread [of a pile]   patte *f* d'éléphant
footing, strip   semelle *f* filante
footlights [theatre]   rampe *f*
footpath   chemin *m* piéton, sentier *m*, voie *f* piétonnière
footpath [along a watercourse]   marchepied *m*
footpath [e.g. on bridge]   trottoir *m*
footpath surfacing   revêtement *m* du trottoir
footprint of a building   emprise *f* d'un bâtiment
footway   piste *f* piétonne, piste *f* pour piétons, trottoir *m*
footway [of bridge, tunnel]   banquette *f*
force   effort *m*, force *f*
force [power]   puissance *f*
force, bending   effort *m* de flexion
force, braking   effort *m*/force *f* de freinage
force, centrifugal   force *f* centrifuge

force, clamping   effort *m* de serrage
force, compressive   effort *m*/force *f* de compression
force, electromotive   force *f* électromotrice
force, friction   force *f* frictive/de frottement
force, lateral   effort *m* latéral
force, longitudinal   effort *m*/force *f* longitudinal
force, motive   force *f* motrice
force, normal   effort *m* normal
force, restoring   force *f* de rappel
force, shear(ing)   effort *m*/force *f* de cisaillement
force, tangential   effort *m* tangentiel
force, tensile   force *f* de traction
force, transmission of   transmission *f* d'effort
force, transverse   force *f* transversale
force, transverse shear   effort *m* tranchant
force, uplift   poussée *f* ascensionnelle
forces, polygon of   polygone *m* des forces
forces, vector sum of [resultant]   résultante *f*
force cup [plumbing]   ventouse *f*
forced   forcé *adj*
forced (air) circulation   circulation *f* (d'air) forcée
forced circulation   circulation *f* accélérée
ford   gué *m*
fore-hearth [of a fireplace]   avant-foyer *m*
forecast   prévision *f*
foreclosure of a mortgage   saisie *f* d'une hypothèque
foreign literature [documentation]   documentation *f* étrangère
foreman   contremaître *m*, maître *m* ouvrier
foreman, (construction)   chef *m* d'équipe
foreman, general   chef *m* de chantier, conducteur *m* de travaux
foreman, shop   chef *m* d'atelier
foreman, site   chef *m* de chantier
forest *adj*   forestier *adj*
forest   forêt *f*
forest ride or track   laie *f*
forest, communal/public   forêt *f* domaniale
forest, deciduous   forêt *f* feuillue
forestry development   exploitation *f* forestière
forge   forge *f*
forge [metal]   forger *vb*
forging   pièce *f* forgée
fork   fourche *f*
forklift truck   chariot *m* élévateur (à fourche)
form, *see* formwork
form   forme *f*
form [document]   formulaire *m*
form [sheet to be filled in]   fiche *f*
form   façonner *vb*
form [a collection, deposit, etc]   former *vb*
form fixing pockets in fresh masonry with a crowbar   battre *vb* le beurre[†]
formaldehyde   formaldéhyde *m*
formation [process of]   formation *f*
formation [subgrade or subsoil]   terrain *m* de fondation
formation [undisturbed ground]   forme *f*, terrain *m* naturel
formation level [of excavations]   plate-forme *f* des terrassements

formed [of concrete]   banché *adj*
formed, cold   formé *adj* à froid
former [precedes noun in French]   ancien *adj*
forming   façonnage *m*
forming [process of]   formation *f*
forming, cold/hot   déformation *f*/façonnage *m* à froid/à chaud
formula   formule *f*
formula, chemical   formule *f* chimique
formula, design   formule *f* de calcul/de dessin
formula, fluctuations [costing, pricing]   formule *f* de révision des prix
formula, pile driving   formule *f* de battage
formwork   coffrage *m*, moule *f*
formwork [curved, for arches, vaults, etc]   cintre *m*
formwork, climbing   coffrage *m* grimpant
formwork, lost or sacrificial   coffrage *m* perdu
formwork, plywood   coffrage *m* en contreplaqué *m*
formwork, re-usable, for floor slabs   table *f* coffrante
formwork, sliding   coffrage *m* glissant
formwork, tunnel   coffrage *m* tunnel
formwork, steel   coffrage *m* métallique
formwork, temporary   coffrage *m* provisoire
formwork, timber   coffrage *m* bois
formwork, travelling   coffrage *m* mobile/roulant
form(work), tunnel   coffrage *m* tunnel
formwork carpenter   charpentier-coffreur *m*, coffreur *m* (pour béton armé), coffreur-boiseur *m*
formwork design, computer-aided   coffrage *m* assisté par ordinateur (CAO)
formwork drawing   plan *m* de coffrage
formwork for casting a wall [i.e. two faced]   banche *f*
formwork inner side   peau *f* de coffrage
formwork rake or taper [to facilitate striking]   dépouille *f*
formwork removal or striking   décoffrage *m*
formwork tie   tirant *m* de coffrage
formwork vibrator   vibreur *m* de coffrage
fort(ress)   fort *m*
forward [send off]   expédier *vb*
forwarding company   compagnie *f* de transport
fossil *adj*   fossile *adj*
fossil   fossile *m*
foul   encrasser *vb*
foul smelling   nauséabond *adj*
foul water sewerage   assainissement *m* des eaux usées
fouled   encrassé *adj*
fouling   encrassement *m*
found (on)   fonder *vb* (sur)
foundation   fondation *f*
foundation [basis]   fondement *m*
foundation [footing]   empattement *m*
foundation [solid, of concrete]   plate-bande *f*
foundation, deep   fondation *f* profonde
foundation, earthquake resistant   fondation *f* parasismique
foundation, improved [ground]   fondation *f* traitée
foundation, long strip   fondation *f* sur semelle continue
foundation, mat   fondation *f* sur radier, radier *m*
foundation, pier   fondation *f* sur puits

foundation, pile(d)   fondation *f* sur pieux/pilotis
foundation, plinth   socle *m* de fondation
foundation, raft   fondation *f* sur radier
foundation, raft [on soft ground]   plate-forme *f* hollandaise
foundation, road   fondation *f* de la chaussée
foundation, shallow   fondation *f* superficielle
foundation, stepped   fondation *f* en redans
foundation, strip   fondation *f* sur semelle
foundation block [e.g. for a pole or post]   bloc *m* de fondation
foundation course [road construction]   couche *f* de fondation
foundation depth   profondeur *f* de la fondation
foundation digging or excavation   fouille *f* de fondation
foundation mat   radier *m*
foundation of wooden piles   pilotis *m*
foundation raft   châssis *m*/radier *m* de fondation
foundation slab   dalle *f* de fondation
foundation slab and walls of a basement   cuvelage *m*
foundation stone, lay the   poser *vb* la première pierre
foundation sub-base [road construction]   sous-couche *f* de fondation
foundation (tie) beam   longrine *f*
foundation trench   tranchée *f* de fondation
foundation work(s)   travaux *m,pl* de fondation
foundations of old buildings found under new   substruction *f*
foundry   fonderie *f*
fountain   jet *m* d'eau *f*, fontaine *f*
fountains   jeux *m,pl* d'eau
four-wheel drive   transmission *f* quatre-roues
fraction   fraction *f*
fraction [grain sizing]   classe *f* granulaire
fracture   cassure *f*, rupture *f*
fracture [point of]   point *m* de rupture
fracture, brittle   rupture *f* fragile
fracture, deformation   cassure *f* ductile
fracture, ductile   cassure *f* ductile
fracture, fatigue   cassure *f*/rupture *f* de fatigue
fracture, fine-grained   cassure *f* à grain fin
fracture elongation   allongement *m* de rupture
fracture grain   grain *m* de la cassure
fracture   casser *vb*
fragility   fragilité *f*
fragment   fragment *m*
fragmental or fragmentary   fragmentaire *adj*
frame   bâti *m*, cadre *m*
frame [door, window]   châssis *m*, dormant *m*
frame [of building]   carcasse *f*
frame [of door]   chambranle *m*
frame [structural]   charpente *f*
frame [support stand]   chevalet *m*
frame [timber or steel]   ossature *f*
frame, clamping   cadre *m* de serrage
frame, counter   contre-cadre *f*
frame, door   bâti *m*/châssis *m* de porte, huisserie *f*
frame, loading   portique *m* de chargement
frame, multi-bay   portiques *m,pl* multiples

frame, multi-storey   cadre *m*/portique *m* à étages
frame, portal   (cadre *m*) portique *m*
frame, single portal   portique *m* simple
frame, steel   ossature *f* métallique
frame, structural   ossature *f* de construction
frame, three-hinge   cadre *m* à trois articulations
frame, window   bâti *m*/châssis *m* de fenêtre, huisserie *f*
frame construction   construction *f* en charpente
frame construction, reinforced concrete   ossature *f* en béton armé
frame for transporting glass   portoir *m*
frame stanchion   béquille *f* de portique
framework   bâti *m*, cadre *m*
framework [of building]   carcasse *f*
framework [structural]   charpente *f*
framework [timber or steel]   ossature *f*
framework, built-in [to accept door, window, etc]   bâti *m* dormant
framework, vertical, with windbracing   palée *f* de contreventement
framework of beams   poutrage *m*
framing   cadre *m*, encadrement *m*
framing [timber or steel]   ossature *f*
framing of roof structure, radiating horizontally from central point   enrayure *f*
framing with verticals stopped under the top member   huisserie *f* à chapeau
free   libre *adj*
free [an object]   dégager *vb*
freelance [self-employed]   travailleur *m* indépendant
free-standing   autonome *adj*, indépendant *adj*
freestone   grès *m* à bâtir/de construction, pierre *f* de taille
freeze   congeler *vb*, geler *vb*
freeze out   séparer *vb* par congélation
freezer   congélateur *m*
freezer, chest   coffre *m* congélateur
freezer, tunnel   tunnel *m* de congélation
freeze-resistant   incongelable *adj*
freezing up [of pipework, etc]   obturation *f* par congélation
freight   fret *m*
freight hall   halle *f* à marchandises
freight terminal, airport   aérogare *f* (de) fret
French curve   pistolet *m*
French drag   brette *f*, chemin *m* de fer, guillaume *m*
French drain   drain *m* de pierraille, drain *m* français, fossé *m* couvert
French flier   marche *f* dansante
French national building research institute   Centre Scientifique et Technique du Bâtiment (CSTB)
French national cartographic institute   Institut Géographique National (IGN)
French national electricity supply company [with GDF]   Electricité de France (EDF)
French national gas supply company [with EDF]   Gaz de France (GDF)
French national railways   Société Nationale des Chemins de Fer Français (SNCF)
French national standard/code   Norme Française (NF)

French national standards authority/institution
Association Française de Normalisation (AFNOR)
French national statistical institute [produces building
cost indices, etc] Institut National de la Statistique des
Études Économiques (INSEE)
French post office La Poste
French society of planners Société Française des
Urbanistes (SFU)
French telephone service France Télécom
French unit of cold [= calorie] frigorie *f*
French window porte-croisée *f*, porte-fenêtre *f*
freon fréon *m*
frequency [electrical, statistical] fréquence *f*
frequency, carrier fréquence *f* porteuse
frequency, natural fréquence *f* naturelle/propre
frequency, resonant fréquence *f* de résonance
frequency, vibration fréquence *f* de vibration
frequency and voltage convertor convertisseur *m* de
fréquence et de tension
frequency curve courbe *f* de fréquence
frequency response réponse *f* en fréquence
fresh [air, concrete, etc] frais *adj*
freshness [of colours] fraîcheur *f*
fretsaw scie *f* à chantourner/à découper
friable friable *adj*
friction friction *f*, frottement *m*
friction, coefficient of coefficient *m* de frottement
friction, internal frottement *m* interne
friction, negative skin frottement *m* négatif
friction, skin frottement *m* latéral/superficiel
friction angle angle *m* de frottement
friction angle, internal angle *m* de frottement interne
friction circle cercle *m* de frottement
fridge[†] frigo[†] *m*, réfrigérateur *m*
frieze [architectural] frise *f*
frog [of railway points] cœur *m* d'aiguille
from above [on pipework drawing] du haut
from below [on pipework drawing] du bas
from top to bottom de fond en comble
front devant *m*, face *f* avant
front [meteorological] front *m*
frontage devanture *f*
frontage, set back [behind another] facade *f* en retrait
fronton fronton *m*
fronton [small, above a door] fronteau *m*
frost gel *m*, gelée *f*
frost [surface deposit, not weather] givre *m*
frost action action *f* du gel
frost blanket [anti-frost layer] couche *f* antigel
frost crack gélivure *f*, givrure *f*
frost deposit dépôt *m* de givre
frost formation givrage *m*
frost heave gonflement *m* dû au gel/par le gel
frost proofing (agent) antigel *m*
frost protection material matériau *m* antigel
frost resistance résistance *f* au gel
frost sensitivity/susceptibility [of stone, etc]
gélivité *f*
frost thermostat thermostat *m* à gel

frost givrer *vb*
frost [glass] dépolir *vb*
frosted [glass] dépoli *adj*
frosting givrage *m*
frost-proof layer [road construction] (sous-)couche *f*
anti-gel
frost-resistant insensible/résistant au gel
frost-riven [e.g. stone] gélif *adj*
frost-sensitive/susceptible [of stone, etc] gélif *adj*
froth écume *f*
froth écumer *vb*
frozen gelé *adj*
frying pan poêle *f*
fuel carburant *m*, combustible *m*
fuel/air ratio rapport *m* air/combustible
fuel, boiler combustible *m* de chauffe
fuel, diesel gas-oil *m*, gazole *m*, mazout *m* diesel
fuel, fossil combustible *m* fossile
fuel, liquid combustible *m* liquide
fuel, low grade combustible *m* pauvre
fuel, solid combustible *m* solide
fuel bowser tonne *f* à carburant
fuel bunker soute *f* à combustible
fuel economy économie *f* de combustible
fuel gas gaz *m* combustible
fuel gas analysis analyse *f* des gaz combustibles
fuel oil fioule *m*, fuel *m* (de chauffage), mazout *m*
fuel oil, domestic mazout *m* domestique
fuel saving économie *f* de combustible
fuel storage stockage *m* de combustible
fuel storeroom soute *f*
fuel tank réserve *f* de carburants, réservoir *m* de
carburant/de combustible
fuel tank [of a vehicle] réservoir *m* à carburant
fulcrum point *m* d'appui
fulcrum [large wedge of wood or stone] orgueil *m*
full plein *adj*
full text [e.g. of a specification] texte *m* intégral
full-load break [electrical] coupure *f* en charge
full-scale *adj* en vraie grandeur
full-scale layout [of axes of a structural framework]
étalon *m*
full size *adj* en vraie grandeur
full-term à terme
fume cupboard hotte *f* de laboratoire
fume hood hotte *f*
fumes fumée *f*, gaz *m* de fumée
function of, as a en fonction de
functional fonctionnel *adj*
functioning fonctionnement *m*
fund subventionner *vb*
fund [debts] consolider *vb*
funds capital *m*, fonds *m*
fungicidal fongicide *adj*
fungicide fongicide *m*
fungus champignon *m*
funicular (railway) funiculaire *m*
funnel [tundish] entonnoir *m*
funnel [for loading] trémie *f*

fur [in a boiler]   tartre *m*
furnace   chambre *f* de combustion, four *m*, fourneau *m*
furnace, annealing   four *m* de recuit
furnace, gas fired   four *m* à gaz
furnace, heating   four *m* de chaudière
furnace, industrial gas   fourneau *m* à gaz industriel
furnace, normalizing   four *m* de normalisation
furnish   meubler *vb*
furnish [provide]   fournir *vb*
furnished   meublé *adj*
furniture   mobilier *m*
furniture [piece of]   meuble *m*
further in, knock [e.g. a nail]   renfoncer *vb*
fuse [for igniting a detonator]   cordeau *m* Bickford
fuse [safety]   fusible *m*
fuse, cartridge   coupe-circuit *m* à cartouche, fusible *m* à

cartouche, fusible-cartouche *m*
fuse, cut-out or safety   fusible *m* de sûreté
fuse, detonating   cordeau *m* détonant
fuse, switched cartridge   coupe-circuit *m* à cartouche
fuse box/cabinet   boîte *f* à fusibles
fuse carriers, modular   coupe-circuit *m* sectionnable
fuse holder/terminal   porte-fusible *m*
fuse [blow the fuse of]   faire *vb* sauter
fuse, blow a   fondre *vb* un fusible
fuseboard   porte-fusibles *m*, table *f* de répartition
fusibility   fusibilité *f*
fusible   fusible *adj*
fusion [of metals, etc]   fusion *f*
fusion, latent heat of   chaleur *f* latente de fusion
fusion face [welding]   bord *m* à souder

G-cramp   bride *f* à capote
gabbro   gabbro *m*
gabion   gabion *m*
gable   pignon *m*, gable *m* [old fashioned word in French]
gable, fire   pignon *m* coupe-feu
gable board   bordure *f* du pignon
gain   gain *m*
gain [in value]   plus-value *f*
gain, casual heat   gain *m* de chaleur occasionnel
gain, conduction (heat)   gain *m* par conduction
gain, heat   gain *m* de chaleur
gain, incidental heat   gains *m,pl* de chaleur variable
gain on disposal   plus-value *f* de cession
gaiter, protective   guêtre *f* de protection
gallery [long room, mining, theatre, etc]   galerie *f*
gallery, inspection   galerie *f* d'inspection/de visite
gallery, service(s)   galerie *f* technique
galvanize   galvaniser *vb*, zinguer *vb*
galvanized   galvanisé *adj*, zingué *adj*
galvanized, hot-dip   galvanisé *adj* à chaud/au bain
galvanizing   galvanisation *f*, zingage *m*
gang [work team]   équipe *f*
ganger   chef *m* d'équipe
gangway   passerelle *f*
gantry   portique *m*
gap   écart *m*, espace *m*, intervalle *m*, vide *m*
gap [in a wall]   percée *f*
gap (between)   écartement *m* (de, entre)
garage [*not* petrol station]   garage *m*
garbage   ordures *f,pl* ménagères
garden   jardin *m*
garden, decorative   jardin *m* d'agrément
garden, formal   jardin *m* à la française
garden, kitchen   potager *m*

garden, landscaped   jardin *m* paysager/à l'anglaise
garden, large   parc *m*
garden, rock   jardin *m* de rocaille, rocaille *f*
garden, roof   jardin *m* en toiture, terrasse-jardin *f*
garden, small   jardinet *m*
garden, terraced   jardin *m* en terrasse
garden, vegetable   potager *m*
garden, water   parterre *m* d'eau
garden shelter   abri *m* jardin
gardener   jardinier *m*
gargoyle   chimère *f*, gargouille *f*
garret   chambre *f* de bonne, comble *m*, galetas *m*
garret [in a mansard roof]   mansarde *f*
gas   gaz *m*
gas, burnt   gaz *m* brûlé
gas, combustion   gaz *m* de combustion
gas, compressed   gaz *m* comprimé
gas, exhaust   gaz *m* extrait/d'échappement/rejeté
gas, flue   gaz *m* de fumée
gas, fuel   gaz *m* combustible
gas, heating   gaz *m* de chauffage
gas, high pressure   gaz *m* comprimé
gas, inert   gaz *m* rare
gas, liquified petroleum (LPG)   gaz *m* de pétrole liquéfié
gas, natural   gaz *m* naturel
gas, oxyacetylene   oxyacétylène *m*
gas, producer   gaz *m* de gazogène
gas, rare   gaz *m* rare
gas, refill [of a lamp]   gaz *m* de remplissage
gas, sewage or sewer   gaz *m* de curage
gas, town   gaz *m* de ville
gas, tracer   gaz *m* traceur
gas, waste   gaz *m* brûlé/de fumée
gas, water   gaz *m* à l'eau

gas blower   soufflerie *f* à gaz
gas bottle   bouteille *f* à gaz/de gaz
gas burner   bec *m* de gaz, brûleur *m*
gas burning/firing   chauffe *f* à/au gaz
gas central heating   chauffage *m* central au gaz
gas cock, main   robinet *m* principal de gaz
gas cooker   cuisinière *f* à gaz, gazinière *f*
gas cooling   refroidissement *m* à gaz
gas cutting   découpage *m* au chalumeau, oxycoupage *m*
gas cutting machine   machine *f* d'oxycoupage
gas cutting torch   chalumeau *m* à gaz
gas cylinder   bouteille *f* à gaz/de gaz, cylindre *m* à gaz
gas density   densité *f* d'un gaz de fumée
gas detector   capteur *m* à gaz
gas escape   déperdition *f* de gaz
gas firing [heating]   chaufferie *f* à gaz
gas fitter   gazier *m*
gas fitter [plumber]   plombier *m*
gas furnace, industrial   fourneau *m* à gaz industriel
gas generator   générateur *m* de gaz
gas geyser [instantaneous bath water heater]   chauffe-bain *m* à gaz
gas governor   régulateur *m* de gaz
gas holder   gazomètre *m*
gas installation   installation *f* au gaz
gas jet burner   brûleur *m* à jet de gaz
gas main   conduite *f* principale du gaz
gas main connection   branchement *m* particulier de gaz
gas meter   compteur *m* de gaz
gas mixture, explosive   mélange *f* de gaz explosive
gas oven   four *m* à gaz
gas pipe from mains valve to consumer meter   tuyau *m* de sortie du robinet chef
gas pipeline   canalisation *f* de gaz, gazoduc *m*
gas producer (plant)   gazogène *m*
gas producing   gazogène *adj*
gas radiator   radiateur *m* à gaz
gas refrigerator   réfrigérateur *m* à gaz
gas service pipe   branchement *m* de gaz
gas stop tap [for a whole building]   robinet *m* chef (de gaz)
gas storage   stockage *m* de gaz
gas test   essai *m* au gaz
gas turbine   turbine *f* à gaz
gas water heater   chauffe-eau *m* à gaz
gas water heater, multi-point   chauffe-eau *m* à gaz à plusieurs points
gas welding   soudage *m* autogène/oxyacétylénique/au gaz
gaseous   gazeux *adj*
gas-lighter   allume-gaz *m*
gasification   gazéification *f*
gasify   gazéifier *vb*
gasket   joint *m*
gasket, foam rubber   joint *m* en caoutchouc mousse
gasket, rubber   joint *m* en caoutchouc
gasket, rubber [in a valve]   portée *f* d'étanchéité en caoutchouc
gasket ring   anneau *m* de joint
gasoline   essence *f*

gasometer holder   gazomètre *m*
gas-tight   étanche *adj*/imperméable *adj* au gaz
gasworks   usine *f* à gaz
gate   porte *f*
gate [field, road]   barrière *f*
gate, barred   porte *f* à claire-voie
gate, carriage   porte *f* charretière/cochère
gate, double   porte *f* à deux battants
gate, lock   vanne *f*
gate, two-leaf, too narrow for vehicles   porte *f* bâtarde
gate lodge   pavillon *m* d'entrée
gate(way)   portail *m*, porte *f*
gateway large enough to admit a vehicle   porte *f* cochère
gauge, off or out of   hors tolérances
gauge   jauge *f*
gauge [indicator]   indicateur *m*
gauge [measuring tool]   calibre *m*
gauge [of a tile or slate, i.e. visible area or height]   pureau *m*
gauge [template or size gauge]   gabarit *m*
gauge, air   indicateur *m* d'air
gauge, Bourdon   tube *m* de Bourdon
gauge, dial [measuring tool]   comparateur *m* à cadran
gauge, differential pressure   manomètre *m* différentiel
gauge, level   indicateur *m*/tube *m* de niveau
gauge, liquid level   jauge *f* de liquide
gauge, loading   gabarit *m* de chargement
gauge, oil level   jauge *f* de niveau d'huile
gauge, oil level [fuel oil]   jauge *f* de mazout
gauge, pressure   indicateur *m*/jauge *f* de pression, manomètre *m*
gauge, rain   pluviomètre *m*
gauge, repeater   indicateur *m* de répétition
gauge, sheet metal   épaisseur *f* de tôle
gauge, spacing   gabarit *m* d'écartement
gauge, steam   jauge *f* de vapeur
gauge, strain   extensomètre *m*, jauge *f* d'allongement/de déformation
gauge, thickness   jauge *f* d'épaisseur
gauge, vacuum   jauge *f* de vide
gauge, water   tube *m* de niveau d'eau
gauge, wind   anémomètre *m*
gauge of sheet metal   épaisseur *f* de tôle
gauge   jauger *vb*
gauging [of cement]   dosage *m*
gear, in   engrené *adj*, en prise
gear, out of or in neutral   au point mort
gear   engrenage *m*
gear, bevel   engrenage *m* conique, renvoi *m* d'angle
gear, reducing   réducteur *m*
gears, train of   train *m* d'engrenages
gear ratio   rapport *m* d'engrenage
gear, throw into   enclencher *vb*
gearbox   boîte *f* de vitesses
gearbox, angle   renvoi *m* d'angle
gel, silica   gel *m* de silice
generate [electricity, steam]   produire *vb*
generating set (emergency or standby)   groupe *m* électrogène (de secours)

generating station   centrale *f* électrique

generating station, steam   centrale *f* thermique

generation [e.g. of steam]   production *f*

generation, on-site   production *f* sur site

generator   alternateur *m*, générateur *m*, groupe *m* électrogène

generator [electrical]   génératrice *f*

generator [portable]   groupe† *m*

generator, acetylene   générateur *m* d'acétylène

generator, direct current   génératrice *f* à courant continu

generator, gas   générateur *m* de gaz

generator, steam   chaudière *f* à vapeur, générateur *m* de vapeur

generator, welding   génératrice *f* pour soudage

generator, wind   aérogénérateur *m*, éolienne *f*

geodesy   géodésie *f*

geodetic mark   signal *m* géodésique

geological   géologique *adj*

geologist   géologue *m/f*

geology   géologie *f*

geology, engineering   géologie *f* de l'ingénieur

geometry   géométrie *f*

geomorphological   géomorphologique *adj*

geomorphology   géomorphologie *f*

geophysical   géophysique *adj*

geophysics   géophysique *f*

geotechnical   géotechnique *adj*

geotechnics   géotechnique *f*

geothermal phenomena, study of   géothermie *f*

geothermic   géothermique *adj*

germ   germe *f*

germ-free   stérile *adj*

germicidal   germicide *adj*

get better   s'améliorer *vb*

get in touch with [a person]   joindre *vb*

get out of plumb   prendre *vb* coup

get rusty   se rouiller *vb*

geyser, bath [instantaneous water heater]   chauffe-bain *m*

geyser, gas [instantaneous bath water heater]   chauffe-bain *m* à gaz

ghost island [road]   espace *m* interdit à la circulation

gib(-headed key)   clavette *f* à talon

gilding   dorure *f*

gimlet   vrille *f*

girder, *see also* beam   poutre *f*

girder [small]   poutrelle *f*

girder, bowstring   arc *m* à tirant

girder, box   poutre *f* caisson

girder, central [of a bridge]   longeron *m*

girder, cross   entretoise *f*

girder, gantry   poutre *f* de roulement

girder, hollow   poutre *f* creuse

girder, lattice   poutre *f* à treillis

girder, longitudinal   longeron *m*

girder, main   poutre *f* maîtresse/principale

girder, open frame   poutre *f* échelle/Vierendeel

girder, plate or solid web   poutre *f* à âme pleine

girder, steel   poutrelle *f* métal

girder, Vierendeel   poutre *f* échelle/Vierendeel

girder, warren   poutre *f* Warren

girder, wooden   poutrelle *f* bois

girt [timber frame structure]   sommier *m*

give (someone) permission (to)   autoriser *vb* (qn à)

give off [heat, etc]   dégager *vb*

give way   s'affaisser *vb*, s'écrouler *vb*

give way [ground]   fuir *vb*

give way [under load, pressure]   céder *vb*

giving way   affaissement *m*, écroulement *m*

glacial   glacial *adj*

glacier   glacier *m*

gland [on electrical equipment]   embout *m*

gland, cable   presse-étoupe *m* de câble

gland, packing/stuffing   chapeau *m* de presse-étoupe

gland packing rope for pumps   bitord *m*

glare   éblouissement *m*

glare factor/index   indice *m* d'éblouissement

glass *adj*   vitré *adj*

glass   verre *m*

glass, bull's eye   verre *m* en cul de bouteille

glass, case hardened   verre *m* trempé

glass, cellular or foam   verre *m* multicellulaire

glass, foam   verre *m* expansé/mousse/multicellulaire, mousse *f* de verre

glass, frosted   verre *m* dépoli

glass, hard [borosilicate, for lamps]   verre *m* dur

glass, laminated   verre *m* feuilleté

glass, lamp   verre *m* de lampe

glass, milk   opaline *f*

glass, opal   verre *m* opale

glass, patterned   verre *m* mousseline

glass, plate   verre *m* à glace

glass, plate or sheet of   glace *f*

glass, protective [used by welders]   verre *m* protecteur

glass, quartz   silice *f*

glass, reeded   verre *m* strié

glass, safety   verre *m* de sécurité

glass, semi-opaque [type of]   verre *m* cathédrale

glass, sheet   verre *m* plat

glass, soft [lime-soda, for lamps]   verre *m* tendre

glass, stained   verre *m* cathédrale/coloré

glass, thick   verre *m* demi-double

glass, toughened   verre *m* trempé

glass, Triplex laminated/safety   triplex *m*, verre *m* triplex

glass, unbreakable   verre *m* incassable

glass, unpolished   verre *m* brut/non poli

glass, white/opal   verre *m* opale

glass, window   verre *m* à vitres

glass, wire reinforced   verre *m* armé

glass, wired   verre *m* grillagé

glasses, protective   lunettes *f,pl* de protection

glass block/brick (paving)   pavé *m* de verre, dalle *f* translucide

glass cutter   coupe-verre *m*, diamant *m*

glass fibre   fibre *f* de verre

glass holder [bathroom fitting]   porte-verre *m*

glass of low quality [cannot be diamond cut]   verre *m* casilleux

glass opaque to infra-red light   verre *m* catathermique
glass pane   vitre *f*
glass paper   papier *m* de verre
glass partition   vitrage *m*
glass roof over platform [station]   marquise *f*
glass slate/tile   panne *f* en verre
glass window, leaded or stained   vitrail *m*
glass with small frosted patterns   mousseline *f*
glass wool (insulation)   laine *f* de verre
glassware   verrerie *f*
glaze coat [paintwork]   lustre *f*
glaze   vitrer *vb*
glaze [paintwork]   lustrer *vb*
glazed   vitré *adj*
glazed area   surface *f* vitrée
glazed cast iron   fonte *f* émaillée
glazed stoneware   grès *m* flambé
glazier   miroitier *m*, vitrier *m*
glazier's hammer   marteau *m* de vitrier
glazier's knife for cutting lead   tailloir *m*
glazing [the action, and the result]   vitrage *m*
glazing, double   double vitrage *m*
glazing, roof   verrière *f*
glazing, secondary   survitrage *m*
glazing, structural silicone   vitrage *m* extérieur collé
glazing bar   fer *m* à vitrage
glazing bar [metal]   petit fer *m*
glazing bar [wood]   petit bois *m*
glazing bars, crossed   croisillon *m*
glazing bead/fillet   parclause *f*, parclose *f*
glazing putty   mastic *m* de vitrier
glazing rods   armature *f*
glazing sprig   pointe *f* de vitrier
glazing T-bar   fer *m* en T/té à vitrage
gloss [paint]   brillant *adj*
gloss   lustre *f*
glossy   lisse *adj*, lustré *adj*
glove   gant *m*
glove, protective   gant *m* de protection, manicle *f*, manique *f*
gloves, rubber   gants *m,pl* en caoutchouc
gloves, work/working   gants *m,pl* de travail
glow   incandescence *f*, lueur *f*
glue   colle *f*
glue, neoprene   colle *f* (au) néoprène
glue, wallpaper   colle *f* à tapisser
glue, wood/woodworking   colle *f* blanche
glue size   colle *f* de pré-encollage
glue   coller *vb*
glued   collé *adj*
gluing   collage *m*
glulam (timber)   bois *m* lamellé-collé
gneiss   gneiss *m*
go dim or pale   pâlir *vb*
go round [rotate]   tourner *vb*
go rusty   rouiller *vb*
go up   monter *vb*
go upstairs   monter *vb* l'escalier
goggles [protective]   lunettes *f,pl* de protection, lunettes-loup/masque *f,pl*

golf clubhouse   pavillon *m* de golf
golf course   golf *m*
good practice, in accordance with   dans les régles *f,pl* de l'art
goods   produits *m,pl*
goods [as in goods and chattels]   biens *m,pl*
goods, industrial   produits *m,pl* industriels
goods vehicle, heavy   poids *m* lourd
goods shed   halle *f* à marchandises
goodwill [of a business]   pas *m* de porte
gorge   canyon *m*, défilé *m*, gorge *f*
gouge   gouge *f*
gouge, rough turning   mèche *f* à bois creuse
government decree or order   décret *m*
governor, gas   régulateur *m* de gaz
grab excavator   pelle *f* en benne preneuse
grab [bucket]   grappin *m*
gradation   classement *m*, échelonnement *m*, gradation *f*
grade [finished level]   niveau *m* de réglage
grade [ground, fill, etc, to required level]   régler *vb*
grade of steel   nuance *f*/qualité *f* d'acier
graded   calibré *adj*
graded [in terms of particle size]   gradué *adj*
graded, well   bien calibré
grader   grader *m*
grader [earth- or roadworks]   niveleuse *f* (à lame)
gradient   gradient *m*, inclinaison *f*, pente *f*
gradient [hillside]   côte *f*
gradient [rise, upwards slope]   montée *f*
gradient, change of [point of]   point *m* d'inflexion
gradient, hydraulic   gradient *m* de hydraulique
gradient, longitudinal   pente *f* longitudinale
gradient, maximum   pente *f* maximale
gradient, point of change of   point *m* d'inflexion
gradient, pressure   gradient *m* de pression
gradient, temperature   gradient *m* thermique/de température
gradient, thermal   gradient *m* thermique
gradients [of a roadway, etc]   dénivellation *f*
grading   classement *m*
grading [by grain or particle size]   classement *m*/composition *f* granulométrique, granulométrie *f*
grading [levelling]   nivellement *m*
grading curve   courbe *f* granulométrique
gradually   par degrés
graduate   licencié *m*
grain, across the   à contre-fil
grain, across the [of wood]   debout *adj*
grain, against the   à contre-fil, à rebours
grain, along or with the   de droit fil
grain [of wood]   fil *m*
grain [particle or texture]   grain *m*
grain, cross or raised [of wood]   rebours *m*
grain of stone at right angles to natural bedding   délit *m*
grain size   diamètre *m* des grains, grosseur *f* de grain
grain size analysis   analyse *f* granulométrique
grain size class or range   classe *f* granulaire
grain size distribution   composition *f* granulométrique, granulométrie *f*

grain size distribution curve   courbe *f* granulométrique
grain size [in soil samples]   dimension *f* de(s) grain(s)
grain size [in steel]   dimension *f* du grain
grain structure [of material]   grain *m*
grain [paint to resemble wood grain]   peigner *vb*
grain [paint or a coating to imitate stone]   pocher *vb*
grandstand   tribune *f*
granite   granit(e) *m*
granite, fine-grained   granulite *m*
grant   prime *f*, subvention *f*
grant (someone) permission (to)   autoriser *vb* (qn à)
grant [rights, etc] to   céder *vb*
granular   granulaire *adj*
granulometric   granulométrique *adj*
graph   abaque *m*, courbe *f*
graph paper   papier *m* millimétré
graph   mettre *vb* sous forme graphique
graphic   graphique *adj*
graph(ic) [representation]   graphique *m*
graphite   graphite *m*
grass   herbe *f*
grass [short]   gazon *m*
grass cutter   tondeuse *f*
grassed   engazonné *adj*
grassed strip   bande *f* engazonnée
grassed surface   surface *f* engazonnée
grassland   prairie *f*
grassy   herbeux *adj*
grate   foyer *m*
grate, rocking   grille *f* oscillante
grate, shaking   grille *f* à secousse
grate, step   grille *f* à gradins
grate, travelling   grille *f* mécanique/roulante
grate, vibrating   grille *f* oscillante
grate area   surface *f* de grille
grate loading   charge *f* de grille
grating   caillebotis *m*, grillage *m*, grille *f*
grating [of a road gulley]   couronnement *m*
grating [to trap rubbish at a pipe entry]   pommelle *f*
gravel   cailloutis *m*, gravier *m*
gravel [ballast]   ballast *m*
gravel, alluvial   gravier *m* d'alluvion
gravel, crushed   gravier *m* concassé
gravel, filter   gravier *m* de filtrage
gravel, fine or pea   gravillon *m*
gravel, fine river   mignonnette *f*
gravel, rolled   gravier *m* roulé
gravel chippings   gravier *m* concassé, gravillons *m,pl* concassés
gravel chippings, fine   gravillon *m* concassé
gravel filter   filtre *m* de graviers
gravel pit   ballastière *f*, carrière *f*, gravière *f*
gravel/leaf guard in a rain water outlet   crapaudine *f*
graveyard   cimetière *m*
gravimetry   gravimétrie *f*
gravity   gravité *f*, pesanteur *f*
gravity, centre of   centre *m* de gravité
gravity, specific   pesanteur *f* spécifique
gravity axis   ligne *f* des centres de gravité

gravity system   système *m* par gravité
grease   graisse *f*
grease nipple   graisseur *m*, raccord *m* de graissage
grease remover   dégraisseur *m*
grease trap   boîte *f*/fosse *f* à graisse, séparateur *m* à graisse
grease   graisser *vb*
greasy [surface]   onctueux *adj*
green pigment [Prussian blue/chrome yellow mixture]   vert *m* anglais
green space [in town or city]   espace *m* vert
greenhouse   serre *f*
greenhouse effect   effet *m* de serre
grid   grille *f*
grid [above a stage]   gril *m*
grid [electrical supply]   réseau *m*
grid [for analysis or drawing]   maillage *m*
grid [maps, surveying]   treillis *m*
grid [pipes, cables, etc]   réseau *m*
grid, structural   grille *f*/trame *f* de structure
grid layout   quadrillage *m*
grid layout, with a   en quadrillage
grid system [of lines, structure, etc]   trame *f*
gridded   quadrillé *adj*
grill [generally within oven in France]   gril *m*
grill, guard   grille *f* de protection
grillage [e.g. for bridge deck analysis]   maillage *m*
grillage [foundations]   grillage *m*
grille   grille *f*
grille [for ventilation]   bouche *f*
grille, air   grille *f* d'aération
grille, air inlet   grille *f* d'entrée d'air
grille, differential   bouche *f* à section variable
grille, discharge or supply   grille *f* de soufflage
grille, extract   bouche *f* d'extraction
grille, fixed-direction   bouche *f* à lames fixes
grille, louvred   bouche *f* d'aération à lames
grille, perforated   grille *f* perforée
grille, protective, in front of an opening   barreaudage *m*
grille, protective, over glazing   garde-verre *m*
grille, radiator   grille *f* de radiateur
grille, recirculation   grille *f* de reprise
grille, return (air)   grille *f* de retour
grille, sill-mounted   grille *f* en tablette de fenêtre
grille, small, in a larger door   guichet *m*
grille, supply   grille *f* de soufflage
grille, supply air   grille *f* de soufflage
grille, transfer   grille *f* de transfert
grille, wall (mounted) supply   grille *f* de soufflage murale
grille differential pressure   perte *f* de charge de bouche d'air
grind   meuler *vb*
grind [into small fine gravel or powder]   égruger *vb*
grind [sharpen]   affûter *vb*
grind (in) [generally with abrasive powder]   roder *vb*
grind to powder   pulvériser *vb*
grinder   meuleuse *f*
grinder [sharpener]   affûteuse *f*

grinder, angle   meuleuse *f*
grinder, floor   meuleuse *f* de sols
grinder, pavement   ponceuse *f* à sol
grinding   affûtage *m*
grinding wheel   meule *f*
grindstone *adj*   meulier *adj*
grindstone   pierre *f* à meuler
grip [hold]   prise *f*
grip [in a verge]   saignée *f*
grip length   longueur *f* de serrage
grip   serrer *vb*
grit arrestor   capte-suie(s) *m*
grit spreader   gravillonneuse *f*
grit/mud bucket in a road gulley   panier *m*
gritblasted   grenaillé *adj*
gritblasting, (steel)   grenaillage *m*
gritstone   meulière *f*
gritter   gravillonneuse *f*
grizzly, vibrating   crible *m* scalpeur
grizzly feeder, vibrating   précribleur *m* vibrant
groove   entaille *f*, rainure *f*, saignée *f*
groove [fluting, striation]   cannelure *f*
groove [in masonry]   tranchée *f*
groove [in woodwork, etc]   feuillure *f*
groove   entailler *vb*, évider *vb*, rainer *vb*
groove, fine, between two mouldings   tarabiscot *m*
grooved [of the edge of a board]   rainé *adj*
groover [for road concrete, asphalt]   rainureuse *f*
grotto   grotte *f*
ground, from or of the   tellurique *adj*
ground   sol *m*, terrain *m*
ground [for paint, rendering]   subjectile *m*
grounds [carpentry]   carcasse *f*, taquets *m,pl*
grounds [park]   parc *m*
ground, boggy   molets *m,pl*
ground, electrical [earth]   terre *f* électrique
ground, high   hauteur *f*
ground, higher/raised [e.g. hills, mountains]   relief *m*
ground, lie of the   site *m*
ground, made   rapport *m* de terre, remblai *m*
ground, natural   terrain *m* naturel
ground, permeable   sol *m* perméable
ground, soft   terrain *m* meuble
ground, swampy   molets *m,pl*
ground, undisturbed   terrain *m* naturel
ground, waste   terrain *m* vague
ground beam   longrine *f*
ground freezing   congélation *f* du sol, gel *m* des sols
ground investigation   reconnaissance *f* de sol
ground investigation by test or trial pits   sondage *m* par tranchée
ground level, at   de plain-pied
ground plate [of a timber frame panel]   sablière *f*
ground sill [of wood, under a prop, etc]   couche *f*, semelle *f*
ground tremor   secousse *f* tellurique
groundwater   eau *f* phréatique/souterraine, nappes *f,pl* phréatiques/souterraines
groundwater, gravitational   eau *f* souterraine en circulation

groundwater, stagnant   eau *f* souterraine stagnante
groundwater, sulphate bearing   eau *f* séléniteuse
groundwater level   niveau *m* de la nappe aquifère, niveau *m* phréatique
groundwater lowering   abaissement de l'eau *f* souterraine
groundwater table   nappe *f* aquifère/d'eau/phréatique
groundwork [carpentry]   taquets *m,pl*
groundworks   terrassement *m*
group [e.g. of houses, piles, trees]   groupe *m*
group replacement [of lamps]   remplacement *m* en groupe
grout   coulis *m* d'injection
grout, clay/cement/etc   coulis *m* d'argile/de ciment/etc
grout blanket [beneath dam]   zone *f* injectée de la fondation
grout course [in penetration surfacing of roads]   couche *f* de liant
grout curtain   diaphragme *m* en béton, rideau *m*/voile *m* d'injection
grout curtain [dam]   écran *m* d'injection, voile *m* d'étanchéité
grout hole   forage *m*/trou *m* d'injection
grout mixer   malaxeur *m* à coulis
grout wall   paroi *f* au coulis
grout   injecter *vb*
grout [tiles]   jointoyer *vb*
grout up with (cement) mortar   injecter *vb* de mortier
grouted into [masonry, etc]   scellé *adj*
grout(ed) zone   zone *f* injectée
grouting   injection *f* (de coulis)
grouting [tiles]   jointoiement *m*
grouting, joint   injection *f* des joints
grouting in   scellement *m*
grow [of plants, trees]   pousser *vb*
grow dim or pale   pâlir *vb*
grow larger   s'agrandir *vb*
grow longer   s'allonger *vb*
growth   croissance *f*
growth, compound   croissance *f* composée
growth (of), inhibit the   retarder *vb* la croissance (de)
growth rate   taux *m* de croissance
growth ring [annual, of a tree]   cerne *m*
groyne   brise-lames *m*, épi *m*
guarantee   garantie *f*
guarantee [security]   caution *f*, cautionnement *m*
guarantee, performance   cautionnement *m*, garantie *f* de bonne fin
guarantee, two-year [on minor work, in France]   biennale *f*, garantie *f* biennale
guarantee, 10-year [on major work, in France]   décennale *f*, garantie *f* décennale
guarantee   garantir *vb*
guarantee deposit   dépôt *m* de garantie
guarantor   caution *f*, garant *m*
guard [against hazard, not a security guard]   protection *f*
guard [safety]   dispositif *m* de protection
guard, bird [grille]   grille *f* anti-volatile
guard, blade [circular saw]   chapeau *m* de scie, coiffe *f*, protège-lame *m*

guard, fire   garde-feu *m*
guard, flame   pare-flamme *m*
guard, hand   garde-main *m*
guard, kick-back [woodworking machine]   écran *m* anti-projection
guard, leg   garde-jambe *m*
guard, machine   protection *f* de machine
guard board   garde-pieds *m*, planche *f* de garde, latte *f* de garde, garde-gravois *m*
guard grill/netting   grille *f* de protection
guard post/stone at corner of a building   bouteroue *f*
guardhouse/guardroom   corps-de-garde *m*
guesthouse   maison *f* d'hôtes
guide   guide *m*
guide [rails] [e.g of a lift or site hoist]   guidage *m*
guide, air   guide *m* d'air
guide, belt   guide *m* de courroie
guide, cage [lift/elevator]   guide *m* de la cabine
guide markings or strips [on road surfaces]   bande *f* de guidage
gulf   golfe *m*
gulley, *see* gully
gully [drainage outlet]   avaloir *f*, bouche *f* d'égout (BE)
gully [gutter]   gouttière *f*
gully [natural feature]   couloir *m*, goulet *m*, ravin *m*
gun, gluing   pistolet *m* à colle
gun, mastic   pistolet *m* à mastiquer
gun, nail   cloueuse *f*
gun for spraying paint, sandblasting, etc   pistolet *m*
gunite   béton *m* projeté, gunite *m*
guniting   gunitage *m*
gusset   raidisseur *m*

gusset (plate)   gousset *m*
gust [of wind]   coup *m* de vent, rafale *f*
gutter [of lead, from roof or terrace]   lancier *m*
gutter [road or street]   cunette *f*, caniveau *m*
gutter [roof], *see also* roof gutter   chéneau *m*
gutter [on building]   gouttière *f*
gutter [triangular section, fixed under the eaves course]   gouttière *f* havraise/nantaise
gutter, box [road or street]   caniveau *m* couvert (en U)
gutter, box [roof]   chéneau *m* encaissé
gutter, built-up, resting on a cornice   chéneau *m* à l'anglaise
gutter, condensation   gorge *f* d'écoulement
gutter, eaves   chéneau *m*
gutter, rain   gouttière *f*
gutter, suspended   gouttière *f* pendante
gutter, triangular section, fixed under the eaves course   gouttière *f* havraise/nantaise
gutter, valley   noue *f*
gutter, valley [folded metal strip]   noquet *m*
gutter, wooden   échenal *m*
gutter bracket/clip   crochet *m*/patte *f* de gouttière
gutter overflow pipe   gueulard *m*
gutter stone [between down pipe and drain]   culière *f*
guy [line]   hauban *m*
guy derrick   derrick *m*, mât *m* de levage, mât *m* inclinable
guy   hauban(n)er *vb*
guying or staying with guys   hauban(n)age *m*
gypsum   gypse *m*, pierre *f* à plâtre
gypsum, ground   plâtre *f* cru
gypsum board   placoplâtre *m*
gypsum plaster   plâtre *f*

H-beam   poutre *f* en H
ha-ha   saut-de-loup *m*
habitat   habitat *m*
habitation, fit for   habitable *adj*
hack [a rendering or surface finish]   repiquer *vb*
hacking [of a rendering, etc]   repiquage *m*
hacksaw   scie *f* à archet/à métaux
hail   grêle *f*
half-hard   mi-dur *adj*
half-hip [roof]   croupe *f* faîtière
half-round [moulding, section]   demi-rond *m*
half-shade   demi-ombre *f*
half-timbering   colombage *m*
hall   hall *m*
hall [as in entrance hall]   vestibule *m*
hall, arrival or departure   hall *m*
hall, dining   réfectoire *m*
hall, entrance   hall *m* (d'entrée)
hall, freight   halle *f* à marchandises

hall, lecture   salle *f* de conférence
halogen   halogène *m*
halogen lamp   lampe *f* à halogène
halogenous   halogène *adj*
halved [of a joint in timber]   mi-bois *adj*
hamlet   hameau *m*
hammer   marteau *m*
hammer, blacksmith's (hand)   marteau *m* de forgeron
hammer, bricklayer's   marteau *m* de maçon
hammer, carpenter's   marteau *m* de charpentier
hammer, chamfering   marteau *m* à bigorner
hammer, chipping   marteau *m* à piquer, marteau-burineur *m*
hammer, claw   marteau *m* à griffe/à panne fendue
hammer, club [small]   massette *f*
hammer, down-(the-)hole   marteau-perforateur *m* hors (du) trou
hammer, drop [pile-driver]   marteau *m* dame, mouton *m* à déclic, singe *m*

hammer, face of a   aire *f* d'un marteau
hammer, flat-face   marteau *m* à planer
hammer, glazier's   marteau *m* de vitrier
hammer, gravity [pile-driver]   mouton *m* sec
hammer, hand   marteau *m* à main
hammer, heavy mason's, with hatched faces   chien *m*
hammer, jack-   marteau *m* piqueur
hammer, large pavior's   épinçoir *m*
hammer, lead   marteau *m* à plomb
hammer, lump [small]   massette *f*
hammer, machinist's   marteau *m* rivoir
hammer, mash   masse *f*
hammer, mason's   têtu *m*, masse *f*
hammer, mason's heavy, with hatched faces   chien *m*
hammer, mason's toothed [for preliminary dressing of soft stone]   laye *f*
hammer, mason's two-pointed (scabbling)   smille *f*
hammer, pile   marteau *m* de battage, mouton *m* (de battage)
hammer, plumber's   marteau *m* à main
hammer, pneumatic   marteau-perforateur *m*, marteau *m* piqueur
hammer, punch   marteau *m* à poinçon
hammer, rebound   scléromètre *m*
hammer, riveting   matoir *m*, rivoir *m*
hammer, scabbling [two-pointed]   smille *f*
hammer, slate   marteau *m* d'ardoisier
hammer, slater's or tiler's [similar to bricklayer's hammer]   martelet *m*
hammer, small   martelet *m*
hammer, small, with one wedge shaped peen parallel to its handle   herminette *f*
hammer, steam [pile-driver]   mouton *m*/sonnette *f* à vapeur
hammer   marteler *vb*
hammer-head(ed) hatchet   hachette *f* à marteau
hammering [finish]   martelage *m*
hand, by   manuellement *adv*
hand over   remettre *vb*
hand over [powers, property sold]   transmettre *vb*
handbasin [small]   lave-mains *m*
handbasin for building-in   vasque *f* à poser
handbook   manuel *m*
handbrake   levier *m* de frein
handcart   chariot *m* à bras
handhole   trou *m* de poing
handing over [keys, etc]   remise *f*
handle   poignée *f*
handle [of a saw]   fût *m*
handle [of a tool or implement]   manche *m*
handle, lever [cupboard or door]   bec *m* de cane, béquille *f*
handle, open [e.g. of a saw]   poignée *f* ouverte
handle, operating   levier *m* de manœuvre
handle, pistol grip   poignée *f* revolver
handle, saw   manche de scie *m*
handle, side grip   poignée *f* latérale
handle, valve   levier *m* de vanne
handle   manier *vb*, manipuler *vb*
handle [a subject]   traiter *vb*

handling   manipulation *f*, manutention *f*
handling, manual/mechanical   manutention *f* manuelle/mécanique
handling cart or trolley   chariot *m* de manutention
handmade or done by hand   exécuté à la main
hand-operated   à (la) main, manuel *adj*
handrail   appui *m*, balustrade *f*, garde-corps *m*, garde-fou *m*, glisse-main *m*, lice/lisse *f*, main *f* courante
handrail [across a window opening]   barre *f* d'appui
handrail, oblong rounded profile of   olive *f*
handrail, stone   tablette *f* d'appui
handrail bracket [built into a wall]   écuyer *m*
handsaw   égoïne *f*, scie *f* à main/égoïne/manuelle
handwheel   volant *m* à main
handwheel, adjusting/adjustment   volant *m* de réglage
handwheel, clamp(ing)   volant *m* de serrage
handwheel, height or rise-and-fall adjusting   volant *m* de réglage en hauteur
handwheel, operating   volant *m* de manœuvre
handwheel, work clamp   volant *m* de serrage
handyman   bricoleur *m*
hang up   suspendre *vb*
hangar   hangar *m*
hanger   étrier *m* de suspension
hanger [suspension bridge]   suspente *f*
hanger, cable   porte-câble *m*
hanger, pipe   collier *m*/support *m* de tuyauterie, étrier *m* de suspension des tubes
hanging   suspendu *adj*
harbour   havre *m*, port *m*
harbour, inner   arrière-port *m*
harbour basin [tidal]   darse *f*
harbour facilities   équipements *m,pl*/installations *f,pl* portuaires
hard   dur *adj*
hard hat   casque *m* de protection
hardboard   panneau *m* de fibres dures
hardboard [trade name]   Isorel *m*
hardboard, perforated   panneau *m* dur perforé
hardcore   blocaille *f*, empierrement *m*, remblai *m*
hardcore, compacted   pierraille *f* compactée
harden   durcir *vb*
harden [cold], **work**   écrouir *vb*
hardened   durci *adj*
hardened [of steel]   écroui *adj*
hardened glass   verre *m* trempé
hardener [for cement, glues]   durcisseur *m*
hardening   durcissement *m*
hardening, age   vieillissement *m*
hardening, work   écrouissage *m*, vieillissement *m* mécanique
hardening by cold working   durcissement *m* par écrouissage
hardening of concrete   durcissement *m* du béton
hardening the surface of soft limestone with fluoro-silicates   fluatation *f*
hardness [metal, water, etc]   dureté *f*
hardness, degree of   degré *m* de dureté
hardness, permanent   dureté *f* permanente

hardness, residual dureté *f* résiduelle
hardness, temporary dureté *f* temporaire
hardness, total dureté *f* totale
hardness measurement mesure *f* de la dureté
hardness measurer, rebound duromètre *m*
hardness number, Brinell chiffre *m* de dureté Brinell
hardness test, Brinell/Rockwell essai *m* de dureté Brinell/Rockwell
hardness of water dureté *f* de l'eau
hardpan [formed on the surface of limestone] calcin *m*
hardstanding [vehicle park] parc *m* de stationnement
hardware ferronnerie *f*, quincaillerie *f*, serrurerie *f*
hardware, computer matériel *m* informatique
hardware, door/window ferrure *f*
hard-wearing inusable *adj*
hardwood bois *m* dur
harmful nocif *adj*
harmful [e.g. gas] nuisible *adj*
harmony with in en rapport avec
harness, body/safety [for working at height] harnais *m* de sécurité
hasp moraillon *m*
hatch trappe *f*
hatch, access trappe *f* d'accès/de visite
hatch, entry/inspection trappe *f* de visite
hatch, escape trappe *f* d'évacuation
hatched hachuré *adj*
hatchet hache *f* à main, hachette *f*
hatchet, hammer-head(ed) hachette *f* à marteau
haul up hisser *vb*
haulier transporteur *m*
haunch [in concrete structure] gousset *m*
haunch (of an arch) aisselle *f* (d'une voûte)
haunches [of an arch] reins *m,pl*
hawk taloche *f*
haze brume *f*
hazel [tree or bush] noisetier *m*
head [leader] chef *m*
head [pressure] charge *f* statique
head [of a nail, hammer] tête *f*
head, countersunk tête *f* fraisée, tête *f* noyée
head, cross-slot [for a screwdriver] pointe *f* cruciforme
head, flat countersunk [of a screw] tête *f* plate
head, hexagonal tête *f* hexagonale
head, pressure hauteur *f* de chute d'eau
head, raised countersunk [of a screw] tête *f* bombée
head, recessed [Phillips/Pozidrive/Supadrive type screw] tête *f* cruciforme
head, recessed [Robertson type screw] tête *f* creuse
head, static hauteur *f* de charge/de la pression statique, charge *f* statique
head, velocity hauteur *f* dynamique
head loss perte *f* de pression
head loss, dynamic chute *f* de pression dynamique
head of a jetty or pier musoir *m*
head of water charge *f* hydraulique, chute *f*, hauteur *f* de chute d'eau
head office siège *m*
header [masonry] boutisse *f*

header [multi-outlet for individual plumbing service pipes] nourrice *f*
header [pipe or duct] collecteur *m*, distributeur *m*
header [of bond stone in a wall] queue *f* [d'une pierre de taille]
head(er), rainwater cuvette *f* de (chéneau)
heading [geology] galerie *f* de reconnaissance
headlamp or -light phare *m*
headquarters siège *m*
headrace (tunnel) [dam] galerie *f* d'amenée
headrace [of a water mill] bief *m* d'amont
headroom dégagement *m*, hauteur *f* libre
headroom (in a staircase or passage) échappée *f*, échappement *m*
health, *see* occupational health
health and safety archive, project dossier *m* adapté
health and safety inspector [French equivalent] inspecteur *m* du travail
health and safety plan plan *m* d'hygiène et de sécurité (PHS), plan *m* de santé et sécurité
health centre infirmerie *f*
heap tas *m*
hearing, threshold of seuil *m* d'audibilité
heart shake [timber defect] cadranure *f*, gélivure *f*, givrure *f*, fente *f* de cœur
hearth âtre *m*, foyer *m*
hearth, back plate or rear wall of a chimney contre-cœur *m*
hearth, raised âtre *m*
hearth back plate plaque *f* de fond
hearth plate plaque *f* d'âtre
hearthstone âtre *m*, foyer *m*
heartwood bois *m* parfait/de cœur, cœur *m* (de bois), duramen *m*
heartwood plank planche *f* de cœur
heat [conduction, engine, etc] thermique *adj*
heat chaleur *f*
heat, convection chaleur *f* de convection
heat, latent, *see* latent heat
heat, mechanical equivalent of équivalent *m* mécanique de la chaleur
heat, radiant chaleur *f* radiante/rayonnante
heat, sensible chaleur *f* sensible
heat, source of foyer *m*/source *f* de chaleur
heat, specific chaleur *f* massique/spécifique
heat, total enthalpie *f* totale
heat, volumetric chaleur *f* volumique
heat, waste chaleur *f* perdue
heat accumulator, reservoir or store accumulateur *m* de chaleur
heat balance bilan *m* thermique
heat capacity capacité *f* calorifique
heat conduction conduction *f* thermique
heat conductivity conductivité *f* thermique
heat consumption dépense *f* de chaleur
heat content enthalpie *f*
heat convection convection *f* de chaleur
heat conveying calorifère *adj*
heat distribution distribution *f* de chaleur

heat economy économie *f* thermique
heat emission émission *f* de chaleur
heat engine moteur *m* thermique
heat exchange échange *m* de chaleur
heat exchanger échangeur *m* thermique/de chaleur
heat exchanger, counter-flow échangeur *m* de chaleur à contre-courants
heat exchanger, cross-flow échangeur *m* à flux croisés/à courants croisés
heat exchanger, direct contact échangeur *m* thermique à contact direct
heat exchanger, gas–air échangeur *m* de chaleur air–gaz
heat exchanger, rotary échangeur *m* rotatif
heat exchanger, waste gas échangeur *m* de chaleur de gaz d'échappement
heat flow écoulement *m* de chaleur
heat flux flux *m* de chaleur
heat gain gain *m* de chaleur
heat gain, casual gain *m* de chaleur occasionnel
heat gain, conduction gain *m* par conduction
heat gain, incidental gains *m,pl* de chaleur variable
heat gain, ventilation charge *f* thermique de ventilation
heat generator générateur *m* de chaleur
heat input apport *m* calorifique/de chaleur, énergie *f* thermique fournie
heat insulating material calorifuge *m*, isolant *m* calorifuge
heat insulation isolation *f* thermique
heat insulator calorifuge *m*
heat loss déperdition *f* calorifique/thermique, perte *f* de chaleur
heat loss, ventilation charge *f* de rafraîchissement
heat loss calculation calcul *m* des déperditions calorifiques/des pertes de chaleur
heat loss coefficient of a habitation (watts/m³/°C) coefficient-G *m* d'un logement
heat loss rate perte *f* de chaleur par unité de temps
heat meter compteur *m* d'énergie thermique/de calories/chaleur
heat of combustion/compression/condensation chaleur *f* de combustion/compression/condensation
heat of hydration chaleur *f* d'hydratation/de prise *f*
heat of mixing/reaction chaleur *f* de mélange/réaction
heat of vaporization chaleur *f* d'évaporation
heat output quantité *f* de chaleur produite
heat output [rating] puissance *f* calorifique
heat output or rate puissance *f* thermique
heat output, rated puissance *f* calorifique nominale
heat output, total/useful puissance *f* thermique totale/utile
heat pump pompe *f* à chaleur
heat radiation rayonnement *m* thermique
heat recovery récupération *f* de chaleur
heat recovery, waste récupération *f* de chaleur
heat recovery device or equipment récupérateur *m* de chaleur
heat recovery system système *m* de récupération de chaleur
heat requirement demande *f* de chaleur

heat resistance résistance *f* à la chaleur
heat reservoir accumulateur *m* de chaleur
heat source source *f* de chaleur
heat transfer transfert *m* de chaleur
heat transfer coefficient coefficient *m* de transfert thermique/de transmission de chaleur
heat transfer cycle cycle *m* de transfert de chaleur
heat transfer fluid fluide *m* caloporteur
heat transfer medium caloporteur *m*
heat transmission transmission *f* de chaleur
heat treatment traitement *m* thermique
heat wheel roue *f* de chaleur
heat chauffer *vb*
heated chauffé *adj*
heated, steam chauffé *adj* à la vapeur
heater réchauffeur *m*
heater, air réchauffeur *m* d'air
heater, back-up radiateur *m* d'appoint
heater, boiler feedwater réchauffeur *m* d'eau d'alimentation de chaudière
heater, convection or convector appareil *m* de chauffage par convection
heater, gas radiant panneau *m* radiant au gaz, radiateur *m* à gaz à rayonnement
heater, gas-fired air réchauffeur *m* d'air à gaz
heater, immersion chauffe-eau *m*, réchauffeur *m* à immersion
heater, instantaneous appareil *m* de chauffage instantané
heater, oil [to heat oil] réchauffeur *m* d'huile
heater, radiant appareil *m* de chauffage par rayonnement
heater, radiant electric radiateur *m* électrique à rayonnement
heater, radiant gas panneau *m* radiant au gaz, radiateur *m* à gaz à rayonnement
heater, radiant strip cordon *m* chauffant, radiateur *m* à plaque
heater, storage appareil *m* de chauffage à accumulation
heater, unit [fan + heating element] aérotherme *f*
heater, unit [wall mounted] convecteur *m* mural
heater, wall radiateur *m* mural
heater, water, *see* water heater
heater battery batterie *f* de chauffage
heater operation fonctionnement *m* du chauffage
heater unit, warm air générateur *m* d'air chaud
heater-sprayer [for bitumen, tar, etc] réchauffeuse-répandeuse *f*
heath lande *f*
heath and moorland garrigue *f*
heating chauffe *f*
heating [system or action of] chauffage *m*
heating, air chauffage *m* d'air
heating, ceiling chauffage *m* par le plafond
heating, central, *see also* central heating chauffage *m* central
heating, coal-fired chauffage *m* au charbon
heating, district chauffage *m* collectif/de quartier/urbain, distribution *f* de chaleur à distance

heating, electric   chauffage *m* électrique

heating, exhaust steam   chauffage *m* à vapeur d'échappement

heating, floor   chauffage *m* par le sol

heating, fresh air   chauffage *m* d'air frais

heating, gas-fired   chauffage *m* au gaz

heating, gas-fired air   chauffage *m* d'air à gaz

heating, high/low pressure steam   chauffage *m* à vapeur (à) haute/basse pression

heating, hot or warm air   chauffage *m* à air chaud

heating, indirect   chaleur *f* indirecte

heating, induction   chauffage *m* par éjecto-convecteur/par induction

heating, infra-red   chauffage *m* au rayons infrarouges

heating, integrated electric [insulation + mechanical ventilation + electric heating]   chauffage *m* électrique intégré (CEI)

heating, intermittent   chauffage *m* intermittent

heating, microbore   chauffage *m* par tuyauteries de très faibles diamètres

heating, off-peak storage   chauffage *m* par accumulation d'énergie hors pointe

heating, oil-fired   chauffage *m* au fuel/au mazout

heating, panel   chauffage *m* à surface rayonnante/par panneau

heating, radiant   chauffage *m* à/par rayonnement

heating, radiant ceiling panel   chauffage *m* au plafond

heating, room   chauffage *m* de(s) locaux

heating, skirting board   plinthe *f* chauffante

heating, small bore   chauffage *m* par tuyauteries de petit diamètre

heating, solar   chauffage *m* solaire

heating, steam   chauffage *m* à (la) vapeur

heating, storage   chauffage *m* à accumulation

heating, supplementary   chauffage *m* d'appoint

heating, thermoelectric   chauffage *m* thermoélectrique

heating, trace   traçage *m* électrique

heating, underfloor   chauffage *m* intégré par le sol, chauffage *m* sous-sol

heating, ventilation and air-conditioning (HVAC)   chauffage *m*, ventilation et air climatisé

heating, wall   chauffage *m* mural/par les murs

heating, window sill   chauffage *m* en allège

heating apparatus/appliance   appareil *m* de chauffage, calorifère *m*, corps *m* de chauffe

heating chamber   chambre *f* de chauffage, étuve *f*

heating economics/economy   économie *f* de chauffage

heating effect   effet *m* calorifique

heating element   élément *m* chauffant/de chauffe, résistance *f*

heating engineer   chauffagiste *m*

heating flue   carneau *m* de chauffage

heating installation   calorifère *m*, installation *f* de chauffage

heating load   charge *f* calorifique

heating method   méthode *f* de chauffe

heating oil, domestic   fuel *m* domestique, fioule *f*, mazout *m* domestique

heating panel, built-in [floor, wall or ceiling]   panneau *m* incorporé

heating panel, buried pipe/tube   panneau *m* à tubes enrobées

heating panel, embedded [floor, wall or ceiling] panneau *m* incorporé

heating plant, central [central boiler]   chaufferie *f* centrale

heating panel, radiant   panneau *m* chauffant

heating period   durée *f* de mise en température, période *f* de chauffage

heating plant   installation *f* de chauffage

heating power   pouvoir *m* calorifique

heating season   saison *m* de chauffage

heating surface, boiler   surface *f* de chauffe d'une chaudière

heating system   système *m* de chauffage

heating system, dry return (steam)   système *m* de chauffage à vapeur avec retour

heating system, hot water   système *m* de chauffage à eau chaude

heating unit, warm air   générateur *m* d'air chaud

heating up   mise *f* en température

heating, turn on the   mettre *vb* le chauffage

heat-insulating   calorifuge *adj*

heatproof   résistant à la chaleur

heat-resistant/resisting   résistant à la chaleur

heat-stained [brown/yellow]   rissolé *adj*

heat-up time   période *f* de réchauffe

heave [of ground]   soulèvement *m*

heave [geologically]   se soulever *vb*

heave [ground, etc]   gonfler *vb*

heave up   hisser *vb*

heaving [of ground]   soulèvement *m*

heaving [of ground, road surface]   gonflement *m*

heavy   lourd *adj*, pesant *adj*

hectare [10,000m^2 = 100 ares = 2.47 acres]   hectare *m* [† = ha]

hedge   haie *f*

hedge trimmer   taille-haies *f*

heel [of a dam]   pied *m*

height [above sea level]   altitude *f*

height [e.g. of a building]   hauteur *f*

height [e.g. of an arch]   montée *f*

height [of a tree, etc]   grandeur *f*

height [of the sun]   élévation *f*

height [summit]   faîte *f*, point *m* culminant

height, capillary rise   hauteur *f* d'ascension capillaire

height, ceiling   hauteur *f* sous plafond

height, clear   hauteur *f* libre

height, clear beneath beams   hauteur *f* sous poutres

height, effective, of water seal in a siphon   garde *f* d'eau

height, floor   hauteur *f* d'étage

height, floor-to-ceiling   hauteur *f* de pièce (sous plafond)

height, floor-to-ceiling [beneath beams]   hauteur *f* sous poutres

height, maximum   hauteur *f* limite

height, overall   hauteur *f* d'encombrement/hors tout

height, slab-to-slab   hauteur *f* dalle à dalle

height, spot [on drawing]   cote *f*

height, storey   hauteur *f* d'étage

height, structure   hauteur *f* de l'ouvrage
height, web   hauteur *f* de l'âme
height between staircase landings   montée *f*
height increase [of building, wall, etc]
   exhaussement *m*
height marker   témoin *m*
**height of a door or window sill above the ground
   surface**   enseuillement *m*
**height of top of well/shaft casing above maximum water
   level**   revanche *f*
height of waterproof flashing above edges of a terrace
   relevé *m* d'étanchéité
height to which lower building rises on a party wall
   héberge *f*
heighten   hausser *vb*, surélever *vb*
heighten [a building, wall, etc]   exhausser *vb*
heightened   surélevé *adj*
heightening   rehaussement *m*
heightening [e.g. of a wall]   surélévation *f*
helical   hélicoïdal *adj*
heliostat   héliostat *m*
helix   hélice *f*
helmet   casque *m*
helmet, full [protective]   casque *m* enveloppant/intégral
helmet, protective or safety   casque *m* de protection
helmet, welding   casque *m* de soudage
helmet diver   scaphandrier *m*
hemp   chanvre *f*
hemp fibre   tresse *f* de filasse
herb   herbe *f*
herbaceous   herbacé *adj*
herbicide   désherbant *m*, herbicide *m*
heritage   patrimoine *f*
hermetic   hermétique *adj*
hermetically sealed   fermée hermétiquement
heterogeneous   hétérogène *adj*
hexagonal/octagonal/etc   à six/huit/etc pans
hidden   non visible
hide [a minor defect, a difference of level]   racheter *vb*
hiding power [of paint]   opacité *f*, pouvoir *m* couvrant
high   élevé *adj*, haut *adj*
high performance   performant *adj*
high power, at   à haute puissance
high pressure   haute pression *adj*
high-efficiency   haut rendement *adj*
high-level [on pipework drawing]   en plafond *adj*
high-return [investment]   performant *adj*
high-speed/-velocity   à haute vitesse *adj*
high-tensile   à haute résistance *adj*
high-tension   haute tension (HT) *adj*
high-strength   à haute résistance *adj*
high-velocity   à haute vitesse *adj*
high-voltage   haute tension (HT) *adj*
highway   routier *adj*
highway [system]   voirie *f*
highway, public   voie *f* publique
highway, two/three/four-lane   route *f* à deux/trois/quatre
   bandes/voies
highway classification   catégorie *f* de route

highway maintenance [work, or department responsible]
   voirie *f*
highway maintenance area   centre *m* d'entretien routier
highway type   type *m* de route
highways and miscellaneous external works   voirie et
   réseaux divers (VRD)
hill   colline *f*, mont *m*
hillock   monticule *m*
hillside   coteau *m*, flanc *m* de coteau, versant *m* de colline
hinder   retarder *vb*
hindrance   obstacle *m*
hinge   charnière *f*
hinge [split]   paumelle *f*
hinge [structural]   articulation *f*
hinge, abutment   articulation *f* aux naissances/de culée
hinge, crown [of a structure]   articulation *f* à la clé
hinge, lift-off   paumelle *f* dégondable/double
hinge, lift-off butt   paumelle *f* à droite
hinge, loose-butt   fiche *f* à vase
hinge, loose-pin   fiche *f* à nœud
hinge, pin   charnière *f* à broche/à fiche
hinge, plastic   rotule *f* plastique
hinge, T/tee   penture *f* anglaise
hinge bracket and pin [of a strap hinge]   gond *m*
hinge knuckle   nœud *m* (de charnière de penture)
hinge pin   boulon *m*/broche *f* de charnière, mamelon *m*
hinge washer [to raise door or shutter]   bague *f* de
   paumelle
hinge   articuler *vb*
hinged   articulé *adj*, pivotant *adj*
hinterland   arrière-pays *m*
hip [roof]   arêtier *m*
hip bath   baignoire-sabot *f*
hip(ped) end [of a roof]   croupe *f*
hire   location *f*
hire [labour]   embaucher *vb*
hit a snag[†]   avoir *vb* un pépin[†]
hoar-frost   gelée *f* blanche, givre *m*
hoarding [around a site]   palissade *f*
hoarding, advertisement   panneau *m* d'affichage,
   panneau *m* publicitaire
hod   oiseau *m*
hod carrier [mason's labourer]   porte-auge *m*
hoist   appareil *m* de levage
hoist, bucket   élévateur *m* à godets
hoist, builder's/site   ascenseur *m*/élévateur *m*/monte-
   charge *m* de chantier
hoist, chain   palan *m* à chaîne
hoist, materials   monte-matériaux *m*
hoist, rope   palan *m* à câble
hoist, sack   monte-sacs *m*
hoist   hisser *vb*
hoisting   levage *m*, montage *m*
hoisting device or equipment   appareil *m* de levage
hold tight   serrer *vb*
hold up [delay]   retarder *vb*
hold up [support]   maintenir *vb*, supporter *vb*
holder [of a card, document, etc]   titulaire *m*
holder, bulb   porte-ampoule *m*

holder, fuse   porte-fusible *f*
holder, label   porte-étiquette *m*
hole   trou *m*
hole [bore-hole, drill-hole]   forage *m*, sondage *m*
hole [in a rafter for peg jointing on to purlins]   brandille *f*
hole [in the ground]   fosse *f*
hole [in timbers for dowel/peg jointing]   enlaçure *f*
hole [in masonry for end of putlog or scaffolding tie]
   boulin *m*
hole [into rock for blasting charge]   fourneau *m*
hole [opening]   pertuis *m*
hole [pre-cast in structure, for services]   réservation *f*
hole [through structure for a rainwater downpipe]
   souillard *m*
hole, cat   chatière *f*
hole, drill(ed)   forage *m*, trou *m* alésé
hole, fixing   trou *m* de fixation
hole, inspection   trou *m* d'homme/de regard
hole, bore hole through or make a hole in   percer *vb*
holiday camp   colonie *f* de vacances
holiday period   période *f* des vacances
hollow *adj*   creux *adj*
hollow   creux *m*
hollow [cut into something]   évidement *m*
hollow [in ground]   cuvette *f*
hollow out   évider *vb*
home [abode]   demeure *f*
home [house]   foyer *m*
home, nursing   clinique *f*
home, old people's   maison *f* de retraite
home, retirement   maison *f* de retraite, résidence *f*
   troisième âge
home, second   résidence *f* secondaire
homogeneous   homogène *adj*
honeycomb [structure, etc]   nid *m* d'abeilles
honeycombing [e.g. in concrete surface]   nid *m* de
   cailloux/de gravillons
hood   capot *m*, chapeau *m*
hood, cooker or fume   hotte *f*
hood, exhaust or extraction   hotte *f* aspirante/
   d'évacuation/d'extraction
hood, fume   hotte *f*
hood, kitchen   hotte *f* de cuisine
hood, lamp [traffic lights, railway signals]   visière *f*
hood, protective, over a circular saw blade   coiffe *f*
hook   crochet *m*
hook, lifting   crochet *m* de levage
hook, pipe   corbeau *m*
hook, reinforcement   crochet *m* d'armature
hook, safety [on a roof]   crochet *m* de sécurité
Hooke's law   loi *f* de Hooke
hooked tie bar   épingle *f*
hoop   cerce *f*
hoop [reinforcement]   frette *f*, spire *f*
hopper   trémie *f*
hopper, feed   benne *f*/trémie *f* d'alimentation
hopper, loading   benne *f* de chargement
hopper, rainwater   cuvette *f* de chéneau
hopper, storage   trémie *f* de stockage

horizon [visual, and geological]   horizon *m*
horizontal   horizontal *adj*
horse [trestle]   tréteau *m*
horse chestnut [tree]   marronier *m* d'Inde
horsepower   cheval-vapeur (CV), puissance *f* en chevaux
hose, fire   tuyau *m* d'incendie
hose, garden   tuyau *m* d'arrosage
hose, metal   tuyau *m* métallique flexible
hose, rubber   tuyau *m* de/en caoutchouc
hose connection, fireman's   raccord *m* pompier/rapide
hose reel   dévidoir *m*
hose reel, fire   robinet *m* d'incendie armé (RIA)
hosepipe   tube *m* flexible, tuyau *m* d'arrosage
hospital   hôpital *m*
hospital, private   clinique *f*
hospital sink   évier *m* d'hôpital
hostelry   hôtellerie *f*
hot   chaud *adj*
hot water heater   calorifère *m* d'eau chaude
hot water heating system   système *m* de chauffage à eau
   chaude
hot water storage   stockage *m* d'eau chaude
hot water system, high/low/medium pressure   système
   m à eau chaude à haute/basse/moyenne pression
hot water tank with immersion heater   cumulus *m*
   électrique
hot, become or get   chauffer *vb*
hot lay [e.g. asphalt]   mettre *vb* en œuvre à chaud
hot-dip galvanized   galvanisé *adj* à chaud/au bain
hot-laid [road materials]   mis *adj* en œuvre à chaud
hotel   hôtel *m*
hotel, large luxury   palace *f*
hothouse   serre *f* chaude
hotplate   plaque *f* (chauffante)
hotplate [on a cooking stove]   plaque *f* (de cuisson)
hours of darkness   heures *f,pl* d'obscurité
house   demeure *f*, maison *f*
house [generally detached, modern]   pavillon *m*
house [one family]   maison *f* individuelle
house, caretaker's   maison *f* de garde/gardien
house, country   château *m*, maison *f* de campagne
house, country [southern France]   bastide *f*
house, detached   maison *f* indépendante/
   individuelle/isolée, pavillon *m*
house, dwelling   maison *f* d'habitation
house, fortified   maison *f* forte
house, half-timbered   maison *f* en colombage
house, isolated country   chartreuse *f*
house, manor   gentilhommière *f*
house, pump   station *f* de pompage
house, semi-detached   maison *f* jumelée
house, single storey   maison *f* à rez-de-chaussée
house, single storey, ground level   maison *f* sur terre-
   plein
house, small detached   villa *m*
house, terrace   maison *f* en bande/rangée
house, town [mansion]   hôtel *m*
house, two-storey   maison *f* à étage
house, weekend   maison *f* d'amis

house agent agent *m* immobilier
house automation [intelligent control systems]
 domotique *f*
house building construction *f* d'habitation
house chimney cheminée *f* de maison
house painter peintre *m* (en bâtiment)
house removal [furniture, etc] déménagement *m*
house [timber, in a housing] enclaver *vb*
household foyer *m*
household equipment équipement *m* ménager
housing logements *m,pl*
housing [case] boîtier *m*, carter *m*
housing [protective] enveloppe *f*
housing, collective habitat *m*/immeuble *m* collectif,
 logement *m*
housing, public [in France] habitation *f* à loyer modéré
 (HLM)
housing density densité *f* des logements
housing estate cité *m*, lotissement *m*
housing into which a blind is raised pavillon *m*
housing priority area zone *f* à urbaniser en priorité
housing project ensemble† *m*
housing starts mises *f,pl* en chantier de logements
Hoyer method of prestressing by pre-tensioned bonded
 wires précontrainte *f* par fils adhérents
hub [e.g. of a fan, impeller] moyeu *m*
hue couleur *f*, teinte *f*
humid humide *adj*
humidification humidification *f*
humidifier humidificateur *m*
humidifier, evaporative humidificateur *m* à évaporation
humidifier, steam humidificateur *m* à vapeur
humidity humidité *f*
humidity, absolute concentration *f* de vapeur
humidity, air humidité *f* de l'air
humidity, degree of teneur *f* en eau
humidity, relative degré *m* hygrométrique, humidité *f*
 relative
humus humus *m*
hunting pompage *m* en régulation
hunting lodge pavillon *m* de chasse
hurricane ouragan *m*
hut cabane *f*
hut, roadman's maison *f* cantonnière

hut, site baraque *f* (de chantier), baraquement *m*
hut, small cabanon *m*
HVAC engineering génie *m* climatique
hydrant prise *f* d'eau
hydrant, cleansing [ground level] bouche *f* de lavage
hydrant, fire [pillar type] borne *f* d'incendie
hydrant, fire [ground level] bouche *f* d'incendie
hydrant, underground bouche *f* d'incendie à clé sous
 trottoir
hydrant cap chapeau *m* de borne-fontaine
hydrate hydrater *vb*
hydration hydratation *f*
hydration, heat of prise *f*, chaleur *f* de
hydraulic hydraulique *adj*
hydraulic depth, mean profondeur *f* hydraulique
 moyenne
hydraulics hydraulique *f*
hydrocarbon hydrocarbure *m*
hydrodynamic hydrodynamique *adj*
hydrogen hydrogène *m*
hydrologist hydrologue *m/f*
hydrology hydrologie *f*
hydrolysis hydrolyse *f*
hydrometer hydromètre *m*
hydrophilic hydrophile *adj*
hydrophobic hydrophobe *adj*
hydrostatic hydrostatique *adj*
hygiene, industrial hygiène *f* du travail, hygiène *f*
 industrielle/professionnelle
hygrograph hygromètre *m* enregistreur
hygrometer hygromètre *m*
hygrometer, aspirated hygromètre *m* à aspiration
hygrometer, dew point hygromètre *m* à point de
 rosée
hygrometer, hair hygromètre *m* à cheveu
hygrometer, recording hygromètre *m* enregistreur
hygrometry hygrométrie
hygroscopic hygroscopique *adj*
hygrostat hygrostat *m*
hyperbolic hyperbolique *adj*
hypothesis hypothèse *f*
hysteresis hystérèse *f*, hystérésis *f*
hysteresis curve courbe *f* d'hystérésis
hysteresis loop boucle *f* d'hystérésis

I-beam poutre *f* en I
ice glace *f*
ice, dry neige *f* carbonique
ice apron or breaker on a bridge pier brise-glace *m*
ice lens lentille *f* de glace
identification mark estampille *f*

identification plate [on a tool, machine, etc] plaque *f*
 signalétique
idle time temps *m* d'arrêt/de pause
idling [of a machine] marche *f* à blanc/vide
igniter allumeur *m*, dispositif *m* d'allumage
ignition allumage *m*

illuminance éclairement *m*
illuminate éclairer *vb*, illuminer *vb*
illuminated illuminé *adj*
illuminated [sign, etc] lumineux *adj*
illumination éclairage *m*, illumination *f*
illumination, background éclairage *m* de fond
illumination, uniformity of uniformité *f* de l'éclairage
illumination measurement mesure *f* de la lumière
image image *f*
image, mirror image *f* symétrique, image-miroir *f*
immediately [i.e. urgently] d'urgence *adv*
immerse immerger *vb*, plonger *vb*
impact choc *m*, coup *m*
impedance [electrical] impédance *f*
impeller roue *f*
impeller [air] propulseur *m*
impeller [centrifugal pump] roue *f* mobile
impeller, fan roue *f* de ventilateur
impeller, pump couronne *f*/roue *f* mobile
impeller blade ailette *f* de ventilateur
imperfection, surface défaut *m* de surface
imperial dome [roof] dôme *m* à bulbe
impermeable étanche *adj*, imperméable *adj*
impermeable to air/gas imperméable à l'air/au gaz
impervious étanche *adj*, imperméable *adj*
implement réaliser *vb*
implementation mise *f* à exécution, mise *f* en œuvre, réalisation *f*
implementation, project réalisation *f* du projet
implements outillage *m*
impost [of an arch] imposte *f*
impregnate (with) imprégner *vb* (de)
impregnation imprégnation *f*
improve [get better] s'améliorer *vb*
improve [make better] améliorer *vb*, perfectionner *vb*
improvement amélioration *f*, perfectionnement *m*
improvement [of buildings, etc] aménagement *m*, mise *f* en valeur
improvement (works) travaux *m,pl* d'amélioration
improvements, make apporter *vb* des améliorations
impurity impureté *f*
in a body en masse
in abeyance en attente
in (actual) practice dans la pratique
in bulk en vrac
in circuit en circuit
in clear [uncoded] en clair
in cutting en déblais
in disrepair dégradé *adj*
in due time à terme
in force [of code of practice, etc] en vigueur
in front of devant *adj*
in gear en prise
in open cut à ciel *m* ouvert
in phase en phase
in place en place
in ruins dégradé *adj*
in spate en crue
in staggered rows en quinconces

in stock en stock
in the background à l'écart
in the form of a network réticulé *adj*
in (the) open country de rase campagne
in the town au bourg
in transit en transit
in use [of code of practice, etc] en vigueur
in working order capable de fonctionner, en état/ordre de marche
inaccurate [instrument] faux *adj*
inadequate insuffisant *adj*
incandescence incandescence *f*
incandescent incandescent *adj*
incident incident *adj*
incinerate incinérer *vb*
incineration incinération *f*
incinerator four *m* d'incinération, incinérateur *m*
incise entamer *vb*
inclination inclinaison *f*
inclination [of a wall about to fall] coup *m*
inclination [slant, banking] dévers *m*
inclination, inwards [opposite of overhang] fruit *m*
inclination, outwards surplomb *m*
inclination, slope pente *f* de talus
incline déverser *vb*
inclined incliné *adj*
inclined [rising] rampant *adj*
inclinometer clinomètre *m*
including comprenant *adj*
inclusion inclusion *f*
inclusive [e.g. of fee, payment] forfaitaire *adj*
incombustible incombustible *adj*
income revenu *m*
income [from an investment] rapport *m*
income [proceeds, yield] produit *m*
income, rental revenus *m,pl* locatifs
income tax impôts *m,pl* sur le revenu
incorporated (in, into) incorporé *adj* (dans)
incorporation incorporation *f*
increase accroissement *m*, augmentation *f*
increase (in) [prices, temperature, wages] hausse *f* (de)
increase in cost [e.g. during course of a contract] plus-value *f*
increase augmenter *vb*
increase [get bigger] s'agrandir *vb*
increase [sth] agrandir *vb*
increase in volume foisonner *vb*
increase the scope of étendre *vb*
increment accroissement *m*, augmentation *f*, incrément *m*
indebtedness endettement *m*
indentation indentation *f*
indentation [e.g. in a coast line] échancrure *f*
independent autonome *adj*, indépendant *adj*, individuel *adj*
indeterminate, statically hyperstatique *adj*
index [cost, etc] indice *m*
index [of a catalogue, etc] index *m*
index, bearing capacity indice *m* portant
index, colour rendering (CRI) indice *f* de rendu des couleurs (IRC)

index, comfort   indice *m* de confort
index, construction costs   Indice du coût de la
   construction (ICC)
index, glare   indice *m* d'éblouissement
index, liquidity   indice *m* de liquidité
index, plasticity   indice *m* de plasticité
index, room   indice *m* du local
index, thermal comfort   indice *m* de confort thermique
index, wear   coefficient *m* d'usure
index card   fiche *f*
index of protection (IP)   indice *f* de protection (IP)
index [costs, prices]   actualiser *vb*
indexation/indexing [of costs, prices, etc]   indexation *f*,
   actualisation *f*
indicate   indiquer *vb*
indication [sign]   indice *m*
indication [of levels, dimensions, etc, on a drawing]
   cotation *f*
indicator   index *m*
indicator [as in wind speed indicator, etc]   indicateur *m*
indicator [mark]   repère *m*
indicator, blow-out [electrical]   voyant *m* de fusion
indicator, draught   indicateur *m* de tirage
indicator, hours run   compteur *m* d'heures
indicator, level, *see* level indicator
indicator, remote   indicateur *m* à distance
indicator, repeater   indicateur *m* de répétition
indicator board   tableau *m* de signalisation
indium   indium *m*
individual *adj*   particulier *adj*
individual [isolated]   isolé *adj*
individually   isolément *adv*
induce   induire *vb*
induced   induit *adj*
induction   induction *f*
induction [of air]   aspiration *f*
induction unit, low pressure   inducteur *m* à basse
   pression
industrial   industriel *adj*
industrial concern [company]   exploitation *f* industrielle
inefficient   inefficace *adj*
inequality   inégalité *f*
inert   inerte *adj*
inertia [e.g. of geometric section]   inertie *f*
inertia, moment of   moment *m* d'inertie
inertia, thermal   inertie *f* thermique
infection   infection *f*
infill [of a framework, with bricks, plaster, etc]   hourdage
   m
infill between ceiling joists [plaster, concave]   auget *m*
infill [a framework, with bricks, plaster, etc]   hourder *vb*
infiltrate   infiltrer *vb*
infiltration [damp, air]   infiltration *f*
inflammability   inflammabilité *f*
inflammable   inflammable *adj*
inflate   gonfler *vb*
inflation   inflation *f*
inflation [physical rather than monetary]   gonflage *m*,
   gonflement *m*

inflexion, point of   point *m* d'inflexion
influence   influence *f*
influence line   ligne *f* d'influence
influx   afflux *m*, entrée *f*
information [piece of]   indication *f*, renseignement *m*
infra-red   infrarouge *adj*
infrastructure   infrastructure *f*
infringement [of rights]   abus *m*
inglenook   coin *m* du feu
inhabit   habiter *vb*
inhabitable [in theory!]   habitable *adj*
inhabitant   habitant *m*
inhabited   habité *adj*
inhale   aspirer *vb*
inhibit the growth (of)   retarder *vb* la croissance (de)
inhibitor   inhibiteur *m*
inhibitor, corrosion   inhibiteur *m* de corrosion
initial [first]   premier *adj*
initial [load, etc]   initial *adj*
initiate [a programme, a project]   commencer *vb*, lancer *vb*
initiate [measures, a method, a policy]   instaurer *vb*
initiate [negotiations]   amorcer *vb*, engager *vb*, ouvrir *vb*
inject   injecter *vb*
injection rod [in a water treatment unit]   canne *f*
   d'injection
injector   injecteur *m*
injector, steam   pompe *f* à jet
injector nozzle   tuyère *f* d'injecteur
injunction   arrêté *m* de suspension
injury [physical/legal]   dommage *m*
injury, physical   dommages *m,pl* corporels
inkjet printer   imprimante *f* à jet d'encre
Inland Revenue [French equivalent]   Direction *f*
   Générale des Impôts; le fisc[†] *m*
inlay   marqueter *vb*
inlet   admission *f*, entrée *f*
inlet [for air]   amenée *f*
inlet [of a culvert]   tête *f* amont
inlet, air   entrée *f* d'air
inlet, drainage [road]   bouche *f* à grille
inlet, dry riser   raccord *m* d'alimentation
inlet, fan   entrée *f* de ventilateur, ouïe *f* d'aspiration de
   ventilateur
inlet, return air   bouche *f* de reprise d'air
inlet, water   adduction *f*, arrivée *f*/entrée *f* d'eau
inlets, distributed   entrées *f,pl* reparties
inn   auberge *f*, hôtellerie *f*
inner *adj*   intérieur *adj*
inner swept path [roads]   bord *m* intérieur de la chaussée
inorganic   inorganique *adj*
input, analogue   entrée *f* analogique
input, energy   apport *m* d'énergie
input, heat   énergie *f* thermique fournie
input peripheral [computer]   périphérique *f* d'entrée
inquire, etc, *see* enquire, etc
inquiry   enquête *f*
inrush of water   venue *f* d'eau
insect   insecte *m*
insensitive to   insensible *adj* à

insensitivity (to)   insensibilité *f* (à)

insert   pièce *f* noyée/scellée

insert [of wood, in masonry, etc, to receive fixings] taquet *m*

insert, channel/concrete [fixing channel]   rail *m* de scellement

insertion   entrée *f*

inside *adj*   dedans *adj*

inside   dedans *m*, intérieur *m*

insist upon   exiger *vb*

insolation   ensoleillement *m*, rayonnement *m* solaire incident

insoluble   insoluble *adj*

inspect   contrôler *vb*, examiner *vb*

inspect [accounts]   revoir *vb*

inspection   contrôle *m*, inspection *f*, révision *f*

inspection, detailed   inspection *f* de détail

inspection, periodic   inspection *f* périodique

inspection, preliminary   surveillance *f* préliminaire

inspection, sanitary   surveillance *f* hygiénique

inspection chamber   chambre *f*/pièce *f*/puits *m* de visite

inspection chamber [small]   regard *m* de visite

inspection cover   trappe *f* de visite

inspection cover/fitting, sealed   tampon *m* hermétique

inspection cover, with an   à regard

inspection door   porte *f* de visite

inspection hatch   trappe *f* de visite

inspection hole   trou *m* d'homme/de regard

inspection pit   fosse *f*

inspection window [e.g. of a boiler]   voyant *m*

inspector   inspecteur *m*

inspector, medical (works) or occupational health médecin *m* inspecteur de travail

install   installer *vb*

install boundary markers   borner *vb*

install rafters   chevronner *vb*

installation, see also plant   installation *f*

installation, electrical   installation *f* électrique

installation, extra-low voltage safety   installation *f* à très basse tension de sécurité/à TBTS

installation, gas   installation *f* au gaz

installation, sanitary   installation *f* sanitaire

installation, wiring   installation *f* électrique

installation cost   coût *m* de l'installation

installation plan   plan *m* d'installation

installer   installateur *m*

installing   installation *f*

instalment   acompte *m*, versement *m*

instalment payment   paiement *m* échelonné

instalment purchase   location *f* vente

instruction   prescription *f*

instructions [printed or written]   instructions *f,pl*

instructions, assembly   notice *f* d'assemblage/de montage

instructions, installation   notice *f*/instructions *f,pl* d'installation

instructions, maintenance   instructions *f,pl* d'entretien

instructions, mounting   instructions *f,pl* d'installation

instructions, operating   consignes *f,pl* d'utilisation, instructions *f,pl* d'exploitation, mode *m* d'emploi

instructions, user or working   mode *m* d'emploi

instrument   instrument *m*, organe *m*

instrument, cable testing   instrument *m* de contrôle de câble

instrument, calibrated   instrument *m* étalonné

instrument, direct reading   instrument *m* à lecture directe

instrument, measuring   appareil *m*/instrument *m* de mesure

instrument, recording   appareil *m* enregistreur, enregistreur *m*

instruments, surveying   instruments *m,pl* topographiques

insufficient   insuffisant *adj*

insulate   calorifuger *vb*, isoler *vb*

insulated   isolé *adj*

insulating   isolant *adj*, isolateur *adj*

insulating, heat   calorifuge *adj*

insulating, see also insulation

insulating [the end of an electrical conductor]   capotage *m*

insulating material, thermal   calorifuge *m*, isolant *m* calorifuge

insulation [electrical]   isolement *m*

insulation [thermal, the process]   calorifugeage *m*

insulation [thermal, acoustic]   isolation *f*

insulation, acoustic   isolation *f* acoustique/phonique/sonore, revêtement *m* acoustique

insulation, blanket type (thermal)   isolant *m* en matelas

insulation, bulk or loose fill   isolant *m* en vrac

insulation, bulk or loose fill (thermal)   isolation *f* thermique en vrac

insulation, cellular thermal   isolation *f* thermique cellulaire

insulation, corkboard thermal   isolation *f* par plaque de liège

insulation, desired/required (level of)   isolement *m* recherché

insulation, double   double isolation *f*

insulation, fibreglass   laine *f* de verre

insulation, foam [for cavity walls]   isolant *m* moussé

insulation, foamed in situ (thermal)   isolation *f* expansée in situ

insulation, half round, preformed pipe   coquille *f*, isolant *m* en coquille

insulation, noise or sound   isolation *f* contre le bruit

insulation, preformed pipe   isolant *m* en coquille

insulation, required (level of)   isolement *m* recherché

insulation, rigid panel/sheet   isolant *m* en panneau rigide

insulation, rock wool   laine *f* de roche

insulation, sound   isolation *f* contre le bruit

insulation, standardized acoustic   isolement *m* acoustique normalisé

insulation, thermal or heat   isolant *m*/isolation *f* thermique

insulation, thermo-setting resin foam thermal   mousse *f* à base d'urée formol

insulation breakdown   défaillance *f* de l'isolement

insulation fault   défaut *m* d'isolement

insulation layer, thermal   couche *f* d'isolation thermique
insulation paper used under sheet metal roofing   papier *m* anglais
insulation thickness   épaisseur *f* isolante
insulator   isolant *m*
insulator [tradesman]   calorifugeur *m*
insulator, thermal   calorifuge *m*
insurance   assurance *f*
insurance, all-risk   assurance *f* tous risques
insurance, civil liability   assurance *f* responsabilité civile
insurance, comprehensive   assurance *f* multirisques/tous risques
insurance, decennial   assurance *f* décennale
insurance, fire   assurance *f* (contre l')incendie
insurance, life   assurance *f* (sur la) vie
insurance, theft   assurance *f* (contre le) vol
insurance against material damage [compulsory for building works in France]   dommage ouvrage *m* (DO)
insurance policy   police *f* d'assurance
insurance policy (joint) covering all parties to the contract   police *f* unique du chantier
insurance taken out by client for building works on a project basis   dommage *m* ouvrage
insure against   garantir *vb* contre
insured [person]   assuré *m*
insurer   assureur *m*
intact   intact *adj*
intake   admission *f*
intake [for air]   amenée *f*
intake, air   prise *f* d'air
intake, bellmouth   trompe *f* d'entrée
intake, fresh air   prise *f* d'air neuf
intake, river   prise *f* en rivière
intake, supply   poste *m* de livraison
intake, water   prise *f* d'eau
intake tunnel [dam]   galerie *f* de prise
integral *adj*   intégré *adj*, intégral *adj*
integral   intégrale *f*
integrate   intégrer
integrated   intégré *adj*
integration   intégration *f*
integrator   intégrateur *m*
integrity   intégrité *f*
intensity   intensité *f*
intensity, current   intensité *f* de courant
intensity, luminous   intensité *f*/puissance *f* lumineuse
intensity, radiation   intensité *f* de rayonnement
intensive use   usage *m* intensif
interaction   interaction *f*
interceptor, petrol   séparateur *m* décanteur
interceptor (drain)   drain *m* d'interception
interchange [highway], *see also* intersection, junction   carrefour *m*, échangeur *m*
interchange, motorway   échangeur *m* autoroutier
interchangeability   interchangeabilité *f*
intercom (system)   interphone *m*
interest   intérêt *m*
interest, compound/simple   intérêt *m* composé/simple
interest rate   taux *m* d'intérêt

interface   interface *f*
interfere [of light rays]   interférer *vb*
interference [of light rays, or electrical signals]   interférence *f*
interference [radio/PA/phone, crackle/crackling]   friture *f*, grésillement *m*
interim   provisoire *adj*
interior, *see also* internal
interior decorator   décorateur *m* d'intérieur, tapissier *m*
interior designer   architecte *m* d'intérieur, en somblier *m*
interlock   enclenchement *m*
interlock, security   couplage *m* de sécurité
interlocking   emboîtement *m*
intermediate   intermédiaire *adj*, mitoyen *adj*
intermittent   intermittent *adj*
intermittent [of a defect]   fugitif *adj*
internal   intérieur *adj*, interne *adj*
internal [of dimensions]   dans œuvre (D/O)
interpretation   interprétation *f*
interpretation of tests, test results   interprétation *f* des essais
interrupt [a circuit]   interrompre *vb*
interrupter [electrical]   interrupteur *m*, rupteur *m*
interruption [of current]   interruption *f*
intersect   se croiser *vb*, recouper *vb*
intersection   croisement *m*, intersection *f*
intersection [road]   carrefour *m*
intersection [surveying]   recoupement *m*
intersection, arm, branch or leg of an   branche *f* de carrefour
intersection, at grade or single level   carrefour *m* à niveau
intersection, cloverleaf   croisement *m* en double huit, carrefour *m*/échangeur *m* en trèfle, saut-de-mouton *m*, trèfle *m*
intersection, grade-separated   carrefour *m* dénivelé
intersection, signal controlled   carrefour *m* à feux
intersection of a minor and major road in a dip   lunette *f*
intersection of setting out   axe *m* repère de piquetage
intersection of two vaults   pénétration *f*
interstitial   interstitiel *adj*
interstitial condensation   condensation *f* interstitielle
interval [of time or space]   intervalle *m*
interval of time   intervalle *m* de temps
interview   entretien *m*
intrados   douelle *f* intérieure, intrados *m*
introduction [to document]   avant-propos *m*
inundation   inondation *f*
invalid   invalide *adj*
invalid [of a clause]   non valable *adj*
invariable   constant *adj*
inventory   inventaire *m*
inventory, resource   inventaire *m* des ressources
inversion   inversion *f*
inversion, temperature   inversion *f* thermique
inversion of polarity   inversion *f* de (la) polarité
invert [of a sewer, etc]   radier *m*
invest   investir *vb*

investigate something   faire *vb* des recherches sur qch
investigation   enquête *f*, recherche *f*, reconnaissance *f*
investigation, field   étude *f* de terrain, reconnaissance *f* de site
investigation, geophysical   étude *f* géophysique
investigation, ground, *see* ground investigation
investigation, laboratory   recherche *f* en laboratoire
investigation, land tenure/ownership   étude *f* foncière
investigation, site   reconnaissance *f* de site
investigation, soil, *see* soil investigation
investigation, X-ray   examen *m* à rayons X, analyse *f* aux rayons X
investment   investissement *m*
investment [in securities]   placement *m*
investment cost   dépenses *f,pl* d'investissement
invitation   appel *m*
invitation to prequalify   appel *m* de candidature
invitation to tender   appel *m* d'offres
invitation to tender, public   avis *m* d'appel public à la concurrence (AAPC)
invite tenders for sth   mettre *vb* qch en adjudication
invoice   facture *f*
invoice clerk   facturier *m*
invoice register   facturier *m*
invoice   facturer *vb*
invoicing   facturation *f*
iodine   iode *m*
ion   ion *m*
ion exchange   échange *m* d'ions
ionization   ionisation *f*
iron   fer *m*
iron, cast, *see also* cast iron   fer *m* coulé/de fonte, fonte *f*
iron, channel   fer *m* (en) U, barre *f* à U
iron, corrugated   tôle *f* ondulée
iron, grey cast   fonte *f* grise
iron, scrap   fer *m* de masse
iron, sheet   fer *m* en tôle
iron, white   fonte *f* blanche
iron, wrought   fer *m* battu/forgé/soudé
iron bar, square   fer *m* carré
iron oxide   oxyde *m* de fer
ironing board   table *f* à repasser

ironmongery   ferronnerie *f*
ironmongery [metal fittings]   quincaillerie *f*
ironwork   ferronnerie *f*
ironwork [piece of]   ferrure *f*
ironwork, ornamental   serrurerie *f* d'art
ironworks   usine *f* métallurgique
irradiation   irradiation *f*
irregular   irrégulier *adj*
irregularity   irrégularité *f*
irremovable   ancré *adj* en place
irrigate   irriguer *vb*
irrigated   irrigué *adj*
irrigation   irrigation *f*
irrigation [by spraying]   aspersion *f*
irrigation, drip   irrigation *f* goutte à goutte
irrigation, sprinkler   irrigation *f* par aspersion
irrigation, subsurface   irrigation *f* souterraine
irrigation, surface   irrigation *f* superficielle/de surface
irrigation channel   fossé *m* d'irrigation
irrigator   asperseur *m*
isenthalpic   isenthalpique *adj*
island [traffic, or small, in water]   îlot *m*
isobar   isobare *f*
isobaric   isobare *adj*
isolated [e.g. a house]   isolé *adj*
isolating device   dispositif *m* de sectionnement
isolation   isolement *m*
isolation [electrical]   sectionnement *m*
isolation [of a community]   enclavement *m*
isolator [electrical]   isolateur *m*
isometric (drawing)   dessin *m* isométrique
isotherm   isotherme *f*
isothermal   isotherme *adj*
isotope   isotope *m*
isotropic   isotrop(iqu)e *adj*
issue (on to)   déboucher *vb* (sur)
item   article *m*
item [in a list]   poste *m*
itemize [in accounts, etc]   détailler *vb*
items covered by the contract   éléments *m,pl* compris dans le contrat
ivy   lierre *m*

jack [lifting]   cric *m*, vérin *m*
jack, double-acting   vérin *m* double-effet
jack, flat   vérin *m* Freyssinet/plat
jack, hydraulic   cric *m*/vérin *m* hydraulique
jack, hydraulic [for steadying a crane, etc]   béquille *f* hydraulique
jack, pneumatic   vérin *m* pneumatique
jack, rack-and-pinion or ratchet-and-pawl   cric *m*, vérin *m* à crémaillère

jack, screw   cric *m*/vérin *m* à vis
jack, stressing   vérin *m* de mise en tension
jack, tooth-and-pinion   cric *m*, vérin *m* à crémaillère
jack   vériner *vb*
jack sth up   soulever *vb* qch au cric
jacket [boiler, tank]   enveloppe *f*
jacket, insulating   jaquette *f* calorifuge
jacket, life   gilet *m* de sauvetage
jacket, steam   chemise *f*/enveloppe *f* de vapeur

jacket, water   chemise *f* d'eau
jacket cover [round a boiler]   jaquette *f*
jacket   chemiser *vb*
jacketed   chemisé *adj*
jackhammer   marteau *m* piqueur
jacking   vérinage *m*
jam   sc bloquer *vb*
jam [a door, machinery, etc]   bloquer *vb*, coincer *vb*
jam on [a brake]   bloquer *vb*
jamb, door [pier or post]   poteau *m* de porte
jamb, door or window [splay]   ébrasement *m*, ébrasure *f*
jamb, door or window [pier or post] pied-droit/piédroit *m*, poteau *m* d'huisserie/de baie/de porte
jamb, hanging   montant *m* charnier/de rive/de ferrage
jamb, window [pier or post]   dormant *m* de fenêtre, poteau *m* de fenêtre
jammed   bloqué *adj*, coincé *adj*
janitor   gardien *m*, portier *m*
jaw [of a chuck]   mors *m*
jaw, fixed/movable [of a vice, etc]   mâchoire *f*/mors *m* fixe/mobile
jemmy   pied-de-biche *m*
jet   jet *m*
jet, discharge   jet *m* de décharge
jet, steam   jet *m* de vapeur
jet throw   portée *f* d'un jet
jetting [excavation]   abattage *m* hydraulique
jetty   appontement *m*, épi *m*, jetée *f*
jetty head   musoir *m*
jib [crane]   bec *m*, flèche *f*
jib, crane with luffing   grue *f* à flèche relevable
jib, lattice   flèche *f* en treillis
jig   gabarit *m*
jigsaw   scie *f* à chantourner, scie *f* anglaise/sauteuse
job   tâche *f*
join   jointure *f*
join [in paintwork]   raccord *m*
join [line of]   joint *m*
join, *see also* connect   joindre *vb*, raccorder *vb*
join [a branch pipe or duct into a main]   piquer *vb*
join by tongue-and-groove   embrever *vb*
join end to end   ab(o)uter *vb*, rabouter *vb*
join side by side   accoler *vb*
join to   raccorder *vb* à
join together   accoupler *vb*
join together [with a binder, a tie, a strap]   lier *vb*
join two pieces of wood   enter *vb*
joiner   menuisier *m*
joiner's bench, small   bidet *m*
joiner's chisel   ciseau *m* de menuisier
joiner's pincers   tenaille *f* de menuisier
joinery   menuiserie *f*
joinery, metal   menuiserie *f* métallique
joinery, panelled [e.g. doors, shutters, etc]   lambris *m*
joinery feature, decorative   habillage *m*
joining   liaison *f*, raccordement *m*
joining [of timbers]   empattement *m*
joining together   accouplement *m*, raccordement *m*
joint *adj*   collectif *adj*

joint   assemblage *m*, joint *m*, jointure *f*
joint [node]   nœud *m*
joint [pipework]   joint *m*, raccord *m*
joint [timber, metal structure]   assemblage *m*
joint, air-tight   joint *m* étanche à l'air
joint, angle   joint *m* d'angle
joint, ball or ball-and-socket   articulation *f*/joint *m* à rotule
joint, bed   joint *m* d'assise/de lit
joint, bevel   assemblage *m* en fausse coupe
joint, bevelled or skew   joint *m* biais
joint, blow(n)   joint *m* soudé au chalumeau
joint, brazed   joint *m* brasé
joint, brazed butt [in lead pipework]   nœud *m* de soudure de jonction
joint, brazed T- [in lead pipework]   nœud *m* (d'empattement)
joint, butt   about(ement) *m*, assemblage *m* par aboutage/d'about, joint *m* abouté/bout à bout/carré/droit/plat
joint, butt-strap   assemblage *m* par couvre-joint
joint, butt-welded   cordon *m* de soudure bout à bout, raccordement *m* par soudure bout à bout
joint, cable   raccord *m* de câble
joint, capillary   joint *m* soudé par capillarité
joint, caulked   joint *m* maté
joint, chamfered   joint *m* en biseau
joint, combed   assemblage *m* à queues droites
joint, composite   joint *m* composé
joint, compression [plumbing]   joint *m* de compression, raccord *m* à compression
joint, cone   joint *m* conique
joint, construction [in concrete]   joint *m*/surface *f* de reprise, joint *m* de construction
joint, contraction   joint *m* de contraction/de retrait
joint, corner   joint *m* à l'angle
joint, double skew notch   assemblage *m* à double épaulement
joint, dovetail   assemblage *m* à queue d'aronde, queue-de-carpe *f*
joint, dovetail halving   assemblage *m* à mi-bois à queue d'aronde
joint, dowelled   enlaçure *f*
joint, dry [without mortar]   liaison *f* à sec
joint, edge   joint *m* de bord
joint, elbow   raccord *m* coude
joint, expansion, *see* expansion joint
joint, field   joint *m* de chantier/de montage
joint, flange(d)   joint *m* à brides/de bride, raccord *m* à brides
joint, flared   joint *m* mandriné
joint, flush [mortar]   joint *m* affleuré/lisse
joint, forked/open mortice-and-tenon   assemblage *m* à enfourchement, enfourchement *m*
joint, halved [in iron/steel]   entaille *f* à mi-fer
joint, halved [in timber]   assemblage *m*/entaille *f* à mi-bois, entablure *f*
joint, halved mortice-and-tenon   paume *f*
joint, halving   assemblage *m* à entaille

joint, high pressure   joint *m* à haute pression
joint, hinged   joint *m* articulé
joint, hollow [mortar]   joint *m* creux
joint, horizontal tongue-and-groove [in cladding]   joint *m* horizontal à recouvrement
joint, keyed [mortar]   joint *m* arrondi/en canal
joint, lap   joint *m* à recouvrement
joint, lap [in reinforcement bar]   recouvrement *m*
joint, lapped   assemblage *m* à clin
joint, lead-caulked [spigot-and-socket pipes]   nœud *m* à embase
joint, lead-caulked or leaded   joint *m* de plomb/coulé à plomb
joint, leak-proof   joint *m* étanche
joint, loose tongue-and-groove   assemblage *m* à fausse languette
joint, mitre(d)   assemblage *m* à onglet
joint, mortar   joint *m* en mortier
joint, mortise   assemblage *m* à emboîtement
joint, mortise-and-tenon   assemblage *m* à tenon et mortaise, tenon *m* simple
joint, movement   joint *m* de déplacement/de dilatation
joint, movement [to allow for unequal loading]   joint *m* de rupture
joint, neoprene   joint *m* en néoprène
joint, O-ring   joint *m* torique
joint, overlapping   joint *m* par recouvrement
joint, pegged   assemblage *m* à clef
joint, pipe   jonction *f* de tuyaux
joint, rebated   joint *m* à emboîtement
joint, recessed [mortar]   joint *m* creux
joint, reducing/reduction   raccord *m* de réduction
joint, riveted   assemblage *m* rivé/à rivets, rivure *f*
joint, rubber   joint *m* caoutchouc
joint, scaffolding   nœud *m* d'échafaudage
joint, scarf   joint *m* biseauté
joint, screw (on)   raccord *m* à visser
joint, screwed   assemblage *m*/joint *m*/raccord *m* vissé
joint, screwed pipe   raccord *m* vissé pour tuyau
joint, settlement   joint *m* de tassement
joint, shrinkage   joint *m* de retrait
joint, single skew notch   assemblage *m* à embrèvement simple
joint, site   joint *m* de chantier/de montage
joint, skew   joint *m* biais
joint, sleeve   raccord *m* manchon
joint, sliding   joint *m* glissant
joint, sliding expansion   compensateur *m* à joints glissants, joint *m* glissant
joint, slip   joint *m* glissant
joint, soldered   joint *m* brasé
joint, spigot [in pipework]   emboîtement *m*
joint, spigot-and-socket   assemblage *m*/raccord *m* à emboîtement, joint *m* d'emboîture/à emboîtement en cloche
joint, splice   joint *m* à recouvrement
joint, standing seam [in sheet metal]   joint *m* debout
joint, straight pipe (sleeve)   raccord *m* manchon

joint, straight scarf(ed)   assemblage *m* à trait de Jupiter droit
joint, sweat(ed)   joint *m* soudé par capillarité
joint, T-/tee-   raccord *m* té, té *m*
joint, toothed [between two pieces of timber]   endentement *m*
joint, universal   charnière *f* universelle
joint, V-shaped edge or end [timber]   assemblage *m* à grain d'orge
joint, vertical loose tongue, between facade panels   joint *m* vertical à pincement
joint, weathered [mortar]   joint *m* creux chanfreiné
joint, weatherproof   joint *m* d'étanchéité
joint, web   joint *m* de l'âme
joint, welded   assemblage *m*/joint *m* soudé
joint, welded socket   joint *m* à emboîtement soudé
joint, welted [in roofing, cladding sheets]   agrafure *f*
joint, wiped solder   brasage *m* à la louche
joints, alternate   joints *m,pl* croisés
joints, staggered   joints *m,pl* en chicane
joints, timber   assemblage *m* de pièces de bois
joint between pipes of different diameter   mamelon *m*
joint filling materials [road or bridge joints]   produits *m,pl* de garnissage
joint grouting   clavage *m*, injection *f* des joints
joint in a road pavement not extending to full depth   faux joint *m*
joint layout [stonework, formwork, tiling, etc]   calepinage *m*
joint layout drawing [for stonework, formwork, tiling, etc]   calepin *m*
joint ring   bague *f* d'étanchéité
joint sealant [mastic paste]   mastic *m* pâteux
joint sealant [road construction]   produit *m* de garnissage
joint sealing, plastic, applied under pressure with a pump   joint *m* à la pompe
joint sealing compound   garniture *f* de joint
joint tape   bande *f* (à joint), sous-joint *m*
joint [join together]   assembler *vb*, raccorder *vb*
joint [masonry]   jointoyer *vb*
joint [timbers]   empatter *vb*
joint notched timbers with a peg   brandir *vb*
jointer [of masonry, tiles]   jointoyeur *m*
jointer [tool for mortar joints]   tire-joint *m*
jointing   assemblage *m*, jointement *m*
jointing, butt   aboutage *m*, aboutement *m*
jointing, end/end-to-end [of timber]   enture *f*
jointing, flange   joint *m* à brides/de bride, assemblage *m* par brides
joint(ing), forked/open mortise-and-tenon   enfourchement *m*
jointing, Klingerite   joint *m* à klingerite
jointing, spigot-and-socket   emboîtement *m*
joint(ing), tongue-and-groove   assemblage *m* à rainures et languettes, embrèvement *m*
jointing end-to-end [of timber]   enture *f*
jointing mastic or paste   lut *m*
jointing of rafters into a ridge   enfourchement *m*

jointing tool [for types of recessed pointing]   fer *m* à
    joints
jointing while bedding   joint *m* fait en montant
jointly liable [e.g. of partners in a joint venture]   solidaire
    adj
joist   poutre *f*, poutrelle *f*, solive *f*
joist [normal, as opposed to trimmer or trimming]   solive
    f courante
joist, angled trimmer   soliveau *m* en empannon
joist, bearing [supporting ends of floor joists]
    lambourde *f*
joist, bridging   lambourde *f*
joist, ceiling   entrait *m*
joist, double floor   doubleau *m*
joist, edge   solive *f* de rive
joist, floor   poutre *f* de plancher
joist, heavy   penne *f*
joist, rolled steel (RSJ)   poutrelle *f* laminée (en acier)
joist, secondary   solive *f*
joist, small trimmer   soliveau *m*
joist, trimmed   solive *f* de remplissage
joist, trimmed [built-in one end, other joist supported]
    boiteuse *f*
joist, trimmer   boiteuse *f*, solive *f* boiteuse
joist, trimming   solive *f* enchevêtrée/porteuse/
    d'enchevêtrure
joist, wall [under internal wall in timber structure]   poutre
    f porte-cloison

joist hanger   étrier *m*
joist shuttering   coffrage *m* de la poutre maîtresse
joist spacing   clair *m*
joisting, secondary   solivage *m*
jolt   secousse *f*
journey   trajet *m*
joystick [computer]   manche *m* à balai
juice [electric current][†]   jus[†] *m*
jump, hydraulic   ressaut *m*
jumper [bar with flattened ends]   barre *f* à mine
junction   jonction *f*, nœud *m*, raccord *m*
junction [pipework]   branchement *m*
junction [road]   carrefour *m*, intersection *f*
junction, cloverleaf   croisement *m* en double huit,
    carrefour *m*/échangeur *m* en trèfle, saut-de-mouton *m*,
    trèfle *m*
junction, signal controlled   carrefour *m* à feux
junction, single oblique [pipework]   branchement *m*
    oblique simple
junction, three-leg or -way [highways]   bifurcation *f*,
    embranchement *m*
junction, three-way [plumbing]   culotte *f* double
junk [worthless]   saleté *f*
justification for further revision of a claim
    réplique *f*
jut out   faire *vb* saillie
jutting out   saillant *adj*, en saillie

k-value (thermal conductivity)   coefficient-k *m*,
    coefficient *m* de transmission thermique
kaolin   kaolin *m*
keep [of a fortress]   donjon *m*
keeper   conservateur *m*
keeper [striking plate] of a lock   gâche *f*
keeping with, in   en rapport avec
kennel   chenil *m*
kentledge   corps *m* mort
kerb   bord *m*, bordure *f* (de trottoir)
kerb, flush   bordure *f* arasée/basse/enterrée
kerb, long paving slab used as   boutisse *f*
kerb, lowered [for passage of vehicles]   bateau *m*
kerb, raised   bordure *f* haute/en saillie
kerb, splayed   bordure *f* inclinée
kerb, vertical faced   bordure *f* droite
kerb channel   bordure *f* caniveau
kerb channel, hollow   bordure *f* creuse
kerb-stone   parement *m*
kerf [when sawing wood]   voie *f* de scie
kerf [width of cut made by cutting torch in steel]
    entaille *f*
kerosene   pétrole *m* lampant

key [in a timber joint or between pieces of masonry]
    adent *m*
key [of a lock or arch, or in a joint]   clef *f*
key [on a drawing, map, etc]   légende *f*
key [on a shaft]   clavette *f*
key [typewriter or computer keyboard]   touche *f*
key, chuck   clef *f* de mandrin
key, flat [used from both sides]   clef *f* bénarde
key, flat part of a   panneton *m*
key, make-and-break   brise-circuit *m*
key, master   passe *f*, passe-partout *m*
key, shear   clef *f* de cisaillement
key, stopcock   clef *f* de robinet
key, sunk   clavette *f* normale/ordinaire
key for operating a butterfly damper   papillon *m*
key map [inset on a drawing]   cartouche *m*
key profile [of a flat key]   chiffre *m*, silhouette *f*
key with a hollow shank   clef *f* forée
key [on to a shaft]   claveter *vb*
keyboard [computer]   clavier *m*
keyhole   trou *m* de serrure
keyhole cover   cache-entrée *m*
keyhole plate   entrée *f* de serrure

keyhole surround plaque *f* d'entrée
keying [e.g. impeller on to shaft] clavetage *m*
keying [fixing with a key] calage *m*, coinçage *m*
keying [roughening] accrochage *m*
keying up [an arch] clavage *m*
keystone clef *f* de voûte, mensole *f*
kiln, lime four *m* à chaux
kiln dry étuver *vb*
kiln drying [of timber] étuvage *m*
kilobar kilobar *m*
kilocalorie kilocalorie *f*
kilogram kilogramme *m*
kilometre kilomètre *m*
kilometres, measure off in kilométrer *vb*
kilometres, measurement in kilométrage *m*
kilometre stones, mark with kilométrer *vb*
kilotonne kilotonne *f*
kilowatt kilowatt *m*
kilowatt hours kilowatts-heures *m,pl*
kindle [a fire] allumer *vb*
kindle [catch fire] s'allumer *vb*
kindled allumé *adj*
kinematic cinématique *adj*
kinetic cinétique *adj*
king-post aiguille *f*, poinçon *m*
kiosk kiosque *m*
kitchen cuisine *f*
kitchen, back arrière-cuisine *f*
kitchen corner coin *m* cuisine
kitchen garden potager *m*
kitchen hood hotte *f* de cuisine
kitchen range [cooker] cuisinière *f*
kitchen sink évier *m*
kitchen unit, prefabricated élément *m* préfabriqué de cuisine
kitchenette cuisinette *f*
Klingerite jointing joint *m* à klingerite
kneading [soil treatment] pétrissage *m*
knee [pipework, etc] coude *m*
kneed coudé *adj*

kneepad genouillère *f*, protège-genoux *m*
knife [e.g. painter's] couteau *m*
knife, draw couteau *m* à deux manches, plane *f*
knife, filling spatule *f*
knife, painter's filling [French regional term] macabé *m*
knife, pruning serpette *f*
knife, putty couteau *m* à mastiquer, spatule *f* de vitrier
knife, riving [of circular saw] couteau *m* diviseur
knife, Stanley or universal couteau *m* universel
knife, stripping spatule *f*
knife, stripping [for wallpaper] couteau *m* décolleur
knife, trimming couteau *m* à araser/à émarger
knife, wallpaper cutting tranchet *m*
knife for cutting lead, glazier's tailloir *m*
knob pomme *f*
knob [of a hand plane] pommeau *m*
knob, door bouton *m* de porte
knock [blow] coup *m*
knock down défaire *vb*
knock further in [e.g. a nail] renfoncer *vb*
knocking down [e.g. of a wall] abattage *m*
knoll monticule *m*
knot [in timber] nœud *m*
knot, arris nœud *m* d'arête
knot, edge nœud *m* de rive
knot, face nœud *m* de face
knot, loose nœud *m* bouchon
knot, sound nœud *m* adhérent/sain
knot, splay nœud *m* plat/tranchant
knot, unsound nœud *m* vicieux
knot cluster nœuds *m,pl* groupés
knot-hole cutter fraise *f* à dénoder
knot-hole moulding machine perceuse *f* à dénoder
knotty [wood] racheux *adj*
knuckle [folded, of a strap hinge] œil *m* roulé
knuckle [of a hinge] charnon *m*, nœud *m*
knuckle, folded, of a strap hinge œil *m* roulé
knurl moleter *vb*
knurled moleté *adj*
krypton krypton *m*

L

L-shaped réparti en équerre
L-shaped cleat [for holding two walls or timbers together] harpon *m*
L-shaped clip or metal cleat équerre *f* d'assemblage
L-wall mur *m* en angle
lab(oratory) labo(ratoire) *m*
label étiquette *f*
label, circuit label *m* de circuit
label holder porte-étiquette *m*
laboratory, research laboratoire *m* de recherches
laboratory, test(ing) laboratoire *m* d'essais

labour main-d'œuvre *f*
labour, materials and percentage overheads work travaux *m,pl* en régie
labour, semi-skilled ouvrier *m* spécialisé
labour, skilled main-d'œuvre *f* qualifiée, ouvrier *m* qualifié
labour, unskilled travail *m* de manœuvre
labour cost charge *f* de main d'œuvre
labour market marché *m* du travail
labour supervisor contrôleur *m* de travail
labourer manœuvre *m*, travailleur *m*

labourer, earthworks   terrassier *m*
labourer, general site   garçon *m* de relais
labourer, mason's   gâcheur *m*, porte-auge *m*
labour(er), unskilled   manœuvre *m*
lack of balance [of rotating element]   balourd *m*
lacquer   laque *f*, peinture *f* laque
ladder   échelle *f*
ladder, extending/extension   échelle *f* coulissante/à
   coulisse
ladder, extending/retractable   échelle *f* escamotable
ladder, fire escape   échelle *f* à incendie/de sauvetage
ladder, fixed   échelle *f* fixe
ladder, folding   échelle *f* pliante
ladder, inspection   échelle *f* de visite
ladder, insulated   échelle *f* isolante
ladder, mechanical   échelle *f* mécanique
ladder, peg   rancher *m*
ladder, retractable [e.g. for loft access]   échelle *f*
   escamotable
ladder, roof   échelle *f* plate/de couvreur
ladder, roof(er's)   échelle *f* plate/de couvreur
ladder hook [to hold top of a ladder]   crochet *m* d'échelle
ladder tower [wheeled]   échelle *f* roulante
ladders in an inverted V, assemble two   éventailler *vb*
lag [delay]   retard *m*
lag [in response]   délai *m*/retard *m* de réponse
lag, thermal   retard *m* thermique
lag, time   décalage *m*
lag [behind]   traîner *vb*
lag [insulate]   calorifuger *vb*, envelopper *vb*
lag in phase   être *vb* en retard de phase
lagged [insulated]   calorifugé *adj*, chemisé *adj*
lagging [insulation]   isolant *m* thermique
lagging [of a boiler]   chemise *f*, enveloppe *f*, garnissage *m*
lagging [of a tank]   enveloppe *f*
lagging [the process]   calorifugeage *m*
lagoon   lagon *m*
laitance   laitance *f*
laitance, emergence of   ressuage *m*
lake   lac *m*, plan *m* d'eau
lake, dam   lac *m* de barrage
lake, ornamental   bassin *m*, pièce *f* d'eau
lake, ox-bow   bras *m* mort
lamella   lamelle *f*
lamellar   lamellaire *adj*, lamelliforme *adj*
laminar   laminaire *adj*
laminate   stratifié *m*
laminated   laminé *adj*, stratifié *adj*
laminated [built up in layers]   contreplaqué *adj*
laminated [in thin sheets]   feuilleté *adj*
laminated [structure]   lamellé *adj*
lamination [the process]   laminage *m*
laminboard   contreplaqué *m* latté, panneau *m* à âme
   lamellée
lamp   lampe *f*
lamp, arc   lampe *f* à arc
lamp, bactericidal or germicidal   lampe *f* germicidale
lamp, ceiling   plafonnier *m*
lamp, dichroic reflector   lampe *f* à réflecteur dichroïque

lamp, discharge   lampe *f* à décharge
lamp, display   lampe *f* à faisceau directionnel
lamp, filament   lampe *f* à incandescence
lamp, fluorescent   lampe *f* fluorescente/à fluorescence
lamp, halogen   lampe *f* à halogène
lamp, hand   baladeuse *f*
lamp, high pressure mercury/sodium (vapour)   lampe *f*
   à vapeur de mercure/sodium à haute pression
lamp, incandescent   lampe *f* à incandescence
lamp, indicating   lampe *f* de signalisation
lamp, inspection   baladeuse *f*
lamp, metal halide   lampe *f* aux halogénures
lamp, metal vapour   lampe *f* à vapeur métallique
lamp, pilot   lampe *f* témoin
lamp, portable inspection   baladeuse *f*
lamp, standard   lampadaire *m*
lamp, table   lampe *f* de table
lamp, vacuum   lampe *f* à vide
lamp, vapour [e.g. mercury or sodium]   lampe *f* à vapeur
lamp, wall   applique *f*
lamp ballast   ballast *m*
lamp bulb, silvered crown   calotte *f* argentée
lamp cap [base]   culot *m*
lamp holder   douille *f*, porte-ampoule *m*
lamp holder, ceiling   douille *f* à plafond
lamp hood [traffic lights, railway signals]   visière *f*
lamp life   durée *f* de vie d'une lampe
lamp (operating) current   courant *m* de fonctionnement
lamp post   lampadaire *m*, poteau *m* de réverbère
lamp power   puissance *f* de la lampe
lamp support, jointed [Anglepoise or similar]
   genouillère *f*
lampshade   abat-jour *m*
land, of or pertaining to   foncier *adj*
land   terrain *m*
land, agricultural   terrain *m* agricole
land, building   terrain *m* constructible
land, cultivated   cultures *f,pl*
land charges [as found by search]   état *m* hypothécaire
land register   cadastre *m*
land registry   bureau *m* des hypothèques, cadastre *m*
land survey   arpentage *m*, levé *m* de terrain
land surveying equipment   matériel *m* de topographie
land surveyor   arpenteur *m*, géomètre *m*, topographe *m*
land surveyor [dealing with larger areas, major
   developments, etc, in France]   géomètre-topographe *m*
land surveyor [dealing with local matters in France]
   géomètre-expert *m*
land tax   impôt *m* foncier
land taxes [in France]   taxes *f* foncières
land under cultivation   cultures *f,pl*
land use (zoning) plan   plan *m* d'occupation des sols
   (POS)
landing [large, serving several rooms]   galerie *f*
landing [large, upstairs]   dégagement *m*
landing [of stair, etc]   palier *m*
landing, ground floor   palier *m* de départ
landing, intermediate   palier *m* de repos
landing, quarter-space   demi-palier *m*

landing board with nosing at top of stairs   plaquette *f*
landing in a spiral staircase   palier *m* circulaire
landing serving several apartments or offices   palier *m* de communication
landing stage   appontement *m*
landing strip   terrain *m* d'atterrissage
landing tread [of staircase]   marche *f* palière
landlord [owner]   propriétaire *m*
landmark   point *m* côté, repère *m*
landscape   paysage *m*
landscaping   aménagement *m* (d'un espace vert), aménagement *m* paysager, paysage *m*
landslide or landslip   éboulement *m* (de terre), glissement *m* (de terrain)
lane   allée *f*
lane [of road]   bande *f*, voie *f*
lane [rural]   chemin *m* rural
lane, acceleration   bande *f*/voie *f* d'accélération
lane, auxiliary traffic [i.e. hard shoulders, etc]   voie *f* supplémentaire
lane, bus stopping   piste *f* d'arrêt
lane, crawler   voie *f* pour véhicules lents
lane, deceleration   bande *f*/voie *f* de décélération
lane, emergency stopping   bande *f* d'arrêt d'urgence (BAU)
lane, general-purpose   voie *f* de circulation normale
lane, storage   voie *f* de stockage
lane, traffic   bande *f*/voie *f* de circulation
lane, tramway   voie *f* de (circulation du) tramway
lane, turning (left/right)   voie *f* de tourne-à-(gauche/droite)
lane, waiting   voie *f* de stockage
lane, weaving   voie *f* d'entrecroisement/d'insertion
lantern [open, above a dome, etc]   lanterne *f*
lantern [skylight]   lanterneau *m*
lap [e.g. of tiles]   chevauchement *m*, recouvrement *m*
lap   recouvrir *vb*
lapped   à recouvrement
lapping   recouvrement *m*
lapse rate, adiabatic   chute *f* de température adiabatique
lap-welded   soudé *adj* à recouvrement
larch   mélèze *m*
larder   cellier *m*, garde-manger *m*
large-scale   à grande échelle
laser   laser *m*
last   dernier *adj*
latch   clenche *f*, fermeture *f*
latch [pivoted type, for a door, gate, etc]   loquet *m*
latch [straight bar, for gate, shutter, etc]   fléau *m*
latch, thumb-piece of a   poucier *m*
latch, tubular [door]   serrure *f* à tubulaire
late   en retard
late, make   retarder *vb*
latent   latent *adj*
latent heat   chaleur *f* latente
latent heat of evaporation/vaporization   chaleur *f* latente d'évaporation
latent heat of fusion/melting   chaleur *f* latente de fusion
later [subsequent]   ultérieur *adj*

lateral   latéral *adj*
lath   bois *m* feuillard, latte *f*
laths [of lath and plaster]   lattis *m*
laths used in wooden trellis-work   boissellerie *m*
lath, counter   contre-latte *f*
lath, notched/toothed, for cupboard shelf supports   crémaillère *f*
lath   latter *vb*
lathe   tour *m*
lathe, wood (turning)   tour *m* à bois/à dégauchir
lathe chuck   mandrin *m*
lathe spindle   broche *f* de tour
latitude   latitude *f*
latrines, site   latrines *f,pl* de chantier
lattice   treillage *m*, treillis *m*
lattice, metal   fer *m* maillé
launching [e.g. of bridge girder]   lançage *m*
launching frame or truss [e.g. for bridge sections]   lanceur *m*
launderette   laverie *f* automatique
laundry (room)   blanchisserie *f*, buanderie *f*
lavatory (WC), *see* toilet, toilets, etc
lavatories, public   toilettes *f,pl*
lavatories, site   latrines *f,pl* de chantier
law   droit *m*, loi *f*
law of conservation of energy   loi *f* de la conservation d'énergie
lawn   gazon *m*, pelouse *f*
lawnmower   tondeuse *f*
lawsuit   litige *m*
lawyer, courtesy title for   maître *m*
lawyer [advocate]   avocat *m*
lawyer [legal adviser, roughly equivalent to solicitor in England and Wales]   conseil *m* juridique
lay   poser *vb*
lay a cable   poser *vb* un câble
lay down   fixer *vb*
lay hot/cold [road material]   mettre *vb* en œuvre à chaud/froid
lay on [paint, etc]   taper *vb*
lay out   aligner *vb*, tracer *vb*
lay out [a road, etc]   construire *vb*
lay out capital   faire *vb* une mise de fonds
lay the foundation stone   poser *vb* la première pierre
lay [pipes, mains]   canaliser *vb*
lay [stone] **at 90° to natural bedding plane**   poser *vb* en délit
layer   film *m*
layer [geological]   banc *m*
layer [of a partition or wall, etc]   feuillet *m*
layer [of bricks, cement, stone, road material]   assise *f*
layer [paint, ground, etc]   couche *f*
layer [pass in welding]   passe *f*
layer [tradesman, layer of cables, pipes, etc]   poseur *m*
layer, anti-frost   couche *f* antigel
layer, asphaltic   feuille *f* de goudron, film *m* bitumineux
layer, bituminous   couche *f* bitumineuse
layer, boundary   couche *f* limite
layer, compressible   couche *f* compressible

layer, corrosion protection   couche *f* antirouille
layer, drainage   couche *f* drainante
layer, drainage [groundworks]   tapis *m* de drainage
layer, dust   couche *f* de poussière
layer, filter [groundworks]   tapis *m* filtrant
layer, filter [road construction]   couche *f* filtrante, sous-
  couche *f* drainante
layer, flexible [road foundations]   tapis *m* bitumineux
layer, frost proof [road construction]   couche *f* antigel
layer, impermeable [e.g. of soil]   couche *f* imperméable
layer, insulating   couche *f* isolante
**layer, lower, of a concrete slab containing its main
  reinforcement**   prédalle *f*
layer, ozone   couche *f* d'ozone
layer, permeable   couche *f* perméable
layer, pipe [machine], *see* pipelayer
layer, pipe [tradesman]   poseur *m* de tuyaux
layer, protective   couche *f* de protection
layer, protective concrete   couche *f* protectrice en béton
layer, road surface   couche *f* de roulement
layer, rust   couche *f* de rouille
layer, slag   lit *m* de scories
layer, stone   empierrement *m*
layer, surface   couche *f* superficielle/de surface
layer, thermal insulation   couche *f* d'isolation thermique
layer, thin, of road construction   tapis *m*
layer, underlying   sous-couche *f*
layer of backfill   couches *f,pl* de remblai
layer of chippings   couche *f* de gravillons
layer of clay   lit *m* d'argile
layer of material   lit *m*
layer of reinforcement   lit *m* d'acier
layer thickness [paint]   épaisseur *f* de couche
layers, successive   couches *f,pl* successives
laying   mise *f* en place, posage *m*, pose *f*
laying depth   profondeur *f* de pose
laying out   traçage *m*
laying out of structural timbers on a full-scale plan
  établissement *m*
layout   disposition *f*
layout [arrangement]   agencement *m*
layout [marking or setting out]   implantation *f*
layout [of a building]   ordonnance *f*
layout [of road, railway, etc]   tracé *m*
layout, full-scale [of axes of a structural framework]
  étalon *m*
layout (drawing)   épure *f*, plan *m*
layout of a network [e.g. of railways, bus routes]
  aménagement *m* d'un réseau
layout plan, *see* plan
leach   délaver *vb*
leached   lessivé *adj*
leaching   délavage *m*, lessivage *m*
lead [electrical]   branchement *m*, connexion *f*
lead [metal]   plomb *m*
lead, black [for darkening lead or zinc work]   plombagine *f*
lead, earth(ing)   conducteur *m* à la masse/de terre
lead, extension   prolongateur *m*, cordon *m* (de)
  prolongateur

lead, flexible   flexible *m*
lead, main earth   collecteur *m* de terre
lead, pencil   mine *f*
lead, red   mine *f* anglaise, minium *m*
lead, refill [pencil]   mine *f*
lead, sheet   plomb *m* en feuille
lead, white   blanc *m* de plomb/de céruse, céruse *f*
lead caulking   calfatage *m* de plomb
lead sheet   feuille *f* de plomb
lead tin solder wire   fil *m* de soudure
lead in phase   être *vb* en avance de phase
lead into [e.g. one space to another]   desservir *vb*
lead to [e.g. water to a drain]   conduire *vb* à
leader   chef *m*
leader [of piling rig]   guide *m* (de battage)
leading [of a leaded light]   résille *f*
leaf [of a door, window, etc]   battant *m*, ouvrant *m*,
  vantail *m*
leaf [of a partition or wall, etc]   feuillet *m*
leaf of a venetian blind   lame *f* de persienne
**leaf of a window with rounded edge fitting into groove
  in the frame**   battant *m* mouton
leak   fuite *f*
leak [the hole itself, in a hydraulic structure, a boiler]
  renard *m*
leak, air/water   fuite *f* d'air/d'eau
leak, duct/pipe   fuite *f* de conduit/tuyau
leak detector, *see* detector
leak   fuir *vb*
leakage   fuite *f*, fuites *f,pl*
leakage [electrical]   déperdition *f* par dispersion
leakage [of electric current]   déviation *f*
leakage coefficient [HVAC]   coefficient *m*
  d'exfiltration
leakage path [HVAC]   cheminement *m* d'exfiltration
leaking [of a boiler]   suintement *m*
lean [of cement or mortar]   maigre *adj*
lean [at an angle]   déverser *vb*
lean (on, against)   s'appuyer *vb* (sur, contre)
lean (sth) (on, against)   appuyer *vb* (qch) (sur, contre)
leaning   dévers *adj*
leaning, inwards [opposite of overhang]   fruit *m*
leaning, outwards   surplomb *m*
lean-to [roof, or building itself]   appentis *m*
lease   bail *m*
lease, building   bail *m* à construction
lease, draft   projet *m* de bail
lease, long term   bail *m* emphytéotique
leasing [with terminal purchase option, in France]   crédit
  m bail
lecture room with stepped seating   amphithéâtre *m*
ledge   saillie *f*
ledge, window   appui *m* de baie/de fenêtre
ledged and braced   à barres et écharpes
ledger [timber frame structure]   sommier *m*
ledger tube [scaffolding]   moise *f*
legal   juridique *adj*
legal action/dispute   litige *m*
legal tender, be   avoir *vb* cours

legend [on a drawing, map, etc] légende *f*

legging [protective gaiter] guêtre *f* de protection

legislation législation *f*

leisure centre base *f* de loisirs

lend prêter *vb*

lender [lending company] financier *m*

lending agency agence *f* de financement

length [duration, space] longueur *f*

length [i.e. a particular length, of pipe, road, etc] tronçon *m*

length, anchorage longueur *f* d'ancrage

length, arc [in welding] longueur *f* de l'arc

length, buckling longueur *f* de flambement

length, lap longueur *f* de recouvrement

length, lap [reinforcement] recouvrement *m* des armatures

length, maximum longueur *f* limite

length, overall longueur *f* hors-tout/totale

length, useful longueur *f* utile

length of grip longueur *f* de serrage

lengthen agrandir *vb* en long, allonger *vb*, prolonger *vb*

lengthening prolongation *f*

lengthwise en longueur *adv*

lens [light or geological]) lentille *f*

lens, ice lentille *f* de glace

lessening diminution *f*

lessor bailleur *m*

let louer *vb*

letter, recorded delivery advise lettre *f* avec avis de réception (AR)

letter, registered lettre *f* recommandée

letter box boîte *f* aux lettres

letting *adj* locatif *adj*

let(ting), furnished location *f* en meublé

level [ground] uni *adj*

level [surface, etc] plan *adj*

level, on the same de plain-pied

level plan *m*

level [height] cote *f*

level [of a building] étage *m*

level [result of making flush] affleurement *m*

level [storey of a building] niveau *m*

level [surveying, etc] niveau *m*

level, at ground de plain-pied

level, background noise intensité *f* de bruit de fond

level, base niveau *m* de base

level, change/difference/variation of dénivellation *f*

level, datum plan *m* de comparaison

level, departure from [of a roadway, etc] dénivellation *f*

level, determination of [surveying] dénivellation *f*

level, finished floor (FFL) niveau *m* du sol fini

level, foundation or sub-grade niveau *m* de la fondation

level, groundwater niveau *m* de la nappe aquifère/phréatique, niveau *m* phréatique

level, high-water niveau *m* haut de l'eau

level, highest/maximum niveau *m* au plus haut

level, lighting niveau *m* d'éclairage

level, loudness/sound niveau *m* sonore

level, loudness or sound intensity niveau *m* d'intensité sonore

level, lowest [of a watercourse] étiage *m*

level, lowest/minimum niveau *m* au plus bas

level, noise, *see* noise level

level, oil niveau *m* d'huile

level, piezometric surface niveau *m* piézométrique

level, plumb niveau *m* de maçon

level, pocket niveau *m* de poche

level, reference [on a drawing] plan *m* de comparaison

level, sea niveau *m* de la mer

level, sound niveau *m* (d'intensité) sonore

level, sound power niveau *m* de puissance sonore

level, sound pressure niveau *m* de pression acoustique

level, spirit niveau *m* à bulle (d'air), niveau *m* à bulle transversale, nivelle *f*

level, (standardized) sound pressure niveau *m* de pression acoustique (normalisé)

level, street rez-de-chaussée *m*

level, surveyor's niveau *m*

level, tailwater [below a dam] niveau *m* aval

level, water, *see* water level

level board [two pegs + a horizontal] chaise *f*

level gauge, liquid/oil/fuel oil jauge *f* (de niveau) de liquide/d'huile/de mazout

level indicator indicateur *m* de niveau

level indicator, float indicateur *m* de niveau à flotteur

level indicator, water indicateur *m* de niveau d'eau

level line marked on walls 1m above finished floor level trait *m* de niveau

level marker repère *m* de nivellement

level marker, reference repère *m* de nivellement

level araser *vb*

level [a surveying instrument] pointer *vb*

level [bring to same level] affleurer *vb*

level [ground, with a thin layer of earth, gravel, etc] régaler *vb*

level, become se niveler *vb*

level again reniveler *vb*

level line along a wall, mark a cingler *vb*

level (out) niveler *vb*

level (with), be affleurer *vb*

levelling [flattening] aplanissement *m*

levelling [making flush] affleurage *m*

levelling [making level, and in surveying] nivellement *m*

levelling [on a map or plan] pointage *m*

levelling [of a lift/site hoist during loading/unloading] isonivelage

levelling, automatic [of a lift cage] nivelage *m* automatique

lever levier *m*

lever, adjusting levier *m* de réglage

lever, balance [of bascule bridge] arrière-bras *m*

lever, brake levier *m* de frein

lever, clamping levier *m* de serrage

lever, control levier *m* de commande/de manœuvre

lever, gear levier *m* de changement de vitesse/de débrayage

lever, hand levier *m* à main, manette *f*

lever, operating   levier *m* de manœuvre
lever, release   levier *m* de déclenchement/de dégagement
lever, reversing   levier *m* de renvoi
lever arm   bras *m* de levier
levy   prélèvement *m*
levy   prélever *vb*
liability   engagement *m*
liability, contingent   engagement *m* hors bilan
liabilities   masse *f* passive, passif *m*
liabilities, contingent   passif *m* éventuel
liabilities, current   passif *m* circulant
liable to flooding   inondable *adj*
lias   lias *m*
library   bibliothèque *f*
licence   permis *m*
licence to trade or engage in a profession   patente *f*
license   patenter *vb*
licensee   licencié *m*
lie of the ground   site *m*
life [length of]   durée *f* de vie
life, effective   durée *f* de vie effective/réelle
life, lamp   durée *f* de vie d'une lampe
life interest [generally in land/property]   viager *m*
life jacket   gilet *m* de sauvetage
lift   ascenseur *m*
lift [of a pump]   refoulement *m*
lift, builder's   élévateur *m*//monte-charge *m* de chantier
lift, drag   téléski *m*
lift, fire fighting or firemen's   ascenseur *m* des pompiers
lift, goods   monte-charge *m*
lift, hydraulic   ascenseur *m* hydraulique
lift, mechanical [ski lift]   remontée *f* mécanique
lift, (mobile) scissor   table *f* élévatrice (mobile)
lift, service [from a kitchen]   monte-plats *m*
lift bank   batterie *f* d'ascenseurs
lift car   cabine *f* d'ascenseur
lift lobby   vestibule *m* d'ascenseur
lift machine/motor room   local *m* technique des ascenseurs
lift pit   cuvette *f* d'ascenseur
lift shaft or well   cage *f*/gaine *f* d'ascenseur
lift   lever *vb*
lift up   soulever *vb*
lifting   levage *m*, soulèvement *m*
lifting up   relèvement *m*
light [colour]   clair *adj*, vif *adj*
light [weight]   léger *adj*
light   lumière *f*
light [clarity]   clarté *f*
light [quality of]   luminosité *f*
light, artificial   lumière *f* artificielle
light, black   lumière *f* noire
light, bulkhead   hublot *m*
light, ceiling   plafonnier *m*
light, flashing   feu *m* clignotant
light, indicating   lampe *f* de signalisation
light, natural   lumière *f* naturelle
light, pavement   dalle *f*/béton *m* translucide, verre *m* dalle
light, pilot   lumière *f* de contrôle

light, pilot [of boiler, gas stove, etc]   veilleuse *f*
light, street   réverbère *m*
light, vault [glass blocks cast in concrete]   béton *m* translucide
light, warning, *see* warning light
lights, stage   herses *f,pl*
lights, traffic   feux *m,pl* de circulation/(de signalisation)
light density   intensité *f* de la lumière
light distribution curve   courbe *f* d'intensité lumineuse
light fitting, self-powered   bloc *m* autonome lumineux
light fitting/fixture, *see also* fixture   luminaire *m*
light fitting/fixture, hanging   luminaire *m* suspendu, suspension *f*
light sensor   détecteur *m* de lumière
light source   source *f* lumineuse
light well   puits *m* de lumière, trémie *f*
light [illuminate]   éclairer *vb*
light [lamp, fire, etc]   allumer *vb*
light up   illuminer *vb*
lightening of paint colour with loss of gloss   farinage *m*
lighter [igniter]   allumeur *m*
lighthouse   phare *m*
lighting   éclairage *m*, illumination *f*, jour *m*
lighting [of a lamp, fire, etc]   allumage *m*
lighting, ambient   éclairage *m* d'ambiance
lighting, effect   éclairage *m* d'effet
lighting, electric   éclairage *m*/lumière *f* électrique
lighting, emergency   éclairage *m* de sécurité/de secours
lighting, motorway   éclairage *m* autoroutier
lighting, overhead   éclairage *m* vertical
lighting, road   éclairage *m* routier/des routes
lighting, roadworks [to guide traffic]   guirlande *f* lumineuse
lighting, safety or security   éclairage *m* de sécurité
lighting, soft   éclairage *m* tamisé
lighting, stage   éclairage *m* de scène
lighting, street   éclairage *m* urbain
lighting, studio   jour *m* d'atelier
lighting, task   éclairage *m* d'appoint
lighting, workplace   éclairage *m* du lieu de travail
lighting circuit   circuit *m* d'éclairage
lighting column (road)   candélabre *m*
lighting design   conception *f* d'éclairage
lighting engineer   éclairagiste *m/f*
lighting equipment   appareils *m,pl* d'éclairage
lighting fixture, recessed   luminaire *m* encastré
lighting installation   installation *f* d'éclairage
lighting level   niveau *m* d'éclairage
lighting mast   mât *m* d'éclairage
lighting track   glissière *f* des lumières
lightning   éclair *m*
lightning conductor   paratonnerre *m*
lightning protection   parafoudre *m*
lightning protection finial   pointe *f* capteur d'énergie
lightning rod   tige *f* de paratonnerre, paratonnerre *m*
lightning surge diverter   parafoudre *m*
lignite   lignite *m*
lime [for mortar, etc]   chaux *f*
lime [tree]   tilleul *m*

lime, fat chaux *f* grasse
lime, hydrated/slaked chaux *f* éteinte/hydratée
lime, hydraulic chaux *f* hydraulique
lime suspended in water lait *m* de chaux
limestone calcaire *m*, pierre *f* calcaire/à chaux *f*
limestone [very soft] tuffeau *m*
limestone, hard calcaire *m* compact
limestone, hard [frost resistant] liais *m*
limestone, oolitic calcaire *m* oolithique
limit limite *f*
limit [in the sense of bounds] cadre *m*
limit, elastic limite *f* d'élasticité
limit, endurance/fatigue limite *f* d'endurance
limit, fatigue limite *f* de fatigue
limit, lower limite *f* inférieure
limit, offset limite *f* d'écoulement, limite *f* élastique
  apparente
limit, plasticity [soil property] limite *f* de plasticité
limit, proportional limite *f* de proportionnalité
limit, temperature limite *f* de température
limit, upper limite *f* supérieure
limit of pour [concrete work] arrêt *m* de coulage
limit of proportionality limite *f* de proportionnalité
limit state état *m* limite
limit state, serviceability état *m* limite de service (ELS),
  état *m* limite d'utilisation
limit state, ultimate (ULS) état *m* limite ultime (ELU)
limited [e.g. access to a motorway, etc] réglementé *adj*
limiter, current limiteur *m* de courant
limiter, draught limiteur *m* de tirage
limiter, flow limiteur *m* de débit
limiter, pressure limiteur *m* de pression
limiter, voltage limiteur *m* de tension
limiting limite *adj*
limits délimitation *f*
line ligne *f*
line [alignment] alignement *m*
line [axis] axe *m*
line [e.g. of posts, trees] cordon *m*
line [e.g. on paper] raie *f*
line [mark] trait *m*
line [of road, railway, etc] tracé *m*
line [of trees, columns, etc] rang *m*
line [on a graph] courbe *f*
line [row] file *f*, rangée *f*
line, assembly chaîne *f* de fabrication
line, branch [main or pipe] conduite *f* secondaire
line, branch [railway] ligne *f* secondaire
line, bricklayer's or builder's cordeau *m*
line, building alignement *m*
**line, carpenter's, bricklayer's or builder's, for marking
  out circles** simbleau *m*
line, condensate tuyauterie *f* d'eau condensé
line, (continuous/discontinuous) white [road marking]
  ligne *f* blanche (continue/discontinue)
line, contour courbe *f* de niveau
line, datum ligne *f* de référence/repère
line, dimension [on a drawing] trait *m* de cote
line, discontinuous [road marking] ligne *f* discontinue

line, dotted ligne *f* hachurée, pointillé *m*
line, feed [pipe] conduite *f* d'alimentation
line, high tension/voltage power ligne *f* à haute tension
line, high tide/water ligne *f*/limite *f* de la marée
line, influence ligne *f* d'influence
line, main (railway) ligne *f* principale (de chemin de fer)
line, overhead (power) ligne *f* aérienne
line, railway ligne *f* de chemin de fer
line, straight droite *f*, ligne *f* droite
line, suction conduite *f* d'aspiration
line, supply [electrical mains] ligne *f* d'alimentation
line, tree limite *f* des arbres
line, very fine painted filet *m* sec
line, water supply conduite *f* d'eau
line of cut [sawing] trait *m* de scie
line of cut marked on face of a stone dérobement *m*
line of flight [of stairs] ligne *f* de foulée
line of flow ligne *f* d'écoulement
line of force ligne *f* de force
line of intersection of two planes of mansard roof ligne
  f de bris
line of rivets file *f* de rivets
line of sight ligne *f* de mire
**line of stones of equal height along the length of a
  facade** cours *m* d'assises
line painter (roads) traceuse *f* de lignes
line [a borehole] tuber *vb*
line [a borehole, shaft, well] cuveler *vb*
line [a conduit, duct, pipe, etc] chemiser *vb*
line [a structure] blinder *vb*
line [contours or outline of paintwork against another
  colour] rechampir *vb*
line [painting, with a lining brush] traîner *vb*
line [with paper, fabric, etc] tapisser *vb*
line up [a machine] dresser *vb*
line up [at same level] niveler *vb*
line up [in line] aligner *vb*
linear linéaire *adj*
linearly linéairement *adv*
lined chemisé *adj*
lining [covering] recouvrement *m*, revêtement *m*
lining [contours or outline of paintwork against another
  colour] rechampissage *m*
lining [of a boiler] garnissage *m*
lining [of a borehole] tubage *m*
lining [of a conduit, duct, pipe, etc] chemisage *m*
lining [of a flue] tubage *m*
lining [of a wall, to improve strength, insulation, etc]
  doublage *m*
lining [of a well, shaft] cuvelage *m*
lining [of an excavation, trench] blindage *m*
lining [shuttering] coffrage *m*
lining [wall] contre-cloison *f*
lining, anti-friction [bearing] foyer *m* d'antifriction
lining, anti-friction [surface] revêtement *m* antifriction
lining, dry doublage *m* sec
lining, internal recouvrement *m* intérieur
lining, refractory revêtement *m* réfractaire
lining, wall contre-cloison *f*

lining material　matériau *m* de revêtement
lining of a well　blindage *m* de puits
lining up [of a machine, etc]　dressage *m*
link　liaison *f*, raccord *m*
link [of a chain]　maillon *m*
link [reinforcement]　cadre *m*
link [shear reinforcement]　épingle *f*
link [structural]　articulation *f*
link, air/rail/road/sea　liaison *f*
　aérienne/ferroviaire/routière/maritime
link, bridge(d)　liaison *f* en pont
link, closed [reinforcement]　cadre *m* fermé
link, inclined [reinforcement]　cadre *m* incliné
link pin [of a chain]　tourillon *m*
link road　bretelle *f* (de raccordement)
link　accoupler *vb*
link [together]　joindre *vb*
link up　relier *vb*
linkage, valve　mécanisme *m* de soupape
linkage [mechanical]　tringlerie *f*
linking　accouplement *m*
linoleum　linoléum *m*
lintel　linteau *m*
lintel [prefabricated lower part containing main
　reinforcement]　prélinteau *m*
lintel, reinforced concrete　linteau *m* de béton armé
lintel, reinforcing [built into a wall above a weak point]
　linteau *m* de décharge
lintel, secondary, built into wall behind main lintel
　arrière-linteau *m*
lintel beam, spanning between two posts or columns
　linçoir *m*, linceau *m*
lintel beam [single or composite]　poitrail *m*
lintel course [of stone]　plate-bande *f*
lip [projecting irregularity in masonry or concrete]
　balèvre *f*
liquefaction　liquéfaction *f*
liquid *adj*　liquide *adj*
liquid　liquide *f*
liquid, cooling　liquide *f* de refroidissement
liquid containing fine particles in suspension
　dispersion *f*, suspension *f*
liquid limit [soil property]　limite *f* de liquidité
liquidation [of a debt]　amortissement *m*
liquidity　liquidité *f*
liquidity index　indice *m* de liquidité
list　liste *f*, nomenclature *f*
list, price　bordereau *m*/liste *f* de prix
list of charges　tarif *m*
list of contents　sommaire *m*, table *f* de matières
listed [as of architectural or historic interest]　classé *adj*
litigation　litige *m*
litre　litre *m*
load, *see also* loading
load [electrical, structural, etc]　charge *f*
load [externally applied]　sollicitation *f*
load, additional　surcharge *f*
load, allowable　charge *f* admissible/de service
load, alternating　charge *f* alternée

load, application of a　application *f* d'une force
load, assumed　charge *f* supposée
load, axial　charge *f* axiale, effort *m* normal
load, balanced　charge *f* équilibrée
load, base or minimum　charge *f* minimum
load, bearing　capacité *f* de charge
load, breaking　charge *f* de rupture
load, buckling　charge *f* de flambement, charge *f*
　d'affaissement
load, collapse　charge *f* d'affaissement
load, compound　charge *f* composée
load, concentrated　charge *f* concentrée
load, connected　charge *f* connectée/de connexion
load, cooling　charge *f* frigorifique/de refroidissement
load, dead　charge *f* constante/morte/permanente
load, dead [self weight]　poids *m* mort/propre
load, design　charge *f* prévisionnelle/de calcul
load, distributed (DL)　charge *f* répartie
load, dynamic　charge *f* dynamique
load, eccentric　charge *f* décentrée/excentrée/excentrique
load, electrical　charge *f* électrique
load, Euler buckling　charge *f* critique d'Euler
load, exceptional　charge *f* exceptionnelle
load, excess or extra　surcharge *f*
load, fire　charge *f* calorifique/d'incendie
load, floor　charge *f* au sol/sur le plancher
load, full　pleine charge *f*
load, full [of a lorry]　plein *m* de camion
load, heat　charge *f* calorifique/thermique
load, imposed　charge *f* variable
load, line　charge *f* linéaire
load, live　charge *f* utile/variable/vive/d'utilisation,
　surcharge *f* mobile
load, maximum　charge *f* limite/maximale
load, maximum [on a road or bridge]　tonnage *m*
　maximum
load, military　charge *f* militaire
load, minimum　charge *f* minimum
load, moving　(sur)charge *f* mobile
load, operating　charge *f* d'exploitation/de service
load, peak　charge *f* maximale/de pointe
load, permanent　charge *f* permanente
load, pile bearing　charge *f* d'un pieu
load, point　charge *f* concentrée/ponctuelle
load, pulsating　charge *f* ondulée
load, safe　charge *f* admissible/de service
load, service　charge *f* d'exploitation/de service
load, snow　surcharge *f* de neige
load, static　charge *f* statique
load, steady　charge *f* constante/permanente
load, sustained　charge *f* de longue durée
load, test or trial　charge *f* d'épreuve/d'essai
load, total　charge *f* totale
load, ultimate　charge *f* limite/à la rupture
load, uniformly distributed (UDL)　charge *f*
　uniformément répartie
load, unit　charge *f* unitaire
load, variable　charge *f* variable, surcharge *f*
load, wind　charge *f*/surcharge *f* de vent

load, working   charge *f* d'exploitation/de service
load at failure   charge de rupture *f*
load case   cas *m* de charge
load case, design   combinaison *f* de cas de charges
load case, first/etc   premier/etc cas *m* de charge
load cell   cellule *f* de contrainte
load distribution   répartition *f* de la charge
load distribution, triangular   répartition *f* triangulaire de charge
load factor   coefficient *m* charge/d'utilisation
load fluctuation   fluctuation *f* de charge
load levelling   nivelage *m* de charge
load per unit area   charge *f* par unité de surface
load platform consisting of thick boards   chantier *m*
load reversal   alternance *f*/renversement *m* de charge
load shedding   délestage *m*, réduction *f* de charge/débit
load table [electrical]   tableau *m* de puissances
load table [structural]   tableau *m* des charges
load test   essai *m* en charge
load variation   variation *f* de charge/débit
load   charger *vb*
load, carry or support a   supporter *vb* une charge
load, take up a   reprendre *vb* une charge
load and transport materials   barder *vb*
loadbearing   porteur *adj*
load-carrying   porteur *adj*
loader [machine]   pelle *f* chargeuse
loader [machine or person]   chargeur *m*
loader, backhoe   chargeuse-pelleteuse *f*, tracto-pelle *f*
loader, belt   chargeuse *f* à bande
loader, bucket   chargeuse *f* à godets
loader, crawler   chargeuse *f* à/sur chenilles
loader, elevating   élévateur-chargeur *m*
loader, front-end   chargeuse *f*
loader, front-end, which can discharge at rear   rétro-chargeuse *f*
loader, low   remorque *f* surbaissée
loader, skid-steer   chargeuse *f* à direction par ripage
loader, wheeled   pelle *f* chargeuse
loading, *see also* load, etc   chargement *m*
loading, constant rate of   augmentation *f* constante de la charge
loading, floor   charge *f* au sol/sur le plancher
loading, grate   charge *f* de grille
loading, wind   surcharge *f* de vent
loading case, *see* load case
loading reduction rule for multi-storey buildings   règle *f* de dégressivité *f* des charges
load-settlement curve   courbe *f* charge-tassement, courbe *f* de tassement/de chargement
loam   terreau *m*
loan   avance *f*, crédit *m*, prêt *m*
loan [borrowed]   emprunt *m*
loan, bank   prêt *m* bancaire
loan, mortgage   prêt *m* hypothécaire
loan, secured   prêt *m* garanti/sur gage
loan at notice   prêt *m* à terme
loan, redeem a   amortir *vb* un emprunt
loan, repay the   rembourser *vb* l'emprunt

lobby   dégagement *m*, sas *m*, vestibule *m*
local   local *adj*
local area network (LAN)   réseau *m* local
locality   endroit *m*, lieu *m*
locate   situer *vb*
locate a fault [electrical], **try to**   rechercher *vb* un dérangement
located   sis *adj*
location   emplacement *f*, situation *f*
location of faults [electrical]   recherche *f* de dérangements
location or site plan   plan *m* d'ensemble/de repérage/de masse/de situation
lock [door, gate, etc]   serrure *f*
lock [e.g. on canal]   écluse *f*
lock, Bramah   serrure *f* batteuse
lock, Bramah safety   serrure *f* de sûreté à pompe
lock, burglar-proof   serrure *f* incrochetable
lock, combination   serrure *f* à combinaisons
lock, cylinder   barillet *m*, serrure *f* à pêne dormant/à barillet
lock, dead   serrure *f* à pêne dormant
lock, double-sided   serrure *f* bénarde
lock, drawback   serrure *f* camarde
lock, drawback, with a bevelled bolt   serrure *f* à bec-de-cane
lock, horizontal [i.e. long edge horizontal]   serrure *f* en longue
lock, mortise   serrure *f* lardée/à mortaiser/encastrée/mortaisée
lock, rim   serrure *f* encloisonnée/en applique/à palastre/à palâtre
lock, safety or security   serrure *f* de sûreté
lock, side through which the bolt passes   têtière *f*
lock, strike plate of a   gâche *f*
lock, spring   serrure *f* à ressort
lock, time   serrure *f* chronométrique/temporisée
lock, tumbler safety   serrure *f* de sûreté à gorges
lock, vertical [i.e. long edge vertical]   serrure *f* en large
lock box or case   coffre *m* de serrure
lock chamber [canal]   sas *m* d'écluse
lock closed with one-and-a-half turns   serrure *f* à tour et demi
lock gate   vanne *f*
lock housing   corps *m* de serrure
lock mortised into the face of a door   serrure *f* entaillée
lock plate   palastre *m*/palâtre *m* (de serrure)
lock rail [of a door]   traverse *f* de serrure
lock wing wall   musoir *m* d'une écluse
lock with a bevelled bolt   serrure *f* à pêne demi-tour
lock   fermer *vb*
lock [wheels, a piece of machinery]   bloquer *vb*
lock up [capital]   immobiliser *vb*
lock up [premises]   fermer *vb* à clef
lockable [with a padlock]   cadenassable *adj*
locked [wheel, machinery, etc]   bloqué *adj*
locker rooms   vestiaires *m,pl*
locking [in position]   blocage *m*
locking [of a door, window, etc]   condamnation *f*

locking accessory   accessoire *m* de fermeture
locking device   dispositif *m* de verrouillage
locking device [insurance terminology]   pointe *f* de
  condamnation
lock-nut   contre-écrou *m*, écrou *m* indesserable/de
  blocage
locksmith   serrurier *m*
locksmithing   serrurerie *f*
lodging   gîte *m*, logement *m*
loess   loess *m*
loft   comble *m*, grenier *m*
loft, rigging [above a stage]   cintre *m*
loft with a ceiling   grenier *m* plafonné
log [large, suitable for sawmilling]   grume *f*
log sawn through-and-through   plot *m*
loggia   loge *f*, loggia *m*
logic   logique *f*
logical   logique *adj*
long   long *adj*
long term [loan, etc]   à long terme
long term, in the   à long terme
longer, grow   s'allonger *vb*
longitude   longitude *f*
longitudinal   longitudinal *adj*
look after [i.e. deal with]   s'occuper de *vb*
loop   boucle *f*
loop, balanced   boucle *f* équilibrée
loop, compensating   boucle *f* de dilatation
loop, control   boucle *f* d'asservissement/de contrôle,
  circuit *m* de régulation
loop, expansion   lyre *f* de dilatation
loop, hysteresis   boucle *f* d'hystérésis
loophole [in a battlement]   meurtrière *f*
loose   desserré *adj*, lâche *adj*
loose [uncompacted, of ground]   meuble *adj*
loose tongue [metal, forced into opposing grooves in
  wood]   languette *f* rapportée
loose, work   prendre *vb* du jeu
loosen   desserrer *vb*, détendre *vb*, dévisser *vb*, lâcher *vb*
loosen [a structure]   étonner *vb*
loosen [sth fixed into masonry, etc]   desceller *vb*
loosen [sth jammed]   décoincer *vb*
looseness   jeu *m*
loosening   desserrage *m*
loosening [of sth fixed into masonry, etc]   descellement *m*
lop [a peak]   écrêter *vb*
lopping [of trees]   élagage *m*
lopping, peak   écrêtage *m*/écrêtement *m* des pointes
lorry   camion *m*, fourgon *m*
lorry, fuel tank(er)   tonne *f* à carburant
lorry, heavy   poids *m* lourd
lorry, tanker   camion-citerne *m*, citerne *f* de transport
lorry, three-way tipper   camion *m* tribenne
lorries, fleet of   parc *m* de camions
lorry depot   gare *f* routière
lorry load   plein *m* de camion
lorry loader, *see* crane, lorry-mounted
lose temper [steel]   détremper *vb*
loss   perte *f*

loss [legal]   dommage *m*
loss [of energy, heat, etc]   déperdition *f*
loss [under insurance policy]   sinistre *m*
loss, chimney or flue gas   perte *f* à la cheminée
loss, conduction   perte *f* par conduction
loss, cooling   perte *f* de froid
loss, earth leakage (current)   perte *f* de courant à la terre
loss, energy   perte *f* d'énergie
loss, entry   perte *f* de charge à l'entrée
loss, evaporation   perte *f* par évaporation
loss, friction(al)   perte *f* par frottement
loss, head   perte *f* de pression
loss, heat   déperdition *f* calorifique/thermique, perte *f* de
  chaleur
loss, percentage   pourcentage *m* de pertes
loss, pipe friction   perte *f* de charge dans un tube
loss, power [electrical]   panne *f* de courant
loss, pressure   perte *f* de charge
loss, radiation   perte *f* par radiation
loss, speed   perte *f* de vitesse
loss, volumetric   perte *f* de volume
loss, water   perte *f* d'eau
loss, weight   perte *f* en poids
loss by conduction/evaporation/radiation   perte *f* par
  conduction/évaporation/radiation
loss of adhesion [of mortar or rendering]   décollement *m*
loss on disposal   moins-value *f* de cession
loss(es), total   pertes *f,pl* totales
lot [at auction]   lot *m*
lot [bid package]   tranche *f* de travaux
lot [plot of land]   lot *m* de terrain, parcelle *f*
loudness   intensité *f* sonore
louvre   volet *m* d'aération
louvre, adjustable   volet *m* d'aération réglable
louvre, ventilation   ouïe *f* d'aération
louvres, adjustable [of a window, blind, etc]   jalousie *f*
low   bas *adj*
low [as in low density, etc]   faible *adj*
lower   inférieur *adj*
lower chord/flange, *see* flange, lower
lower part   pied *m*, partie *f* inférieure
lower [water level, prices, etc]   baisser *vb*
lower on a rope   affaler *vb*
lower the level of [a wall]   déraser *vb*
lowering   abaissement *m*
lowering [of a wall, stonework, etc, to required level]
  dérasement *m*
lowering [of price]   diminution *f*
lowest [tender]   moins-disant *adj*
low-tension, *see* low-voltage
low-voltage   basse tension *f*
low-voltage cabinet   armoire *f* basse tension
low-voltage distribution board or panel   tableau *m*
  général basse tension (TGBT)
lubricant   lubrifiant *m*
lubricate   graisser *vb*, lubrifier *vb*
lubricating   lubrifiant *adj*
lubricating cup   godet *m* graisseur
lubrication   lubrification *f*

lubrication, forced   lubrification *f* forcée
lubrication chart   tableau *m* de graissage
lubrication system   système *m* de lubrification
lubricator   graisseur *m*
luffer board   abat-son *m*
luffing jib, crane with   grue *f* à flèche relevable
lug   ergot *m*
lug, fixing [ear]   oreille *f*
luggage rack [on vehicle]   porte-bagages *m*
lukewarm   tiède *adj*
lumen   lumen *m*
lumen depreciation   chute *f* de flux lumineux
lumen maintenance factor   facteur *m* de dépréciation

luminaire   luminaire *m*
luminance   luminance *f*
luminescence   luminescence *f*
luminescent   luminescent *adj*
luminosity   luminosité *f*
luminous   lumineux *adj*
luminous efficacy   efficacité *f* lumineuse
lump [bump]   bosse *f*
lump [in lime]   pigeon *m*
lump sum   forfait *m*, somme *f* forfaitaire
lump together   bloquer *vb*
lux   lux *m*

macadam   macadam *m*
macadam, bitumen grouted or bitumen penetration   macadam *m* traité en pénétration
macadam, coated/dry-bound   macadam *m* enrobé/vibré
macadam, grouted   macadam *m* en pénétration
macadam, rolled   macadam *m* cylindré
macadam, waterbound   macadam *m* à l'eau
macadamize   macadamiser *vb*
macerator   macérateur *m*
macerator [plumbing]   broyeur *m*
machine   machine *f*
machine, arc welding   machine *f* pour soudage à l'arc
machine, bending   machine *f* à cintrer
machine, boring   foreuse *f*
machine, calculating   machine *f* à calculer
machine, chain morticing   fraiseuse *f* à chaîne, mortaiseuse *f* à chaîne
machine, cleaning   nettoyeuse *f*
machine, cold pavement milling   raboteuse *f* (routière) à froid
machine, compacting   engin *m* de compactage
machine, dishwashing   lave-vaisselle *m*, machine *f* à laver la vaisselle
machine, drilling   foreuse *f*
machine, facsimile   télécopieur *m*
machine, finishing [road construction]   surfaceuse *f*
machine, gas cutting   machine *f* d'oxycoupage
machine, knot-hole moulding   perceuse *f* à dénoder
machine, milling   fraiseuse *f*
machine, morticing   machine *f* à mortaiser
machine, office   machine *f* de bureau
machine, pipe laying   pose-tubes *m*
machine, pipe threading   machine *f* à fileter des tubes
machine, planing   machine *f* à raboter, raboteuse *f*
machine, plate shearing   cisaille *f* à tôle
machine, (pneumatic) pipe jacking   pousse-tube *m* (pneumatique)
machine, punching   poinçonneuse *f*

machine, refrigerating   machine *f* frigorifique/à froid
machine, riveting   riveteuse *f*
machine, rotary   machine *f* tournante
machine, self-propelled   engin *m* automoteur
machine, sharpening   affûteuse *f*
machine, shearing   cisaille *f*, découpeuse *f*
machine, slipform paving   machine *f*/surfaceuse *f*/vibreuse-surfaceuse *f* à coffrage glissant
machine, soil stabilizing   engin *m* stabilisateur/de traitement des sols
machine, spindle moulding   toupie *f*
machine, tunnel boring   taupe *f*, tunnelier *m*
machine, tunnelling   tunnelier *m*
machine, universal woodworking   machine *f* à bois polyvalente
machine, veneer slicing   machine *f* à trancher le bois
machine, washing   lave-linge *m*, machine *f* à laver
machine, welding   machine *f* à souder
machine, woodworking   machine *f* à bois, machine-outil *f* à (travailler le) bois
machine guard   protection *f* de machine
machine part   élément *m* de machine
machine plant   engin *m*
machine shop   atelier *m* de constructions mécaniques
machine tool   machine-outil *f*
machine   usiner *vb*
machining   travail *m*, usinage *m*
machining allowance   tolérance *f* d'usinage
machinist's hammer   marteau *m* rivoir
made ground   rapport *m* de terre, remblai *m*
made to order   fait sur commande
magnesia   magnésie *f*
magnesium   magnésium *m*, ciment *m* magnésien
magnet   aimant *m*
magnitude   ampleur *f*
mahogany   acajou *m*
main   principal *adj*
main, buried   réseau *m* enterré

main, condenser conduite *f* de condenseur
main, delivery [from a pump] conduite *f* de refoulement
main, distribution conduite *f* principale de distribution
main, district heating conduite *f* principale de chauffage à distance
main, dry falling colonne *f* descendante
main, gas conduite *f* principale du gaz
main, ring réseau *m* primaire en boucle
main, rising conduite *f* montante
main, rising [from a borehole] colonne *f*/conduite *f* de refoulement
main, steam conduite *f* de vapeur
main, trunk conduite *f* principale
main, water conduite *f* principale d'eau
mains [pipework] conduite *f*
mains, electrical canalisation *f* électrique
mains, feeder câble *m* principale
mains, town canalisation *f* de ville
mainframe [computer] unité *f* centrale
maintain entretenir *vb*, maintenir *vb*
maintained entretenu *adj*
maintenance entretien *m*, maintenance *f*, maintien *m*
maintenance, computer-assisted planned maintenance *f* assistée par ordinateur (MAO)
maintenance, corrective/preventive maintenance *f* corrective/préventive
maintenance area centre *m* d'entretien
maintenance area, road centre *m* d'entretien routier
maintenance factor, lumen facteur *m* de dépréciation
maintenance instructions instructions *f,pl* d'entretien
maintenance service service *m* d'entretien
maintenance work of a minor nature corvée *f*
maisonette appartement *m* (en) duplex
major majeur *adj*
make [brand] marque *f*
make [of an electrical switch] enclenchement *m*
make produire *vb*
make [in a calculation] faire *vb*
make a hole in percer *vb*
make an opening [for a window, etc] percer *vb*
make clear préciser *vb*
make enquiries demander *vb* renseignements
make fast [tie] amarrer *vb*
make flush affleurer *vb*
make good [a minor defect, a difference of level] racheter *vb*
make good [around an element fixed in masonry, etc] calfeutrer *vb*
make improvements apporter *vb* des améliorations
make late retarder *vb*
make level araser *vb*
make out an invoice établir *vb* une facture
make over [a lease] céder *vb*
make notes prendre *vb* des notes
make repairs faire *vb* des réparations
make rusty rouiller *vb*
make unfit for its purpose rendre *vb* impropre à sa destination
make use of utiliser *vb*

make-and-break device [electrical] autorupteur *m*
make-and-break key brise-circuit *m*
maker constructeur *m*, fabricant *m*
maker's nameplate plaque *f* du constructeur
making [a window, i.e. piercing for] percement *m*
making flush affleurage *m*
making good, final [of surfaces] ragréage *m*, ragrément *m*
making over [of property, etc] transmission *f*
male [connection, thread, etc] mâle *adj*
mall [avenue] mail *m*
malleability malléabilité *f*
malleable malléable *adj*
mallet, wood maillet *m* (à tête rectangulaire)
manage diriger *vb*, gérer *vb*
management conduite *f*, direction *f*, gestion *f*
management, building gestion *f* d'immeuble
management, energy gestion *f* de l'énergie
management, project direction *f* de projet, gestion *f* du projet
management, property gestion *f* du patrimoine
management company compagnie *f*
management contractor, responsible for co-ordination of all the works entreprise *f* pilote
management of the works [by the *maître d'œuvre*] maîtrise *f* d'œuvre
management staff [member of] cadre *m*
manager directeur *m*, gérant *m*
manager, construction maître *m* de chantier
manager, project chef *m*/gestionnaire *m* de projet, maître *m* d'œuvre, maître *m* de travaux, projeteur *m*
manager, site chef *m*/maître de chantier
manager, technical directeur *m* technique
manager, works chef *m* d'exploitation
manageress directrice *f*
mandate mandat *m*
mandatory obligatoire *adj*
manganese manganèse *m*
manhole chambre *f* de visite, cheminée *f*/puits *m* d'accès, trou *m* d'homme/de visite
manhole [for drain or sewer] regard *m*
manhole [small], **inspection** regard *m* de visite
manhole, cable puits *m* à câbles
manhole, with a à regard
manhole cover couvercle *m* de trou d'homme, plaque *f*/tampon *m* d'égout
manifold collecteur *m*
manifold, exhaust collecteur *m* d'échappement
manifold, supply collecteur *m* de distribution
manoeuvre manœuvre *m*
manoeuvre manœuvrer *vb*
manometer manomètre *m*
manometer, differential manomètre *m* différentiel
manometer, inclined tube manomètre *m* (à tube) incliné
manometric manométrique *adj*
manor château *m*, manoir *m*
manpower effectifs *m,pl*, main-d'œuvre *f*
mansion, private hôtel *m* particulier

mantelpiece or mantelshelf   manteau *m* (de cheminée), tablette *f* (de cheminée)
manual *adj*   à la main, manuel *adj*
manual   manuel *m*
manual, operating   manuel *m* de service/d'exploitation
manual override   déverrouillage *m* manuel
manually   manuellement *adv*
manufacture   fabrication *f*, fabrique *f*, usinage *m*
manufacture   fabriquer *vb*, usiner *vb*
manufactured   fabriqué *adj*
manufacturer   fabricant *m*
map   carte *f*
map, contour   carte *f* avec courbes de niveau
map, grid   carte *f* quadrillée
map, relief   carte *f* en relief
map, skeleton   canevas *m*
map, thermal   carte *f* thermique
maple   érable *m*
mapping agency of France [Ordnance Survey in the UK]   Service *m* Cartographique de l'État
marble   marbre *m*
marble, breccia   marbre *m* brèche
marble, Carrara   carrare *m*
marble of a single colour   marbre *m* simple
marble of several different colours   marbre *m* composé
marble [paintwork]   brécher *vb*
marbling [paintwork]   bréchage *m*, faux-marbre *m*
margin   marge *f*
margin [amount]   écart *m*
margin, gross   marge *f* brute
margin, profit   marge *f* bénéficiaire
margin, safety   marge *f* de sécurité
maritime engineering   génie *m* maritime
mark, *see also* marker   repère *m*
mark [line]   trait *m*
mark, high tide/water   ligne *f*/limite *f* de la marée
mark   marquer *vb*, pointer *vb*
mark a level line along a wall   cingler *vb*
mark off or out   tracer *vb*
mark out   repérer *vb*
mark out [fixing position by outlining]   piquer *vb*
mark out [land boundaries]   borner *vb*
mark out [with pegs or stakes]   bornoyer *vb*
mark out [with survey poles]   baliser *vb*
mark with a chalk line   ligner *vb*
mark with kilometre stones   kilométrer *vb*
marked   marqué *adj*
marker   marqueur *m*, repère *m*
marker, boundary or height   témoin *m*
marker, boundary [may be stone, plastic, etc]   borne *f*
marker, distance   borne *f*
marker, reference level   repère *m* de nivellement
marker stake/pole [civil works]   balise *f*
market   marché *m*
market, covered   halle *f*
market prices/rates   cours *m* du marché
market study   étude *f* de marché
marking   repérage *m*
marking [e.g. white lining of roads]   marquage *m*

marking [of line of cut with a chalk line]   lignage *m*
marking, edge   marquage *m* latéral
marking, embedded [on roads]   marquage *m* scellé
marking, painted [on roads]   marquage *m* peint
marking, road   marquage *m* routier
marking off or out   traçage *m*
marking out [boundaries of land]   bornage *m*
marking out [layout of buildings, roads, etc]   implantation *f*
marking system for cables/wiring   système *m* de repérage pour filerie
marking with kilometre stones   kilométrage *m*
marl   marne *f*
marly   marneux *adj*
marquetry   marqueterie *f*
marsh   marais *m*, marécage *m*
marshalling yard   chantier *m*/gare *f* de triage
marshalling yard signal box   poste *m* de butte
mask [shield]   masque *m*
mask [used in painting]   cache *f*
mask, arc-welding   masque *m* de soudage à l'arc
mask, breathing   masque *m* respiratoire
mask, welder's   masque *m* de soudeur
masking [for painting]   masquage *m*
mason   maçon *m*
mason building rough/rubble masonry   limousin(ant) *m*
mason, stone   tailleur *m* de pierre
mason's chisel, narrow   ognette *f*
mason's hammer   masse *f*, têtu *m*
mason's hammer, heavy, with hatched faces   chien *m*
mason's labourer   gâcheur *m*, porte-auge *m*
mason's labourer [French regional expression]   garçon-maçon *m*
mason's tool with lines of teeth [for surfacing stone by scraping]   chemin *m* de fer
mason's toothed chisel   gradine *f*
mason's toothed hammer [for preliminary dressing of soft stone]   laye *f*
mason's two-pointed (scabbling) hammer   smille *f*
mason's wedge [for splitting stone]   tranche *f*
masonry *adj*   maçonné *adj*
masonry [also plastering and rendering]   maçonnerie *f*
masonry [bricks and mortar]   bâtisse *f*
masonry [the work]   maçonnage *m*
masonry, non-structural, and other minor works   ouvrages *m,pl* légers/menus
masonry, projecting irregularity in   balèvre *f*
masonry, rough or rubble   limousinage *m*, limousinerie *f*
masonry bay or panel   pan *m* de maçonnerie
masonry consisting of two or more materials [e.g. brick and stone]   maçonnerie *f* composite
masonry partition   cloison *f* maçonnée
masonry slab with waterproof surface [lab bench, kitchen slab, etc]   paillasse *f*
masonry wall   mur *m* en maçonnerie
mass   masse *f*
mass [of a structure, etc]   massif *m*
mass, centre of   centre *m* de gravité

massing [of buildings, etc] masse *f*
massive massif *adj*
mass-produced fabriqué *adj* en série
mast mât *m*, poteau *m*, pylône *m*
mast, guyed/stayed mât *m* haubané
mast, lighting mât *m* d'éclairage
mast, trellis mât *m* en treillis
master key passe *f*, passe-partout *m*
master plan plan *m* directeur
master maîtriser *vb*
mastic mastic *m*
mastic, boiler mastic *m* de chaudière
mastic, resin based mastic *m* acrylique
mastic, sealing mastic *m* d'étanchéité
mastic gun pistolet *m* à mastiquer
mastic paste [joint sealant] mastic *m* pâteux
mat [blinding] béton *m* de propreté
matchboard planche *f* profilée
matchboarding lambris *m*
material matériau *m*, matière *f*
material, absorbent or absorption matière *f* absorbante
material, binding [road or groundworks] agglomérant *m*
material, borrow matériau *m* d'emprunt
material, cohesive matériau *m* cohérent
material, combustible matière *f* combustible
material, excavated déblais *m,pl*, matériau *m* de déblai
material, fill matériau *m* de remplissage/de remblai
material, filter matière *f* de filtrage
material, impermeable matériau *m* imperméable
material, inflammable matière *f* inflammable
material, insulating matière *f* isolante
material, lining matériau *m* de revêtement
material, loose matériau *m* meuble
material, non-conducting non-conducteur *m*
material, raw matière *f* première
material, riddled or sieved criblures *f,pl*
material, soluble matière *f* soluble
material, toxic substance *f* toxique
material damage insurance [compulsory for building works in France] dommage ouvrage *m* (DO)
materials [supply of, as opposed to labour] fourniture *f*
materials, salvaged matériaux *m,pl* de récupération
matrix matrice *f*
matrix, diagonal/square matrice *f* diagonale/carrée
matt mat *adj*
matt [make dull] amatir *vb*
matter [substance] matière *f*
matter, cohesive éléments *m* cohérents
matter, colouring matière *f* colorante
matter, dissolved matières *f,pl* en solution
matter, volatile matière *f* volatile
matting [of gilding] matage *m*
mattock pioche *f*
mattress (blinding) béton *m* de propreté
mattress, fascine fascine *f*, tapis *m* de branchage
maturity échéance *f*
maximize porter *vb* au maximum
maximum *adj* limite *adj*, maximal *adj*, maximum *adj*
maximum maximum *m*

maximum allowable/permissible maximum *m* admissible/permissible
maximum, reach a plafonner *vb*
mayor maire *m*
meadow prairie *f*, pré *m*
mean *adj* moyen *adj*
mean [average] moyenne *f*
mean, arithmetic(al)/geometric(al) moyenne *f* arithmétique/géométrique
mean, numerical moyenne *f* numérique
means [of doing sth] moyen *m*
means of applying pressure moyen *m* de pression
means of transport moyen *m* de transport
meander [of a watercourse] méandre *m*
meaning sens *m*
measure [unit of, or dimension] mesure *f*
measure [for liquids, powders] doseur *m*
measure [rod of unit length] pige *f*
measure, folding mètre *m* pliant
measure, tape mesure *f*/mètre *m* à ruban
measures [arrangements] dispositions *f,pl*
measures [steps] mesures *f,pl*
measures, protective mesures *f,pl* de protection
measure [gauge] jauger *vb*
measure [land] arpenter *vb*
measure [length, and in quantity surveying] métrer *vb*
measure off in kilometres kilométrer *vb*
measure out [a dosage] doser *vb*
measure (sth) mesurer *vb* (qch), prendre *vb* les dimensions/le mesure (de qch)
measurement [dimension] dimension *f*, mesure *f*
measurement [of length, in metres] métrage *m*
measurement [of quantities and work done] métré *m*
measurement [the process] mesurage *m*, mensuration *f*
measurement [with compasses/dividers] compassement *m*
measurement, consumption mesure *f* de la consommation
measurement, flow mesure *f* de débit
measurement, moisture mesure *f* d'humidité
measurement, noise mesure *f* du bruit
measurement, speed mesure *f* de la vitesse
measurement, unit of unité *f* de mesure
measurement between interior faces mesure *f* dans œuvre (D/O)
measurement in kilometres kilométrage *m*
measurement of air quantity mesure *f* de la quantité d'air
measurement of building, preliminary [before works] avant-métré *m*
measurement of discharge mesure *f* de débit/décharge
measurement of efficiency mesure *f* du rendement
measurement of hardness/illumination/pressure/etc mesure *f* de la dureté/lumière/pression/etc
measurement of quantities métré *m* quantitatif
measurement of settlement mesure *f* de tassement
measuring mesurage *m*
measuring [of length, in metres] métrage *m*
measuring [with compasses/dividers] compassement *m*

measuring device/equipment appareil *m*/dispositif *m* de mesure
measuring point point *m* de mesure
measuring range champ *m* de mesure
measuring rod pige *f*
measuring tape, 10 metre décamètre *m*
measuring technique technique *f* de mesure
mechanical mécanique *adj*
mechanically controlled à commande mécanique
mechanics, fluid mécanique *m* des fluides
mechanics, rock mécanique *m* des roches
mechanism dispositif *m*, mécanisme *m*
mechanism, alarm mécanisme *m* d'alarme
mechanism, control mécanisme *m* de contrôle
mechanize mécaniser *vb*, motoriser *vb*
medium *adj* moyen *adj*
medium hard mi-dur *adj*
medium agent *m*
medium [e.g. air] milieu *m*
medium [instrument] organe *m*
medium [of paint] liant *m*
medium, absorbent/absorption matière *f* absorbante
medium, combustion comburant *m*
medium, dampening agent *m* humectant
medium, diluting substance *f* diluante
medium, filter matière *f* filtrante
medium, refrigerating agent *m* de refroidissement, réfrigérant *m*
meeting [individuals] rencontre *f*
meeting [several people] réunion *f*
meeting notes procès-verbal *m* (PV)
meeting room salle *f* de réunion(s)/de séances
melt [of metal] coulée *f*
melt (down, away) fondre *vb*
melt the ice on déglacer *vb*
melting *adj* fondant *adj*
melting [of metals, etc] fusion *f*
melting, latent heat of chaleur *f* latente de fusion
melting or fading of one colour into another fondu *m*
member barre *f*, élément *m*, membre *m*
member, compressed/compression barre *f*/pièce *f* comprimée, élément *m* comprimé
member, lattice or truss barre *f* de treillis
member, longitudinal longeron *m*
member, prefabricated élément *m* préfabriqué
member, small vertical, in a roof truss jambette *f*
member, structural élément *m* de construction/de structure
member, tension barre *f*/pièce *f* tendue, élément *m* tendu, tirant *m*
membrane membrane *f*, peau *f*
membrane, damp-proof chape *f* d'étanchéité
membrane, waterproof, *see* waterproof membrane
memo note *f* de service
memory [computer] mémoire *f*
memory, magnetic tape [computer] mémoire *f* à bande magnétique
memory, random access (RAM) mémoire *f* vive
memory, read only (ROM) mémoire *f* morte

men and materials, resources of ressources *f,pl* en hommes et en matériel
mend réparer *vb*
merchant, builder's commerçant *m* en matériaux de construction
merchant, scrap metal ferrailleur *m*
merchant, timber marchand *m* de bois
mercury mercure *m*
mere étang *m*
merged fusionné *adj*
merger [companies, etc] fusion *f*
Merulius lacrymans [dry rot fungus] mérule *m*
mesh [i.e. individual opening in a mesh] maille *f*
mesh, welded wire fil *m* de treillis soudé
mesh, wire, *see* wire mesh
metal *adj* métallique *adj*
metal métal *m*
metal, base métal *m* de base
metal, expanded métal *m* déployé
metal, ferrous métal *m* ferreux
metal, light alliage *m*/métal *m* léger
metal, non-ferrous métal *m* non ferreux
metal, parent métal *m* de base
metal, soft métal *m* tendre
metal, thin, straight strip or sheet of feuillard *m*
metal, weld métal *m* déposé
metal [a road] caillouter *vb*
metalled [road] empierré *adj*
metallic métallique *adj*
metallization métallisation *f*
metallography métallographie *f*
metallurgy métallurgie *f*
metalwork serrurerie *f*
metamorphic [of rock] métamorphique *adj*
metamorphism [of rock] métamorphisme *m*
meteorology météorologie *f*
meter [electricity, water] compteur *m*
meter, consolidation œdomètre *m*
meter, current ampèremètre *m*
meter, current [paddle type] moulinet *m*
meter, domestic water compteur *m* d'eau domestique
meter, earth resistance tellurohmètre *m*
meter, electricity compteur *m* électrique/d'électricité
meter, flow débitmètre *m*
meter, gas compteur *m* de gaz
meter, heat compteur *m* d'énergie thermique/de calories/de chaleur
meter, hourly compteur *m* d'heures
meter, local compteur *m* divisionnaire
meter, main compteur *m* général
meter, oil compteur *m* d'huile
meter, principal compteur *m* général
meter, sound level sonomètre *m*
meter, universal test multimètre *m*, voltampèremètre *m*
meter, vane water compteur *m* d'eau à turbine à aube, hydromètre *m* à ailettes
meter, Venturi compteur *m* (à) Venturi
meter, water compteur *m* d'eau
meter cupboard coffret *m* de compteur

meter pit   fosse *f* à compteur
meter room   local *m* comptage
methane   méthane *m*
method   méthode *f*, mode *m*
method, balancing   méthode *f* d'équilibrage
method, calculation   méthode *f* de calcul
method, cleaning   méthode *f* de nettoyage
method, construction   méthode *f* de construction
method, cooling   méthode *f* de refroidissement
method, defrost(ing)   méthode *f* de dégivrage
method, discharging   méthode *f* de déchargement
method, finite element   méthode *f* des éléments finis
method, firing [of a boiler]   méthode *f* de chargement
method, graphical   méthode *f* graphique
method, heating   méthode *f* de chauffe
method, loading   méthode *f* de chargement
method, manufacturing   procédé *m* de fabrication
method, recording   méthode *f* d'enregistrement
method, standard   procédé *m* normalisé
method, static regain [for duct sizing]   méthode *f* de regain statique
method, test(ing)   méthode *f* d'essai
method, unloading   méthode *f* de déchargement
method, velocity reduction   méthode *f* de regain statique
method, work(ing)   méthode *f* de travail
method of operation   mode *m* de fonctionnement
method of payment   mode *m* de paiement
methodology   méthodologie *f*
metre   mètre *m*
metre, cubic   mètre *m* cube
metre, linear   mètre *m* linéaire
metre, square   mètre *m* carré
metric   métrique *adj*
metric tonne   tonne *f*
mezzanine (floor)   entresol *m*, mezzanine *f*
mica   mica *m*
microbore heating   chauffage *m* par tuyauteries de très faibles diamètres
microclimate   microclimat *m*
microcrack   microfissure *f*
micro-filter   filtre *m* ultra-fin
microfissure   microfissure *f*
micrograph   micrographie *f*
micrometer   micromètre *m*
micron   micron *m*
micro-pile   micro-pieu *m*
microprocessor   microprocesseur *m*
micro-section   coupe *f* micrographique
microstructure   microstructure *f*
micro-switch   microrupteur *m*
microwave   micro-onde *f*
midden pit [sump for liquid manure]   fosse *f* à purin
middle   *adj*   moyen *adj*
middle   centre *m*, milieu *m*
middle third   noyau *m* central, tiers-central *m*
mid-plane   plan *m* moyen
mid-span   mi-travée *adj*
mid-span, moment at   moment *m* en travée
migrate [of water in a permeable medium]   migrer *vb*

mildew   moisi *m*, moisissure *f*
milestone [strictly = kilometre stone]   borne *f* kilométrique
milking parlour   salle *f* de traite
mill, *see also* watermill, windmill   moulin *m*
mill, hammer   broyeur *m* à marteau
mill, rolling   laminoir *m*
mill scale   calamine *f*, peau *f*/scorie *f* de laminage
mill state [steel]   état *m* brut de laminage
mill [a log to produce square edged timber]   aviver *vb*
mill   fraiser *vb*, moleter *vb*
milled   moleté *adj*
millimetre   millimètre *m*
milling   fraisage *m*
millings   limaille *f* (d'acier, de fer)
millstone   meule *f*, meulière *f*
millstream   bief *m* (du moulin)
mine   mine *f*
mine, open-cast   mine *f* à ciel ouvert
mine shaft   puits *m* de mine
mineral   *adj*   minéral *adj*
mineral   minéral *m*
mini-digger or -excavator   micro-pelle *f*
minimize   réduire *vb* au minimum
minimum   *adj*   minimal *adj*, minimum *adj*
minimum   minimum *m*
mining   *adj*   minier *adj*
mining   abattage *m*
mining, opencast   exploitation *f* à ciel ouvert
mining engineer [from l'École des Mines]   ingénieur *m* des mines
mini-pile   picot *m*
mini-trunking [electrical]   goulotte *f*, moulure *f*
minor   mineur *adj*
minute [of time, or note of sth]   minute *f*
minute   minuter *vb*
minutes [of a meeting]   compte *m* rendu, procès-verbal *m* (PV)
minutes-taker   rapporteur *m*
mirror   glace *f*, miroir *m*
mirror backing panel   parquet *m* de glace
mirror image   image *f* symétrique, image-miroir *f*
mirror, polish a   doucir *vb* une glace/un miroir
mirrors, polishing of   douci *m*
misappropriation   dilapidation *f*
miscellaneous   divers *adj*
misconnect   mal coupler *vb*
mission [task]   mission *f*
mist   brouillard *m*, brume *f*
mist [on glass, etc]   buée *f*
misuse   abus *m*
mitre   onglet *m*
mitre box   boîte *f* à coupe(s), boîte *f* à onglets/d'onglet
mitre cut   coupe *f* biaise
mitre cut [at other than 45°]   faux onglet *m*
mitre cut [to form a right angle]   coup *m* d'onglet *m*
mitre square   équerre *f* d'onglet *m*
mitt [protective]   mitaine *f*
mitten [protective]   moufle *f*

mix   malaxer *vb*, mélanger *vb*
mix [adding water to]   gâcher *vb*
mix [in correct proportions]   doser *vb*
mix [several colours of paint together]   fondre *vb*
mixed [in the sense of combined or composite]   mixte *adj*
mixer   malaxeur *m*, mélangeur *m*
mixer, batch   malaxeur *m* discontinu
mixer, concrete   bétonnière *f*
mixer, grout   malaxeur *m* à coulis
mixer, pan   malaxeur *m* plat
mixer, pulverizing   pulvérisateur-mélangeur *m*
mixer, revolving drum   mélangeur *m* à tambour tournant
mixer, tilting   bétonnière *f* à cuve basculante
mixing [of mortar, etc]   gâchage *m*
mixing, heat of   chaleur *f* de mélange
mix-in-place [soil stabilization method]   mélange *f* en place
mixture   mélange *f*
mixture, explosive gas   mélange *f* de gaz explosive
mixture, freezing   mélange *f* frigorifique
mixture, gas–air   mélange *f* air–gaz
mixture, well-graded [e.g. of road materials]   mélange *f* bien graduée
mnemonic   mnémonique *f*
moat [of a fortress]   douve *f*, fossé *m*
mobilization   mobilisation *f*
mock-up   maquette *f*
mode of payment   mode *m* de paiement
model *adj*   modèle *adj*
model   modèle *m*
model [architectural, mock-up]   maquette *f*
model, scale   modèle *m* réduit
model-maker   maquettiste *m/f*
modem   modem *m*, modulateur-démodulateur *m*
modification   modification *f*
modified   modifié *adj*
modify   modifier *vb*
modillion [architectural]   modillon *m*
modular   modulaire *adj*
modulation, full   modulation *f* complète
module [in design, layout, etc]   module *m*
module, data display   module *m* d'affichage de données
modulus   module *m*
modulus of compressibility   module *m* de compressibilité
modulus of elasticity   coefficient *m*/module *m* d'élasticité
modulus of rigidity   module *m* de rigidité
modulus, section   module *m* de résistance
modulus, shear   module *m* de cisaillement/d'élasticité transversale
modulus, Young's   module *m* d'élasticité/de Young
Mohr's circle [of stress]   cercle *m* de Mohr
moist   moite *adj*
moisten   humecter *vb*, mouiller *vb*
moistening   mouillage *m*
moisture, atmospheric   humidité *f* de l'air
moisture, available   eau *f* disponible
moisture balance   équilibre *m* d'humidité
moisture content   degré *m* d'humidité, teneur *f* en humidité

moisture content, equilibrium   teneur *f* en eau d'équilibre
moisture (content) of air   degré *m* hygrométrique
mole [breakwater]   môle *m*
mole [tunnelling]   tunnelier *m*
mole drainage pipe-layer   pose-drains *m*
mole-drain   drain-taupe *m*
Mollier diagram   diagramme *m* de Mollier
molybdenum   molybdène *m*
moment   moment *m*
moment [of forces]   couple *m*
moment, bending   moment *m* de flexion/fléchissant
moment, fixed-end or restraint   moment *m* d'encastrement
moment, overturning   couple *m*/moment *m* de renversement
moment, panel or span   moment *m* en travée
moment, restoring or resisting   moment *m* résistant
moment, statical   moment *m* statique
moment, support   moment *m* sur appuis
moment, torsional/twisting   couple *m*/moment *m* de torsion
moment, turning   couple *m*/moment *m* de rotation
moment about the major/minor axis   moment *m* par rapport au grand/petit axe
moment at mid-span   moment *m* en travée
moment at supports   moment *m* sur appuis
moment diagram   aire *f* des moments
moment of area, first/second   moment *m* statique/d'inertie
moment of inertia   moment *m* d'inertie
moment of inertia, polar   moment *m* d'inertie polaire
momentum   moment *m*, quantité *f* de mouvement
monitor   contrôler *vb*
monitoring   mesures *f,pl* automatiques
monkey [piling]   mouton *m*
monkey wrench   clef *f* anglaise, pince-étau *f*
monolith [made from a single piece of stone]   monolithe *m*
monolithic   monolithe *adj*
monorail   monorail *m*
monthly   mensuel *adj*
monument   monument *m*
monument [fixed datum point]   repère *m* fixe
monument, historic   monument *m* historique
moor(land)   lande *f*
mooring   poste *m* d'amarrage
mooring [the process]   amarrage *m*
mop [a surface, to dry it]   tamponner *vb*
moraine   moraine *f*
mortar   mortier *m*
mortar [for brick/block laying]   mortier *m* normal/ordinaire
mortar, activated [treated to improve workability]   mortier *m* activé
mortar, adhesive, tile fixing   mortier *m* colle
mortar, bastard/lime and cement [for coating]   mortier *m* bâtard
mortar, bedding/bonding   mortier *m* de pose/de reprise
mortar, cement   mortier *m* de ciment, mortier *m* fin gris

mortar, fat mix   mortier *m* gras
mortar, jointing   coulis *m*
mortar, lime   mortier *m* de chaux, mortier *m* fin blanc
mortar, quick-setting   mortier *m* à prise rapide
mortar, rendering   mortier *m* d'enduit
mortar, slow-setting   mortier *m* à prise lente
mortar, wet   gâchis *m*
mortar bed [to receive cut stone, etc]   bain *m* de mortier
mortar of clay and straw   bauge *f*
mortar pan/trough/tub   auge *f*/bac *m* à mortier
mortar skin, outer   croûte *f* de mortier
mortar strength   résistance *f* du mortier
mortaring in [fixing]   scellement *m*
mortgage   crédit *m*, hypothèque *f*
mortgage [house]   emprunt-logement *m*
mortgage, foreclosure of a   saisie *f* d'une hypothèque
mortgage charge   privilège *m* d'hypothèque
mortgage loan   prêt *m* hypothécaire
mortgage registrar [registrar of land charges]   conservateur *m* des hypothèques
mortgage   engager *vb*
mortise   mortaise *f*
mortise, side of a, or material removed to form a tenon   épaulement *m*
mortise in a piece of metal   empênage *m*
mortise   mortaiser *vb*
mortiser, *see* mortising machine
mortising   mortaisage *m*
mortising chain   chaîne *f* à mortaiser
mortising chain, endless   chaîne *f* à mortaiser sans fin
mortising machine   machine *f* à mortaiser
mortising machine, chain   fraiseuse *f* à chaîne, mortaiseuse *f* à chaîne
mosaic   mosaïque *f*
motel   motel *m*
motion, direction of   sens *m* de déplacement
motion, setting sth in   déclenchement *m*
motor   moteur *m*
motor, alternating current (a.c.)   moteur *m* à courant alternatif
motor, direct current (d.c.)   moteur *m* à courant continu
motor, electric   moteur *m* électrique
motor, lifting   moteur *m* de levage
motor, shunt   moteur *m* shunt
motor, single-phase   moteur *m* monophasé
motor, squirrel cage   moteur *m* à cage d'écureuil
motor, stand-by   moteur *m* de secours/de réserve
motor, three-phase   moteur *m* (à courant) triphasé
motor bed-plate   socle *m* de moteur
motor protection   protection *f* de machine
motor rating   puissance *f* (nominale) du moteur
motor winding   bobinage *m*/enroulement *m* de moteur
motor-blower [for road cleaning]   motosouffleur *m*
motor-driven   actionné par moteur
motorgrader   motorgrader *m*, niveleuse *f* automotrice
motorize   motoriser *vb*
motorized   automoteur *adj*
motorscraper   motorscraper *m*
motorway   autoroute *f*

motorway, rural   autoroute *f* de rase campagne
motorway, toll   autoroute *f* à péage
motorway, urban   autoroute *f* urbaine
motorway fork   branche *f* d'autoroute
motorway interchange, *see also* intersection   intersection *f* d'autoroute, échangeur *m* autoroutier
motorway maintenance area   centre *m* d'entretien autoroutier
mould [for casting or moulding]   moule *f*
mould [form]   forme *f*
mould [mildew]   moisi *m*, moisissure *f*
mould [soil]   humus *m*
mould   mouler *vb*
mould [mildew]   moisir *vb*
moulded   moulé *adj*, mouluré *adj*
moulder   mouleur *m*
moulder, spindle   toupie *f*
moulding   moulure *f*
moulding, circular   cordon *m*
moulding, concave circular   gorge *f*
moulding, cover [e.g. between wall and door frame]   calfeutrement *m*
moulding, drip   jet *m* d'eau
moulding, flat [used as a cover strip]   demi-baguette *f*
moulding, flat portion of a   contre-champ *m*
moulding, half-round   demi-rond *m*
moulding, matching profile of a   contre-profil *m*
moulding, ogee   talon *m*
moulding, ogee [concave above, convex below]   doucine *f*
moulding, ogee [convex above, concave below]   doucine *f* renversée
moulding, ovolo [on a capital]   échine *f*
moulding, plane, separating two curved mouldings   réglet *m*
moulding, quarter- or half-round   baguette *f*
moulding, quarter-round   bouvement *m*, quart *m* de rond
moulding, reversed   contre-profil *m*
moulding, round   boudin *m*
moulding, torus   tondin *m*, tore *m*
moulding applied to emphasize an architrave   contre-chambranle *f*
mouldy, go [of wood]   chancir *vb*
mount   mettre *vb* en place
mount flush (with)   encastrer *vb* (dans)
mountain   montagne *f*
mounting, anti-vibration   plots *m,pl* anti-vibratiles, support *m* anti-vibration
mounting, surface   montage *m* en saillie
mounting base   socle *m*
mounting bracket   patte *f* de fixation
mounting frame   châssis *m*
mouse [computer]   souris *f*
mousetrap   souricière *f*
mouth [of a plane]   lumière *f*
mouth [of a river]   bouche *f*, embouchure *f*
mouth [pipe, well, etc]   orifice *m*
mouth [vent or inlet]   bouche *f*
mouth, river   embouchure *f*

movable   amovible *adj*, mobile *adj*
move   déplacer *vb*
movement   mouvement *m*
movement, backwards   recul *m*
movement of a part of a structure, small sideways   ripage *m*
moving   roulant *adj*
moving [of heavy materials, on site]   bardage *m*
moving house   déménagement *m*
muck [excavated material]   déblais *m,pl*
muck from tunnelling   marin *m*
muck out [a tunnel]   mariner *vb*
mucking-out stage of tunnel works   marinage *m*
mud   boue *f*, gâchis *m*
mud [alluvium]   limon *m*
muddy   boueux *adj*, gâcheux *adj*
mud-flap [vehicle]   bavette *f* garde-boue
mud/grit bucket in a road gulley   panier *m*
muffled   feutré *adj*

muffler   silencieux *m*
mullion [both horizontals and verticals]   meneau *m*
multi-cell [e.g. of bridge deck]   multicellulaire *adj*
multi-disciplinary   pluridisciplinaire *adj*
multi-disciplinary design and consulting firm   bureau *m* d'études
multi-layer [e.g. of paint system]   multicouche *adj*
multi-level   multiniveau *adj*
multimeter   multimètre *m*, voltampèremètre
multiplier   multiplicateur *m*
multi-plier [adjustable wrench]   pince *f* multiprise
multi-purpose   à usages multiples, polyvalent *adj*
multi-purpose appliance/unit   appareil *m* polyvalent
multi-zone [e.g. air-conditioning system]   multizone *adj*
muntin [of a door]   petit montant *m*
music pavilion   pavillon *m* de musique
mustiness   moisissure *f*
mutual   réciproque *adj*
mutual agreement, by   de gré à gré

nail   clou *m*
nail, clout   clou *m* calot(t)in/à tête de diamant, pointe *f* à tête *f* large
nail, cut   clou *m* découpé, clou *m* étampé
nail, duckbill   patte *f* à chambranle
nail, felt   clou *m* à tête large
nail, lost-head   clou *m* à tête perdue/à tête d'homme
nail, roofing   pointe *f* à papier bitumé
nail, round-head   clou *m* à tête plate
nail, slate   clou *m* à ardoise
nail, wire   pointe *f* à tête *f* conique, pointe *f* de Paris
nail, wooden   cheville *f* de bois
nail gun   cloueuse *f*
nail puller   arrache-clous *m*, pied *m* de biche, tire-clous *m*
nail punch   chasse-clou *m*, chasse-goupille *m*, chasse-pointe *m*
nail with small head   pointe *f*
nail   clouer *vb*
nailing   clouage *m*
nailing, secret   clouage *m* invisible
name   désignation *f*
name board, contractor's   plaque *f* d'entreprise
nameplate [on a tool, machine, etc]   plaque *f* signalétique
nameplate, maker's   plaque *f* du constructeur
narrow   étroit *adj*
narrow [get or grow narrower]   se resserrer *vb*
narrow [make narrower]   réduire *vb* la largeur de, rétrécir *vb*
narrow-gauge [of railways]   à voie étroite
narrowing [e.g. between hearth and mantlepiece]   rétrécissement *m*

narrowing of ends of timbers abutting same roof post   déjoutement *m*
narrows [of a river]   pertuis *m*
national engineering council [France]   Conseil *m* National des Ingénieurs de France (CNIF)
national federation of building contractors [France]   Fédération Nationale du Bâtiment (FNB)
national federation of civil engineering contractors [France]   Fédération Nationale des Travaux Publics (FNTP)
national forest authority in France [similar to British Forestry Commission]   Conservation des Eaux et Forêts
national homes improvement agency [France]   Agence Nationale pour l'Amélioration de l'Habitat (ANAH)
national job creation agency [France]   Agence Nationale pour l'Emploi (ANPE)
national level survey [France]   Niveau *m* Général de la France (NGF), nivellement *m* national
national mining and geological research institution [France]   Bureau *m* de Recherches Géologiques et Minières Français (BRGM)
national planning commission [France]   Commissariat *m* Général du Plan (CGP)
national planning regulations [France]   Règles *f,pl* Nationales d'Urbanisme
national railways [France]   Société Nationale des Chemins de Fer Français (SNCF)
national railways [Belgium]   Société Nationale des Chemins de Fer Belges (SNCB)
national standard   norme *f* nationale
national standard, French   norme *f* française (NF)

national statistical institute [responsible for official building cost indices, etc, in France]   Institut National de la Statistique des Études Économiques (INSEE)
national telephone service [in France]   France Télécom
natural   naturel *adj*
natural surroundings   milieu *m* naturel
natural ground   terrain *m* naturel
nature [character]   caractère *f*
nature reserve/sanctuary   réserve *f* naturelle
nauseating   nauséabond *adj*
nave   nef *f*
navvy   terrassier *m*
navvy pick [two-pointed]   bigot *m*
nearness   proximité *f*
necking [of a test piece]   striction *f*
needle   aiguille *f*
needle, Proctor plasticity   aiguille *f* de Proctor
needle scaler   marteau *m* à décalaminer
negative   négatif *adj*
negative [photographic]   cliché *m*
negotiate   négocier *vb*, traiter *vb*
negotiated [e.g. of contracts]   gré à gré
negotiator   négociateur *m*
neighbourhood of, in the   à proximité de
neon sign   enseigne *f* au néon
neoprene [synthetic rubber]   néoprène *m*
net [price, weight, etc]   net *adj*
net   filet *m*
net, safety   filet *m* de recueil, filet *m*/voile *m* de sécurité
nett *adj, see* net
netting, guard   grille *f* de protection
netting, wire, *see* wire netting
nettle   ortie *f*
network [e.g. of services]   réseau *m*
network, communication   réseau *m* de communication
network, delivery   réseau *m* de distribution
network, distribution   réseau *m* de distribution
network, drainage   réseau *m* d'évacuation/de drains
network, pipe   réseau *m* de canalisations
network, primary road   voirie *f* principale
network, road   réseau *m* routier, voirie *f*
network, rural road   voirie *f* rurale
network, supply   réseau *m* d'alimentation
network analysis (PERT) programme   planning *m* pert
network of grid lines   trame *f*
network system [with loops]   réseau *m* ramifié
neutral [e.g. axis, pressure, colour, conductor]   neutre *adj*
neutral, common   neutre *m* commun
neutral axis   axe *m* neutre
neutral conductor   conducteur *m*/fil *m* neutre
neutral point   point *m* neutre
neutral wire   fil *m* neutre
newel   pilastre *m*
newel [solid, of a spiral staircase]   noyau *m*
newel discontinuous between floors   noyau *m* suspendu
newel of a spiral staircase   noyau *m* d'escalier
newel of a spiral staircase with moulded-in handrail   noyau *m* à corde

newel of open tread spiral staircase   noyau *m* vide/à jour
newel post   noyau *m* de bois, pilastre *m*
next to   contigu *adj* à
nib [of a beam, a plain tile]   talon *m*
nibbler   grignoteur *m*
niche   niche *f*
nickel   nickel *m*
night set-back   ralenti *m* de nuit
night shift   poste *m* de nuit
nipple   nipple *m*
nipple [plumbing]   mamelon *m*
nipple, barrel   nipple *m* double
nipple, grease   raccord *m* de graissage
nipple, reducing   embout *m* de réduction
nitrate   nitrate *m*
nitric oxide   oxyde *m* nitrique
nitrogen   azote *m*
node [in a structural frame, a network]   nœud *m*
noise   bruit *m*
noise, air   bruit *m* d'air
noise, airborne   bruit *m* aérien
noise, ambient   bruit *m* d'ambiance
noise, break-out   bruit *m* émis
noise, duct break-out   bruit *m* émis par les conduits
noise, impact   bruit *m* d'impact
noise, predominantly high frequency   bruits *m,pl* à dominantes aiguës
noise, traffic   bruit *m* de la circulation
noise absorber   piège *m* à son
noise control   contrôle *m* acoustique
noise intensity   intensité *f* de bruit
noise level   niveau *m* sonore/de bruit
noise level, background   intensité *f* de bruit de fond
noise measurement   mesure *f* du bruit
noise nuisance   nuisance *f* sonore
noise reduction   réduction *f* de bruit
noise threshold level   seuil *m* du son
noiseless   silencieux *adj*
noisy   bruyant *adj*
no-load   à vide
no-load, run on or at   marcher *vb* à vide
nominal   nominal *adj*
nomogram   abaque *m*, nomographie *f*
non-cohesive   non-cohérent *adj*, pulvérulent *adj*
non-conducting [electrically]   isolant *adj*
non-drip   sans goutte
non-ferrous   non-ferreux *adj*
non-inflammable   ininflammable *adj*
non-metallic   non-métallique *adj*
non-skid   antidérapant *adj*
non-standard [not from standard production]   hors série
non-standard [not in accordance with standards]   hors normes
non-trade   hors exploitation
non-yielding   rigide *adj*
normal   normal *adj*
normalize   normaliser *vb*
north, northerly, northern, northward   nord *adj*

north   nord *m*
north-east   nord-est *m*
north-west   nord-ouest *m*
nose [of a latch or tool]   bec *m*
nosing [of a stair tread]   astragale *m*, nez *m* de marche
not included in estimate   hors devis
not recommended   déconseillé *adj*
not to scale (NTS)   non à l'échelle
notarial   notarial *adj*
notarial practice/office   étude *f* notariale
notarized document   acte *f* authentique
notary [functions overlap those of solicitors in England and Wales]   notaire *m*
notation   notation *f*
notch   échancrure *f*, entaille *f*
notch [e.g. in a flat key]   encoche *f*
notch [in masonry]   tranchée *f*
notch effect   effet *m* d'entaille
notch in a principal rafter to accept end of roof post   engueulement *m*
notch sensitivity   sensibilité à l'effet d'entaille
notch to receive shoulders of a tenon   épaulement *m*
notch   entailler *vb*, gruger *vb*
notch [edge of a piece of timber]   délarder *vb*
notching [of edge of a piece of timber]   délardement *m*
notching in a roof post for ends of braces, etc   dégueulement *m*
note   note *f*
note [summary]   bordereau *m*
notes of a meeting   procès-verbal *m* (PV)
notes, make or take   prendre *vb* des notes
notice, advance   préavis *m*
notice, preliminary or prior [of start of works]   avis *m* préalable
notice board   tableau *m* d'annonces
noxious   délétère *adj*, nocif *adj*
nozzle   buse *f*, tuyère *f*

nozzle [controllable]   ajutage *m*
nozzle [of a pipe or tube]   bec *m*, embout *m*
nozzle, anti-splash tap   brise-jet *m*
nozzle, ejector   buse *f* d'éjecteur
nozzle, injector   tuyère *f* d'injecteur
nozzle, inlet   jet *m* d'entrée
nozzle, outlet   buse *f* de soufflage, jet *m* de décharge
nozzle, spray   buse *f* d'injecteur
nozzle tip   pointe *f* de jet
nuclear   nucléaire *adj*
number [digit, figure]   chiffre *m*
number [in a series, e.g. telephone number, etc]   numéro *m*
number [quantity]   nombre *m*
number, Brinell hardness   chiffre *m* de dureté Brinell
number, even/odd/whole   nombre *m* pair/impair/entier
number, serial   numéro *m* de série
numbering   chiffrage *m*
numerical   numérique *adj*
numerical mean   moyenne *f* numérique
nursery [generally trees]   pépinière *f*
nurseryman   pépiniériste *m*
nursing home   clinique *f*
nut [for bolt]   écrou *m*
nut, blind   écrou *m* borgne
nut, cage   écrou *m* cage
nut, castellated   écrou *m* crénelé
nut, collar   écrou *m* à collet
nut, fixing   écrou *m* de fixation
nut, gland   écrou *m* du presse-étoupe
nut, gland [of a tap, valve, etc]   presse-étoupe *m*
nut, hexagonal   écrou *m* hexagonal/à six pans
nut, lock   écrou *m* indesserrable/de blocage
nut, sleeve   manchon *m* taraudé
nut, square   écrou *m* carré
nut, wing   écrou *m* papillon/à oreilles

O-ring joint   joint *m* torique
oak [tree, wood]   chêne *m*
obligation, contractual   obligation *f* contractuelle
obligation to give way [road traffic]   obligation *f* d'attente
oblique   en biais, oblique *adj*
obliquity   obliquité *f*
obscure [dark]   obscur *adj*
observation   observation *f*
observation [surveying]   coup *m* de lunette *f*
observe   surveiller *vb*
obsolete   désuet *adj*
obstruct   obstruer *vb*
obstruct [a pipe, a road]   boucher *vb*

obstruct [a road]   encombrer *vb*
obstruct [action, plans, progress]   entraver *vb*
obstructed   bouché/encombré/obstrué *adj*
obstruction   obstruction *f*
obstruction [barrier]   barrage *m*
obstruction [blockage, of a road, pipe, etc]   bouchon *m*
obstruction [congestion]   encombrement *m*
obstruction [clogging, of a pipe]   engorgement *m*
obstruction [of or to progress]   obstacle *m*
obstruction lighting [roadworks]   guirlande *f* lumineuse
occupancy detector   détecteur *m* de présence
occupancy rate   degré *m* d'occupation
occupant   occupant *m*
occupation [of land, premises]   occupation *f*

occupation under licence   occupation *f* précaire
occupational health   hygiène *f* du travail
occupational health inspector   médecin *m* inspecteur du travail
occupational health service   service *m* médical du travail
occupational health specialist   médecin *m* du travail
occupier   occupant *m*
occupy [place, time, job, etc]   occuper *vb*
octagonal   octogonal *adj*
odometer   odomètre *m*
odour   odeur *f*
odourless   inodore *adj*
oedometer   œdomètre *m*
off or out of gauge   hors tolérances
offcut   chute *f*
offcut [from trimming the edges of something]   rognure *f*
offer   offre *f*
offer   proposer *vb*
offer up [stone in masonry construction]   barder *vb*
office [building or room]   bureau *m*
office [insurance company, etc]   agence *f*
office [of lawyer, etc]   cabinet *m*
office, notarial   étude *f* notariale
office, open-plan   bureau *m* paysagé/en espace fonctionnel
office, registered   siège *m* social
office block   immeuble *m* de bureaux
office chair   siège *m* de bureau
office machine   machine *f* de bureau
offices   locaux *m,pl*
official *adj*   officiel *adj*
***Official Gazette**   Journal Officiel* (Le JO)
off-peak period   période *f* hors pointe
off-peak periods [for electrical tariffs]   heures *f,pl* creuses
offset [cranking]   coudage *m*
offset [of a pipe, flue, etc]   dévoiement *m*
offset between faces of adjoining walls   retraite *f*
offset   désaxer *vb*
offset [displace]   décaler *vb*
ogee, *see also* moulding, ogee   accolade *f*
ogive   ogive *f*
ohm   ohm *m*
Ohm's law   loi *f* d'Ohm
ohmmeter   ohmmètre *m*
oil   huile *f*
oil, burning   huile *f* combustible
oil, crude   pétrole *m* brut
oil, diesel   gas-oil *m*, gazole *m*, mazout *m* diesel
oil, domestic fuel/heating   fioule *f*, fuel *m*/mazout *m* domestique
oil, flushing   huile *f* de lavage
oil, fuel   fioule *m*, fuel *m* (de chauffage), mazout *m*
oil, gas   gas-oil *m*
oil, heating   fuel *m* (de chauffage), fioule *m*, mazout *m* (chauffage)
oil, light   huile *f* légère
oil, linseed   huile *f* de lin
oil, paraffin   pétrole *m* lampant
oil, refrigerant   huile *f* à frigorigène

oil burner nozzle or jet   gicleur *m* de brûleur à fuel
oil burning/firing   chauffe *f* à/au fuel
oil collector, separator   déshuileur *m*
oil drain(age)   vidange *f* d'huile
oil filler cap   bouchon *m* de remplissage d'huile
oil filler pipe   chargeur *m* d'huile
oil gas   gaz *m* d'huile
oil tank   réservoir *m* de mazout
oil tank [domestic]   cuve *f* à mazout
oil   graisser *vb*, lubrifier *vb*
oil-can   burette *f*
oil-free   exempt d'huile, sans huile
oilstone   pierre *f* à huile, affiloir *m*
oil-tight   étanche *adj* à l'huile
oily [surface]   onctueux *adj*
old [used, worn]   usagé *adj*
old [when following noun]   ancien *adj*
olive [in plumbing joint]   olive *f*
olive grove   oliveraie *f*
olive tree   olivier *m*
on edge   de chant, sur champ
on embankment   en remblai
on fire, be   brûler *vb*
on one floor/level   de plain-pied
on period   période *f* de fonctionnement
on site   à pied d'œuvre, sur le tas
on the banks of   sur les bords de
on the same level   de plain-pied
one-dimensional   monodimensionnel *adj*
one-level   plain-pied *adj*
one-piece   monobloc *adj*
one-pipe system   système *m* monotube
one-way (road, street)   à sens unique
ooze [wet mud]   limon *m*
oozing   suintement *m*
opacity   opacité *f*
opacity [of paint]   pouvoir *m* couvrant
opacity meter [for opacity or hiding power of paint]   opacimètre
opaque   opaque *adj*
opaque, render   opacifier *vb*
open   libre *adj*
open [of a door, a valve, etc]   ouvert *adj*
open [of a view]   dégagé *adj*
open [to the sky]   à ciel ouvert
open [a window, valve, etc]   ouvrir *vb*
open a circuit   couper *vb* un circuit
open (on to)   déboucher *vb* (sur)
open out [a hole]   agrandir *vb*
open out [e.g. the end of a pipe]   fraiser *vb*
open the flood-gates or sluices   lever *vb* les vannes
opencast [mining, etc]   à ciel ouvert
opening *adj*   ouvrant *adj*
opening   bouche *f*, orifice *m*, ouverture *f*
opening [architectural]   baie *f*
opening [hole]   pertuis *m*
opening [the action]   ouverture *f*
opening, exhaust   bouche *f* d'évacuation, orifice *m* de sortie/de soufflage

opening, inlet orifice *m*/ouverture *f* d'entrée
opening, louvred persienne *f*
opening, louvred [in a bell tower] abat-son *m*
opening, outlet orifice *m* de soufflage/de sortie
opening, small, in a partition or wall guichet *m*
opening, small, to admit light donne-jour *m*
opening, supply (air) bouche *f* de soufflage (d'air)
opening, vent [built into a partition or wall] ventouse *f*
opening, window ouverture *f* de fenêtre
opening in a floor for vertical access or circulation trémie *f*
opening in a wall to provide light [e.g. on a stair] rayère *f*
opening in a wall in accordance with a particular agreement or easement jour *m* de servitude
opening in dock or retaining wall for escape of water [larger than a weephole] dalot *m*
opening light [window] ouvrant *m*
opening overlooking a neighbour to which the latter cannot object jour *m* de souffrance
opening size, nominal mesh [of a sieve] ouverture *f* nominale [d'un tamis]
opening with sliding cover for flue/chimney-sweep's brush coulisse *f* de ramonage
open-plan paysagé *adj*, sans cloisons
openwork construction, of ajouré *adj* [opposite of *plein*]
operate marcher *vb*
operate [a machine] manœuvrer *vb*
operate [a quarry, a railway, etc] exploiter *vb*
operate [sth] faire *vb* fonctionner
operate in parallel/series marcher *vb* en parallèle/série
operated locally manœuvré à pied d'œuvre
operating *adj* de fonctionnement
operating characteristics caractéristiques *f,pl* de fonctionnement
operating conditions conditions *f,pl* de fonctionnement, mode *m* opératoire
operating current courant *m* de fonctionnement
operating curve courbe *f* d'efficacité
operating cycle cycle *m* opératoire
operating data bases *f,pl* d'exploitation
operating period période *f* de fonctionnement
operating point point *m* de fonctionnement
operating record or log registre *m* d'exploitation
operating time durée *f* de marche
operation [e.g. of a railway] exploitation *f*
operation opération *f*
operation [manoeuvre] manoeuvre *m*
operation [process] processus *m*
operation [running] fonctionnement *m*
operation, continuous fonctionnement *m* continu, marche *f* continue
operation, cycling/cyclic opération *f* cyclique/périodique
operation, method of mode *m* de fonctionnement
operation, no-load [of a machine] marche *f* à blanc/vide
operation, quiet fonctionnement *m* silencieux
operation, series opération *f* en cascade
operation, short-term service *m* de courte durée
operation, silent fonctionnement *m* silencieux
operation, smooth allure *f* régulière

operator [of a machine] conducteur *m* d'engin
operator, plant conducteur *m* d'engin
opposite direction, in the en sens contraire
optical fibre fibre *f* optique
optimum *adj* optimal *adj*
optimum optimum *m*
option variante *f*
optional facultatif *adj*
orange peel [paint defect] peau *f* d'orange
orchard verger *m*
orchestra pit fosse *f* d'orchestre
order [architectural, for goods, etc] ordre *m*
order [for goods or services] commande *f*
order [instruction] prescription *f*
order, repeat commande *f* ultérieure
order not included in the contract commande *f* hors marché
order of size ordre *m* de grandeur
order reference référence *f* de commande
order [goods or services] commander *vb*
order, official [administrative] arrêté *m*, ordonnance *f*, ordre *m* officiel
order, put in ranger *vb*
ordinary commun *adj*
Ordnance Survey (OS) [near French equivalent] Institut Géographique National (IGN)
organic organique *adj*
organization chart organigramme *m*
orientate orienter *vb*
orientation [e.g. of a crane boom] orientation *f*
orifice orifice *m*
original *adj* original *adj*
original original *m*
original [of a drawing, estimate, etc] minute *f*
original spare part [genuine] pièce *f* d'origine
ornament in the shape of an olive olive *f*
ornament(ation) ornement *m*
oscillate osciller *vb*
oscillation oscillation *f*
oscillogram oscillogramme *m*
oscillograph oscillographe *m*
oscilloscope oscilloscope *m*
osmosis osmose *f*
osmotic osmotique *adj*
Otto cycle cycle *m* d'Otto
out of circuit hors (de) circuit
out of gauge [e.g. of load] hors gabarit
out of order [of a machine] dérangé *adj*
out of plumb, *see also* out-of-plumb hors d'aplomb, en porte-à-faux
out of plumb, be porter *vb* à faux, surplomber *vb*
out of plumb, get prendre *vb* coup
out of stock épuisé *adj*
out of the way à l'écart
out of true [bent, distorted] gauchi *adj*
out of true [not vertical] hors d'aplomb
out of true [twisted] tordu *adj*
out of true, run tourner *vb* à faux
outbuilding annexe *f*, dépendance *f*

outcome résultat *m*
outcrop [geological] affleurement *m*
outcrop [geologically] affleurer *vb*
outer extérieur *adj*
outfall débouché *m*
outfall [of sewer] décharge *f*
outfall, sea rejet *m* en mer
outfall, sewage décharge *m* des eaux usées
outfall drain évacuateur *m*
outflow écoulement *m*
outhouse appentis *m*
outlay, capital mise *f* de fonds
outlet bouche *f*, issue *f*
outlet [commercial or opening] débouché *m*
outlet [drainage] exutoire *m*
outlet [of a culvert] tête *f* aval
outlet [of a lake, reservoir, etc] émissaire *m*
outlet, air échappement *m*/sortie *f* d'air
outlet, bottom vidange *f* de fond
outlet, cable sortie *f* de câbles
outlet, ceiling ouverture *f* du plafond
outlet, dry riser prise *f* d'incendie
outlet, exhaust orifice *m* de sortie/de soufflage
outlet, fan sortie *f* de ventilateur
outlet, key-operated bouche *f* à clé
outlet, recirculation bouche *f* de reprise
outlet, slotted bouche *f* linéaire, orifice *m* fendu
outlet, sluicing [of a dam] orifice *m* de chasse
outlet, socket, *see* socket outlet
outlet, supply (air) bouche *f* de soufflage (d'air)
outlet, vaned bouche *f* à ailettes
outlet, water départ *m*/sortie *f* d'eau
outlet opening orifice *m* de soufflage/de sortie
outline [drawn] contour *m*, esquisse *f*, ligne *f*, tracé *m*
outline [of a building, gauge, template] gabarit *m*
outline [written] ébauche *f*, esquisse *f*, schéma *m*
outline, rough ébauche *f*
outline drawing plan *m*
outline sketch ébauche *f*
outlining [drawing a line round] contournement *m*
outlook vue *f*
outlook, unobstructed/unrestricted vue *f* imprenable
out-of-date périmé *adj*
out-of-plumb dévers *m*
output production *f*, rendement *m*
output [of a motor, pump, fan, etc] débit *m*
output, analogue sortie *f* analogique
output, computer sortie *f* d'ordinateur
output, condenser rendement *m* de condenseur
output, heat quantité *f* de chaleur produite
output, maximum/peak rendement *m* maximal
output, minimum puissance *f* minimale
output, rated heat puissance *f* calorifique nominale
output, thermal puissance *f* thermique
output, useful heat puissance *f* thermique utile
output of a pipe refoulement *m*
output peripheral [computer] périphérique *f* de sortie
outside *adj* dehors *adj*, en dehors, extérieur *adj*, hors de
outside dehors *m*

outstand encorbellement *m*
outstation poste *m* secondaire
oval [shaped] ovale *adj*
oven four *m*
oven, coke four *m* à coke
oven, drying étuve *f* de séchage, four *m* à sécher
oven, gas four *m* à gaz
oven, microwave four *m* à micro-ondes
oven, side [beside open fireplace] potager *m*
oven dried séché *adj* à l'étuve
overall [of measurements] hors-œuvre *adj* (H/O)
overall dimension/size dimension *f* hors tout
overall layout [of a plan] masse *f*
overall layout plan plan *m* (de) masse, plan *m* général d'implantation
overburden terrain *m* de couverture/de recouvrement
overcharge [price too high] surfaire *vb*
over-compaction surcompactage *m*
over-consolidated surconsolidé *adj*
overcooling refroidissement *m* excessif
overdesigning surdimensionnement *m*
overdraft solde *m* débiteur
overdrawn [bank account] découvert *adj*
overdue en retard
overflow [of liquid] débordement *m*, déversement *m*
overflow [pipe, chute] déversoir *m*
overflow, stormwater déversoir *m* d'eau de pluie
overflow déborder *vb*
overhang surplomb *m*, encorbellement *m*, porte-à-faux *m*
overhang [of a wall] contre-fruit *m*
overhang surplomber *vb*
overhanging en porte-à-faux, en surplomb
overhaul [equipment, plant] reviser/réviser *vb*
overhauling revision/révision *f*
overhead aérien *adj*
overheads frais *m,pl* généraux
overheat surchauffer *vb*, échauffer *vb*
overheat [of a bearing, etc] chauffer *vb*
overheated surchauffé *adj*
overheating échauffement *m*, surchauffe *f*
overlap chevaucher *vb*, recouvrir *vb*
overlap(ping) chevauchement *m*, chevauchure *f*, recouvrement *m*
overlap(ping) [of tiles, boarding, etc] enchevauchure *f*
overlapping of (reinforcement) bars recouvrement *m* d'armatures/des barres
overload surcharge *f*
overload [cut-out, trip] disjoncteur *m* à maximum
overload capacity capacité *f* de surcharge
overload cut-out device/switch interrupteur *m* de surcharge
overload protection protection *f* contre les surcharges
overload surcharger *vb*
overloading surcharge *f*
overmodulate surmoduler *vb*
over-occupation surpeuplement *m*
overpass autopont *m*, passage *m* supérieur (PS)
overpass [motorway] pont *m* autoroutier
overpopulation surpeuplement *m*

overshoot   dépasser *vb*
oversize [e.g. of load]   hors gabarit
oversize [in thickness]   surépaisseur *f*
oversize [particles not passing through a sieve]   refus *m*
   (supérieur)
oversizing   surdimensionnement *m*
oversleeve [protective]   manchette *f*
overstress   dépassement *m* de contrainte, surcontrainte *f*
over-stretching   surtension *f*
overtighten [a bolt]   bloquer *vb*
overtightened   bloqué *adj*
overtime   heures *f,pl* supplémentaires, surtemps *m*
overturn [vehicle]   verser *vb*
overturning   renversement *m*
overturning moment   couple *m*/moment *m* de
   renversement
overturning stability   stabilité *f* au renversement
over-voltage   surtension *f*
over-voltage protection   protection *f* contre les
   surtensions
owner   propriétaire *m*

owner [client for building work]   maître *m* d'ouvrage
owner, bare   nu-propriétaire *m*
ownership, joint   copropriété *f*, indivision *f*
ownership, joint [of intervening wall, hedge, etc]
   mitoyenneté *f*
oxidation   oxydation *f*
oxide, iron   oxyde *m* de fer
oxidize   oxyder *vb*
oxidized   oxydé *adj*
oxidizer   agent *m* oxydant
oxyacetylene cutting   découpage *m* autogène
oxyacetylene gas   oxyacétylène *m*
oxyacetylene torch   chalumeau *m* oxyacétylénique
oxyacetylene welding   soudage *m*
   autogène/oxyacétylénique, soudure *f* oxyacétylénique
oxygen   oxygène *m*
oxygen, dissolved   oxygène *m* dissous
oxygen cutting   découpage *m* à l'oxygène, oxycoupage *m*
oxyhydrogen welding   soudage *m* au chalumeau
   oxhydrique
ozone   ozone *m*

PABX (private automatic branch exchange)
   autocommutateur *m*
pace   allure *f*
pack [a gland]   garnir *vb*
pack down   tasser *vb*
package   conditionner *vb*
package, bid/contract/work   lot *m*
package dealer   ensemblier *m*
packaged [of a unit, equipment]   autonome *adj*,
   monobloc *adj*
packaged unit [HVAC]   appareil *m* totalement équipé
packaging   emballage *m*
packer   emballeur *m*
packing [for a gland or stuffing-box]   garniture *f* de
   presse-étoupe
packing [gland, tap, valve, etc]   garniture *f*
packing [of a gland]   garnissage *m*
packing [of a joint]   fourrure *f*
packing [of a plate cooling tower]   plaque *f* de
   remplissage
packing, asbestos   garniture *f* d'amiante
packing, high pressure   joint *m* à haute pression
packing case   cadre *m*, caisse *f*
packing down   tassement *m*
packing rope for pump glands   bitord *m*
packing-box   presse-étoupe *m*
pad [e.g. for spreading varnish]   tampon *m*
padded   feutré *adj*
padlock   cadenas *m*
page [e.g. in drawing numbers]   planche *f*

paint   peinture *f*
paint, anti-corrosion or anti-rust   peinture *f* anti-rouille
paint, asphaltic or bitumastic   peinture *f* bitumineuse
paint, bituminous   enduit *m* bitumeux
paint, emulsion   émulsion *m*, peinture *f* émulsion/à
   dispersion
paint, enamel   peinture *f* émail
paint, gloss   peinture *f* brillante
paint, intumescent   peinture *f* intumescente
paint, latex   peinture *f* au latex
paint, lead   peinture *f* au plomb
paint, oil   peinture *f* à l'huile
paint, oil-based   peinture *f* glycérophtalique
paint, preliminary coat of   fond *m*
paint, rust proofing   peinture *f* anti-rouille
paint, semi-transparent oil   glacis *m*
paint, thin coat applied to a surface prior to filling
   déguisage *m*
paint, thinnest possible coat of   frottis *m*
paint defect – peeling   pelage *m*
paint defect – surface cracking   craquelure *f*
paint defect due to painting on top of paint not yet dry
   crocodilage *m*, peau *f* de crapaud/crocodile
paint defect known as orange peel   peau *f* d'orange
paint kettle   camion *m*
paint roller   rouleau *m* à peindre
paint spray pistol   pistolet *m* à peinture
paint spraying   peinture *f* au pistolet/par pulvérisation
paint spraying booth   cabine *f* de peinture
paint stripper   décapant *m*

paint stripping by scraping   grattage *m*
paint   peindre *vb*, peinturer *vb*
paint white   blanchir *vb*
paintbrush   pinceau *m*
paintbrush, flat   pinceau *m* plat, queue-de-morue *f*
paintbrush, round   pinceau *m* rond
paintbrush used for wood-graining   ta(u)pette *f*,
  veinette *f*
painted   peint *adj*
painted area replacing finger plate on a door   anglaise *f*
painter, house   peintre *m* (en bâtiment)
painter specializing in fillets   fil(et)eur *m*
painter specializing in varnish work   vernisseur *m*
painter's cradle [chair]   sellette *f*
painter's curved knife for cutting heavy wallpaper
  serpette *f*
painter's filler applied in a single coat   enduit *m* non
  repassé
painter's filler applied in several layers to produce
  smooth surface   enduit *m* repassé
painter's filler for plaster   enduit *m* gras
painter's filler for wood or old base   enduit *m* maigre
painter's filling knife [French regional term]   macabé *m*
painter's shave hook   violon *m*
painting   peinture *f*, travaux *m,pl* de peinture
painting [the action, or its result]   peinturage *m*
painting, brush   peinture *f* à la brosse
painting, roller   peinture *f* au rouleau
painting white [and its results]   blanchiment *m*
pair [couple]   couple *f*
pair, twisted   paire *f* torsadée
pair of compasses   compas *m*
pair of pliers   pince *f*, pinces *f,pl*
pair of scales   balance *f*
pair of steps   escabeau *m*, triquet *m*
paired   jumelé *adj*
paired [e.g. of twin columns, arches]   géminé *adj*
paired [in pairs]   en paires
pale [colour]   pâle *adj*
pale, go, grow or turn   pâlir *vb*
palisade   palissade *f*
pallet   palette *f*
palm tree   palme *f*
pamphlet   brochure *f*
pan [to collect condensate, etc]   cuvette *f*
pan, mortar   bac *m* à mortier
pan, WC   cuvette *f*, siège *m* à l'anglaise
pane [of glass]   carreau *m*, vitre *f*
pane, window, *see* window pane
pane [individual] **in a glazed panel**   dé *m*
panel [board]   tableau *m*
panel [flat face of a building, a roof]   pan *m*
panel [of a door, etc]   panneau *m*
panel [section or bay]   travée *f*
panel, alarm   tableau *m* des alarmes
panel, central control   tableau *m* général de
  commande
panel, changeover   tableau *m* de commutation

panel, control   panneau *m* de régulation, tableau *m* de
  régulation/de commande/de contrôle
panel, electrical   tableau *m* électrique
panel, end   travée *f* de rive
panel, glazed   verrière *f*
panel, large glazed   panneau *m* de verrière
panel, lowest and projecting, of a *porte cochère*
  tablier *m*
panel, opening   panneau *m* ouvrant
panel, plain and without a border   panneau *m* à glace
panel, plain decorative, in a wall or panel   table *f*
panel, plain decorative, raised/recessed   table *f*
  saillante/défoncée
panel, precast   panneau *m* préfabriqué
panel, radiant   panneau *m* rayonnant
panel, radiant heating   panneau *m* chauffant
panel, recessed plain decorative   table *f* défoncée
panel, remote control   tableau *m* de télécommande
panel, roof   pan *m* de comble
panel, sandwich [generally with insulating core]
  panneau *m* sandwich
panel, sliding   panneau *m* à glissière
panel construction   construction *f* à panneaux
panel edge 30–50mm wide [joinery]   plate-bande *f*
panel flush with its frame   lambris *m*/panneau *m*
  arasé
panel forming a chamfer between two walls   pan *m*
  coupé
panel heating   chauffage *m* à panneau rayonnant/par
  paneau
panel heating, wall/ceiling   chauffage *m* mural/au
  plafond
panel pin   pointe *f*
panel proud of one side of its surrounding frame
  lambris *m* à table saillante
panel   lambrisser *vb*
panel [in wood]   boiser *vb*
panelled   lambrissé *adj*
panelling [of a room or space]   lambrissage *m*
panelling [tongue and groove]   lambris *m*
panelling, wooden   boiserie *f*
panelling on the sloping side of a curb or mansard roof
  lambris *m* de comble
pantile   panne *f*, tuile *f* flamande/en S/creuse sans
  emboîtement
pantograph   pantographe *m*
pantry   garde-manger *m*
paper [document]   pièce *f*
paper, backing or lining   papier *m* d'apprêt
paper, bitumen   papier *m* bitumé
paper, filter   papier *m* filtrant
paper, glass   papier *m* de verre
paper, graph   papier *m* millimétré
paper, roofing   carton *m* bitumé
paper, roofing [waterproof]   pare-pluie *m*
paper, squared   papier *m* quadrillé
paper, tracing   papier-calque *m*
paper, waste   papier *m* de rebut

paper, waterproof [used in roofs or facades]
   pare-pluie *m*
paper, wind barrier   pare-vent *m*
paper hanging   pose *f* de papier peint
paper mill   usine *f* de pâte à papier
paper (with wallpaper)   tapisser *vb* (de papier peint)
paperhanger [tradesman]   tapissier *m*
paperhanger's brush   balai *m* à encoller
paperhanger's ceiling prop   porte-lé *m*
paperhanger's scissors   ciseaux *m,pl* de tapissier
parabola   parabole *f*
parabolic   parabolique *adj*
paraffin oil   pétrole *m* lampant
parallel (to)   parallèle *adj* (à)
parallel, operate or run in   marcher *vb* en parallèle
parallelogram of forces/velocities   parallélogramme *m*
   des forces/vitesses
parameter   paramètre *m*
parapet   garde-corps *m*, garde-fou *m*, parapet *m*
parapet on which one can lean   mur *m* d'appui
parcel [of land]   parcelle *f* [de terrain]
parcel [package]   colis *m*, paquet *m*
parcel, bordering   parcelle *f* riveraine
parcel out [land]   lotir *vb*
parcel out [supplies, etc]   aménager *vb*
pare [a piece of wood to required thickness]   doler *vb*
pare (down)   rogner *vb*
pare down [to make sth fit]   délarder *vb*
Paris airport authority   Aéroports de Paris (ADP)
Paris region express transport   Réseau Express
   Régional (RER)
Paris transport authority   Régie Autonome des
   Transports Parisiens (RATP)
parish   paroisse *f*
park   jardin *m*, parc *m*
parking, *see also* carpark
parking, basement   parking *m* en sous-sol
parking area, aircraft   aire *f* de stationnement
parking area or lot   aire *f* de parking, aire *f*/parc *m*/zone *f*
   de stationnement, parking *m*
parking bay or space   place *f* de stationnement
parking in the basement   parking *m* en sous-sol
parking lane   zone *f* de stationnement
parking meter   parcomètre *m*
parquet, chevron pattern   parquet *m* en chevrons
parquet, fan pattern   parquet *m* en éventail
parquet, framed inlaid panel   parquet *m* à la française/à
   compartiments/d'assemblage
parquet, framed panel   parquet *m* en feuilles
parquet, herringbone pattern [mitred ends]   parquet *m* à
   fougères/en fougère, parquet *m* à/en point de Hongroie
parquet, herringbone pattern [square ends]   parquet *m* à
   bâtons rompus
parquet, mosaic   parquet *m* (en) mosaïque
parquet floor support batten   lambourde *f*
parquet floor(ing)   parquet *m*
parquet flooring, chevron pattern   parquet *m* en
   chevrons

parquet layer [craftsman]   parqueteur *m*
parsonage   presbytère *m*
part   pièce *f*
part [constituent]   composant *m*
part [fraction, portion]   partie *f*
part, rear   arrière *m*
part, replacement or spare   pièce *f* de rechange
part by volume/weight   partie *f* en volume/en poids
part of a structure built into a neighbouring property
   enhachement *m*
part of building projecting in front of facade   avant-
   corps *m*
part [rope, cable, etc, under load]   céder *vb*
part [ground, etc]   fendre *vb*
partial [load, etc]   partiel *adj*
partial rebuilding of damaged work   reprise *f*
particle   grain *m*, particule *f*
particle, dust   grain *m* de poussière
particle, suspended   particule *f* en suspension
particles, aerosol   particules *f,pl* d'aérosols
particles, airborne   particules *f,pl* aériennes
particles, dust   particules *f,pl* de poussières
particle shape   forme *f* des grains
particle shape category   classe *f* de forme des grains
particle shape gauge [for road material]   pied *m* à
   coulisse
particle shape test   essai *m* de forme des grains
particle size   dimension *f* brute des particules, dimension
   f des grains *m,pl*, taille *f* de particule
particle size analysis   analyse *f* granulométrique
particle size distribution   composition *f*
   granulométrique, granulométrie *f*, granularité *f*
particle size distribution curve   courbe *f*
   granulométrique
particles of aggregate   granulats *m,pl*
particular   particulier *adj*
particulars   renseignements *m,pl*
parting [rope, etc]   rupture *f*
partition   paroi *f*, séparation *f*
partition [of uprights and plaster, brick or similar infill]
   galandage *m*
partition, fire [division]   cloison *f* coupe-feu
partition, framed   cloison *f* en charpente/en treillis, mur
   m ossaturé
partition, (timber) framed   cloison *f* en charpente
partition, glass   vitrage *m*
partition, loadbearing   cloison *f* porteuse, paroi *f*
   porteuse
partition, masonry   cloison *f* maçonnée
partition, sliding   cloison *f* mobile
partition, small [e.g. between flues, or a division within a
   duct]   languette *f*
partition, stud   cloison *f* en treillis
partition, stud [lath and plaster]   cloison *f* lattée et
   plâtrée
partition, wood   pan *m* de bois
partition (wall)   cloison *f*
partition   cloisonner *vb*

partitioning cloisonnage *m*, cloisonnement *m*
partner associé *m*
partnership partenariat *m*
party [as in party wall] mitoyen *adj*
party, interested ayant droit *m*
party involved intervenant *m*
party line ligne *f* à postes groupés
party line telephone poste *m* groupé
party wall mur *m* mitoyen
party wall, jointly owned space on each side of pied *m* d'aile
pass [between mountains] col *m*, défilé *m*
pass [in welding] passe *f*
pass on [property] transmettre *vb*
passable passable *adj*
passage conduit *m*
passage(way) corridor *m*, couloir *m*, passage *m*, dégagement *m*
passage(way), underground passage *m* souterrain, souterrain *m*
passenger passager *m*
passenger concourse hall *m* des passagers
passenger terminal, airport aérogare *f* (de) passagers
passivation passivation *f*
paste pâte *f*
paste, emulsion [for wallpaper] colle *f* à dispersion
paste, Teflon pâte *f* de téflon
paste, wallpaper colle *f* de pâte
pasting machine [for wallpaper] machine *f* à encoller
patch of land bout *m* de terrain
patch réparer *vb*
patent brevet *m*
patented breveté *adj*
paternoster patenôtre *f*
path allée *f*, chemin *m*, sentier *m*
path, air flow voie *f* d'écoulement d'air
path, inner swept [roads] bord *m* intérieur de la chaussée
patina patine *f*
patio [enclosed courtyard] patio *m*
pattern dessin *m*, disposition *f*
pattern, flow configuration *f* d'écoulement
pave [with flagstones or slabs] daller *vb*
pave a road with pebbles caillouter *vb*
paved dallé *adj*, pavé *adj*
pavement trottoir *m*
pavement [and the action of laying it] dallage *m*
pavement [of a road] chaussée *f*
pavement [of a road or runway] corps *m* de chaussée, superstructure *f*
pavement, concrete revêtement *m* bétonné
pavement, flexible chaussée *f* souple
pavement, rigid chaussée *f* rigide, revêtement *m* bétonné
pavement milling machine, cold raboteuse *f* (routière) à froid
pavement milling machine reprofileur *m* (à froid), reprofileuse *f*

paver [highways] épandeuse *f*, finisseur *m* routier
paver, asphalt finisseur *m* à asphalte
paver, road finisseur *m* routier
paver, slipform machine *f*/surfaceuse *f*/vibreuse-surfaceuse *f* à coffrage glissant
paving [the process] pavage *m*
paving [and the action of laying it] dallage *m*
paving, concrete block pavé *m* de béton
paving, large sett pavés *m.pl* échantillons
paving, mosaic sett pavé *m* mosaïque
paving, radial-sett pavage *m* mosaïque
paving machine, slipform, *see* slipform paver
paving sett pavé *m*
paving sett, large/small pavé *m* échantillon/mosaïque
paving setts, herringbone pattern of arête *f* de poisson
paving slab of thin sandstone platine *f*
paving slab, long, used as kerb boutisse *f*
paving slab, small dallot *m*
paving slab, stone dalle *f*, lause/lauze *f*
paving slabs, heavy slate orneaux *m,pl*
paving stone pavé *m*
paving stone 60–80 mm thick écale *f*
paving train, slipform machine *f* à coffrage glissant
pavior dalleur *m*
pavior's rammer or punner demoiselle *f*, hie *f*
pawl cliquet *m*
pay salaire *m*
pay payer *vb*
pay [an account] régler *vb*
pay cash payer *vb* (au) comptant
pay off a loan amortir *vb* en emprunt
payable when due payable à l'échéance
paying off [mortgage] dégagement *m*
payment paiement *m*
payment [of an account] règlement *m*
payment, advance paiement *m* d'avance
payment, cash paiement *m* comptant/en liquide(s)
payment, down acompte *m*
payment, instalment paiement *m* échelonné
payment, method or mode of mode *m* de paiement
payment, monthly [e.g. salary] mensualité *f*
payment, partial paiement *m* partiel
payment, stage(d) paiement *m* échelonné
payment by cheque paiement *m* par chèque
payment date échéance *f*
payment in advance paiement *m* d'avance
payment in full paiement *m* comptant/intégral
payment on account acompte *m*
peak [as in peak hours, etc] pointe *f*
peak [of a hill] sommet *m*
peak day jour *m* de pointe
peak demand demande *f* de pointe
peak hours ·heures *f,pl* pointes/de pointe
peak load charge *f* maximale/de pointe
peak load [output] débit *m* maximum
peak lopping or smoothing écrêtage *m*/écrêtement *m* des pointes
peak period période *f* de pointe

peal of bells   carillon *m*

pear tree   poirier *m*

peat   sol *m* tourbeux, tourbe *f*

pebble   caillou *m*, galet *m*, petit galet *m*/gravier *m*

pebbledash   cailloutage *m*, crépi *m* moucheté

pedestal   embase *f*, piédestal *m*, socle *m*

pedestal of a bearing   chevalet *m*

pedestrian *adj*   piéton *adj*, piétonnier *adj*

pedestrian   piéton *m*

pedestrian area/plaza/square   piazza *f*

pedestrian opening in large gate   guichet *m*

pedestrian way   piste *f* piétonne

pedestrianized   piétonné

pedestrianized street   rue *f* piétonne

pediment   fronton *m*

peel [a log to produce veneer, e.g. for plywood]
   dérouler *vb*

peel off [of paint, etc, by itself]   s'écailler *vb*

peeling [of paint layers]   pelage *m*

peen [of a hammer]   panne *f*

peen   marteler *vb*

peg   broche *f*

peg [e.g. for setting out]   piquet *m*

peg [for marking a layout or alignment]   taquet *m*

peg [used in jointing]   fiche *f*

peg of a peg ladder   ranche *f*

peg joint notched timbers   brandir *vb*

peg (out)   bornoyer *vb*, piqueter *vb*

peg out   implanter *vb*

pegging out [for setting out, surveying]   piquetage *m*

pegging out [boundaries]   bornage *m*

pelmet   bandeau *m*

pelmet [solid]   lambrequin *m*

pen, ruling   tire-ligne *m*

penalty, delay   pénalité *f* de retard

penalty clause   clause *f* pénale

pencil   crayon *m*

pencil, bricklayer's   crayon *m* de maçon

pencil, clutch   porte-mine *m*

pendentive   pendentif *m*, trompe *f*

pendentive, small   trompillon *m*

pending   en attente

penetration   pénétration *f*

penetration, depth of   (profondeur *f* de) pénétration *f*

penetration needle [for soil resistance measurement]
   aiguille *f* de Proctor

penetrometer   pénétromètre *m*

peninsula   péninsule *f*, presqu'île *f*

pension [income]   rente *f*

penstock [penning gate]   vanne *f* de tête d'eau

penstock [pressure pipe of a dam]   conduite *f* forcée

penthouse   auvent *m*, étage *m* hors-toit

penthouse roof   appentis *m*

per capita [per head]   par personne/tête

per cent   pour cent

per hour   à l'heure

percentage   pourcentage *m*

percentage by weight   pourcentage *m* du poids

percentage of [proportion]   teneur *f* en

percolate   infiltrer *vb*

percolation   filtration *f*, infiltration *f*

percolation of water to the surface of setting concrete
   remontée *f* d'eau

percolation path   chemin *m* d'infiltration

percussion   percussion *f*

perfect   parfaire *vb*

perforated   ajouré *adj*, perforé *adj*

perforated plate [of a plumb line]   chas *m*

perforation   perforation *f*

perforation [to allow light to enter]   ajour *m*

performance   performance *f*

performance [output, yield]   rendement *m*

performance, acoustic   effet *m* acoustique

performance, actual/practical   performance *f* pratique

performance, coefficient of   facteur *m* de
   fonctionnement

performance, design/predicted   performance *f* prévue

performance, erratic   irrégularité *f*

performance, target   performance *f* prévue

performance bond   cautionnement *m*, garantie *f* de bonne
   fin

performance characteristic curve   courbe *f* de
   fonctionnement

performance coefficient   coefficient *m* de performance

pergola   pergola *f*, treille *f* à l'italienne

perimeter   circonférence *f*, périmètre *m*

perimeter, wetted   périmètre *m* mouillé

period *adj*   d'époque

period [of time]   période *f*

period, actual working   durée *f* effective de service

period, contract   délai *m* d'exécution, durée *f* de contrat

period, curing   durée *f* de prise

period, delivery [time for delivery]   délai *m* de
   livraison

period, exposure   temps *m* d'exposition

period, holiday   période *f* de(s) vacances

period, off-peak   période *f* hors pointe

period, peak   période *f* de pointe

period, reheat   période *f* de réchauffe

period, rent-free   franchise *f* de loyer

period, rest   période *f* de repos

period, retention   période *f* de rétention

period, running   période *f* de fonctionnement

period, shut-down   période *f* d'arrêt

period, sunshine   période *f* d'ensoleillement

period, training   période *f* d'instruction

periods, off-peak [for electrical tariffs]   heures *f,pl*
   creuses

periods, slack   heures *f,pl* creuses

period of sunshine   période *f* d'ensoleillement

period of time [for completion of task, payment, etc]
   délai *m*

periodic   périodique *adj*

peripheral   périphérique *adj*

periphery [e.g. surrounding a building]   pourtour *m*

perlite   perlite *m*

permanent permanent *adj*
permanent [colour] fixe *adj*
permanent [definitive] définitif *adj*
permanent way [railway] voie *f* ferrée
permeability perméabilité *f*
permeability, degree of degré *m* de perméabilité
permeability coefficient coefficient *m* de perméabilité
permeable perméable *adj*
permeable material matériau *m* perméable
permeate (with) imprégner *vb* (de)
permissible admissible *adj*
permission, *see* permit
permission, planning permis *m* de construire
permission (for or to) autorisation *f* (de)
permission (to someone to), give or grant autoriser *vb*
  (qn à)
permit licence *f*, permis *m*
permit, building permis *m* de construire
permit, demolition permis *m* de démolir
permit, scrub clearance permis *m* de défrichage
permit, work permis *m* de travail
permit (for or to) autorisation *f* (de)
perpendicular normal *adj*
perpendicular (to) perpendiculaire *adj* (à)
person particulier *m*
person responsible for letting of the contract
  personne *f* responsable du marché (PRM)
perspective *adj* perspectif *adj*
perspective perspective *f*
perspective, axonometric perspective *f* axonométrique
perspective, bird's eye view perspective *f* cavalière
perspective (drawing) dessin *m* en perspective
PERT programming and network analysis planning *m*
  pert
pervibration [internal vibration of concrete]
  pervibration *f*
pervious perméable *adj*
petrol essence *f*
petroleum [not petrol] pétrole *m*
petroleum, crude pétrole *m* brut
pH value valeur *f* du pH
pH indicator pH-mètre *m*
phase [electrical] phase *f*
phase [stage] étape *f*, période *f*, phase *f*
phase angle angle *m* de phase
phase conductor conducteur *m* de phase
phase difference or shift décalage *m* (de phase),
  déphasage *m*
phase displacement déplacement *m* de phase
phased progressif *adj*
phasing phasage *m*
phosphating phosphatisation *f*
phosphorous phosphore *m*
photo[†] photo[†] *f*, photographie *f*
photo reconnaissance reconnaissance *f* photographique
photocopier photocopieur *m*
photocopy photocopie *f*
photo-electric cell cellule *f* photo-électrique

photogrammetry photogrammétrie *f*
photograph photographie *f*
photograph, X-ray photographie *f* aux rayons X
photography photographie *f*
photograph(y), aerial photographie *f* aérienne
photographer photographe *m*
photo-theodolite photothéodolite *m*
pick pic *m*, pioche *f*
pick [a lock] crocheter *vb*
pick out trier *vb*
pick out [a colour] rehausser *vb*
pick out [joints in masonry] dégarnir *vb*
pick up [a signal] capter *vb*
pick up [an error] relever *vb*
pick up [improve] s'améliorer *vb*
pick up [something dropped] ramasser *vb*
pick up speed prendre *vb* de la vitesse
pickaxe pioche *f*
picket [stake, peg] piquet *m*
picking off cement/plaster rendering/coating
  hachement *m*
pickle [metal] blanchir *vb*, décaper *vb*, dérocher *vb*
pickle with acid mordre *vb*
pickling [of metal] décapage *m*, dérochage *m*
pickup, acoustic capteur *m* acoustique
pickup truck pick-up *m*
picture (of) image *f* (de)
picture frame cadre *m*
piece [fragment, object] pièce *f*
piece, added or additional ajoute *m*
piece, binding [timber structure] moise *f*
piece, branch raccord *m* de branchement
piece, connecting raccord *m*
piece, extension rallonge *f*
piece, make-up [for extending pipe] tube *m* de rallonge
piece, make-up [of wood, in joinery] alaise *f*
piece, strengthening renfort *m*
piece, transformation pièce *f* de transformation
piece, transition [ductwork] pièce *f* de réduction
piece of the wrong length [e.g. of wood] courçon *m*
piece of turf morceau *m* de gazon
piece of wall [whole height] pan *m* de mur
piece of wood reinforcing cob walling palançon *m*
piece of wood used for tightening a bow saw garrot *m*
piece of wood used to disguise a crack, etc flipeau/
  flipot *m*
piece worker tâcheron *m*
pier [breakwater] môle *m*
pier [jetty, quay] jetée *f*
pier [quay or wharf] quai *m*
pier [length of wall between two openings] trumeau *m*
pier [of a bridge] pile *f*, poteau *m*
pier [of an arch, bridge, window] pied-droit/piédroit *m*
pier [on piles] estacade *f*
pier, abutment arc *m* boutant
pier, central part of [bridge] fût *m*
pier, corner [to stiffen junction of two walls] chaîne *f*
  d'angle

pier, set back, supporting an archway or lintel course
   fausse-alette *f*

pier, supporting [between arches or parts of a wall]
   jambage *m*

pier in a wall [of cut stone, etc]   chaîne *f*

pier in a wall [reinforcing]   dosseret *m*

pierce   percer *vb*

pierce with many holes   cribler *vb*

pierced   percé *adj*

piercing   percement *m*

pierhead   musoir *m*

piezometer   piézomètre *f*

pigment   pigment *m*

pigsty   porcherie *f*

pilaster   pilastre *m*

pile [for foundations]   pieu *m*, pilot *m*

pile [of sth]   empilage *m*, tas *m*

pile [stack]   pile *f*

pile, batter   pieu *m* incliné

pile, bored   pieu *m* foré

pile, cast in situ   pieu *m* moulé dans le sol

pile, continuous flight auger   pieu *m* à la tarière
   continue

pile, drilled cast in situ   pieu *m* moulé foré

pile, driven   pieu *m* battu/foncé

pile, driven cast in situ   pieu *m* moulé battu

pile, end-bearing   pieu *m* portant en pointe/résistant à la
   pointe

pile, fender   pieu *m* d'accostage

pile, friction   pieu *m* flottant

pile, in situ   pieu *m* in situ

pile, refusal of a   refus *m* d'un pieu

pile, reinforced concrete   pieu *m* en béton armé

pile, sand   pieu *m* à sable

pile, screw   pieu *m* à vis

pile, sheet, *see* sheet pile

pile, spread base of a   patte *f* d'éléphant

pile, wooden   pilot *m*

piles, contiguous   pieux *m,pl* contigus

piles, king   pieux *m,pl* berlinois

pile bearing load   charge *f* d'un pieu

pile cap   casque *m*/tête *f* de pieu

pile cap [footing for masonry]   plate-forme *f*

pile cap (beam)   longrine *f*

pile cluster   groupe *m* de pieux

pile driver, *see also* piling rig   bélier *m*, hie *f*, sonnette
   f/mouton *m* (de battage)

pile driving   battage *m* de pieux, pilotage *m*

pile driving by vibration   vibro-fonçage *m*

pile driving formula   formule *f* de battage

pile driving in sand with the aid of water injection
   lançage *m*

pile (driving) record   carnet *m* de battage

pile driving resistance   résistance *f* du pieu au battage

pile group   faisceau *m* de pieux *m,pl*, pilotis *m*

pile helmet   casque *m* de battage, coiffe *f* de pieu

pile loading test   essai *m* de chargement d'un pieu

pile shoe   sabot *m*

pile tip   pointe *f* du pieu

pile vibratory driver   vibro-fonceur *m*

pile regularly to work out volume [e.g. sand]
   emmétrer *vb*

piling [i.e. several piles]   pilotis *m*

piling, sheet, *see* sheet piling

piling of earth   mise *f* en cavalier

piling rig, *see also* pile driver   appareil *m*/engin *m* de
   battage

piling vibrator   vibrateur *m* de fonçage et d'arrachage,
   vibro-fonceur *m*

pillar   colonne *f*, pilier *m*

pillar [in underground workings]   pilier *m*

pillar, half [applied to face of a wall]   demi-pilier *m*

pillar, small   pédicule *m*

pillar at end of party wall, visible from the street
   jambe *f* étrière

pilot hole   avant-trou *m*

pilot light   veilleuse *f*

pin *adj* [of a joint]   articulé *adj*

pin   boulon *m*, goujon *m*

pin [of electrical plug, fluorescent tube, etc]   broche *f*

pin [e.g. to retain a doorknob]   goupille *f*

pin [of a hinge]   broche *f*, fiche *f*, mamelon *m*

pin [of lift-off hinge]   lacet *m*

pin [used in jointing]   cheville *f*, fiche *f*

pin, hinge   broche *f* à charnière

pin, split   goupille *f* fendue

pin base or cap [fluorescent tube]   culot *m* à broches

pincers   tenailles *f,pl*

pincers, joiner's   tenaille *f* de menuisier

pincers, large   tricoises *f,pl*

pinch bar   pied-de-biche *m*, pince *f*

pine [tree, timber]   pin *m*

pine wood [forest]   pinède *f*

pin-joint [mechanism or structural]   articulation *f*

pinnacle [on a building, buttress]   pinacle *m*

pipe   buse *f*, canal *m*, canalisation *f*, conduite *f*, tube *m*,
   tuyau *m*

pipe, admission   amenée *f*

pipe, asbestos cement   tuyau *m* en fibrociment

pipe, balance or balancing   tuyau *m* d'équilibrage

pipe, bitumen coated steel   tuyau *m* en acier revêtu de
   bitume

pipe, bleed   tuyau *m* soutirage

pipe, blow-off   tuyau *m* de purge

pipe, branch   tuyau *m* de branchement

pipe, breather   reniflard *m*

pipe, buried   conduite *f* enterrée

pipe, butt welded   tuyau *m* soudé en bout

pipe, bypass   by-pass *m*, conduite *f*/tube *m*/tuyau *m* de
   dérivation

pipe, casing [in borehole, etc]   tube *m* de revêtement

pipe, cast iron/clay/concrete   tuyau *m* en
   fonte/céramique/béton

pipe, circulation   tuyau *m* de circulation

pipe, clay [piece of]   poterie *f*

pipe, concrete   buse *f*

pipe, connecting tuyau *m* de raccordement

pipe, cooling tuyau *m* de refroidissement

pipe, cooling water conduite *f* d'eau de refroidissement

pipe, copper/steel/etc tube *m*/tuyau *m* en cuivre/acier/etc

pipe, delivery [e.g. from a pump] tuyau *m* de refoulement

pipe, delivery [to an appliance, a building] conduite *f* d'alimentation

pipe, discharge tuyau *m* de décharge

pipe, distributor tuyau *m* de distribution

pipe, drain, *see* drain pipe

pipe, drain down or draining tube *m*/tuyau *m* de vidange

pipe, drainage tube *m*/tuyau *m* d'évacuation/de drainage

pipe, drip pipette *f*

pipe, earthenware tuyau *m* en grès

pipe, exhaust tuyau *m* de décharge

pipe, exhaust [vehicle] pot *m*/tuyau *m* d'échappement

pipe, expansion tube *m* de compensation/de rallonge, tuyau *m* de trop-plein

pipe, extension tube *m* de rallonge

pipe, feed tuyau *m* d'alimentation

pipe, filling tuyau *m* de remplissage

pipe, filter [drainage] tuyau *m* filtrant/perforé

pipe, finned tuyau *m* à ailettes

pipe, flanged tuyau *m* à brides

pipe, flexible tube *m*/tuyau *m* flexible/souple

pipe, flexible metal tuyau *m* métallique flexible

pipe, flue conduit *m* de fumée, tuyau *m* de cheminée

pipe, galvanized tuyau *m* galvanisé

pipe, gas conduite *f*/tuyau *m* de gaz

pipe, gas service branchement *m* de gaz

pipe, heating caloduc *m*, tuyau *m* de chauffage

pipe, horizontal distribution [between meter and appliance connections] ceinture *f* principale

pipe, horizontal distribution [from rising main] ceinture *f* d'appartement/d'étage

pipe, inlet tuyau *m* d'entrée

pipe, leaking fuite *f* de tuyau

pipe, main condensate conduit *m* principal de condensation

pipe, malleable iron tuyau *m* en fonte malléable

pipe, outlet tuyau *m* de décharge

pipe, overflow trop-plein *m*, tuyau *m* de trop-plein

pipe, perforated [drainage] tuyau *m* filtrant/perforé

pipe, plastic/polyethylene/rubber tuyau *m* de/en plastique/polyéthylène/caoutchouc

pipe, prefabricated tuyau *m* préfabriqué

pipe, pressure conduite *f* forcée

pipe, prestressed concrete tuyau *m* en béton précontraint

pipe, rainwater tube *m*/tuyau *m* de descente (pluviale)

pipe, refrigerating tuyau *m* de refroidissement

pipe, return conduite *f*/tuyau *m* de retour

pipe, riser conduite *f* montante

pipe, screwed/threaded tube *m* fileté/taraudé

pipe, service conduite *f* de distribution/de raccordement

pipe, sewer [large] égout *m*

pipe, sewer [small] tuyau *m* d'égout

pipe, slotted [drainage] tuyau *m* fendu

pipe, smoke tube *m* de fumée

pipe, socketless tube *m* sans emboîtement

pipe, soil tube *m*/tuyau *m* d'égout

pipe, spun iron, with socket-and-spigot joints fonte *f* salubre

pipe, steam tuyau *m* de vapeur

pipe, suction conduite *f* d'aspiration

pipe, supply amenée *f*, conduite *f*/tuyau *m* d'alimentation

pipe, swan neck col *m* de cygne

pipe, tremie [concreting] tube *m* de bétonnage

pipe, tremie [funnel] entonnoir *m*

pipe, vacuum tuyau *m* à vide

pipe, vent tube *m*/tuyau *m* d'évent

pipe, vent [emergency, safety] tube *m* de sécurité/sûreté

pipe, vent(ilation) tuyau *m* d'aération/de ventilation, ventilation *f*, ventilation *f* secondaire

pipe, vertical colonne *f*

pipe, vertical waste, below a WC chute *f*

pipe, waste tuyau *m* de dégagement

pipe, waste [of bath, shower, etc] écoulement *m*

pipe, waste tuyau *m* de décharge/d'échappement/d'évacuation

pipe, water supply conduite *f*/tuyau *m* (d'alimentation) d'eau

pipe, welded tube *m* soudé

pipe, wrought iron tuyau *m* en fer forgé

pipe bend coude *m* de tube

pipe bender cintreuse *f* de tube

pipe chase [in masonry, concrete work] fourreau *m*

pipe cleaner, flexible furet *m*

pipe clip [collar] collier *m* (d'attache), gâche *f*

pipe coil serpentin *m*

pipe connection, screwed raccord *m* vissé pour tuyau

pipe cutter coupe-tubes *m*, coupe-tuyaux *m*

pipe end bout *m* de tube

pipe fitter or layer poseur *m* de tuyaux

pipe fitting appareillage *m* pour tube

pipe fitting, three-way tuyau *m* à trois voies

pipe fitting, four-way tuyau *m* en croix

pipe flange bride *f* de tube

pipe hanger collier *m*/support *m* de tuyauterie, étrier *m* de suspension des tubes

pipe hook corbeau *m*

pipe insulation, half round, preformed coquille *f*, isolant *m* en coquille

pipe jacking machine (pneumatic) pousse-tube *m* (pneumatique)

pipe joint, *see* coupling, or joint, pipe

pipe junction fitting culotte *f*

pipe layer [tradesman] poseur *m* de tuyaux

pipe leak fuite *f* de tuyau

pipe network réseau *m* de canalisations

pipe reducer   raccord *m* de réduction
pipe sleeve [in concrete, masonry, etc]   fourreau *m*, réservation *f*
pipe socket   manchon *m* de tuyau
pipe support   support *m* de tube/tuyau
pipe system   système *m* de tuyauteries
pipe thread   filetage *m* de tubes
pipe threading machine   machine *f* à fileter des tubes
pipe trench   tranchée *f* pour conduites
pipe union   raccord *m* union
pipe wrapping   revêtement *m* des tubes
pipe wrench   clef *f* de tuyau, pince *f* multiprise, serre-tubes *m*
pipe-bending   cintrage *m* des tubes
pipelayer [machine]   pose-tubes *m*, tracteur *m* poseur de canalisations
pipelayer, mole drainage   pose-drains *m*
pipelaying tractor   tracteur *m* poseur de canalisations
pipeline   pipe-line *m*
pipeline [for oil products]   oléoduc *m*
pipeline, gas   canalisation *f* de gaz, gazoduc *m*
pipeline, pressure   canalisation *f* sous pression, conduite *f* forcée
pipework   canalisation *f*, tuyauterie *f*
pipework, balancing   tuyauterie *f* d'équilibr(ag)e
pipework, buried   tuyauterie *f* enterrée
pipework, condensate   tuyauterie *f* d'eau condensé
pipework, drainage   canalisation *f* d'assainissement
pipework, flow (feed)   canalisation *f* d'alimentation
pipework, foul water drain   tuyauterie *f* des eaux usées
pipework, heating   tuyauterie *f* de chauffage
pipework, plastic [for pressurized plumbing and heating systems]   hydrocâblage *m*
pipework, return   canalisation *f* de retour
pipework, steam   tuyauterie *f* de vapeur
pipework, suction   tuyauterie *f* d'aspiration
pipework, ventilating   tuyauterie *f* d'aération
pipework unit (prefabricated) for kitchen and bathroom   bloc-eau *m*
piping, *see also* pipe, pipework   canalisation *f*, tuyauterie *f*
pit   fosse *f*, fouille *f*, trou *m*
pit [of a theatre or concert hall]   parterre *m*
pit [mine]   (puits *m* de) mine *f*
pit, ash   cendrier *m*, fosse *f* à cendres
pit, blow-down   fosse *f* de vidange
pit, cable   puits *m* à câbles
pit, gravel   ballastière *f*, carrière *f*, gravière *f*
pit, inspection   fosse *f*
pit, masonry lined   bassin *m*
pit, meter   fosse *f* à compteur
pit, midden [sump for liquid manure]   fosse *f* à purin
pit, sand   carrière *f* de sable, sablière *f*
pit, seepage   puits *m* d'infiltration
pit for liquid agricultural manure   fosse *f* à purin
pitch (tar)   brai *m*
pitch [of gear teeth, a spiral, etc]   pas *m*

pitch [spacing]   distance *f* d'espacement, espacement *m*
pitch diameter [of a gear wheel]   diamètre *m* primitif
pitch of a roof   chute *f* d'un toit
pitch, rivet   écartement *m*/pas *m* des rivets
pitching, stone   hérisson *m*, enrochement *m*, masque *f* en enrochement, perré *m*
pitching stone   pierre *f* du hérisson
pitot tube   tube *m* de Pitot
pit-prop   étançon *m*
pitted [of metal]   piqué *adj*
pitting   corrosion *f* ponctuelle
pitting [of metal]   piquage *m*
pitting [on metal surface]   piqûre *f*
pivot   pivot *m*
pivot   pivoter *vb*
pivoting *adj*   pivotant *adj*
pivoting   pivotement *m*
place   endroit *m*, point *m*
place [e.g a square in a town]   place *f*
place   placer *vb*, poser *vb*
place [an investement, a person]   placer *vb*
place end-to-end   rabouter *vb*
placer spreader   précribleur répartiteur *m*
placing [location]   emplacement *f*
placing [the action]   mise *f* en place
plain [unadorned]   simple *adj*
plain   plaine *f*
plain, alluvial   zone *f* d'alluvionnement
plain, flood   zone *f* d'inondation
plan [drawing]   dessin *m*, plan *m*, projection *f* horizontale
plan [design, or intention]   projet *m*
plan [of a building]   ordonnance *f*
plan [of property parcels in a commune]   cadastre *m*
plan [of road, railway, etc]   tracé *m*
plan, area development   plan *m* de secteur
plan, building   plan *m* de construction
plan, commissioning   plan *m* de mise en route
plan, detailed health and safety   plan *m* particulier de santé et sécurité
plan, dimensioned, of a plot of land   mesurage *m*
plan, drainage   plan *m* d'évacuation des eaux
plan, emergency (response)   plan *m* d'intervention en cas d'urgence
plan, general transport/traffic   plan *m* général des transports
plan, health and safety   plan *m* d'hygiène et de sécurité (PHS), plan *m* de santé et sécurité
plan, installation   plan *m* d'installation
plan, land use or zoning   plan *m* d'occupation des sols (POS)
plan, layout   plan *m* d'implantation/de situation
plan, location or site   plan *m* d'ensemble/de masse/de repérage/de situation
plan, local [showing permissible land uses]   plan *m* d'occupation des sols (POS)
plan, master   plan *m* directeur
plan, modernization   plan *m* de modernisation

plan, overall layout plan *m* (de) masse *f*, plan *m* général d'implantation
plan, plot layout plan *m* parcellaire
plan, quality assurance plan *m* d'assurance (de la) qualité (PAQ)
plan, site facilities or layout plan *m* d'installation du chantier
plan, study plan *m* d'études
plan, subdivision [showing layout of plots] plan *m* parcellaire
plan, zoning plan *m* d'occupation des sols (POS)
plan drawing dessin *m* des plans
plan of a ZAC [*zone d'aménagement concerté*] plan *m* d'aménagement de zone (PAZ)
plan view vue *f* en plan
plan [a house, an estate, etc] concevoir *vb*, dresser *vb* les plans de
plan [intend to] avoir *vb* l'intention de, projeter *vb* de
plan [programme] planifier *vb*
plan [work out, a project, etc] élaborer *vb*, préparer *vb*
plan to do sth projeter *vb* de faire qch
plane [surface, etc] plan *adj*
plane [for timber, etc] rabot *m*
plane [of a surface] nu *m*
plane [surface] plan *m*
plane [tree] platane *m*
plane, bedding plan *m* de stratification
plane, cleavage plan *m* de clivage
plane, compass rabot *m* cintré/à semelle cintré
plane, fault plan *m* de faille
plane, grooving guimbarde *f*, rabot *m* à languette
plane, hand rabot *m* à main
plane, inclined plan *m* incliné
plane, jack demi-varlope *f*, rabot *m* à dégrossir, riflard *m*
plane, moulding mouchette *f*, rabot *m* à moulures
plane, ogee/moulding doucine *f*
plane, plough bouvet *m*
plane, rabbet/rebate guillaume *m*
plane, router guimbarde *f*
plane, shear plan *m* de cisaillement
plane, sliding plan *m* de glissement
plane, slip plan *m* de glissement
plane, smoothing rabot *m* plat
plane, sole of a plan *m* d'un rabot
plane, toothing rabot *m* à coller/à dents/denté
plane, trying rabot *m* long, varlope *f*
plane, web plan *m* de l'âme
plane, work(ing) plan *m* utile/de travail
plane handle corne *f*/nez *m*/pommeau *m* d'un rabot
plane iron [blade] fer *m* (de rabot)
plane of cleavage of a stone bed lit *m*
plane of sliding plan *m* de glissement
plane of web plan *m* de l'âme
plane table planche *f* de topographe
plane raboter *vb*
plane [make even a wood or metal surface] planer *vb*
plane [flooring] replanir *vb*

planed (off) raboté *adj*
planer machine *f* à raboter, raboteuse *f*
planer [for road surfaces] aplanisseuse *f*
planer, cold [roadworks] fraiseuse *f*/raboteuse *f* (routière) à froid
planer, surface [for straightening wood] dégauchisseuse *f*, machine *f* à dresser/à corroyer
planer mark on wood surface onde *f*
planimetry planimétrie *f*
planing rabotage *m*
planing [making even a wood or metal surface] planage *m*
planing machine machine *f* à raboter, raboteuse *f*
planing of flooring replanissage *m*
planing tools outillage *m* à façonner
plank, *see also* board, timber, wood
plank 30–60mm thick ais *m*
plank 65×170mm bast(a)ing *m*
plank less than 54mm thick planche *f*
plank, heartwood planche *f* de cœur
plank, rough [for casing a well, timbering a tunnel] chon *m*
plank, square-edged planche *f* avivée/équarrie/à arêtes vives
plank, trimmed planche *f* avivée/équarrie/à arêtes vives
plank, untrimmed planche *f* en grume/non équarrie
plank, waney-edged planche *f* en grume/non équarrie
platform, plank platelage *m*
planking planchéiage *m*, platelage *m*
planking and strutting blindage *m*
planking between templates [arch/vault construction] couchis *m* de voûte
planner [of projects] projeteur *m*
planner [of work, rather than urban, town, etc] planificateur *m*
planner, town [not officially recognized title in France] urbaniste *m/f*
planners, French society of Société *f* Française des Urbanistes (SFU)
planning [sequencing] planification *f*
planning, housing and highways department [in France], *see also* country planning etc l'Équipement[†] *m*
planning, town urbanisme *m*
planning application demande *f* de permis de construire
planning application documents dossier *m* de permis de construire
planning by-law or regulation règle *f* d'urbanisme
planning certificate, town or urban certificat *m* d'urbanisme
planning department/office [town planning] administration *f*/service *m* d'urbanisme
planning department/office [sequencing work] bureau *m*/service *m* de planification
planning stage, at (the) en projet
plant [botanical] herbe *f*, plante *f*
plant [fixed] *see also* installation centrale *f*, installation *f*
plant [equipment] équipements *m,pl*, matériel *m*

plant [factory rather than equipment]   usine *f*

plant, air-conditioning   centrale *f* de climatisation, installation *f* de climatisation (d'air)/de conditionnement d'air

plant, ash handling   installation *f* du transport des cendres

plant, asphalt   centrale *f*/poste *m* d'enrobage

plant, asphalt, batch type   poste *m* d'enrobage à mélange discontinu

plant, asphalt drum mix(ing)   tambour *m* sécheur enrobeur

plant, asphalt recycling   poste *m* de recyclage de chaussée bitumineuse

plant, asphaltic concrete batching   centrale *f* d'enrobage

plant, batching [concrete]   centrale *f* à béton, usine *f* à béton/de bétonnage

plant, boiler   chaufferie *f*, installation *f* des chaudières

plant, boilermaking   chaudronnerie *f*

plant, central boiler   centrale *f* de chauffage

plant, central heating   installation *f* de chauffage central

plant, central heating [central boiler]   chaufferie *f* centrale

plant, chilled water   installation *f* d'eau glacée

plant, chlorination   poste *m* de chloration

plant, cleaning   station *f* d'épuration

plant, coating   poste *m* d'enrobage

plant, compacting   engins *m,pl* de compactage

plant, concrete recycling   poste *m* de recyclage de béton résiduel

plant, cooling   installation *f* de réfrigération

plant, crushing [for aggregates, rock, etc]   installation *f* de concassage

plant, district heating   centrale *f*/installation *f* de chauffage urbain

plant, dry batching   centrale *f* de malaxage à sec

plant, drying   installation *f* de séchage

plant, dual duct   installation *f* à deux conduits

plant, duplicate   installation *f* de réserve

plant, dust extraction/removal   installation *f* de dépoussiérage

plant, earth-moving   engins *m,pl*/matériel *m* de terrassement

plant, evaporating   installation *f* d'évaporation

plant, factory   équipement *m* d'usine

plant, factory-assembled   matériel *m* assemblé en usine

plant, filter/filtration   installation *f* de filtrage

plant, freezing   installation *f* de congélation

plant, ice-making   installation *f* de production de glace

plant, gas producer   gazogène *m*

plant, heating   installation *f* de chauffage

plant, high-velocity   installation *f* à haute vitesse

plant, hot air   installation *f* de chauffage à air chaud

plant, incinerator   installation *f*/usine *f* d'incinération

plant, low-pressure gas   installation *f* à gaz à basse pression

plant, low-velocity   installation *f* à faible vitesse

plant, machine   engin *m*

plant, modular air-conditioning   centrale *f* modulaire de climatisation

plant, open air/outdoor   installation *f* en plein air

plant, piece of   engin *m*

plant, pickling [metal]   atelier *m* de décapage

plant, power   centrale *f* de production d'énergie

plant, purification   station *f* d'épuration

plant, recovery   installation *f* de récupération

plant, recycling   installation *f* de recyclage

plant, refuse collecting   installation *f* de ramassage des ordures

plant, road retreatment   train *m* de reconditionnement/retraitement de chaussée

plant, screening [for granular materials]   installation *f* de criblage, station *f* de tamisage

plant, screening [for sewage, etc]   installation *f* de filtrage

plant, separate heating   installation *f* de chauffage annexe

plant, separating   installation *f* de triage

plant, sewage treatment   station *f* de traitement des eaux usées

plant, single-duct air-conditioning   installation *f* de climatisation à un conduit

plant, site   matériel *m* de chantier

plant, sludge digestion   digesteur *m*

plant, sludge pressing   filtre-presse *m*

plant, softening [water]   installation *f* d'adoucissement

plant, standby   installation *f* de réserve

plant, suction   installation *f* d'aspiration

plant, tidal power   usine *f* marémotrice

plant, treatment   poste *m* de conditionnement, station *f* de traitement

plant, vacuum   installation *f* à vide

plant, variable volume   installation *f* à volume variable

plant, ventilating   installation *f* d'aération/de ventilation

plant, washing   installation *f* de lavage

plant, waste collecting   installation *f* de ramassage des ordures

plant, water softening   installation *f* d'adoucissement

plant, water supply   installation *f* d'alimentation en eau

plant, water treatment   poste *m* de conditionnement d'eau, usine *f* de traitement des eaux

plant cost   coût *m* de centrale

plant operator   conducteur *m* d'engin

plant room   local *m* technique/de centrale

plant semi-trailer   semi-remorque *f* porte-engins

plant [grass, flowers, etc, or a stake]   planter *vb*

plaque [on a wall]   plaque *f*

plaster   plâtre *m*

plaster, alabaster   plâtre *m* d'albâtre

plaster, alum [type of]   ciment *m* anglais

plaster, alum/borax [used for stucco work]   plâtre *m* aluné/boraté

plaster, anhydrous   plâtre *m* anhydre

plaster, coating   plâtre *m* à enduit

plaster, fair-faced enduit *m* lisse de plâtre
plaster, fibrous staff *m*
plaster, finely sieved [for fine mouldings, etc] plâtre *m* au sas
plaster, finish(ing) plâtre *m* de finition
plaster, gypsum plâtre *m*
plaster, trowelled plâtre *m* lissé
plaster debris from demolition plâtras *m*
plaster infill [concave] **between ceiling joists** auget *m*
plaster skim coat(ing) enduit *m* fin
plaster work, moulded pigeonnage *m*
plaster plâtrer *vb*
plasterboard panneau *m*/plaque *f* de plâtre, Placoplâtre *m* [trade name]
plasterer plafonneur *m*, plâtrier *m*
plasterer's hawk taloche *f*
plasterer's tool, small patte *f* d'oie
plasterer's tool for hacking plaster hacheteau *m*
plasterer's trough auge *f*
plastering plâtrage *m*
plasterwork légers ouvrages *m,pl*
plasterwork, rough hourdage *m*
plastic plastique *f*
plastic limit [soil property] limite *f* de plasticité
plastic range domaine *m* plastique
plasticity plasticité *f*
plasticity [of concrete] fluidité *f*
plasticity chart diagramme *m* de plasticité
plasticity index indice *m* de plasticité
plasticity limit [soil property] limite *f* de plasticité
plasticity needle, Proctor aiguille *f* de Proctor
plasticize [i.e. to add a plasticizer] plastifier *vb*
plasticizer [e.g. for concrete] plastifiant *m*
plasticizer [for asphalt, bitumen, concrete] fluidifiant *m*, fondant *m*
plate plaque *f*
plate [circular] disque *m*
plate [of metal] tôle *f*
plate [slab] dalle *f*
plate, anchor plaque *f* d'ancrage
plate, base plaque *f* d'assise/de base
plate, bearing plaque *f* d'appui
plate, bottom [of a wall panel] sablière *f*
plate, check plaque *f* d'arrêt
plate, chequer plaque *f* striée, tôle *f* gaufrée/striée
plate, cover plaque *f* de protection
plate, end [of a pipe, etc] tampon *m*
plate, fascia plaque *f* de recouvrement
plate, foot [of wood, under a prop, etc] couche *f*, semelle *f*
plate, ground [of a timber frame panel] sablière *f*
plate, heavy or thick tôle *f* forte
plate, metal platine *f*
plate, orifice diaphragme *m*, plaque *f* d'obturation
plate, perforated plaque *f* perforée
plate, reinforcing or stiffening plaque *f*/tôle *f* de renfort
plate, splice éclisse *f*, plate-bande *f*

plate, thick tôle *f* forte
plate, top [of a timber frame panel] sablière *f* supérieure/de forme
plate, vibrating [for compacting] patin *m* vibrant
plate, wall, *see* wall plate
plate preventing wrong shape key entering a lock garde *f* de serrure
plate to cover unused hole in a sanitary appliance mascaron *m*
plateau plateau *m*, relief *m* tabulaire
platform plate-forme *f*
platform [e.g. of station] quai *m*
platform, aerial engin *m* élévateur à nacelle
platform, aerial access/work élévateur *m* à nacelle
platform, board plate-forme *f* de madriers, platelage *m*
platform, control plate-forme *f* de commande
platform, drilling plate-forme *f* de forage
platform, elevating plate-forme *f* élévatrice
platform, lifting plate-forme *f* de levage
platform, loading [railway] perron *m* de chargement
platform, observation plate-forme *f* d'observation
platform, scaffolding board plate-forme *f* de madriers, platelage *m*
platform, scissor lift (work) plate-forme *f* à ciseaux/compas
platform, station quai *m* de gare
platform, work estrade *f* de travail
platform, work(ing) passerelle *f*/plancher *m*/plate-forme *f* de travail
platform, work(ing) [on scaffolding] plateau *m* d'échafaudage
platform roof, glass [station] marquise *f*
plating, cadmium/chromium/nickel cadmiage *m*/chromage *m*/nickelage *m*
play [clearance] jeu *m*
plenum plénum *m*
plenum chamber chambre *f* de mise en charge/en pression
pliers tenailles *f,pl*
pliers, combination (cutting) pince *f* universelle
pliers, pair of pince *f*, pinces *f,pl*
pliers, round nose pince *f* à becs ronds
pliers, steelfixer's pinces *f,pl* à ligatures
plinth embase *f*, socle *m*
plinth [basement] soubassement *m*
plinth [of a column] plinthe *f*
plot [graph] graphique *m*
plot [of land] parcelle *f* [de terrain]
plot, bordering parcelle *f* riveraine
plot, building lot *m*, terrain *m* à bâtir
plot boundary limite *f* du terrain
plot layout plan plan *m* parcellaire
plot ratio coefficient *m* d'occupation du sol (COS)
plot [a curve, etc] tracer *vb*
plotter [computer] traceur *m*
plotter, flat-bed traceur *m* à plat
plotter, thermal traceur *m* thermique
plot(ting) [of curves, data] tracé *m*

plough   labourer *vb*

plough, trenching   défonceuse *f*

plug [for blocking or stopping up sth]   bouchon *m*, tampon *m*

plug [electrical]   prise *f* électrique mâle

plug [for fixing, e.g. in a wall]   tampon *m*

plug [of a bath, etc]   bonde *f*

plug, brazed lead [blocking the end of a lead pipe]   nœud *m* de tamponnage

plug, connector   fiche *f* de connexion

plug, drain   robinet *m* de vidange

plug, earthed   fiche *f* (mâle) de sécurité

plug, electric [not socket]   fiche *f*

plug, extension   fiche *f* (mâle) de prolongateur

plug, fusible   bouchon *m* fusible

plug, lockable   bouchon *m* cadenassable

plug, screwed   bouchon *m* à vis

plug, spark   bougie *f*

plug, spigot   bouchon *m* mâle

plug, three-phase   fiche *f* (mâle) pour triphasé

plug, three-pin   fiche *f* (mâle) à trois broches

plug, wall [fixing]   cheville *f*, tampon *m*

plug, wall [socket outlet]   bouchon *m* de prise/de contact

plug gap   écartement *m* des électrodes

plug hole [of a bath, etc]   bonde *f*

plug [a hole, a pipe, etc]   boucher *vb*

plug [a leak, a hole]   colmater *vb*

plug [a wall]   tamponner *vb*

plug in [electrically]   brancher *vb*

plug up [a hole, etc]   taper *vb*

plugged [openings, holes]   bouché *adj*

plugged [wall, etc]   tamponné *adj*

plugging [of leaks, holes]   colmatage *m*

plugging and filling prior to wallpapering   bordage *m*

plum tree   prunier *m*

plumb [i.e. vertical]   à plomb, aplomb, d'aplomb

plumb [of a wall, etc]   droit *m*

plumb, be out of   porter *vb* à faux, surplomber *vb*

plumb, get out of   prendre *vb* coup

plumb, out of   dévers *adj*, en porte-à-faux

plumb bob   fil *m* à plomb

plumb line   fil *m* à plomb

plumb with   à l'aplomb de

plumbed with   fait à l'aplomb de

plumber   plombier *m*

plumber's chain tongs   serre-tubes *m*

plumber's hammer   marteau *m* à main

plumber's mate [French regional expression]   garçon-plombier *m*

plumber's rasp   écouenne *f*

plumber's tool for expanding/bellmouthing pipe ends   toupie *f*

plumber's union [joint]   raccord *m* union

plumber's vice   serre-tubes *m*

plumbing   plomberie *f*

plume [smoke, steam, etc]   panache *m*

plummet   fil *m* à plomb

plunger [router]   berceau *m*

plus [sign]   plus *m*

pluviometry   pluviométrie *f*

ply [veneer of plywood]   pli *m*

plywood   (bois *m*) contreplaqué *m*

plywood, moulded   contreplaqué *m* moulé

plywood construction   contre-placage *m*

plywood with lightweight core   contreplaqué *m* sandwich

pneumatic   pneumatique *adj*

pneumatically controlled   à commande pneumatique

pneumatics   pneumatique *f*

pocket   réservation *f*

pocket [for building a fixing into a wall, etc]   auget *m*

pocket calculator   calculatrice *f* de poche

point [in a process, programme, of an idea, etc]   point *m*

point [sharp end]   pointe *f*

point, anchorage   point *m* d'ancrage

point, application   point *m* d'application

point, balance   point *m* d'équilibre

point, boiling   point *m* d'ébullition

point, branch-off   point *m* de dérivation

point, break-even   seuil *m* de rentabilité

point, breaking   limite *f* de rupture

point, burning   point *m* d'inflammation

point, compressive yield   limite *f* d'écrasement, limite *f* élastique à la compression

point, contact   point *m* de contact

point, crossover   point *m* d'intersection

point, cut-in   point *m* d'enclenchement

point, cut-off   point *m* de coupure

point, cut-out   point *m* de déclenchement

point, datum [for setting out, etc]   point *m* de repère

point, delivery   poste *m* de livraison

point, departure   point *m* de départ

point, fixed   point *m* fixe

point, fixing   point *m* de fixation

point, flash   point *m* d'éclair/d'inflammation

point, freezing   point *m* de congélation

point, high/low   point *m* haut/bas

point, highest [of a hill, tree]   sommet *m*

point, measuring   point *m* de mesure

point, melting   point *m* de fusion

point, operating   point *m* de fonctionnement

point, reference   point *m* côté/de référence, repère *m*

point, support   point *m* d'appui

point, triangulation   point *m* géodésique

point, turning   tournant *m*

point, yield   limite *f* élastique/d'écoulement

point of application   point *m* d'application

point of bearing   point *m* d'appui

point of change of gradient   point *m* d'inflexion

point of contraflexure or inflexion   point *m* d'inflexion

point of entry/exit [for traffic]   point *m* d'entrée/de sortie

point of intersection   point *m* d'intersection

point of intersection of two arc(he)s   tiers-point *m*

point of meeting of glazing bars of a fanlight   trompillon *m*

point [a surveying instrument]   pointer *vb*
point [a wall]   bloquer *vb*
point [masonry]   jointoyer *vb*
pointed   pointu *adj*
pointer   aiguille *f*, index *m*
pointer [masonry, tiles]   jointoyeur *m*
pointing [masonry]   joint *m*
pointing [brickwork, tiles, the action and its result]   jointoiement *m*
pointing, flush   joint *m* affleuré/plat/plein
pointing, hollow   joint *m* creux
pointing, recessed   joint *m* creux
pointing, verge [of a roof]   ruellée *f*
pointing, weathered   joint *m* creux chanfreiné
pointing of a ridge [mortar or plaster]   (r)embarrure *f*
points [railway]   aiguillages *m,pl*
points blade   lame *f* d'aiguille
points cable   transmission *f* funiculaire
points rail   aiguille *f*
poisonous   vénéneux *adj*
Poisson's ratio   coefficient *m*/module *m* de Poisson
poker [fire iron]   pique-feu *m*, tisonnier *m*
poker [for a boiler]   ringard *m*
poker, vibrating [for concrete]   aiguille *f* vibrante/de pervibration
poker vibrator   pervibrateur *m*
polarity   polarité *f*
polarity, change/inversion/reversal of   inversion *f* de (la) polarité
polarize   polariser *vb*
pole   mât *m*, poteau *m*, pylône *m*
pole [electrical, magnetic]   pôle *m*
pole, electricity supply   poteau *m* électrique
pole, gin   écoperche *f*
pole, marker [civil works]   balise *f*
pole, negative/positive [electrical]   pôle *m* négatif/positif
pole, ranging   jalon *m*
pole derrick   derrick *m*, mât *m* de levage
police station   gendarmerie *f*
policy [company]   politique *f*
policy [of insurance]   police *f*
policy decision   décision *f* de principe
polish   polir *vb*
polish [stone]   riper *vb*
polish [a mirror]   doucir *vb*
polished   poli *adj*
polishing [of plate glass, to produce parallel faces]   doucissage *m*
polishing of mirrors   douci *m*
pollutant *adj*   polluant *adj*
pollutant   polluant *m*
pollute   polluer *vb*
pollution   contamination *f*, pollution *f*
pollution, air or atmospheric   pollution *f* d'air/de l'air
pollution, degree of   degré *m* de pollution
polyester   polyester *m*
polyethylene   polyéthylène *m*

polygon of forces   polygone *m* des forces
polymer   polymère *m*
polymerization   polymérisation *f*
polyphase   polyphasé *adj*
polystyrene   polystyrène *m*
polystyrene, expanded/extruded   polystyrène *m* expansé/extrudé
polystyrene, expanded (Styropor)   styropor *m*
polyvinyl chloride (PVC)   polychlorure *m* de vinyl (PVC ou PCV)
pond   bassin *m*, étang *m*, mare *f*
pond [lake]   plan *m* d'eau
pond [in garden or park]   pièce *f* d'eau
pontoon   ponton *m*
pool   bassin *m*, étang *m*
pool, swimming   piscine *f*
poplar   peuplier *m*
poplar plantation   peupleraie *f*
population density   densité *f* de la population
porch   porche *m*
porch roof [without posts]   auvent *m*
pore [e.g. of stone]   pore *m*
porosity   porosité *f*
porous   poreux *adj*
port [both opening and harbour]   port *m*
port, blow-out   orifice *m* d'échappement
port, intake   orifice *m* d'admission
port/harbour facilities   équipements *m,pl*/installations *f,pl* portuaires
portable   portatif *adj*, transportable *adj*
portal [e.g. of tunnel]   portail *m*
portal action   action *f* de portique
portal frame   cadre *m* portique, ossature *f* en portique, portique *m*
portal frame, single   portique *m* simple
portcullis   sarrasine *f*
porter [of a building]   portier *m*
portico   portique *m*
position   emplacement *f*, situation *f*, station *f*
position [location, or job]   position *f*
position [place]   endroit *m*
position, burning [of a lamp]   position *f* de fonctionnement
position   mettre *vb* en place
positive   positif *adj*
possession   jouissance *f*
possible   éventuel *adj*
post   piquet *m*
post [of a fence or sheet pile wall]   palis *m*
post [of gate, roof]   montant *m*
post [roof, truss]   poinçon *m*
post [small vertical strut]   potelet *m*
post [timber structure]   poteau *m*
post, banister   balustre *m*, barreau *m*
post, corner   poteau *m* cornier/d'angle
post, crown   poinçon *m*
post, hanging [of a carriage gateway, etc]   chardonnet *m*
post, principal   poteau *m* de refend

post, roof poinçon *m*

post, small vertical potelet *m*

post, support(ing) poteau *m* de décharge

post, window or door, *see* jamb

post driver engin *m* pour poser les poteaux

post office bureau *m* de poste

Post Office, The [in the UK] Poste *f*, La [in France]

post (wind)mill moulin *m* (à vent) sur pile

post [a letter, parcel] mettre *vb* à la poste

post-tension [external prestressing] précontrainte *f* externe

post-tension(ing) post-contrainte *f*, post-tension *f*

post-treatment traitement *m* ultérieur

postage (costs) frais *m,pl* d'envoi

postage paid franc de port, port payé

postal order chèque *m* postal

poster placard *m*

postern [of a fortress] poterne *f*

postern-gate poterne *f*

postpone ajourner *vb*, différer *vb*, remettre *vb*, reporter *vb*

postponement ajournement *m*

pot, glue pot *m* à colle

pot bearing [rubber] appareil *m* d'appui à pot

pot burner brûleur *m* à coupelle/à pot

potability check [for drinking water] contrôle *m* de potabilité

potential *adj* potentiel *adj*

potential [electrical, etc] potentiel *m*

potential, change in changement *m* de potentiel

potential drop chute *f* de potentiel

potentiometer potentiomètre *m*

pothole [in the ground] cloche *f* d'effondrement, marmite *f* d'érosion, poche *f*

pothole [in a road, pathway, etc] flache *f*, nid *m* de poule

pottery ware [terracotta] poterie *f*

pour, limit of [concrete work] arrêt *m* de coulage

pour [concrete] couler *vb*

pour [concrete, into formwork or shuttering] bancher *vb*

pour [liquids] verser *vb*

pour away vider *vb*

pour out or forth verser *vb*

pouring [of concrete, asphalt road material, etc] coulage *m*

pouring [of concrete using formwork or shuttering] banchage *m*

pouring out [of liquid, etc] versement *m*

powder poudre *f*

powder, grind to pulvériser *vb*

powder coated thermolaqué *adj*

power pouvoir *m*

power [of a machine, motor, etc] puissance *f*

power [electric] courant *m*

power, at full à toute puissance

power, at high à haute puissance

power, acoustic intensité *f* acoustique

power, electric, *see* electric power

power, rated puissance *f* nominale

power, reactive puissance *f* réactive

power, tidal énergie *f* marémotrice

power, unit of unité *f* de puissance

power, usable or useful effet *m* utile

power, wind énergié *f* éolienne

power factor facteur *m* de puissance

power factor correction compensation *f* de facteur de puissance

power failure or loss [electrical] panne *f* de courant

power generated or absorbed, electrical charge *f*

power generation, electrical production *f* de puissance électrique

power line, high tension/voltage ligne *f* à haute tension

power line, overhead ligne *f* aérienne

power point prise *f* de courant

power requirement besoin *m* d'énergie, énergie *f* nécessaire

power source source *f* d'électricité

power station centrale *f* (électrique/thermique)

power station, atomic centrale *f* atomique

power station, combined heat and centrale *f* combinée chaleur force/de chaleur/force

power station, electric/hydro-electric centrale *f* hydro-électrique/hydraulique

power station, nuclear centrale *f* atomique/nucléaire

power station, thermal centrale *f* thermique

power station, tidal usine *f* marémotrice

power supply alimentation *f* en électricité

power supply, electric [amount] puissance *f* électrique

power switch [disconnector] délesteur *m*

powerhouse centrale *f*

powerhouse [electric] centrale *f* électrique

pozzolan(a) pouzzolane *m*

practical fonctionnel *adj*, pratique *adj*

practice [actual, common] pratique *f*

practice, in (actual) dans la pratique

practice, put something into mettre *vb* qch en pratique

practice for building works, code of [in France] document *m* technique unifié (DTU)

practise [a trade, skill or profession] exercer *vb*, pratiquer *vb*

pre-assembled prémonté *adj*

pre-assembly préassemblage *m*

prebore préforer *vb*

prebored préforé *adj*

precast *adj* préfabriqué *adj*

precast préfabriquer *vb*

precast in the factory/on site préfabriquer *vb* en atelier/sur chantier

precasting préfabrication *f*

precasting yard aire *f* de préfabrication

precedent condition condition *f* suspensive

precedents [of a legal case] jurisprudence *f*

precincts [e.g. of a cathedral] pourtour *m*

precipitation précipitation *f*

precipitation, daily quantité *f* journalière de pluie

precision justesse *f*, précision *f*

pre-coated préenrobé *adj*

precompaction précompactage *m*

precompression précompression *f*
pre-cooler prérefroidisseur *m*
pre-cooling prérefroidissement *m*
pre-design avant-projet *m*
prediction prévision *f*
predrill préforer *vb*
predrilled préforé *adj*
pre-emptive right [generally to acquire, in France] droit *m* de préemption
prefabricate (in the factory/on site) préfabriquer *vb* (en atelier/sur chantier)
prefabricated préfabriqué *adj*
prefabricated in the factory/on site préfabriqué *adj* en atelier/sur chantier
prefabricated lower part of a lintel containing main reinforcement prélinteau *m*
prefabrication préfabrication *f*
prefabrication of elements fitting together préfabrication *f* fermée
prefabrication of elements to be combined on site with traditional construction préfabrication *f* ouverte
prefabrication of lightweight elements préfabrication *f* légère
prefabrication of major elements préfabrication *f* lourde
prefilter préfiltre *m*
preheat préchauffer *vb*
preheater préchauffeur *m*
preheat(ing) préchauffage *m*
preheating, air préchauffage *m* d'air
pre-insulated préisolé *adj*
preliminary préliminaire *adj*
preload(ing) préchargement *m*
premises bâtisse *f*, biens *m,pl* bâtis, lieux *m,pl*, locaux *m,pl*
premises, rear [of a building] arrière-corps *m*
premium [insurance] cotisation *f*, prime *f*
premixed prédosé *adj*
prepack conditionner *vb*
preparation préparation *f*
preparatory step [towards sth] acheminement *m*
prepare [fit or lay out] aménager *vb*
prepare (for) amorcer *vb*
prequalification candidature *f*
prequalify, invitation to appel *m* de candidature
presbytery presbytère *m*
present [i.e. current] actuel *adj*
preservation conservation *f*, maintien *m*
pre-set *adj* préréglé *adj*
pre-set prérégler *vb*
pre-setting préréglage *m*
press presse *f*
press, veneering presse *f* à plaquer
press a mortise into a tenon emmortaiser *vb*
press down (on) peser (sur) *vb*
press excess paint out of a brush dégorger *vb*
press [metals] emboutir *vb*
press-button, *see* button, press

pressed [generally of metal sheet] embouti *adj*
pressostat, air pressostat *m* de manque d'air
pressure, high haute pression
pressure tight étanche *adj* à la pression
pressure compression *f*, impression *f*, pression *f*
pressure, acoustic pression *f* acoustique
pressure, active earth poussée *f* des terres
pressure, ambient pression *f* ambiante
pressure, arch poussée *f* de voûte
pressure, atmospheric pression *f* atmosphérique
pressure, axial pression *f* axiale
pressure, back contre-pression *f*
pressure, balance pression *f* d'équilibre
pressure, barometric pression *f* barométrique
pressure, boiler pression *f* dans la chaudière
pressure, burner pression *f* dans le brûleur
pressure, control pression *f* de contrôle
pressure, critical pression *f* critique
pressure, delivery pression *f* de refoulement
pressure, design or nominal working pression *f* nominale de travail
pressure, differential pression *f* différentielle
pressure, discharge pression *f* de décharge
pressure, distribution pression *f* d'alimentation
pressure, dynamic pression *f* dynamique
pressure, earth pression *f* des terres
pressure, equalizing pression *f* de compensation
pressure, excess surpression *f*
pressure, gas pression *f* de gaz
pressure, grille differential perte *f* de charge de bouche d'air
pressure, horizontal poussée *f*
pressure, hydraulic pression *f* hydraulique
pressure, hydraulic [total] charge *f*
pressure, low basse pression *f*
pressure, maximum safe pression *f* limite de sécurité
pressure, nominal working pression *f* nominale de travail
pressure, oil pression *f* d'huile
pressure, operating pression *f* de marche/de service/normale de travail
pressure, partial pression *f* partielle
pressure, passive earth butée *f* des terres
pressure, pore water pression *f* d'eau interstitielle
pressure, reduced pression *f* réduite
pressure, rock contrainte *f* dans la roche/initiale dans le massif
pressure, static pression *f* statique
pressure, steam or vapour pression *f* de vapeur
pressure, suction pression *f* d'aspiration
pressure, surface pression *f* superficielle
pressure, test pression *f* d'épreuve/d'essai
pressure, test [of a boiler] timbre *m*
pressure, uplift sous-pression *f*, poussée *f* ascensionnelle
pressure, vapour, *see* vapour pressure
pressure, velocity pression *f* de vitesse
pressure, water pression *f* de l'eau
pressure, wind poussée *f*/pression *f* du vent

pressure, working   pression *f* de marche/de service/normale de travail

pressure booster   surpresseur *m*

pressure change   changement *m* de pression

pressure compensator   compensateur *m* de pression

pressure control valve   soupape *f* de surpression

pressure decrease or reduction   réduction *f* de pression

pressure distribution   distribution *f* de pression

pressure drop   chute *f*/perte *f* de pression

pressure drop across a filter   pression *f* filtrante différentielle

pressure gauge, differential   manomètre *m* différentiel

pressure head   hauteur *f* manométrique/de pression/de charge/de refoulement

pressure loss   chute *f*/perte *f* de pression, perte *f* de charge

pressure measurement   mesure *f* de la pression

pressure meter   pressiomètre *m*

pressure pipeline   canalisation *f* sous pression

pressure reducing point   point *m* de détente

pressure reducing valved   étendeur *m*, détendeur *m* de pression, soupape *f* réductrice de pression

pressure reducing valve, spring-loaded   détendeur *m* à ressort, soupape *f* de réduction à ressort

pressure reducing valve, steam   détendeur *m* de vapeur

pressure reducing valve, two stage   détendeur *m* à double détente

pressure reduction   détente *f*, réduction *f* de pression

pressure regulating valve   contrôleur *m*/réducteur *m* de pression

pressure regulator   régulateur *m* de pression

pressure relief valve   robinet *m* décompresseur, soupape *f* de sécurité

pressure rise   augmentation *f* de pression

pressure sensing device   détecteur *m* de pression

pressure switch   déclencheur *m*, manostat *m*, pressostat *m*

pressure tank   réservoir *m* d'air anti-bélier

pressure test   épreuve *f*/essai *m* de pression

pressure transducer   capteur *m* de pression

pressure treatment [for timber]   traitement *m* autoclave

pressure tunnel [dam]   galerie *f* en charge

pressure vessel   réservoir *m* sous pression

pressure, be under   être *vb* sous pression

pressure, resist   contre-buter *vb*

pressure-void ratio curve   courbe *f* pression-indice des vides

pressurization   pressurisation *f*

pressurize   mettre *vb* en pression

pressurized   en surpression *adj*, sous pression *adj*

prestress   mettre *vb* en précontrainte, précontraindre *vb*

prestressed   précontraint *adj*

prestress(ing)   précontrainte *f*

prestressing, external [post-tension]   précontrainte *f* externe

prestressing by pre-tensioned bonded wires (Hoyer method)   précontrainte *f* par fils adhérents

prestressing by tendons   précontrainte *f* par câbles

pretension [internal prestressing]   précontrainte *f* interne

pretension(ing)   prétension *f*

pretreated   prétraité *adj*

prevention, accident   prévention *f* des accidents

prevention, rust   protection *f* anti-rouille

previous owner   ancien propriétaire *m*

price   coût *m*, prix *m*

price [of base metals, securities]   cours *m*

price, asking   prix *m* demandé/de départ

price, basic   prix *m* de base

price, bottom or floor   prix *m* plancher

price, cash   prix *m* au comptant

price, ceiling   prix *m* plafond

price, closing   cours *m* de clôture, dernier cours *m*

price, contract [specific contract]   montant *m* du marché

price, contract, inclusive or lump sum   prix *m* à forfait, prix *m* forfaitaire

price, cost   coût *m*/prix *m* de revient, prix *m* coûtant

price, estimated   prix *m* calculé

price, factory   prix *m* de fabrique

price, factory (gate)   prix *m* départ usine

price, firm/fixed   forfait *m*, prix *m* à forfait, prix *m* fixe/forfaitaire

price, list   prix *m* de catalogue

price, lump sum   prix *m* forfaitaire/à forfait

price, marked   prix *m* marqué

price, market   prix *m* courant

price, opening   cours *m* d'ouverture

price, purchase   prix *m* d'achat

price, retail   prix *m* de détail

price, selling   prix *m* de vente

price, set   prix *m* fixe

price, trade   prix *m* net

price, unit   prix *m* unitaire

price, updated   prix *m* mis à jour

price, wholesale   prix *m* de gros/en gros

prices, market   cours *m* du marché

price book, builder's   série *f* des prix du bâtiment

price list   bordereau *m*/liste *f* de prix, tarif *m*

price revision   revision/révision *f* des prix

price revision formula   formule *f* de révision des prix

price schedule   bordereau *m* de prix

price, fix/quote/set a   établir *vb* un prix

price for, fix a   tarifer *vb*

primary distributor [road]   voie *f* de distribution, réseau *m* primaire

prime [a pump]   amorcer *vb*

prime [a surface]   apprêter *vb*, primer *vb*

prime [a surface, with paint or varnish]   nourrir *vb*

prime [for painting]   imprimer *vb*

primer [first coat on absorbent substrate]   couche *f* d'impression

primer [paint]   couche *f* de fond/d'impression

primer, thick white [prior to painting]   réparure *f*

priming [for decorating]   apprêt *m*

priming [of a boiler, or painting]   primage *m*

priming [of a pump]   amorçage *m*, amorce *f*

priming [painting]   impression *f*, primage *m*

principle   principe *m*

principles, basic   éléments *m,pl*
principle of modular construction   principe *m* de la construction modulaire
principle of superposition   principe *m* de superposition
print [of a drawing, etc]   tirage *m*
print   imprimer *vb*
printer [computer]   imprimante *f*
printer, inkjet   imprimante *f* à jet d'encre
prior notice (of commencement of works)   avis *m* préalable
priory   prieuré *m*
private   privatif *adj*, privé *adj*
private areas [of buildings]   parties *f,pl* privatives
private individual   particulier *m*
probability   probabilité *f*
probability calculation   calcul *m* de probabilités
probe   sonde *f*
probe, temperature   sonde *f* de température
probe, vibrating   vibrolance *m*
probe [to test, etc]   sonder *vb*
problem   problème *m*
problem, design   problème *m* de dessin/de conception
procedure   opération *f*, procédure *f*, processus *m*
procedure, execution   méthode *f* d'exécution
procedure, starting   opération *f* de mise en marche
proceedings [legal]   procès *m*
process   processus *m*
process [administrative, legal]   procédure *f*
process [industrial, physical]   procédé *m*
process [legal action]   procès *m*
process, activated sludge   procédé *m* des boues *f,pl* activées
process, air-conditioning   procédé *m* de conditionnement d'air
process, chemical treatment   procédé *m* de traitement chimique
process, defrosting   procédé *m* de dégivrage
process, drying   procédé *m* de séchage
process, freezing   procédé *m* de congélation
process, gradual   gradation *f*
process, manufacturing   procédé *m* de fabrication
process, mixing   processus *m* de mélange
process, standard   procédé *m* normalisé
process, treatment   procédé *m* de traitement
process server, court   huissier *m*
process   traiter *vb*
process an order   donner *vb* suite à une commande
processing   traitement *m*
processor, central [of a computer]   unité *f* centrale
processor (unit), central   processeur *m* central
Proctor compaction test   essai *m* de compactage Proctor
Proctor curve   courbe *f* Proctor
Proctor density   densité *f* Proctor
Proctor plasticity needle   aiguille *f* de Proctor
produce   produire *vb*
product   production *f*, produit *m*
product, alternative   produit *m* de substitution
product, clean(s)ing   produit *m* d'entretien

product, combustion   produit *m* de combustion
product, licensed   produit *m* sous licence
products, industrial   produits *m,pl* industriels
products, rolling mill   produits *m,pl* laminés
products, semi-finished   produits *m,pl* semi-finis
production   production *f*, fabrication *f*
production schedule   programme *m* de fabrication
productivity   productivité *f*, rendement *m*
professional *adj*   professionnel *adj*
profile [metal section]   profilé *m*
profile [section]   coupe *f*, profil *m*, section *f*
profile [of a moulding, cornice, etc]   modénature *f*
profile [of an architectural element]   galbe *m*
profile, cambered [of a road]   profil *m* en forme de toit *m*, profil *m* en travers en toit
profile, matching [of a moulding]   contre-profil *m*
profile, soil   coupe *f* de sondage
profile, velocity   profil *m* des vitesses
profile board [in excavations]   chevalet *m* pour tirer au cordeau
profile of a flat key   silhouette *f*, chiffre *m*
profiler, pavement   fraiseuse *f* routière
profit   bénéfice *f*, produit *m*, profit *m*, rapport *m*
profit, pre-tax   bénéfice *f* avant impôt
profit, retained   bénéfice *f* mise en réserve
profit, trading   bénéfice *f* d'exploitation
profit margin   marge *f* bénéficiaire
profitability   rentabilité *f*
program [computer]   logiciel *m*
program, spreadsheet   tableur *m*
programmable   programmable *adj*
programme   phasage *m*, planning *m*, programme *m*
programme [calendar, timetable]   calendrier *m*
programme [plan]   plan *m*
programme, bar chart   planning *m* à barres
programme, building   programme *m* de construction
programme, construction/works   phasage *m* des travaux, planning *m* d'exécution des travaux
programme, design   phasage *m* des études
programme, manufacturing   programme *m* de fabrication
programme, network analysis (PERT)   planning *m* pert
programme, site   calendrier *m* du chantier
programme, study   plan *m* d'études
programmer [for electrical equipment, etc]   programmateur *m*
programming, PERT, and network analysis   planning *m* pert
progress (of the works)   avancement *m* (des travaux)
prohibited area   zone *f* interdite
project   projet *m*
project [undertaking]   entreprise *f*, opération *f*
project, building   projet *m* de bâtiment/de construction
project, construction   grands travaux *m,pl*
project, road   projet *m* routier
project content   programme *m*
project director   directeur *m* de travaux
project health and safety archive   dossier *m* adapté

project implementation réalisation *f* du projet

project management direction *f* de projet, gestion *f* du projet

project manager chef *m*/gestionnaire *m* de projet, maître *m* d'œuvre, directeur *m* de travaux, projeteur *m*

project faire *vb* ressaut/saillie

project [predict] prévoir *vb*

projecting en saillie, saillant *adj*

projection saillie *f*

projection [drawing] projection *f*

projection [from/above a surface] ressaut *m*

projection in front of facade avant-corps *m*

prolong [in space or time] prolonger *vb*

prolongation prolongation *f*

promise engagement *m*

promise [to do something] s'engager *vb*

promoter [of development/construction schemes] promoteur *m*

promotion promotion *f*

proofing [against water ingress, etc] étanchement *m*

proof-reading revision/révision *f*

prop butée *f*, cale *f*, étai *m*, jambe *f* de force, soutien *m*, support *m*

prop [to side of excavation] buton *m*

prop [generally small, temporary] chandelle *f*

prop [sensibly vertical] accore *m*

prop [stay] pied *m* d'appui

prop, adjustable telescopic [Acrow type] chandelle *f* à crémaillère, étrésillon *m* à vérin

prop, inclined étançon *m*

prop, push-pull étai *m* tirant-poussant

prop, roundwood rondin *m*

prop, shuttering chandelle *f*

props used during underpinning chevalement *m*

prop accorer *vb*, étançonner *vb*

prop [side of excavation] butonner *vb*

prop prior to underpinning chevaler *vb*

prop (up) appuyer *vb*, buter *vb*, étayer *vb*

propane propane *m*

propeller hélice *f*

propeller, twin-bladed [wind generator] hélice *f* bipale

property [as in property sale, etc] immobilier *adj*

property propriété *f*

property [attribute] caractéristique *f*, propriété *f*

property [goods, possesions] biens *m,pl*

property [land] domaine *m*

property [real estate] immobilier *m*

property, bordering propriété *f* riveraine

property, freehold propriété *f* foncière perpétuelle

property, mechanical propriété *f* mécanique

property, unencumbered propriété *f* allégée

property damage dommages *m,pl* matériels

property developer promoteur *m* immobilier/de construction

property gazette indicateur *m* immobilier

property management gestion *f* du patrimoine

property valuer expert *m* immobilier

proportion proportion *f*

proportion system of Le Corbusier modulor *m*

proportion correctly [mixtures] doser *vb*

proportional proportionnel

proportionality, limit of limite *f* de proportionnalité

proportioning [in mixtures] dosage *m*

proportioning [size] dimensionnement *m*

proportioning unit [for mixing] groupe *m* de dosage

proposal offre *f*, proposition *f*

propose proposer *vb*

propping étaiement *m*, étançonnement *m*, étayage *m*

propping [side of excavation] butonnage *m*

propping to facilitate underpinning enchevelament *m*

proprietary system méthode *f* brevetée

proprietor propriétaire *m*

proprietor, riparian riverain *m*

proscenium proscenium *m*

proscenium arch manteau *m* d'Arlequin

prospect [view] prospect *m*

protect [shelter] abriter *vb*

protect [safeguard] sauvegarder *vb*

protect (from, against) protéger *vb* (de, contre)

protected abrité *adj*, protégé *adj*, sauvegardé *adj*

protected site [in France] site *m* classé

protection [mask] masque *m*

protection (from, against) protection *f* (de, contre)

protection, cathodic protection *f* cathodique

protection, degree of degré *m* de protection

protection, earth mise *f* à (la) terre, protection *f* de terre

protection, fire protection *f* incendie/contre l'incendie/les incendies

protection, flood protection *f* contre les crues

protection, index of (IP) indice *f* de protection (IP)

protection, motor protection *f* de machine

protection, overload protection *f* contre les surcharges

protection, over-voltage protection *f* contre les surtensions

protection, reinforced concrete masque *m* en béton armé

protection, rust protection *f* anti-rouille

protection, surface [road construction] couche *f* de protection

protection, thermistor protection *f* à la thermistance

protection against corrosion protection *f* contre la corrosion

protection against liquids protection *f* contre les liquides

protection against solid bodies protection *f* contre les corps solides

protective protecteur *adj*

protective conductor conducteur *m* de protection

protective device or equipment, collective équipement *m* de protection collective

protective earthing mise *f* à (la) terre, protection *f* de terre

protective equipment, personal équipement *m* de protection individuelle (EPI)

protective film film *m* protecteur

protective gaiter guêtre *f* de protection

protective glasses   lunettes *f,pl* de protection
protective goggles   lunettes-loup/-masque *f,pl*
protective hood over a circular saw blade   coiffe *f*
protective (over)sleeve   manchette *f* de protection
protectors, ear   coquilles *f,pl* de protection d'oreille,
  serre-nuque *m*, serre-tête *m* antibruit
protectors, eye   lunettes *f,pl* de protection
prototype   prototype *m*
protuberance in continuity of an arch or vault profile
  jarret *m*
prove [i.e. test]   mettre *vb* à l'épreuve
provide   fournir *vb*
provide access to   desservir *vb*
provide access to [e.g. one room to another]
  commander *vb*
provide regular transport services to   desservir *vb*
provide with a coping [a wall]   chaperonner *vb*
provider of services [e.g. of design or construction]
  prestataire *m*
providing [supplying]   fourniture *f*
province, within his/her   de son domaine/ressort
province   province *f*
provincial   provincial *adj*
provision [reserve or supply]   approvisionnement *m*,
  provision *f*
provision [of goods, supplies, etc]   fourniture *f*
provision of roads and main services to a building site
  viabilité *f*
provision of transport services   desserte *f*
provisional   provisoire *adj*, transitoire *adj*
provisional acceptance [of works]   réception *f* provisoire
proximity   proximité *f*
prune [trees, etc]   tailler *vb*
pruning [trees, etc]   taille *f*
pruning, gentle   taille *f* douce
pruning, hard   taille *f* sévère
pruning, safety   taille *f* de sécurité
pruning knife   serpette *f*
Prussian blue   bleu *m* de Prusse
pry bar   pied *m* de biche, pied-de-biche *m*
psychrometer   psychromètre *m*
psychrometer, aspiration   psychromètre *m* à aspiration
psychrometer, sling   psychromètre *m* fronde/à rotation
psychrometric chart   diagramme *m* psychrométrique/de
  Carrier
public   communal *adj*
public address (PA) system [provision with]
  sonorisation *f* (générale)
public convenience/lavatories   toilettes *f,pl*
public domain [i.e. not copyright]   domaine *m* public
public forest   forêt *f* domaniale
public health engineer   ingénieur *m* sanitaire
public health (PH) engineering   génie *m* sanitaire,
  hydraulique *f* urbaine, technique *f* d'assainissement
public highway   voie *f* publique
public notary   notaire *m* public
public sector procurement regulations [in France]
  Code *m* des Marchés Publics

public servant   fonctionnaire *m*
public services   services *m,pl* publics
public transport   transports *m,pl* (en commun)
puddle [line with, a pond, basin, etc]   corroyer *vb*
pug   hourder *vb*
pugging   hourdage *m*
pugging [of a timber frame panel]   hourdis *m*
pull   effort *m* de traction
pull [of a magnet]   attraction *f*
pull down   défaire *vb*
pull out [e.g. sheet piles]   arracher *vb*
pull to pieces   défaire *vb*
pull up [hoist]   hisser *vb*
pulley   poulie *f*
pulley, drive   poulie *f* motrice/d'entraînement
pulley, loose   poulie *f* folle
pulley, return   poulie *f* de renvoi *m*
pulley block   palan *m*
pulley wheel   réa *m*
pulling   traction *f*
pulling out [e.g. sheet piles]   arrachage *m*
pulsating   pulsatoire *adj*
pulse counting   comptage *m* d'impulsions
pulverize   pulvériser *vb*
pulverized fuel ash (PFA)   cendres *f,pl* volantes
pumice   ponce *f*
pumice stone   pierre *f* ponce
pump   pompe *f*
pump, auxiliary   pompe *f* de secours
pump, axial-flow   pompe *f* hélice
pump, boiler feed   pompe *f* d'alimentation de chaudière
pump, booster   pompe *f* de surpression
pump, centrifugal   pompe *f* centrifuge
pump, centrifugal high pressure   pompe *f* centrifuge à
  haute pression
pump, centrifugal multi-stage   pompe *f* centrifuge
  multicellulaire
pump, centrifugal single stage   pompe *f* à simple effet
pump, circulating   pompe *f* de circulation
pump, circulating [heating system]   accélérateur *m*,
  circulateur *m*
pump, concrete   pompe *f* à béton
pump, cooling water circulating   pompe *f* de circulation
  d'eau de refroidissement
pump, diaphragm   pompe *f* à diaphragme
pump, drainage   pompe *f* de relevage/de drainage
pump, emergency   pompe *f* de secours
pump, feed   pompe *f* d'alimentation
pump, fire   pompe *f* à incendie
pump, gear   pompe *f* à engrenage
pump, grout   pompe *f* d'injection
pump, hand   pompe *f* manuelle/à main
pump, heat   pompe *f* à chaleur
pump, high pressure (centrifugal)   pompe *f* (centrifuge)
  à haute pression
pump, horizontal   pompe *f* horizontale
pump, hot water heating   pompe *f* du chauffage à l'eau
  chaude

pump, hydraulic   pompe *f* hydraulique
pump, mixer/mixing   pompe *f* mélangeuse
pump, mobile concrete   pompe *f* automobile à béton
pump, motor   moto-pompe *f*
pump, multi-stage   pompe *f* à plusieurs étages
pump, oil   pompe *f* à huile
pump, oil feed   pompe *f* à mazout/d'amenée d'huile
pump, petrol   distributeur *m* d'essence
pump, piston/reciprocating   pompe *f* à piston
pump, pressure test   pompe *f* d'épreuve
pump, proportioning   pompe *f* de dosage
pump, rotary   pompe *f* rotative
pump, run   pompe *f* de service
pump, self-priming   pompe *f* à amorçage automatique
pump, sewage   pompe *f* des eaux d'égout
pump, single stage centrifugal   pompe *f* à simple effet
pump, sludge   pompe *f* à matières épaisses
pump, slurry   pompe *f* à boue
pump, standby   pompe *f* de réserve
pump, submersible   pompe *f* immergée/immersible/submersible
pump, suction   pompe *f* aspirante
pump, twin   pompe *f* jumelle
pump, vacuum   pompe *f* à vide
pump, wind   éolienne *f*
pump body   corps *m* de pompe
pump casing   carter *m* de pompe
pump characteristics   caractéristiques *f,pl* d'une pompe
pump house   station *f* de pompage
pump set, electric   groupe *m* électro-pompe
pump (up, out, etc)   pomper *vb*
pump up [inflate]   gonfler *vb*
pumping   pompage *m*
pumping [of drainage]   refoulement *m*
pumping cylinder   cylindre *m* de pompage
pumping station   station *f* de relevage/de pompage
pumping station, booster   station *f* de pompage intermédiaire
punch   mandrin *m*
punch [drift]   poinçon *m*
punch, centre   pointeau *m*
punch, hole   perforeuse *f*
punch, marking   matoir *m*
punch, metal   poinçonneuse *f*
punch, nail   chasse-clou *m*, chasse-goupille *m*, chasse-pointe *m*
punch   poinçonner *vb*
punched card   carte *f* perforée
punching   poinçonnement *m*
punching machine   poinçonneuse *f*
punner   batte *f*
purchase   achat *m*, acquisition *f*
purchase, hire/instalment   achat *m* à tempérament, location *f* vente
purchase   acquérir *vb*
purchase compulsorily   exproprier *vb*
purchaser   acheteur *m*, acquéreur *m*
purge   purger *vb*

purging, air   purge *f* d'air
purification   assainissement *m*, épuration *f*, purification *f*
purification, air   épuration *f* de l'air
purification, natural   épuration *f* naturelle
purification, water   épuration *f* de l'eau, traitement *m* des eaux potables
purifier [air, water]   épurateur *m*
purifier, air   épurateur *m* d'air
purify   assainir *vb*, épurer *vb*, purifier *vb*
purifying, *see* purification
purity   pureté *f*
purity, degree of   degré *m* d'épuration
purlin   filière *f*, panne *f*, ventrière *f*
purlin, lattice   panne *f* en treillis
purlin, ridge   panne *f* faîtière
purlin in a pyramidal roof   lierne *f*
purlin jointed into rather than laid upon principal rafter   panne *f* à lierne
purpose   but *m*
purpose, fit/unfit for its   propre/impropre *adj* à sa destination
push   pousser *vb*
push block [carpenter's or joiner's]   poussoir *m*
pushbutton   bouton *m* de poussoir, poussoir *m*
pushing   poussée *f*
push-launching [e.g. of bridge girder]   poussage *m*
put [an object or person]   placer *vb*
put away   ranger *vb*
put back [postpone]   remettre *vb*
put back by [a clock]   retarder *vb* de
put back in place   ranger *vb*, remettre *vb*, replacer *vb*
put down [pay]   verser *vb*
put in order   ranger *vb*
put into store or storage   entreposer *vb*
put off [a decision]   remettre *vb* à plus tard
put off [defer]   reporter *vb*
put on top of   superposer *vb* sur
put out [extinguish]   éteindre *vb*
put out [sub-contract]   donner *vb* à un sous-traitant
put right   mettre *vb* en ordre
put right a mistake   rectifier *vb* une erreur
put (sth) right   remédier *vb* (à qch)
put sth into practice   mettre *vb* qch en pratique
put to the test   mettre *vb* à l'essai
put up [a building]   établir *vb*
put up [a notice, etc]   afficher *vb*
put up some capital   faire *vb* une mise de fonds
putlog   boulin *m*
putting away   rangement *m*
putting back [of sth into its place]   remise *f*
putting into service   mise *f* en service
putty   mastic *m*
putty, glazing   mastic *m* de vitrier
putty, linseed oil   mastic *m* lin naturel
putty knife   couteau *m* à mastiquer, spatule *f* de vitrier
pycnometer or pyknometer   picnomètre/pyknomètre *m*
pylon   pylône *m*
pyrometer   canne *f* thermoélectrique, pyromètre *m*

quake, tectonic/volcanic tremblement *m* tectonique/volcanique

qualification qualification *f*

qualified [as or to do something] qualifié *adj*

qualify qualifier *vb*

quality qualité *f*

quality, air qualité *f* de l'air

quality, fuel gas qualité *f* du gaz combustible

quality, indoor air qualité *f* d'air intérieur

quality, steam qualité *f* de la vapeur

quality, water qualité *f* d'eau/de l'eau

quality assurance (QA) assurance *f* qualité (AQ)

quality assurance plan plan *m* d'assurance (de la) qualité (PAQ)

quality control (QC) contrôle *m* (de) qualité

quality control office for building design bureau *m* de contrôle

quantify quantifier *vb*

quantity quantité *f*

quantity [total measured, to be multiplied by rate] timbre *m*

quantity, approximate quantité *f* approximative

quantity of flow débit *m*

quantity per day quantité *f* par jour

quantity survey métré *m* quantitatif

quantity surveying métrage *m*, métré *m*

quantity surveyor [near French equivalent] économiste *m/f*, expert *m* métreur, métreur *m*

quantities, bill of devis *m* quantitatif, tableau *m* des quantités des matériaux

quantities, measurement of métré *m* quantitatif

quarry carrière *f*

quarry, sand carrière *f* de sable, sablière *f*

quarry run [ungraded material] tout-venant *adj*

quarry stone moellon *m*

quarry stone, squared and rusticated moellon *m* d'appareil

quarry waste déchets *m,pl* (de) carrière

quarryman carrier *m*

quarter [of a town, e.g. commercial] quartier *m*

quarter [of year] trimestre *m*

quarter [one-fourth] quart *m*

quarters [living] logement *m*

quarter-sawing [of timber from the log] débit *m* sur maille

quartz quartz *m*

quay quai *m*

quay, berthing quai *m* d'amarrage

quench tremper *vb*

quenching trempe *f*

quick-acting à action rapide

quick-drying à séchage rapide

quicklime chaux *f* anhydre/vive

quick-release à déclenchement rapide

quicksand sable *m* boulant/mouvant, sables *m,pl* boulants

quit [premises] vider *vb*

quoin coin *m*, pierre *f* d'angle

quoin bond [at corners of stonework] appareil *m* en besace

quotation/quote [price] estimation *f*

quote devis *m*

quote a price établir *vb* un prix

rabbet rainure *f*

rack [toothed] **of mountain railway** crémaillère *f*

racking [raking] **back** bout *m* en escalier/en attente

radial radial *adj*

radian radian(t) *m*

radiating radiant *adj*, rayonnant *adj*

radiate rayonner *vb*

radiation [of heat] rayonnement *m*

radiation, background rayonnement *m* de base

radiation, duct rayonnement *m* des conduits d'air

radiation, heat or thermal rayonnement *m* thermique

radiation, incident solar rayonnement *m* solaire incident

radiation, infra-red rayonnement *m* infrarouge

radiation, ionizing radiation *f* ionisante, rayonnement *m* ionisant

radiation, low temperature rayonnement *m* à basse température

radiation, solar rayonnement *m* solaire

radiation, thermal rayonnement *m* thermique

radiation, ultraviolet rayonnement *m* ultraviolet

radiation, visible [light] rayonnement *m* visible

radiator corps *m* de chauffe, radiateur *m*

radiator, back-up radiateur *m* d'appoint

radiator, cast radiateur *m* en fonte

radiator, column radiateur *m* à colonnes

radiator, electric radiateur *m* électrique

radiator, finned-tube radiateur *m* à tubes à ailettes
radiator, gas radiateur *m* à gaz
radiator, high pressure radiateur *m* à haute pression
radiator, panel (steel) radiateur *m* panneau/à panneaux (en acier)
radiator, recessed radiateur *m* encastré
radiator, skirting board radiateur *m* sur plinthe
radiator, steel panel radiateur *m* panneau en acier
radiography radiographie *f*
radius rayon *m*
radius, bending rayon *m* de courbure
radius, crane working portée *f* de grue
radius, transition rayon *m* de raccordement
radius, turning rayon *m* de giration
radius, working rayon *m* d'action
radius of bending or curvature rayon *m* de courbure
radius of gyration rayon *m* de giration
radius of influence rayon *m* d'influence
raft [base] radier *m*
raft, brushwood [road foundation on peat] tapis *m* de branchage
raft, foundation châssis *m*/radier *m* de fondation
rafter chevron *m*
rafter, common chevron *m* intermédiaire
rafter, hip arêtier *m*, chevron *m* d'arête, empannon *m*
rafter, jack empannon *m* de long pan/de croupe
rafter, principal arbalétrier *m*, chevron *m* arbalétrier, maître-chevron *m*
rafter, valley empannon *m* à noulet, noue *f*, noulet *m*
rafter end queue-de-vache *f*
rafter end or head tête *f* de chevron
rail *adj* [of railways] ferroviaire *adj*
rail [of a door or shutter] emboîture *f*, traverse *f*
rail [of boarded door or shutter] barre *f*
rail [of post-and-rail fencing] lice/lisse *f*
rail [railway track] rail *m*
rail, banister glisse-main *m*, main *f* courante
rail, barrier [level crossing] lisse *f*
rail, bottom [of a door] traverse *f* inférieure
rail, check or guard [at points] contre-rail *m*
rail, conductor rail *m* de contact
rail, curtain tringle *f* (de rideau)
rail, flat-bottomed rail *m* à base plate
rail, guard [handrail] garde-corps *m*, garde-fou *m*, parapet *m*, rambarde *f*
rail, guard [at points] contre-rail *m*
rail, guide rail *m* de guidage
rail, head [of a timber frame panel] sablière *f* supérieure
rail, live or third rail *m* conducteur
rail, lock [of a door] traverse *f* de serrure
rail, middle [of a barrier] sous-lisse *f*
rail, sill [timber structure] entretoise *f*/lisse *f* d'appui
rail, slide glissière *f*
rail, top [of a barrier, balustrade] lice/lisse *f*
rail, top [of a door] traverse *f* supérieure
rail clip [railway track] plaque *f* de serrage
rail head [of rail] champignon *m*
rail or sleeper clip crapaud *m* de rail
rail-brake rail-frein *m*

railing grille *f*
railing [handrail] balustrade *f*
railway *adj* ferroviaire *adj*
railway chemin *m* de fer
railway, elevated/overhead chemin *m* de fer aérien/surélevé, métro(politain) *m* aérien
railway, rack chemin *m* de fer à crémaillère
railway, single/double track chemin *m* de fer à simple/double voie
railway, underground chemin *m* de fer souterrain, métro(politain) *m*
railway, urban chemin *m* de fer urbain
railway line ligne *f* de chemin de fer
railway line, branch ligne *f* secondaire (de chemin de fer)
railway line, main ligne *f* principale (de chemin de fer)
railway network réseau *m* ferroviaire
railway right-of-way (delimited) on highway terre-plein *m* spécial
railway station gare *f*
railway station, underground station *f* métro
railway track voie *f* ferrée
railway tunnel tunnel *m* ferroviaire
rain pluie *f*
rain pleuvoir *vb*
rainfall précipitation *f*
rainfall [quantity] pluviométrie *f*
rainfall, annual quantité *f* annuelle de pluie
rainfall, daily quantité *f* journalière de pluie
rainfall chart courbe *f* pluviométrique
rainfall intensity intensité *f* des précipitations
rainstorm trombe *f* d'eau
rainwater eau *f* de pluie, eaux *f,pl* pluviales (EP)
rainwater channel on top of cornice goulotte *f*
rainwater downpipe descente *f* eaux pluviales
rainwater downpipe, hole through structure for a souillard *m*
rainwater goods [galvanized metal] zinguerie *f*
rainwater outlet [from a roof gutter] moignon *m*
rainwater outlet grating [to prevent blockage] araignée
rainwater shoe dauphin *m*
rainwater soakaway [pebble-filled] bétoire *f*
rainwater stack chute *f* eaux pluviales (chute EP)
raise [a building, etc] bâtir *vb*, construire *vb*
raise [a monument] ériger *vb*, élever *vb*
raise [increase] augmenter *vb*, relever *vb*
raise [a question or subject, lift sth up] soulever *vb*
raise [heighten, increase height of] exhausser *vb*, hausser *vb*, rehausser *vb*, relever *vb*, surélever *vb*
raise [lift up] lever *vb*
raise [wages, prices, taxes, etc] augmenter *vb*
raise funds se procurer *vb* les fonds
raised en relief *adj*
raised [domed] bombé *adj*
raised [heightened] exhaussé *adj*, rehaussé *adj*, surélevé *adj*
raising [heightening] rehaussement *m*, surélévation *f*
raising [increasing] relèvement *m*
rake râteau *m*

rake [for a boiler]   ringard *m* à crochet
rake out [joints in masonry]   dégarnir *vb*
raking [of a gravel path or drive]   repassage *m*
ram   vérin *m*
ram [hydraulic, etc]   bélier *m*
ram, heavy beam used as a, in demolition work
   bélier *m*
ram, hydraulic   bélier *m*/vérin *m* hydraulique
ram [earth, etc]   damer *vb*, pilonner *vb*
ram [paving stones, etc]   bliner *vb*
ram down   bourrer *vb*, tasser *vb*
ram sand down into joints between slabs or stones
   ficher *vb*
rammer   dame *f*, dameur *m*, engin *m* de damage
rammer [tool]   pilon *m*
rammer, frog   grenouille *f*
rammer, hand   batte *f*
rammer, pavior's   demoiselle *f*, hie *f*
rammer, power   dameuse *f* mécanique
rammer, vibrating   dame *f* vibrante, pilonneuse *f*
   vibrante
ramming   pilonnage *m*
ramming down   bourrage *m*, tassement *m*
ramp   rampe *f*
ramp [regular and with even slope]   glacis *m*
ramp, access   rampe *f* d'accès
ramp, moving   rampe *f* roulante
rampart   rempart *m*
random access memory (RAM)   mémoire *f* vive
random rubble [masonry], *see also* rubble stone   opus *m*
   incertum
random rubble construction   blocage *m*
range [area, duration]   étendue *f*
range [between limits]   variation *f*
range [distance, scope]   portée *f*
range [field]   champ *m*
range [of colours, products, etc]   gamme *f*
range [of prices]   fourchette *f*
range [scale, e.g. of prices, salaries]   échelle *f*
range [surveying]   alignement *m*, direction *f*
range [within a series or table]   plage *f*
range, crane   portée *f* de grue
range, daily   variation *f* journalière/quotidienne
range, monthly   variation *f* mensuelle
range, tidal   marnage *m*
range of accessories   gamme *f* d'accessoires
range of adjustment   plage *f* de réglage
range of error   plage *f* d'erreur
range of measurement   champ *m*
range of scatter   domaine *m* de dispersion
range of temperature, etc   domaine *m* de températures,
   etc
Rankine cycle   cycle *m* de Rankine
rapidity   vitesse *f*
rapids [in a watercourse]   rapides *m,pl*
rasp   râpe *f*
rasp, plumber's   écouenne *f*
rasp, wood   râpe *f* à bois
rasp   râper *vb*

ratchet   cliquet *m*, rochet *m*
rate   taux *m*
rate [pace, rhythm]   cadence *f*
rate [price]   tarif *m*
rate, adiabatic lapse   chute *f* de température adiabatique
rate, air change   taux *m* de renouvellement d'air/de l'air
rate, bank   taux *m* officiel d'escompte
rate, basic   tarif *m* de base
rate, charging [electrical]   taux *m* de charge
rate, consumption   taux *m* de consommation
rate, cooling   vitesse *f* de refroidissement
rate, decay [of decrease]   taux *m* de décroissance
rate, decay [of deterioration]   vitesse *f* de détérioration
rate, discount   taux *m* d'escompte
rate, evaporation   taux *m* d'évaporation
rate, exchange [currency]   cours *m*, taux *m* de change
rate, feed   vitesse *f* d'alimentation
rate, filtration   vitesse *f* de filtrage
rate, flat [charge]   tarif *m* forfaitaire/uniforme
rate, flow, *see* flow rate
rate, fresh air   taux *m* d'air frais
rate, growth   taux *m* de croissance
rate, interest   taux *m* d'intérêt
rate, lending   taux *m* de prêt
rate, loading   vitesse *f* de charge(ment)
rate, mass flow   débit *m*
rate, mixing   dosage *m* de mélange
rate, occupancy   degré *m* d'occupation
rate, outside air   taux *m* d'air frais
rate, strain   vitesse *f* de déformation
rate, two-part [electricity tariff]   tarif *m* binôme
rate, wage   taux *m* des salaires
rates, market   cours *m* du marché
rate of increase   vitesse *f* d'augmentation
rate of spread [of tars, bitumen, etc]   dosage *m*
rate for, fix   tarifer *vb*
rates [local taxes]   impôts *m,pl* locaux
ratify   sanctionner *vb*
rating [index]   indice *m*
rating, contact   pouvoir *m* de coupure
rating, motor   puissance *f* nominale du moteur
rating, nominal   intensité *f* nominale
rating, one hour   puissance *f* nominale à l'heure
rating, thermal   intensité *f* thermique
ratio   proportion *f*, rapport *m*
ratio, compression   rapport *m*/taux *m* de compression
ratio, cooling efficiency   coefficient *m* d'effet frigorifique
ratio, energy efficiency   rendement *m* énergétique
ratio, entrainment   taux *m* d'entraînement
ratio, induction   taux *m* d'induction
ratio, mixing   rapport *m* de mélange
ratio, quality–price   rapport *m* qualité–prix
ratio, shrinkage   indice *m* de retrait
ratio, uniformity [lighting]   facteur *m* d'uniformité
ratio, void   indice *m* des vides
ratio, water–cement   rapport *m* eau–ciment
ravine   ravin *m*
raw [untreated, e.g. water]   brut *adj*
raze   raser *vb*

raze to the ground   abattre *vb* à la terre
reach   portée *f*
reach, within   à portée
reach [arrive at]   atteindre *vb*
reach [rise to]   monter *vb* à
reach [spread to]   s'étendre *vb* à
reach [gain access to]   accéder *vb* à
reach a ceiling/maximum   plafonner *vb*
reaction   réaction *f*
reaction, chain   réaction *f* en chaîne
reaction, heat of   chaleur *f* de réaction
reaction, support   réaction *f* d'appui
reactive   réactif *adj*
reactor [nuclear power station]   réacteur *m*
reactor core   cœur *m* du réacteur
reactor vessel   cuve *f* du réacteur
read and approved [in contract documents]   lu et
    approuvé
read only memory (ROM)   mémoire *f* morte
read [proofs]   revoir *vb*
read a meter   relever *vb* un compteur
reading [of a meter, instrument]   lecture *f*
reading, remote [e.g. of meters]   relevé *m* à distance
readjustment   rectification *f*
ready for use   prêt *adj* à l'emploi/à l'usage/d'emploi
ready money   argent *m* comptant
ready reckoner   barème *m*
ready-mix concrete lorry, *see also* concrete   camion *m*
    malaxeur
reagent   réactif *m*
real time   temps *m* réel
real-estate   biens *m,pl* immeubles/immobiliers,
   , immobilier *m*
real-estate agent   agent *m* immobilier
real-estate dealer   marchand *m* de biens
reallocation of land holdings to improve use [in France]
    remembrement *m*
ream   aléser *vb*
reamer   alésoir *m*, fraise *f*
reaming   alésage *m*
reappraisal   contre-expertise *f*
reappraise   reviser/réviser *vb*
rear *adj*   arrière *adj*
rear premises [of a building]   arrière-corps *m*
rear wall of a chimney hearth   contre-cœur *m*
rebar[†]   acier *m* d'armature
rebate [financial]   escompte *m*
rebate [in a panel or surface]   défoncement *m*
rebate [in woodwork, etc]   feuillure *f*
rebatten [a roof]   relatter *vb*
rebattening [of a roof]   remaniage *m*
rebound, Schmidt   rebondissement *m* Schmidt
rebound hardness measurer   duromètre *m*
rebuild   refaire *vb*
rebuilding   reconstruction *f*, réfection *f*
recalculation   vérification *f*
receipt [acknowledgement of]   récépissé *m*
receipt [action or fact of]   réception *f*
receipt [proof of payment]   quittance *f*, reçu *m*

receiver   récepteur *m*
receiver [of fluid, gas]   récipient *m*
receiver [tank]   collecteur *m*
receiver, vacuum   collecteur *m*/récipient *m* à vide
receptacle   vase *m*
reception [room]   réception *f*
reception area   aire *f* de réception
recess [alcove]   alcôve *f*, enfoncement *m*
recess [cavity, groove]   évidement *m*
recess [groove, rebate]   feuillure *f*
recess for sluice gates   enclave *f*
recessed   enfoncé *adj*
recharge [battery, groundwater, etc]   recharger *vb*
rechargeable   rechargeable *adj*
recharging   recharge *f*
recheck [of measurements, readings, etc]   contrôle *m* des
    mesures
reciprocal   réciproque *adj*
recirculate   recirculer *vb*
recirculation   recirculationn *f*
reckoning [numerically]   chiffrage *m*
reclaimer, bucket-wheel   engin *m* de reprise à roue-pelle
reclaimer, road   régénérateur *m* de chaussée
recoat [of renderings, finishes]   repasser *vb*
recommend   conseiller *vb*, préconiser *vb*, recommander *vb*
recommendation   recommandation *f*
recompact [geotechnics]   recompacter *vb*
recompression   recompression *f*
reconnaissance, photo   reconnaissance *f* photographique
reconnaissance, site   étude *f* de reconnaissance du site
reconnect   relier *vb*
reconnect [electrically]   rebrancher *vb*
reconnect [mechanically, electrically]   réaccoupler *vb*
reconstituted [e.g. stone]   reconstitué *adj*
reconstruct   reconstruire *vb*
reconstruction   reconstruction *f*
reconstruction [e.g. of something damaged or worn]
    reconstitution *f*
re-cooling   post-refroidissement *m*
record, as-built, of work subsequently covered up
    attachement *m*
record of evidence   procès-verbal *m* (PV)
record of finished work   dossier *m* des ouvrages exécutés
    (DOE)
record of work completed and payments made and
    outstanding   situation *f*
records   archives *f,pl*
records, as-built   dossier *m* des ouvrages exécutés (DOE)
record [a minute, a judgement, etc]   minuter *vb*
record [a reading]   enregistrer *vb*
recorder   enregistreur *m*
recorder, data   enregistreur *m* de données
recorder, drum chart   tambour *m* enregistreur
recorder, flow   enregistreur *m* de débit
recorder, volume   enregistreur *m* de volume
recording *adj*   enregistreur *adj*
recording   enregistrement *m*
recoup [a loss]   récupérer *vb*
re-cover [a roof, etc]   recouvrir *vb*

recover [debts, etc]   recouvrer *vb*
recover [waste, etc]   récupérer *vb*
recover damages   obtenir *vb* des dommages-intérêts
recoverable   récupérable *adj*
recovered [of used materials]   récupéré *adj*
recovering   recouvrement *m*
recovery   récupération *f*
recovery, (waste) heat   récupération *f* de chaleur
recovery installation or plant   installation *f* de récupération
recreation area   aire *f* récréative
recreation centre   base *f* de loisirs
recreation of former state [drawn or modelled]   restitution *f*
rectangle   rectangle *m*
rectangular   à angles droits, rectangulaire *adj*
rectification   correction *f*, rectification *f*
rectifier [electrical]   clapet *m*, rectificateur *m*, redresseur *m*
rectifier, current   rectificateur *m* de courants
rectify   rectifier *vb*
rectify [an error]   redresser *vb*
rectifying [electrical]   redressement *m*
rectilinear   rectiligne *adj*
rectory   presbytère *m*
recuperation   récupération *f*
re-cut [stone]   retailler *vb*
re-cutting [of the edge of paving slabs/stones]   repassage *m*
recycle   recycler *vb*
recycler, (asphalt)   recycleur *m* (d'asphalte)
recycling   recyclage *m*
redeem a loan   amortir *vb* un emprunt
redemption date [of a loan]   échéance *f*
redemption, debt   amortissement *m* d'une dette
redress [weathered stonework to give it a new appearance]   regratter *vb*
reduce   réduire vb
reduce [price, etc]   rabattre *vb*
reduce [thickness of a piece of wood]   affaiblir *vb*
reduce size [to make sth fit]   délarder *vb*
reduce speed   ralentir *vb*
reduce the expenses/costs   rogner *vb* les dépenses/frais
reducer   réducteur *m*
reducer, male–female [plumbing]   réduction mâle–femelle
reducer, pipe   raccord *m* de réduction
reducer, pressure   détendeur *m*, réducteur *m* de pression
reduction   diminution *f*
reduction [in cost or price]   abattement *m*, rabais *m*, remise *f*
reduction [in scale of a map, drawing, etc]   réduction *f*
reduction in (cross-)section   réduction *f* de section
reduction in thickness due to dressing   recoupement *m*
reduction of area [of a test piece]   striction *f*
reduction of noise   réduction *f* de bruit
reduction of pressure   réduction *f* de pression
redundant [of a system able to redistribute loads]   surabondant *adj*
redundant [out of work]   licencié *adj*

redwood [North American]   séquoia *m*
redwood [Scandinavian]   pin *m* du nord
reel   bobine *f*, touret *m*
re-erect [something fallen or lying down]   redresser *vb*
re-examine   reviser/réviser *vb*
re-examine [accounts, a document]   revoir *vb*
reface [masonry]   ravaler *vb*
refacing work   ravalement *m*
refectory   réfectoire *m*
refer (a matter) (to a court)   saisir *vb* (un tribunal) (d'une affaire)
reference   référence *f*
reference [of a matter to an authority, etc]   renvoi *m*
reference, order(ing)   référence *f* de commande
reference axis   axe *m* de référence
reference conditions   conditions *f,pl* standards/de référence
reference level [on a drawing]   plan *m* de comparaison
reference level marker   repère *m* de nivellement
reference point   point *m* côté/de référence, repère *m*
refill   recharge *f*
refill [a surface for second coat of paint]   renduire *vb*
refine   raffiner *vb*
refinery   usine *f* de raffinage
refixing   repose *f*
reflectance   pouvoir *m* réflecteur
reflecting or reflective stud   plot *m* (rétro-) réfléchissant/réflectorisé
reflection factor [lighting]   facteur *m* de réflexion
reflectivity   pouvoir *m* réfléchissant, réflectivité *f*
reflector   réflecteur *m*, réverbère *m*
refractor   réfracteur *m*
refractory   réfractaire *adj*
refresh   rafraîchir *vb*
refreshing   rafraîchissement *m*
refrigerant   fluide *m* frigorigène, frigorigène *m*, réfrigérant *m*
refrigerate   frigorifier *vb*, réfrigérer *vb*
refrigerated   réfrigéré *adj*
refrigeration   froid *m*, réfrigération *f*, refroidissement *m*
refrigeration, compression   réfrigération *f* par compression
refrigeration demand   demande *f* de froid
refrigeration engineering   technique *f*/technologie *f* du froid
refrigeration installation or plant   installation *f* frigorifique/de refroidissement/de réfrigération
refrigeration system, compression   système *m* de réfrigération à compression
refrigeration system, direct expansion   système *m* de réfrigération à détente directe
refrigerator   frigo[+] *m*, réfrigérateur *m*
refrigerator, domestic   réfrigérateur *m* ménager
refrigerator, gas   réfrigérateur *m* à gaz
refurbish, *see also* rehabilitate   remettre *vb* à neuf
refurbish [e.g. joinery, masonry]   rafraîchir *vb*
refurbished   rénové *adj*
refurbishment [e.g. of shop interior]   lifting *m*
refurbishment (works)   travaux *m,pl* de réhabilitation

refusal of a pile   refus *m* d'un pieu
refuse   déchets *m,pl*, détritus *m*, ordures *f,pl*
refuse, household   ordures *f,pl* ménagères
refuse collecting installation or plant   installation *f* de ramassage des ordures
refuse disposal or treatment   traitement *m* des ordures ménagères
regeneration   régénération *f*
region   région *f*
regional development [planning]   aménagement *m* du territoire
regional development agency [in France]   Délégation *f* à l'Aménagement du Territoire et à l'Action Régionale (DATAR)
register   registre *m*
register, cadastral   cadastre *m*
register [a letter, parcel]   recommander *vb*
register [indicate]   indiquer *vb*
register [property in the cadastral record]   cadastrer *vb*
registrar   conservateur *m*
registrar of mortgages and land charges   conservateur *m* des hypothèques
registrar or clerk of the court   greffier *m*
registration [of mortgages, etc]   inscription *f*
registration [of post]   recommandation *f*
registration duty   droit *m* d'enregistrement
regression coefficient   coefficient *m* de régression
regression equation   équation *f* de régression
regressive   régressif *adj*
regulate   réguler *vb*
regulate [a machine, a system]   régler *vb*
regulating unit   ensemble *m* de régulation
regulating valve, pressure   contrôleur *m* de pression
regulation [adjustment]   réglage *m*, régulation *f*
regulation [of flow of a watercourse]   régularisation *f*
regulation [rule]   prescription *f*
regulation, automatic   régulation *f* automatique
regulation, hand/manual   réglage *m* manuel, régulation *f* manuelle
regulation, planning   règle *f* d'urbanisme
regulation of flow   régulation *f* de(s) débit(s)
regulation system   système *m* de régulation
regulations   code *m*, règlement *m*, réglementation *f*
regulations, building or construction [in France]   Code *m* de la Construction
regulations, development [in France]   Code *m* de l'Environnement
regulations, high rise building [in France]   Codes *m,pl* pour Structures Élevées
regulations, public sector procurement [in France]   Code *m* des Marchés Publics
regulations, technical [e.g. building regulations]   prescriptions *f* techniques
regulations, town planning [in France]   Code *m* d'Urbanisme
regulator   régleur *m*, régulateur *m*
regulator [control device]   contrôleur *m*
regulator, air pressure   régulateur *m* de la pression d'air

regulator, back pressure   régulateur *m* de pression d'aspiration
regulator, delivery/discharge   régulateur *m* de débit
regulator, expansion   régulateur *m* d'expansion
regulator, high pressure   régulateur *m* à haute pression
regulator, low pressure   régulateur *m* de basse pression
regulator, nozzle   régulateur *m* de jet
regulator, output   régulateur *m* de débit
regulator, pressure   régulateur *m* de pression
regulator, speed   régulateur *m* de vitesse
regulator, variable voltage   régulateur *m* par tension variable
regulator, voltage   régulateur *m* de tension
rehabilitate   réhabiliter *vb*, rénover *vb*, restaurer *vb*
rehabilitated   rénové *adj*, restauré *adj*
rehabilitation   réhabilitation *f*
rehabilitation (works)   travaux *m,pl* de réhabilitation
rehandle [materials]   remanier *vb*
rehang [a door or shutter on pin hinges]   regonder *vb*
reheat   réchauffer *vb*
reheater   post-réchauffeur *m*, réchauffeur *m*
reheating   réchauffage *m*
rehouse   reloger *vb*
re-ignition   réallumage *m*
reimburse   rembourser *vb*
reinforce   renforcer *vb*
reinforce [i.e. fit metal reinforcement]   ferrailler *vb*
reinforced [of a door, etc]   blindé *adj*
reinforced concrete, *see also* reinforcement   béton *m* armé
reinforced concrete construction   construction *f* en béton armé
reinforced concrete design   étude *f* de béton armé
reinforced concrete facing or protection   masque *m* en béton armé
reinforced concrete footing   semelle *f* armée
reinforced concrete frame construction   ossature *f* en béton armé
reinforced concrete pile   pieu *m* en béton armé
reinforced concrete slab   dalle *f* armée
reinforced concrete structure   ossature *f* béton armé
Reinforced Earth Company [trade name]   Terre *f* Armée
reinforcement [for concrete: the material]   acier *m*/fer *m* à béton/d'armature, armature *f*
reinforcement [for concrete: design, material and installation of]   ferraillage *m*
reinforcement [general term]   renforcement *m*
reinforcement [supplementary, around holes, etc]   renfort *m*
reinforcement, area of compression/tension   section *f* d'acier comprimé/tendu
reinforcement, compression/negative   armature *f* supérieure/de compression
reinforcement, hoop [rectangular]   cadre *m* d'armature
reinforcement, layer of   lit *m* d'acier
reinforcement, longitudinal   armature *f* longitudinale
reinforcement, main   armature *f* principale
reinforcement, mesh   treillis *m* d'armature

reinforcement, positive   armature *f* inférieure/tendue/de traction

reinforcement, pre-stressing   armature *f* de précontrainte

reinforcement, projecting   crosse *f* (d'armature)

reinforcement, ribbed (steel) bar   acier *m* crénelé/nervuré/haute adhérence, armature *f* crénelée, barre *f* à haute adhérence

reinforcement, round (bar)   fer *m* rond, rond *m* à béton

reinforcement, tensile/tension   armature *f* inférieure/tendue/de traction

reinforcement, transverse   armature *f* transversale

reinforcement, vertical   armature *f* verticale

reinforcement, welded fabric or mesh   (armature *f* en) grillage *m* soudé, armature *f* en treillis soudé, treillis *m* soudé

reinforcement bar   barre *f* de renforcement

reinforcement (bar), deformed round   rond *m* à béton crénelé

reinforcement bar, main   acier *m* porteur

reinforcement bar, round   fer *m* rond, rond *m* à béton

reinforcement bar, straight   barre *f* droite

reinforcement bars, bottom/top   armature *f* inférieure/supérieure

reinforcement bars, connecting   armature *f* de liaison

reinforcement bars, overlapping of   recouvrement *m* d'armatures/des barres

reinforcement bars, shape of   forme *f* des barres

reinforcement cage   cage *f* (d'armature)

reinforcement concreter   cimentier *m* en béton armé

reinforcement coupler   coupleur *m* d'armatures

reinforcement cover   enrobage *m* des aciers/des armatures

reinforcement drawing   plan *m* de ferraillage

reinforcement hook   crochet *m* d'armature

reinforcement schedule   bordereau *m* d'armatures

reinforcement spacer   cale *f* à béton, distancier *m*

reinforcement spacing   écartement *m*/entredistance *f* des armatures

reinforcement starter bars, *see* starter bars

reinforcement steel, *see also* reinforcement   acier *m* à béton, armature *f*

reject   rebut *m*

rejects   pièces *f,pl* de rebut

reject [goods, etc]   rebuter *vb*

reject sth   mettre *vb* qch au rebut

rejoint [stonework]   reficher *vb*

relative [density, position, size, value, etc]   relatif *adj*

relax [stress, tension, etc]   détendre *vb*

relaxation   relaxation *f*

relay   relai(s) *m*

relay, alarm or control   relai(s) *m* de contrôle

relay, call   relai(s) *m* d'appel

relay, changeover   relai(s) *m* inverseur

relay, positioning   relai(s) *m* de positionnement

relay, (primary) protection   relai(s) *m* de protection (primaire)

relay, protective   relai(s) *m* de protection

relay, safety   relai(s) *m* de sécurité/de protection

relay, thermal   relai(s) *m* thermique

relay, time delay   relai(s) *m* temporisé

relay [individual paving slabs]   repiquer *vb*

relay [paving, etc]   remanier *vb*

relaying   repose *f*

relaying [of individual paving slabs]   repiquage *m*

release   déclenchement *m*

release [unscrewing]   dévissage *m*

release [of a mortgage or charge]   mainlevée *f*

release [of gas, steam, etc]   dégagement *m*

release [undoing, of a clamp, nut, handbrake, etc]   desserrage *m*

release agent [for concrete formwork]   démoulant *m*, produit *m* de démoulage

release agent [oil]   huile *f* de décoffrage

release system   système *m* de déclenchement

release [a clamp, nut, handbrake, etc]   desserrer *vb*

release [heat, steam, etc]   dégager *vb*

release [let sth go]   lâcher *vb*

release [set off, start]   déclencher *vb*

re-level   reniveler *vb*

reliability   fiabilité *f*, solidité *f*

reliable   fiable *adj*, solide *adj*

relic   vestige *m*

relief [projecting from a surface]   relief *m*

relieve [i.e. to take over from]   relayer *vb*

reload   recharger *vb*

relocate   déplacer *vb*

remainder   résidu *m*, reste *m*

remains   restes *m,pl*, vestiges *m,pl*

remains of existing buildings   restes *m,pl* d'ouvrage

remake   refaire *vb*

remake [dismantle and reassemble joinery]   retailler *vb*

remark   observation *f*

remedy   remède *m*

remedy (sth)   remédier *vb* (à qch)

remittance [of money]   remise *f*

remnant   vestige *m*

remote   éloigné *adj*

remote [as in remote control, etc]   à distance *adj*

remote indicator   indicateur *m* à distance

remote reading [e.g. of meters]   relevé *m* à distance

remote(ly) controlled or operated   manœuvré *adj* à distance

removable   amovible *adj*, démontable *adj*

removal   évacuation *f*

removal [elimination]   élimination *f*

removal [taking out or down]   démontage *m*

removal, asbestos   désamiantage *m*

removal, ash   évacuation *f* des cendres

removal, house [furniture, etc]   déménagement *m*

removal, waste   élimination *f*/évacuation *f* des déchets

removal of builder's/building rubbish/rubble   élimination *f* des décombres, évacuation *f* des gravats/des décombres

removal of fill [from a hole, an excavation, etc]   débourrage *m*

removal of overburden in open-cast mining   découverte *f*

removal of overlap from edge of wallpaper   arasement *m*

removal of projecting parts of a construction   abattage *m*
removal of rocks   dérochage *m*
removal of surface layer   décapage *m*
removal of test pieces   prélèvement *m* d'échantillons
removal of waste   élimination *f* des déchets
remove   enlever *vb*, retirer *vb*
remove [edges, stains]   ôter *vb*
remove [moving house]   déménager *vb*
remove [pollutants, waste]   éliminer *vb*
remove [sth fixed into masonry, etc]   desceller *vb*
remove a thin layer from [timber or stone]   démaigrir *vb*
remove cover strips, beadings, decorations, etc
   déshabiller *vb*
remove ground around foundations   déchausser *vb*
remove part of a moulding to make a 90° joint
   ébarber *vb*
remove rendering   décrépir *vb*
remove rocks   dérocher *vb*
remove soft parts of dimension stone   ébousiner *vb*
remove the surface layer   décaper *vb*
remover, household   déménageur *m*
remover, road marking   effaceuse *f* (de marquage
   routier)
remuneration   rémunération *f*
renewal, urban   rénovation *f* urbaine
render   revêtement *m*
render [a facade, stonework, etc]   ravaler *vb*
render [a wall, etc]   crépir *vb*
render [heavily to bring surface to desired plane]   garnir *vb*
render opaque   opacifier *vb*
render (with)   enduire *vb* (de)
render with a large, flat trowel   ratisser *vb*
renderer   cimentier *m*, cimentier-ravaleur-applicateur *m*,
   façadier *m*
rendering   enduit *m* hourdé, revêtement *m*
rendering [application of]   enduisage *m*
rendering [of a facade, stonework, etc]   ravalement *m*
rendering [of a wall, etc]   crépissage *m*
rendering [plaster, mortar]   enduit *m*
rendering, alpine finish   enduit *m* tyrolien/au balai,
   mouchetis *m*, crépi *m* tyrolien
rendering, bastard [with a half-and-half lime/cement
   binder]   enduit *m* bâtard
rendering, brushed finish   enduit *m* au balai
rendering, cement   enduit *m* de ciment
rendering, external   enduit *m* extérieur
rendering, polymer-based, with mineral filler   enduit *m*
   de parement plastique
rendering, preliminary   renformis *m*
rendering, reinforced   enduit *m* armé
rendering, rough or roughcast   crépi *m*
rendering, single layer   monocouche *f*
rendering, smooth finish   enduit *m* lisse/lissé
rendering, surface preparation for, by light picking
   hachement *m*
rendering, Tyrolean finish   crépi *m* tyrolien, enduit *m*
   moucheté/tyrolien/au balai, mouchetis *m*
rendering applied in a thick coat to masonry facings
   enduit *m* garnissant

rendering brushed to remove roughness before
   complete setting   enduit *m* brossé
rendering bush-hammered three weeks after setting
   enduit *m* bouchardé
rendering finished with a devil float two to four hours
   after application   enduit *m* gratté
rendering finished with a felt covered trowel   enduit *m*
   feutré
rendering levelled with a rule   enduit *m* dressé
rendering polished after two to six days with a brick or
   rotating sander   enduit *m* grésé/poncé
rendering polished with a fine abrasive disc   enduit *m*
   poli
rendering raked with a metal comb or toothed trowel
   two to six hours after application   enduit *m*
   peigné/raclé
rendering sprayed with water and sponged some hours
   after application   enduit *m* lavé
rendering thrown at a surface to produce a rustic effect
   enduit *m* jeté
rendering treated mechanically after hardening   enduit
   m ravalé
rendering trowelled to produce a rustic effect   enduit *m*
   à la truelle
rendering trowelled with a float to draw laitance to
   surface   enduit *m* frotté
rendering used to bring a surface to desired facing
   plane   garnissage *m*
rendering with a large, flat trowel   ratissage *m*
rendering with a ribbed or scored finish   enduit *m*
   ribbé/strié
rendering with a smooth finish   enduit *m* lisse/lissé
rendering with a special finish applied with a roller
   enduit *m* tramé
rendering with floated or trowelled finish   enduit *m*
   taloché
rendering with mineral filler, polymer-based   enduit *m*
   de parement plastique
rendering, apply preliminary   renformir *vb*
renew   remplacer *vb*, renouveler *vb*
renewal   renouvellement *m*, rénovation *f*
renewal, urban   rénovation *f* urbaine
renovate   réaménager *vb*, réhabiliter *vb*, rénover *vb*
renovated   rénové *adj*
renovation   réaménagement *m*, rénovation *f*
renovation, needing   à remettre *vb* en état
renovation, urban   rénovation *f* urbaine
rent   loyer *m*
rent   louer *vb*
rental/renting   location *f*
rental [system, e.g. electricity supply]   abonnement *m*
rented   loué *adj*
re-open [a road]   débarrer *vb*
repair   réfection *f*, réparation *f*, réparer *vb*
repair [of paintwork, plaster, tiling, etc]   raccord *m*
repair damaged edge(s)   aviver *vb*
repair of masonry by replacement of individual brick
   or stones   lancis *m*
repair rendering   recrépir *vb*

repair shop   atelier *m* de réparation(s)
repay   rembourser *vb*
repay the loan   rembourser *vb* l'emprunt *m*
repeat [of a pattern, e.g. wallpaper]   raccord *m*
repeat   répéter *vb*
repeat [an error]   reproduire *vb*
replace   remplacer *vb*, renouveler *vb*, replacer *vb*
replace a degraded brick or stone in masonry
   relancer *vb*
replacement   remplacement *m*
replacement [repair, of individual roof slates or tiles]
   recherche *f*
replacement of individual bricks or stones in existing
   masonry   lancis *m*
replacement part   pièce *f* de rechange
replacing a degraded brick or stone in masonry
   relancis *m*
replaster   replâtrer *vb*
replastering [the action, and its result]   replâtrage *m*
reply   réponse *f*
re-point   rafraîchir *vb* les joints
re-pointing [the action, and its result]   rejointoiement *m*
report   compte-rendu *m*, procès-verbal *m* (PV), rapport *m*
report [of facts, etc]   exposé *m*
report, monthly   rapport *m* mensuel
report, test   compte-rendu *m* d'essai, rapport *m* d'essais
report, weekly   rapport *m* hebdomadaire
representation   représentation *f*
representative   représentant *m*
representative [of somebody, e.g client]   mandataire *m/f*
reproduce   reproduire *vb*
request (for)   demande *f* (de)
require   exiger *vb*
required   demandé *adj*
requirement   besoin *m*, condition *f* requise, exigence *f*
requirement, air   quantité *f* d'air nécessaire
requirement, energy   énergie *f* nécessaire
requirement, power   besoin *m* d'énergie, énergie *f*
   nécessaire
requirement, space   place *f* nécessaire
requirements, space [dimensions, for equipment]   cotes
   f,pl d'encombrement
requirements, table of technical   tableau *m* de
   contraintes
requirements, technical   exigence *f,pl* techniques
re-render   recrépir *vb*
re-route   réacheminer *vb*
resale   revente *f*
resaw [a log, to square edged timber]   aviver *vb*
research   recherche *f*
research centre   centre *m* de recherches
research laboratory   laboratoire *m* de recherches
research into sth, do   faire *vb* des recherches sur qch
resell   revendre *vb*
reservation, central [roads]   berme *f* centrale, séparateur
   m, terre-plein *m* central
reservation, raised (central)   séparateur *m* en saillie
reserve [provision]   provision *f*
reserve [stock, or as in nature reserve]   réserve *f*

reserve of water   réserve *f* d'eau
reservoir   bassin *m* de rétention
reservoir [dammed]   barrage-réservoir *m*
reservoir [small]   étang *m*
reservoir [structural rather than dammed]   réservoir *m*
reservoir, balancing   réservoir *m* d'arrêt
reservoir, cylindrical   ballon *m*
reservoir, heat   accumulateur *m* de chaleur
reservoir, hydraulic   réservoir *m* hydraulique
reservoir, storage   réservoir *m* d'accumulation
reset [equipment]   recaler *vb*
reshaping   remaniage *m*
residence   demeure *f*, résidence *f*
residence [place of]   domicile *m*
residence, gentleman's   maison *f* de maître
residence, principal/secondary   résidence *f*
   principale/secondaire
resident [living along a road or river]   riverain *m*
residential   résidentiel *adj*
residual   résiduel *adj*
residual current device (RCD)   appareil *m* différentiel
residual current switch   interrupteur *m* différentiel
residue   résidu *m*
resilience   résilience *f*
resilient [of a material]   résilient *adj*
resin   résine *f*
resin pocket [timber defect]   poche *f* de résine
resin-bonded   collé *adj* à la résine synthétique
resinous [of timber]   résineux *adj*
resist pressure/thrust   contre-buter *vb*
resistance   résistance *f*
resistance, air   résistance *f* de l'air
resistance, air flow   résistance *f* aéraulique/au passage de
   l'air
resistance, coefficient of   coefficient *m* de résistance
resistance, cold   résistance *f* au froid
resistance, conductor   résistance *f* d'un conducteur
resistance, corrosion   résistance *f* à la corrosion
resistance, electrical   résistance *f* électrique
resistance, fatigue   résistance *f* à la fatigue
resistance, film   résistance *f* thermique superficielle
resistance, fire, *see* fire resistance
resistance, flame   tenue *f* à la flamme
resistance, flow   résistance *f* à l'écoulement
resistance, flow [frictional]   résistance *f* par friction
resistance, frictional [arising from friction]   résistance *f*
   de frottement
resistance, frictional [to friction]   résistance *f* au
   frottement
resistance, frost   résistance *f* au gel
resistance, heat   résistance *f* à la chaleur
resistance, impact   résistance *f* au choc
resistance, insulation [electrical]   résistance *f*
   d'isolement
resistance, limited   tenue *f* limitée
resistance, mechanical   résistance *f* mécanique
resistance, penetration   résistance *f* à la pénétration
resistance, required (period of) fire   durée *f* coupe-feu
   réglementaire

resistance, shear(ing)　résistance *f* au cisaillement
resistance, sliding　résistance *f* au glissement
resistance, spread of flame　pare-flammes *m*
resistance, starting　résistance *f* de démarrage
resistance, starting [the apparatus]　rhéostat *m* de démarrage
resistance, thermal　résistance *f* thermique
resistance, weather　résistance *f* aux intempéries
resistance, weathering [of soils, stone, etc]　résistance *f* à l'altération
resistance to penetration of the point of a pile　effet *m* de pointe
resistivity　résistivité *f*
resistivity, thermal　résistivité *f* thermique
resistor　résistance *f*
re-smooth [a surface]　radoucir *vb*
resonance　résonance *f*
resort　station *f*
resource　ressource *f*
resources, natural　ressources *f,pl* naturelles
resources, water　ressources *f,pl* en eau
resources of men and materials　ressources *f,pl* en hommes et en matériel
resource exploitation　mobilisation *f* de la ressource/des ressources
resource inventory　inventaire *m* des ressources
respirator　appareil *m* filtrant
respite [time allowed]　délai *m*
response　réponse *f*
response, control　réponse *f* de contrôle
response time　temps *m* de réponse
responsibility　responsabilité *f*
responsibility, within his/her　de son ressort *m*
responsible　responsable *adj*
rest, tool　porte-outil *m*
rest area [roadside]　aire *f* de repos
rest period　période *f* de repos
rest (sth) (on, against)　appuyer *vb* (qch) (sur, contre)
restart of concrete pouring　reprise *f* de bétonnage
restoration　réfection *f*, restauration *f*
restoration [e.g. of something damaged or worn]　reconstitution *f*
restoration [of a building]　restitution *f*
restoration [of a facade, stonework, etc]　ravalement *m*
restoration [of masonry, joinery]　rafraîchissement *m*
restoration, for　à restaurer *adj*
restoration (works)　travaux *m,pl* de restauration
restore　réhabiliter *vb*, rénover *vb*, restaurer *vb*
restore [a facade, stonework, etc]　ravaler *vb*
restored　rénové *adj*, restauré *adj*
restorer　restaurateur *m*
restrained [built in]　encastré *adj*
restraining [by building in]　encastrement *m*
restriction [constriction]　étranglement *m*
result　résultat *m*
result [final, overall]　bilan *m*
results, scatter of　dispersion *f* des résultats
resultant　résultant *adj*
resultant　résultante *f*

resulting　résultant *adj*
resurface [a road]　refaire *vb* la surface de
resurface [a facade, stonework, etc]　ravaler *vb*
resurface [re-render]　recrépir *vb*
resurfacing [of a facade, stonework, etc]　ravalement *m*
re-survey [of land, etc]　réarpentage *m*
re-survey [of condition, causes, etc]　contre-expertise *f*
retailer　commerçant *m*
retain　soutenir *vb*
retaining smoke particles　fumivore *adj*
retarder　retardeur *m*
retarder [railway]　rail-frein *m*
retarder, vapour　retardeur *m* de vapeur
retarding agent [for concrete]　retardateur *m* de prise
retention [e.g. of water]　rétention *f*
retention [support]　soutènement *m*
retention [until final acceptance of works]　retenue *f* de garantie
retention period　période *f* de rétention
retest　contre-essai *m*
retile [roof]　remanier *vb*
retirement　retraite *f*
retractable　escamotable *adj*
retreat　retraite *f*
retreatment plant/train, road　train *m* de reconditionnement/retraitement de chaussée
return [financial]　rapport *m*
return [of goods, etc]　renvoi *m*
return (on) [an investment]　rendement *m* (de)
re-use　remploi *m*, réemploi *m*, réutilisation *f*
re-use　réutiliser *vb*
reveal [of a door, window]　embrasure *f*, tableau *m*
revenue [financial]　rapport *m*
revenue stamp　timbre *m* fiscal
reverberation　réverbération *f*
reverberation chamber　chambre *f* réverbérante
reverberation time　période *f*/temps *m* de réverbération
reversal　renversement *m*
reversal [electrical]　inversion *f*
reversal of polarity　inversion *f* de (la) polarité
reverse [current]　inverser *vb*
reversible　réversible *adj*
reversing　renversement *m*
reversionary interest [in property]　nue-propriété *f*
revetment [drystone]　perré *m*
revetment, paved　perré *m* maçonné
revetment of heavy stone around a bridge pier　contre-garde *f*
revise [accounts, a document]　revoir *vb*
revise [proofs, documents]　reviser/réviser *vb*
revision [of accounts, a document]　revision/révision *f*
revolution [turn]　révolution *f*, tour *m*
revolutions per minute (rpm)　tours *m,pl* par minute
revolve　tourner *vb*
revolve [about a point]　pivoter *vb*
rheology　rhéologie *f*
rheostat　rhéostat *m*
rheostat, field　rhéostat *m* de champ
rheostat, starting　rhéostat *m* de démarrage

rib [e.g. below reinforced concrete arch or vault]
  nervure *f*
rib arch [of a vault]   nervure *f*
rib springing from corner of a vault   tierceron *m*
rib   nervurer *vb*
ribbed   nervuré *adj*
rich [of a plaster or cement rich in binder]   gras *adj*
riddle   crible *m*, tamis *m*
riddle [screen]   claie *f*
riddle   cribler *vb*, tamiser *vb*
riddled   criblé *adj*, tamisé *adj*
riddling   criblage *m*, tamisage *m*
ride [in a forest]   laie *f*
rider [amending a contract]   avenant *m*
ridge [as in ridge tile, etc]   faîtière *adj*
ridge [geology]   ligne *f* de crête
ridge [of a mountain]   arête *f*, crête *f*
ridge [roof]   faîtage *m*, faîte *m*, ligne *f* de faîtage
ridge beam   faîte *m*, madrier *m* de faîtage, poutre *f* de
  faîte
ridge board or purlin   panne *f* faîtière
ridge capping [lead, of a slate roof]   brifier *m*
ridge capping [tile or metal]   enfaîtement *m*, faîtage *m*
ridge course [of roof tiles or slates]   rang *m* de faîtage
ridge end tile, decorative   faîteau *m*
ridge [ground for drainage purposes]   butter *vb*
ridge [i.e. cover ridge of a roof]   enfaîter *vb*
right [accurate]   exact *adj*, juste *adj*
right [correctly]   correctement *adv*, juste *adv*
right   droit *m*
right, pre-emptive [generally to acquire, in France]   droit
  m de préemption
right of enjoyment   jouissance *f*
right, put   mettre *vb* en ordre
right, put a mistake   rectifier *vb*
righting [setting upright]   redressement *m*,
  relèvement *m*
rigid   rigide *adj*
rigidity   raideur *f*, rigidité *f*
rigidity, modulus of   module *m* de rigidité
rigidity, torsional   rigidité *f* à la torsion
rim   bord *m*
rim [of a wheel]   bourrelet *m*, jante *f*
ring   anneau *m*, bague *f*, rond *m*
ring [figure or shape]   cercle *m*
ring, clamping   collier *m* de serrage
ring, curtain   boucle *f* de rideau
ring, elliptical   anneau *m* elliptique
ring, focusing [e.g. of a surveying telescope]   bague *f* de
  mise au point
ring, gasket   anneau *m* de joint
ring, growth [annual, of a tree]   cerne *m*
ring, packing   anneau *m* de garniture
ring, sealing   anneau *m*/bague *f* d'étanchéité
ring, shaft   anneau *m* de puits
ring, split   anneau *m* brisé
ring, supporting   anneau *m* de support
ring holding pipe against a wall, metal   ceinture *f*
ring of strip metal covering a joint   ceinture *f*

ring to which upper ends of rafters in a conical roof are
  fixed   gobelet *m*
ring with a tail for fixing to/into wood, masonry, etc
  piton *m*
ringing [of a bell]   sonnerie *f*
ring-shake [timber defect]   roulure *f*
rinse   rincer *vb*
rip [timber]   refendre *vb*
riparian   riverain *adj*
riparian proprietor   riverain *m*
ripper [groundworks]   rippeur *m*
ripper [rock breaker]   défonceuse *f* portée
ripple [electrical]   ondulation *f*
rip-rap   enrochement *m* (de protection), perré *m*
ripsaw   scie *f* à refendre
rise [ascent, climb]   montée *f*
rise [of water level in a river]   crue *f*
rise [slope]   côte *f*, pente *f*
rise [small hill]   hauteur *f*
rise in   hausse *f* de, élévation *f* de
rise in [prices, temperature, wages]   augmentation *f* de
rise in [wages]   relèvement *m* de
rise (of an arch)   flèche *f* (d'un arc)
rise   monter *vb*
rise [wages, prices, taxes, etc]   augmenter *vb*
riser [of stair]   contremarche *f*, montant *m*
riser [vertical duct or pipe]   colonne *f* (montante)
riser, communications   gaine *f* informatique
riser, dry   colonne *f* sèche
riser, pressurized   colonne *f* en charge
riser, service(s)   gaine *f* technique
riser, supply   colonne *f* montante d'alimentation
riser, wet   colonne *f* humide/en charge
rising [ground, tide]   montant *adj*
rising [prices, pressure, temperature, wages]   en hausse
rising, *see* rise
risk   risque *f*
risk   risquer *vb*
rive [wood, slate, etc]   fendre *vb*
river *adj*   fluvial *adj*
river   cours *m* d'eau, rivière *f*
river [with direct outflow to the sea]   fleuve *m*
river, navigable   voie *f* navigable
river bed   lit *m* de fleuve/rivière
riverside *adj*   bord de rivière, au
rivet   rivet *m*
rivet, button or round head   rivet *m* à tête ronde
rivet, countersunk   rivet *m* à tête fraisée
rivet head, flat countersunk   rivure *f* fraisée
rivet head as set during riveting process   rivure *f*
rivet pitch or spacing   écartement *m*/pas *m* des rivets
rivet over   bouteroller *vb*
rivet (together)   river *vb*
rivet-snap   bouterolle *f*
riveted, hot/cold   rivé à chaud/froid
riveter [worker]   riveur *m*
riveting   rivetage *m*
riveting [the result]   rivure *f*
riveting, staggered   rivure *f* en quinconce

riveting die bouterolle *f*
riveting machine riveteuse *f*
riveting tool riveuse *f*
riving knife [of circular saw] couteau *m* diviseur
rivulet ruisseau *m*, ruisselet *m*
road *adj* routier *adj*
road chaussée *f*, chemin *m*, route *f*, voie *f*
road, access or approach chemin *m*/route *f*/voie *f*
    d'accès
road, access and service voie *f* de desserte
road, county chemin *m* départemental, route *f*
    départementale
road, delivery accès *m* livraison
road, earth piste *f*
road, embanked route *f* en remblai
road, entry/exit slip bretelle *f* d'accès/de sortie
road, farm chemin *m* d'exploitation
road, forest chemin *m* forestier/d'exploitation
road, high-speed voie *f* (à circulation) rapide
road, local distributor (collector) voie *f* de collecte
road, loop rue *f* en boucle
road, main or major route *f* principale
road, major or trunk route *f* nationale/à grande
    circulation
road, metalled chemin *m* empierré/ferré
road, minor route *f*/voie *f* secondaire
road, minor access/service voie *f* secondaire de desserte
road, one-way chaussée *f*/route *f* à sens unique
road, ring boulevard *m* périphérique, périphérique *m*,
    rocade *f*
road, rural chemin *m* rural
road, secondary chemin *m* départemental, route *f*
    départementale, route *f* secondaire
road, slip voie *f* d'insertion
road, trunk route *f* nationale
road, two/three/four lane route *f* à deux/trois/quatre
    bandes/voies
roads, provided with [of a site for building] viabilisé *adj*
road along a dike or embankment levée *f*
road area including footways and verges plate-forme *f*
road boundary between public and private domains
    alignement *m*
road centre-line axe *m* de chaussée
road construction [the complete pavement] corps *m* de
    chaussée
road for delivery of equipment/plant accès *m* livraison
    matériel
road for delivery of materials accès *m* livraison
    matériaux
road marking marquage *m* routier
road marking eraser/remover effaceuse *f* (de marquage
    routier)
road metal caillasse *f*, écales *f,pl*, pierraille *f*
road network or system réseau *m* routier, voirie *f*
road network, primary voirie *f* principale
road network, rural voirie *f* rurale
road on embankment/in cutting route *f* en
    remblai/déblai
road paved with setts chaussée *f* pavée

road safety prévention *f*/sécurité *f* routière
road system réseau *m* routier, voirie *f*
roads with a grid layout rues *f,pl* en quadrillage
roadbase, flexible couche *f* de base souple
roadbase, rigid couche *f* de base rigide
roadheader abatteuse *f*
roadman cantonnier *m*
roadman's hut maison *f* cantonnière
roadway chaussée *f*
roadway, plank chemin *m* en madriers
roadway of a bridge aire *f* d'un pont
roadworks travaux *m,pl* de voirie, voirie *f*
roadworks lighting [to guide traffic] guirlande *f*
    lumineuse
robust solide *adj*
robustness solidité *f*
rocaille [decoration in rococo architecture] rocaille *f*
rock [block of stone] rocher *m*
rock [in general] roche *f*
rock, crushed pierre *f* cassée/concassée, roche *f* naturelle
    concassée
rock, eruptive roche *f* éruptive
rock, hard roches *f,pl* compactes
rock, metamorphic roche *f* métamorphique
rock, natural roche *f* naturelle
rock, parent roche-mère *f*
rock, sedimentary roche *f* sédimentaire/stratifiée
rock, solid [aggregate, road material] matériau *m*
    rocheux compact
rock, sound rocher *m* intact
rocks, decomposed/weathered roches *f,pl* altérées
rocks, removal of dérochage *m*
rock bolt boulon *m* d'ancrage
rock breaker brise-roc(he) *m*, dérocheuse *f*
rock face, steep falaise *f*
rock waste décharge *m* de roche
rocker *adj* [of a switch] à bascule
rockery rocaille *f*
rockfall chute *f* de pierre(s), éboulement *m*
rockfill enrochement *m*, remblai *m* pierreux
Rockwell hardness test essai *m* de dureté Rockwell
rocky rocheux *adj*
rod [metal] barre *f*, tige *f*
rod [curtain, etc] tringle *f*
rod [wood] baguette *f*, perche *f*
rod, anchoring tirant *m* d'ancrage
rod, connecting [not locomotive] fer *m* de liaison
rod, control tringle *f* de commande
rod, curtain barre *f* à rideau, tringle *f* (de rideau)
rod, earth(ing) piquet *m*/prise *f* de terre
rod, filler [welding] baguette *f* d'apport
rod, levelling [surveying] mire *f*, tige *f* de nivellement
rod, measuring pige *f*
rod, ranging jalon *m*
rod, square steel, used for reinforcing plaster work
    fanton *m*
rod, stair tringle *f* d'escalier
rod, surveyor's, *see also* surveyor's rod mire *f*, jalon *m*
rod, threaded tige *f* filetée

rod, wooden   perche *f*
rod man [surveying]   porte-mire *m*
roll [e.g.of wallpaper]   rouleau *m*
roll crusher   broyeur *m* à cylindres
roll [metal]   laminer *vb*
roll [to compact]   compacter *vb* au rouleau *m*, cylindrer
   vb, rouler *vb*
roll up [e.g. carpet, cable]   enrouler *vb*
rolled [compacted]   cylindré *adj*
rolled [e.g. steel]   laminé *adj*
rolled, cold/hot   laminé *adj* à froid/chaud
roller   cylindre *m*
roller [compactor]   compacteur *m*
roller [metal]   galet *m*
roller [of a painter's roller]   manchon *m*
roller [painter's, road, etc]   rouleau *m*
roller [wood, for moving heavy loads]   roule *f*
roller, conveyor   rouleau *m* transporteur
roller, cylinder   rouleau *m* cylindre
roller, double-drum vibratory   rouleau *m* vibrant à
   double bille
roller, feed   rouleau *m* entraîneur
roller, grid   rouleau *m* à grille
roller, paint   rouleau *m* à peindre
roller, pedestrian-controlled (vibratory)   rouleau *m*
   (vibrant) à guidage à main
roller, pneumatic-/rubber-tyred   compacteur *m* à
   pneu(matique)s, rouleau *m* (compacteur) à pneu(s)
roller, road   rouleau *m* compresseur
roller, seam [for wallpaper]   rouleau *m* de colleur,
   roulette *f*
roller, self-propelled single drum   rouleau *m* monobille
   automoteur
roller, sheepsfoot   rouleau *m* pied de mouton/à pieds de
   mouton
roller, smooth-wheeled   compacteur *m* à jante *f* lisse,
   rouleau *m* lisse
roller, stippling   rouleau *m* à pochoir
roller, tamping   rouleau *m* dameur
roller, tamping foot   rouleau *m* à pieds dameurs
roller, tapered foot   rouleau *m* à pieds coniques
roller, (towed) tamping   rouleau *m* vibrant (tracté)
roller, (towed) vibrating/vibratory   rouleau *m* vibrant
   (tracté)
roller, tyred   rouleau *m* (compacteur) à pneu(s)
roller, vibrating/vibratory   cylindre *m*/rouleau *m* vibrant
roller covered with pyramidal points for surface
  finishing   boucharde *f* d'enduiseur
roller track   chemin *m* de roulement
rolling   roulant *adj*
rolling [of steel]   laminage *m*
rolling [to compact]   cylindrage *m*
rolling, cold/hot   laminage *m* à froid/chaud
rolling, compaction by   compactage *m* par cylindrage
rolling, final   cylindrage *m* définitif
rolling mill   laminoir *m*
rolling stock [railway]   matériel *m* roulant
roof, *see also* roofing   toit *m*, toiture *f*
roof, concrete shell   coque *f* en béton

roof, conical/turret   toit *m* conique/en poivrière
roof, flat   toit *m* plat/en terrasse
roof, gable   comble *m* sur pignon, toit *m* à pignon/en
   pente/en dos d'âne/en selle
roof, glass, over platform [station]   marquise *f*
roof, glass/glazed   verrière *f*
roof, hipped   toit *m* à quatre arêtiers/en croupe, toiture *f* à
   quatre pentes
roof, hipped-gable   toit *m* à croupe faîtière ·
roof, lean-to, *see also* roof, mono-pitch   toit *m*/toiture *f* en
   appentis
roof, mansard   mansarde *f*, toit *m*/toiture *f* à la mansarde
roof, mansard [structure, roof space]   comble *m*
   brisé/mansardé
roof, mansard, upper part of a   faux-comble *m*,
   terrasson *m*
roof, mission or Spanish tiled   toiture *f* romaine
roof, mono-pitch   toit *m* à un versant/à un pan/en
   appentis, toiture *f* à une seule pente
roof, mortared edge of a tiled   parement *m*
roof, nearly flat   terrasse *f*, terrasson *m*
roof, pantile(d)   toiture *f* flamande
roof, penthouse   appentis *m*
roof, pitched   toiture *f* inclinée
roof, plain tiled   toit *m* de tuiles à talon
roof, prestressed concrete shell   coque *f* précontrainte
roof, rafter   toit *m* à deux pans/à chevrons/à deux
   versants
roof, ridge   toit *m* à deux égouts/à deux pentes
roof, sawtooth   toit *m*/toiture *f* à redans/redents, toit *m* en
   shed/en dents de scie
roof, slated   toit *m* d'ardoises
roof, space-frame   couverture *f* industrielle
roof, span   comble *m* à deux pentes/à deux versants
roof, sun   toiture-terrasse *f*
roof, suspended [hung from cables]   couverture *f*
   suspendue
roof, temporary, to shelter repairs/works   parapluie *m*
roof, tent   toit *m* en tente
roof, thatched   toit *m* de chaume
roof, thin shell concrete   voile *m* mince
roof, upper part of a mansard   faux-comble *m*,
   terrasson *m*
roof batten   latte *f* volige
roof boarding/sheathing   voligeage *m*
roof cladding/covering   couverture *f*
roof glazing   verrière *f*
roof gutter [boxed metal]   chéneau *m* encaissé/à
   l'encaissement
roof gutter [metal channel in wooden trough]   chéneau *m*
   à l'anglaise
roof gutter [rounded metal, on metal feet]   gouttière *f* à
   l'anglaise
roof ladder   échelle *f* plate/de couvreur
roof panel   pan *m* de comble
roof post   poinçon *m*
roof rack [on a vehicle]   porte-bagages *m*
roof sheathing board   bardeau *m*, volige *f*
roof slater/tiler   ardoisier *m*

roof space comble *m*/combles *m,pl*
roof space, conical, of a tower or spire flèche *f*
roof space, convertible combles *m,pl* aménageables
roof space too low to be occupied faux comble *m*
roof space storey étage *m* lambrissé
roof (space) with convex slopes comble *m* à l'impériale
roof structure, timber comble *m*, charpente *f* de comble
roof structure and covering toiture *f*
roof terrace plate-forme *f*, toiture-terrasse *f*
roof timbers combles *m,pl*
roof valley noulet *m*
roof vent [smoke vent] trappe *f* de désenfumage
roof ventilator chatière *f*, ventilateur *m* de toiture
roof couvrir *vb*
roofed(-in) area terrain *m* couvert
roofer couvreur *m*
roofer's ladder échelle *f* plate/de couvreur
roofing couverture *f*, toiture *f*
roofing, thatched couverture *f* de chaume
roofing, tiled couverture *f* en tuiles
roofing bracket console *f* de couvreur
roofing nail pointe *f* à papier bitumé
roofing paper carton *m* bitumé
roofing paper [waterproof] pare-pluie *m*
rooflight fenêtre *f* de toit, vitrage *m*
rooflight [raised] lanterneau *m*
rooflight [Velux, patented] Velux *m*
rooftop faîte *m*
room pièce *f*
room [amount of space] place *f*
room [bedroom, or hotel] chambre *f*
room [large, or formal] salle *f*
room [lawyer's or doctors, or small] cabinet *m*
room [office, study] bureau *m*
room [premises, and as in plantroom] local *m*
room, auction salle *f* des ventes
room, clean chambre *f* propre, salle *f* blanche
room, cold armoire *f* frigorifique, chambre *f* frigorifique/froide
room, conference salle *f* de conférence
room, consulting cabinet *m* de consultation
room, control régie *f*, salle *f* de commande/de contrôle
room, dining salle *f* à manger
room, drawing salon *m*
room, drawing [formal] salle *f*/salon *m* de réception
room, dressing cabinet *m* de toilette, dressing *m*
room, equipment local *m* technique, salle *f* des appareils
room, first aid infirmerie *f*
room, games salle *f* de jeux
room, guest chambre *f* d'amis
room, kitchen-dining cuisine-salle *f* à manger
room, laundry blanchisserie *f*, buanderie *f*
room, linen lingerie *f*
room, living living *m*, salle *f* de séjour, séjour *m*
room, lumber débarras *m*
room, meeting salle *f* de réunion(s)/de séances
room, occupied local *m* occupé, pièce *f* occupée
room, plant local *m* technique/de centrale

room, rest salle *f* de repos/de récupération
room, shower salle *f* d'eau/de douches
room, sick infirmerie *f*
room, sitting salon *m*
room, small cabinet *m*
room, smoking fumoir *m*
room, store, *see also* storeroom local *m* de stockage
room, test local *m* d'essai
room, utility buanderie *f*
room, waiting salle *f* d'attente
rooms, changing vestiaires *m,pl*
room characteristic caractéristique *f* du local
room constant [reverberation] constante *f* de réverbération
root [of a plant, tree, and mathematical] racine *f*
root, cube/square racine *f* cubique/carrée
root, weld base *f* de la soudure
root crack fissure *f* à la base
root mean square (RMS) value valeur *f* effective
root of title origine *f* de propriété
rooter [earthworks] défonceuse *f* tractée
rope câble *m*, corde *f*
rope, manilla manille *f*
rope, steel-wire câble *m* d'acier
rope, wire câble *m* métallique
ropeway téléphérique *f*
ropeway, aerial blondin *m*, câble *m* transporteur, transporteur *m* aérien à câble
ropeway, cabin télécabine *f*
rose [of a door handle or bell push] rosette *f*
rosewood palissandre *m*
rot pourriture *f*
rot [timber] carie *f*
rot, dry/wet, *see* dry/wet rot
rot pocket or rotten knot hidden in sound wood malandre *f*
rot [organic matter] décomposer *vb*
rot se décomposer *vb*
rotary rotatif *adj*
rotate tourner *vb*
rotating rotatif *adj*
rotation rotation *f*
rotation [i.e. one turn] tour *m*
rotation [e.g. in shift working] roulement *m*
rotations per minute (rpm) tours *m,pl* par minute
rotation, anti-clockwise rotation *f* à gauche
rotation, centre of centre *m* de rotation
rotation, clockwise rotation *f* à droite
rotation, direction of sens *m* de rotation
rotor rotor *m*
rotor, twin-bladed [wind generator] hélice *f* bipale
rot-proof imputrescible *adj*
rotting décomposition *f*
rotunda rotonde *f*
rough [approximate] approximatif *adj*
rough [of a surface] rude *adj*, rugueux *adj*
rough [uneven, of ground] accidenté *adj*
rough [unworked, without finishes] brut *adj*
rough hew [wood, stone] bûcher *vb*

rough in [before final coat of plaster or rendering]
  gobeter *vb*
rough out [a plan, etc]   ébaucher *vb*
rough out [trim to size, of woodwork]   dégrossir *vb*
rough plane   dégauchir *vb*
roughcast [rendering]   crépi *m* (moucheté/tyrolien),
  hourdis *m*, enduit *m* moucheté
roughcast, Tyrolean   crépi *m* tyrolien, enduit *m* au balai
roughcast [render]   hourder *vb*
roughening [to provide a key]   accrochage *m*
roughing out [to size]   dégrossissage *m*
roughing out [sketching]   ébauche *f*
roughing-in [plastering or rendering]   gobetage *m*
roughing-in [plaster or render]   gobetis *m*
roughness   rugosité *f*
roughness coefficient   coefficient *m* de rugosité
round   arrondi *adj*, rond *adj*
round [e.g. moulding or steel bar]   rond *m*
round (off) [a corner or edge, a number]   arrondir *vb*
round off [a job, etc]   terminer *vb*
round off the corner of [e.g. of exposed timbers]
  carderonner *vb*
roundabout   carrefour *m* giratoire, rond-point *m*
rounded [domed]   bombé *adj*
rounded [of a sand or gravel particle]   roulé *adj*
rounded (off)   arrondi *adj*
round-edged   à bords arrondis
roundwood   bois *m* de brin/de grume, bois *m* grume,
  brin *m*
route [of road, railway, etc]   tracé *m*
router   défonceuse *f*, toupie *f*
routing [of a despatch of goods]   acheminement *m*
row   rangée *f*
row [e.g. of posts, trees]   cordon *m*
row [of tiles or slates]   rang *m*
row of paving slabs of equal size   range *f* de pavés
rows, in staggered   en quinconces
rows, set in   ranger *vb*
rub   frotter *vb*
rub down   poncer *vb*
rubber *adj*   en caoutchouc
rubber   caoutchouc *m*
rubbing   frottement *m*
rubbing down and touching up [of paintwork]   touche *f*
  et frottis *m*
rubbing down with a cloth or felt and a fine abrasive
  chiffonnage *m*
rubbish [earth, etc]   éboulis *m*
rubbish [from buildings]   décombres *m,pl*
rubbish [junk]   saleté *f*
rubbish [waste]   ordures *f,pl*
rubbish, demolition   gravats *m,pl*, gravois *m*
rubbish, domestic/household   ordures *f,pl* ménagères
rubbish collection, (domestic)   ramassage *m* d'ordures
  (ménagères)
rubbish compactor, (domestic)   compacteur *m* d'ordures
  (ménagères)
rubbish disposal, domestic   traitement *m* d'ordures
  ménagères

rubble   blocaille *f*, pierraille *f*
rubble [from buildings]   décombres *m,pl*
rubble, small round piece of   tête *f* de chat
rubble masonry, *see also* random rubble   limousinage *m*,
  limousinerie *f*
rubble stone   moellon *m* brut
rubble stone, roughly squared   moellon *m* ébousiné
rubble stone, tooled, with sharp edges   moellon *m*
  piqué
rubble stone laid on edge   moellon *m* en coupe
rubble stone laid on its bedding plane   moellon *m*
  plat
rubble stone placed to take concentrated load   libage *m*
rubble stone sawn to dimensions at the quarry
  moellon *m* préscié
rubble stone too irregular to be squared   moellon *m*
  bloqué
rubble stone tooled with a mason's hammer   moellon *m*
  smillé
ruin   ruine *f*
ruin   ruiner *vb*
ruin(s), fall into   tomber *vb* en ruines
ruinous [building, etc]   délabré *adj*
ruinous [expensive]   ruineux *adj*
rule   règle *f*
rule [metre]   mètre *m*
rule, empirical   règle *f* empirique
rule, folding   mètre *m* pliant
rule following a power float to ensure flatness of slab
  lissoir *m*
rule in squares   quadriller *vb*
ruler   régle *f* graduée
ruler, scale   échelle *f* de réduction
rules   règlement *m*
ruling   ordonnance *f*
run [of excess paint]   coulure *f*
run [a business]   administrer *vb*
run [a computer program]   exécuter *vb*
run [a machine]   faire *vb* fonctionner/marcher
run [a quarry, a railway, etc]   exploiter *vb*
run [flow, e.g. a river]   couler *vb*
run [machine]   fonctionner *vb*, marcher *vb*
run [motor]   tourner *vb*
run [paint]   couler *vb*
run a pipe into   amener *vb* un tuyau dans
run a plane over sth   passer *vb* qch au rabot
run across [i.e. one side to another]   traverser *vb*
run along [beside]   longer *vb*
run away [liquids]   s'écouler *vb*
run down [accumulator or battery]   se décharger *vb*
run dry [pump]   se désamorcer *vb*
run dry [spring, well, river, stream]   se dessécher *vb*, se
  tarir *vb*
run hot   chauffer *vb*
run in parallel/series   marcher *vb* en parallèle/série
run into debt   s'endetter *vb*
run off [rainwater, etc]   s'écouler *vb*, ruisseler *vb*
run on or at no load   marcher *vb* à vide
run out [supplies, stocks]   s'épuiser *vb*

run out of true   avoir *vb* du balourd, tourner *vb* à faux
run over [overflow]   déborder *vb*
run to spoil [unsuitable excavated material]   mettre *vb* en dépôt
run true   tourner *vb* sans balourd
rung [of a ladder]   échelon *m*, traverse *f*
runner [of a drawer]   coulisse *f*, coulisseau *m*
running *adj*   courant *adj*
running, *see also* operation   fonctionnement *m*
running [an enterprise, railway, etc]   exploitation *f*
running [e.g. of an installation]   conduite *f*
running [of a house, shop, etc]   tenue *f*
running [of equipment, machines]   fonctionnement *m*, marche *f*
running, no-load [of a machine]   marche *f* à blanc/vide
running in   mise *f* au point
running light or on no-load [of a machine]   marche *f* à blanc/vide
running-board [vehicle]   marchepied *m*
run-off   écoulement *m* (des eaux), ruissellement *m*

run-off, surface water   eaux *m,pl* de ruissellement/de surface *f*
run-off coefficient   coefficient *m* de ruissellement
run-up [sports ground, for jumping or throwing]   aire *f* d'élan
runway, airfield/airport   piste *f* (d'atterrissage)
runway, crane   chemin *m* de la grue
rupture   cassure *f*, rupture *f*
rupture   casser *vb*
rural   champêtre *adj*
rust   rouille *f*
rust   se corroder *vb*, se rouiller *vb*
rust [sth]   corroder *vb*, rouiller *vb*
rust coloured   rubigineux *adj*
rust protective agent   moyen *m* de protection contre la rouille
rustproof   inoxydable *adj*
rustproof [of a treatment]   antirouille *adj*
rustic   campagnard *adj*, champêtre *adj*, rustique *adj*
rusty   rouillé *adj*

S-trap   siphon *m* en S
S-bend   coude *m* de renvoi
sack   sac *m*
sack hoist   monte-sacs *m*
saddle [between mountains]   passe *f*
saddle [geology]   anticlinal *m*
saddle (piece) [flashing above chimney, etc]   besace *f*
safety   sécurité *f*, sûreté *f*
safety, fire, *see* fire safety
safety, road   prévention *f*/sécurité *f* routière
safety, site   sécurité *f* du travail
safety against local buckling   sécurité *f* au voilement
safety arrestor [of lift or hoist]   parachute *m*
safety barrier [highways]   glissière *f* de sécurité
safety barrier, double/single sided   glissière *f* double/simple de sécurité
safety barrier, steel [highways]   glissière *f* métallique de sécurité
safety belt   ceinture *f* de sécurité
safety boot   botte *f* de sécurité, chaussure *f* (avec coquille) de sécurité
safety coordination   coordination *f* de la sécurité
safety curtain [in a theatre]   rideau *m* de fer
safety cut-out   disjoncteur *m* de sécurité, dispositif *m* de sécurité par coupure
safety cut-out [pressure switch operated]   pressostat *m* de sécurité
safety cut-out, high/low pressure   pressostat *m* de sécurité haute/basse pression
safety device   dispositif *m* de sécurité/de protection, sécurité *f*

safety device, automatic   dispositif *m* automatique de sécurité
safety device, low water   sécurité *f* de manque d'eau
safety factor   coefficient *m*/facteur *m* de sécurité
safety fence [road]   élément *m* de glissement
safety fuse   fusible *m* [de sûreté]
safety glass   verre *m* de sécurité
safety glass (Triplex)   verre *m* triplex
safety harness [for working at height]   baudrier *m*/harnais *m* de sécurité
safety helmet   casque *m* (de protection)
safety hook [of a roof]   crochet *m* de sécurité
safety hoop [on a fixed ladder]   crinoline *f*
safety interlock   couplage *m* de sécurité
safety lighting   éclairage *m* de sécurité
safety margin   marge *f* de sécurité
safety net   filet *m* de recueil/de sécurité, voile *m* de sécurité
safety plug   bouchon *m* fusible
safety rail   garde-corps *m*, garde-fou *m*
safety requirement   condition *f* de sécurité
safety valve   soupape *f* de sécurité/de sûreté
safety valve, spring-loaded   soupape *f* de sûreté à ressort
safety warnings   signalisation *f* des risques
sag [amount of]   flèche *f*
sag   s'affaisser *vb*, fléchir *vb*, plier *vb*
sagging [of a beam, etc]   affaissement *m*, courbure *f*, fléchissement *m*, flexion *f*
salary   salaire *m*
salary, monthly   mensualité *f*
salary slip   bordereau *m* de salaire

sale   vente *f*
sale, auction   adjudication *f*, vente *f* aux enchères
sale, bill of   acte *f* de vente
sale, contract of   contrat *m* de vente
sale, deed of   acte *f* authentique de vente, acte *f* finale de vente
sale, firm   vente *f* ferme
sale, outright   vente *f* à forfait
sale agreement, preliminary [to buy and sell]   compromis *m* de vente
sale agreement, preliminary [to sell]   promesse *f* de vente
sale by auction   vente *f* aux enchères
sale contract   contrat *m* de vente
sale on approval   vente *f* à l'essai
sale on deferred payment terms   vente *f* à tempérament
saleroom   salle *f* des ventes
salerooms [generally for auctions]   hôtel *m* des ventes
salesman, technical   agent *m* technico-commercial
salient   saillant *adj*
salient angle   angle *m* saillant
saline   salin *adj*
salinity   salinité *f*
salt-glazed ware   grès *m*
saltpetre   salpêtre *m*
salvage [waste, etc]   récupérer *vb*
salvaged [of used materials]   récupéré *adj*
same time, at the   ensemble *adv*
sample   échantillon *m*, éprouvette *f*
sample, borehole or core   carotte *f*
sample, control   échantillon *m* prélevé par hasard
sample, core   carotte *f*, échantillon *m* carotté
sample, disturbed/undisturbed   échantillon *m* remanié/non remanié
sample, soil   échantillon *m* de sol
sample, test   échantillon *m* d'essai/pour essai
sample, up to   conforme à l'échantillon
sample, water   échantillon *m* d'eau
sampler   appareil *m* de prise d'échantillon(s), échantillonneur *m*
sampler [for deep cores]   carottier *m*
sampling   échantillonnage *m*, prélèvement *m*/prise *f* d'échantillon
sampling method, cylindrical mould [soil testing]   méthode *f* du cylindre
sand   sable *m*
sand [very fine grain]   sablon *m*
sand, blown   sable *m* éolien
sand, coarse   grève *f*, sable *m* grossier
sand, compacted   sable *m* compacté
sand, fine   sable *m* fin
sand, natural   sable *m* naturel
sand, river   sable *m* alluvionnaire/de rivière
sand, shell   sable *m* coquiller
sand, soft   sable *m* doux
sand box   boîte *f* de sable
sand dune   dune *f*
sand pit/quarry   carrière *f* de sable, sablière *f*
sand trap [sewage plant]   dessableur *m*

sand trapping [sewage treatment]   dessablage *m*
sand   poncer *vb*
sandbank   relais *m*
sandblasted   sablé *adj*
sandblasted, lightly   gommé *adj*
sandblast   décaper *vb* au jet de sable, sabler *vb*
sandblast, lightly   gommer *vb*
sandblasting   décapage *m* au (jet de) sable, sablage *m*
sandblasting, light   gommage *m*
sander   ponceuse *f*
sander, belt   ponceuse *f* à bande
sander, orbital   ponceuse *f* vibrante
sanding   ponçage *m*
sanding disc, paper   disque *m* de papier abrasif
sanding/sandpaper block   bloc *m* à poncer, cale *f*
sandpaper   papier *m* de verre
sandstone   grès *m*
sandy   sableux *adj*
sandy beach or sea shore   grève *f*
sanitary   sanitaire *adj*
sanitation   assainissement *m*
sapwood   aubier *m*, bois *m* imparfait
sarking   voligeage *m*
sarking board   volige *f*
sash, double [of a window]   contre-châssis *m*
sash, inner [of a double window]   contre-fenêtre *f*
sash bar   fer *m* à vitrage
sash cord, seven-strand   septain *m*
satellite [building]   satellite *m*
satellite dish [aerial]   parabole *f*
satellite terminal [airport]   aérogare *f* satellite
satin [paint finish]   satiné *adj*
saturated   saturé *adj*
saturated [ground]   gorgé d'eau
saturation   saturation *f*
saturation, degree of   degré *m* de saturation
saturation line   ligne *f* de saturation
saturation point   point *m* de saturation
saturation vapour density [HVAC]   densité *f* de saturation
save [a building, etc]   sauver *vb*
save [computer data]   conserver *vb*
save [money, labour, materials]   économiser *vb*
save [time]   gagner *vb*
saving   économie *f*
saving, energy   économie *f*/maîtrise *f* d'énergie
saw, *see also* chainsaw, handsaw, ripsaw, etc   scie *f*
saw, abrasive (wheel) cut-off   tronçonneuse *f*
saw, bow   scie *f* à travers
saw, circular   scie *f* circulaire
saw, compass or keyhole   scie *f* à guichet
saw, concrete   scie *f* à béton
saw, crosscut   scie *f* à tronçonner, tronçonneuse *f*
saw, crosscut [double ended]   scie *f* à deux mains, scie *f* passe-partout
saw, edger   déligneuse *f*, scie *f* de délignage
saw, frame   scie *f* multiple
saw, hole   scie *f* trépan
saw, masonry   scie *f* de maçon

saw, panel   scie *f* à panneaux
saw, pavement   scie *f* à sol
saw, rock cutter or cutting   scie *f* à rocher
saw, scroll   scie *f* à chantourner
saw, tenon   scie *f* à dosseret/à dos/d'encadreur
saw bench   plateau *m* de sciage
saw carriage   chariot *m* de scie
saw chain   chaîne *f* de la scie
saw cut [in wood]   saignée *f*
saw guard [of a saw bench]   chapeau *m* de scie
saw guide [of a chainsaw]   traverse *f* de la scie
saw handle   manche *m* de scie
saw mark [timber surface defect]   trait *m* de scie
saw rasp   râpe-scie *f*
saw setting tool   pince *f* à avoyer, tourne-à-gauche *m*
saw   scier *vb*
saw [a log to square edged timber]   aviver *vb*
sawdust   sciure *f*
sawing   sciage *m*
sawing, through-and-through   sciage *m* en plots
saw-line or -cut   trait *m* de scie
sawmill   scierie *f*
sawtooth   dent *f* de scie
scabble (with a scabbling hammer)   smiller *vb*
scabbled [of stone dressed with many short, parallel cuts]   smillé *adj*
scabbled stone finish   parement *m* smillé
scabbling [with a scabbling hammer, the action, or its result]   smillage *m*
scaffolding [and erection of scaffolding]   échafaudage *m*
scaffolding, cantilever or outrigger   échafaudage *m* cantilever/en porte-à-faux/en encorbellement
scaffolding, flying   balancelle *f*, échafaudage *m* volant
scaffolding, frame or square   échafaudage *m* sur cadre
scaffolding, ground   échafaudage *m* de/sur pied
scaffolding, horizontal   plancher *m* de service
scaffolding, ladder   échafaudage *m* à échelles/sur échelles
scaffolding, ladder jack   échafaudage *m* par échelles et taquets/sur taquets d'échelles
scaffolding, mobile or tower [wheeled]   échafaudage *m* roulant
scaffolding, projecting or suspended   échafaudage *m* suspendu
scaffolding, steel tube or tubular steel   échafaudage *m* métallique tubulaire/en tubes d'acier
scaffolding, trestle   échafaudage *m* sur tréteaux
scaffolding, truss-out   échafaudage *m* à chaise/sur consoles
scaffolding board platform   plate-forme *f* de madriers *m,pl*, platelage *m*
scaffolding ledger   sommier *m* d'échafaudage
scaffolding pole   étamperche *f*, poinçon *m* d'échafaudage, perche *f*, pointier *m*
scaffolding pole [vertical]   échasse *f* (d'échafaudage)
scaffolding pole [wood or metal, may have pulley at upper end]   écoperche *f*
scaffolding tie   boulin *m*
scald   ébouillanter *vb*

scale [deposit]   encroûtement *m*
scale [in a boiler or pipes]   incrustation *f*, tartre *m*
scale [of a map, ruler, thermometer, etc]   échelle *f*
scale [of prices, salaries, etc]   barème *m*
scale, boiler   dépôt *m* d'une chaudière, incrustation *f* de chaudière
scale, change of   changement *m* d'échelle
scale, drawn to   à l'échelle
scale, full   en grandeur *f* nature
scale, hardness   échelle *f* de dureté
scale, mill   calamine *f*
scale, not to (NTS)   non à l'échelle
scale formation   entartrage *m*, formation *f* d'incrustation
scale of, drawn to a   rapporté à l'échelle de
scale of charges   barème *m*/tableau *m* des tarifs
scale removal   désincrustation *f*
scale [a boiler, etc]   décrasser *vb*, détartrer *vb*, écailler *vb*
scale down [a drawing]   réduire *vb* l'échelle de
scale off   s'écailler *vb*
scale up [a drawing]   augmenter *vb* proportionnellement
scale up [an activity]   augmenter *vb*
scaler, needle   marteau *m* à décalaminer
scales [pair of]   balance *f*
scaling [a boiler]   piquage *m*
scantling [timber size]   équarrissage *m*
scarcity [of goods, etc]   pénurie *f*
scarf [a weld]   amorcer *vb*
scarfing [of a weld]   amorçage *m*
scarifier   scarificateur *m*
scarp   escarpement *m*
scarp [of a fortress]   escarpe *f*
scatter, range of   domaine *m* de dispersion
scatter of results   dispersion *f* des résultats
scatter [light]   disperser *vb*
scatter(ing) [of light]   dispersion *f*
scavenge   balayer *vb*
scavenging   balayage *m*
scenery   paysage *m*, vue *f*
schedule   nomenclature *f*
schedule [of prices, salaries, etc]   barème *m*
schedule, bending   nomenclature *f* d'/des aciers
schedule, construction   planning *m* général
schedule, price   bordereau *m* de prix
schedule, production   programme *m* de fabrication
schedule of works   bordereau *m* des travaux
scheme   projet *m*
scheme stage design report   avant-projet *m* sommaire (APS), dossier *m* de la solution d'ensemble préconisée
schist   schiste *m*
Schmidt rebound   rebondissement *m* Schmidt
school [buildings], *see also* building, school   école *f*
school, riding   manège *m*
scissor lift, (mobile)   table *f* élévatrice (mobile)
scissor lift (work) platform   nacelle *f* à ciseaux, plate-forme *f* à ciseaux/compas
scissors, paperhanger's   ciseaux *m,pl* de tapissier
sclerometer   scléromètre *m*
scope   contenu *m*, ressort *m*
scope, within his/her   de son ressort *m*

score [incise]   entamer *vb*
scotia [architectural moulding]   scotie *f*
scour [by water]   affouillement *m*
scour protection   risberme *f*
scour [by water flow]   affouiller *vb*
scour [clean]   décrasser *vb*
scouring [cleaning]   décrassage *m*
scrap (metal)   ferraille *f*
scrap sth   mettre *vb* qch au rebut
scrape   gratter *vb*, racler *vb*
scrape [stone]   riper *vb*
scraper   décapeuse *f*, grattoir *m*, raclette *f*, racloir *m*
scraper [earthworks]   racleur *m*, scrape(u)r *m*
scraper [for stone]   ripe *f*
scraper [ground surfaces]   machine *f* décapeuse
scraper, elevating   décapeuse *f* élévatrice
scraper, foot [metal]   décrottoir *m*
scraper, motor   décapeuse *f* automotrice
scraping   décrassage *m*
scraping down to bare plaster before repapering
   grattage *m* à vif
scraping of stone to smooth or polish it   ripage *m*
scratch [on wood or metal]   rayure *f*
scratch   gratter *vb*, rayer *vb*
screed   chape *f*
screed, cement   chape *f* en ciment
screed, concrete compression   chape *f* d'étanchéité
screed, floating   chape *f* flottante
screed, monolithic   chape *f* incorporée
screed(er), vibrating   règle *f* (surfaceuse) vibrante
screen [computer, wall, etc]   écran *m*
screen [for sewage]   dégrilleur *m*
screen [grille]   grille *f*
screen [riddle]   claie *f*, crible *m*, tamis *m*
screen, anti-dazzle/-glare [roads]   dispositif *m* antiphare,
   écran *m* anti-éblouissant
screen, classifying   crible *m* de classement
screen, dewatering   grille *f* essoreuse
screen, (fine) wire mesh   tamis *m* en toile métallique
screen, fire   écran *m* de cheminée, garde-feu *m*
screen, gravel   crible *m* à gravier
screen, intake   grille *f* d'entrée
screen, mesh   tamis *m* à mailles
screen, revolving   trommel *m*
screen, rotary   crible *m* rotatif
screen, scalping   crible *m* scalpeur
screen, shaking   crible *m* à secousses
screen, sizing   crible *m* classeur
screen, suction   crépine *f* d'aspiration
screen, vibrating   crible *m* vibrant à balourd
screen, welding   écran *m* de soudeur
screen, wind [*not* windscreen]   paravent *m*
screen [of e.g. smoke, trees, etc]   rideau *m*
screen [electrically]   blinder *vb*
screen [grade]   cribler *vb*, tamiser *vb*
screened [electrically]   blindé *adj*
screening [electrical]   blindage *m*
screening [grading]   criblage *m*, tamisage *m*
screening [sewage]   dégrillage *m*

screening off floating matter from surface of sewage
   écrémage *m*
screw   vis *f*
screw [with bolt head]   tire-fond *m*
screw, adjusting   vis *f* de rectification/de rappel/de
   réglage
screw, clamping   vis *f* de blocage/de serrage
screw, coach   tire-fond *m*
screw, earthing   vis *f* de mise à la terre
screw, fixing or locking   vis *f* de blocage
screw, grub   vis *f* sans tête
screw, hexagon head   vis *f* à tête hexagonale
screw, levelling   vis *f* calante
screw, securing   vis *f* de fixation
screw, self-tapping   vis *f* (auto)taraudeuse
screw, square-headed   vis *f* à quatre pans
screw, stud   goujon *m*
screw, vice   vis *f*
screw, wood   vis *f* à bois
screw cutting machine   machine *f* à fileter
screw [fix with a]   visser *vb*
screw down   serrer *vb* à vis
screw home   serrer *vb* à bloc, visser *vb* à bloc/à fond
screwdriver   tournevis *m*
screwdriver, cross-point   tournevis *m* cruciforme
screwdriver, pump/spiral ratchet/yankee   tournevis *m* à
   spirale et à rochet
screwdriver bit [for Phillips/Pozidrive/Supadrive/recessed
   head screws]   pointe *f* cruciforme
screwdriver-drill   perceuse-visseuse *f*
screw-head cover   cache-vis *m*
scribe [wood]   rouanner *vb*
scriber   trusquin *m*
scriber for wood   roanne/rouanne *f*
scrim   bande *f* à joint
scrub [bushes]   brousailles *f,pl*
scrub   brosser *vb*
scrub [gas]   épurer *vb*
scrubber [washer]   laveur *m*
scrubber, air   épurateur *m* d'air
scrubbing [of a gas]   épuration *f*
scrubbing method   méthode *f* de nettoyage
scrutiny   vérification *f*
scullery   arrière-cuisine *f*, lavoir *m*
scum   écume *f*
sea level   niveau *m* de la mer
seal   scellement *m*
seal [commercial]   label *m*
seal [on a document]   cachet *m*, sceau *m*
seal [on a parcel, a meter]   plomb *m*
seal [to prevent escape of air, water, etc]   bourrage *m*
seal [to render air- or watertight]   dispositif *m*
   d'étanchéité
seal [washer]   rondelle *f* d'étanchéité
seal [water in a siphon]   garde *f* d'eau
seal, air-tight   joint *m* étanche à l'air
seal, bituminous   étanchement *m* bitumineux
seal, clay   bouchon *m*/scellement *m* d'argile, joint *m* de
   terre glaise

seal, foam   joint *m* mousse
seal, labyrinth   joint *m* à labyrinthe
seal, leak-proof   joint *m* étanche
seal, neoprene   joint *m* en néoprène
seal, rubber   joint *m* caoutchouc
seal, slurry [road construction]   coulis *m* bitumineux
seal, water, *see* water seal
seal, weatherproof   joint *m* d'étanchéité
seal contact [in a valve]   contact *m* d'étanchéité
seal   sceller *vb*
seal [a surface, with paint or varnish]   nourrir *vb*
seal [hermetically]   fermer *vb* hermétiquement
seal [with a lead seal]   plomber *vb*
seal (off) [a leak, leaking joint, etc]   colmater *vb*
seal off [close or plug]   boucher *vb*
seal with mastic or paste   luter *vb*
sealed   scellé *adj*
sealed [water- or airtight]   étanche *adj*, hermétique *adj*
sealed, hermetically   fermé hermétiquement
sealing   scellement *m*
sealing [of documents]   cachetage *m*
sealing [with lead]   plombage *m*
sealing off   étanchement *m*
sealing profile, extruded   profilé *m* extrudé d'étanchéité
sealing run [in welding]   cordon *m* d'étanchéité
seam   joint *m* à recouvrement
seam [geological]   filon *m*
seam [welded]   cordon *m*
seam [welded or brazed]   soudure *f*
seam, butt welded   ligne *f* de soudure bout à bout
seam, longitudinal   cordon *m* longitudinal
seam, welded   cordon *m* de soudure, soudure *f*
seam joint, standing [in sheet metal]   joint *m* debout
seam [sheet metal]   agrafer *vb*
seamed [of metal sheeting]   agrafé *adj*
seamless   sans soudure
search for   rechercher *vb*
seashore   front *m* de mer
seaside *adj*   au bord de (la) mer
season [of the year]   temps *m*
seasonal   saisonnier *adj*
seasonal changeover   changement *m* (de température) été-hiver
seat   siège *m*
seat, toilet   siège *m* de cabinet
seat of the fire   foyer *m* de l'incendie
seating [of a bearing]   chaise *f*
seating [of a valve or tap]   siège *m*
seating [of the end of a prop or strut]   assise *f*
seating, anchorage   recul *m* à l'ancrage
seating, cover   siège *m* de trapillon
seating, trapdoor   siège *m* de trapillon
seating, valve   siège *m* de soupape/de vanne/d'obturateur
secluded   retiré *adj*
second fixings   finitions *f,pl*
second moment of area   moment *m* d'inertie
second piece sawn from a log after the slab   contre-dosse *f*
secondary   secondaire *m*

section [cross-section, or structural]   section *f*
section [drawing]   coupe *f*
section [length of wall, roof plane, etc]   pan *m*
section [length, stretch]   tronçon *m*
section [profile]   profil *m*
section, angle iron   cornière *f*
section, built-up   section *f* composée, profil *m* composé
section, bulb angle   cornière *f* à boudin
section, change in   changement *m* de section
section, circular hollow (CHS)   profil *m* creux rond
section, closed [of which some parts are inaccessible]   profil *m* fermé
section, cold rolled   profil *m* à froid
section, compound   section *f* composée, profil *m* composé
section, cracked   section *f* fissurée
section, cross, *see* cross-section
section, deceleration [roads]   zone *f* de décélération
section, depth of   hauteur *f* du profil
section, equal-leg angle   cornière *f* à ailes égales
section, half-round   demi-rond *m*
section, hollow   section *f* creuse
section, homogeneous   section *f* homogène
section, light   profil *m* léger
section, long(itudinal)   coupe *f* longitudinale, coupe *f*/profil *m* en long
section, metal   profilé *m*
section, net   section *f* nette
section, open [all of whose faces are accessible]   profil *m* ouvert
section, polished   coupe *f* polie
section, rectangular hollow   profil *m* creux rectangulaire
section, reduction in   réduction *f* de section
section, rolled   profilé *m*
section, round hollow (RHS)   profil *m* creux rond
section, round-edged angle   cornière *f* à angles arrondis
section, solid   section *f* pleine
section, square hollow (SHS)   profil *m* creux carré
section, standard   profil *m* normal(isé)
section, steel   acier *m* profilé, profilé *m* métallique
section, storage [roads]   zone *f* de stockage
section, T-   fer *m*/profilé *m* en T/té
section, thin   coupe *f* mince
section, transformed   section *f* homogénéisée
section, transverse   coupe *f* en travers
section, typical   coupe *f* type, coupe-type *f*
section, unequal-leg (angle)   cornière *f* à ailes inégales
section, usable/useful   section *f* utile
section, waiting [roads]   zone *f* de stockage
section (drawing)   dessin *m* en coupe
section line [drawing]   indication *f* de coupe
section modulus   module *m* de résistance
section X–X [on a drawing]   coupe *f* (suivant) X–X
sectional   sectionnel *adj*
section(al) (area), gross   section *f* brute
sector   secteur *m*
sector, blanked   secteur *m* d'obturation
sector, private/public   secteur *m* privé/public
secure (to)   fixer *vb* (à, sur)
secured (to)   fixé *adj* (à, sur)

securing bolt   boulon *m* de fixation
security, *see also* safety
security [for loan]   garantie *f*
security [safety]   sécurité *f*
security [stock exchange]   titre *m*
securities [stock exchange]   valeurs *f,pl*
security, collateral   nantissement *m*
security checkpoint or control point   poste *m* de
  contrôle sécurité
security interlock   couplage *m* de sécurité
security lighting   éclairage *m* de sécurité
security measures [e.g on site]   gardiennage *m*
security of tenure   droit *m* de renouvellement
security van   fourgon *m*
sediment   boue *f*, dépôt *m*, sédiment *m*
sediment, build-up or deposit of   formation *f* de
  sédiments
sediment, muddy or silt   sédiment *m* limoneux
sediment basin   dessableur *m*
sedimentary   sédimentaire *adj*
sedimentation   sédimentation *f*
sedimentation [of sewage]   décantation *f*
sedimentation [silting]   envasement *m*
sedimentation basin   bassin *m* de décantation,
  décanteur *m*
sedimentation tank   bac *m*/bassin *m* de décantation
see [cross-reference in document]   voir
seep   percoler *vb*
seep into   infiltrer *vb*
seepage   infiltration *f*, percolation *f*
seepage [leaking]   fuite *f*
seepage [oozing, sweating]   suintement *m*
seepage path   trajet *m* d'infiltration
seepage pit   puits *m* d'infiltration
segment [of an arch]   claveau *m*, voussoir *m*
segment [of time]   plage *f*
segment, hinge   voussoir *m* d'articulation
segment, match-cast   voussoir *m* conjugué
segregation   ségrégation *f*
seismic   s(é)ismique *adj*
seismograph   s(é)ismographe *m*
seismography   s(é)ismographie *f*
seismological   s(é)ismologique *adj*
seismology   s(é)ismologie *f*
seize [e.g. of a bearing]   se bloquer *vb*, se gripper *vb*
seize (up) [bearing, machine]   gripper *vb*
seized [bearing, machinery, etc]   bloqué *adj*
seizure of real property   saisie *f* immobilière
selected   sélectionné *adj*
selector, floor [lift/elevator]   sélecteur *m* d'étage
self-acting   automatique *adj*
self-adhesive   autocollant *adj*
self-cleaning   autonettoyant *adj*
self-cleansing [of drains or sewers]   autocurage *m*
self-contained   autonome *adj*
self-erecting [e.g. tower crane]   automontable *adj*
self-financing   autofinancement *m*
self-levelling   autonivelant *adj*
self-locking   autoblocage *m*

self-propelled   automoteur *adj*
self-service   libre service *m*
self-supporting   autoportant *adj*, autoporteur *adj*,
  autostable *adj*
seller   vendeur *m*
semi-automatic   semi-automatique *adj*
semi-circular   semi-circulaire *adj*
semi-column   demi-ceint *m*
semiconductor   semi-conducteur *m*
semi-opaque   demi-opaque *adj*
semi-trailer   semi-remorque *f*
semi-trailer, flat bed   semi-remorque *f* plateau/plate-
  forme
semi-trailer, plant   semi-remorque *f* porte-engins
semi-trailer, tipping   semi-benne *f*, semi-remorque *f* à
  benne
semi-trailer tractor unit   tracteur *m* de semi-remorque
send off [despatch]   expédier *vb*
sender   expéditeur *m*
sensitive   sensible *adj*
sensitiveness/sensitivity (to)   sensibilité *f* (à)
sensor   capteur *m* (de mesure), élément *m* sensible
sensor, dew point   élément *m* sensible au point de rosée
sensor, humidity   élément *m* sensible à l'humidité
sensor, remote   capteur *m* à distance
sensor, solar   capteur *m* de rayonnement solaire
separate   individuel *adj*, séparé *adj*
separate trades [contracting]   lots *m,pl* séparés
separate   séparer *vb*
separate [sort]   trier *vb*
separately   isolément *adv*
separating   séparatif *adj*
separation   séparation *f*
separation [fire, or between buildings]   isolement *m*
separation (between) [distance]   écartement *m* (de, entre)
separator   séparateur *m*
separator [in equipment]   cloison *m*
separator [reinforcement]   écarteur *m*
separator, air   séparateur *m* d'air
separator, centrifugal (air)   séparateur *m* (d'air)
  centrifuge
separator, dust   séparateur *m* de poussières
separator, expansion   séparateur *m* à expansion
separator, oil   séparateur *m* d'huile
separator, raised traffic   séparateur *m* en saillie
separator, steam   épurateur *m* de vapeur, séparateur *m*
  d'eau de condensation
separator, water   séparateur *m* d'eau
septic tank   fosse *f* septique
sequence   ordre *m*, séquence *f*
sequence, assembly   ordre *m* de montague
sequence, control   séquence *f* de contrôle
sequence, starting   séquence *f* de mise en route
sequence control   régulation *f* séquentielle
sequence operation   asservissement *m*
serial number   numéro *m* de série
series [electrical, mathematical, etc]   série *f*
series [of doors, columns, etc]   enfilade *f*
series operation   opération *f* en cascade

series, operate or run in   marcher *vb* en série
serve with regular transport   desservir *vb*
service   service *m*
service, maintenance   service *m* d'entretien
services [as provided by architect, etc]   services *m,pl*
services [e.g. of design or construction]   prestations *f,pl*
services, electrical   installation *f* électrique
services, engineering   services *m,pl* d'ingénierie
services, statutory   services *m,pl* concédés
service charges included   charges *f,pl* comprises
service tree [and its hard wood]   cormier *m*
services channel   caniveau *m* technique
service(s) riser   gaine *f* technique
services sleeve [hole in concrete, masonry, etc]
   réservation *f*
service [equipment, plant]   reviser/réviser *vb*
service [maintain]   entretenir *vb*
serviceability [of equipment]   disponibilité *f* technique,
   état *m* satisfaisant (de fonctionnement)
serviceability limit state   état *m* limite de service (ELS),
   état *m* limite d'utilisation
serviceable   utile *adj*
serviced [of a site for building]   viabilisé *adj*
serviced [maintained]   entretenu *adj*
servicing   entretien *m*, maintenance *f*
servo-control   servocommande *f*
servo-motor   servomoteur *m*
set [predetermined]   fixe *adj*
set back (from)   en retrait (de)
set on edge   posé sur champ
set [of drawings, tools, etc]   jeu *m*
set [e.g. of drawings]   série *f*
set [of equipment, etc]   batterie *f*
set [of saw teeth]   chasse *f*, voie *f*
set, permanent   déviation *f* permanente
set, permanent [lengthening]   allongement *m* permanent
set of shelves   rayonnage *m*
set of spanners   jeu *m* de clés/clefs
set of tools   outillage *m*
set point   point *m* de consigne, valeur *f* de référence
set point, temperature   point *m* de consigne de
   température
set square   équerre *f* (à dessin)
set   fixer *vb*
set [a blade, tool, etc]   affiler *vb*
set [a clock, equipment, a thermostat, etc]   régler *vb*
set [concrete, glue]   prendre *vb*
set [glass in a frame]   sertir *vb*
set [saw teeth]   avoyer *vb*
set [sharpen, a saw]   affûter *vb*
set a price   établir *vb* un prix
set back [a clock, a programme]   retarder *vb*
set back from [i.e. build set back from]   bâtir
   vb/construire *vb* en retrait
set back [recess]   renfoncer *vb*
set fire to   allumer *vb*
set in rows   ranger *vb*
set out [peg or mark out]   tracer *vb*, implanter *vb*
set to level [top of a wall, etc]   araser *vb*

sett, large/small paving   pavé *m* échantillon/mosaïque
sett, paving   pavé *m*
sett paving [of a road], *see also* paving   chaussée *f* pavée
setter out [of timber structural frameworks]   gâcheur *m*
setting [surroundings]   cadre *m*
setting [of a clock, equipment, a thermostat, etc]
   réglage *m*
setting [e.g. of a building]   cadre *m* à étages
setting [of concrete, glue, etc]   prise *f*
setting, cut-out   position *f* de coupure
setting, remote   réglage *m* à distance
setting, rough   réglage *m* approximatif
setting back [in space, e.g. behind a building line]
   reculement *m*
setting heat [of concrete]   chaleur *f* de prise
setting in motion [releasing]   déclenchement *m*
setting out   traçage *m*
setting out [layout of buildings, roads, etc]   implantation *f*
setting out [of a road, etc]   tracé *m*
setting out datum point   point *m* de repère
setting out drawing or plan   plan *m* d'implantation
setting out line   axe *m* d'implantation
setting out of lines of slates or tiles   lignage *m*
setting out peg   point *m* de repère
setting out point   point *m* d'implantation
setting sth in motion   déclenchement *m*
setting to work [of equipment]   mise *f* en service
setting up of survey poles   balisage *m*
setting upright   relèvement *m*
settle [sink, subside]   s'affaisser *vb*
settle [foundation]   prendre *vb* coup
settle [a question, finally]   trancher *vb*
settle [an argument, an account, details, etc]   régler *vb*
settle [compact or sink down]   se tasser *vb*
settle [dust, etc]   se déposer *vb*
settle [in a country]   s'établir *vb*
settle [take up residence]   s'installer *vb*
settle, cause to [earth, a road, a structure]   affaisser *vb*
settle (on) [a date, a programme, etc]   régler *vb*
settlement [of an account]   règlement *m*
settlement [of foundations, etc]   affaissement *m*,
   enfoncement *m*
settlement [of ground, earth]   tassement *m*
settlement, consolidation   tassement *m* de consolidation
settlement, differential   tassement *m* différentiel
settlement, unequal [of supports, etc]   dénivellation *f*
settlement analysis   étude *f* de tassement
settlement date   échéance *f*
settlement measurement   mesure *f* de tassement
settlement of account [closure]   arrêté *m* de compte
settlement of (legal) disputes   règlement *m* des litiges
settlement of supports   tassement *m* des appuis
settlement tendency [of ground or fill]   sensibilité *f* au
   tassement
settlement–load curve   courbe *f* de tassement/de
   chargement/efforts–déformations
settlement–time curve   courbe *f* de tassement en fonction
   du temps
sewage   eaux *f,pl* usées (EU), eaux, *f,pl* d'égout

sewage, domestic   eaux *f,pl* usées domestiques
sewage, industrial   eaux *f,pl* usées industrielles
sewage, raw   eaux *f,pl* d'égout brutes
sewage, urban [combined soil, waste and rainwater]   effluent *m* urbain
sewage degreasing   écumage *m*
sewage disposal   assainissement *m* (des eaux usées), évacuation *f* des eaux d'égouts
sewage disposal plant, *see* sewage treatment plant
sewage farm   champs *m,pl* d'épandage
sewage tank, sealed/septic   fosse *f* étanche/septique
sewage treatment   épuration *f*/traitement *m* des eaux usées
sewage treatment plant   station *f* d'épuration/de traitement des eaux usées, usine *f* à purification des eaux d'égouts
sewage water   eau *f* d'égout
sewer   égout *m*
sewer, combined   égout *m* mixte/à canalisation unique
sewer, main   collecteur *m*, égout *m* collecteur/principal
sewer, main collector   collecteur *m* principal, grand collecteur *m*
sewer, outfall   égout *m* de décharge
sewer, separate   égout *m* séparatif
sewer, stormwater   égout *m* pluvial, collecteur *m* (eaux) pluvial(es), collecteur *m*/conduite *f* des eaux pluviales
sewer, sub-main   collecteur *m* secondaire
sewer, trunk   grand collecteur *m*
sewer connection   branchement *m* d'égout
sewer main   égout *m* collecteur
sewerage   assainissement *m*
sewerage, foul water   assainissement *m* des eaux usées
sewerage system   système *m* du tout à l'égout, réseau *m* d'assainissement
sewer(age) system, combined/separate   réseau *m* unitaire/séparatif
shackle   manille *f*
shade   ombre *f*
shade [blind]   store *m*
shade [colour]   couleur *f*
shade [of colour]   teinte *f*, ton *m*
shade card or chart   carte *f* de coloris
shade [from light]   ombrager *vb*
shaded [from light]   ombragé *adj*
shading coefficient   coefficient *m* de protection du soleil
shading device   dispositif *m* de pare-soleil
shadow   ombre *f*
shadow cast by one body on another   ombre *f* portée
shaft [duct]   gaine *f*
shaft [for foundations or underground access]   puits *m*
shaft [for stairs, lift]   cage *f*
shaft [of a bolt, rivet, etc]   tige *f*
shaft [of a column]   pilier *m*
shaft [of a manhole]   cheminée *f* verticale
shaft [of a pile]   fût *m*
shaft [rotating]   arbre *m*
shaft, access   cheminée *f* d'accès
shaft, air   cheminée *f* d'aération

shaft, communications   gaine *f* informatique
shaft, drive/driving   arbre *m* de commande
shaft, elevator or lift   cage *f*/gaine *f* d'ascenseur
shaft, mine   puits *m* de mine
shaft, pressure [dam]   puits *m* en charge
shaft, pumping out   puits *m* d'exhaure
shaft, service(s)   cage *f*/gaine *f* technique
shaft, transmission   arbre *m* de transmission *f*
shaft, ventilation   cheminée *f*/conduit *m*/gaine *f*/puits *m* d'aération/de ventilation, trémie *f* (d'aération)
shake [timber defect]   gerçure *f*
shake [a structure]   étonner *vb*
shaking   secousse *f*
shaking table [for testing concrete spread]   table *f* à secousse
shale   ardoise *f*, schiste *m* argileux
shale, burnt colliery   schiste *m* brûlé
shale, clay   argile *f* schisteuse, schiste *m* argileux
shale, expanded   schiste *m* expansé
shallow   peu profond *adj*
shallowness   peu *m* de profondeur
shank [of a bolt, rivet, etc]   tige *f*
shank [of a rivet]   fût *m*
shape   forme *f*
shape, particle   forme *f* des grains
shape category or class   classe *f* de forme
shape coefficient   coefficient *m* de forme
shape of bars [reinforcement]   forme *f* des barres
shape   contourner *vb*, façonner *vb*
shaped   profilé *adj*
shaped [e.g. U-shaped]   en forme de
shaped, cold   façonné à froid
shaping   contournement *m*
shaping [of brick, stone, etc]   taille *f*
shaping [of stone, timber, etc]   façonnage *m*
shaping, cold/hot   déformation *f*/façonnage *m* à froid/chaud
share [in a business, etc]   intérêt *m*
share [in a company]   action *f*
share capital   capital *m* social
share certificate   action *f*
shared   collectif *adj*
shareholder   actionnaire *m/f*
sharing out   répartition *f*
sharp [blade, tool, etc]   bien affilé, bien aiguisé
sharp [i.e. cutting]   tranchant *adj*
sharp [of a point]   aigu *adj*
sharp [of an outline]   net *adj*
sharp [of contrast, difference]   prononcé *adj*
sharp [pointed]   pointu *adj*
sharpen [a pencil]   tailler *vb*
sharpen [blade, tool, etc]   affiler *vb*, affûter *vb*, aiguiser *vb*
sharpen [put an edge on]   aviver *vb*
sharpener [oilstone]   affiloir *m*
sharpener [person]   affûteur *m*
sharpener [whetstone]   aiguisoir *m*
sharpening   affilage *m*, affûtage *m*, aiguisage *m*
sharpening machine   affûteuse *f*

sharpening stone   pierre *f* à affûter
sharpness [of reflection]   netteté *f*
sharpness angle [cutting tool]   angle *m* d'affûtage/du bec
shatter   briser *vb*
shatter [glass]   éclater *vb*
shave hook, painter's   violon *m*
shave [a piece of wood to required thickness]   doler *vb*
shave hook, triangular   ébardoir *m*
shaving   copeau *m*
shavings [wood turning]   tournures *f,pl*
shear *adj*   de cisaillement
shear   cisaillement *m*
shear, punching   poinçonnement *m*
shear angle   angle *m* de cisaillement
shear blade   lame *f* de cisaille
shear box   boîte *f* de cisaillement
shear centre   centre *m* de cisaillement
shear connector [composite construction]   goujon *m*
shear crack   fissure *f* de cisaillement
shear diagram   diagramme *m* de l'effort tranchant
shear flow   flux *m* de cisaillement
shear key   clef *f* de cisaillement
shear modulus   module *m* d'élasticité transversale/de cisaillement
shear plane   plan *m* de cisaillement
shear zone   zone *f* de rupture par cisaillement
shear   cisailler *vb*
shearing *adj*   de cisaillement
shearing   cisaillement *m*
shearing [metal, etc]   cisaillage *m*
shearlegs or shears   bigues *f,pl*, chèvre *f*
shears [for cutting]   cisaille *f*, cisailles *f,pl*
shears, bolt-cutting   cisaille *f* américaine/coupe-boulons
shears, hand   cisaille *f* à main
shears, trimming [for metal]   ébarbeur *m*
sheath   couverture *f*
sheath [of a cable]   gaine *f*
sheath [clad or lag, e.g. a boiler]   chemiser *vb*, envelopper *vb*
sheathed   chemisé *adj*
sheathed [e.g. cable]   sous gaine *adj*
sheathing [cladding or lagging]   chemise *f*, enveloppe *f*
sheathing [cladding or lagging – the action]   chemisage *m*
sheathing, roof   voligeage *m*
sheave   réa *m*
shed   remise *f*
shed [generally open-sided]   hangar *m*
shed [lean-to]   appentis *m*
shed, goods   halle *f* à marchandises
shed, open   auvent *m*
sheepfold or sheep pen   bergerie *f*
sheet   feuille *f*
sheet [e.g. in drawing numbers]   planche *f*
sheet [form, to be filled in]   fiche *f*
sheet [of metal]   plaque *f*
sheet, balance   bilan *m*
sheet, corrugated iron or steel   tôle *f* ondulée
sheet, embossed   tôle *f* gaufrée
sheet, flat metal   tôle *f* plate

sheet, galvanized steel   tôle *f* (d'acier) galvanisée/zinguée
sheet, metal   tôle *f*
sheet, perforated   plaque *f* perforée
sheet, rubber   feuille *f* de caoutchouc
sheet, thin   feuillet *m*
sheet, tinned steel   fer *m* blanc
sheet, waterproof sheet [tarpaulin]   bâche *f*
sheet, work   fiche *f* de travail
sheet, zinc   tôle *f* en zinc
sheet lead   feuille *f* de plomb, plomb *m* en feuille
sheet metal   feuillard *m*, tôle *f*
sheet (metal), embossed   tôle *f* gaufrée
sheet (metal), enamelled   tôle *f* émaillée
sheet (metal), gauge of   épaisseur *f* de tôle
sheet (metal), rolled   tôle *f* laminée
sheet (metal), thin   tôle *f* fine/mince
sheet metal cover/plug, removable   virole *f*
sheet pile   palplanche *f*
sheet pile, flat web   palfeuille *f*
sheet pile, trench   palplanche *f* légère
sheet pile caisson   batardeau *m*
sheet pile cut-off   parafouille *f* en palplanches
sheet pile driving   battage *m* de palplanches
sheet pile extractor   arracheur *m* de palplanches
sheet pile screen   écran *m*/rideau *m* de palplanches
sheet pile wall   palée *f*, écran *m*/paroi *f*/rideau *m* de palplanches
sheet piling cut-off   cloison *f* en palanches
sheet piling   palée *f*, rideau *m* de palplanches
sheet rubber   feuille *f* de caoutchouc
sheet [an excavation]   blinder *vb*
sheeting [of an excavation]   blindage *m*
sheeting, bitumen asbestos   feutre *m* bituminé chargé d'amiante
sheeting, bitumenized   bitume *m* armé
sheeting, corrugated   plaque *f* ondulée
sheeting, polyethylene   feuille *f* de polyéthylène
shelf   rayon *m*, tablette *f*
shelf [in the ground]   ressaut *m*
shelf [wooden board less than 54mm thick]   planche *f*
shelf or shelf unit   étagère *f*
shelves, set of   rayonnage *m*
shell [e.g. surround to column]   coque *f*
shell [of a building]   cage *f*, carcasse *f*
shell, upstream/downstream [of a dam]   masque *m* amont/aval
shell limestone   coquillart *m*
shell roof, concrete   coque *f* en béton
shell roof, prestressed concrete   coque *f* précontrainte
shellac   laque *f*
shelter   abri *m*
shelter, bus/passenger/queue   aubette *f*
shelter belt   barrière *f* naturelle
shelter   abriter *vb*
sheltered from water   hors d'eau
shield [tunnelling]   bouclier *m*
shield   abriter *vb*
shielding   abri *m*
shielding class [HVAC]   classe *f* d'abri/de protection

shift [work team]   équipe *f* (de travail)
shift, day/night   poste *m* de jour/nuit
shift work   travail *m* par relais
shift   déplacer *vb*
shift the zero [of an instrument]   décaler *vb*
shifts, work in   se relayer *vb*
shifting [of heavy materials, on site]   bardage *m*
shine   lustre *f*
shingle [roofing, cladding]   aissante *f*, aisseau *m*, bardeau *m*, essente *f*, tavaillon *m*
shingle [small stones]   galets *m,pl*
shingle, asphalt   bardeau *m* d'asphalte
shingle [a roof, a wall]   essenter *vb*
shipping [sending off]   expédition *f*
shock   choc *m*, coup *m*, secousse *f*
shock, electric   choc *m*/décharge *f*/secousse *f* électrique
shock, thermal   choc *m* thermique
shock absorber   amortisseur *m*
shoe [at foot of downpipe]   gargouille *f*
shoe [scaffolding base]   patin *m*
shoe [socket to prevent crushing]   sabot *m*
shoe scraper [metal]   décrottoir *m*
shoe with heat resistant sole   chaussure *f* à semelle anti-chaleur
shop   magasin *m*
shops   commerces *m,pl*
shop front   devanture *f* de magasin
shop steward   délégué *m*/représentant *m* syndical
shop window   montre *f*, vitrine *f*
shopkeeper   commerçant *m*
shopping centre   centre *m* commercial, zone *f* d'activité commerciale (ZAC)
shopping district   quartier *m* commerçant
shopping facilities   commerces *m,pl*
shopping town   bourg *m* commerces
shore [of a lake, sea]   bord *m*, rive *f*
shore [of an ocean, sea]   côte *f*
shore [prop]   étai *m*
shore [prop, generally small, temporary]   chandelle *f*
shore [sensibly vertical]   accore *m*
shore, inclined   étançon *m*
shore [an excavation]   blinder *vb*, boiser *vb*
shore (up)   étançonner *vb*, appuyer *vb*, étayer *vb*
shored [of an excavation]   blindé *adj*
shoring   étaiement *m*, étançonnement *m*, étayage *m*
shoring [of an excavation]   blindage *m*
shoring/propping to facilitate under-pinning
enchevalement *m*
short circuit   court-circuit *m*
short circuit   court-circuiter *vb*, mettre *vb* en court-circuit
short circuiting   mise *f* en court-circuit
short cut   raccourci *m*, traverse *f*
short term [loan, etc]   à court terme
short term, in the   à court terme
shortage [of goods, money, etc]   pénurie *f*
shortening   raccourcissement *m*
shortening, elastic   raccourcissement *m* élastique
short-term   à court terme, de courte durée

shotblast   grenailler *vb*
shotblasted   grenaillé *adj*
shotcrete   béton *m* projeté, ciment *m* projeté
shotcrete or shotcreting   gunitage *m*
shotcreting unit   dispositif *m* de projection de béton
shot-fire [a nail or wall plug]   spiter *vb*
shot-firer   boutefeu *m*
shoulder [stopping lane]   bande *f* d'arrêt
shoulder [verge of a roadway]   accolement *m*, accotement *m*, bande *f* latérale
shoulder [of a tenon]   arasement *m*
shoulder, hard/soft   accotement *m* stabilisé/non-stabilisé
shoulder, hard [emergency stop lane]   bande *f* d'arrêt d'urgence (BAU)
shovel   bêche *f*, pelle *f*
shovel, drag   pelle *f* équipée en rétro
shovel, face   pelle *f* (équipée) en butte
shovel, front   pelle *f* (équipée) en butte
shovel, hydraulic   pelle *f* hydraulique
shovel, loading or tractor   pelle *f* chargeuse
shovel, mechanical/power   excavateur *m*, pelle *f* mécanique, pelleteuse *f*
shovelful   pelletée *f*
show [time, temperature]   marquer *vb*
shower   douche *f*
shower [of rain]   averse *f*
shower, sudden [of rain]   giboulée *f*
shower base, *see* shower tray
shower bath   baignoire-douche *f*
shower cabinet   cabine *f* de douche
shower fittings   robinetterie *f* de douche
shower head or nozzle   pomme(au) *f* de douche
shower head [hand held]   douchette *f*
shower room   salle *f* d'eau
shower screen   pare-douche *m*
shower tray   bac *m* de douche, pédiluve *f*, receveur *m* à douches/de douche, socle *m* receveur
shrink   rétrécir *vb*
shrinkage   contraction *f*, retrait *m*, rétrécissement *m*
shrinkage distortion   déformation *f* de retrait
shrinkage ratio   indice *m* de retrait
shroud [electrically]   blinder *vb*
shrouded [electrically]   blindé *adj*
shrouding [electrical]   blindage *m*
shunt [electrical]   shunt *m*
shunt circuit   circuit *m* dérivé
shunt [current]   dériver *vb*
shunt [electrically]   shunter *vb*
shunting   shuntage *m*
shunting [of electrical current]   dérivation *f*
shunt-wound   excité en dérivation
shut *adj*   fermé *adj*
shut   fermer *vb*
shut [door, window]   joindre *vb*
shut down [lid over]   rabattre *vb*
shut down (or off) the steam   couper *vb* la vapeur
shut off [electricity, gas]   couper *vb*, fermer *vb*
shut-down period   période *f* d'arrêt
shut-down switch   arrêt *m*, interrupteur *m* automatique

shut-off device   dispositif *m* d'arrêt (automatique)
shutter   volet *m*
shutter [exterior, solid, of a window or door]   contrevent *m*
shutter [photographic]   obturateur *m*
shutter, folding   volet *m* brisé
shutter, internal window   sourdière *f*
shutter, louvred or slatted   persienne *f*
shutter, roller   volet *m* coulissant/roulant
shutter, roller [steel]   fermeture *f* à l'anglaise
shutter board [formwork]   planche *f* de coffrage
shutter boards or boarding [formwork]   bois *m* de coffrage
shutter bolt   espagnolette *f*
shutter catch   arrêt *m* de volet
shutter retaining clip   crochet *m* de contrevent/volet
shuttered [of concrete]   banché *adj*
shuttered [of a socket outlet]   avec éclipse
shuttering   coffrage *m*, moule *f*
shuttering, bottom   fond *m* de coffrage
shuttering, column   coffrage *m* du poteau
shuttering, main joist or beam   coffrage *m* de la poutre maîtresse
shuttering, permanent [precast concrete plank(s)]   prédalle *f*
shuttering for casting a wall [i.e. two face]   banche *f*
shuttering prop or strut   chandelle *f*
shutting (down, off, up)   fermeture *f*
shutting stile   montant *m* de serrure
siccative   siccatif *adj*
siccative [as used in paint]   siccatif *m*
sick bay/room   infirmerie *f*
sick building syndrome   syndrome *m* de bâtiment *m* malsain
side   côté *m*
side [of a hill, roof, etc]   versant *m*
side [of a plane]   joue *f*
side [of angular object, building, etc]   pan *m*
side [sloping, e.g. of an embankment, hill, etc]   flanc *m*
side, high/low pressure   côté *m* haute/basse pression
side, long [of a building, a roof plane]   long-pan *m*
side connection [between two ducts or pipes]   empattement *m*
side of a valley   versant *m* de vallée
sides [of a vault]   reins *m.pl*
side-tipping wagon   girafe *f*
siding [railway]   voie *f* de garage/de triage
siding [weatherboard]   clin *m*
sideboard   buffet *m*
sieve   crible *m*, tamis *m*
sieve, test   tamis *m* de contrôle
sieve   cribler *vb*, passer *vb* au tamis, tamiser *vb*
sieve opening size   maille *f* d'un tamis
sieved   tamisé *adj*
sieving   criblage *m*, tamisage *m*
sift, sifted, etc, *see* sieve, sieving, etc
sight glass   verre-regard *m*, viseur *m*, voyant *m*
sighting [surveying]   coup *m* de lunette, visée *f*
sighting telescope [surveying]   lunette *f* viseur/de visée
sighting-board [surveying]   voyant *m*

sign [indication]   indice *m*
sign [board, notice]   panneau *m*
sign [indication, symbol]   signe *m*
sign, advance direction   signal *m* de présignalisation
sign, call   indicatif *m*
sign, conventional   symbole *m* de représentation, signe *m* conventionnel
sign, direction   signal *m* direction
sign, illuminated   enseigne *f* lumineuse
sign, illuminated (road)   signal *m* lumineux
sign, road or traffic   indication *f* routière, panneau *m* indicateur/de signalisation, signal *m* routier
sign, speed restriction   signal *m* de ralentissement/de limitation de vitesse
sign, stop   panneau *m* stop
sign, street (name)   plaque *f*
sign convention   convention *f* de signe
sign [a letter, etc]   signer *vb*
sign a contract   passer *vb* un contrat
sign in/on [on arrival at work]   pointer *vb* en arrivant
sign off/out [on leaving work]   pointer *vb* en partant
sign on [labour]   embaucher *vb*
signage [putting up signs, and the signs]   signalisation *f*
signal   signal *m*
signal, alarm   signal *m* d'alarme
signal box   cabine *f*/poste *m* d'aiguillage
signal box, marshalling yard   poste *m* de butte
signal gantry   portique *m* à signaux/de signalisation
signal light, flashing   feu *m* clignotant
signature   signature *f*
signing [e.g. of contract]   passation *f*
signpost   poteau *m* indicateur
silencer   pot *m* d'échappement, silencieux *m*
silencing   atténuation *f* des sons/du son
silent   silencieux *adj*
silica   silice *f*
silica gel   gel *m* de silice, silicagel *m*
silicate   silicate *m*
silicon   silicium *m*
silicon carbide   carborundum *m*, carbure *m* de silicium
silicone   silicone *f*
silky   soyeux *adj*
sill [of a dam]   seuil *m* déversoir
sill [of a door]   seuil *m*
sill, ground [of wood, under a prop, etc]   couche *f*, semelle *f*
sill, window   appui *m* (de fenêtre)/(de baie), banquette *f*/tablette *f* de fenêtre, pièce *f* d'appui, rebord *m*
silo   silo *m*
silo, cement   silo *m* à ciment
silt   limon *m*
silt sediment   sédiment *m* limoneux
silt (up)   se déposer *vb*
siltation or silting   alluvionnement *m*, envasement *m*
siltstone   grès *m* fin
silty   limoneux *adj*
simply supported   sur appui libre
simulation   simulation *f*
simultaneous   simultané *adj*

184

simultaneously   simultanément *adv*
single   simple *adj*, unique *adj*
single-   mono-
single-cell [e.g. of bridge deck]   monocellulaire *adj*
single-layer [e.g. of paint system]   monocouche *adj*
single-phase   monophasé *adj*
single-phase(d)   en monophasé
single-pole   unipolaire *adj*
single-stage   mono-étagé *adj*
single-storey [house]   plain-pied *adj*
sink [stone, porcelain, etc]   évier *m*
sink, cleaner's   évier *m* agent de ménage
sink, galvanized iron   auge *f*
sink, hospital   évier *m* d'hôpital
sink, kitchen   évier *m*
sink hole   entonnoir *m*
sink [a borehole, a well]   forer *vb*
sink [a pile, shaft, well]   foncer *vb*
sink [dig, e.g. a well]   creuser *vb*
sink, cause to [foundations, wall, etc]   affaisser *vb*
sink (down) [ground, a structure, etc]   s'affaisser *vb*, se tasser *vb*
sink in [penetrate]   pénétrer *vb*
sink-hole [geology]   bétoire *f*, cratère *m* d'effondrement, doline *f*
sinking   abaissement *m*
sinking [a borehole, a well]   forage *m*
sinking [a well or shaft]   fonçage *m*
sinking [of foundations, etc]   enfoncement *m*
sinking (in) [of foundations, walls, etc]   affaissement *m*
sinking, unequal [of supports, etc]   dénivellation *f*
sintering   frittage *m*
siphon   siphon *m*
siphon   siphonner *vb*
siphon basket   panier *m* siphon
siphon trap   clôture *f* à siphon
Siporex (lightweight concrete)   béton *m* Siporex, siporex *m*
site, on   à pied d'œuvre, sur le tas
site, within the   dans l'enceinte *f* du chantier
site [area or plot of land]   terrain *m*
site [location, position]   emplacement *f*
site [of a house, town, etc]   position *f*
site [of installation of a piece of equipment]   lieu *m* de mise en place
site [where works are taking or to take place]   chantier *m*
site, archaeological   site *m*
site, building   chantier *m*/lieu *m* de construction
site, camp(ing)   camping *m*, terrain *m* de camping
site, construction   chantier *m*, chantier *m*/lieu *m* de construction
site, housing   lotissement *m*, terrain *m* à bâtir
site, mobile or temporary work   chantier *m* mobile
site agent, *see also* site manager   conducteur *m* de travaux, contrôleur *m* de chantier
site agent [engineer]   ingénieur *m* d'exécution
site arrangements [for installation]   disposition *f* des lieux
site boundary   bordure *f* du terrain

site canteen   cantine *f*
site clearance   aménagement *m* du site
site closing down   repliement *m*
site connection or joint   joint *m* de chantier
site development   aménagement *m* du terrain
site engineer   ingénieur *m* de chantier
site equipment   matériel *m* de chantier
site facilities   installations *f,pl* de chantier
site facilities plan   plan *m* d'installation du chantier
site fence   clôture *f* de chantier
site for dumping earth, provisional   relais *m*
site foreman   chef *m* de chantier
site foreman [of structural timber works]   gâcheur *m*
site hoist, *see* hoist
site installations   installations *f,pl* de chantier
site investigation   campagne *f* de reconnaissance, reconnaissance *f* de site/de sol
site latrines/lavatories   latrines *f,pl* de chantier
site layout plan   plan *m* d'implantation de chantier/d'installation du chantier
site manager, *see also* site agent   chef *m*/maître *m*/gestionnaire *m* de chantier
Site of Special Scientific Interest (SSSI) [UK]   Zone *f* Naturelle d'Intérêt Écologique, Faunistique et Floristique (ZNIEFF) [France]
site plan   plan *m* d'ensemble/de situation/de masse
site plant   matériel *m* de chantier
site preparation [provision of basic facilities]   aménagement *m* du chantier
site reconnaissance   étude *f* de reconnaissance du site
site safety   sécurité *f* du travail
site servicing [provision of basic services]   viabilisation *f*
site supervisor, *see* site agent, site manager
site survey   étude *f* (de reconnaissance) du site
site   implanter *vb*, placer *vb*, situer *vb*
siting   localisation *f*
situate   situer *vb*
situated   sis *adj*, situé *adj*
situation [location]   emplacement *f*, position *f*, situation *f*
situation [job]   emploi *m*, position *f*
size [dimension]   dimension *f*
size [for decorating]   apprêt *m*, enduit *m*
size [for wallpapering]   colle *f* de pré-encollage
size [general, and height]   grandeur *f*
size [glue]   colle *f*
size [measure or measurement]   mesure *f*
size [of an object]   taille *f*
size [weight, thickness, volume]   grosseur *f*
size, particle   taille *f* de particule
size, raw   dimension *f* brute
size, overall   dimension *f* hors tout
size bracket or range [grain sizing]   classe *f* granulaire
size [adjust to size]   mettre *vb* à dimension
size [a surface]   apprêter *vb*
size [with glue]   encoller *vb*
sizing [for decorating]   apprêt *m*, enduit *m*
sizing [of ducts, pipes, structural members, etc]   dimensionnement *m*
sizing with glue   encollage *m*

skeleton   ossature *f*
sketch   croquis *m*, esquisse *f*, schéma *m*, tracé *m*
sketch, first   jet *m* premier
sketch, outline   ébauche *f*
sketch, preliminary   esquisse *f*
sketch, preliminary design   avant-projet *m* sommaire
  (APS)
sketch, rough   croquis *m*, ébauche *f*, schéma *m* de
  principe
sketch design   avant-projet *m*
sketch design study   étude *f* d'avant-projet
sketch   dessiner *vb*, ébaucher *vb*, tracer *vb*
sketching out   ébauche *f*
skew   biais *m*
skewed, on the skew   en biais
skidder   débardeur *m*/tracteur *m* forestier, tracteur *m*
  débardeur
skin   peau *f*
skin, external/internal   paroi *f* extérieure/intérieure
skin, mortar outer   croûte *f* de mortier
skip   benne *f*
skip, loading   benne *f*/caisse *f* de chargement, chargeur *m*
skip, revolving   benne *f* à bascule
skirt   contourner *vb*
skirting 220–250mm high   stylobate *m*
skirting (board)   plinthe *f*
skirting round   contournement *m*
sky lift   engin *m* élévateur à nacelle
skylight   châssis *m* vitré, claire-voie *f*
skylight [dormer]   lucarne *f*
skylight [hinged]   châssis *m* tabatière, tabatière *f*
skylight [raised]   coupole *f* vitrée, lanterneau *m*
slab   plaque *f*
slab [cut from the outside of a log]   dosse *f*, dosseau *m*
slab [cut stone, etc]   tranche *f*
slab [slate]   table *f*
slab [heavy, large]   bloc *m*
slab [reinforced concrete, stone, etc]   dalle *f*
slab, base or foundation   dalle *f* de fondation, radier *m*
slab, composite   dalle *f* avec prédalle
slab, concrete   dalle *f* de/en béton
slab, flat [e.g. a worktop]   tablette *f*
slab, flat, on headed columns   plancher *m* champignon
slab, floating   dalle *f* flottante
slab, floor   dalle *f* de plancher
slab, foundation   dalle *f* de fondation
slab, ground   dallage *m*, dalle *f* de fond
slab, hollow   dalle *f* fléchie
slab, hollow pot floor   dalle *f* en corps creux
slab, one way   dalle *f* travaillant dans un sens
slab, ribbed   dalle *f* nervurée
slab, solid   dalle *f* pleine
slab, stone   dalle *f*
slab, transition   dalle *f* de transition
slab, waffle   dalle *f* caissonnée
slab floor, hollow (block)   plancher *m* à corps creux
slab of masonry with waterproof surface [lab bench,
  kitchen slab, etc]   paillasse *f*
slab on grade   dallage *m*

slab panel   panneau *m* de dalle
slab [a pavement]   daller *vb*
slack [loose]   lâche *adj*
slack [of a nut]   desserré *adj*
slack [fine coal]   fines *f,pl* de charbon
slacken   détendre *vb*, lâcher *vb*
slacken [a nut, etc]   desserrer *vb*
slackening   desserrage *m*
slag   scorie *f*
slag [for use as aggregate]   mâchefer *m*
slag [for use as aggregate or in slag cement]   laitier *m*
slag, blastfurnace   laitier *m* de haut-fourneau
slag, crushed blastfurnace   laitier *m* de haut-fourneau
  concassé
slag, expanded blastfurnace   laitier *m* expansé, mousse *f*
  de laitier
slag, ground granulated blastfurnace (ggbfs)   sable *m*
  de laitier concassé
slag, steelworks   laitier *m* d'aciérie
slag inclusion   inclusion *f* de laitier
slag layer   lit *m* de scories
slag wool   laine *f* de laitier/de scories
slake [lime]   déliter *vb*
slaked lime   chaux *f* éteinte/hydratée
slaking [lime]   délitage *m*
slant   biais *m*
slant   déverser *vb*
slantwise [on the slant]   en biais, en sifflet *adv*
slat   latte *f*
slate   ardoise *f*
slate, edge of a   chef *m*
slate, glass   panne *f* en verre
slate, roofing   ardoise *f*
slate, small sized   cartelette *f*
slate, stone   lause/lauze *f*
slate clip   crochet *m* d'ardoise
slate cut at an angle adjacent to ridge or valley
  approche *f*
slate cutter   coupe-ardoises *m*
slater   ardoisier *m*, couvreur *m* en ardoise
slater's hammer [similar to bricklayer's hammer]
  marteau *m* d'ardoisier, martelet *m*
slater's iron   coupe-ardoises *m*
slatted opening   persienne *f*
slaughterhouse   abattoir *m*
sledgehammer   mail *m*, marteau *m* à deux mains/à
  frapper devant, masse *f*
sledgehammer, heavily toothed   chien *m* taillé
sleeper [of railway track]   traverse *f*
sleeper tower [temporary support]   camarteau *m*
sleeping policeman   dos *m* d'âne
sleeve   manche *f*, manchette *f*, manchon *m*
sleeve [cast or two-piece, for pipe repair]   coulisse *f*
sleeve [sheath]   gaine *f*
sleeve, cable protection   gaine *f* de câbles
sleeve, connecting   manchon *m* de raccordement
sleeve, flanged   manchon *m* à brides
sleeve, flexible   manchette *f* flexible/souple
sleeve, inlet   manchon *m* d'entrée

sleeve, intake   manchon *m* d'entrée
sleeve, pipe [in masonry, concrete work]   fourreau *m*, réservation *f*
sleeve, protective   fourreau *m*, manchette *f* de protection
sleeve, reducing   manchon *m*
sleeved [e.g. of concrete piles]   chemisé *adj*
sleeving   buse *f*
slenderness (ratio)   élancement *m*
slewing [of a crane]   pivotement *m*
slice   tranche *f*
slide   glissière *f*
slide   glisser *vb*
slide chair [railway points]   coussinet *m* de glissement
slide rule   règle *f* à calcul
slide rule, circular   cercle *m* à calculer
slide(way)   coulisse *f*, coulisseau *m*
sliding *adj*   roulant *adj*
sliding   glissement *m*
sliding pickup [electrical]   collecteur *m*
sling [for lifting]   élingue *f*
sling [for lifting]   brayer *vb*
slinging [for lifting]   brayage *m*
slip [note, summary]   bordereau *m*
slip, rotational [geotechnics]   glissement *m* circulaire
slip   glisser *vb*
slip   glissement *m*
slipform   coffrage *m* glissant
slipform paver   machine *f*/surfaceuse *f*/vibreuse-surfaceuse *f* à coffrage glissant
slipping   glissement *m*
slipshod   lâche *adj*
slit   fissure *f*
slope   pente *f*
slope [banking, and of a wall, etc]   dévers *m*
slope [embankment, excavation]   talus *m*
slope [gradient]   inclinaison *f*
slope [hillside]   côte *f*, coteau *m*
slope [regular]   glacis *m*
slope [the ground or roof, not the angle]   versant *m*
slope [upwards]   montée *f*
slope [bank a road, slant a wall]   déverser *vb*
slope [side of earthworks]   taluter *vb*
slope, cut(ting)   talus *m* en déblai
slope, down   pente *f* descendante
slope, embankment   talus *m* en remblai
slope, flange   inclinaison *f* des ailes
slope, gentle   talus *m* à pente faible
slope, grassed   talus *m* herbacé
slope, lower, of a curb or mansard roof   brisis *m*
slope, reverse   contre-pente *f*
slope, steep   talus *m* à forte pente
slope, up   pente *f* ascendante
slope angle   angle *m* de talus
slope failure   glissement *m* de pente
slope inclination   pente *f* de talus
sloper [earthworks]   taluteuse *f*
sloping   dévers(é) *adj*, rampant *adj*
sloping part of a building [stairs, roof, vault, etc]   rampant *m*

slot [e.g. in a flat key]   encoche *f*
slot [e.g. in a screw head]   fente *f*
slot [groove, channel]   cannelure *f*, rainure *f*
slot [groove, channel or notch]   entaille *f*
slot, air inlet   fente *f* de prise d'air
slot, dovetail   entaille *f* à queue d'aronde
slot, ventilating/ventilation   fente *f* d'aération
slot [cut a slot in]   rainer *vb*
slotted   fendu *adj*
slotted strap holding a fanlight in open position   coulisse *f* de vasistas
slotted tube to guide key into a lock   canon *m*
slovenly [work]   lâché *adj*
slow (down)   ralentir *vb*
slow-acting   temporisé *adj*
sludge [sewage]   boues *f,pl*
sludge, activated   boues *f,pl* activées
sludge digestion plant   digesteur *m*
sludge drying bed   lit *m* de séchage (des boues)
sludge pressing plant   filtre-presse *m*
sludge process, activated   procédé *m* des boues *f,pl* activées
sludge recovery   récupération *f* des boues
sludge settling tank   silo *m* de décantation de boues
sluice   pertuis *m*
sluice-gate   bonde *f*, pale *f*, vanne *f*
slump   affaissement *m*
slump   s'affaisser *vb*
slump test cone   cône *m* d'Abrams
slurry   coulis *m*
slurry [for drilling]   boue *f* de forage
slurry, bitumen [road construction]   coulis *m* goudronneux
slurry, cement   lait *m* de ciment
slurry, tar   coulis *m* goudronneux
slurry seal [road construction]   coulis *m* bitumineux
smash   briser *vb*
smell   odeur *f*
smelling, evil or foul   nauséabond *adj*
smith [ironworker]   forgeron *m*
smithy   forge *f*
smoke   fumée *f*
smoke bomb   bombe *f* fumigène
smoke clearance   désenfumage *m*
smoke deflector   déflecteur *m* de fumée
smoke density   densité *f* d'un gaz de fumée, opacité *f* de fumée
smoke exhaust outlet   sortie *f* d'évacuation de fumée
smoke extract system, mechanical   système *m* mécanique de désenfumage
smoke extract system, natural   système *m* naturel de désenfumage
smoke flue   conduit *m* de fumée
smoke vent   exutoire *m*
smoke vent, emergency   cheminée *f* d'appel
smoke-tight   étanche *adj* à la fumée
smooth   lisse *adj*
smooth [ground]   uni *adj*
smooth, very [of paint or varnish]   tendu *adj*

smooth [a joint, a groove, in wood]   recaler *vb*
smooth [a surface by applying an agent]   abreuver *vb*
smooth [down or off]   adoucir *vb*
smooth [joinery, timber]   blanchir *vb*
smooth [timber with a jack plane]   rifler *vb*
smooth [wood]   planer *vb*
smooth again [a surface]   radoucir *vb*
smooth a plaster surface prior to decoration   égrener *vb*
smooth by rubbing [a surface]   gréser *vb*
smooth cut edge of glass   gréser *vb*, gruger *vb*
smooth off [a peak]   écrêter *vb*
smoothed edge of cut stone around the face   ciselure *f*
smoothing   lissage *m*, planage *m*
smoothing [a surface, the cut edge of glass, by rubbing]   grésage *m*
smoothing, peak   écrêtage *m*/écrêtement *m* des pointes
smoothing (down or off)   adoucissement *m*
smoothing of plaster prior to decoration   égrenage *m*
smoothing of the surface of a slab   surfaçage *m*
smoothing off peaks   écrêtement *m* des pointes
smoothness   lisse *m*
snag, hit a[†]   avoir *vb* un pépin[†]
snap action   action *f* à déclic
snap head as set during riveting   rivure *f* sphérique
snow   neige *f*
snow guard, rail or trap [on a roof]   garde-neige *m*
snow load   surcharge *m* de neige
soak   inonder *vb*, tremper *vb*
soak [masonry before rendering]   abreuver *vb*
soak (with)   imprégner *vb* (de)
soak in or up   absorber *vb*
soakaway   puits *m* perdu/d'infiltration
soakaway for rainwater [pebble-filled]   bétoire *f*
soaked with   imbibé de
soaking   trempe *f*
soaking of timber in water   flottage *m*
soap dish or holder   porte-savon *m*
soap distributor   distributeur *m* de savon
social security charge or contribution   cotisation *f* de la sécurité sociale
Society of Architects [statutory body in France]   Ordre *m* des Architectes
Society of Land Surveyors   Ordre *m* des Géomètres-Experts [in France]
sociologist   sociologue *m*
sociology   sociologie *f*
socket [electrical], *see* socket outlet
socket [shoe on end of structural member]   sabot *m*
socket, bayonet   culot *m* à baïonnette, douille *f* à baïonnette
socket, bulb   douille *f*
socket, connecting   manchon *m* de raccordement
socket, electrical plug   socle *m* de prise de courant
socket, extension   fiche *f* (femelle) de prolongateur
socket, female plug   fiche *f* femelle
socket, floor, with hinged lid [electrical]   boîte *f* de parquet à couvercle pivotant
socket, hinged floor [electrical]   boîte *f* de parquet pivotante

socket, pipe   manchon *m* de tuyau
socket, plug, *see* socket outlet
socket, razor   prise *f* (de courant pour) rasoir
socket, screwed   manchon *m* vissé
socket, threaded   manchon *m* fileté
sockets, trailing [electrical]   bloc *m* multiprise (à cordon)
socket box, floor   boîte *f* de parquet
socket end   embout *m* femelle
socket fitting, ceiling   douille *f* à plafond
socket fitting, floor   boîte *f* de parquet
socket fitting, fluorescent lamp   douille *f* pour lampe fluorescente
socket forming tool [plumbing]   évaseur *m*
socket joint, welded   joint *m* à emboîtement soudé
socket outlet   prise *f* (de courant) femelle, prise *f* (électrique femelle), socle *m*, socle *m* (de prise) de courant
socket outlet, double (plug)   socle *m* de deux prises de courant
socket outlet, earthed   prise *f* de courant de sécurité
socket outlet, floor mounted   socle *m* de prise de sol
socket outlet, flush mounted   socle *m* de prise à encastrer
socket outlet, panel mounting   socle *m* de tableau
socket outlet, surface mounted   socle *m* (en) saillie
socket outlet, three-pin   socle *m* de prises de courant à trois contacts
socket outlet box   boîtier *m* de prises de courant
sod [piece of turf]   morceau *m*/plaque *f* de gazon
soffit   sous-face *f*
soffit [architectural]   soffite *m*
soffit [of an arch or tunnel]   cintre *m*
soffit [of an arch or vault]   douelle *f*, intrados *m*
soft [of a verge, ground, etc]   meuble *adj*
soft layer on face of hard stone in its bedding plane   moye *f*
soft metal seal over a screw in a meter casing   cache-vis *m*
soft shoulder   accotement *m* non-stabilisé
soften   ramollir *vb*
soften [steel]   détremper *vb*
soften [edges, colours, water, etc]   adoucir *vb*
softener, water   adoucisseur *m* d'eau
softening   ramollissement *m*
softening [of edges, colours, water, etc]   adoucissement *m*
softwood [timber]   bois *m* tendre
softwood [trees]   essences *f,pl* conifères
soil, from or of the   tellurique *adj*
soil   sol *m*, terre *f*
soil, coarse-grained   sol *m* grenu
soil, cohesive   sol *m* cohérent
soil, fine-grained   sol *m* fin
soil, non-cohesive   sol *m* pulvérulent
soil, organic [topsoil]   sol *m* vivant, terre *f* végétale
soil, undisturbed   sol *m* non remanié
soil, unsuitable   sol *m* impropre
soil analysis   analyse *f* du sol
soil and waste stack   tuyau *m* de chute unique/de descente d'eaux ménagères
soil compaction   compactage *m* de sol

soil **deformation properties**   deformabilité *f* d'un sol
soil **filtration**   filtration *f* par le sol
soil **horizon**   couche *f* de sol
soil **investigation by test or trial pits**   sondage *m* par
   tranchée
soil **investigation/survey**   étude *f* pédologique/du sol,
   reconnaissance *f*/essais *m,pl* de sol
soil **moisture**   humidité *f* du sol
soil **profile**   coupe *f* de sondage
soil **science**   pédologie *f*
soil **scientist**   pédologiste *m/f*, pédologue *m/f*
soil **stabilizer**   stabilisateur *m* de sols
soil **stabilizing machine**   engin *m* stabilisateur/de
   traitement des sols
soil **stack**   chute *f* eaux vannes (chute EV), tuyau *m* de
   chute
soil **structure**   structure *f* du sol
soil **survey,** *see* soil investigation
soil **treatment**   traitement *m* des sols
soil-cement   sol-ciment *m*
soiling   salissure *f*
sol **air temperature**   température *f* fictive extérieure
solar   solaire *adj*
solar **cell**   cellule *f* solaire, photopile *f*
solar **flat plate collector**   capteur *m* plan
solar **(heat) gain**   apports *m,pl* solaires/de chaleur solaire
solar **time**   temps *m* solaire
solarium   solarium *m*
solder   étain *m* à braser, soudure *f*
solder **fitting** [pipe jointing]   raccord *m* montage soudé
solder **wire, lead tin**   fil *m* de soudure
solder **wire, tin based**   fil *m* d'étain
solder   braser *vb*, souder *vb*
solder, soft   métal *m* d'apport de brasage
solder, hard   souder *vb* au cuivre
solder, soft   souder *vb* à l'étain
soldering   soudage *m*, soudure *f*
soldering, hard   soudure *f* au cuivre
soldering, lead   brasure *f* à plomb, soudure *f* au plomb
soldering, soft   soudure *f* à l'étain
soldering **equipment**   matériel *m* de brasage
soldering **gun**   pistolet *m* à souder (électrique)
soldering **iron**   fer *m* à souder/à braser, soudoir *m*
soldering **iron, high-speed**   fer *m* à souder instantané
sole **of a plane**   plan *m*/semelle *f*/talon *m* d'un rabot
sole **piece or plate** [of wood, under a prop, etc]   chaise *f*,
   couche *f*, semelle *f*
sole **plate** [of a timber frame panel]   sablière *f*
solenoid   solénoïde *m*
Solétanche rotary diaphragm wall rig [trade name]
   Hydrofraise *m*
solid   solide *adj*
solid [e.g. solid timber, etc]   massif *adj*
solid [without openings]   plein *adj*
solid **block or mass** [of concrete, masonry, etc]   massif *m*
solid **cap**   couronnement *m* massif
solid **webbed**   à âme pleine
solidification   solidification *f*
solidify [by freezing]   se geler *vb*

solidity   solidité *f*
solubility   solubilité *f*
soluble [problem, substance]   soluble *adj*
soluble, make   solubiliser *vb*
solution   solution *f*
solution [result of dissolving]   dissolution *f*
solution, compromise   solution *f* de compromis
solvent *adj*   dissolvant *adj*
solvent   dissolvant *m*, solvant *m*
soot   suie *f*
soot **arrestor**   capte-suie(s) *m*
sooted up [of a spark plug]   encrassé *adj*
sort   trier *vb*
sort [grade]   calibrer *vb*
sort **out** [a problem]   régler *vb*
sort **through** [e.g. mail, drawings, etc]   dépouiller *vb*
sorted   calibré *adj*
sorting   classement *m*
sound [acoustic]   acoustique *adj*, sonore *adj*
sound [of a building, a business, etc]   solide *adj*
sound [of timber]   sain *adj*
sound   son *m*
sound [noise]   bruit *m*
sound, airborne   bruit *m* aérien
sound, speed of   vitesse *f* du son
sound, structure-borne   vibration *f*
sound **absorber**   amortisseur *m*
sound **attenuation**   affaiblissement *m* phonique,
   atténuation *f* des sons/du son
sound **attenuator**   dispositif *m* d'insonorisation
sound **control**   contrôle *m* acoustique
sound **deadening**   amortissement *m* de bruit
sound **insulation**   isolation *f* contre le bruit
sound **intensity**   intensité *f* sonore
sound **intensity level**   niveau *m* d'intensité sonore
sound **level**   niveau *m* sonore
sound **power**   puissance *f* acoustique
sound **pressure level, standardized**   niveau *m* de
   pression acoustique normalisé
sound **source**   source *f* sonore
sound **spectrum**   spectre *m* sonore
sound **transmission coefficient**   coefficient *m* de
   transmission du son
sound **wave**   onde *f* sonore
sound [to investigate at depth]   sonder *vb*
sounding [to measure depth of water]   sondage *m*
sounding **board**   abat-voix *m*
sounding **rod or probe**   tige *f* de sondage
soundness   solidité *f*
soundproof *adj*   insonore *adj*
soundproof   insonoriser *vb*, isoler *vb*
soundproofing *adj*   isolant *adj*
soundproofing   insonorisation *f*
soundproofing [materials, devices]   éléments *m,pl*
   d'insonorisation
source, alternative energy   source *f* alternative d'énergie
source, (electrical) power   source *f* d'électricité
source, energy   source *f* d'énergie
source, error   source *f* d'erreur

source, heat foyer *m*/source *f* de chaleur
source, light source *f* lumineuse
source, power source *f* d'électricité
source, sound source *f* sonore
source, waste energy source *f* d'énergie perdue
south, southern, southerly, southward sud *adj*
south sud *m*
south-east sud-est *m*
south-west sud-ouest *m*
spa station *f*
space espace *m*, place *f*
space [apart] intervalle *m*
space, contained (vertical) cage *f*
space, cupboard rangement *m*
space, empty vide *m*
space, flat aire *f*
space, floor surface *f* couverte/des étages
space, green [in town or city] espace *m* vert
space, habitable surface *f* habitable
space, occupied local *m* occupé
space, open espace *m* libre
space, required place *f* nécessaire
space, storage [cupboards, shelving] rangement *m*
space, storage or waiting [road] zone *f* de stockage
space, usable surface *f* utilisable
space (between) écartement *m* (de, entre)
space between adjacent buildings or parts of buildings isolement *m*
space created by setting back of part of a building retraite *f*
space in front of a building [in regulations] prospect *m*
space requirement place *f* nécessaire
space requirements [dimensions, for equipment] cotes *f,pl* d'encombrement
space under a suspended floor vide *m* sanitaire
space-frame treillis *m* tridimensionnel
spaced at espacé *adj* de
spacer dispositif *m* d'écartement, écarteur *m*, espaceur *m*
spacer [in equipment, or reinforcement] entretoise *f*
spacer, bolted [timber structure] aiguille *f*
spacer, reinforcement cale *f* à béton, distancier *m*
spacer, tile croisillon *m*
spacing distance *f* d'espacement, écart *m*, espacement *m*
spacing [between, joists, rafters, etc] travée *f*
spacing [pitch] pas *m*
spacing, fin écartement *m*/pas *m* d'ailettes
spacing, reinforcement écartement *m*/entredistance *f* des armatures *f,pl*
spacing, rivet écartement *m*/pas *m* des rivets
spacing, support distance *f* entre les supports
spacing (between) écartement *m* (de, entre), entredistance *f* (de)
spacing between centre lines [e.g. of beams] entraxe *f*
spacing between joists clair *m*
spacing gauge or template gabarit *m* d'écartement
spacing (out) échelonnement *m*
spacing piece entretoise *f*
spade bêche *f*
spall écailler *vb*

spalled stone finish parement *m* smillé
spalling écaillage *m*
spalling of stone edge or surface épaufrure *f*
span [distance between supports] portée *f*
span [part between supports] travée *f*
span, central/centre travée *f* centrale
span, clear portée *f* entre nus/libre
span, end travée *f* de rive
span, lift(ing) [bridge] travée *f* levante/mobile
span, main portée *f* principale
span, side travée *f* latérale
span arch of a bridge, main arche *f* maîtresse
span of an arch archée *f*
span of an arch between springs ouverture *f* d'un arc
span (over) enjamber *vb*, franchir *vb*, traverser *vb*
spandrel [of door of cathedral] tympan *m*
spanner, *see also* wrench clef *f*
spanner, adjustable/shifting clef *f* à crémaillère/à molette
spanner, open clef *f* plate
spanner, open-ended clef *f* à fourches
spanner, ring clef *f* polygonale/à œil
spanner, socket clef *f* à tube
spanners, set of jeu *m* de clés
spanning in two directions [slab] continu *adj* en deux directions [dalle]
spar chevron *m*
spare *adj* de réserve
spare (part) pièce *f* détachée/de rechange
spare part, guaranteed genuine pièce *f* d'origine
spark étincelle *f*
spate, in en crue
spatula spatule *f*
special spécial *adj*
special design type *m* spécial
special dispensation dérogation *f*
specialist spécialiste *m*
specific spécifique
specific gravity pesanteur *f*/poids *m* spécifique
specific heat chaleur *f* massique/spécifique
specification cahier *m* des charges, descriptif *m*, spécification *f*
specification, detailed cahier *m* des charges détaillé
specification, technical cahier *m* des charges/clauses techniques
specifications, detailed technical spécifications *f,pl* techniques détaillées
specify préciser *vb*, spécifier *vb*
specimen échantillon *m*
spectacles lunettes *f,pl* (à branches)
spectral energy distribution répartition *f* d'énergie spectrale
spectrographic spectrographique *adj*
spectrum spectre *m*
spectrum, sound spectre *m* sonore
speed allure *f*, vitesse *f*
speed [of a motor] régime *m*
speed, at full or top à toute allure, à plein régime
speed, average or mean moyenne *f*, vitesse *f* moyenne

speed, critical vitesse *f* critique
speed, fan vitesse *f* du ventilateur
speed, maximum vitesse *f* limite/maximum
speed, tip vitesse *f* périphérique
speed control, automatic réglage *m* de vitesse automatique
speed decrease or reduction ralentissement *m*
speed indicator indicateur *m* de vitesse
speed loss perte *f* de vitesse
speed of sound vitesse *f* du son
speed variator variateur *m* de vitesse
speedometer indicateur *m* de vitesse
sphere sphère *f*
spherical sphérique *adj*
spherical vault, small calotte *f*
spigot, conical, formed in a pipe end emboîture *f* conique
spigot, with a shoulder, formed in a pipe end emboîture *f* à épaulement
spigot joint [in pipework] emboîtement *m*
spike pointe *f* longue
spike [long headless nail] broche *f*
spike on railings pique *f*
spike together brocher *vb*
spiked hook on top of a wall or fence épi *m*
spiked ironwork on top of a wall or fence artichaut *m*, chardon *m*, hérisson *m*
spillway déversoir *m*, évacuateur *m* de crue
spillway, emergency évacuateur *m* de secours
spindle arbre *m*
spindle [grindstone, pump, etc] axe *m*
spindle [of a valve, tap] tige *f*
spindle, lathe broche *f* de tour
spine arête *f*
spiral *adj* hélicoïdal *adj*
spiral spirale *f*
spiral [staircase] tournant *adj*
spiral grain [timber defect] fibre *f* torse, fil *m* tors
spiral staircase, *see* staircase, spiral
spire [conical or pyramidal] flèche *f*
spire (of a bell tower) aiguille *f* (de clocher)
spirit, white white spirit *m*
spirits, methylated alcool *m* à brûler
spit [of land] langue *f*, presqu'île *f*
splashback [above kitchen worktop, lab bench] dosseret *m*
splice enture *f*, joint *m*
splice [in a rope, cable, conductor, etc] épissure *f*
splice [in reinforcing bar, a cable, wire, etc] ligature *f*
splice, web joint *m* de l'âme
splice plate éclisse *f*, plate-bande *f*
splicing assemblage *m* par couvre-joint
splinter [of wood] écharde *f*
splinter [of wood, stone] éclat *m*
split *adj* fendu *adj*
split fente *f*
split [in plaster, etc] lézarde *f*
split [in timber] gerce *f*
split pin clavette *f* fendue, goupille *f* (fendue)
split [e.g. slate] refendre *vb*

split [lengthwise] fendre *vb*
split [stone in its bedding plane] déliter *vb*
split [wood] éclater *vb*
split into sheets feuilleter *vb*
splitter [in ductwork] aubage *m* directeur, baffle *m*
splitting refend *m*
spoil [excavated material] déblais *m,pl*
spoil, drill(ing) débris *m* de forage
spoil bank [roadworks] cavalier *m*
spoil draught coupe-tirage *m*
spoil abîmer *vb*, détériorer *vb*
spoil completely ruiner *vb*
spokeshave vastringue *f*, wabstringue *f*
sponge éponge *f*
sports field or ground terrain *m* de sport(s)
sports stadium stade *m*
spot [locality] lieu *m*
spot [of dirt, mark] tache *f*
spot [place] endroit *m*
spot cash argent *m* comptant
spot height [on drawing] cote *f*
spotlight projecteur *m*
spotlight [theatre] projecteur *m* de scène
spotted with holes [of coatings, rendering, varnish, etc] piqué *adj*
sprawl [of a town, etc] étendue *f*
spray chamber chambre *f* de pulvérisation
spray curtain rideau *m* de pulvérisation
spray gun pistolet *m*
spray painting peinture *f* au pistolet/par pulvérisation
spray pistol, paint pistolet *m* à peinture
spray [a liquid] atomiser *vb*
spray [a liquid, paint] pulvériser *vb*
spray (with copper sulphate) sulfater *vb*
sprayer atomiseur *m*, pulvérisateur *m*
sprayer [for bitumen, road emulsion, etc] répandeuse *f*
spray(er), hand pulvérisateur *m* à main
sprayer, tar- goudronneuse *f*
spraying [of concrete, mortar, rendering] projection *f*
spraying, metal métallisation *f*
spraying with textile + glue/binder [for acoustic damping] flocage *m*
spread [of fire] propagation *f*
spread [a river, a stain] s'élargir *vb*
spread [distribute: loads, sand, chippings, etc] répandre *vb*
spread [extend, stretch: a town, branches, etc] s'étendre *vb*
spread [fire, water] s'épandre *vb*
spread [flood, river, water] se répandre *vb*
spread [diffuse: heat, light, news, etc] diffuser *vb*
spread [ideas, news, plans, etc] communiquer *vb*
spread [lay out: a map, a carpet, etc] étendre *vb*
spread [paint] étendre *vb*
spread [sewage, fertilizer] épandre *vb*
spread gravel or chippings on tar gravillonner *vb*
spread out [e.g. gravel, objects] étaler *vb*
spread out [e.g. payments, over time] échelonner *vb*
spread to [e.g. fire] atteindre *vb*

spreader [for fertilizer, sewage, etc] épandeuse *f*
spreader [machine] engin *m* d'épandage
spreader [road works] répandeuse *f*
spreader, chippings gravillonneuse *f*
spreader, hot-mix (tar) goudronneuse *f*
spreading [diffusion] diffusion *f*
spreading [of chippings, tar, etc] répandage *m*
spreading [of paint, varnish] étendage *m*
spreading [of sewage on land] épandage *m*
spreading a thin layer of earth, etc, to level ground
  régalage *m*
spreadsheet program tableur *m*
spring [mechanical] ressort *m*
spring [of an arch or vault] naissance *f*, retombée *f*
spring [of timber] voilement *m* longitudinal de rive
spring [of water] fontaine *f*, résurgence *f*, source *f*
spring, circular ressort *m* annulaire
spring, constantly flowing source *f* à débit *m* permanent
spring, leaf ressort *m* à lames
spring, return ressort *m* de rappel
spring leaf [*not* botanic] lame *f* de ressort
springer or springing stone [of an arch] coussinet *m*,
  sommier *m*, tas *m* de charge
springing line [of an arch] ligne *f* de naissance
sprinkle [water] arroser *vb*
sprinkler [irrigation or watering] arroseur *m*, asperseur *m*
sprinkler [fire protection] sprinkle(u)r *m*
sprinkler, automatic extincteur *m* automatique
sprinkler, boom arroseur *m* à flèche
sprinkler, revolving/rotating arroseur *m* rotatif
sprinkler head diffuseur *m* (d'extincteur automatique
  d'incendie)
sprinkler installation or system [fire protection]
  installation *f* d'extinction par sprinkle(u)rs
sprinkler system, dry système *m* d'extinction à sec
sprinkler system, fire système *m* d'extinction
  automatique
sprinkling [irrigation, watering] arrosage *m*, aspersion *f*
sprinkling [fire protection] sprinklage *m*
spruce [tree, wood] épicéa *m*
spyhole [in an entrance door] judas *m*
spyhole in a larger door guichet *m*
squall [of wind] coup *m* de vent, rafale *f*
square carré *adj*
square [at right angles] à angle droit, d'équerre
square [as in a town] place *f*
square [figure, second power] carré *m*
square [for marking right angles] équerre *f*
square [on a grid] mail *m*
square, set équerre *f* à dessin
square root racine *f* carrée
square [a corner] couper *vb* à angle droit
square [a number, or make square] carrer *vb*
square [books] balancer *vb*
square [stone] dresser *vb*, équarrir *vb*
square [timber] araser *vb*, équarrir *vb*
square up with [pay debts, an account] regler *vb* les
  comptes avec
squared équarri *adj*

square-edged à vive arête, à angles vifs
square-edged angle (section) cornière *f* à angles vifs
squaring [of timber, stone] équarrissage *m*
squeegee raclette *f*
squinch trompe *f*
squinch, small trompillon *m*
stability stabilité *f*
stability, dimensional stabilité *f* dimensionnelle
stability, flame stabilité *f* de la flamme
stability, overturning stabilité *f* au renversement
stability against overturning stabilité *f* au renversement
stability analysis étude *f* de stabilité
stabilization stabilisation *f*
stabilization, cement stabilisation *f* au ciment
stabilization, soil stabilisation *f* du/d'un sol, traitement *m*
  des sols
stabilization, subsoil stabilisation *f* du sous-sol
stabilization basin bassin *m* de stabilisation
stabilize stabiliser *vb*
stabilized stabilisé *adj*
stabilized earth terre *f* stabilisée
stabilizer [compound or product] produit *m* stabilisant,
  stabilisant *m*, stabilisateur *m*
stabilizer, draught régulateur *m* de tirage
stabilizing stabilisant *adj*
stabilizing agent, *see* stabilizer
stable *adj* constant *adj*, stable *adj*
stable écurie *f*
stable position [of a body such as a beam] assiette *f*
stack [chimney] tuyau *m* de cheminée
stack [of wood, etc] empilage *m*, pile *f*
stack, rainwater chute *f* eaux pluviales (chute EP)
stack, vent colonne *f* de ventilation
stack, waste chute *f* eaux usées (chute EU)
stadium, (sports) stade *m*
staff, levelling [surveying] mire *f*, tige *f* de nivellement
staff, surveyor's mire *f*, jalon *m*
staff representative délégué *m* du personnel,
  représentant *m* des travailleurs
stage [of a process, work, etc] étape *f*, phase *f*
stage [of a pump, a compressor] étage *m*
stage [of a theatre] scène *f*
stage [point in a process, programme, etc] point *m*
stage, hanging [scaffolding] échafaudage *m* suspendu
stage lighting éclairage *m* de scène
stage lights herses *f,pl*
stagger [positional] décalage *m*
stagger [in position] décaler *vb*
staggered en quinconce
staggered [in rows] en quinconces
staggering [spacing] échelonnement *m*
stain [mark] tache *f*
stain [tint] teinte *f*
stain [mark] tacher *vb*
stain [tint] teinter *vb*
stains indicating rot in timber rougeurs *m,pl*
stained brown/yellow by heat rissolé *adj*
stainless [as of steel] inoxydable *adj*
stair [of stairs] marche *f*

stair clearance   échappée *f*, échappement *m*

stair nose   nez *m* de marche

stair rod   tringle *f* d'escalier

stair string   limon *m* (d'escalier)

stair well   trémie *f* d'escalier

stair(s), *see also* staircase   escalier *m*

stairs, flight of   volée *f* d'escalier

stairs, steep flight of narrow   échelle *f* de meunier

staircase   escalier *m*

staircase, close or housed string (straight)   escalier *m* à la française/à limon (droit)

staircase, cut string   escalier *m* à crémaillère/à l'anglaise

staircase, escape   escalier *m* d'évacuation/de secours

staircase, external   escalier *m* hors d'œuvre

staircase, hidden   escalier *m* dérobé

staircase, open spiral   escalier *m* en escargot

staircase, open string   escalier *m* à crémaillère/à l'anglaise

staircase, open tread   escalier *m* de meunier/en échelle

staircase, quarter-turn with winders   escalier *m* à quartier tournant

staircase, secret   escalier *m* dérobé

staircase, service   escalier *m* de service

staircase, shallow, with open treads   escalier *m* à échelle de meunier

staircase, skeleton   escalier *m* de meunier/en échelle

staircase, spiral   colimaçon *m*, escalier *m* à vis/en colimaçon/en hélice/en limaçon/tournant, escargot *m*, limaçon *m*

staircase, spiral, with open newel   escalier *m* à vis sans noyau, escalier *m* tournant à noyau *m* creux

staircase, spiral, with solid newel   escalier *m* à vis à noyau, escalier *m* tournant à noyau *m* plein

staircase, winding, *see also* staircase, spiral, etc   escalier *m* tournant

staircase between walls   escalier *m* encloisonné

staircase tread width   emmarchement *m*

staircase with large going and relatively small rise   pas *m* d'âne

staircase without risers   échelle *f* de meunier/en échelle

stairway   escalier *m*

stairwell   cage *f*/jour *m* d'escalier

stairwell, width of   jour *m*

stake [e.g. for setting out]   piquet *m*

stake [in a business, etc]   intérêt *m*

stake, marker [civil works]   balise *f*

stake stabilizing foot of angled prop or shore   accul *m*

staking [e.g. for setting out]   piquetage *m*

stalk   tige *f*

stalls, front or orchestra [of a theatre]   orchestre *m*

stamp   timbre *m*

stamp [commercial]   label *m*

stamp [e.g. on approved drawing]   visa *m*

stamp, excise   timbre *m* fiscal

stamp, official   timbre *m* officiel

stamp, revenue   timbre *m* fiscal

stamp, rubber   tampon *m*

stamp duty   droit *m* de timbre

stamp [a document or letter]   tamponner *vb*

stamp [metals]   emboutir *vb*

stamped [letter, document, etc]   tamponné *adj*

stanchion, *see also* column, post   poteau *m*

stanchion, frame   béquille *f* de portique

stanchion, lattice   poteau *m* en treillis

stanchion head   tête *f* de poteau

stand, drill   support *m* de perceuse

stand [support]   chevalet *m*

stand [trestle]   tréteau *m*

stand [building, monument, peak, etc]   se dresser *vb*

standard   normal *adj*, standard *adj*

standard [code of practice]   norme *f*

standard [e.g. example, trial]   étalon *m*

standard, current   norme *f* en vigueur

standard, national   norme *f* nationale

standard air   air *m* normal

standard air, weight of   poids *m* de l'air normal

standard atmosphere   atmosphère *f* normale

standard colour code   code *m* de couleurs normalisées

standard deviation   déviation *f* standard, écart *m* type

standard test piece   éprouvette *f* normalisée

standard unit [equipment]   appareil *m* normal

standard unit [module]   module *m*

standardization   étalonnage *m*, étalonnement *m*

standardize [an instrument, etc]   étalonner *vb*

standardized   étalonné *adj*

standby   de réserve, de secours

standby equipment   équipement *m* de réserve

standby generating set   groupe *m* électrogène de secours

standing charge   abonnement *m*

standpipe   tuyau *m* de prise d'eau

staple [U-shaped]   clou *m* en U, crampon *m*

staple [for fixing electrical wiring]   (clou *m*) cavalier *m*

staple   agrafe *f*

staple   agrafer *vb*

staple for building into masonry [e.g. step iron]   cavalier *m* à deux scellements

stapled   agrafé *adj*

stapler   agrafeuse *f*

star connection [electrical]   étoile *f*

star voltage   tension *f* étoilée

star-delta   étoile-triangle *adj*

start, *see also* starting

start   démarrer *vb*

start [a generator]   amorcer *vb*

start [of pump, motor]   s'amorcer *vb*

start (up) [a machine, etc]   mettre *vb* en marche/en route

start from cold   démarrer *vb* à froid

starter   commutateur *m* de démarrage, démarreur *m*

starter [for a lamp]   starter *m*

starter, automatic   démarreur *m* automatique, dispositif *m* de démarrage automatique

starter, direct-on-line (DOL)   démarreur *m* à opération directe

starter, part-winding   démarreur *m* sur fraction d'enroulement

starter, single-phase   démarreur *m* monophasé

starter bar [reinforcement]   acier *m*/barre *f* en attente, attente *f*, crosse *f* (d'armature)

starter bars   attentes *f,pl* de ferraillage, fer *m* de reprise
starter bars, fine   soie *f*
starter bars, small, in reinforced concrete   chevelu *m*
starter button   bouton *m* de commande/de démarrage
starter device or gear   dispositif *m* de démarrage
starting [activating, releasing]   déclenchement *m*
starting (up) [of machines, equipment]   démarrage *m*, mise *f* en marche/en route
starting up [of work]   ouverture *f*
starting, full-load   mise *f* en marche à pleine charge
starting, no-load   démarrage *m* à vide
starting, star-delta   démarrage *m* delta-étoile
starting current   courant *m* de démarrage
starting point   point *m* de départ
starting procedure   opération *f* de mise en marche
starting time   temps *m* de mise en marche
starting torque   couple *m* de démarrage
state [condition]   état *m*
state, existing [of a building, site, etc]   état *m* des lieux
state, normal   état *m* normal
state control (of)   régie *f* (de)
state of equilibrium   état *m* d'équilibre
state of stress   état *m* de contrainte/de tension
state-owned company or organization [France]   régie *f* (de l'État)
statement [of facts, etc]   exposé *m*, exposition *f*
statement, written   procès-verbal *m* (PV)
statement of account [balance in an account]   relevé *m* de compte
statement of account [balance of unpaid accounts]   relevé *m* de comptes
static   statique *adj*
static [interference]   parasites *m,pl*
static regain method [for duct sizing]   méthode *f* de regain statique
statics   statique *f*
statical moment   moment *m* statique
statically determinate   isostatique *adj*
statically indeterminate   hyperstatique *adj*
station   gare *f*
station [place, position]   station *f*
station, atomic power   centrale *f* atomique
station, balancing   poste *m* d'équilibrage
station, booster pumping   station *f* de pompage intermédiaire
station, bus   gare *f* routière
station, combined heat and power (CHP)   centrale *f* combinée chaleur force/de chaleur/force
station, electric/hydro-electric power   centrale *f* hydro-électrique/hydraulique
station, filling   station-service *f*
station, fire   caserne *f* des pompiers
station, goods/passenger   gare *f* de marchandises/de voyageurs
station, nuclear power   centrale *f* atomique/nucléaire
station, petrol   station-service *f*
station, power, *see* power station
station, pumping   station *f* de relevage/de pompage
station, research   laboratoire *m* de recherches

station, service [vehicular]   station-service *f*
station, survey   signal *m* géodésique
station, thermal power   centrale *f* thermique
station, transformer   poste *m* de transformation
station, underground (railway)   station *f* métro
station hall or concourse   hall *m* de gare
stationary plant mixing [soil stabilization method]   mélange *f* en centrale
statistics   statistique *f*
stator   stator *m*
statute   ordonnance *f*
statutory auditor   Commissaire *m* aux Comptes
statutory representative   contrôleur *m* des travaux, fonctionnaire *m* délégué
statutory services or utilities   services *m,pl* concédés
statutory undertaker   concessionnaire *m*
stay [at/in a place]   séjour *m*
stay [door or window]   entrebâilleur *m*
stay [prop]   pied *m* d'appui
stay [prop, generally small, temporary]   chandelle *f*
stay [support]   portant *m*, soutien *m*
stay [tension member]   barre *f* tendue
stay [wooden, sensibly vertical]   accore *m*
stay, notched/drilled, for holding rooflight open   crémaillère *f*
stay [brace]   entretoiser *vb*
stay [support]   appuyer *vb*
stay [with guys]   hauban(n)er *vb*
staying/stay ropes   hauban(n)age *m*
steady   constant *adj*, stable *adj*
steady state (condition)   état *m* stationnaire, état *m* (de régime) permanent
steady state, non-   régime *m* variable
steam   vapeur *f*
steam, dry   vapeur *f* sèche
steam, exhaust or waste   vapeur *f* d'échappement
steam, flash   autovaporisation *f*
steam, live   vapeur *f* vive
steam, low/high pressure   vapeur *f* à basse/haute pression
steam, saturated   vapeur *f* saturée
steam, superheated   vapeur *f* surchauffée
steam, waste   vapeur *f* d'échappement
steam curing [of concrete]   étuvage *m*
steam generating station   centrale *f* thermique
steam heated   chauffé *adj* à la vapeur
steam humidifier   humidificateur *m* à vapeur
steam tables   tables *f,pl* de vapeur
steam-proof/steam-tight   étanche *adj* à la vapeur
steamroller   rouleau *m* compresseur
steel *adj*   en acier, métallique *adj*
steel *adj* [relating to the steel industries]   sidérurgique *adj*
steel   acier *m*
steel, alloy   acier *m* allié
steel, austenitic   acier *m* austenitique
steel, bar   fer *m* marchand
steel, basic   acier *m* basique
steel, carbon   acier *m* au carbone

steel, carbon (0.6–0.7%C)   acier *m* dur
steel, case-hardened   acier *m* cémenté
steel, cast   acier *m* coulé/fondu/moulé, fonte *f* d'acier
steel, cold drawn   acier *m* étiré à froid
steel, cold hardened or worked   acier *m* écroui
steel, compression [reinforcement]   acier *m* comprimé
steel, corrosion-resistant or stainless   acier *m*
  inox(ydable)
steel, crude   acier *m* brut
steel, deformed or ribbed [reinforcement]   acier *m*
  crénelé/haute adhérence
steel, fine grain   acier *m* à grain fin
steel, forged   acier *m* forgé
steel, galvanized   acier *m* galvanisé/zingué
steel, grade of   nuance *f*/qualité *f* d'acier
steel, high-alloy   acier *m* hautement allié
steel, high-carbon   acier *m* à haute teneur en carbone
steel, high-speed   acier *m* rapide
steel, high-strength   acier *m* à haute résistance/à
  résistance élevée
steel, low-alloy   acier *m* faiblement allié
steel, low-carbon   acier *m* à bas carbone/à faible teneur
  de carbone
steel, mild (0.15–0.3%C)   acier *m* doux
steel, open-hearth   acier *m* Martin
steel, oxygen refined   acier *m* à l'oxygène
steel, reinforcement/reinforcing, *see also* reinforcement
  acier *m* à béton, armature *f*
steel, rolled   acier *m* laminé
steel, semi-hard (0.4–0.6%C)   acier *m* demi dur
steel, semi-mild   acier *m* demi doux
steel, stainless   acier *m* inox(ydable)
steel, structural   acier *m* de construction
steel, tempered   acier *m* trempé
steel, zinc coated   acier *m* zingué
steel bar providing extra support to glazing   vergette *f*
steel bender, *see* steelfixer
steel deck or decking plate [of a bridge]   tablier *m*
  métallique
steel fabricator   entreprise *f* de constructions métalliques
steel formwork   coffrage *m* métallique
steel frame building   bâtiment *m* à ossature métallique
steel structure   ossature *f* métallique
steel grit   grenaille *f*
steel grit blasting   grenaillage *m*
steel mill   usine *f* sidérurgique
steel panel, bay or wall   pan *m* de fer
steel plate   tôle *f* d'acier
steel reinforcement bar, *see also* reinforcement   barre *f*
  d'armature en acier
steel sheet   tôle *f* d'acier
steel sheet, corrugated   tôle *f* ondulée
steel sheet, galvanized   tôle *f* (d'acier) galvanisée/zinguée
steel sheet, tinned   fer *m* blanc
steel strip up to 160mm wide   fers *m,pl* plats
steel strip wider than 160mm   fers *m,pl* larges
steel structure   ossature *f* métallique
steel superstructure   superstructure *f* métallique
steel wool   laine *f* d'acier, laine *f*/paille *f* de fer

steelfixer   ferrailleur *m*, treillageur *m*
steelfixer's flat tongs [for bending fine reinforcement]
  béquette *f*
steelfixer's pliers   pinces *f,pl* à ligatures
steelworks   aciérie *f*, usine *f* sidérurgique
steelyard   balance *f* romaine
steep [slope]   raide *adj*
steep   tremper *vb*
steeple   clocher *m*
stem   tige *f*
stemming [blasting holes]   bourrage *m*
stencil, lettering   trace-lettres *m*
stencil (plate)   pochoir *m*
stencil brush   pochon *m*
stencil for circles/ellipses   trace-circles/ellipses *m*
stencilling   peinture *f* au pochoir
step [action, measure]   démarche *f*
step [in a gable, an embankment, etc]   redan *m*, redent *m*
step [in metallic roofing, a gutter, etc]   ressaut *m*
step [of a process, work, etc]   étape *f*
step [of stair]   marche *f*, pas *m*
step [of terracing, or control system]   gradin *m*
step, balanced/dancing   marche *f* dansante
step, block   marche *f* pleine
step, bottom   marche *f* de départ/du bas
step, skewed and of varying width   marche *f* biaise
step, straight   marche *f* droite
step, top   dernière marche *f*, marche *f* palière
step, varying width [winder]   marche *f* tournante
steps   escalier *m*
steps [in front of a house]   perron *m*
steps [measures]   mesures *f,pl*
steps, administrative   démarches *f,pl* administratives
steps, flight of   escalier *m*
steps, front [of a building]   perron *m*
steps, pair of, *see also* stepladder   triquet *m*
step iron [of a ladder]   échelon *m*
step [wall, embankment]   recouper *vb*
stepladder   échelle *f* double/pliante, escabeau *m*
stepped [e.g. a gable, an embankment]   redenté *adj*, en
  redan
stepped [spaced out, staggered]   échelonné *adj*
stepped in section   étagé *adj* en section
stepping [of an embankment]   recoupement *m*
stereogram/stereograph [image]   stéréogramme *m*
stereograph [instrument]   stéréographe *m*
stereometer   stéréomètre *m*
stereometric camera   appareil *m* de stéréométrie
stereo-photography   stéréophotographie *f*
stereoscope [instrument]   stéréoscope *m*
stereotomy   stéréotomie *f*
sterile   stérile *adj*
sterilizer   appareil *m* de stérilisation, étuve *f* de
  désinfection
stick [machines, etc]   coincer *vb*
stick (to)   adhérer *vb* (à)
stickiness   adhésivité *f*
sticking   adhésion *f*
stiff [rigid]   rigide *adj*

stiffen   raidir *vb*
stiffener   raidisseur *m*, renfort *m*
stiffener, web   raidisseur *m* d'âme
stiffness   raideur *f*, rigidité *f*
stiffness, modulus of   module *m* d'élasticité
stile [of door, window]   montant *m*
stile, closing   montant *m* de battement
stile, corner   poteau *m* cornier/d'angle
stile, hanging   montant *m* charnier/de ferrage/de rive
stile, meeting/shutting   montant *m* de serrure
stilling basin [of a dam]   bassin *m* d'amortissement/de
   dissipation/de tranquilisation, chambre *f*
   d'amortissement
stillson wrench   clef *f* de tuyau
Stirling cycle   cycle *m* de Stirling
stirrer   agitateur *m*
stirrup [reinforcement]   cadre *m*, cadre *m*
   d'armature/fermé, étrier *m*
stirrup, inclined [reinforcement]   cadre *m* incliné
stirrup bracket   support *m* à étrier
stock [of a plane]   fût *m*
stock [of equipment, etc]   parc *m*
stock [reserve]   réserve *f*
stock [supply]   provision *f*
stock, in   en stock
stock with   meubler *vb* de
stocking [holding]   stockage *m*
stock-list   inventaire *m*
stock-list, according to   suivant inventaire
stockpile [of materials]   réserve *f* de matériaux
stockpile [space for material(s)]   aire *f* de stockage
stockpile   mettre *vb* en dépôt
stockpiling   mise *f* en dépôt
stoker   chargeur *m* de foyer
stoker, chain grate   foyer *m* à grille (mécanique)
stoking [of a boiler]   chauffe *f*
stone   pierre *f*
stone [40–60mm]   caillou *m*
stone, arch   claveau *m*, voussoir *m*
stone, artificial   pierre *f* artificielle
stone, boundary   pierre *f* de bornage
stone, breast [under a window or door sill]   parpaing *m*
   d'appui
stone, building   pierre *f* à bâtir
stone, coping   tablette *f*
stone, corner   pierre *f* de refend
stone, crushed   cailloutis *m*, pierre *f* cassée/concassée
stone, dressed   pierre *f* appareillée
stone, exposed   pierre *f* apparente
stone, final dressing of, before polishing   préparation *f*
   au poli
stone, natural   pierre *f* naturelle
stone, pitching   pierre *f* du hérisson
stone, polygonal facing   soleil *m*
stone, pumice   pierre *f* ponce
stone, quarry   moellon *m*
stone, reconstituted   pierre *f* reconstituée
stone, reconstituted, based on crushed basalt
   basaltine *f*

stone, riven   pierre *f* tranchée
stone, rubble, *see* rubble stone
stone, sharpening   pierre *f* à affûter
stone, small pieces of building   carottin *m*
stone, soft, from the Paris area   lambourde *f*
stone, thin slab or sheet of   pierre *f* pelliculaire
stone, toothing   harpe *f*, pierre *f* d'attente
stones, broken   pierraille *f*
stones, loose   rocaille *f*
stone (chippings) content   teneur *f* en gravillons
stone cut parallel its bedding plane   passe *f*
stone cutter or finisher   tailleur *m* de pierre(s)
stone depth perpendicular to wall face   queue-de-
   moellon *f*
stone elements of flat or curved arch   clavage *m*
stone exposed on both faces of a wall   parpaing *m*
stone finish, spalled   parement *m* smillé
stone sink   pierre *f* d'évier
stone that breaks up when one tries to dress it   poul *m*
stone-fall   chute *f* de pierre(s)
stonemason   tailleur *m* de pierre(s)
stoneware   faïence *f*, grès *m*
stoneware, glazed   grès *m* flambé
stonework   maçonnerie *f*
stonework, exposed   pierre *f* apparente
stop   butoir *m*, ergot *m*
stop [halt]   arrêt *m*
stop, depth   blocage *m*/butée *f* de profondeur
stop, door   butoir *m* de/d'une porte
stop end [for a pipe]   raccord *m* bouchon
stop end [in formwork]   coffrage *m* d'arrêt
stop rod [for door, shutter, etc, in non-rebated opening]
   battement *m*
stop   arrêter *vb*
stop the flow [of a liquid]   étancher *vb*
stop up [cracks, joints, etc]   calfeutrer *vb*, colmater *vb*
stop up [a hole, etc]   taper *vb*
stop up [a hole, a pipe, etc]   boucher *vb*
stopcock   robinet *m* d'arrêt/de fermeture
stopcock key   clef *f* de robinet
stoping   abattage *m* en gradins
stoppage [blockage]   engorgement *m*, obstruction *f*
stoppage [of machinery]   interruption *f* (de marche)
stoppage [of machinery, work, etc]   arrêt *m*
stopped up   bouché *adj*, obturé *adj*
stopper   bouchon *m*, obturateur *m*, tampon *m*
stopping [of machinery, work, etc]   arrêt *m*
stopping [parking of vehicles]   stationnement *m*
stopping [with lead]   plombage *m*
stopping of flow of a liquid   étanchement *m*
stopping up [of holes, cracks or joints]   masticage *m*
storage   entreposage *m*, stockage *m*
storage [of wheeled vehicles]   garage *m*
storage, annual   emmagasinement *m* annuel
storage, cold [the process]   conservation *f* par le froid
storage, dead [water]   eau *f* morte
storage, heat   accumulation *f* de chaleur, stockage *m*
   thermique/de chaleur
storage, rubble   stockage *m* de décombres

storage, seasonal emmagasinement *m* saisonnier	**straightenable** dépliable *adj*
storage, thermal entreposage *m* thermique, stockage *m* thermique	**straightening** redressement *m*
	straightening [e.g. of a road] rectification *f*
storage, waste stockage *m* de déchets	**straightening** [of a component, a piece of material] dressage *m*
storage, water accumulation *f*/emmagasinement *m* d'eau	
storage, water [by damming] retenue *f* d'eau	**straightening** [of a piece of wood, metal, etc] dégauchissage *m*
storage area aire *f* de stockage, zone *f* de stockage et d'entreposage	**strain** [deformation] déformation *f*
	strain, plane déformation *f* plane
storage bin silo *m* de stockage	**strain, shear** déformation *f* de cisaillement
storage calorifier ballon *m* à chauffage mixte	**strain, yield** déformation *f* à la limite d'écoulement
storage capacity, effective water eau *f* utilisable	**strain at failure** déformation *f* à la rupture
storage space [cupboards, shelving] rangement *m*	**strain control** contrôle *m* des déformations
storage unit élément *m* de rangement	**strain ellipse** ellipse *f* des déformations
storage yard parc *m*	**strain rate** vitesse *f* de déformation
storage, put into entreposer *vb*	**strained beyond the elastic limit** déformé *adj* au delà de la limite élastique
store [depot] parc *m*	
store [depot, warehouse] entrepôt *m*	**strainer** [at foot of suction pipe] crépine *f*
store [stock or reserve] provision *f*, réserve *f*	**strainer** [e.g. for wire fencing] tendeur *m*
store, cold entrepôt *m* frigorifique	**strainer** [for fluids] filtre *m* à tamis
store, department grand magasin *m*	**strainer** [for sewage] dégrilleur *m*
store, heat accumulateur *m* de chaleur	**strainer, drinking water** crépine *f* eau potable
store [documents] conserver *vb*	**strainer, emergency water** crépine *f* incendie
store [heat, electricity, etc] accumuler *vb*, emmagasiner *vb*	**strainer, inlet** crépine *f* d'entrée
	strainer, pipeline filtre *m* de la canalisation
store [in computer memory] mémoriser *vb*	**strainer, suction** crépine *f*
store [put into] entreposer *vb*	**strainer check valve** clapet *m* crépine
storehouse, storeroom réserve *f*	**strand** [of a cable] brin *m*, toron *m*
storekeeper magasinier *m*	**strand** [of rope] cordon *m*
storeroom [for food and wine] cellier *m*	**strand, cable** toron *m* de câble
storey étage *m*	**strap** [connection] lien *m*
storey, attic étage *m* d'attique	**strap** [leather, for carrying glass] bricole *f*
storey, basement étage *m* en sous-sol	**strap** [of a strap hinge] penture *f*
storey height hauteur *f* d'étage	**strategy, control** stratégie *f* de contrôle
storey in the roof space étage *m* lambrissé	**strategy, fire protection** stratégie *f* de protection *f* contre l'incendie
storey of uniform height étage *m* carré	
storing, *see* storage	**stratification** stratification *f*
storm orage *m*, tempête *f*	**stratification, air** stratification *f* de l'air
stormwater *see* rainwater	**stratified** stratifié *adj*
stormwater drainage assainissement *m* pluvial, canalisation *f* des eaux de pluie	**stratify** stratifier *vb*
	stratigraphy stratigraphie *f*
stormwater overflow déversoir *m* d'eau de pluie	**stratum** couche *f*, strate *f*
stormwater sewer égout *m* pluviale, collecteur *m* (des eaux) pluviale(s), conduite *f* des eaux pluviales	**straw** paille *f*
	streak [line] trait *m*
	stream cours *m* d'eau, ruisseau *m*
stove poêle *m*	**stream** [current or of traffic] courant *m*
stove [cooking] poêle *m* de cuisine	**stream** [current, of water] courant *m* d'eau
stove [portable] réchaud *m*	**stream** [of liquids] s'écouler *vb*
stove, cooking cuisinière *f*, poêle *m* de cuisine	**stream, converging/diverging** [traffic] courant *m* convergent/divergent
stove, gas fourneau *m*/poêle *m* à gaz	
stove, heating poêle *m* de chauffage	**stream, mountain** torrent *m*
stove, heating [enclosed in masonry] fourneau *m* de construction	**stream, through-traffic** courant *m* principal (de trafic)
	stream, with the au fil de l'eau
stove, oil poêle *m* à mazout	**stream(bed), dry** ruisseau *m* à sec
stove, portable réchaud *m*	**streamline** ligne *f* de courant
stove, slow-burning poêle *m* à feu continu	**streamlined** aux lignes *f,pl* aérodynamiques, profilé *adj*
stove, wood(-burning) poêle *m* à bois	
stow (away) ranger *vb*	**street** rue *f*
straight droit *adj*, rectiligne *adj*	**street, pedestrianized** rue *f* piétonne
straighten [an alignment, etc] rectifier *vb*	**street, shopping** rue *f* commerçante
straighten [piece of wood, metal] dégauchir *vb*	
straighten out or up redresser *vb*	

street drain   caniveau *m*
street furniture   mobilier *m* urbain
street gully   bouche *f* d'égout (BE)
street lamp/light   réverbère *m*
street level   rez-de-chaussée *m*
street (name) sign   plaque *f*
strength   force *f*, propriété *f* mécanique, résistance *f*
strength, adhesive   adhérence *f*, force *f* de collage
strength, bending   résistance *f* à la flexion
strength, bonding   facteur *m* d'adhérence
strength, breaking   limite *f* de rupture, résistance *f* à la rupture
strength, buckling   résistance *f* au flambement
strength, bursting   résistance *f* à l'éclatement
strength, cleavage   résistance *f* au clivage
strength, compressive   résistance *f* à la compression
strength, crushing   résistance *f* à l'écrasement
strength, cube   résistance *f* sur cube
strength, dielectric   rigidité *f* diélectrique
strength, fatigue   résistance *f* aux efforts alternés
strength, impact   résistance *f* au choc
strength, residual   résistance *f* résiduelle
strength, shear   résistance *f* au cisaillement
strength, tensile   résistance *f* à la traction/en traction
strength, yield   limite *f* élastique
strength class   classe *f* de résistance
strength of bond   force *f* de liaison
strengthen   renforcer *vb*
strengthen [a building, etc]   conforter *vb* [un bâtiment, etc]
strengthen [cracked or unstable work]   reprendre *vb*
strengthen [foundations]   consolider *vb*
strengthen [ground with piles]   palifier *vb*
strengthen [reinforce]   renforcer *vb*
strengthening   renforcement *m*
strengthening [of foundations]   consolidation *f*
strengthening [by replacing blocks of material, e.g. stone]   reprise *f* par encastrement/incrustement
stress   contrainte *f*, effort *m*, taux *m* de travail
stress, allowable   contrainte *f* admissible
stress, alternating   contrainte *f*/sollicitation *f* alternée, effort *m* alterné/pulsatif/pulsatoire
stress, applied   contrainte *f* appliquée
stress, axial   contrainte *f* normale, effort *m* axial/centré
stress, bearing   pression *f*/réaction *f* d'appui
stress, bending   contrainte *f* de flexion
stress, bond   contrainte *f* d'adhérence
stress, breaking or tearing   contrainte *f* de rupture
stress, buckling or collapse   contrainte *f* d'affaissement
stress, compression/compressive   contrainte *f* de compression, effort *m* compression
stress, creep   contrainte *f* de fluage
stress, critical   contrainte *f* critique
stress, effective   contrainte *f* effective
stress, extreme fibre   contrainte *f* au bord
stress, failure   contrainte *f* de rupture
stress, fatigue [repeated]   effort *m* de fatigue
stress, flexural   contrainte *f* de flexion
stress, horizontal   contrainte *f* horizontale

stress, internal   contrainte *f* propre, tension *f* interne
stress, maximum   effort *m* maximum
stress, normal   contrainte *f* normale
stress, permissible   effort *m* admissible, contrainte *f* permissible
stress, principal   contrainte *f* principale
stress, pulsating   effort *m* pulsatif/pulsatoire
stress, repeated (fatigue)   effort *m* de fatigue
stress, residual   contrainte *f*/tension *f* résiduelle, effort *m* résiduel
stress, secondary   contrainte *f*/effort *m* secondaire
stress, shear   contrainte *f* de cisaillement
stress, shear(ing)   contrainte *f* tangentielle
stress, shrinkage   contrainte *f* de retrait
stress, state of   état *m* de contrainte/tension
stress, static   effort *m* statique
stress, temperature   contrainte *f* thermique
stress, tensile   contrainte *f*/effort *m* de traction/tension, tension *f*, sollicitation *f* en traction
stress, thermal   tension *f* d'origine thermique, contrainte *f* thermique
stress, torsional   effort *m* de torsion
stress, triaxial   contrainte *f* triaxiale
stress, ultimate   contrainte *f* limite, limite *f* de rupture
stress, ultimate tensile   effort *m* limite de traction
stress, working   contrainte *f* de service/permissible
stress, yield   contrainte *f* à la limite d'écoulement
stress analysis   détermination *f* des efforts, vérification *f* des contraintes
stress concentration   concentration *f* des contraintes
stress crack, thermal   crique *f* due aux tensions thermiques
stress diagram   diagramme *m* des contraintes
stress dispersion/dissipation   dispersion *f*/dissipation *f* des contraintes
stress distribution   distribution *f*/répartition *f* des contraintes
stress loss   chute *f* de tension
stress relief   libération *f* de tension
stress relief by annealing   recuit *m* de normalisation/de détente
stress   solliciter *vb*
stress–strain curve   courbe *f* contrainte–allongement/contrainte–déformation
stress–strain diagram   diagramme *m* contrainte–déformation/effort–déformation
stretch [of ground]   bande *f*, étendue *f*
stretch [of road or route]   tronçon *m*
stretch [of time]   période
stretch [become longer]   s'allonger *vb*
stretch [become wider]   s'élargir *vb*
stretch [elastically or plastically]   étirer *vb*
stretch [extend, make wider or longer]   tendre *vb*
stretch [funds, resources]   faire *vb* durer
stretcher   panneresse *f*
stretcher, cut stone   carreau *m*
stretcher, stonework   carreau *m*
stretching [lengthening]   allongement *m*
stretching [widening]   élargissement *m*

strike [withdrawal of labour]   grève *f*
strike [geological]   direction *f*
strike plate of a lock   gâche *f*
strike [an arc]   amorcer *vb*
strike [arch or vault formwork]   décintrer *vb*
strike [concrete formwork]   décoffrer *vb*
striking [of an arc]   amorçage *m*
striking face [of a hammer]   tête *f* de frappe
striking of arch or vault formwork   décintrage *m*
striking of concrete formwork   décoffrage *m*
string [of a staircase]   limon *m*
string [used for lining]   cordeau *m*
string, close   limon *m* à la française
string, cut or open   limon *m* à crémaillère/à l'anglaise
string, false, applied to edge of a staircase   fausse crémaillère *f*
string, outer   limon *m* apparent
string, stair   limon *m* (d'escalier)
string, stone [part which supports treads]   parpaing *m* d'échiffre
string, wall   contre-limon *m*, faux limon *m*
string, wooden spiral staircase   courbe *f* rampante
string piece [in staircase]   sablière *f*
string wall   échiffre *f*
stringer [staircase], *see* string
stringer [structural, not staircase]   longeron *m*
strip [e.g. of wallpaper]   lé *m*
strip [of a material or slab]   bande *f*
strip [of a material]   barrette *f*
strip [of wood]   baguette *f*, tasseau *m*
strip 60–80mm wide of flooring   lame *f*
strip, bimetallic   bilame *f*
strip, connecting   barrette *f* de raccordement
strip, cover   bande *f* de recouvrement, calfeutrement *m*, couvre-joint *m*
strip, cover [over joint or crack]   parclause *f*, parclose *f*
strip, cover [wiring]   baguette *f*
strip, draught   bande *f* de calfeutrement, bourrelet *m*, colmatage *m*
strip, dust   bande *f* de poussière
strip, earthing terminal   bornier *m* de terre
strip, edging   bande *f* de chant
strip, edging [on a door]   couvre-chant *m*
strip, grassed   bande *f* engazonnée
strip, sealing   garniture *f* d'étanchéité
strip, small, of metal or plastic   lamelle *f*
strip, terminal   barrette *f* de connexion
strip, thin, flat metallic   méplat *m*
strip, thin, of wood or metal   lame *f*
strip, threshold   bande *f*, barre *f* de seuil
strip of road surface next to the kerb forming gutter   contre-bordure *f*
strip of wood to be machined for flooring   lame *f* de parquet
strip [a cable or wire of insulation]   dénuder *vb*
strip [a cable, rendering from a wall, etc]   dépouiller *vb*
strip [empty, a house, etc]   vider *vb*
strip [paint]   décaper *vb*
strip [rendering from a wall]   plumer *vb*

strip, resilient [under a partition, to allow for structural movement]   semelle *f* résiliente
strip of wood used to disguise a crack, etc   flipeau/flipot *m*
strip down [a machine etc]   démonter *vb*
strip off paint with acid   mordre *vb*
strip the thread of   arracher *vb* le filet de
striper [road marking]   traceuse *f* de lignes
stripper [for glued-on surfacings, carpets, etc]   décolleuse *f*
stripper, paint   décapant *m*
stripper, steam [for wallpaper]   décolleuse *f* (à vapeur)
stripper(s), wire   dénude-fil *m*, pince *f* à dénuder
stripping [paint, topsoil]   décapage *m*
stripping of paint or wallpaper by scraping   grattage *m*
stroke [of a jack, a piston, etc]   course *f*
strong   fort *adj*, résistant *adj*, solide *adj*
strongroom   chambre *f* forte
structural masonry [main walls, etc]   gros ouvrages *m,pl*
structural shape   profil *m*
structural steel workshop   atelier *m* de constructions métalliques
structurally participating precast concrete plank   prédalle *f* participante
structure   construction *f*, ouvrage *m*, structure *f*
structure [framework]   ossature *f*
structure, accessory/auxiliary   ouvrage *m* accessoire
structure, honeycomb   structure *f* en nid d'abeilles
structure, internal   structure *f* interne
structure, loadbearing   construction *f* porteuse
structure, permanent   construction *f* en dur
structure, reinforced concrete   ossature *f* béton armé
structure, road   structure *f* de chaussée
structure, soil   structure *f* du sol
structure, steel   ossature *f* métallique
structure, supporting   ouvrage *m* de soutènement, système *m* porteur
structure, temporary   construction *f* provisoire
structure, transfer   structure *f* de transfert
structures, analysis of   calcul *m* de structures
structures, engineering   ouvrages *m,pl* d'art
structures, existing [when starting construction of new]   restes *m,pl* d'ouvrages
structure and envelope, building   gros œuvre *m*
structure resisting ground pressure   soutènement *m*
structure rising above main body of a building   pavillon *m*
structure supporting a floor   gîtage *m*
strut, *see also* prop, stay   barre *f* comprimée, cale *f*, contre-fiche *f*, diagonale *f*, jambe *f* de force, portant *m*
strut [between joists]   étrésillon *m*
strut [supporting a wall]   bracon *m*
strut [small vertical]   potelet *m*
strut [solid, between floor joists]   lierne *f*
strut [steel roof]   bielle *f*
strut, shuttering   chandelle *f*
strut, small vertical   potelet *m*
strut, vertical, in a roof truss   poteau *m*
strut of a beam   décharge *m* d'une poutre
strut supporting a purlin or principal rafter   bielle *f*

struts used during underpinning   chevalement *m*

strut, *see also* prop, stay   contre-ficher *vb*, entretoiser *vb*,
étayer *vb*, étrésillonner *vb*

strut prior to underpinning   chevaler *vb*

strutted [of an excavation]   blindé *adj*

strutting [the action]   étrésillonnement *m*

strutting, herringbone [between timber floor joists]
croix *f* de Saint-André

strutting between joists   étrésillonnage *m*

strutting used during underpinning   chevalement *m*

stubble [of crops]   chaume *m*

stucco   crépi *m*, stuc *m*

stuck [jammed]   coincé *adj*

stud   clou *m*

stud [generally threaded]   boulon *m*

stud [timber structure]   colombe *f*, montant *m*, poteau *m*,
tournisse *fm*

stud, corner   poteau *m* cornier/d'angle

stud, intermediate   poteau *m* de remplissage

stud, reflecting or reflective   plot *m* (rétro-)
réfléchissant/réflectorisé

stud, small vertical   potelet *m*

stud, support(ing)   poteau *m* de décharge

stud bolt   goujon *m*, prisonnier *m*, tige *f* filetée

stud gun   pistolet *m* de scellement à cartouches
explosives

stud timber with nails [as key for abutting masonry]
larder *vb*

studwork [timber]   colombage *m*

study   étude *f*

study [room]   bureau *m*, cabinet *m* d'étude

study, case   étude *f* de cas

study, design   étude *f* de/du projet

study, environmental   étude *f* d'impact

study, feasibility   avant-projet *m*, étude *f* d'avant-
projet/de faisabilité

study, final design   avant-projet *m* détaillé (APD)

study, hydrological   étude *f* hydrologique/de bassins
hydrologiques

study, impact   étude *f* d'impact

study, market   étude *f* de marché

study, pilot or preparatory   avant-projet *m*, étude *f*
d'avant-projet

study, pre-design   étude *f* d'avant-projet

study, pre-feasibility   étude *f* de pré-faisibilité

study, pre-investment   étude *f* de pré-investissement

study, preliminary   avant-projet *m*, étude *f*
préliminaire/d'avant-projet, reconnaissance *f*

study, preliminary design   avant-projet *m* sommaire (APS)

study, preparatory   étude *f* d'avant-projet

study of geothermal phenomena   géothermie *f*

study plan or programme   plan *m* d'études

studies, design   études *f,pl* de conception

study   étudier *vb*, faire *vb* des études

stuff [a gland]   garnir *vb*

stuff (full)   bourrer *vb*

stuffing [of a gland, etc]   bourrage *m*, garnissage *m*

stuffing-box   presse-étoupe *m*

stump [of a tree]   souche *f*

sturdiness   solidité *f*

sturdy   solide *adj*

stylobate [architectural]   stylobate *m*

sub-base   sous-couche *f*

sub-base (course) [road construction]   couche *f* de
fondation/de forme, sous-couche *f* de base

sub-contract   sous-traiter *vb*

(sub-)contract, labour-only   louage *m*

sub-contracting   sous-traitance *f*

sub-contractor   entreprise *f* sous-traitante, sous-traitant *m*

sub-contractor, labour-only   entreprise *f* intervenante,
tâcheron *m*

sub-contractor, minor works   tâcheron *m*

subdivision plan [showing layout of plots]   plan *m*
parcellaire

subdued [light]   tamisé *adj*

sub-floor   sous-plancher *m*

sub-frame   contre-cadre *f*

sub-frame [of an opening, prefabricated or cast in situ]
précadre *m*

sub-grade [formation]   terrain *m*/sol *m* de fondation

sub-grade [of a roadway]   sous-couche *f*

subject   sujet *m*

subject [e.g of a report, study]   objet *m*

subject to tax(ation)   imposable *adj*

sublease or sublet [land]   sous-affermer *vb*

sublease or sublet [property]   sous-louer *vb*

subleasing or subletting   sous-location *f*

sublet [a contract]   sous-traiter *vb*

submerge   submerger *vb*

sub-meter   compteur *m* divisionnaire

submission   soumission *f*

submit a bid or tender   faire *vb* une soumission

subscription   abonnement *m*

subscription [social security]   cotisation *f*

subsequent   ultérieur *adj*

subside   s'affaisser *vb*

subside [ground, a structure, etc]   se tasser *vb*

subside, cause to   affaisser *vb*

subsidence   abaissement *m*, effondrement *m*, subsidence
f, tassement *m*

subsidence [ground]   affaissement *m*

subsidence, mining   subsidence *f* minière

subsidence, unequal [of supports, etc]   dénivellation *f*

subsidiary (company)   filiale *f*

subsidize   subventionner *vb*

subsidy   prime *f*, subvention *f*

subsoil   sol *m*/terrain *m* de fondation, sous-sol *m*

substance, flammable   substance *f* inflammable

substance, oxidizing   comburant *m*, substance *f* comburante

substation   sous-centrale *f*, sous-station *f*

substitute   produit *m* de substitution

substitute (for)   substituer *vb* (à)

substitution (of, for)   substitution *f* (de, à)

substrate [for paint, rendering]   subjectile *m*

substratum   sous-couche *f*

substratum, impermeable   substratum *m* imperméable

substructure   infrastructure *f*, substructure *f*

substructure [road construction]   couche *f* de base

subtenant   sous-locataire *m/f*
subterranean   souterrain *adj*, subterrané *adj*
suburb   banlieue *f*, faubourg *m*
suburban line (railway)   voie *f* de banlieue
subway   passage *m* inférieur (PI)/souterrain, souterrain *m*
succeed [of business, project, etc]   réussir *vb*
succession [of doors, columns, etc]   enfilade *f*
successive layers   couches *f,pl* successives
suck in or up [water]   aspirer *vb*
suction   aspiration *f*, succion *f*
suction disc or cup [plumbing]   ventouse *f*
suction head or lift   hauteur *f* d'aspiration
suction port   orifice *m* d'aspiration
suction side [of a fan, etc]   côté *m* d'aspiration
suffix [mathematical]   indice *m*
sugar soap [French equivalent]   lessive *f* St Marc
suit   convenir *vb* à
suitable   convenant *adj*
sulphate   sulfate *m*
sulphate bearing groundwater   eau *f* séléniteuse
sulphur   soufre *m*
sulphur dioxide   anhydride *m* sulfureux
sum [amount of money]   montant *m*
sum [of figures, money]   somme *f*
sum, lump   forfait *m*, somme *f* forfaitaire
summary   relevé *m*, résumé *m*, sommaire *m*
summary pre-project [scheme stage design/report]   avant-projet *m* sommaire (APS)
summerhouse   belvédère *f*, pavillon *m*/kiosque *m* de jardin
summit [highest point]   point *m* culminant
summit [of a roof, hill]   faîte *f*, sommet *m*
summons [writ]   assignation *f*
sump   puisard *m*
sump [of a machine]   pot *m* de décantation
sump [pit]   fosse *f*
sump, drain(age)   puisard *m* de relevage
sump, firemen's   puisard *m* incendie
sump for liquid manure   fosse *f* à purin
sun   soleil *m*
sun blind   abat-jour *m*, jalousie *f*
sun lounge   véranda *f*
sun roof   toiture-terrasse *f*
sun terrace   solarium *m*
sunbeam   rayon *m* de soleil
sundial   cadran *m* solaire
sundial blade   gnomon *m*
sunken   enfoncé *adj*
sunny   ensoleillé *adj*
sunshade   brise-soleil *m*, pare-soleil *m*
sunshine period   période *f* d'ensoleillement
supercooling   surfusion *f*
super-elevation [of a road]   dévers *m*
superheat   surchauffe *f*
superheat, degree of   degré *m* de surchauffe
superheat   surchauffer *vb*
superheated   surchauffé *adj*
superheater   surchauffeur *m*
superheating   surchauffe *f*

superposition, principle of   principe *m* de superposition
supersaturate   sursaturer *vb*
supersaturated   sursaturé
supersaturation   sursaturation *f*
superseded   périmé *adj*
superstructure   superstructure *f*
supervise   contrôler *vb*, surveiller *vb*
supervising authority   autorité *f* de surveillance
supervision   contrôle *m*, surveillance *f*
supervision, constant/continuous   surveillance *f* permanente
supervision, general work   contrôle *m* général des travaux (CGT)
supervision of the works   direction *f* des travaux
supplier   fournisseur *m*
supply [of gas, water, electricity, etc]   alimentation *f*
supply [of goods]   approvisionnement *m*, fourniture *f*
supply, air   apport *m* d'air, soufflage *m*
supply, electricity/power, *see* power supply
supply, emergency   alimentation *f* de secours/de sécurité
supply, gas   alimentation *f* en gaz
supply, stabilized [electrical]   alimentation *f* stabilisée
supply, uninterruptible   alimentation *f* sans coupure
supply, water, *see* water supply
supply of air   apport *m* d'air
supply of energy   apport *m* d'énergie
supply [furnish]   fournir *vb*
supply [with electric power, water, gas, etc]   alimenter *vb*
supply-air system   système *m* de soufflage d'air
supplying [furnishing]   fourniture *f*
support, *see also* supporting   appui *m*, soutènement *m*, soutien *m*, support *m*
support, bearing   support *m* de soutien
support, built-in   appui *m* encastré
support, pin-joint(ed)   appui *m* articulé
support, pipe   support *m* de tube/tuyau
support, simple   appui *m* libre
support bracket   patte *f* de support
support frame [fixed into masonry, to support body of a chimney, etc]   chaise *f*
support leg [for trailer, mobile crane, etc]   béquille *f*
support moment   moment *m* sur appuis
support point   point *m* d'appui
support reaction   réaction *f* d'appui
support tower [for cableway]   pylône *m*
support   appuyer *vb*, étayer *vb*, maintenir *vb*, soutenir *vb*, supporter *vb*
support [equipment]   empatter *vb*
support a load   supporter *vb* une charge
support with props   accorer *vb*
support with props or shoring   étançonner *vb*
supported (on, against)   appuyé *adj* (sur, contre)
supported on four sides   appuyé *adj* sur quatre côtés
surcharge (loading)   surcharge *f*
surface *adj*   superficiel *adj*
surface   superficie *f*, surface *f*
surface [flat space]   aire *f*
surface, bearing   surface *f* portante/d'appui
surface, bearing [of a joist or beam]   tablette *f*

surface, contact   surface *f* de contact
surface, cooling   surface *f* de refroidissement
surface, datum   surface *f* de référence
surface, formed [of concrete]   parement *m*
surface, grassed   surface *f* engazonnée
surface, heating   surface *f* chauffante/de chauff(ag)e
surface, inner (wall)   paroi *f* interne
surface, internal [an arch or vault]   intrados *m*
surface, loose [of ground]   foisonnement *m*, surface *f* foisonnée
surface, machined   surface *f* usinée
surface, plane   surface *f* plane
surface, radiant   surface *f* de rayonnement
surface, radiator   surface *f* de rayonnement
surface, rough   surface *f* rugueuse
surface, running   couche *f* de roulement
surface, shear   surface *f* de cisaillement
surface, sliding/slip   surface *f* de glissement
surface, water   surface *f* de l'eau
surface, work   plan *m* de travail
surface layer, removal of   décapage *m*
surface layer, road   couche *f* de roulement
surface preparation for rendering by light picking hachement *m*
surface tension   tension *f* superficielle
surface water run-off   eaux *f,pl* de ruissellement/de surface
surface waterproofing compound   hydrofuge *m* (de surface)
surface   faire *vb* surface
surface [a facade, stonework, etc]   ravaler *vb*
surface [a road]   revêtir *vb*
surface [polish]   polir *vb*
surface [prepare for painting, etc]   apprêter *vb*
surface [rough plane]   dégauchir *vb*
surfacer or surface planer   machine *f* à corroyer/à dresser
surfacing [smoothing of a slab]   surfaçage *m*
surfacing [filler]   enduit *m* de ragréage
surfacing [of a facade, stonework, etc]   ravalement *m*
surfacing [of wood]   dégauchissage *m*
surfacing [preparation for painting, etc]   apprêtage *m*
surfacing [rendering, cladding, etc]   revêtement *m*
surfacing [roads, etc]   finition *f* de surface, revêtement *m*
surfacing, anti-skid   revêtement *m* antidérapant
surfacing, asphaltic/bituminous   revêtement *m* bitumineux
surfacing, bituminous bound   couche *f* de surface bitumineuse
surfacing, carriageway   revêtement *m* de chaussée
surfacing, footpath   revêtement *m* du trottoir
surfacing, multi-layer waterproof   multicouche[†] *f*
surfacing, pavement (footpath)   revêtement *m* du trottoir
surfacing, road   revêtement *m* de chaussée
surfacing, watertight [of a dam]   masque *f* étanche
surfacing machine   machine *f* à dresser/à corroyer
surfacing resembling stone [of plaster, lime and sundry aggregates]   métalline *f*

surfacing with chippings or grit spread on tar gravillonnage *m*
surge, lightning   choc *m* de foudre
surge, voltage   onde *f* de surtension
surge diverter, lightning   parafoudre *m*
surge tank [for pressure regulation]   cheminée *f* d'équilibre, réservoir *m* d'égalisation de pression
surge tank [anti-water hammer]   réservoir *m* d'air anti-bélier
surgery   infirmerie *f*
surging [in a pipe]   contre-foulement *m*
surmount   surmonter *vb*
surmounted (by)   surmonté *adj* (de)
surplus   surplus *m*
surplus (of)   excès *m* (de)
surplus (over)   excédent *m* (sur)
surrounded (by) [e.g. walls]   ceint *adj* (de)
surrounding *adj*   environnant *adj*
surroundings   ambiance *f*, cadre *m*, environnement *m*, environs *m,pl*, milieu *m*
surroundings [approaches]   abords *m,pl*
surroundings, natural   milieu *m* naturel
survey   levé *m*, reconnaissance *f*
survey [independent, of work, materials, etc]   expertise *f*
survey [of quantities]   métré *m*
survey, aerial   levé *m* aérien
survey, cadastral   levé *m* cadastral
survey, dimensional, of a structure   relevé *m* de l'ouvrage
survey, field   étude *f* de terrain, levé *m* sur le terrain
survey, ground   levé *m* terrestre/de terrain
survey, infra-red   examen *m* infrarouge
survey, land, *see* land survey
survey, land tenure/ownership   étude *f* foncière
survey, photogrammetric aerial   levé *m* aéro-photogrammétrique
survey, preliminary   reconnaissance *f*
survey, site   étude *f* de reconnaissance du site
survey, soil, *see* soil investigation
survey, topographic(al)   étude *f*/levé *m* topographique
survey data   indications *f,pl* topographiques
survey mark   indication *f* topographique
survey pole   balise *f*
survey poles, setting up of   balisage *m*
survey station   signal *m* géodésique
survey   faire *vb* le levé (topographique) de
survey [a building]   examiner *vb*
survey [for land registration purposes]   cadastrer *vb*
survey [in quantity surveying]   métrer *vb*
survey [land]   arpenter *vb*
survey [work, materials, etc, independently]   expertiser *vb*
survey, carry out a [topological]   lever *vb* un plan
surveying   géodésie *f*, topographie *f*
surveying [levels]   nivellement *m*
surveying equipment   matériel *m* de topographie
surveying instruments   instruments *m,pl* topographiques
surveyor [independent expert]   expert *m*

surveyor, land arpenteur *m*, (expert-)géomètre *m*, topographe *m*

surveyor, quantity [near French equivalents] économiste *m/f*, expert *m* métreur, métreur *m*

surveyor, topographical topographe *m*

surveyor's level niveau *m*

surveyor's rod or staff mire *f*, jalon *m*

surveyor's rod with sliding target mire *f* à glissière

surveyor's square équerre *f* d'arpenteur

surveyor's telescope lunette *f* de visée, lunette *f* viseur

susceptibility (to) sensibilité *f* (à)

suspend suspendre *vb*

suspended suspendu *adj*

suspended [in a liquid] flottant *adj*

suspender [e.g. wire or rod for fibrous plaster work] suspente *f*

suspension [equipment supporting a lift cage] suspente *f*

suspension points points *m,pl* de suspension

swage [metals] emboutir *vb*

swallow-hole bétoire *f*

swamp marais *m*

swampy ground molets *m,pl*

swan-neck cou *m* de cygne, tuyau *m* coudé

swan-necked à col de cygne

sward gazon *m*

sweat [a pipe joint] ressuer *vb*

sweat [develop surface droplets] ressuer *vb*

sweat [soft solder] souder *vb* à l'étain

sweating [e.g. of a damp wall] suintement *m*

sweating of fresh concrete [laitance] ressuage *m*

sweep, chimney ramoneur *m*

sweep [a chimney or flue] ramoner *vb*

sweep [ground, a floor, etc] balayer *vb*

sweep (across) [of light] balayer *vb*

sweeper, road [machine] balayeuse *f*

sweeping [of a chimney or flue] ramonage *m*

sweeping (out, up) balayage *m*

swell [increase in volume] foisonner *vb*

swell [inflate] gonfler *vb*

swelling [bump] bosse *f*

swelling [inflating] gonflage *m*

swelling [inflation] gonflement *m*

swelling of a baluster or column panse *f*

swelling of plasterwork due to saltpetre enflure *f*

swept path [of a vehicle] aire *f* de traînée

swimming bath or pool piscine *f*

swing or swinging [bridge, crane, etc] basculant *adj*

swirl remous *m*

swirl tourbillonner *vb*

switch [electrical] interrupteur *m*, inter[†] *m*

switch, automatic time interrupteur *m* automatique de réglage

switch, changeover commutateur *m*, interrupteur-inverseur *m*, inverseur *m* (de courant)

switch, commutator commutateur *m*

switch, control interrupteur *m* de contrôle

switch, cord-operated interrupteur *m* à tirette

switch, daylight operated interrupteur *m* crépusculaire

switch, door contacteur *m* de porte

switch, double pole interrupteur *m* à deux pôles, contacteur *m* bipolaire

switch, electric time minuterie *f*

switch, end-of-travel interrupteur *m* de fin de course

switch, float interrupteur *m* à flotteur

switch, flow [cut-out] disjoncteur *m* d'écoulement

switch, foot interrupteur *m* au pied

switch, isolating sectionneur *m*

switch, knife interrupteur *m* à couteau

switch, light-sensitive interrupteur *m* crépusculaire

switch, liquid level interrupteur *m* de mise à niveau de liquide

switch, magnetic interrupteur *m* magnétique

switch, main commutateur *m*/interrupteur *m* principal

switch, manual override [on a time switch] interrupteur *m* de marche forcée

switch, mercury interrupteur *m* à mercure

switch, micro- microrupteur *m*

switch, overload interrupteur *m* de surcharge

switch, override interrupteur *m* dérogation

switch, plug interrupteur *m* à fiche

switch, power [disconnector] délesteur *m*

switch, pressure déclencheur *m*, manostat *m*, pressostat *m*

switch, pull interrupteur *m* à tirette

switch, pushbutton interrupteur *m* à poussoir

switch, remote interrupteur *m* à distance, télérupteur *m*

switch, residual current interrupteur *m* différentiel

switch, rocker interrupteur *m* à bascule

switch, rotary/rotating interrupteur *m* rotatif

switch, safety interrupteur *m* de sécurité

switch, selector commutateur *m*

switch, shut-down arrêt *m*, interrupteur *m* automatique

switch, single-pole interrupteur *m* unipolaire

switch, starting interrupteur *m* de démarrage

switch, stop arrêt *m*

switch, thermal interrupteur *m* thermique

switch, time, *see* time switch

switch, time-lag interrupteur *m* temporisé

switch, two-way commutateur *m* de va-et-vient, interrupteur *m* va-et-vient

switch box boîte *f* de commande

switch off couper *vb*, déclencher *vb*

switch off [electrical circuit] mettre *vb* hors circuit/tension

switch off the current couper *vb* le courant

switch off the light(ing) couper *vb* la lumière

switch on enclencher *vb*

switch on [electrical circuit] mettre *vb* en circuit/sous tension

switch on [electrically] allumer *vb*, brancher *vb*

switch on [electricity, gas, etc] ouvrir *vb*

switch on [the light, lighting] allumer *vb*

switchboard tableau *m*, tableau *m* de distribution/de commande

switchboard, electrical tableau *m* électrique

switchboard, high voltage tableau *m* HT

switchboard, telephone standard *m*, table *f*

switched off [out of circuit]   hors circuit
switchgear   dispositif *m* de couplage
switching, local group   sélecteur *m* local groupé
switching off [of lighting]   extinction *f*
switching off or out   fermeture *f*
switching on   ouverture *f*
switching on [of electricity, lighting]   allumage *m*
switching period   période *f* de mise en circuit
swivel   tournant *adj*
swivel joint   joint *m* à rotule
swivel pin [of a gate]   tourillon *m*
swivel [sth]   faire *vb* pivoter/tourner
swivel   pivoter *vb*, tourner *vb*
swivelling [bridge, crane, etc]   basculant *adj*
swivelling   pivotement *m*
syenite   syénite *f*
symbol   signe *m*, symbole *m*
symbol, conventional   symbole *m* de représentation, signe *m* conventionnel
symmetrical   symétrique *adj*
symmetry   symétrie *f*
symptom   indice *m*, signe *m*
synchronization   synchronisation *f*
syncline   synclinal *m*
synthetic rubber (neoprene)   néoprène *m*
system   système *m*
system [administrative]   régime *m*
system [pipes, cables etc]   réseau *m*
system, addressable fire detecting   localisateur *m* d'incendie
system, administrative   régime *m*
system, alarm   système *m* d'alarme
system, all air   système *m* tout air
system, automatic   système *m* automatique
system, back-up   réseau *m* de réserve
system, changeover   système *m* de commutation
system, combined/separate drainage   système *m* d'assainissement unitaire/séparatif
system, communication   système *m* de communication
system, compression refrigeration   système *m* de réfrigération à compression
system, computer-based   système *m* piloté par ordinateur
system, construction   système *m* de construction
system, defrosting   système *m* de dégivrage
system, detection   système *m* de détection
system, diffusion-absorption   système *m* à absorption-diffusion
system, (direct) cooling   système *m* de refroidissement (direct)

system, direct expansion   système *m* à détente directe/d'expansion directe
system, direct expansion refrigeration   système *m* de réfrigération à détente directe
system, direct vent   système *m* de mise à l'atmosphère
system, distribution, *see* distribution system
system, drainage   ouvrage *m* d'assainissement, système *m* de drainage, réseau *m* d'évacuation/de drains
system, dry sprinkler   système *m* d'extinction à sec
system, dual duct   système *m* à deux conduits
system, duct   réseau *m* de conduits, système *m* de canalisation
system, ductwork   réseau *m* de distribution d'air
system, earthing   prise *f* de terre
system, exhaust   système *m* d'extraction d'air
system, feedback   système *m* de contre-réaction
system, fire detection   détection *f* incendie
system, fire sprinkler   système *m* d'extinction automatique
system, gravity   système *m* par gravité
system, heating/hot water/etc, *see* heating/hot water/etc
system, lubrication   système *m* de lubrification
system, mechanical smoke extract   système *m* mécanique de désenfumage
system, metering [in mixing plant]   doseur *m*
system, radiant cooling   système *m* de rafraîchissement par rayonnement
system, patented   méthode *f* brevetée
system, single-pipe   système *m* monotube
system, sliding   système *m* à glissière
system, supply   réseau *m* d'alimentation
system, supporting   système *m* porteur
system, total energy   système *m* d'énergie totale
system, transmission   organes *m,pl* de transmission
system, variable air volume (VAV)   système *m* à volume d'air variable (VAV)
system, ventilating   installation *f*/système *m* d'aération
system, water supply, *see* water supply
system, waterproofing   système *m* d'étanchéité *f*/d'imperméabilisation
system, wiring   système *m* de câblage
system breakdown   écroulement *m* d'un système
system for marking cables/wiring   système *m* de repérage pour filerie
system losses   pertes *f,pl* dans le réseau
systems analysis   analyse *f* fonctionnelle
systems analyst   analyste *m* fonctionnel
systems engineer   ingénieur *m* système
systems program   logiciel *m* de base

T

T-bar   fer *m* (en) T/té
T-beam   poutre *f* (en) T/té
T-beam (reinforced)   poutre *f* en T/té (en béton armé)
T-bevel   fausse équerre *f*
T-hinge   penture *f* anglaise
T-iron or section   fer *m*/profilé *m* en T/té
T-junction   jonction *f* en T/té
T-section   profilé *m* en T/té
T-square   té *m*
T/tee with cleaning eye   té *m* de purge
T/tee, reducing   té *m* de réduction
T/tee, regulating   té *m* de réglage
table [design]   abaque *m*
table [list, figures, etc]   barème *m*, tableau *m*
table, bedside   chevet *m*, table *f* de nuit
table, coffee or occasional   table *f* basse
table, design   abaque *m*
table, dressing   coiffeuse *f*, table *f* de toilette
table, drop-leaf   table *f* à abattants
table, extending   table *f* à rallonges
table, gate-legged   table *f* anglaise
table, paste/pasting   table *f* à encoller
table, plane   planche *f* de topographe
table, water, *see* water table
table of contents   table *f* de matières
table of technical requirements   tableau *m* de contraintes
tablet [of stone]   table *f*
tacheometer/tachymeter   tachéomètre *m*
tachometer   tachymètre *m*
tack   petit clou *m*, pointe *f*
tack [for wood, carpets, etc]   broquette *f*
tack [short nail]   semence *f*
tack, round-headed   bossette *f*
tack, upholstery   pointe *f* à damas
tackle [pulley block]   palan *m*
tailboard lift   hayon *m* élévateur
tailings   résidus *m,pl* (de mine)
tailrace   canal *m* de fuite
tailrace [of a water mill]   bief *m* d'aval
tailrace tunnel   galerie *f* de fuite
tailstock   poupée *f* mobile
tailwater   eau *f* aval
tailwater level [below a dam]   niveau *m* aval
take away   enlever *vb*
take care/charge of [deal with]   s'occuper *vb* de
take core samples   prélever *vb* des carottes
take fire   s'allumer *vb*
take notes   prendre *vb* des notes
take off its hinges [a door, shutter, etc]   dégonder *vb*
take off/out   retirer *vb*
take on [staff]   embaucher *vb*, engager *vb*
take out [an insurance policy]   souscrire *vb*

take samples   prélever *vb* des échantillons
take the strain off a beam   soulager *vb* une poutre
take up a load   reprendre *vb* une charge
take-off piece   tubulure *f* de dérivation/de raccordement
talus   éboulis *m*
talus cone   cône *m* d'éboulis
tambour [of a column, under a cupola or dome, etc]   tambour *m*
tamp (down)   bliner *vb*, damer *vb*, pilonner *vb*, tasser *vb*
tamper   dame *f*, pilon *m*
tamper, vibratory   dame *f*/pilonneuse *f* vibrante
tamper with   altérer *vb*
tamping (down)   damage *m*, pilonnage *m*, tassement *m*
tangential   tangentiel *adj*
tank   bac *m*, bâche *f*, cuve *f*, réservoir *m*
tank [of expansion unit]   vase *m*
tank, additional, back-up or reserve   réservoir *m* supplémentaire
tank, aeration   bassin *m* d'aération
tank, balance   réservoir *m* d'équilibre
tank, bitumen, with spraying equipment   citerne *f* à bitume avec rampe de répandage, citerne *f* thermique répandeuse de liants bitumineux
tank, blow-down   réservoir *m* de décompression
tank, brine   bac *m* à saumure
tank, coagulation   bassin *m* de coagulation
tank, collecting   réservoir *m* collecteur
tank, cooling water   réservoir *m* d'eau de refroidissement, rose *m* d'expansion fermé
tank, copper   citerne *f* de cuivre
tank, cylindrical   ballon *m*
tank, diaphragm [central heating system]   cumulus *m*
tank, dip   cuve *f* à immersion
tank, discharge   réservoir *m* de décharge
tank, drain   puisard *m*
tank, equalizer   réservoir *m* égalisateur
tank, expansion   cumulus *m*, réservoir *m* de dilatation, vase *m* d'expansion/de dilatation
tank, feed   bâche *f* alimentaire/d'alimentation, réservoir-nourrice *m*
tank, filter   réservoir *m* à filtration/d'épuration
tank, firemen's water   puisard *m* incendie
tank, flash   réservoir *m* pour autovaporisation
tank, flushing   réservoir *m* de chasse d'eau
tank, fuel   réserve *f* de carburants, réservoir *m* de carburant/de combustible
tank, fuel [of a vehicle]   réservoir *m* à carburant
tank, header   réservoir *m* en charge
tank, (hot) water   ballon *m* d'eau (chaude)
tank, membrane expansion   vase *m* de dilatation à membrane
tank, mixing   réservoir *m* de mélange

tank, oil   réservoir *m* de mazout
tank, oil [domestic]   cuve *f* à mazout
tank, (rain)water   citerne *f*
tank, reserve   réservoir *m* supplémentaire/de secours
tank, sealed sewage   fosse *f* étanche
tank, sedimentation   bac *m*/bassin *m* de décantation
tank, septic   fosse *f* septique
tank, steel   réservoir *m* en acier
tank, storage   citerne *f*/réservoir *m* de stockage
tank, suction   bâche *f* d'aspiration
tank, supply   réservoir *m* d'alimentation
tank, surge, *see* surge tank
tank, water   bac *m* à eau, citerne *f*
tank [i.e. to make waterproof]   aveugler *vb*
tap, *see also* valve, etc   robinet *m*
tap [electrical], *see* tap-off, tapping
tap [for tapping off water]   prise *f* d'eau
tap [for threading]   taraud *m*
tap, bleed(er)   robinet *m* de purge
tap, calibrated [used prior to advent of flow meters]
   robinet *m* jauge *f*
tap, combination [with hot and cold controls]
   mélangeur *m*
tap, drain   robinet *m* purgeur/de vidange
tap, draw-off   robinet *m* de puisage
tap, gas   robinet *m* de gaz
tap, gas stop [for a whole building]   robinet *m* chef (de
   gaz)
tap, main   robinet *m* principal
tap, mixer [twin tap, single spout]   robinet *m* mélangeur
tap, mixer [monobloc, lever operated]   mitigeur *m*
tap, stop   robinet *m* d'arrêt
tap, swan-neck   col *m* de cygne
tap, swivel   robinet *m* à bec orientable
tap, waste   robinet *m* purgeur
tap, water   robinet *m* d'eau
tap nozzle, anti-splash   brise-jet *m*
tap washer   clapet *m*, joint *m*/rondelle *f* de robinet
tap wrench   tourne-à-gauche *m*
tap [a thread]   tarauder *vb*
tap [electrical supply]   soutirer *vb*
tap [pick up electrical current]   capter *vb*
tap [strike lightly]   taper *vb*
tap off   brancher *vb*
tap-off (point) [electrical]   prise *f* de soutirage
tape, 10 metre [measuring]   décamètre *m*
tape, adhesive   ruban *m* adhésif
tape, cable   câble-ruban *m*
tape, fixing   ruban *m* de fixation
tape, insulating   ruban *m* isolant
tape, masking   papier *m* cache
tape, Teflon   ruban *m* de téflon, téflon *m* en ruban
tape measure   mesure *f* à ruban, mètre *m* (à) ruban
taper [highways]   biseau *m*
taper [pipework]   cône *m*
taper of formwork [to facilitate striking]   dépouille *f*
tapered   effilé *adj*
tapering [of a column, upwards]   contracture *f*
tapering [decreasing, reduction]   diminution *f*

tapering *adj* [to a point]   en pointe
tapped hole   trou *m* fileté/taraudé
tapping [a thread]   taraudage *m*
tapping [electrical supply]   soutirage *m*
tapping [for a sensor, a current take-off]   prise *f*
tapping [hydraulic, electrical]   branchement *m*
tapping [shunting of electrical current]   dérivation *f*
tapping, flowmeter   prise *f* de débit-mètre
tapping point [into a main pipe]   piquage *m*
tapping into a main pipe [action of]   piquage *m*
tar   goudron *m*
tar, coal [protective or waterproof coating]   coaltar *m*
tar, cold   goudron *m* fluide
tar, dense   goudron-bitume *m* (GB)
tar spraying machine   goudronneuse *f*
tar-concrete   béton *m* au goudron
target [surveying]   cible *f*, voyant *m*
target date   échéance *f*
target performance   performance *f* prévue
tariff   tarif *m*
tariff, maximum demand   tarif *m* de pointe
tariff, off-peak   tarif *m* (des) heures creuses
tar-like deposit in a flue or chimney   bistre *m*
tarmac [airport runway]   aire *f* d'envol, piste *f*
tarmacadam   macadam *m* goudronné, tarmacadam *m*
tarnish   tacher *vb*, ternir *vb*
tarnishing [metal surfaces]   ternissement *m*
tarpaulin   bâche *f*
tarred   goudronné *adj*
tar-sprayer   goudronneuse *f*
task [professional]   mission *f*
tax   impôt *m*, taxe *f*
tax [duty]   droit *m*
tax, income   impôts *m,pl* sur le revenu
tax, land/real estate   impôt *m* foncier
tax, letting [in France]   droit *m* de bail
tax, residential [in France]   taxe *f* d'habitation
tax, value added (VAT)   taxe *f* sur la valeur ajoutée
   (TVA)
taxes, land/real estate [in France]   taxes *f* foncières
tax code/rules [in France]   Code *m* Général des Impôts
   (CGI)
tax excluded   hors taxe (HT)
tax included   toutes taxes comprises (TTC)
tax office   centre *m*/hôtel *m* des impôts
tax on professional activity [in France]   taxe *f*
   professionnelle
tax on turnover   impôt *m* sur le chiffre d'affaires
tax situation   fiscalité *f*
tax [sth]   taxer *vb*
tax [somebody]   imposer *vb*
taxable   imposable *adj*
taxation   imposition *f*
tax-free [excluding VAT, etc]   hors taxe *adj* (HT)
taxiway [airfield]   voie *f* de circulation
teak   teck *m*
team   équipe *f*
tear [in fabric, etc]   déchirure *f*
tear [a fabric, etc]   déchirer *vb*

tear down [dismantle]   défaire *vb*
tear out/up [a plant, a tree]   arracher *vb*
tear up [old road surface, etc]   défoncer *vb*
technical   technique *adj*
technical advisory note [issued by CSTB in France]   avis *m* technique
technical appraisal [independent, of work, materials, etc]   expertise *f*
technical consultancy office [generally contractor owned]   bureau *m* d'études techniques (BET)
technical literature   documentation *f* technique
technical salesman   agent *m* technico-commercial
technical term   terme *m* de métier
technician   technicien *m*
technique   technique *f*
technology of chimneys, flues, fireplaces, stoves, etc   fumisterie *f*
tectonic quake   tremblement *m* tectonique
tee-, *see* T-
teeth of a toothed drag   brettelure *f*
Teflon tape   ruban *m* de téflon, téflon *m*
Teflon paste   pâte *f* de téflon
telecommunications   télécommunications *f,pl*
telefax   télécopie *f*
telephone *adj*   téléphonique *adj*
telephone   téléphone *m*
telephone, internal   interphone *m*
telephone exchange   central *m* téléphonique
telephone exchange, automatic   central *m* automatique
telephone exchange, local automatic   central *m* automatique urbain
telephone extension   poste *m* (téléphonique)
telephone handset   combiné *m* (téléphonique)
telephone network   réseau *m* téléphonique
telephone receiver   récepteur[†] *m*
telephone switchboard   standard *m*, table *f*
telescope [of a theodolite]   lunette *f*
telescope, sighting or surveyor's   lunette *f* viseur/de visée
television set   poste *m* de télévision, téléviseur *m*, télévision *f*
telltale *adj*   témoin *adj*
telltale   témoin *m*
temper [of steel]   coefficient *m* de dureté
temper [steel], lose   détremper *vb*
temperature [stress, etc]   thermique *adj*
temperature   température *f*
temperature, air or ambient   température *f* ambiante
temperature, apparent   température *f* apparente
temperature, average or mean   moyenne *f* des températures
temperature, balance point   température *f* d'équilibre
temperature, black body   température *f* de brillance
temperature, black body equivalent   température *f* équivalente de corps noir
temperature, boiling   température *f* d'ébullition
temperature, boundary layer   température *f* de la couche limite
temperature, combustion   température *f* de combustion

temperature, condensing   température *f* de condensation
temperature, continuous working   température *f* continue de service
temperature, critical   température *f* critique
temperature, delivery   température *f* délivrée
temperature, design   température *f* imposée
temperature, discharge   température *f* de décharge/de refoulement
temperature, dry bulb   température *f* sèche/au thermomètre
temperature, equilibrium   température *f* d'équilibre
temperature, evaporation   température *f* d'évaporation
temperature, flame   température *f* de la flamme
temperature, flow   température *f* de départ
temperature, flue (gas)   température *f* de fumée
temperature, flue or waste gas   température *f* des gaz brûlés
temperature, freezing   température *f* de congélation
temperature, globe   température *f* au thermomètre globe
temperature, ignition   température *f* d'allumage/d'ignition
temperature, initial   température *f* initiale
temperature, injection   température *f* d'injection
temperature, mean   moyenne *f* des températures
temperature, mixing   température *f* de mélange
temperature, operating   température *f* de fonctionnement/de service
temperature, outlet   température *f* de sortie
temperature, outside   température *f* extérieure
temperature, outside air   température *f* de l'air extérieur
temperature, radiant   température *f* radiante
temperature, return   température *f* de retour
temperature, room   température *f* intérieure du local
temperature, skin   température *f* superficielle
temperature, smoke   température *f* de la fumée
temperature, sol air   température *f* fictive extérieure
temperature, supply   température *f* d'alimentation
temperature, surface   température *f* de surface
temperature, wall   température *f* de paroi
temperature, waste gas   température *f* des gaz brûlés
temperature, wet bulb   température *f* humide
temperature, working   température *f* d'utilisation/de travail
temperature correction   correction *f* de température
temperature difference   écart *m* de température
temperature distribution   distribution *f* de la température
temperature drop   chute *f* de température
temperature gradient   gradient *m* de température
temperature inversion   inversion *f* thermique
temperature map   carte *f* thermique
temperature probe   sonde *f* de température
temperature range   domaine *m* de températures
temperature regulation system   système *m* de régulation de température
temperature regulator   thermorégulateur *m*
temperature rise   échauffement *m*, élévation *f* de la température
temperature rise, maximum   échauffement *m* limite

temperature sensor élément *m* sensible à la température
temperature set point point *m* de consigne de température
temperature variation variation *f* de température
tempering [of steel] revenu *m*
template calibre *m*, gabarit *m*
template [for an arch] cintre *m*
template, curved [for archways, etc] cerce *f*, cerceau *m*
template, spacing gabarit *m* d'écartement
template for circles/ellipses trace-circles/ellipses *m*
temporary provisoire *adj*, temporaire *adj*
tenant locataire *m*
tenant, co- or joint colocataire *m/f*
tender offre *f*, soumission *f*
tender, best offre *f* mieux-disant
tender, by par voie d'adjudication
tender, invitation to appel *m* d'offres
tender, lowest offre *f* moins-disant
tender, negotiated adjudication *f* libre, marché *m* de gré à gré/négocié
tender, open adjudication *f* publique
tender, public invitation to avis *m* d'appel public à la concurrence (AAPC)
tender, restricted adjudication *f* restreinte
tender analysis assistance assistance *f* au dépouillement des offres
tender awards or results résultats *m,pl* des marchés
tender documents dossier *m* de consultation (des entreprises) (DCE), dossier *m* d'appel d'offres
tender documents, preparation of établissement *m* des dossiers d'appel d'offres
tender invitation, public appel *m* d'offres public, avis *m* d'appel public à la concurrence (AAPC)
tender invitation, restricted appel *m* d'offres restreint
tender, put sth out to mettre *vb* qch en adjudication
tender, submit a faire *vb* une soumission, soumissionner *vb*
tenders, invite mettre *vb* en adjudication
tenderer soumissionnaire *m*
tenderer, successful adjudicataire *m*, bénéficiaire *m* de l'adjudication, entreprise *f* adjudicataire, titulaire *m* du marché
tenderers, invited candidats *m,pl* sélectionnés
tendering [i.e. putting out to] adjudication *f*
tendering [i.e. submitting] soumission *f*
tendering, competitive marché *m* par adjudication/sur appel d'offres
tendering, limited appel *m* d'offres restreint
tendering, open adjudication *f* publique
tendering, restricted adjudication *f* restreinte
tendon [stressing] câble *m*
tenon tenon *m*
tenon [large] mors *m*
tenon saw, back of a dosseret *m*
tenon empatter *vb*
tenoner fraiseuse *f* à queue, tenonneuse *f*
tenoning empattement *m*
tensiometer extensomètre *m*, tensiomètre *m*
tension tension *f*, traction *f*

tension, high haute tension *f*
tension, surface tension *f* superficielle
tensioner tendeur *m*
tensioning mise *f* en tension
tensor tenseur *m*
tenure jouissance *f*
tenure, security of droit *m* de renouvellement
ten-year *adj* décennal *adj*
ten-year civil liability [on major work, in France] responsabilité *f* civile décennale
ten-year guarantee décennale *f*, garantie *f* décennale
tepid tiède *adj*
term terme *m*
term [lapse of time] échéance *f*
terms [of a contract] conditions *f,pl*
terms of delivery/supply conditions *f,pl* de livraison
terminal [airport] terminal *m*
terminal [of an accumulator/battery] pôle *m*
terminal, airport freight/passenger aérogare *f* (de) fret/passagers
terminal, directional supply air bouche *f* de soufflage orientable
terminal, electrical borne *f*
terminal, fuse porte-fusible *m*
terminal, spare borne *f* de réserve
terminal block bloc *m* de jonction
terminal shield cache-borne *m*
terminal strip barrette *f* de connexion
terminal strip, earthing bornier *m* de terre
terminal tag on an electrical lead cosse *f*
terminate terminer *vb*
terminology terminologie *f*
termite termite *m*
terrace terrasse *f*
terrace [horizontal strip of ground] terre-plein *m*
terrace, roof plate-forme *f*, toiture-terrasse *f*
terrace, sun solarium *m*
terracotta ware poterie *f*
terrain terrain *m*
terrain classes [HVAC] classification *f* des terrains
terrain effects effets *m,pl* du terrain
terrazzo granito *m*, terrazzo *m*
territory territoire *f*
test, *see also* testing épreuve *f*, essai *m* (de contrôle), expérience *f*
test, acceptance essai *m* de recette
test, ball hardness billage *m*, essai *m* billage
test, bend(ing) essai *m* de pliage
test, boiler (output) essai *m* de rendement de chaudière
test, bond (strength) essai *m* d'adhérence
test, breaking essai *m* de cassure
test, Brinell hardness essai *m* de dureté Brinell
test, buckling essai *m* (de) flambement
test, bulging essai *m* d'élargissement
test, check contre-essai *m*
test, comparative essai *m* comparatif
test, compression essai *m* (de) compression
test, compression [for soils] essai *m* de compressibilité
test, consolidation essai *m* de consolidation

test, constant loading rate   essai *m* à vitesse de chargement constante

test, constant rate of increase of strain   essai *m* à vitesse de déformation constante

test, control   essai *m* de contrôle/de vérification

test, creep   essai *m* de fluage

test, crushing   essai *m* d'écrasement

test, crushing [e.g. concrete]   essai *m* de résistance à la compression

test, ductility   essai *m* de ductilité

test, duplicate   contre-épreuve *f*, contre-essai *m*

test, endurance/fatigue   essai *m* d'endurance/fatigue

test, field   essai *m* en situ

test, flattening   essai *m* d'écrasement

test, flue gas   mesure *f* des gaz de fumée

test, freezing or frost sensitivity   essai *m* de gélivité

test, full-scale   essai *m* grandeur nature/en vraie grandeur

test, gas   essai *m* au gaz

test, hardness   essai *m* de dureté

test, impact   essai *m* au/de choc

test, in situ   essai *m* en place/in situ

test, laboratory   essai *m* de/en laboratoire

test, leak or leakage   épreuve *f*/essai *m* d'étanchéité

test, load   essai *m* en charge

test, materials   essai *m* des matériaux

test, methylene blue   essai *m* au bleu de méthylène

test, model   essai *m* sur modèle reduit

test, moisture–density   essai *m* de compactage

test, non-destructive   essai *m* non destructif

test, notch bend   essai *m* de pliage sur éprouvette entaillée

test, notch impact bending   essai *m* de résilience

test, particle shape   essai *m* de forme des grains

test, penetration   essai *m* au pénétromètre/de pénétration

test, permeability   essai *m* de perméabilité

test, pile loading   essai *m* de chargement d'un pieu

test, plate bearing   essai *m* (de chargement) à la plaque

test, preliminary   essai *m* préliminaire

test, pressure   épreuve *f*/essai *m* de pression

test, Proctor compaction   essai *m* de compactage Proctor

test, pumping   essai *m* de pompage

test, punching   essai *m* de poinçonnage

test, repeated notched bar impact bending   essai *m* de fatigue par choc sur le barreau entaillé

test, Rockwell hardness   essai *m* de dureté Rockwell

test, routine   essai *m* courant

test, shear   essai *m* de cisaillement

test, sieving   tamisage *m* de contrôle

test, single   essai *m* isolé

test, slump   essai *m* d'affaissement (au cône d'Abrams)/d'Abrams/de fluidité

test, smoke [opacity]   essai *m* d'opacité de fumée

test, smoke [tightness]   essai *m* d'étanchéité à la fumée

test, soot   mesure *f* de la teneur en suies

test, static loading   essai *m* de chargement statique

test, tensile/tension   essai *m* de traction

test, transverse bending   essai *m* de pliage en travers

test, triaxial compression   essai *m* de compression triaxial

test, ultrasonic   essai *m* aux ultrasons

tests, acceptance, on samples   recette *f*

test apparatus   appareil *m* d'essai

test apparatus, dynamic penetration   appareil *m* de sondage à pénétration dynamique

test bar, notched   barreau *m* entaillé

test bench or stand   banc *m* d'essai

test certificate   certificat *m* d'épreuve/d'essai

test chamber   chambre *f* d'essai

test core [ground investigation]   sondage *m* carotté

test cube [concrete]   éprouvette *f* cubique

test equipment [for trials]   appareillage *m* d'essai

test equipment [for verification]   appareillage *m* de contrôle

test hole   sondage *m*

test piece, standard   éprouvette *f* normalisée

test pit   puits *m* de reconnaissance, sondage *m* par tranchée

test pit, machine dug   puits *m* de reconnaissance à la pelle (mécanique), sondage *m* à la pelle

test pits, ground investigation by   sondage *m* par tranchée

test piece/sample/specimen   échantillon *m*, éprouvette *f* (d'essai)/(pour essai)

test piece [bar]   barreau *m* d'essai

test piece, duplicate   contre-éprouvette *f*

test piece, random   échantillon *m* prélevé par hasard

test piece, standard   éprouvette *f* normalisée

test piece, tensile   éprouvette *f* de traction

test pieces, removal or selection of   prélèvement *m* d'échantillons

test report   compte-rendu *m* d'essai, rapport *m* d'essais

test result   résultat *m* d'essai

test results, evaluation or interpretation of   interprétation *f* des essais

test room   local *m* d'essai

test run   marche *f* d'essai

test sample or specimen, *see* test piece

test sieve   tamis *m* de contrôle

test site   aire *f* d'essai

test tube   éprouvette *f*

test   éprouver *vb*, faire *vb* des essais/un essai

test [check, verify]   contrôler *vb*

test [try out]   essayer *vb*

tester, continuity [electrical]   détecteur *m* de tension, vérificateur *m* de continuité

tester, earth (continuity)   détecteur *m* de contact à la terre

tester, high voltage   vérificateur *m* de haute tension

tester, tong   tenaille *f* d'essai

testing   revision/révision *f*, vérification *f*

test(ing) laboratory   laboratoire *m* d'essais

test(ing) method   méthode *f* d'essai

testing of materials   essai *m* des matériaux

tetrapod   tétrapode *m*

texture   texture *f*

thatch   chaume *m*

thatcher   couvreur *m* en chaume

thaw   dégeler *vb*, déglacer *vb*, fondre *vb*

thawing *adj*   fondant *adj*
thawing   décongélation *f*
theft, insurance against   assurance *f* contre le vol
theodolite   théodolite *m*
theodolite, set up a   mettre *vb* un théodolite en station
theorem   théorème *m*
theoretic(al)   théorique *adj*
theoretically   théoriquement *adv*
theory   théorie *f*
thermal [insulation, stress, etc]   thermique *adj*
thermal insulation, *see* insulation
thermistor   thermistance *f*
thermistor protection   protection *f* à la thermistance
thermocouple   thermocouple *m*
thermodynamics   thermodynamique *f*
thermoelectric   thermoélectrique *adj*
thermograph   enregistreur *m* de température,
   thermomètre *m* enregistreur
thermo-hygrograph/recording thermo-hygrometer
   thermo-hygromètre *m* enregistreur
thermometer   thermomètre *m*
thermometer, alcohol   thermomètre *m* à alcool
thermometer, bi-metallic   thermomètre *m*
   bilame/bimétallique
thermometer, dial   thermomètre *m* à cadran
thermometer, dry bulb   thermomètre *m* sec
thermometer, globe   thermomètre *m* globe
thermometer, mercury   thermomètre *m* à mercure
thermometer, plunger   thermomètre *m* à plongeur
thermometer, recording   thermomètre *m* enregistreur
thermometer, remote indicating   thermomètre *m*
   indicateur à distance
thermometer, remote reading   téléthermomètre *m*,
   thermomètre *m* à lecture à distance
thermometer, resistance   thermomètre *m* à résistance
thermometer, thermocouple   canne *f* thermoélectrique
thermometer, wet bulb   thermomètre *m* mouillé/à bulbe
   mouillé
thermometer bulb   bulbe *m* de thermomètre
thermometer well   puits *m* de thermomètre
thermoplastic *adj*   thermoplastique *adj*
thermoplastic   thermoplastique *m*
thermo-setting   thermodurcissable *adj*
thermostat   thermostat *m*
thermostat, chilled water   thermostat *m* d'arrêt eau
   glacée
thermostat, cooling water   thermostat *m* de l'eau de
   refroidissement
thermostat, freeze-up protection   thermostat *m* de
   protection contre le bouchage
thermostat, frost   thermostat *m* à gel
thermostat, high limit   thermostat *m* de sûreté
thermostat, high oil temperature   thermostat *m* d'huile
   (sécurité haute)
thermostat, immersion   thermostat *m* plongeur/à
   immersion
thermostat, low oil temperature   thermostat *m* d'huile
   (sécurité basse)
thermostat, radiator   thermostat *m* de radiateur

thermostat, room   thermostat *m* d'ambiance
thermostatic   thermostatique *adj*
thick [of paint or varnish]   englué *adj*
thicket   taillis *m*
thickness   épaisseur *f*
thickness, boundary layer   épaisseur *f* de couche limite
thickness, coating   épaisseur *f* de couche
thickness, flange [of a beam]   épaisseur *f* d'aile
thickness, plate   épaisseur *f* de tôle
thickness, throat   épaisseur *f* du cordon
thickness, wall   épaisseur *f* de paroi/de mur
thickness, web [of a beam]   épaisseur *f* de l'âme
thickness of a wall at a door or window opening
   jouée *f*
thickness of insulation   épaisseur *f* isolante
thicknesser   raboteuse *f*
thicknessing table (moving)   table *f* de rabotage (mobile)
thin   mince *adj*
thin [of a column, of paint containing little oil]   maigre *adj*
thin down [timber or stone]   démaigrir *vb*
thinner   diluant *m*
thinness [of a rendering, paintwork, etc]   maigreur *m*
thinning [of a plank, etc]   affinage *m*
thinning down [of timber, stone]   démaigrissement *m*
thin-walled   à paroi mince
thixotropic   thixotropique *adj*
thoroughfare, main   artère *f* principale
thoroughly   à fond
thread [of a screw]   filet *m*, filetage *m*, pas *m*
thread, connecting   filetage *m* de raccordement
thread, external   filet *m*/filetage *m* extérieur
thread, female   filet *m* intérieur, filet *m*/filetage *m* femelle
thread, internal   filet *m*/filetage *m* intérieur, filet *m*
   femelle
thread, left-handed   filet *m*/pas *m* à gauche, filet *m*
   renversé
thread, long   filet *m* allongé
thread, male   filet *m* extérieur/mâle, filetage *m* mâle
thread, pipe   filetage *m* de tubes
thread, right-handed   filet *m*/pas *m* à droit
thread, screw   filetage *m* de vis
thread, taper   filetage *m* conique
thread, Whitworth   filet *m* Whitworth
thread, wood screw   filetage *m* pour bois
thread chaser [wood turning]   peigne *m* à fileter (le bois)
thread [cut or form a thread]   fileter *vb*, tarauder *vb*
threaded   taraudé *adj*
threaded [i.e. to be screwed]   à visser
threading [of a screw]   filetage *m*
three-dimensional   tridimensionnel *adj*
three-phase   triphasé *adj*
threshold [of a door]   pas *m*, seuil *m*
threshold, break-even   seuil *m* de rentabilité
threshold level, noise   seuil *m* du son
threshold of hearing   seuil *m* d'audibilité
threshold strip   bande *f*/barre *f* de seuil
throat [of a fireplace]   avaloir *m*, chambre *f* à fumée
throat, narrowest part of [of a chimney]   gorge *f*
throat restrictor [fireplace chimney]   trappe *f* de fumée

throttle   étrangleur *m*
throttle   étrangler *vb*
through-and-through sawing   sciage *m* en plots
through-and-through sawn log   plot *m*
throughput   débit *m*
throw aside [goods]   rebuter *vb*
throw into gear   enclencher *vb*
throw something away   mettre *vb* qch au rebut
throw-out   rebut *m*
**throwing excavated material directly to the surface in a
  shallow trench**   jet *m* sur berge
**throwing excavated material up on to a bench in a
  trench wall**   jet *m* sur banquette
thrust   charge *f* axiale, poussée *f*
thrust [of the ground]   butée *f*
thrust, axial   poussée *f* axiale
thrust boring [e.g. for pipes under roads]   fonçage *m*
  horizontal
thrust   pousser *vb*
thrust, resist   contre-buter *vb*
thumb-piece of a door/gate latch   poucier *m*
thunderstorm   orage *m*
thyristor   thyristor *m* à té
tick (off) [on or from a list]   pointer *vb*
tide   marée *f*
tide, ebb   marée *f* descendante
tide, flood   marée *f* montante
tide, high   marée *f* haute, pleine mer *m*
tide, low   basses eaux *f,pl*, marée *f* basse
tidy (away or up)   ranger *vb*
tie   tirant *m*
tie [structural]   lien *m*
tie [to hold reinforcement]   attache *f*, ligature *f*
tie, cable   collier *m*
tie, closed [reinforcement]   cadre *m* fermé
tie, corner   lien *m* d'angle
tie, diagonal   moise *f* en écharpe
tie, diagonal [roof]   écharpe *f*, poinçon *m* rampant
tie, diagonal [windbrace]   entretoise *f* de contreventement
tie, dovetail or fishtail   aronde *f*
tie, formwork   tirant *m* de coffrage
tie, metal   fenton *m*
tie, peripheral   chaînage *m* périphérique
tie, wall   agrafe *f*, lien *m*
ties [for walls]   chaînage *m*
tie member   tirant *m*
tie (rod)   entretoise *f*
tie together [with a ring beam, etc]   chaîner *vb*
tie (up)   attacher *vb*
tie up [a deal]   conclure *vb*
tie up [a parcel]   ficeler *vb*
tie up [capital]   immobiliser *vb*
tie-back   tirant *m* d'ancrage
tie-bar   tige *f* d'ancrage
tie-plate   traverse *f* de liaison
tierce-point   tiers-point *m*
tierceron [rib of a vault]   tierceron *m*
tie-rod [steel roof]   entrait *m*
tie-rod [across a truss or building]   tirant *m*

tiers or tiered rows, set or lay out in   étager *vb*
tight   serré *adj*
tight fit   adjustement *m*/ajustage *m* serré
tighten   serrer *vb*
tighten [restrictions, etc]   renforcer *vb*
tighten [tension]   tendre *vb*
tighten (up) [a screw, nut]   resserrer *vb*
tighten home   serrer *vb* à bloc
tightening   serrage *m*
tile [roofing]   tuile *f*
tile, cork   carreau *m* de liège
tile, corner, hip or valley   tuile *f* cornière
tile, decorative ridge   crête[†] *f*
tile, decorative ridge end   faîteau *m*
tile, edge   tuile *f* de rive
tile, filler   bouchon *m*
tile, flat or plain   tuile *f* plate (ordinaire)
tile, floor   carreau *m*
tile, floor [hexagonal, quarry]   tommette *f*
tile, glass   panne *f* en verre
tile, hip   tuile *f* cornière, tuile *f* faîtière d'about
tile, interlocking   tuile *f* mécanique/à emboîtement
tile, interlocking ridge   tuile *f* faîtière mécanique
tile, make-up   bouchon *m*
tile, nib(bed)   tuile *f* crochet
tile, over   tuile *f* mâle/de dessus
tile, plain   tuile *f* plate (ordinaire)
tile, plain [convex]   tuile *f* coffine
tile, plain, with rounded end   tuile *f* plate écaille
tile, plain ridge   faîtière *f* simple
tile, quarry   tommette *f*
tile, ridge   faîtière *f*, tuile *f* arêtière/faîtière/de croupe
tile, ridge course   tuile *f* faîtière de dernier rang
tile, semi-circular ridge   enfaîteau *m*
tile, Spanish   tige *f* de botte, tuile *f* canal/creuse/ronde
tile, Spanish [broad type used in Provence]   tuile *f*
  sarrasine
tile, spigot-and-socket ridge   faîtière *f* à bourrelet
tile, thermoplastic   dalle *f* thermoplastique
tile, under   tuile *f* femelle/de dessous
tile, valley   noue *f*, tuile *f* cornière
tile, ventilation   chatière *f*
tile, vinyl asbestos   dalle *f* thermoplastique
tile, wall   carreau *m* (de revêtement)
tile, wall-capping [special French]   tuile *f* chaperonne
tiles, wall   faïence *f*
tile cutter   coupe-tuiles *m*
tile fixing mortar [adhesive]   mortier *m* colle
tile kiln   tuilerie *f*
tile roofing   couverture *f* en tuiles
tile works   tuilerie *f*
tile [a floor]   daller *vb*
tile [a floor, a wall]   carreler *vb*
tiled   carrelé *adj*
tiler [floor or wall]   carreleur *m*, poseur *m* de carreaux
tiler [roofs]   couvreur *m*
tiler, roof   ardoisier *m*
tiler's hammer [similar to bricklayer's hammer]
  martelet *m*

tiling   carrelage *m*
tiling batten   liteau *m*
tiltdozer   bouteur *m* inclinable
timber, *see also* board, plank, wood   bois *m*
timber, adzed   bois *m* refait
timber, building or framing   bois *m* de charpente
timber, building or square(d)   bois *m* d'œuvre
timber, constructional, long or structural   bois *m* de construction
timber, end grain   bois *m* de bout
timber, exposed   bois *m* apparent, charpente *f* apparente
timber, glue/laminated   bois *m* lamellé-collé, bois *m* lamellé
timber, impregnated   bois *m* imprégné
timber, long   bois *m* de long
timber, mature (standing)   futaie *f* haute
timber, planed   bois *m* raboté
timber, resawn   bois *m* de sciage
timber, rough or undressed   bois *m* en grume
timber, round   bois *m* rond, grume *f*
timber, sawn   bois *m* avivé/de sciage
timber, semi-dry   bois *m* d'entrée
timber, side grain   bois *m* de fil
timber, square(d)   bois *m* équarri/d'équarrissage
timber, square-edged   avivés *m,pl*
timber, treated [impregnated]   bois *m* imprégné/traité
timber, twisted [after squaring]   bois *m* gauche
timber, waney-edged   bois *m* flache
timber, worm-eaten   bois *m* mouliné
timber dog   clameau *m* (à deux points)
timber frame   charpente *f* en bois
timber frame construction   construction *f* en bois
timber frame panel   pan *m* de bois
timber frame partition or wall   cloison *f* en charpente
timber framing   ossature *f* en bois
timber in log form   bois *m* de brin
timber joints   assemblage *m* de pièces de bois
timber member, curved structural   guitare *f*
timber merchant   marchand *m* de bois
timber (panel) floor(ing)   parquet *m*
timbers, roof   combles *m,pl*
timber [an excavation]   blinder *vb*, boiser *vb*
timbered [of an excavation]   blindé *adj*
timberer [specializing in timbering excavations]   boiseur *m*
timbering [of an excavation]   soutènement *m* de bois
timbering [of excavations, action and result]   blindage *m*, boisage *m*
timbering [propping or shoring]   étayage *m*
time [length of]   temps *m*
time, curing   durée *f* de prise
time, dead   temps *m* mort
time, exposure   temps *m* d'exposition
time, extension of   délai *m*, prolongation *f* de temps
time, idle   temps *m* d'arrêt/de pause
time, interval of   intervalle *m* de temps
time, response   temps *m* de réponse
time, reverberation   période *f*/temps *m* de réverbération
time, running-up   temps *m* de mise en marche

time, setting [concrete, glue]   temps *m* de prise
time, solar   temps *m* solaire
time, starting   temps *m* de mise en marche
time, unit of   unité *f* de temps
time allowed [e.g. for completion]   délai *m*
time between failures, mean   temps *m* moyen entre pannes
time delay   temporisation *f*
time delay [e.g. of a time lock]   délai *m* de temporisation
time interval   intervalle *m* de temps
time lag [delay]   retard *m*, temps *m* de retard
time lag [separation]   décalage *m*
time switch   interrupteur *m* horaire/à minuterie, minuterie *f*
time switch, automatic   interrupteur *m* automatique de réglage
time switch, control   minuterie *f* à commande/de contrôle
time switch, electric   minuterie *f*
time to repair, mean   durée *f* de réparation
timekeeper   pointeau *m*, pointeur *m*
timer, seven-day   horloge *f* hebdomadaire
timing device, automatic   minuterie *f*
tin   étain *m*
tin   étamer *vb*
tinning   étamage *m*
tin-plate   fer-blanc *m*
tint   teinte *f*
tint   teinter *vb*
tinting   teintage *m*
tip [point]   pointe *f*
tip [dump]   décharge *f*
tip [extremity]   bout *m*, extrémité *f*
tip of a drill   mouche *f*
tip or point of a traffic island   musoir *m*
tip [unload]   décharger *vb*
tipper, three-way   tribenne *f*
tipper (lorry), three-way   camion *m* tribenne
tipper (lorry/truck)   camion *m* benne
tipping semi-trailer   semi-benne *f*, semi-remorque *f* à benne
tipping trailer   remorque *f* basculante/benne
titanium dioxide [white pigment]   blanc *m* de titane
title [deeds]   titre *m*
title, root of   origine de propriété *f*
title block [of drawing]   cartouche *m*
to above/below [on pipework drawing]   vers le haut/bas
to be discussed   à débattre
to measure   sur mesure
to one side   à l'écart
to scale [of a model or drawing]   à l'échelle
toe, downstream/upstream [of a dam]   pied *m* aval/amont
toe (or foot) of a slope   pied *m* de talus
toeboard   garde-pieds *m*
together, linked [e.g. of partners in a joint venture]   conjoint *adj*
toggle action   action *f* à déclic
toilet (WC)   cabinet *m* d'aisances, waters *m,pl*, WC *m,pl*

toilets (WCs)   cabinets *m,pl*, commodités *m,pl*, lavabos *m,pl*, lieux *m,pl* (d'aisances), toilettes *f,pl*
toilets [public convenience]   toilettes *f,pl*
toilet bowl or pan   cuvette *f* de cabinet/WC
toilet facilities   installations *f,pl* sanitaires
toilet flush(ing) cistern   chasse *f* d'eau, réservoir *m* de chasse (d'eau)
toilet paper   papier *m* hygiénique
toilet roll holder   porte-rouleau *m*
toilet seat   siège *m* de cabinet
toilet seat [lifting flap]   abattant *m*
tolerance   tolérance *f*
tolerance [margin]   marge *f* de tolérance
toll   péage *m*
toll booth [bridge, motorway]   poste *m* de péage
tone [of colour]   ton *m*
tone down [contrast, a colour]   adoucir *vb*
tong tester   tenaille *f* d'essai
tongs   tenaille *f*, tenailles *f,pl*
tongs, flat [for bending fine reinforcement]   béquette *f*
tongs, plumber's chain   serre-tubes *m*
tongue [in a mitred joint]   pigeon *m*
tongue [in tongue-and-groove joint]   languette *f*
tongue [of land]   langue *f*
tongue [weather-moulding] **on the edge of a casement**   languette *f*, mouton *m*
tongue [the edge of a board]   langueter *vb*
tongue-and-groove   bouvet *m*
tongue-and-groove joint, assemble a   affourcher *vb*
tongued-and-grooved   bouveté *adj*, rainuré et languetté *adj*
tongued-and-grooved edges   coupe *f* embrevée
toning down [contrast, a colour]   adoucissement *m*
tonne, metric   tonne *f*
tool   outil *m*
tool, bending [reinforcement]   griffe *f* à couder
tool, caulking   ciseau *m* à mater
tool, crimping   pince *f* de sertissage
tool, drilling   fraise *f*
tool, hand   outil *m* à main
tool, hollowing [wood turning]   alésoir *m*
tool, machine   machine-outil *f*
tool, parting [wood turning]   burin *m*/ciseau *m* de tourneur
tool, pneumatic(ally operated)   outil *m* pneumatique
tool, riveting   riveuse *f*
tool, saw setting   tourne-à-gauche *m*
tool, smoothing   lissoir *m*
tools, hand   outillage *m* à main
tool for boring peg holes [timber structures]   laceret *m*
tool for bush hammering   boucharde *f*
tool for driving out tre(e)nails   repoussoir *m*
tool for grooving a substrate before applying rendering or stucco   chien *m*
tool for ramming packing into a spigot-and-socket joint   cordoir *m*
tool for smoothing cut edges of glass   grégeoir *m*, grugeoir *m*
tool rest   porte-outil *m*

tools, planing   outillage *m* à façonner
tools, set of   outillage *m*
tool stone surface with knurled wheel or hammer   mouliner *vb*
tool-box   boîte *f*/caisse *f* à outils
tooling stone surface with knurled wheel or hammer   moulinage *m*
toothed drag   brette *f*, chemin *m* de fer, guillaume *m*
toothed drag, small, for dressing plastered surfaces   rabotin *f*
toothed drag, teeth of a   brettelure *f*
toothing [for subsequent brick- or stonework]   harpe *f* (d'attente)
toothing [key between old and new masonry]   arrachement *m*
toothing [of masonry]   appareil *m* en besace
top of, at the   au sommet de, en haut de
top [of stairs, a wall, a ladder]   haut *m*
top [of a hill, ladder, tree]   sommet *m*
top [of a list, table]   tête *f*
top [of a mountain, roof, tree]   faîte *m*
top [side or surface]   dessus *m*
top by [surmount with]   couronner *vb* de, surmonter *vb* de
top out [a building]   terminer *vb* le gros œuvre
top up [battery, tank, etc]   ramener *vb* à niveau, reniveler *vb*
top up with   remettre *vb* de, remplir *vb* avec
top-iron [of a plane]   contre-fer *m*
topographic(al)   topographique *adj*
topography   topographie *f*
topped (by)   surmonté *adj* (de)
topsoil   sol *m* cultivable/vivant, terre *f* végétale/de couverture
topsoil stripping   décapage *m* de terre végétale
torch [for heating, brazing, etc]   chalumeau *m*
torch, brazing   chalumeau-braseur *m*, chalumeau-soudeur *m*, torche *f* à braser
torch, cutting   chalumeau *m* (oxycoupeur), chalumeau-coupeur *m*
torch, gas cutting   chalumeau *m* à gaz
torch, oxyacetylene   chalumeau *m* oxyacétylénique
torque   couple *m*, couple *m* de rotation/de torsion
torque, starting   couple *m* de démarrage
torrent   torrent *m*
torsion   torsion *f*
total *adj*   total *adj*
total   total *m*
total [in figures]   chiffre *m*
total [sum, amount]   somme *f*
total [sum, of money]   montant *m*
totally   entièrement *adv*
touch up [paintwork, masonry, pointing, etc]   regarnir *vb*
touch up [paintwork, photos]   retoucher *vb*
touching up [paintwork, masonry, pointing, etc]   regarnissage *m*
touching up [paintwork, photos]   retouche *f*
tough   dur *adj*, solide *adj*
toughen   durcir *vb*
toughness   solidité *f*

toughness [of metals]   ténacité f
tow [a barge]   haler vb
tow [a vehicle]   remorquer vb
tow [equipment, a trailer]   tracter vb
towed [equipment]   tracté adj
towed scarifier   défonceuse f tractée
towel rail   porte-serviette m
towel rail, heated   chauffe-serviettes m, porte-serviette m
  chauffant
tower   tour f
tower, bell   campanile m
tower, bell or church   clocher m
tower, control   tour f de contrôle
tower, corner   tour f d'angle
tower, flanking   tour f de flanquement
tower, high-tension/-voltage   pylône m ligne haute
  tension
tower, support [for cableway]   pylône m
tower, watch   tour f de guet
tower, water   château m d'eau
tower block [flats/apartments]   tour f d'habitation
tower up [building, monument, peak, etc]   se dresser vb
town, right in the   en pleine ville
town   ville f
town [large, or nucleus of]   cité f
town, small   bourg m
town hall   hôtel m de ville, mairie f
town or urban planning certificate   certificat m
  d'urbanisme
town planner   urbaniste m/f
town planning adj   urbanistique adj
town planning   urbanisme m
town planning regulations [in France]   Code m
  d'Urbanisme
towpath   banquette f/chemin m de halage, marchepied m
toxic   toxique adj
toxic material or substance   substance f toxique
toxicity   toxicité f
trace   calquer vb
trace heating   traçage m électrique
trace [a drawing, etc]   décalquer vb
trace the outline of   contourner vb
tracer [in a drawing office]   calqueur m, traceur m
tracing   traçage m
tracing paper   (papier-) calque m
track   chemin m, voie f
track, crane   voie f de grue
track, cycle   piste f cyclable
track, farm   chemin m/sentier m d'exploitation f
track, forest   chemin m forestier, laie f
track, light(ing)   glissière f des lumières
track, plank   chemin m en madriers
track, railway   voie f ferrée
track, roller   chemin m de roulement
tracks [of caterpillar vehicle]   chenilles f,pl
tracked [of caterpillar vehicle]   à/sur chenilles
traction   traction f
traction, electric   traction f électrique
tractor   tracteur m

tractor, crawler   chenillard m
tractor, off-highway   tracteur m de chantier
tractor, pipelaying   tracteur m poseur de canalisations
tractor, track-laying   tracteur m à chenilles
tractor, wheeled   tracteur m sur pneu(matique)s
tractor unit, semi-trailer   tracteur m de semi-remorque
tractor-scraper   avant-train tracteur m, décapeuse f (avec
  avant-train moteur)
trade adj   professionnel adj
trade   commerce m
trade association   association f professionnelle
trade journal   périodique f/revue f professionnelle
trade union   syndicat m
trade union representative   délégué m syndical,
  représentant m syndical
trade-in   reprise f
trademark   marque f (de fabrique), nom m de marque
trademark, registered   marque f déposée
tradesman   artisan m
traffic   circulation f, trafic m
traffic, heavy   trafic m lourd
traffic, intercity   trafic m interurbain
traffic, light   trafic m très faible
traffic, local   trafic m intérieur/local
traffic, medium-heavy or moderate   trafic m moyen
traffic, motor vehicle   circulation f automobile
traffic, one-way   circulation f uni-directionnelle
traffic, rail   trafic m ferroviaire
traffic, road   circulation f routière
traffic, through   circulation f principale, trafic m de
  transit
traffic, urban   trafic m urbain
traffic area   surface f de circulation
traffic direction   sens m de circulation
traffic island   îlot m
traffic island [generally triangular]   pointe f de cœur
traffic island, channelizing   îlot m séparateur
traffic island, tip or point of a   musoir m
traffic island with direction signs   îlot m directionnel
traffic lane, see lane, etc
traffic load(ing)   sollicitation f du trafic
traffic noise   bruit m de la circulation
traffic plan, general   plan m général des transports
traffic regulations   réglementation f de la circulation
traffic roundabout   rond-point m
traffic route [major]   axe m de circulation
traffic separator,.raised   séparateur m en saillie
trafficable   circulable adj
trafficked, lightly   à faible circulation
trailer   remorque f
trailer, bottom-dump   remorque f à déversement par le
  fond
trailer, flat   plateau m
trailer, flat-bed (semi-)   (semi-)remorque f plateau/plate-
  forme
trailer, plant (semi-)   (semi-)remorque f porte-engins
trailer, semi-   semi-remorque f
trailer, tipping (semi-)   (semi-)remorque f
  basculante/benne

train [group of wagons]   rame *f*
train, fast   rapide *m*
train, freight/passenger   train *m* de marchandises/de voyageurs
train, underground   rame *f* de métro
train of gears   train *m* d'engrenages
train [a surveying instrument]   pointer *vb*
train [in a skill]   former *vb*
trained   formé *adj*
training   formation *f*
training, professional   formation *f* professionnelle
training and supervision   encadrement *m*
training period   période *f* d'instruction
trammel [beam compass]   compas *m* à verge
tram(way)   tramway *m*
tramway lane   voie *f* de (circulation du) tramway
transaction   marché *m*
transceiver   émetteur-récepteur *m*
transducer   transducteur *m*
transducer, pressure   capteur *m* de pression
transfer   transfert *m*
transfer   transférer *vb*
transfer [a lease, etc]   céder *vb*
transform (into)   transformer *vb* (en)
transformation piece   pièce *f* de transformation
transformer   transformateur *m*
transformer, bell   transformateur *m* pour sonnerie
transformer, rotary   convertisseur *m* rotatif
transformer, single-phase   transformateur *m* monophasé
transformer, step-up   transformateur *m* élévateur
transformer station   poste *m* de transformation
transient   transitoire *adj*
transient [electrical]   pointe *f* de tension
transistor   transistor *m*
transit mixer (truck)   camion *m* malaxeur
transition [roads]   raccordement *m*
transition, crest [road vertical alignment]   angle *m* saillant
transition, gradient   changement *m* de pente
transition, valley [road vertical alignment]   angle *m* rentrant
transition, vertical   raccordement *m* vertical
transition curve [roads]   courbe *f* de raccordement
transition curve, circular   raccordement *m* circulaire
transition curve, spiral   clothoïde *f*
transition region   domaine *m* de régime transitoire
transitional   transitoire *adj*
transitory   transitoire *adj*
translation [e.g. of a crane]   translation *f*
transmission [of drive, heat, sound, signals]   transmission *f*
transmission factor [lighting]   facteur *m* de transmission
transmission of force   transmission *f* d'effort
transmission of heat   transmission *f* de chaleur
transmission line, electrical [as opposed to distribution line]   artère† *f*, ligne *f* de transmission
transmission system   organes *m,pl* de transmission
transmissivity   transmissivité *f*
transmit   transmettre *vb*

transmittance (U-value), thermal   coefficient-U *m*
transmitter   émetteur *m*, transmetteur *m*
transom   meneau *m* horizontal, traverse *f* (d'imposte *f*)
transparency [drawing]   contre-calque *f*
transparency [of air, water, etc]   transparence *f*
transparency [photographic]   diapositive *f*
transparent [air, glass, etc]   transparent *adj*
transport [of heavy materials, on site]   bardage *m*
transport   transport *m*
transport, ash   transport *m* des cendres
transport, public   transports *m,pl* (en commun)
transport, suburban   transport *m* de banlieue
transport, vertical   appareil *m* élévateur
transport company   compagnie *f* de transport
transport facilities   facilités *f,pl* de transport
transport plan, general   plan *m* général des transports
transport   transporter *vb*
transportation   transport *m*
transverse   transversal *adj*
transversely   en travers
trap [plumbing]   siphon *m*
trap, bottle   siphon *m* à culotte démontable
trap, condensing or steam   purgeur *m* de vapeur
trap, drain   siphon *m*
trap, floor   siphon *m* de sol
trap, grease   boîte *f*/fosse *f* à graisse, séparateur *m* à graisse
trap, oil   siphon *m* d'huile
trap, S-   siphon *m* en S
trap, siphon   clôture *f* à siphon
trap, steam   bouteille *f* de purge
trapdoor   trappe *f*
trapdoor [in a stage]   trapillon *m*
trapezoidal [of plain tiles for pyramidal roofs]   gironné *adj*
travel [of a tool, etc]   course *f*
travel limiting stop [of a lift/elevator]   butée *f*
travelling   roulant *adj*
travolator   tapis *m* roulant, trottoir *m* roulant
tray, cable   chemin *m* de câbles
tray, drip   bac *m* de récupération, égouttoir *m*
tray, overflow [beneath a head tank]   terrasson *m* de sécurité
tray fixable to ladder rungs   taquet *m* d'échelle
tre(e)nail   fenton *m*
tread [of a stair]   giron *m*, marche *f*
tread width of a staircase   emmarchement *m*
treat   traiter *vb*
treated   traité *adj*
treated aesthetically or architecturally   ouvragé *adj*
treatment   traitement *m*
treatment [purification]   purification *f*
treatment, after- or post-   traitement *m* ultérieur
treatment, anti-skid surface   revêtement *m* antidérapant
treatment, boiler water   traitement *m* des eaux de chaudière
treatment, dip [for timber]   trempage *m*
treatment, feed water   épuration *f* de l'eau d'alimentation

treatment, heat   traitement *m* thermique
treatment, primary   traitement *m* premier
treatment, refuse   traitement *m* des ordures ménagères
treatment, secondary   traitement *m* ultérieur
treatment, sewage   épuration *f* des eaux usées
treatment, surface   traitement *m* de surface
treatment, surface [filler, coating]   enduit *m* superficiel
treatment, surface [finish]   finition *f*
treatment, waste water   traitement *m* des eaux d'égouts
treatment, water, *see* water treatment
treatment process, (chemical)   procédé *m* de traitement
   (chimique)
tree   arbre *m*
trees [cluster, plantation, etc]   futaie *f*
tree, deciduous   arbre *m* à feuillage caduc
tree feller   abatteur *m*
tree felling permission/permit   permis *m* d'abattage
tree guard   ceinture *f* de protection d'arbre
tree line   limite *f* des arbres
tree nursery   pépinière *f*
tree planting   plantation *f* d'arbres
tree pruning   élagage *m*
trellis   treillage *m*, treillis *m*
tremor, earth/ground   secousse *f* tellurique
trench   fossé *m*, rigole *f*, tranchée *f*
trench, cable   tranchée *f* à câbles
trench, drainage   saignée *f*, tranchée *f* drainante
trench, foundation   tranchée *f* de fondation
trench, open   tranchée *f* ouverte
trench, pipe   tranchée *f* pour conduites
trench, shallow, to carry away rainwater   goulotte *f*
trench backfilling   remplissage *m* de tranchée
trench beneath a dam   parafouille *f*
trench bottom   fond *m* de fouille
trench cutting   travaux *m,pl* de fouilles en tranchées
trench excavation   fouille *f* en rigole
trench   creuser *vb* une tranchée, fouiller *vb*
trencher [bucket-wheel]   excavateur *m* à godets
trencher/trench excavator [machine]   trancheuse *f*
trencher/trench excavator [worker]   trancheur *m*
trenching   fouille *f* (en rigole)/(en tranchée), tranchée *f*,
   travaux *m,pl* de fouilles en tranchées
trestle   chevalet *m*, palée *f*, tréteau *m*
trial [test]   essai *m*
trial, boiler (output)   essai *m* de rendement de chaudière
trial, field   essai *m* en place/in situ
trial, full scale   essai *m* grandeur nature, essai *m* en vraie
   grandeur
trial hole   sondage *m*
trial pit   puits *m* de reconnaissance, sondage *m* par
   tranchée
trial pit, machine dug   puits *m* de reconnaissance à la
   pelle (mécanique), sondage *m* à la pelle
trial run   marche *f* d'essai
triangulation   triangulation *f*
triangulation, general   canevas *m* d'ensemble
triangulation, skeleton   canevas *m*
triangulation point or station   point *m* géodésique
triaxial   triaxial *adj*

tributary [of a stream, etc]   affluent *adj*
tributary [stream, etc]   affluent *m*
trickle [i.e. small flow]   filet *m*
trigger lever   levier *m* de déclenchement
trim   habillage *m*
trim [cover strip]   couvre-joint *m*
trim saw   scie *f* de délignage
trim [a pile]   recéper *vb*
trim [edges]   rogner *vb*
trim [stone, etc]   dégauchir *vb*
trim [with a saw]   déligner *vb*
trim edges [of joinery]   dégraisser *vb*
trim the edges of   rogner *vb*
trim to level   araser *vb*
trimmer [earthworks]   décapeuse *f* niveleuse, niveleur *m*
trimmer [round an opening, timber frame construction]
   chevêtre *m*
trimmer [supporting beams across a chimney, roof
   opening, etc]   linçoir *m*
trimmer, edge [for lawns]   coupe-bordure *m*
trimmer, edge [plumber's tool]   débordoir *m*
trimmer, hedge   taille-haies *f*
trimming [e.g. a pile, a tree]   recépage *m*
trimming [of stone]   dégauchissage *m*
trimming [of wood, a block of stone, to size]
   abattement *m*
trimming [structure round opening in a floor]
   enchevêtrure *f*
trip [disengage, switch off]   déclencher *vb*
triple-pole   tripolaire *adj*
tripod   pied *m* à trois branches, trépied *m*
trolley   chariot *m*
trough [geological]   synclinal *m*
trough, mortar   auge *f*/bac *m* à mortier
trough, plasterer's   auge *f*
trowel   truelle *f*
trowel, brick(laying)   truelle *f* à mortier
trowel, bricklayer's   briqueteuse *f*
trowel, laying on   taloche *f*
trowel, machine [for smoothing concrete, etc]   lisseuse *f*
trowel, trapezoidal   grelichonne *f*
truck   chariot *m*
truck [lorry]   camion *m*
truck, heavy   poids *m* lourd
truck, industrial   chariot *m* de manutention
truck, light   camionnette *f*
trucks, fleet of   parc *m* de camions
truckmixer   autobétonnière *f*, camion *m* malaxeur, toupie
   f à béton
truckmixer with concrete pump   malaxeur *m* pompe à
   béton, toupie-pompe *f* (à béton)
true [measure, report]   exact *adj*
true [plane]   plan *adj*
true [straight]   droit *adj*
true, out of [bent, distorted]   gauchi *adj*
true, out of [not vertical]   hors d'aplomb
true, out of [twisted]   tordu *adj*
true [a surface]   rectifier *vb*, redresser *vb*
true, run   tourner *vb* sans balourd

true, run out of   avoir *vb* du balourd, tourner *vb* à faux
true up [a machine]   dresser *vb*
true up [a piece of wood, stone]   dégauchir *vb*
true up [make or bring to same level]   niveler *vb*
truing [a surface]   rectification *f*, redressement *m*
truing up   dressage *m*
truing up [of a piece of wood, stone]   dégauchissage *m*
trunk [of a tree]   tronc *m*
trunking, dado/skirting   goulotte *f*
trunking, mini- [electrical]   moulure *f*
trunking, perimeter   plinthe *f* technique
truss   ferme *f*
truss [girder]   poutre *f* à treillis
truss, double pitch   ferme *f* anglaise
truss, half   demi-ferme *f*
truss, mansard   ferme *f* à la mansarde
truss, north light or sawtooth   ferme *f* en shed
truss, queen   comble *m* à plancher suspendu
truss, roof   charpente *f* de comble, ferme *f* (à treillis)
truss, sawtooth   ferme *f* en shed
truss, scissors   ferme *f* à entrait brisé/en écharpes
truss, solid web   ferme *f* à âme pleine
truss, steel roof   ferme *f* en acier
truss post [timber roof]   poinçon *m*
truss [a beam]   renforcer *vb*
trussing [of a beam]   renforcement *m*
try [attempt, test or trial]   essai
try out [experiment with]   essayer *vb*
try out [test]   expérimenter *vb*
try out   mettre *vb* à l'essai
try to do sth   essayer *vb*/tâcher *vb* de faire qch
try to locate a fault   rechercher *vb* un dérangement
tub   cuve *f*
tub, mortar   bac *m* à mortier
tube   buse *f*, tube *m*
tube[†] [underground railway]   métro(politain)[†] *m*
tube, arc [lighting]   brûleur *m*
tube, Bourdon   tube *m* de Bourdon
tube, condenser   tube *m* de condenseur
tube, connecting   tube *m* de connexion
tube, drain   tube *m* de vidange
tube, drainage   tube *m* d'évacuation
tube, finned   tube *m* ailetté/à ailettes
tube, fluorescent   lampe *f* fluorescente, tube *m* fluorescent
tube, inner [of tyre]   chambre *f* d'air
tube, ledger [scaffolding]   moise *f*
tube, pitot   tube *m* de Pitot
tube, screwed or threaded   tube *m* fileté/taraudé
tube, seamless (steel)   tube *m* (en acier) sans soudure
tube, spiral fin   tuyau *m* à ailettes hélicoïdales
tube, steel/copper/etc   tube *m* d'/en acier, de/en cuivre/etc
tube, vacuum   tuyau *m* à vide
tube, welded   tube *m* soudé
tube station   station *f* de métro
tubing [pipes, etc]   tubes *m,pl*
tubing [of a bore-hole, etc, the action and the result]   tubage *m*
tubular   tubulaire *adj*

tufa   tuf *m*
tumble down [a structure]   s'écrouler *vb*
tundish (funnel)   entonnoir *m*
tungsten   tungstène *m*
tungsten carbide   carbure *m* de tungstène
tuning   réglage *m*
tuning [running in]   mise *f* au point
tunnel   tunnel *m*
tunnel [gallery, in a dam, a mine]   galerie *f*
tunnel [underground passage]   souterrain *m*
tunnel, cooling   tunnel *m* de réfrigération
tunnel, diversion   galerie *f* de dérivation
tunnel, inlet [dam]   galerie *f* d'amenée
tunnel, intake [dam]   galerie *f* de prise
tunnel, outlet [dam]   galerie *f* de fuite/de vidange
tunnel, railway   tunnel *m* ferroviaire
tunnel, road   tunnel *m* routier
tunnel, undermining   sape *f*
tunnel, wind   tunnel *m* aérodynamique/d'essai aéraulique
tunnel boring machine   taupe *f*, tunnelier *m*
tunnel form(work)   coffrage *m* tunnel
tunnelling machine   tunnelier *m*
turbidity   turbidité *f*
turbine   turbine *f*
turbine, gas   turbine *f* à gaz
turbine, steam   turbine *f* à vapeur
turbo-blower   turbosoufflante *f*
turbo-charged   à turbocompresseur
turbo-compressor   turbo-compresseur *m*
turbo-compressor, centrifugal   turbo-compresseur *m* radial
turbo-generator   turbo-générateur *m*
turbulence   turbulence *f*
turbulent   turbulent *adj*
turf [in situ]   gazon *m*
turf, piece of   morceau *m*/plaque *f* de gazon
turf   gazonner *vb*
turfing   gazonnement *m*
turn [action of changing direction]   virage *m*
turn [change direction]   virer *vb*
turn [rotation]   rotation *f*, tour *m*
turn   tourner *vb*
turn [on a lathe]   travailler *vb* au tour
turn (a piece of cut stone) to lie on another face   donner *vb*/faire *vb* quartier *m* (à une pierre de taille)
turn off the current   couper *vb* le courant
turn off or out the light(ing)   couper *vb* la lumière
turn off the water   couper *vb* l'eau
turn off tight [a tap or valve]   serrer *vb*
turn on [an electrical circuit]   mettre *vb* en circuit/sous tension
turn on [a tap, electricity, gas, etc]   ouvrir *vb*
turn on [the light, lighting]   allumer *vb*
turn round [an object]   tourner *vb*
turn pale   pâlir *vb*
turnadozer   bouteur *m* à pneus
turnbuckle   lanterne *f* de serrage, tendeur *m* à la lanterne
turning [about a point]   pivotement *m*
turning [fork in a road]   embranchement *m*

turning [side road]   route *f* latérale
turnings   limaille *f* (d'acier, de fer)
turnkey [projects]   clef en main *adj*
turnover   chiffre *m* d'affaires
turns per minute   tours *m,pl* par minute
turret   tourelle *f*
turret [architecture]   donjon *m*
turret, pepper-pot   poivrière *f*
twin-bladed   bipale *adj*
twin(ned)   jumelé *adj*
twist [as in twist drill]   hélicoïdal *adj*
twist [of a road]   contour *m*, tournant *m*, virage *m*
twist [of timber]   dévers *m*, gauchissement *m*
twist [the action]   torsion *f*
twist [out of required plane]   gauchis *m*
twist [sth]   contourner *vb*, (se) déformer *vb*, tordre *vb*
twist [a road]   serpenter *vb*
twist [cable, moulding, reinforcement bar]   torsader *vb*

twisted   gauche *adj*, tordu *adj*, torsadé *adj*
twisting [of a road]   virage *m*
two-pole   bipolaire *adj*
two-way *adj*   va-et-vient
two-way [of a slab]   continu *adj* en deux directions
   [dalle]
two-year guarantee   biennale *f*, garantie *f* biennale
tying [to]   amarrage *m*
tympanum   tympan *m*
type   type *m*
type (out) [on a keyboard]   taper *vb* à la machine
typewriter   machine *f* à écrire
typewriter key   touche *f*
typical   type *adj*, typique *adj*
tyre, *see also* inner tube   pneu *m*
tyre, compacting   pneu *m* compacteur
tyre valve   valve *f*
tyre valve cap   capuchon *m* de valve

U-bend   courbe *f* de retour
U-bend [pipework]   coude *m*/tube *m* en U
U-shaped valley   vallée *f* en fond de bateau/en auge
U-value (thermal transmittance)   coefficient-U *m*
ultimate [e.g. at failure]   limite *adj*, ultime *adj*
ultimate limit state (ULS)   état *m* limite ultime (ELU)
ultraviolet   ultraviolet *adj*
unbar   débarrer *vb*
unblock [a hole, pipe, etc]   déboucher *vb*
unblock [a pipe, sink, etc]   dégorger *vb*
unblock [a road, traffic]   dégager *vb*
unblock [clear a way]   désencombrer *vb*
unbolt [a door, etc]   déverrouiller *vb*
unbolting   déverrouillage *m*
unbreakable   incassable *adj*
unburnt   imbrûlé *adj*
uncoded   en clair
unconform(abil)ity [geology]   discordance *f*
uncovered   découvert *adj*
undamaged   intact *adj*
under   sous *adv*
under construction   en construction
under cover   à l'abri
under pressure   sous pression
undercoat   couche *f* de fond/d'impression, sous-couche *f*
undercoat [on non-absorbent substrate]   couche *f*
   primaire
undercoat [primer]   apprêt *m*
undercoat [reinforcing a primer or first undercoat]
   couche *f* de renforcement
undercut [in welding]   caniveau *m*
undercutting   excavation *f*

undercutting used for sinking a caisson   havage *m*
underfloor draught pipe to a fireplace   ventouse *f*
underfloor space   vide *m* sanitaire
underground   souterrain *adj*, subterrané *adj*
underground [e.g. services]   mis *adj* sous terre
undergrounding [e.g. of power lines]   enfouissement *m*
undergrowth   brousailles *f,pl*, sous-bois *m*
underlying   sous-jacent *adj*
undermine   saper *vb*
undermine [with water]   affouiller *vb*
underpass   passage *m* inférieur (PI)
underpin   reprendre *vb* en sous-œuvre
underpinning *adj*   en sous-œuvre
underpinning   reprise *f* en sous-œuvre, substruction *f*
underpinning, props or struts used during
   chevalement *m*
underpinning, shoring/propping to facilitate
   enchevalement *m*
underpinning, step-by-step excavation during   fouille *f*
   en rigole
underpinning beam [temporary support]   tête *f* de
   chevalement
underside   arase *f* inférieure, dessous *m*, sous-face *f*
undersize   dimension *f* inférieure à la normale
undersize [in a sieving test]   tamisat *m* inférieur
undertaking [task]   entreprise *f*, œuvre *f*
undervaluation   sous-estimation *f*, sous-évaluation *f*
underwriter   assureur *m*
undisturbed [e.g. sample, in geotechnics]   intact *adj*, non
   remanié *adj*
undrinkable   non potable *adj*
undulating [of land, countryside]   vallonné *adj*

unearth   déterrer *vb*
unenforceable [of regulation, etc]   inapplicable *adj*
uneven   inégal *adj*, irrégulier *adj*
unevenness [of a surface, road, etc]   inégalité *f*
unfit for its purpose   impropre *adj* à sa destination
unfix   détacher *vb*, enlever *vb*
unfix [a lifting cable]   débrider *vb*
unfixed   non-fixé *adj*
unfreeze   dégeler *vb*
unfurnished   non-meublé *adj*
ungraded   tout-venant *adj*
unhealthy   malsain *adj*
unhinge [a door, shutter, etc]   dégonder *vb*
unhook [a lifting cable]   débrider *vb*
unidirectional   unidirectionnel *adj*
uniform   uniforme *adj*
uniformity   uniformité *f*
uniformity, coefficient of   coefficient *m* d'uniformité
uniformity of illumination   uniformité *f* de l'éclairage
uniformity ratio [lighting]   facteur *m* d'uniformité
uniformly distributed   uniformément réparti *adj*
uniformly distributed load (UDL)   charge *f*
   uniformément répartie
uninhabited   inhabité *adj*
union *adj*   syndical *adj*
union [pipe joint]   raccord *m* union, union *f*
union [sleeve]   manchon *m*
union, pipe or plumber's [joint]   raccord *m* union
union, trade   syndicat *m*
union representative or steward   représentant *m*
   syndical
uniphase   monophasé *adj*
unit [mathematical, physical]   unité *f*
unit [e.g. of built-in furniture]   élément *m*
unit, air-cooled condensing   appareil *m*/unité *m* de
   condensation refroidi par air
unit, breaking [for aggregates]   groupe *m* de broyage
unit, central processor   processeur *m* central
unit, cladding   élément *m* de façade
unit, cold aggregate feed   prédoseur *m* des enrobés
unit, condensing   groupe *m* condenseur
unit, constant volume   boîte *f* de détente à volume
   constant
unit, cooking top   table *f* de cuisson
unit, cooling   bloc *m* refroidisseur, groupe *m* de
   réfrigération
unit, crosswall   élément *m* mural transversal
unit, dewatering   groupe *m* d'essorage
unit, direct expansion coil   batterie *f* à/de détente directe
unit, drilling   foreuse *f*
unit, drying   groupe *m* sécheur
unit, dwelling   unité *f* d'habitation
unit, feeding   groupe *m* d'alimentation
unit, induction   éjecto-convecteur *m*, inducteur *m*
unit, low pressure induction   inducteur *m* à basse
   pression
unit, mixing   groupe *m* malaxeur
unit, precast   pièce *f* préfabriquée
unit, prefabricated door   bloc-porte *m*

unit, prefabricated kitchen   élément *m* préfabriqué de
   cuisine
unit, prefabricated pipework [kitchen/bathroom]   bloc-
   eau *m*
unit, prefabricated window   bloc-croisée *m*
unit, screening [for granular materials]   groupe *m*
   classeur/de criblage
unit, sealed condensing   groupe *m* compresseur-
   condenseur hermétique
unit, shelf   étagère *f*
unit, shotcreting   dispositif *m* de projection de béton
unit, slave   unité *f* esclave
unit, standard [equipment]   appareil *m* normal
unit, standard [module]   module *m*
unit, storage   élément *m* de rangement
unit, variable volume   appareil *m* à volume variable,
   boîte *f* de détente à volume variable
unit, ventilating   appareil *m*/unité *f* de ventilation
unit, wall   élément *m* suspendu
unit, warm air heater/heating   générateur *m* d'air
   chaud
unit, weighing [for mixing materials]   groupe *m* doseur
units, wall   mural[†] *m*
unit heater [fan + heating element]   aérotherme *f*
unit heater [wall mounted]   convecteur *m* mural
unit of measurement   unité *f* de mesure
unit of power   unité *f* de puissance
unit of time   unité *f* de temps
unit weight   poids *m* spécifique
unit weight, dry   densité *f* sèche, poids *m* volumique sec
unkey   décaler *vb*
unlatch [door]   déclencher *vb*
unlevel part   faux-niveau *m*
unload   décharger *vb*
unload [timber, from a barge, ship]   débarder *vb*
unloaded [no-load]   à vide
unloading   décharge *m*
unloading area   aire *f* de déchargement
unloading bay   aire *f*/surface *f* de déchargement
unloading device/mechanism   dispositif *m* de
   délestage
unlocking   déverrouillage *m*
unmade [of a road]   non-goudronné *adj*
unpeg [remove pegs]   décheviller *vb*
unplug [a duct, pipe, etc]   détamponner *vb*
unpolished   mat *adj*
unprime [a pump, siphon]   désamorcer *vb*
unscrew   dévisser *vb*
unscrewed, come   se dévisser *vb*
unscrewing   dévissage *m*
unseal or remove seal from   desceller *vb*
unsling   débrayer *vb*
unsorted   tout-venant *adj*
unsound [work, etc]   défectueux *adj*
unsoundness   défectuosité *f*
unstable   labile *adj*
unsteadiness   irrégularité *f*
unsteady   irrégulier *adj*
unsteady [wall, structure, etc]   chancelant *adj*

untreated [without finishes]   brut *adj*
untrue   faux *adj*
untrue [apparently vertical but in fact out of plumb]   faux aplomb
untrue, run [of rotating element]   avoir *vb* du balourd
unweathered [e.g. sample, in geotechnics]   intact *adj*
unwieldy   pesant *adj*
unwind   dérouler *vb*
unworked [rough]   brut *adj*
update [costs, prices]   actualiser *vb*
updating [of costs, prices]   actualisation *f*
upheaval [geological]   soulèvement *m*
upholding   maintien *m*
upholster   tapisser *vb*
upholsterer   tapissier *m*
upholstery   tapisserie *f*
upholstery fabric   tissu *m* d'ameublement
upholstery tack   pointe *f* à damas
upkeep   entretien *m*, tenue *f*
uplift   soulèvement *m*
uplift [hydraulic]   poussée *f* d'Archimède
upper   supérieur *adj*
upper circle [of a theatre]   deuxième balcon *m*
upper part   partie *f* supérieure
upper part [cap, of a bearing, a pile]   chapeau *m*
upright   debout *adv*, droit *adj*
upright   droit *m*
upright [of ladder, scaffolding, etc]   montant *m*
upside down   sens dessus dessous
upstairs   en haut
upstream   en amont

upstream [direction]   à contre-fil de l'eau
upthrust [geological]   soulèvement *m*
up-to-date   mis *adj* à jour
upwind   en amont
urban   urbain *adj*
urbanize   urbaniser *vb*
urgent   urgent *adj*
urgently   d'urgence
urinal   urinoir *m*
usable   utile *adj*, utilisable *adj*
usage   consommation *f*
use   emploi *m*, utilisation *f*
use, directions for   mode *m* d'emploi
use, ready for   prêt *adj* à l'emploi/à l'usage/ d'emploi
use   se servir *vb* de, utiliser *vb*
use up   épuiser *vb*
used [secondhand]   d'occasion *adj*
used [worn]   usagé *adj*
useful   utile *adj*
usefulness   utilité *f*
user   usager *m*
usual   normal *adj*
usufruct   usufruit *m*
usufructuary   usufruitier *m*
utilization factor [lighting]   coefficient *m*/facteur *m* d'utilisation
utility (company)   entreprise *f* de service public
utilities, statutory   services *m,pl* concédés
utility services company   concessionnaire *m*
utilize   utiliser *vb*

V-belt   courroie *f* trapézoïdale
V-belt drive   entraînement *m* à courroie trapézoïdale
V-shaped edge or end joint [timber]   assemblage *m* à grain d'orge
vacant   vide *adj*
vacant possession, with   libre *adj* d'occupation
vacate [premises, a property]   quitter *vb*
vacate [empty]   vider *vb*
vacuum   vide *m*
vacuum, high   vide *m* poussé
vacuum cleaner, all purpose [wet and dry]   aspirateur *m* universel
vacuum cleaner, cylinder   aspirateur-chariot *m*, aspirateur-traîneau *m*
vacuum cleaner, upright   aspirateur *m* balai
vale [small valley]   vallon *m*
valley   vallée *f*
valley [small]   vallon *m*
valley [between roofs]   cornière *f*
valley, drowned   vallée *f* enfoncée/submergée

valley, dry   vallée *f* morte/sèche
valley, hanging   vallée *f* suspendue
valley, river   vallée *f* fluviale
valley, roof   noulet *m*
valley, synclinal   vallée *f* évasée/en berceau
valley, U-shaped   vallée *f* en auge/en fond de bateau
valuation   estimation *f*, évaluation *f*
valuation [independent, of work, materials, etc]   expertise *f*
value   prix *m*, valeur *f*
value, actual   valeur *f* réelle
value, approximate   valeur *f* approximative
value, basic   valeur *f* de base
value, bearing capacity   indice *m* portant
value, calibrated   valeur *f* étalonnée
value, calorific, *see* calorific value
value, contract   valeur *f* de contrat
value, default   valeur *f* de défaut
value, desired   valeur *f* désirée
value, drop/fall in   moins-value *f*

value, empirical   valeur *f* empirique
value, estimated [cost]   coût *m* estimé/estimatif
value, limit   valeur *f* extrême/limite
value, limiting   valeur *f* limite
value, mean   valeur *f* moyenne
value, measured   valeur *f* mesurée
value, nominal   valeur *f* nominale
value, numerical   valeur *f* numérique
value, peak   valeur *f* crête/de pointe (maximale)/de
   pic/maximum
value, pH   valeur *f* du pH
value, reference   valeur *f* de référence
value, root mean square (RMS)   valeur *f* effective
value, saturation   valeur *f* de saturation
value, set   valeur *f* de consigne
value, theoretical   valeur *f* théorique
value, threshold limit   valeur *f* limite admissible
value, typical   valeur *f* type
value added tax (VAT)   taxe *f* sur la valeur ajoutée
   (TVA)
value [work, materials, etc, independently]   expertiser *vb*
value-for-money   rapport *m* qualité-prix
valuer   expert *m*
valuer, property   expert *m* immobilier
valve [for drawing off water]   prise *f* de eau
valve [generally flap type]   clapet *m*
valve [generally large, and with a turning or sliding gate]
   vanne *f*
valve [generally screw down type]   robinet *m*
valve [generally spring-loaded or operated]   soupape *f*
valve [of a machine, a tyre]   valve *f*
valve, air inlet   clapet *m* d'entrée d'air
valve, air purge/vent   purgeur *m* d'air
valve, air release   soupape *f* d'échappement d'air
valve, anti-siphon   soupape *f* anti-siphon (casse-vide)
valve, anti-vacuum   soupape *f* casse-vide
valve, automatic air (AAV)   purgeur *m* d'air
   automatique
valve, automatic (spring-loaded)   soupape *f* à ressort
valve, back-pressure   vanne *f* de pression d'aspiration
valve, balancing   robinet *m* d'équilibrage
valve, balancing (overflow)   soupape *f* de trop-plein
valve, ball (check)   clapet *m* à bille, soupape *f* à
   boulet
valve, bellows   vanne *f* à soufflet
valve, bleed   vanne *f* de soutirage
valve, bleeder or blow-off   purgeur *m*, soupape *f* de
   purge
valve, blowdown   robinet *m* de décharge
valve, butterfly   robinet *m*/vanne *f* (à) papillon, vanne *f*
valve, bypass   soupape *f*/vanne *f* de dérivation
valve, change-over   soupape *f* directionnelle
valve, check   clapet *m*, clapet *m* anti-retour/de retenue/de
   non-retour
valve, circuit control   vanne *f* de réglage de circuit
valve, cock   robinet *m* à boisseau
valve, control   soupape *f* de contrôle/de régulation
valve, control/modulating   robinet *m* modulateur
valve, control/regulating   vanne *f* de réglage

valve, diaphragm   robinet *m* à membrane, soupape *f* à
   diaphragme
valve, diaphragm control   soupape *f* à diaphragme
valve, direct flow   robinet *m* à passage direct
valve, disc   clapet *m* à membrane, vanne *f* à disque
valve, discharge/relief   soupape *f* de décharge
valve, diverter   robinet *m*/vanne *f* de dérivation
valve, double shut-off   soupape *f* à double arrêt
valve, double-seated   robinet *m* à double siège
valve, drain   robinet *m*/soupape *f*/vanne *f* de vidange
valve, draw-off   robinet *m* de puisage
valve, electromagnetically operated   robinet *m* électro-
   magnétique
valve, emergency relief   vanne *f* de sécurité et de
   décharge
valve, emergency stop   robinet *m* d'arrêt d'urgence
valve, emergency water   vanne *f* de secours
valve, exhaust   purgeur *m*, soupape *f*
   d'échappement/d'évacuation
valve, expansion   détendeur *m*, soupape *f* de détente
valve, extract(ion)   vanne *f* d'extraction
valve, feed check   robinet *m* d'alimentation
valve, fire   robinet *m* d'incendie
valve, fireman's   raccord *m* pompier/rapide
valve, flapper   clapet *m* fléchissant, soupape *f* à languettes
valve, float   flotteur *m*, robinet *m*/soupape *f* à flotteur
valve, flow   clapet *m* de refoulement
valve, flow control   vanne *f* de régulation de débit
valve, (flow) regulating   robinet *m* de réglage
valve, flush   robinet *m* de chasse
valve, foot   clapet *m* d'aspiration, pédibulateur *m*
valve, full-way   vanne *f* à passage direct
valve, gas   robinet *m* de gaz
valve, gas appliance isolating   robinet *m* de commande
   d'appareil à gaz
valve, gate   robinet-vanne *f*, vanne *f* d'arrêt
valve, globe   robinet *m* à soupape, robinet *m* d'arrêt
   sphérique
valve, high-lift   soupape *f* à haute levée
valve, inlet   soupape *f* d'admission
valve, isolating   organe *m* de coupure, robinet *m*/vanne *f*
   d'isolement
valve, isolating [in an emergency water pipe system]
   contre-barrage *f*
valve, isolating stop   vanne *f* de sectionnement
valve, key-operated   bouche *f*/robinet *m* à clé
valve, main   robinet *m* principal
valve, main control   vanne *f* principale de contrôle
valve, main gas   robinet *m* principal de gaz
valve, main stop   robinet *m* d'arrêt général, soupape *f*
   d'arrêt principale
valve, mixing   vanne *f* mélangeuse/de mélange
valve, mixing [monobloc with single lever]   mitigeur *m*
valve, mixing [twin tap, single spout]   mélangeur *m*
valve, modulating   robinet *m*/soupape *f* de modulation
valve, motorized   soupape *f* à moteur, vanne *f* motorisée
valve, multi-way   soupape *f* à plusieurs voies
valve, needle   clapet-pointeau *m*, robinet *m*/soupape *f* à
   pointeau

valve, non-return (flap)   clapet *m* anti-retour, clapet *m*/vanne *f* de non-retour, soupape *f* de retenue
valve, outlet   robinet *m* d'écoulement
valve, overflow   vanne *f* de trop plein
valve, plug   obturateur *m*, robinet *m* à boule
valve, pressure, etc, *see* pressure, etc
valve, purge   robinet *m* de purge
valve, quick action/acting   soupape *f* à action rapide
valve, quick release   soupape *f* à ouverture rapide
valve, radiator   valve *f* de radiateur
valve, reducing   soupape *f* réductrice/de réduction
valve, refluxclapet *m* anti-retour/de non-retour, soupape *f* de retenue
valve, refrigeration   vanne *f* à frigorigène
valve, regulating   régulateur *m*, vanne *f* régulatrice/de réglage/de régulation
valve, relief   décompresseur *m*, soupape *f* de décharge, vanne *f* de sûreté
valve, safety   soupape *f* de sécurité/de sûreté
valve, shut-off   soupape *f* d'arrêt, vanne *f* de fermeture
valve, sluice-   robinet-vanne *f*, vanne *f* de communication, vannelle *f*, ventelle *f*
valve, sluice- [in a canal lock]   vanne *f* d'écluse
valve, solenoid   robinet *m* solénoïde
valve, solenoid control   électrovanne *f*
valve, spring-loaded pressure reducing   détendeur *m* à ressort, soupape *f* de réduction à ressort
valve, spring-loaded safety   soupape *f* de sûreté à ressort
valve, steam throttle   registre *m* de vapeur
valve, stop   soupape *f*/vanne *f* d'arrêt, vanne *f* de barrage/de sectionnement
valve, strainer check   clapet *m* crépine
valve, suction   clapet *m* d'aspiration
valve, thermostatic   vanne *f* thermostatique
valve, thermostatic control   soupape *f* de contrôle thermostatique
valve, thermostatic expansion   détendeur *m* thermostatique
valve, three-way   robinet *m*/soupape *f*/vanne *f* à trois voies
valve, throttle   régulateur *m*, soupape *f* d'étranglement
valve, two-stage pressure reducing   détendeur *m* à double détente
valve, two-way   vanne *f* deux voies
valve, tyre   valve *f*
valve, water pressure   soupape *f* de pression d'eau
valve, weight-loaded   soupape *f* à contrepoids
valve body   corps *m* de vanne
valve cap, tyre   capuchon *m* de valve
valve chamber   chambre *f* de vannes
valve handle   levier *m* de vanne
valve key   clef *f* de vanne
valve linkage   mécanisme *m* de soupape
valve seat(ing)   siège *m* de soupape/de vanne/d'obturateur
valve spindle   tige *f* de soupape
van   camionnette *f*
van [large]   fourgon *m*
van, light   fourgonnette *f*

vanadium   vanadium *m*
vandal resistant   antivandale *adj*
vane [of a fan]   aube *f*
vane, directional   aube *f* directrice
vane, guide   aube *f* directrice, déflecteur *m*, vanne *f*/pale *f* de guidage
vane, turning   aube *f* directrice
vane, wind   girouette *f*
vanes, variable inlet   aubages *m,pl* d'entrée à inclinaison variable
vaporization   vaporisation *f*
vaporization, flash   vaporisation *f* instantanée
vaporization, (latent heat) of   chaleur *f* (latente) d'évaporation
vapour   vapeur *f*
vapour barrier   barrière *f* de vapeur, coupe-vapeur *m*, écran *m*/protection *f* pare-vapeur, pare-vapeur *m*
vapour density   densité *f* de vapeur
vapour density, saturation   densité *f* de saturation
vapour permeability   perméance *f* à la vapeur
vapour pressure   pression *f* de vapeur
vapour pressure diagram   diagramme *m* de pression de vapeur
vapour retarder   retardeur *m* de vapeur
vapour-tight   étanche *adj* aux vapeurs
variable *adj*   variable *adj*
variable   variable *f*
variable, control   variable *f* de contrôle
variable, controlled   variable *f* régulée
variable, dependent   variable *f* dépendante
variable, independent   variable *f* aléatoire
variable dimension   dimension *f* variable
variation   variation *f*
variation [difference]   écart *m*
variation, allowable   écart *m* admissible, tolérance *f*
variation, annual   variation *f* annuelle
variation, daily   variation *f* journalière/quotidienne
variation, dimensional [in relation to size specified]   écart *m*
variation, flow   variation *f* de débit
variation, load   variation *f* de charge
variation, monthly   variation *f* mensuelle
variation, output   variation *f* de débit
variation, permissible   écart *m* admissible, tolérance *f*
variation, yield   variation *f* de débit
variation in level [of supports, etc]   dénivellation *f*
variation of pressure/speed/temperature/etc   variation *f* de pression/vitesse/température/etc
variator   variateur *m*
varied/varying   varié *adj*
varnish   vernis *m*
varnish   vernir *vb*
varnished   verni *adj*
varnishing   vernissage *m*
vary   varier *vb*
VAT (value added tax)   taxe *f* sur la valeur ajoutée (TVA)
vat   bac *m*, cuve *f*
vault [arch]   voûte *f*

vault, archway   voûte *f*
vault, barrel or tunnel   tonnelle *f*, voûte *f* en berceau
vault, cloister(ed)   voûte *f* en arc de cloître
vault, cradle   berceau *m*
vault, cross   voûte *f* d'arêtes
vault, flat   voûte *f* plate
vault, groin(ed)   voûte *f* d'arêtes *f,pl*
vault, lightweight brick or tile   voûte *f* sarrasine
vault, rib(bed)   voûte *f* sur croisée d'ogives
vault, secondary, piercing a barrel vault   lunette *f*
vault, sloping, beneath treads of a spiral staircase
   coquille *f*
vault, small spherical   calotte *f*
vault in form of quarter sphere   cul-de-four *m*
vault in the form of an arch   arche *f*
vault light [glass blocks cast in concrete]   béton *m*
   translucide
vault springing from two parallel walls   berceau *m*
vault [make sth vault shaped]   cintrer *vb*
vaulted   cintré *adj*, voûté *adj*
vector sum of forces [resultant]   résultante *f*
vegetable   potager *adj*
vehicle   véhicule *m*
vehicle [of paint]   liant *m*
vein [geological]   filon *m*
vein [in wood or stone]   veine *f*
veining [in wood or stone]   veinage *m*
velocity   vitesse *f*
velocity, air   vitesse *f* de l'air
velocity, average/mean   vitesse *f* moyenne
velocity, centre-line   vitesse *f* axiale
velocity, convection   vitesse *f* de convection
velocity, critical   vitesse *f* critique
velocity, discharge   vitesse *f* de décharge
velocity, exhaust gas   vitesse *f* des gaz d'échappement
velocity, exit   vitesse *f* de sortie
velocity, face   vitesse *f* frontale
velocity, inlet   vitesse *f* d'entrée
velocity, outlet   vitesse *f* d'échappement de sortie
velocity, terminal   vitesse *f* terminale
velocity contour   profil *m* de vitesse
velocity head   hauteur *f* dynamique
velocity profile   profil *m* des vitesses
velocity reduction method   méthode *f* de regain statique
velometer   indicateur *m* de vitesse
Velux [patented rooflight]   Velux *m*
vendor   vendeur *m*
veneer   feuille *f* de placage, placage *m*
veneer [of plywood]   pli *m*
veneer   plaquer *vb*
veneer, face   placage *m* extérieur
veneer(ing) [of wood, the action and the result]   placage
   m
veneering press   presse *f* à plaquer
venetian [of a blind]   vénitien *adj*
venetian blind   store *m* vénitien
venetian blind leaf/blade   lame *f* de persienne
vent   évent *m*
vent [air]   trou *m* d'aération

vent, air   orifice *m* d'aération
vent, discharge   évent *m* de décharge
vent, emergency smoke   cheminée *f* d'appel
vent, exhaust   bouche *f* d'évacuation
vent, openable   bouche *f* d'évacuation ouvrable
vent, roof [smoke vent]   trappe *f* de désenfumage
vent, smoke   exutoire *m*, trappe *f* de désenfumage
vent opening [built into a partition or wall]   ventouse *f*
vent stack   colonne *f* de ventilation, colonne *f* des
   ventilations secondaires, ventilation *f*
vent(ing), air   purge *f* d'air
ventilate   aérer *vb*, ventiler *vb*
ventilated   aéré *adj*
ventilating [system, etc]   aérateur *adj*
ventilating, *see also* ventilation
ventilating equipment   installation *f* d'aération/de
   ventilation
ventilating system   installation *f*/système *m* d'aération
ventilation   aération *f*, ventilation *f*
ventilation, controlled   ventilation *f* contrôlée
ventilation, cross   ventilation *f* transversale
ventilation, exhaust   ventilation *f* par extraction
ventilation, fire   désenfumage *m*
ventilation, high pressure   ventilation *f* à haute pression
ventilation, mechanical   ventilation *f* mécanique
ventilation, natural   aération *f* aléatoire, ventilation *f*
   naturelle
ventilation, pressurized   ventilation *f* à surpression
ventilation, transverse   ventilation *f* transversale
ventilation rate (air changes per hour)   renouvellements
   m,pl d'air à l'heure
ventilator   aérateur *m*, bouche *f* d'aération, ventilateur *m*
ventilator, ceiling   bouche *f* de plafond
ventilator, roof   chatière *f*, ventilateur *m* de toiture
venture, joint   groupement *m* d'entreprises, partenariat *m*
veranda   véranda *f*
veranda, small, on columns   méniane *f*
verge [road]   accolement *m*, accotement *m*, bas-côté *m*,
   bord *m*
verge [roof]   avant-toit *m*
verge, grass   bande *f* engazonnée/non stabilisée
verge, paved [road]   bande *f* stabilisée
verge, soft   accotement *m* meuble/non stabilisé, bande *f*
   non stabilisée
verge pointing [of a roof]   ruellée *f*
verification [of measurements, readings, etc]   contrôle *m*
   des mesures
verification of compliance with facade projection and
   height restrictions   recolement *m*
verify   vérifier *vb*
verify compliance with facade projection and height
   restrictions   recoler *vb*
vermiculations [decoration on stonework]
   vermiculures *f,pl*
vermiculite   vermiculite *f*
vernier   vernier *m*
vertex [of an angle, a curve]   sommet *m*
vertical *adj*   à plomb, aplomb, d'aplomb, montant,
   vertical *adj*

vertical [of ladder, scaffolding, window/door jamb, etc] montant *m*

vertical [welding]   ascendant *m* [soudage]

vessel [receptacle], *see also* tank   cuve *f*, récipient *m*, vase *m*

vessel, collecting   réservoir *m* collecteur

vessel, containment [of a reactor]   enceinte *f* du réacteur

vestibule   vestibule *m*

viability, economic   rentabilité *f* économique

viaduct   passage *m* supérieur (PS), viaduc *m*

vibrate   vibrer *vb*

vibrating, *see also* vibration, vibratory   vibrant *adj*, vibratoire *adj*

vibration   vibration *f*

vibration [of concrete to compact it]   vibrage *m*

vibration, forced   vibration *f* forcée

vibration, free   vibration *f* libre

vibration, frequency of   fréquence *f* de vibration

vibration, internal, of concrete   pervibration *f*

vibration absorber/damper   amortisseur *m* de vibrations

vibrator   vibr(at)eur *m*

vibrator, concrete   dame *f*/vibrateur *m*/vibreur *m* à béton

vibrator, formwork   vibreur *m* de coffrage

vibrator, immersion/poker   pervibrateur *m*

vibrator, piling   vibrateur *m* de fonçage et d'arrachage, vibro-fonceur *m*

vibrator, plate [for compacting]   patin *m* vibrant

vibrator, poker   pervibrateur *m*

vibrator-finisher [road works]   vibreuse-finisseuse *f*

vibratory   vibratoire *adj*

vibro-compactor   vibro-compacteur *m*

vibroflotation   vibro-flottation *f*

vibro-hammer   vibro-fonceur *m*

vibro-mould equipment for precast concrete   vibro-mouleuse *f*

vicarage   presbytère *m*

Vicat needle   aiguille *f* de Vicat

vice   étau *m*

vice, front   presse *f* d'établi

vice, hand   tenaille *f* à vis

vice, plumber's   serre-tubes *m*

vice, plumber's chain   serre-tubes *m*

vice, small   étau *m* à agrafe

vice, tail   presse *f* parisienne/arrière de l'établi

vicinity   environs *m,pl*

view   prospect *m*, vue *f*

view, bird's eye [plan/drawing]   plan *m* cavalier

view, (cross-)sectional   vue *f* transversale/en coupe

view, exploded   vue *f* éclatée

view, field of   champ *m* visuel

view, front   élévation *f*/vue *f* de face

view, plan   vue *f* en plan

view, side   élévation *f*/vue *f* de côté, vue *f* latérale

view, unobstructed/unrestricted   vue *f* imprenable

view from above   vue *f* de dessus

viewpoint   point *m* de vue

viewpoint [at a beauty spot]   belvédère *f*

villa   pavillon *m*, villa *m*

village   village *m*

village [small, hamlet]   hameau *m*

vine   vigne *f*

vineyard   vignoble *m*

virginia creeper   vigne *f* vierge

viscometer   viscomètre *m*

viscosity   viscosité *f*

visibility [highways]   visibilité *f*

visibility distance   distance *f* de visibilité

visible   apparent *adj*, visible *adj*

vision   vision *f*

visual   visuel *adj*

visual display unit (VDU)   console *f* de visualisation

voice and data cabling   câblage *m* voix données image

void   vide *m*

void [concrete defect]   nid *m*

void, ceiling   vide *m* sous plafond

void ratio   indice *m* des vides

volatile   volatil *adj*

volcanic quake   tremblement *m* volcanique

volt   volt *m*

voltage   tension *f*, voltage *m*

voltage, breakdown [insulation]   tension *f* de claquage

voltage, grid [of a valve]   tension *f* de grille

voltage, grid [supply]   tension *f* du secteur, voltage *m* du réseau

voltage, insulation   tension *f* d'isolement

voltage, load   tension *f* en charge

voltage, low/high   basse/haute tension *f*

voltage, mains   tension *f* du secteur, voltage *m* du réseau

voltage, medium   tension *f* moyenne

voltage, no-load   tension *f* à vide

voltage, operating [of a lamp]   tension *f* de fonctionnement

voltage, starting   tension *f* d'allumage

voltage, supply   tension *f* du secteur, voltage *m* du réseau

voltage change   changement *m* de tension

voltage detector [electrical]   vérificateur *m* de tension

voltage drop   chute *f* de tension

voltage measurement   mesure *f* de la tension

voltmeter/voltage meter   voltmètre *m*

volume   volume *m*

volume [capacity]   capacité *f*

volume [contents]   contenance *f*

volume, change in   changement *m* de volume

volume, gross   volume *m* brut

volume, habitable [floor area × ceiling height]   volume *m* habitable

volume, net or useful   volume *m* utile

volume, part by   partie *f* en volume

volume measurement [of solids]   cubature *f*

volume range   plage *f* de volume

volume to be transported   volume *m* à transporter

volume, increase in   foisonner *vb*

volumetric   volumétrique *adj*

volume(tric) flow rate   débit *m* volumique

volumetric loss   perte *f* de volume

volute [spiral ornamentation]   volute *f*

vortex   tourbillon *m*

voussoir   claveau *m*, voussoir *m*

voussoir on each side of keystone   contre-clef *f*

voussoirs and keystone of an arch   bandage *m*

wage salaire *m*
wage earner salarié *m*
wage rate taux *m* des salaires
waggon chariot *m*
wainscot boiser *vb*
wainscot(ing) boiserie *f*, lambris *m*
waiting attente *f*
waiver request demande *f* de dérogation
walk allée *f*
walk-out grève *f*
walkway coursive *f*, passerelle *f*, trottoir *m*
walkway [for moving materials on site] plancher *m* de service
walkway, access/service chemin *m* de service
walkway, inspection passerelle *f* de visite
walkway, moving tapis *m* roulant
walkway, service passerelle *f* de service
walkway between sloping roof and gutter banquette *f*
walkway grid grille *f* de circulation
wall mur *m*
wall [enclosure] enceinte *f*
wall [face of] paroi *f*
wall, baffle/screen mur *m* d'écran
wall, Berlin [strutting system for deep excavations] paroi *f* berlinoise
wall, blockwork mur *m* en parpaing/en agglomérés
wall, breakwater [bridge pier protection] arc *m* de radier
wall, cavity cloison *f* de doublage, mur *m* creux
wall, curtain cloison *f* sèche, mur *m* rideau
wall, cut-off [dam] mur *m* parafouille
wall, defensive [e.g. round a town] muraille *f*
wall, diaphragm paroi *f* étanche
wall, diaphragm (cast in situ) paroi *f* moulée (dans le sol)
wall, dividing mur *m* de cloison/de séparation
wall, dock mur *m* bajoyer
wall, drystone mur *m* de pierres sèches
wall, dwarf or low [with coping to carry railings, etc] mur *m* bahut
wall, elbow height accotoir *m*, accoudoir *m*
wall, enclosing mur *m* de clôture
wall, end face of a tête *f* de mur
wall, external mur *m* extérieur
wall, fire (division) mur *m* coupe-feu, cloison *m* coupe-feu/pare-feu
wall, framed cloison *f* en charpente/en treillis
wall, gable (end) mur *m* pignon
wall, gravity retaining mur-poids *m* de soutènement
wall, grout paroi *f* au coulis
wall, hollow mur *m* à double paroi
wall, interior or internal cloison *f*
wall, L- mur *m* en angle
wall, leaning mur *m* déversé

wall, lining contre-mur *m*
wall, loadbearing cloison *f* porteuse, gros mur *m*, mur *m* porteur, paroi *f* porteuse
wall, loadbearing separating mur *m* de refend
wall, low bahut *m*, muret *m*, murette *f*
wall, low [with coping to carry railings, etc] mur *m* bahut
wall, low, on which one can lean mur *m* d'appui
wall, masonry mur *m* en maçonnerie
wall, party mur *m* mitoyen
wall, partition cloison *f*, mur *m* de séparation
wall, piece or length of [whole height] pan *m* de mur
wall, quay or wharf mur *m* bajoyer/de quai
wall, reinforced concrete paroi *f* en béton armé
wall, retaining mur *m* de soutènement
wall, revetment mur *m* de chemise
wall, river turcie *f*
wall, separating mur *m* de séparation
wall, separating [between properties, not necessarily a party wall] mur *m* séparatif
wall, separating [within a building, load-bearing] mur *m* de refend
wall, set back portion of a décrochement *m*
wall, sheet pile palée *f*, paroi *f*/rideau *m* de palplanches
wall, shuttered [cast-in-place] mur *m* banché
wall, side [at right angles to a bridge abutment] mur *m* en retour
wall, side [of doorway, arch, etc] pied-droit/piédroit *m*
wall, side, of a canal lock mur *m* bajoyer
wall, slurry paroi *f* au coulis
wall, small supporting [for a sink, a mantlepiece, etc] jambage *m*
wall, stud mur *m* ossaturé/de bois
wall, stud [lath and plaster] cloison *f* lattée et plâtrée
wall, supporting mur *m* d'appui
wall, thin (concrete) voile *m* (de béton)
wall, thin reinforced/unreinforced voile *m* armé/non armé
wall, (timber) framed cloison *f* en charpente
wall, wharf mur *m* bajoyer/de quai
wall, wing [at an angle to a bridge abutment] aile *f* de pont, mur *m* en aile
wall, wing, of a lock musoir *m* d'une écluse
wall at the back of a chimney hearth contre-cœur *m*
wall built against another contre-mur *m*
wall finish, picked [joints raked out of stonework] parement *m* piqué
wall finishing in marble or mosaic applique *f*
wall finishing specialist ravaleur *m*
wall friction angle angle *m* de friction sur paroi
wall panelling [thin stone or wood panel, etc] lambris *m*
wall plate [carrying feet of rafters] plate-forme *f*, solive *f* de ferme

wall plate [of a wall panel]   sablière *f*
wall plate [supporting a timber frame panel]   sablière *f* de
  chambrée
wall plate [timber]   lisse *f* d'assise
wall rib [of an arch]   formeret *m*
wall supported by archways   mur *m* en décharge
wall supporting chimney pipe(s)   mur *m* dosseret
wall supporting first steps of a staircase   mur *m*
  d'échiffre
wall surrounding a flue   mur *m* d'enveloppe
wall unit   élément *m* suspendu
wall units   mural[†] *m*
wall with simulated arches or openings   mur *m* orbe
walling, curtain   murs *m,pl* rideaux
wallpaper   papier *m* peint/de tenture, tapisserie *f*
wallpaper, embossed   papier *m* gaufré
wallpaper, tapestry   tapisserie *f*
wallpaper cutting knife   tranchet *m*
wallpaper paste   colle *f* de pâte
wallpaper pasting table   table *f* à encoller
wallpaper seam roller   rouleau *m* de colleur, roulette *f*
wallpaper stripper, steam   décolleuse *f* à vapeur
wallpaper stripping   grattage *m*
wallpaper   tapisser *vb*
wallpapering   pose *f* de papier peint
walnut [tree, wood]   noyer *m*
wane [on the edge of sawn timber]   flache *f*
waney-edged [of timber]   flacheux *adj*
waney-edged board/timber   bois *m* flache, chon *m*,
  contredosse *f*, redos *m*
ward [of a hospital]   pavillon *m*
warden   conservateur *m*
wardrobe   armoire *f* (de chambre à coucher), garde-robe *f*
warehouse   dépôt *m*, entrepôt *m*, hangar *m*, magasin *m*
warehouse, dockside   dock *m*
warm   chaud *adj*
warm air curtain   rideau *m* d'air chaud
warm air heater/heating unit   générateur *m* d'air chaud
warm, become or get   chauffer *vb*
warm sth   chauffer *vb* qch
warm (up) sth   réchauffer *vb* qch
warming (up)   réchauffage *m*
warn   avertir *vb*
warning [of a lamp]   témoin *adj*
warning   alerte *f*, avertissement *m*
warning device   dispositif *m* d'alerte/d'avertissement
warning light/signal   voyant *m* (d'avertissement)
warning light [flashing or winking]   clignotant *m*
warp   se déformer *vb*, se gauchir *vb*
warp [sth]   contourner *vb*, déformer *vb*
warp [wood]   déjeter *vb*, déverser *vb*, voiler *vb*
warped [wood]   déjeté *adj*, voilé *adj*
warping   déformation *f*, gauchissement *m*
warp(ing) [of wood]   dévers *m*, voilement *m*
warrant   garantir *vb*
wash stand   lave-mains *m*, table *f* de toilette, toilette *f*
wash tub   lavoir *m*
wash   laver *vb*
wash [clean]   nettoyer *vb*

wash [rinse]   rincer *vb*
wash away [banks, foundations, etc]   affouiller *vb*
wash down [paintwork, walls, etc]   lessiver *vb*
wash out [colours]   délaver *vb*
washable   lavable *adj*
wash-basin   cuvette *f*, lavabo *m*
washer   rond *m*, rondelle *f*
washer [ring]   anneau *m*, bague *f*
washer, air   laveur *m* d'air
washer, bellows   rondelle *f* Belleville
washer, lock [square section]   rondelle *f* Grower
washer, lock, external/internal tooth   rondelle *f* à
  denture extérieure/intérieure
washer, packing   rondelle *f* de garniture
washer, sealing   rondelle *f* d'étanchéité
washer, spring   anneau *m*/rondelle *f* à ressort
washer, tap   clapet *m*, joint *m*/rondelle *f* de robinet
washer, thrust   rondelle *f* de butée
washer between boiler or radiator sections   nipple *m*
wash-house   buanderie *f*, lavoir *m*
washing   lavage *m*
washing away [erosion]   affouillement *m*
washing down [paintwork]   lessivage *m*
washing machine   lave-linge *m*, machine *f* à laver
washing out [of colours]   délavage *m*
washroom   buanderie *f*, lavoir *m*
waste   gaspillage *m*
waste [rubbish]   déchets *m,pl*, détritus *m*
waste, demolition   mouchette *f*
waste, quarry   déchets *m,pl* (de) carrière
waste, reclaimable   détritus *m* récupérable
waste, rock   décharge *m* de roche
waste(s), recyclable   déchets *m,pl* recyclables
waste bank [excavations]   cavalier *m*
waste crusher   broyeur *m* à ordures
waste disposal   élimination *f* des déchets
waste removal   élimination *f*/évacuation *f* des déchets
waste stone from dressing stone to dimension   recoupe *f*
waste storage   stockage *m* de déchets
waste water   eaux *f,pl* usées (EU)
waste water, domestic   eaux *f,pl* ménagères/usées (EU)
waste water treatment   traitement *m* des eaux d'égouts
waste   gaspiller *vb*
watch tower   tour *f* de guet
watchman   gardien *m*
water   eau *f*
water, absorbed   eau *f* de rétention
water, body of   masse *f* d'eau
water, boiler feed   eau *f* d'alimentation de chaudière
water, bound   eau *f* fixée/liée
water, brackish   eau *f* saumâtre
water, capillary   eau *f* capillaire
water, chilled   eau *f* glacée
water, condensation   eau *f* de condensation
water, condenser   eau *f* de condenseur
water, cooling   eau *f* de refroidissement
water, depth of   hauteur *f* d'eau
water, distilled   eau *f* distillée
water, drinking   eau *f* potable

water, expanse of   masse *f* d'eau
water, filtered   eau *f* filtrée
water, fire-fighting   eau *f* pour incendie/pour la lutte contre l'incendie
water, foul   eaux *f,pl* vannes (EV)
water, free   eau *f* libre/de gravité
water, fresh   eau *f* douce/fraîche
water, hard   eau *f* dure
water, head of   charge *f* hydraulique, chute *f*
water, heating   eau *f* de chauffage
water, hot, *see* hot water etc
water, iced   eau *f* glacée
water, infiltration   eau *f* d'infiltration
water, inrush of   venue *f* d'eau
water, loss of   perte *f* d'eau
water, mains   eau *f* de distribution/de ville
water, make-up   eau *f* d'appoint
water, make-up [in treatment unit]   eau *f* de dilution
water, mixing [for mortar, concrete, etc]   eau *f* de gâchage
water, perched   nappe *f* suspendue
water, pore   eau *f* interstitielle
water, process   eau *f* de traitement industriel
water, raw   eau *f* brute
water, recirculated   eau *f* de retour
water, reserve of   réserve *f* d'eau
water, retained   eau *f* de rétention
water, return   eaux *f,pl* de retour
water, running   eau *f* courante
water, salt   eau *f* salée
water, seepage   eau *f* de percolation
water, sewage   eau *f* d'égout
water, soft   eau *f* douce
water, spray   eau *f* pulvérisée
water, stagnant   eau *f* stagnante, eaux *f,pl* mortes
water, surface   eau *f*/eaux *f,pl* de ruissellement, eau *f* superficielle/de surface
water, tap   eau *f* de distribution/de ville
water, treated   eau *f* traitée
water, untreated   eau *f* brute/non traitée
water, usable   eau *f* utilisable
water, WC waste   eaux *f,pl* vannes (EV)
water, waste, *see* waste water
water, well   eau *f* de puits
water analysis, chemical   analyse *f* chimique de l'eau
water check [e.g. fillet between roof slope and wall]   filet *m*
water classification   classification *f* des eaux
water closet (WC), *see* toilet, or lavatory
water content   teneur *f* en eau
water content [of concrete]   dosage *m* en eau
water cooler, drinking   refroidisseur *m* d'eau potable
water cooling installation   installation *f* à refroidissement par eau
(water) cooling jacket   chemise *f* réfrigérante (d'eau)
water curtain [fire barrier]   rideau *m* d'eau
water demand   besoin *m* en eau
water depth [draught]   tirant *m* d'eau
water discharge channel   canal *m* de rejet

water discharge density [irrigation]   densité *f* d'arrosage
water distribution, branched parallel rising main system of   distribution *f* en chandelle
water distribution, drop system of   distribution *f* en parapluie
water diviner   sourcier *m*/sourcière *f*
water extraction [from the ground]   puisage *m*
water feature   pièce *f* d'eau
water filtration   filtration *f* de l'eau
water hammer   choc *m* de l'eau, coup *m* de bélier
water hammer damper   antibélier *m*
water hardness   dureté *f* de l'eau
water heater   chauffe-eau *m*, réchauffeur *m* d'eau
water heater, bath [instantaneous]   chauffe-bain *m*
water heater, gas   chauffe-eau *m* à gaz
water heater, instantaneous   chauffe-eau *m* instantané
water heater, multi-point gas   chauffe-eau *m* à gaz à plusieurs points
water level   niveau *m* d'eau/de l'eau
water level, change in   changement *m* de niveau d'eau
water level, low/high   niveau *m* bas/haut de l'eau
water level, upper   niveau *m* d'eau supérieur
water level electrode   électrode *f* de niveau
water level indicator   indicateur *m* de niveau d'eau
water main   conduite *f* principale d'eau
water main connection   branchement *m* d'eau général
water of hydration   eau *f* d'hydratation
water requirement(s)   besoin *m* en eau
water reserve   réserve *f* d'eau
water run-off   écoulement *m* des eaux
water seal, *see also* siphon   clôture *f* à eau, fermeture *f* à eaux
water seal [in a siphon]   fermeture *f* hydraulique
water seal height in a siphon   garde *f* d'eau
water softener   adoucisseur *m* d'eau
water softener, domestic   adoucisseur *m* domestique
water softening   adoucissement *m* d'eau
water softening installation or plant   installation *f* d'adoucissement
water sterilization with sodium hypochlorite   verdunisation *f*
water storage   accumulation *f*/emmagasinement *m* d'eau
water storage [by damming]   retenue *f* d'eau
water storage capacity, effective   eau *f* utilisable
water storage heater   chauffe-eau *m* à accumulation
water storage level, highest/maximum [in a dam]   niveau *m* de la retenue
water storage tank   réservoir *m*
water supply   adduction *f* (d'eau), alimentation *f* en eau, distribution *f* d'eau
water supply, domestic   alimentation *f* en eau domestique
water supply, drinking   distribution *f* d'eau potable
water supply, emergency   alimentation *f* en eau de secours
water supply, emergency [well or sump]   puisard *m* incendie
water supply, gravity   adduction *f* gravitaire
water supply, hot   service *m* d'(alimentation en) eau chaude

water supply, hot/cold   alimentation *f* en eau chaude/froide

water supply, pumped   adduction *f* par refoulement

water supply, rural   distribution *f* d'eau rurale

water supply (company)   service *m* des eaux

water supply installation or plant   installation *f* d'alimentation en eau

water supply pipe   tuyau *m* d'alimentation d'eau

water supply system   réseau *m* d'alimentation en eau

water supply system of branched parallel rising mains   distribution *f* en chandelle

water system   circuit *m* d'eau

water table, *see also* groundwater   nappe *f* aquifère/d'eau/phréatique

water table, lowering of a   rabattement *m* d'une nappe d'eau

water table, perched   nappe *f* d'eau suspendue

water table, phreatic   nappe *f* phréatique

water tank   bac *m* à eau, citerne *f*

water tank (hot)   ballon *m* d'eau (chaude)

water tanker/bowser   tonne *f* à eau

water thermostat [in a boiler]   aquastat *m*

water tower   château *m* d'eau

water treatment   épuration *f* de l'eau, traitement *m* de l'eau/d'eau/des eaux potables

water treatment, feed   épuration *f* de l'eau d'alimentation

water treatment plant   poste *m* de conditionnement d'eau, usine *f* de traitement des eaux

water vapour   vapeur *f* d'eau

water vapour, saturated   vapeur *f* de saturation

water   arroser *vb*

water, add [to plaster, etc, for mixing]   gâcher *vb*

water-bearing [ground]   aquifère *adj*

water-cooled   à refroidissement par eau, refroidi par eau

watercourse   cours *m* d'eau

waterfall   cascade *f*, cataracte *f*, chute *f* d'eau

watering   arrosage *m*

watering, compaction by   compactage *m* par arrosage

watering can   arrosoir *m*

water-jet cutting   découpe *f* au jet d'eau

waterlogged   imbibé/imprégné d'eau

waterlogged [earth, ground]   hydromorphe *adj*

waterlogged land   terre *f* hydromorphe

waterlogging [of earth, ground]   hydromorphie *f*

watermill   moulin *m* à eau

waterproof   étanche *adj* (à l'eau), hydrofuge *adj*, imperméable *adj*

waterproof membrane   écran *m* étanche

waterproof membrane [bridge deck]   chape *f* d'étanchéité

waterproof membrane [dam]   feuille *f* d'étanchéité

waterproof membrane, reinforced bitumen   chape *f* de bitume armé

waterproof surfacing, multi-layer   multicouche[†] *f*

waterproof   étancher *vb*, imperméabiliser *vb*

waterproofing   étanchéisation *f*, imperméabilisation *f*

waterproofing [of a road bridge]   chape *f*/dispositifs *m,pl* d'étanchéité

waterproofing, multi-layered   étanchéité *f* multicouche(s)

waterproofing agent   imperméabilisant *m*

waterproofing compound [for mortar, concrete]   hydrofuge *m* (de masse)

waterproofing compound for surface application   hydrofuge *m* (de surface)

waterproofing system   système *m* d'étanchéité *f*/ d'imperméabilisation

waterproofness   étanchéité *f*

water-repellent   hydrofuge *adj*

water-resisting [e.g. of a canal lining]   inafouillable *adj*

watershed   ligne *f* de faîte/de partage des eaux

watertight   étanche *adj* (à l'eau), hermétique *adj*, imperméable *adj*

watertight against water jet at any angle   étanche *adj* à la lance

watertight when immersed   étanche *adj* à l'immersion

watertightness   étanchéité *f*

waterway   cours *m* d'eau

waterway, navigable   cours *m* d'eau navigable

waterworks   station *f* de pompage

watt   watt *m*

wattage   puissance *f*

wattmeter   wattmètre *m*

wave   vague *f*

wave [physics, i.e. sound, light, etc]   onde *f*

wave, sound   onde *f* sonore

wavelength   longueur *f* d'onde

wax   cire *f*

wax   cirer *vb*

way   voie *f*

way [of doing something]   moyen *m*

way in   entrée *f*

way (through)   passage *m*

WC (water closet), *see* toilet, or lavatory

weak   faible *adj*

weaken   affaiblir *vb*

weaken [a cement or mortar mix]   amaigrir *vb*, appauvrir *vb*

wear   usure *f*

wear [of a road, etc]   dégradation *f*

wear, frictional   usure *f* par frottement

wear and tear   usure *f*

wear and tear [of a house, rented property, etc]   dégradation *f*

wear and tear, fair   usure *f* normale

wear and tear allowance   amortissement *m* pour dépréciation

wear away   éroder *vb*

wear away or out   user *vb*

wear index   coefficient *m* d'usure

wearing away   attrition *f*, usure *f*

wearing course, bituminous   revêtement *m* bitumineux/hydrocarboné, couche *f* de roulement/de surface hydrocarbonée

wearing course [road construction]   couche *f* d'usure/de roulement

weather   temps *m*

weather, bad/inclement   intempérie *f*

weather conditions   conditions *f,pl* climatiques

weather data   données *f,pl* météorologiques
weather forecast   météo† *f*, prévision *f* météorologique
weather moulding [tongue, on edge of a casement]   languette *f*, mouton *m*
weather rebate [semi-circular, in the edge of a casement]   gueule-de-loup *f*, noix *f*
weather rebate in a jamb   contre-noix *f*
weather report   bulletin *m* météo(rologique)
weather strip [draught excluder]   bourrelet *m*
weather strip [on a window]   coupe-froid *m*
weather strip [on edge of outer leaf of a double door or window]   battement *m*
weather strip(ping)   bande *f* de calfeutrement
weather vane   girouette *f*
weather [geotechnics]   altérer *vb*
weather [crumble]   s'effriter *vb*
weather-board [on a door, window]   auvent *m*, jet *m* d'eau
weatherboard [cladding]   bardeau *m*, clin *m*
weatherboarding [cladding]   planches *f,pl* à recouvrement
weathercock   girouette *f*
weathered [earth, ground, rock]   altéré *adj*
weathering   dégradation *f*
weathering [geotechnics]   altération *f*
weathering [of stone]   désintégration *f*, effritement *m*
weathering, powdery, of stone surface   effleurissement *m*
weathering resistance [of soils, stone, etc]   résistance *f* à l'altération
weatherproof/weathertight   étanche *adj*, résistant *adj* aux intempéries
weatherproof [of electrical equipment]   étanche *adj*
weather-resistant   résistant *adj* aux intempéries
weathertight   résistant *adj* aux intempéries
weaving section [of road]   section *f* d'entrecroisement
weaving [of traffic]   entrecroisement *m*, insertion *f*
web [e.g. of girder]   âme *f*
web splice   joint *m* de l'âme
web plane   plan *m* de l'âme
web plate (of a composite beam)   âme *f* (d'une poutre composée)
web stiffener   raidisseur *m* d'âme
wedge   cale *f*, clavette *f*, coin *m*, taquet *m*
wedge, hardwood   coin *m* de bois dur
wedge, mason's [for splitting stone]   tranche *f*
wedge for jamming a tenon in an oversize mortise   rossignol *m*
wedge jammed into saw cut to prevent saw binding   bondieu *m*
wedge under foot of a timber strut or stay   détente *f*
wedge used to tighten props or shoring   coin *m* de serrage
wedge in   enclaver *vb*
wedge (up)   caler *vb*
wedging   calage *m*, coinçage *m*
wedging in   enclavement *m*
weed   mauvaise herbe *f*
weephole [of a retaining wall]   barbacane *f*
weephole [vertical slot in a wall]   chantepleure *f*

weephole, condensation [in joinery]   trou *m* de buée
weeping [e.g. of a joint, a damp wall]   suintement *m*
weigh batcher [concrete]   dispositif *m* de pesage, doseur *m*
weigh   peser *vb*
weigh down [causing deflection]   affaisser *vb*
weigh down on   peser *vb* sur
weighbridge   balance *f* à bascule, pont-bascule *m*
weighbridge [railway]   bascule *f* à wagon
weigher, aggregates/cement   bascule *f* à agrégats/ciment
weigher, belt   tapis *m* peseur
weighing   pesage *m*
weighing unit [for mixing materials]   groupe *m* doseur
weight   poids *m*
weight [grams/m²]   grammage *m*
weight [relative]   pesanteur *f*
weight, atomic   poids *m* atomique
weight, balance   contrepoids *m*
weight, dead   poids *m* mort
weight, dry [of material]   poids *m* de matériau sec
weight, dry unit   densité *f* sèche, poids *m* volumique sec
weight, gross   poids *m* brut
weight, live   poids *m* utile
weight, molecular   poids *m* moléculaire
weight, net   poids *m* net/utile
weight, part by   partie *f* en poids
weight, percentage by   pourcentage *m* du poids
weight, self   poids *m* propre
weight of standard air   poids *m* de l'air normal
weighty   pesant *adj*
weir   barrage *m*, seuil *m* en rivière
weir [of a dam]   déversoir *m*
weir, control   barrage *m* de retenue/de régulation
weld   cordon *m* (de soudure), soudure *f*
weld, arc   soudure *f* à l'arc
weld, butt   soudure *f* bout à bout
weld, continuous   soudure *f* continue
weld, field or site   soudure *f* de chantier
weld, fillet   cordon *m* d'angle, soudure *f* en/d'angle
weld, flat   soudure *f* à plat
weld, hand/manual   soudure *f* à la main
weld, lap   soudure *f* à recouvrement
weld, multi-layer   soudure *f* en plusieurs passes
weld, plug   soudure *f* en bouchon
weld, seal   soudure *f* d'étanchéité
weld, side   cordon *m* latéral
weld, tack   soudure *f* de pointage
weld cracking   fissuration *f* dans la soudure
weld metal   métal *m* déposé
weld root   base *f* de la soudure
weld   souder *vb*
weldability   soudabilité *f*
weldable   soudable *adj*
welder   soudeur *m*
welder's mask   masque *m* de soudeur
welding   soudage *m*, soudure *f*
welding (work)   travaux *m,pl* de soudage
welding, acetylene   soudage *m* autogène/oxyacétylénique
welding, arc   soudage *m* à l'arc
welding, bronze   soude-brasage *m*

welding, butt   soudage *m* bout à bout, soudage *m* en bout
welding, carbon arc   soudage *m* à l'arc au charbon
welding, deposit   rechargement *m*
welding, discontinuous/intermittent   soudage *m* discontinu
welding, electric (arc)   soudure *f* électrique (à l'arc)
welding, electric resistance   soudure *f* électrique par résistance
welding, fillet   soudage *m* d'angle
welding, fusion   soudage *m* par fusion
welding, gas   soudage *m* au gaz/autogène
welding, hand   soudage *m* manuel
welding, oxyacetylene   soudage *m* autogène, soudage *m*/soudure *f* oxyacétylénique
welding, oxyhydrogen   soudage *m* au chalumeau oxhydrique
welding, resistance   soudage *m* par résistance
welding, spot   soudure *f* par points
welding, stitch or tack   pointage *m*, soudage *m* par points
welding, stud   soudage *m* de goujons
welding blowpipe   chalumeau-soudeur *m*
welding cable   câble *m* de soudage
welding filler rod   baguette *f* d'apport
welding flux   flux *m*/poudre *f* à souder
welding generator   génératrice *f* pour soudage
welding helmet   casque *m* de soudage
welding jig   montage *m* de soudage
welding machine   machine *f* à souder
welding machine, (arc)   machine *f* pour soudage à l'arc
welding rod   baguette *f* à souder/de soudure/pour soudage, fil *m* à souder
welding screen   écran *m* de soudeur
welding sequence   ordre *m* de soudage
welding set   poste *m* de soudage
welding shield, hand-held   écran *m* à main
welding shop   atelier *m* de soudage
welding torch   chalumeau *m* (soudeur)
well [for stairs, lift]   cage *f*
well [water source]   puits *m*
well, absorbing/drain   puits *m* drainant/filtrant
well, artesian   puits *m* artésien
well, drain   puits *m* drainant/filtrant
well, elevator/lift   cage *f*/gaine *f* d'ascenseur
well, firemen's   puisard *m* incendie
well, hot   source *f* de chaleur souterraine
well, open   puits *m* à ciel ouvert
well, open [of a staircase]   noyau *m* évidé
well, relief   puits *m* de décharge
well, ventilation   puits *m* d'aération/de ventilation, trémie *f* (d'aération)
well, water   puits *m* à eau
well casing or lining   blindage *m* de puits
well of a staircase, width of   jour *m*
well-graded/-sorted   bien calibré
well-lit   lumineux *adj*
west, western, westerly, westward   ouest *adj*
west   ouest *m*
wet   humide *adj*, mouillé *adj*
wet bulb depression   différence *f* psychrométrique

wet bulb temperature   température *f* humide
wet bulb thermometer   thermomètre *m* mouillé/à bulbe mouillé
wet rot   carie *f* humide
wet   mouiller *vb*
wetness   humidité *f*
wharf   appontement *m*, quai *m*
wharf wall   mur *m* de quai
wheel, *see also* handwheel   roue *f*
wheel [handwheel]   volant *m*
wheel, adjusting/adjustment   volant *m* de réglage
wheel, clamp(ing)   volant *m* de serrage
wheel, gear   pignon *m*
wheel, heat   roue *f* de chaleur
wheel, pulley   réa *m*
wheel, spare   roue *f* de secours
wheel, steering   volant *m*
wheel, thermal   échangeur *m* rotatif
wheelbarrow   brouette *f*
whet [blade, tool, etc]   affiler *vb*
whetstone   pierre *f* à aiguiser
whirlpool   tourbillon *m* d'eau
whirlpool bath   baignoire *f* à brassage
white painting   blanchiment *m*
white spirit   white spirit *m*
white, paint   blanchir *vb*
whitewash   badigeon *m* (blanc), blanc *m* de chaux
whitewash, lime/cement   lait *m* de chaux/ciment
whitewash   badigeonner *vb*
whitewashing   badigeonnage *m*
wholesale   en gros
wholesaler   grossiste *m*
wide   large *adj*
widened   élargi *adj*
widening   élargissement *m*
width   largeur *f*
width [of sheet material, e.g. carpet]   lé *m*
width, base   épaisseur *f* à la base
width, clear   largeur *f* libre/utile
width, crest [of a dam]   épaisseur *f* en crête
width, duct   largeur *f* de canal
width, flange   largeur *f* d'aile
width, net or usable   largeur *f* utile
width, nominal   largeur *f* nominale
width, overall   largeur *f* totale/hors tout
width, staircase tread   emmarchement *m*
width, stair(well)   jour *m*
width, usable   largeur *f* utile
width of wallpaper between edgings   lé *m*
width usable by vehicles   largeur *f* roulable
wild cherry [tree, wood]   merisier *m*
willow   saule *m*
winch   cabestan *m*, treuil *m*
wind   vent *m*
wind power   énergie *f* éolienne
wind screen [*not* windscreen]   paravent *m*
wind shield   brise-vent *m*
wind tunnel   tunnel *m* aérodynamique/d'essai aéraulique
wind [e.g. a road]   serpenter *vb*

wind round   contourner *vb*
wind (up)   enrouler *vb*
windbrace   contrevent *m*
windbrace, diagonal   entretoise *f* de contreventement
windbracing   contreventement *m*
windbracing [in a metal or timber wall panel]   guette *f*
windbracing, crossed [timber frame structure]   croix *f*
windbracing, vertical framework with   palée *f* de contreventement
windbreak   abat-vent *m*, brise-vent *m*, coupe-vent *m*
winder [stair tread of varying width]   marche *f* gironnée/dansante/tournante
winding [staircase]   tournant *adj*
winding [electrical]   bobinage *m*, enroulement *m*
winding, motor   bobinage *m*/enroulement *m* de moteur
winding, transformer   bobinage *m* de transformateur
winding staircase, *see* staircase
windlass   cabestan *m*, treuil *m*
windmill   éolienne *f*, moulin *m* à vent
(wind)mill, smock   moulin *m* (à vent) hollandais
window   fenêtre *f*
window [small, at mezzanine level]   mezzanine *f*
window, attic   lucarne *f*
window, basement [small, generally barred]   soupirail *m*
window, bay   baie *f* vitrée, bretèche *f*, fenêtre *f* en baie/en saillie
window, bow   bretèche *f*, fenêtre *f* en saillie
window, bull's eye   œil-de-bœuf *m*
window, casement   croisée *f*, fenêtre *f* à battants
window, circular   oculus *m*, œil *m*
window, dormer   lucarne *f*
window, double casement   fenêtre *f* à deux vantaux
window, double glazed   double fenêtre *f*
window, folding   fenêtre *f* pli(ss)ante/en accordéon
window, French   porte-croisée *f*, porte-fenêtre *f*
window, hinged   fenêtre *f* à charnières
window, horizontally pivoted   fenêtre *f* basculante/à axe horizontal
window, inspection [e.g. of a boiler]   voyant *m*
window, inward opening casement   fenêtre *f* à la française
window, lattice   fenêtre *f* treillagée
window, leaded glass   vitrail *m*
window, linked leaves, each pivoting about a vertically sliding axis at one edge   fenêtre *f* à l'australienne
window, louvred   châssis *m* d'aération
window, mansard [in the lower slope of the roof]   fenêtre *f* en mansarde
window, mansard dormer   fenêtre *f* mansardée
window, mullion   fenêtre *f* à meneau
window, open, in a party wall   jour *m* de coutume
window, oriel   oriel *m*
window, outward opening casement   fenêtre *f* à l'anglaise
window, picture   baie *f* vitrée
window, sash   châssis *m* (à guillotine), fenêtre *f* à guillotine/à coulisse
window, secondary, placed against another   contre-fenêtre *f*

window, shop   devanture *f* de magasin, montre *f*, vitrine *f*
window, single casement   fenêtre *f* à un vantail
window, single leaf, inwards opening, with a vertically sliding axis at its lower edge   fenêtre *f* à la canadienne
window, single leaf, outwards opening, with a vertically sliding axis near its upper edge   fenêtre *f* à l'italienne
window, single-glazed   fenêtre *f* simple/à simple vitrage
window, sliding   fenêtre *f* coulissante
window, small, providing very little light   faux-jour *m*
window, stained glass   verrière *f*, vitrail *m*
window, storm   contre-fenêtre *f*
window, three-leaf articulated   châssis *m*/fenêtre *f* à l'australienne
window, vertically pivoted   fenêtre *f* pivotante
window casement with a tongue [*mouton*] on its mating edge   battant *m* mouton
window cleaner   laveur *m* de carreaux
window cleaners' cradle   nacelle *f* de nettoyage
window hardware   ferrure *f*
window head [rail] [timber structure]   traverse *f* dormante
window in a wall in accord with a particular agreement or easement   jour *m* de servitude
window jamb [pier or post]   dormant *m*/poteau *m* de fenêtre, poteau *m* d'huisserie/de baie
window jamb [splay]   ébrasement *m*, ébrasure *f*
window ledge or sill   appui *m* (de fenêtre)/(de baie), banquette *f*/tablette *f* de fenêtre, pièce *f* d'appui, rebord *m*
window opening   ouverture *f* de fenêtre *f*
window opening, splayed inside   abat-jour *m*
window overlooking a neighbour to which the latter cannot object   jour *m* de souffrance
window pane   carreau *m* (de vitre), vitre *f*
window retaining strap, limiting opening   compas *f* de sécurité
window unit (prefabricated)   bloc-croisée *m*
window wider than it is tall   mezzanine *f*
window with openable glass louvres   fenêtre *f* à jalousies
windows   vitrage *m*
windscreen [*not* wind screen]   pare-brise *m*
windsock   manche *f* à air
wine-growing or producing   viticole *adj*
wing [of a building]   aile *f*, pavillon *m*
wing, small [of a building]   ailette *f*
wing-screw   vis *f* à ailettes
wink [of a signal light]   clignoter *vb*
winking [light]   clignotant *adj*
wire   fil *m* (métallique)
wire, annealed   fil *m* recuit
wire, balancing   câble *m* de compensation
wire, barbed   barbelé *m*, fil *m* de fer barbelé
wire, call   fil *m* d'appel
wire, connecting/connection   fil *m* de liaison
wire, curtain [plastic covered spring]   spirotringle *f*, tringle *f* à ressort
wire, draw   tire-fils *m*
wire, drawn   fil *m* tréfilé
wire, earth(ing)   fil *m* de terre/de masse
wire, feed   fil *m* d'amenée

wire, fuse   fil *m* fusible
wire, lead tin solder   fil *m* de soudure
wire, live   fil *m* de sous tension
wire, return   fil *m* de retour
wire, span   hauban *m*
wire, steel   fil *m* d'acier/de fer
wires [electrical transmission line]   ligne *f*
wires, bonded   fils *m,pl* adhérents
wire brush   brosse *f* métallique
wire brush, circular   brosse *f* circulaire
wire cutters   coupe-fil *m*
wire drawing   tréfilerie *f*
wire edge on a cutting tool   morfil *m*
wire fencing   grillage *m*
wire gauze   toile *f* métallique
wire mesh   grillage *m*
wire mesh, steel   treillis *m* de fil d'acier
wire mesh, welded   fil *m* de treillis soudé
wire mesh screen [fine]   tamis *m* en toile métallique
wire nail   pointe *f* à tête *f* conique, pointe *f* de Paris
wire netting   grillage *m*, fer *m* maillé, treillis *m* métallique
wire stripper(s)   dénude-fil *m*, pince *f* à dénuder
wire wool   paille *f* de fer
wire up [a building]   faire *vb* l'installation électrique de
wire up [a circuit]   installer *vb*
wire up [a piece of equipment]   monter *vb*
wiring   câblage *m*
wiring, circuit   câblage *m* de circuit
wiring, house   câblage *m* électrique intérieur
wiring, permanent   câblage *m* permanent
wiring, remote control   câblage *m* de télécommande
wiring, surface   câbles *m,pl* apparents
wiring accessory   accessoire *m* de câblage
wiring diagram   schéma *m* de câblage/de montage
wiring duct, auxiliary [communications wiring, etc]   colonne *f* pilote
wiring installation   installation *f* électrique
wiring marking system   système *m* de repérage pour filerie
wiring system   système *m* de câblage
wiring up [of a circuit]   installation *f*
wiring up [of a piece of equipment]   montage *m*
with or along the grain   de droit fil
with the stream   au fil de l'eau
withdraw   retirer *vb*
withdrawable   retirable *adj*
within his/her province/responsibility/scope   de son ressort *m*
witness   témoin *m*
wood *adj*, *see also* wooden   à/de/en bois
wood, *see also* board, plank, timber, etc   bois *m*
wood, fungal breakdown of   échauffure *f*
wood, glued-laminated   bois *m* lamellé-collé, lamellé-collé *m*
wood, imitation [painted]   faux-bois *m*
wood, impregnated   bois *m* imprégné
wood, improved   bois *m* amélioré
wood, laminated   bois *m* lamellé, (bois *m*) lamellé-collé *m*
wood, painted imitation   faux-bois *m*

wood, plastic   bois *m* synthétique
wood, sawn   bois *m* avivé
wood, semi-dry   bois *m* d'entrée
wood, thin strip of   lame *f*
wood, treated [impregnated]   bois *m* imprégné
wood, worm-eaten   bois *m* mouliné
wood cut parallel to the grain   bois *m* de refend
wood grain on a lengthwise cut face   maille *f* du bois
wood with grain disturbed by knots   bois *m* tranché
wooded   arboré *adj*, boisé
wooden   à/de/en bois *adj*
woodland   pays *m* boisé
woodshed   bûcher *m*
woodwool   copeaux *m,pl*, laine *f* de bois
woodwork   boiserie *f*, menuiserie *f*
woodworker's bench   établi *m* de menuisier
woodworking machine   machine *f* à bois, machine-outil *f* à (travailler le) bois
woodworking machine, universal   machine *f* à bois polyvalente
wool, glass (insulation)   laine *f* de verre
wool, mineral   laine *f* minérale/de minéral
wool, slag   laine *f* de laitier/de scories
wool, wire   paille *f* de fer
word processing   traitement *m* de texte
word processor   machine *f* à/de traitement de texte
work   travail *m*
work [book, drawing, painting, etc]   œuvre *f*
work [i.e. something built, done, made]   ouvrage *m*
work [result of building/construction]   ouvrage *m*
work, at [e.g. forces]   en jeu
work, description/record of finished   dossier *m* des ouvrages exécutés (DOE)
work, design   études *f,pl* (de conception), travail *m* d'étude
work, major civil engineering   ouvrage *m* d'art
work, minor conversion/maintenance   corvée *f*
work, refacing   ravalement *m*
work, shift   travail *m* par relais
work, structural timber or steel   charpenterie *f*
work, unskilled   travail *m* de manœuvre
work, welding   travaux *m,pl* de soudage
work, wrought iron   ferronnerie *f*
work fanning out from one centre   ouvrage *m* en éventail
work hardening   vieillissement *m* mécanique
work hardening [cold]   écrouissage *m*
work only [excluding materials]   façon *f*
work package   lot *m*
work plan/programme/schedule   plan *m* de travail
work platform   estrade *f*/plancher *m* de travail
work platform, aerial   élévateur *m* à nacelle
work platform, elevating   plate-forme *f*/table *f* élévatrice
work platform, telescopic boom   nacelle *f* à bras/flèche télescopique
work section   tranche *f* de travaux
work sheet   fiche *f* de travail
work site area   emprise *f* du chantier
work station   poste *m* de travail

work to be covered up, description of   constat *m*
work with materials supplied free   travail *m* à façon
work   travailler *vb*
work [a machine]   manœuvrer *vb*
work [form or shape]   façonner *vb*
work [function]   marcher *vb*
work harden [cold]   écrouir *vb*
work in fits and starts   travailler *vb* bâtons rompus
work in shifts   se relayer *vb*
work loose   prendre *vb* du jeu
work out [amounts, numbers, etc]   chiffrer *vb*
work round   contourner *vb*
workability [concrete]   maniabilité *f*
workability [of concrete, material]   ouvrabilité *f*
workable [of a machine, etc]   capable *adj* de fonctionner
workable [of a material]   ouvrable *adj*
workbench   banc *m*, établi *m*
worked   travaillé *adj*
worked, cold   écroui *adj*
worker   ouvrier *m*, travailleur *m*
worker, construction   ouvrier *m* de construction
worker, piece   tâcheron *m*
worker, self-employed   travailleur *m* indépendant
worker, semi-skilled   ouvrier *m* spécialisé
worker, skilled   ouvrier *m* qualifié
worker, unskilled   manœuvre *m*
worker consultation   consultation *f* des travailleurs
worker information   information *f* des travailleurs
worker participation, equal   participation *f* équilibrée des travailleurs
worker recreation area   zone *f* sociale réservée aux travailleurs
worker training   formation *f* des travailleurs
workers' representative   représentant *m* des travailleurs
workforce   main-d'œuvre *f*
working *adj* [of equipment, etc]   utilisable *adj*
working [mine, quarry]   abattage *m*
working [of a mine, railway, etc]   exploitation *f*
working [of apparatus, a company, etc]   fonctionnement *m*
working [of equipment]   marche *f*
working [of stone, timber, etc]   façonnage *m*
working, cold   écrouissage *m*
working, opencast   exploitation *f* à ciel ouvert
working area, clean   zone *f* de travail blanche
working area [radius] of crane   aire *f* de balayage de la grue
working condition [of equipment, etc]   état *m* de fonctionnement
work(ing) day [of the week]   jour *m* ouvrable/de travail
work(ing) hours   heures *f,pl* de travail
working order, in   en état de marche, en ordre de marche
working party   groupe *m* de travail
working period, actual   durée *f* effective de service
workman   ouvrier *m*
workman, jobbing   tâcheron *m*
workmanship   exécution *f*, travail *m*
workmanship [of a craftsman]   maîtrise *f*
workmanship, bad   malfaçon *f*
workmanship, bad [piece of]   loup *m*

workmanship, defect due to bad   malfaçon *f*
work(manship), faulty   malfaçon *m*
workplace lighting   éclairage *m* du lieu de travail
works   travaux *m,pl*
works, acceptance of the, and final account   réception *f* des travaux (RDT)
works, accessory/auxiliary   ouvrage *m* accessoire
works, associated   ouvrages *m,pl* annexes
works, building and civil engineering   bâtiment *m* et travaux publics *m,pl* (BTP)
works, civil or (main) structural   gros œuvre *m*
works, clerk or inspector of   surveillant *m* de travaux
works, completion of the   achèvement *m* des travaux
works, demolition   travaux *m,pl* de démolition
works, description of   descriptif *m*, description *f* de travaux
works, design (and supervision) of the   étude *f* (et supervision) des travaux
works, dismantling   travaux *m,pl* de démantèlement
works, drainage   ouvrage *m* de drainage/ d'assainissement
works, excavation   travaux *m,pl* de creusement/ terrassement
works, finishing   travaux *m,pl* de finissage
works, finishing [closure, services, partitions, finishes]   parachèvement *m*, second œuvre *m*
works, foundation   travaux *m,pl* de fondation
works, main structural (contract)   gros œuvre *m*
works, minor [non-structural masonry, etc]   ouvrages *m,pl* légers/menus
works, programme of the   phasage *m* des travaux
works, progress of the   avancement *m* des travaux
works, rehabilitation   travaux *m,pl* de réhabilitation
works, restoration   travaux *m,pl* de restauration
works, roofing   travaux *m,pl* de toiture
works, schedule of   bordereau *m* des travaux
works, secondary [e.g. finishes, fitting out]   second œuvre *m*
works, structural   gros œuvre *m*
works, (sub)contractor for small   tâcheron *m*
works, supervision of the   direction *f* des travaux
works, supporting [provision of support]   travaux *m,pl* de soutènement
work(s), underground   travaux *m,pl* souterrains
works doctor or medical officer   médecin *m* d'entreprise/du travail
workshop   atelier *m*
workshop, structural steel   atelier *m* de constructions métalliques
workshop certificate   certificat *m* d'usine
worktop   plan *m* de travail
worm-eaten   vermoulu *adj*
worm-eaten state [of wood]   vermoulure *f*
worn   usagé *adj*
worn away or out [by use]   usé *adj*
woven   tissé *adj*
wrapping, pipe   revêtement *m* des tubes
wreath [of a handrail]   quartier *m* tournant
wrecking bar   pied-de-biche *m*

wrench, *see also* spanner   clef *f*
wrench, adjustable/shifting   clef *f* à crémaillère/à molette
wrench, all-purpose   pince *f* universelle
wrench, chain [plumbing]   clef *f* à chaîne
wrench, monkey   clef *f* anglaise/à molette, pince-étau *f*
wrench, open ended   clef *f* à fourches
wrench, pipe   clef *f* à tubes/de tuyau, pince *f* multiprise, serre-tubes *m*
wrench, ring   clef *f* polygonale
wrench, screw   clef *f* anglaise
wrench, strap [plumbing]   clef *f* à sangle
wrench, tap   tourne-à-gauche *m*

wrench, torque   clef *f* à choc
wrinkling [paint defect]   ridage *m*
wrist band [protection for heavy work]   poignet *m* de force
writ [summons]   assignation *f*
write a cheque   tirer *vb* un chèque
write off   amortir *vb*
wrong direction, in the   à contre-sens
wrong way round   sens devant derrière
wrong way up   sens dessus dessous
wrought   travaillé *adj*
wrought iron   fer *m* battu/forgé/soudé
wrought iron work   ferronnerie *f*

X-ray examination or investigation   examen *m* à rayons X, analyse *f* aux rayons X
X-ray equipment   appareil *m* radiographique

X-ray photograph   photographie *f* aux rayons X
xenon   xénon *m*

Y-connection   couplage *m* en étoile
Y-junction [plumbing, ductwork]   culotte *f* (simple)
yard [court]   cour *f*
year of manufacture   année *f* de fabrication
yearly   annuellement *adv*
yew   if *m*
yield   produit *m*
yield [financial]   rapport *m*
yield [output]   débit *m*
yield (of) [an investment]   rendement *m* (de)

yield point   limite *f* élastique/d'écoulement
yield point, compressive   limite *f* d'écrasement, limite *f* élastique à la compression
yield [collapse]   céder *vb*
yield [give way]   s'écouler *vb*
yield plastically   déformer *vb* de manière plastique
yielding   écoulement *m*
Young's modulus   module *m* d'élasticité/de Young
yttrium   yttrium *m*

zebra crossing [UK]   passage *m* clouté
zeolite   zéolite *m*
zinc   zinc *m*
zinc coated   zingué *adj*
zinc coating   galvanisation *f*, zingage *m*
zinc sheet   tôle *f* en zinc
zinc white   blanc *m* de neige

zone   aire *f*, zone *f*
zone, borrow   zone *f* d'emprunt
zone, buffer   zone *f* butoir
zone, comfort   zone *f* de confort
zone, compression   zone *f* comprimée
zone, danger(ous)   zone *f* dangereuse
zone, dead   zone *f* morte

zone, deceleration zone *f* de décélération
zone, failure zone *f* de rupture
zone, fire zone *f* d'incendie
zone, freezing zone *f* de pénétration du gel *m*
zone, grout(ed) zone *f* injectée
zone, hardened zone *f* trempée
zone, local control zone *f* de contrôle local
zone, multi-activity zone *f* d'interventions multiples
zone, neutral zone *f* neutre

zone, occupied zone *f* d'occupée/d'occupation
zone, shear zone *f* de rupture par cisaillement
zone, tension zone *f* tendue
zone, transition domaine *m* de régime transitoire, zone *f* de transition
zone allowed for finishes, services etc réservation *f*
zoning zonage *m*
zoning plan plan *m* d'occupation des sols (POS)

PART TWO

Français/Anglais
French/English

à action rapide   quick-acting
à ailettes   finned
à âme pleine   solid webbed
à angle droit   square [at right angles]
à angles droits   rectangular
à angle(s) vif(s)   square-edged
à barres et écharpes   ledged and braced
à bascule   rocker *adj* [of a switch]
à bon marché   cheaply
à bords arrondis   round-edged
à chemise d'air   air-lagged
à chenilles   tracked [of caterpillar vehicle]
à ciel couvert   covered [from the sky]
à ciel ouvert   open [to the sky], opencast [mining, etc]
à col de cygne   swan-necked
à commande mécanique   mechanically controlled
à commande pneumatique   pneumatically controlled
à contre-courant   counterflow *adj*
à contre-fil   across or against the grain, against the current [of water]
à contre-fil de l'eau   upstream
à contre-sens   in the wrong direction
à courants croisés   cross-flow *adj*
à court terme   in the short term, short term [loan, etc]
à débattre   to be discussed
à déclenchement rapide   quick-release *adj*
à distance   remote [as in remote control, etc]
à encastrer   flush mounting
à fond   thoroughly
à gauche   anti-clockwise
à grain fin   fine-grained
à grande échelle   large-scale
à haute puissance   at high power
à haute résistance   high-strength, high-tensile
à haute vitesse   high-speed, high-velocity
à joint fermé   close-jointed [rock, etc]
à l'abri   under cover
à l'abri de   sheltered from
à l'aplomb de   plumb with
à l'écart   in the background, out of the way, to one side
à l'échelle   to scale [of a model or drawing]
à l'heure   per hour
à la main   hand-operated, manual
à long terme   long term [loan, etc]
à pans coupés   with corners cut off
à plat joint   butted together [boards or planks, side-to-side]
à pied d'œuvre   on site
à plein régime   at full or top speed
à portée   within reach
à rebord   with a raised edge
à rebours   against the grain, backwards
à recouvrement   lapped *adj*
à refroidissement par eau   water-cooled

à regard   with an inspection cover or manhole
à remettre *vb* **en état**   needing renovation
à sens unique   one-way [road, street]
à six/huit/etc pans   hexagonal/octagonal/etc
à terme   full-term, in due time
à texture fermée/serrée   close-grained or -textured [rock, etc]
à toute allure   at full or top speed
à toute puissance   at full power
à usages multiples   multi-purpose
à turbocompresseur   turbo-charged
à vide   no-load or unloaded
à vide, marcher *vb* **à**   run unloaded or at no-load
à visser   threaded [i.e. to be screwed]
à vive arête   square-edged
à voie étroite   narrow-gauge [of railways]
à volet   with cover flap [of a socket outlet]
abaissement *m*   decrease, drawdown, falling, lowering, sinking, subsidence
abaissement *m* **de courant**   decrease in current [electrical]
abaissement *m* **de l'eau souterraine**   groundwater lowering
abaissement *m* **de la température**   fall in temperature
abaissement *m* **du point de rosée**   dew point depression
abandonné *adj*   abandoned, derelict, disused
abaque *m*   abacus [slab on top of a column], design chart/diagram/table, diagram, graph, nomogram   .
abat-jour *m*   lampshade, sun blind, window opening [splayed inside]
abat-son *m*   louvred opening [in a bell tower], luffer board
abattage *m*   cutting down or felling [of trees], demolition or knocking down [e.g. of a wall], mining or working of a mine or quarry, removal of projecting parts of a construction
abattage en gradins   stoping
abattage, front *m* **d'**   vertical face of trench excavation
abattage *m* **à l'explosif**   blasting [ground working]
abattage *m* **de retombée**   angled cut in stone arch springer or abutment
abattage *m***, fouille** *f* **en**   excavation by removal of rectangular blocks
abattage *m* **hydraulique**   jetting [excavation]
abatteur *m*   tree feller
abattant *m*   toilet seat [lifting flap]
abattement *m*   reduction [in cost or price], trimming [of wood, a block of stone, to size]
abatteuse *f*   boomheader, roadheader
abattis *m*   brushwood
abattis, faire *vb* **un**   clear fell
abattre *vb*   clench/clinch [a nail], cut down or fell [a tree]
abattre *vb* **à la terre**   raze to the ground

abat-vent *m*   cowl [on top of a chimney], windbreak

abat-voix *m*   sounding board

abattoir *m*   abattoir, slaughterhouse

abîmer *vb*   damage, spoil

ablation *f*   ablation

abonnement *m*   rental [system, e.g. electricity supply], standing charge, subscription

abord *m*   access or approach [to land]

abordable *adj*   accessible or approachable [easily]

abords *m,pl*   surroundings [approaches]

aboucher *vb*   butt together [two planks, pipes]

about *m*   butt [end], end or extremity [of a piece of timber or steel, a wall]

aboutage *m*   butt jointing

ab(o)utement *m*   butt joint or jointing

ab(o)uter *vb*   butt [place end-to-end], join end to end

aboutir *vb*, **ne pas**   fail [negotiations]

Abrams, cône *m* **d'**   slump test cone

Abrams, essai *m* **d'**   slump test [concrete]

abrasif *adj*   abrasive *adj*

abrasif *m*   abrasive *n*

abrasion *f*   abrasion

abreuver *vb*   drench or soak [masonry before rendering], smooth [a surface by applying an agent]

abri *m*   cover, shelter, shielding

abri *m* **jardin**   garden shelter

abri *m* **voiture**   carport

abri, à l'   under cover

abri de, à l'   sheltered from

abrité *adj*   protected

abriter *vb*   protect, shelter, shield

abscisse *f*   abscissa

abside *f*   apse [of a cathedral, church]

absorbant *adj*   absorbent *adj*

absorbant *m*   absorbent

absorber *vb*   absorb, soak in or up

absorption *f*   absorption

absorptivité *f* **acoustique**   acoustic absorptivity

abus *m*   misuse, infringement [of rights]

abuter *vb, voir* **ab(o)uter**

acajou *m*   mahogany

acanthe *m*   acanthus [plant or architectural decoration]

accélérateur *adj*   accelerating

accélérateur *m*   accelerator [in a heating or ventilation system], circulating pump [heating system]

accélérateur *m* **de prise**   concrete accelerator

accélération *f*   acceleration

accélération, voie *f* **de**   acceleration lane

acceptation *f*   acceptance

accès *m*   access

accès facile, d' *adj*   easily accessible

accès, zone *f* **d'**   access point

accès *m* **de nettoyage**   cleaning access

accès *m* **en pente**   sloping access

accès *m* **livraison**   delivery road

accès *m* **livraison matériaux**   road for delivery of materials

accès *m* **livraison matériel**   road for delivery of equipment/plant

accès *m* **pour les pompiers**   fire brigade access

accès *m* **réglementé**   limited access

accessibilité *f*   accessibility

accessible *adj*   accessible

accession *f*   accession [to ownership]

accessoire *adj*   accessory *adj*

accessoire *m*   attachment, accessory

accessoire *m* **de câblage**   wiring accessory

accessoire *m* **de finition**   finishing accessory

accessoire *m* **de fermeture**   locking accessory

accessoire *m* **de levage**   lifting gear

accessoire *m* **de pose**   fixing accessory

accessoires *m,pl*   fittings [as in pipe fittings]

accident *m*   accident [mishap]

accolade *f*   ogee

accolement *m*   shoulder [verge of a roadway], verge [road]

accoler *vb*   join side by side

accord *m*   agreement [about sth]

accore *m*   prop, shore or stay [wooden, sensibly vertical]

accorer *vb*   prop, support with props

accotement *m*   shoulder [of a roadway], verge [of a roadway]

accotement *m* **meuble**   soft verge

accotement *m* **non-stabilisé**   soft shoulder or verge

accotement *m* **stabilisé**   hard shoulder

accotoir *m*   elbow height wall or balustrade

accoudoir *m*   elbow height wall or balustrade

accouplement *m*   connecting electrically, coupling, joining together, linking

accouplement *m* **à plateaux**   flange coupling

accouplement *m* **de réduction**   reducing coupler/coupling

accouplement *m* **direct**   direct coupling

accoupler *vb*   connect electrically, couple, join together, link

accrochage *m*   adhesion [of layers, e.g. paint, rendering, road metal], keying, or roughening to provide a key

accrochage, émulsion *f* **d'**   bitumen(-based) emulsion, tack coat

accroissement *m*   accretion, increase, increment

accul *m*   stake stabilizing foot of angled prop or shore

accumulateur *m*   accumulator or battery [electrical]

accumulateur *m* **de chaleur**   heat accumulator, reservoir or store

accumulateur *m* **de matières**   bulk materials store, bunker

accumulation *f* **d'eau**   water storage

accumulation *f* **de chaleur**   heat storage

accumuler *vb*   store [heat, electricity, etc]

acétylène *m*   acetylene

acétylène, générateur *m* **d'**   acetylene generator

acétylène *m* **dissous**   dissolved acetylene

achat *m*   purchase

acheminement *m*   preparatory step [towards sth], routing [of a despatch of goods]

acheteur *m*   buyer, purchaser

achèvement *m*   completion [e.g. of a section of work]

achèvement *m* **dans les délais voulus**   completion on schedule, on time

achèvement *m* **des travaux**   completion of the works
achever *vb*   end [work, a task], finish [complete, e.g. a building, a task]
acide *adj*   acid *adj*
acide *m*   acid
acier *m*   reinforcement (bar), steel
acier *m* **à bas carbone**   low-carbon steel
acier *m* **à béton**   reinforcement/reinforcing steel
acier *m* **à faible teneur de carbone**   low-carbon steel
acier *m* **à grain fin**   fine grain steel
acier *m* **à haute résistance**   high-strength steel
acier *m* **à haute teneur en carbone**   high-carbon steel
acier *m* **à l'oxygène**   oxygen refined steel
acier *m* **à résistance élevée**   high-strength steel
acier *m* **allié**   alloy steel
acier *m* **au carbone**   carbon steel
acier *m* **austenitique**   austenitic steel
acier *m* **basique**   basic steel
acier *m* **brut**   crude steel
acier *m* **cémenté**   case-hardened steel
acier *m* **comprimé**   compression steel [reinforcement]
acier *m* **coulé**   cast steel
acier *m* **crénelé**   deformed or ribbed (steel) bar [reinforcement]
acier *m* **d'armature**   rebar[†], reinforcement, reinforcing steel
acier *m* **de construction**   structural steel
acier *m* **de répartition**   distribution bar [reinforcement]
acier *m* **demi doux**   semi-mild steel
acier *m* **demi dur**   semi-hard steel (0.4–0.6%C)
acier *m* **doux**   mild steel (0.15–0.3%C)
acier *m* **dur**   carbon steel (0.6–0.7%C)
acier *m* **écroui**   cold hardened or worked steel
acier *m* **en attente**   starter bar
acier *m* **étiré à froid**   cold drawn steel
acier *m* **faiblement allié**   low-alloy steel
acier *m* **fondu**   cast steel
acier *m* **forgé**   forged steel
acier *m* **galvanisé**   galvanized steel
acier *m* **haute adhérence**   deformed or ribbed (steel) bar [reinforcement]
acier *m* **hautement allié**   high-alloy steel
acier *m* **inox(ydable)**   corrosion-resistant or stainless steel
acier *m* **laminé**   rolled steel
acier *m* **lisse**   plain (steel) bar
acier *m* **Martin**   open-hearth steel
acier *m* **moulé**   cast steel
acier *m* **nervuré**   ribbed (steel) bar [reinforcement]
acier *m* **porteur**   main bar [reinforcement]
acier *m* **profilé**   sectional steel, steel section
acier *m* **rapide**   high speed steel
acier *m* **trempé**   tempered steel
acier *m* **zingué** *m*   galvanized or zinc-coated steel
aciérie *f*   steelworks
acompte *m*   deposit, down payment, instalment, payment on account
acousticien *m*   acoustician
acoustique *adj*   acoustic(al), sound *adj*
acoustique *f*   acoustics

acquéreur *m*   buyer, purchaser
acquérir *vb*   acquire, buy, purchase
acquisition *f*   acquisition, purchase
acrotère *m*   acroterion
acrylique *adj*   acrylic
acte *f*   deed, document
acte *f* **authentique**   notarized document [signed before a notary]
acte *f* **authentique de vente**   conveyance, deed of sale
acte *f* **de vente**   bill of sale
acte *f* **finale de vente**   conveyance, deed of sale
acte *f* **sous seing privé**   document signed under hand
actif *m*   assets [of a company]
actif *m* **circulant**   circulating assets
actif *m* **immobilisé**   fixed assets
actif *m* **net**   net assets
action rapide, à   quick-acting
action *f*   act(ion), share [in a company], share certificate
action *f* **à déclic**   snap or toggle action
action *f* **de contrôle**   control action
action *f* **de portique**   portal action
action *f* **du gel**   frost action
action *f* **oxydante**   oxidation action
actionnaire *m/f*   shareholder
actionné par moteur   motor-driven
actionner *vb*   drive
actionneur *m*   actuator, driver [device]
activité *f*   activity
actualisation *f*   indexation/indexing or updating [of costs, prices, etc]
actualiser *vb*   index/update [costs, prices, etc]
actuel *adj*   current [e.g. price], present [e.g state of affairs]
actuellement *adv*   currently, at the moment
adaptateur *m*   adapter [person]
adaptation *f*   adaptation, adjustment, attachment [for electric drill, etc]
adapté *adj* **(à)**   adapted or fitted (for)
adapter *vb*   adapt, adjust, fit
adapteur *m*   adaptor [plumbing]
additif *m*   additive
addition *f*   adding (up)
addition *f* **à**   adding to
additionnel *adj*   additional, extra
adduction *f*   water inlet
adduction *f* **(d'eau)**   water supply
adduction *f* **gravitaire**   gravity water supply
adduction *f* **par refoulement**   pumped water supply
adent *m*   dovetail, key [in a timber joint or between pieces of masonry]
adhérence *f*   adhesion, adhesive strength, bond
adhérent *adj*   adherent, adhesive
adhérent, nœud *m*   sound knot [in timber]
adhérer *vb*   adhere
adhérer *vb* **(à)**   cling or stick (to)
adhésif *m*   adhesive
adhésion *f*   adhesion, sticking
adhésivité *f*   adhesiveness, stickiness
adiabatique *adj*   adiabatic

adjacent *adj* (**à**)   adjacent (to)
adjonction *f*   addition [= fact of adding]
adjudicataire *m*   successful bidder or tenderer
adjudication *f*   adjudication, allocation or awarding, auction sale, tendering [i.e. putting out to tender]
adjudication, marché *m* **par**   competitive tender contract
adjudication, mettre *vb* **qch en**   invite tenders for sth
adjudication, par voie d'   by auction or tender
adjudication *f* **libre**   negotiated tender
adjudication *f* **publique**   open tender(ing)
adjudication *f* **restreinte**   restricted tender(ing)
adjuvant *m*   additive
adjuvant *m* (**du béton**)   admixture (concrete)
adjuvant *m* **entraîneur d'air**   air-entraining additive or agent
administration *f* **d'urbanisme**   planning department/office [town planning]
administrer *vb*   run [a business]
admis *adj*   allowed [allowable]
admissible *adj*   allowable, permissible
admission *f*   admission, inlet, intake
adossé *adj*   back to back
adossé *adj* (**à qch**)   backing (on to sth)
adossement *m*   backing on to *n*
adoucir *vb*   ease [a curve], smooth [down or off], soften [edges, colour, water, etc], tone down [contrast, a colour]
adoucissement *m*   smoothing (down or off), softening [of edges, colours, water, etc], toning down [contrast, a colour]
adoucissement *m* **d'eau**   water softening
adoucisseur *m* **d'eau**   water softener
adoucisseur *m* **domestique**   domestic water softener
adsorber *vb*   adsorb
adsorption *f*   adsorption
aérien *adj*   aerial [overhead]
aérateur *adj*   ventilating [system, etc]
aérateur *m*   air-entraining agent [for concrete], fan or ventilator
aérateur-climatiseur *m*   room air-conditioner
aération *f*   ventilation
aération *f* **aléatoire**   natural ventilation
aéré *adj*   aerated, ventilated
aérer *vb*   aerate, ventilate
aérien *adj*   airborne, overhead
aérien, transporteur *m*   overhead cable transporter
aérobie *adj*   aerobic
aérodynamique *adj*   aerodynamic
aérogare *f* (**de**) **fret**   airport freight terminal
aérogare *f* (**de**) **passagers**   airport passenger terminal
aérogare *f* **satellite**   satellite terminal [airport]
aérogénérateur *m*   aerogenerator, wind generator
aéroport *m*   airport
Aéroports de Paris (ADP)   Paris airport authority
aérosol *m*   aerosol
aérotherme *f*   unit heater [fan + heating element]
affaiblir *vb*   attenuate or weaken, reduce [thickness of a piece of wood]
affaiblissement *m*   attenuation
affaiblissement *m* **acoustique**   acoustic attenuation

affaiblissement *m* **diaphonique**   crosstalk attenuation
affaiblissement *m* **phonique**   acoustic or sound attenuation
affaire *f*   bargain, business, deal
affaissement *m*   collapse [floor, etc], deflection [beam, etc], giving way, sagging [of a beam, etc], settlement or sinking in [of foundations, walls, etc], slump, subsidence [ground]
affaissement (au cône d'Abrams), essai *m* **d'**   slump test
affaisser *vb*   bend or weigh [sth] down [causing deflection], cave in [under load], cause to collapse, cause to deflect, cause to settle [earth, a road, a structure], cause to sink or subside [foundations, wall, etc]
s'affaisser *vb*   cave in, collapse, fail [collapse], give way, sink (down) [ground, a structure, etc], settle [sink, subside]
affaler *vb*   lower on a rope
affermir *vb*   compact [ground, to avoid later settlement]
affichage *m*   display [computer]
afficher *vb*   display or put up [a notice, etc]
affilage *m*   sharpening
affiler *vb*   set, sharpen or whet [a blade, tool, etc]
affiloir *m*   oilstone, sharpener
affinage *m*   fining down or thinning [of a plank, etc]
affinement *m*   fining down or thinning [of a plank, etc]
affleur *adj*   flush [level]
affleurage *m*   levelling, making flush
affleurement *m*   evenness [result of levelling, making flush], level [result of making flush], outcrop [geological]
affleurer *vb*   be even, flush or level (with), outcrop [geologically]
affleurer *vb*   level [bring to same level], make flush
affluence *f*   flow [of open water]
affluent *adj*   affluent or tributary [of a stream, etc]
affluent *m*   tributary [stream, etc]
afflux *m*   influx
affouillement *m*   erosion [of ground, by flow of water], scour or washing away [by water]
affouiller *vb*   erode [ground, by flow of water], scour or undermine [by flow of water], wash away [banks, foundations, etc]
affourcher *vb*   assemble a tongue-and-groove joint
affûtage *m*   grinding, sharpening
affûter *vb*   grind or sharpen [blade, tool, etc], set [sharpen, a saw]
affûteur *m*   sharpener [person]
affûteuse *f*   grinder, sharpening machine
agencé *adj* (**en**)   fitted out (as)
agence *f*   agency, office(s)
Agence de l'Environnement et de la Maîtrise de l'Énergie (ADEME)   environment and energy economy agency [France]
agence *f* **de financement**   lending agency
agence *f* **immobilière**   estate agency
Agence *f* **Nationale pour l'Amélioration de l'Habitat (ANAH)**   national homes improvement agency [France]

Agence *f* **Nationale pour l'Emploi (ANPE)**   national job creation agancy [France]

agencement *m*   arrangement, layout

agencer *vb*   adjust [a machine], arrange, fit up/together

agent *m*   agent, medium

agent *m* **de fixation**   fixing agent

agent *m* **de refroidissement**   cooling or refrigerating medium

agent *m* **humectant**   dampening medium

agent *m* **immobilier**   real-estate agent

agent *m* **oxydant**   oxidizer

agglomération *f*   built-up or urban area

aggloméré *m*   agglomerate, conglomerate [stone], chipboard, particle board

aggloméré *m* **de béton**   hollow concrete block

agglomérant *m*   binding material [road or groundworks]

agios *m*   bank charges

agitateur *m*   agitator, stirrer

agitation *f*   agitation

agrafé *adj*   seamed [of metal sheeting], stapled

agrafe *f*   cramp or dog [holding stones together in masonry], retaining clip [in inclined metal frame glazing], staple, wall tie

agrafer *vb*   clip (together), fix together with a dog or cramp, seam [sheet metal], staple

agrafeuse *f*   stapler

agrafure *f*   welted joint [in roofing, cladding sheets]

agrandir *vb*   enlarge, increase [sth], open out [a hole]

s'agrandir *vb*   expand, grow larger, increase [get bigger]

agrandir *vb* **en large**   widen

agrandir *vb* **en long**   lengthen

agrandissement *m*   enlargement, extension

agréer *vb*   accept, approve

agrégats *m,pl*   aggregate [for concrete]

agrégats *m,pl* **concassés**   crushed aggregate

agrégats *m,pl* **enrobés**   premixed aggregate

agrégats *m,pl* **légers**   lightweight aggregate

agrégats *m,pl* **roulés**   river aggregate

agrégation *f*   binder [fine sand or crushed stone], binding [e.g. of road material]

agrément *m*   acceptance, approval, consent

Agrément Technique Européen (ATE)   European Technical Agrément (approval)

agressif *adj*   aggressive

agressivité *f*   aggressiveness or aggressivity

agricole *adj*   agricultural

agriculteur *m*   farmer

agronome *m*   agriculturalist

agronomie *f*   agriculture

aigu *adj*   sharp [of a point]

aiguillages *m,pl*   points [railway]

aiguille *f*   bolted connector or spacer in timber structure, king-post, thin and headless nail, needle, pointer, points rail

aiguille *f* **(de clocher)**   spire (of a bell tower)

aiguille *f* **de Proctor**   penetration needle [for soil resistance measurement], Proctor plasticity needle

aiguille *f* **de Vicat**   Vicat needle

aiguille *f* **vibrante**   vibrating poker [for concrete]

aiguisage *m*   sharpening

aiguiser *vb*   sharpen [blade, tool, etc]

aiguisoir *m*   sharpener [whetstone]

aile *f*   flange [of beam, angle iron], wing [of a building]

aile *f* **de pont**   wing wall [at an angle to a bridge abutment]

aile *f* **inférieure/supérieure**   lower/upper chord or flange [of a truss, girder]

aile *f* **saillante**   projecting flange

ailette *f*   blade [of a fan, turbine], fin [cooling, radiating, etc], small wing [of a building]

ailette *f* **de refroidissement**   cooling fin

ailette *f* **de ventilateur**   fan or impeller blade

ailettes, à *adj*   finned

ailettes, tube *m* **à**   finned tube

ailettes, vis *f* **à**   wing-screw

aimant *m*   magnet

air *m*   air

air, entraîneur *m* **d'**   air entraining agent [for concrete]

air *m* **additionnel**   make-up air

air *m* **ambiant**   ambient air

air *m* **chargé de poussière**   dust-laden air

air *m* **comburant**   combustion air

air *m* **comprimé**   compressed air

air *m* **d'appoint**   make-up air

air *m* **d'aspiration**   aspirated air

air *m* **de circulation**   recirculated air

air *m* **de combustion**   combustion air

air *m* **de compensation**   make-up air

air *m* **de refroidissement**   cooling air

air *m* **de refoulement**   discharge air

air *m* **de reprise**   recirculated or return air

air *m* **de soufflage**   outlet air

air *m* **en excès**   excess air

air *m* **entraîné**   entrained air

air *m* **évacué/extrait**   exhaust air

air *m* **exhalé**   exhaled air

air *m* **expulsé/rejeté**   exhaust air

air *m* **extérieur**   fresh, outdoor or outside air

air *m* **filtré**   clean [filtered] air

air *m* **frais**   fresh air

air *m* **inclus**   entrapped air

air *m* **intérieur**   indoor air

air *m* **neuf**   primary air

air *m* **normal**   standard air

air *m* **parasite**   infiltrated air

air *m* **primaire**   primary air

air *m* **pulsé**   forced air

air *m* **pur**   clean air

air *m* **rafraîchissant**   cooling air

air *m* **recirculé**   recirculated or return air

air *m* **recyclé**   recirculated or return air

air *m* **repris**   return air

air *m* **secondaire**   secondary air

air *m* **supplémentaire**   make-up air

air *m* **vicié**   vitiated air

airain *m*   bronze

aire *f*   area, flat space, floor, surface [flat space], zone

aire *f* **d'élan**   run-up [sports ground, for jumping or throwing]

aire *f* **d'envol**   tarmac [airport runway]

aire *f* **d'essai**   test site

aire *f* **d'un marteau**   face of a hammer

aire *f* **d'un pont**   floor or roadway of a bridge

aire *f* **de balayage de la grue**   working area [radius] of crane

aire *f* **de captage**   catchment basin

aire *f* **de chargement**   loading area/bay

aire *f* **de déchargement**   unloading area/bay

aire *f* **de drainage**   drainage area or basin

aire *f* **de la section**   cross-sectional area

aire *f* **de parking**   parking area or lot

aire *f* **de préfabrication**   precasting yard

aire *f* **de réception**   reception area

aire *f* **de repos**   rest area [roadside]

aire *f* **de service(s)**   service area (motorway)

aire *f* **de stationnement**   apron or aircraft parking area, parking area or lot

aire *f* **de stockage**   stockpile [space for material(s)], storage area

aire *f* **de traînée**   swept path [of a vehicle]

aire *f* **des moments**   moment diagram

aire *f* **récréative**   recreation area

aire *f* **transversale**   cross-sectional area

ais *m*   board or plank 30–60mm thick

aissante *f*/**aisseau** *m*   shingle [wood]

aisselier *m*   angle or corner brace

aisselle *f* **(d'une voûte)**   haunch (of an arch)

ajour *m*   perforation [to allow light to enter]

ajouré *adj*   of openwork construction [opposite of *plein*], perforated

ajournement *m*   adjournment, postponement

ajourner *vb*   postpone

ajout *m*   secondary constituent of cement

ajoute *m*   added or additional piece

ajouter *vb*   add

ajustage *m*   fitting [adjustment]

ajustage *m* **à jeu**   loose fit

ajustage *m* **facile/lâche**   loose fit

ajustage *m* **serré**   tight fit

ajustement *m*   fit

ajustement *m* **précis**   close fit

ajustement *m* **serré**   tight fit

ajuster *vb*   fit [adapt, adjust]

ajuster *vb* **qch à**   fit sth to

ajutage *m*   nozzle [controllable]

alaise *f*   flashing [metal, around a chimney or roof opening], make-up piece [of wood, in joinery]

alarme *f*   alarm

alarme *f* **d'incendie**   fire alarm

alarme *f* **intempestive**   false alarm

alcali *m*   alkali

alcalin *adj*   alkaline

alcalinité *f*   alkalinity

alcool *m*   alcohol

alcool *m* **à brûler**   methylated spirits

alcôve *f*   alcove, recess

alerte *f*   warning

alésage *m*   bore [of a hole, tube, etc], boring (out), reaming

aléser *vb*   ream

alésoir *m*   hollowing tool [wood turning], reamer

algicide *m*   algicide

algues *f,pl*   algae

algorithme *m* **de contrôle**   control algorithm

alidade *f*   alidade or range [surveying]

alignement *m*   alignment or line, bringing into line, boundary alignment, building line, road boundary between public and private domains

alignement *m* **droit**   straight alignment

aligner *vb*   align or line up, draw up, lay out

alimentateur *m*   feeder

alimentateur *m* **à balourd**   vibrating feeder

alimentateur *m* **à chariot**   reciprocating feeder

alimentateur *m* **à courroie**   belt feeder

alimentateur *m* **à mouvement alternatif**   reciprocating feeder

alimentateur *m* **à plateau**   plate feeder

alimentateur-doseur *m*   cold feeder

alimentation *f*   supply [of gas, water, electricity, etc]

alimentation, vitesse *f* **d'**   feed rate

alimentation *f* **de chaudière**   boiler feed

alimentation *f* **de secours**   emergency supply

alimentation *f* **de sécurité**   emergency supply

alimentation *f* **en eau (chaude/froide)**   (hot/cold) water supply

alimentation *f* **en eau de secours**   emergency water supply

alimentation *f* **en eau domestique**   domestic water supply

alimentation *f* **en électricité**   power supply

alimentation *f* **en gaz**   gas supply

alimentation *f* **sans coupure**   uninterruptible supply

alimentation *f* **stabilisée**   stabilized supply [electrical]

alimenter *vb*   feed or supply [with electric power, water, gas, etc]

allée *f*   drive [lined with trees], lane, path, walk

allège *f*   breast [wall under window ledge]

alliage *m*   alloy

alliage *m* **ferreux**   ferrous alloy

alliage *m* **forgé**   wrought alloy

alliage *m* **léger**   light alloy or metal

allier *vb*   alloy

allongement *m*   elongation, stretching [lengthening]

allongement *m* **avant striction**   elongation before reduction in cross-section

allongement *m* **de rupture**   fracture elongation

allongement *m* **longitudinal**   linear expansion

allongement *m* **permanent**   permanent set [lengthening]

allonger *vb*   elongate, lengthen

s'allonger *vb*   extend [lengthen itself], grow longer, stretch [become longer]

allouer *vb*   allocate or apportion [share, etc], allow [an expense, etc], award [e.g. a sum as damages]

allumage *m*   ignition, lighting [of a lamp, fire, etc], switching on [of electricity, lighting]

allumé *adj*   alight, kindled

allume-feu *m*   fire-lighter

allume-gaz *m*   gas-lighter

allumer *vb*   kindle [a fire], light [lamp, fire, etc], set fire to, switch on [electrically], switch or turn on [the light, lighting]

s'allumer *vb*   catch alight, kindle [catch fire], take fire

allumeur *m*   igniter, lighter

allure *f*   pace or speed

allure, à toute   at full or top speed

allure *f* **poussée**   boost

allure *f* **régulière**   smooth running

alluvial *adj*   alluvial

alluvionnaire *adj*   alluvial

alluvionnement *m*   siltation, silting, silting up

alluvions *f,pl*   alluvium

altérabilité *f*   alterability, degradability

altération *f*   alteration, deterioration, weathering [geotechnics]

altération *f* **bactérienne**   bacterial decay

altéré *adj*   altered, weathered [earth, ground, rock]

altérer *vb*   tamper with, weather [geotechnics]

alternance *f* **de charge**   load reversal

alternateur *m*   alternator, convertor [d.c. to a.c], generator

altimètre *m*   altimeter

altitraceur *m*   altitude recorder

altitude *f*   altitude, height [above sea level]

alu[†] *m*   aluminium

aluminate *m*   aluminate

alumine *f*   alumina

aluminium *m*   aluminium

alvéole *f*   cavity [in masonry, stone]

alvéole *f* **technique**   service duct

amaigrir *vb*   weaken [a mortar or cement mix]

amarrage *m*   mooring [the process], tying [to]

ambiance *f*   surroundings

ambiant *adj*   ambient

amarrer *vb*   brace [a wall], fix [a mast or derrick, by cables, etc], make fast [tie]

amatir *vb*   matt [make dull, a finish]

âme *f*   core [of a cable, wall, etc], web [e.g. of girder]

âme *f* **(d'une poutre composée)**   web plate (of a composite beam)

âme pleine, à   solid webbed

âme *f* **massive**   solid core

amélioration *f*   appreciation [of value of property, etc], betterment, improvement

améliorations, apporter *vb* **des**   make improvements

améliorer *vb*   improve [make better]

s'améliorer *vb*   improve [get better], pick up [improve]

aménageable en   convertible into

aménagé *adj*   equipped

aménagé *adj* **(en)**   converted (into)

amenage *m*   carriage [bringing]

aménagement *m*   adaptation, conversion, development [building, construction], fitting out, improvement [of buildings, etc]

aménagement *m* **(d'un espace vert)**   landscaping

aménagement *m* **d'un réseau**   layout of a network [e.g. of railways, bus routes]

aménagement *m* **du chantier**   site preparation [provision of basic facilities]

aménagement *m* **du site**   site clearance

aménagement *m* **du terrain**   site development

aménagement *m* **du territoire**   regional development [planning]

aménagement *m* **paysager**   landscaping

aménagements *m,pl*   amenities, fittings [furnishings, e.g. office, house]

aménager *vb*   distribute or parcel out [supplies, etc], equip, fit out or up [shop, etc], prepare [fit or lay out]

aménager *vb* **(en)**   convert (into) [building, etc]

amenée *f*   admission/supply pipe, inlet/intake [for air]

amener *vb* **un tuyau dans**   run a pipe into

amiante *m*   asbestos

amiante-ciment *m*   asbestos cement

ammoniac *m*   ammonia

ammoniaque *f*   ammonia solution

amont, en   upstream

amont *m*   upstream part

amonte, tête *f*   inlet [of a culvert]

amorçage *m*   building up [magnetic field], excitation [of a generator], priming [of a pump], scarfing [of a weld], striking [of an arc]

amorce *f*   beginning [e.g. of road or tunnel construction], blasting cap or detonator charge, priming [of a pump]

amorce *f* **de crique**   incipient crack

amorce *f* **de fissure**   incipient crack

amorcer *vb*   begin, excite or start [a generator], initiate [negotiations], prepare (for), prime [a pump], scarf [a weld], strike [an arc], build up [of magnetic field]

s'amorcer *vb*   energize [of generator], start [of pump, motor]

amortir *vb*   amortize, depreciate or write off, attenuate

amortir *vb* **un emprunt**   pay off or redeem a loan

amortissement *m*   amortization [of a loan], attenuation [e.g. of oscillations], depreciation [amount written off], liquidation [of a debt]

amortissement *m* **d'une dette**   debt redemption

amortissement *m* **de bruit**   sound deadening

amortissement *m* **dégressif**   declining balance depreciation

amortissement *m* **linéaire**   linear depreciation

amortissement *m* **pour dépréciation**   wear and tear allowance

amortisseur *m*   damper, shock or sound absorber

amortisseur *m* **de vibrations**   vibration absorber or damper

amovible *adj*   movable, removable

ampérage *m*   current [electrical]

ampère   ampere

ampèremètre *m*   ammeter, current meter

ampèremètre *m* **enregistreur**   recording ammeter

amphithéâtre *m*   amphitheatre, lecture room with stepped seating

ampleur *f*   magnitude

ampliation *f*   certified copy, counterpart [copy]

amplificateur *m*   amplifier

amplification *f*   amplification
amplifier *vb*   amplify
amplitude *f*   amplitude
ampoule *f*   bulb [lamp]
ampoule *f* **à baïonette**   bayonet cap bulb
ampoule *f* **(à culot) à vis**   screw cap bulb
ampoule *f* **dépolie**   frosted bulb
anaérobie *adj*   anaerobic
analogie *f*   analogy
analogique *adj*   analogical
analogue *adj*   analogue, analogous *adj*
analogue *m*   analogue
analyse *f*   analysis
analyse *f* **aux rayons X**   X-ray examination or investigation
analyse *f* **(chimique) de l'eau**   (chemical) water analysis
analyse *f* **coûts–avantages**   cost–benefit analysis
analyse *f* **de contrôle**   check analysis
analyse *f* **de coulée**   melt analysis
analyse *f* **de produit**   random analysis [on finished material]
analyse *f* **des coûts**   analysis or breakdown of costs
analyse *f* **des gaz brûlés**   waste gas analysis
analyse *f* **des gaz combustibles**   fuel gas analysis
analyse *f* **des réseaux**   network analysis
analyse *f* **dimensionnelle**   dimensional analysis
analyse *f* **du sol**   soil analysis
analyse *f* **fonctionnelle**   systems analysis
analyse *f* **granulométrique**   grain size or sieve analysis
analyse *f* **thermique**   thermal analysis
analyser *vb*   analyse
analyseur *m*   analyser
analyseur *m* **acoustique**   acoustic analyser
analyseur *m* **de gaz de fumée**   chemical flue gas analyser
analyseur *m* **infrarouge**   infra-red analyser
analyste *m* **fonctionnel**   systems analyst
ancien *adj*   former [when preceding noun in French], ancient or old [when following noun in French]
ancien propriétaire *m*   previous owner
ancrage *m*   anchor(age)
ancrage, boulon *m* **d'**   anchor bolt
ancrage *m*, **massif** *m* **d'**   anchor(age) block
ancrage, plaque *f* **d'**   anchor plate
ancrage *m* **mort**   dead anchorage
ancré *adj* **en place**   fixed in place, irremovable
ancre *f*   anchor [fixing]
ancrer *vb*   anchor (in place), brace
anémomètre *m*   anemometer, wind gauge
anémomètre *m* **à fil chaud**   hot wire anemometer
anémomètre *m* **à moulinet**   revolving vane anemometer
anémomètre *m* **à palette**   deflecting vane anemometer
anémomètre *m* **enregistreur**   recording anemometer
anglais, appareil *m*   English bond
anglais, vert *m*   green pigment [Prussian blue/chrome yellow mixture]
anglaise *f*   door panel painted darker than surround, painted area replacing finger plate on a door
anglaise, chéneau *m* **à l'**   metal roof gutter [channel in wooden trough]

anglaise, cour *f*   area below street level giving light to basement windows
anglaise, escalier *m* **à l'**   cut or open string staircase
anglaise, fenêtre *f* **à l'**   outward opening window
anglaise, fermeture *f* **à l'**   roller shutter [steel]
anglaise, gouttière *f* **à l'**   metal roof gutter [rounded, on metal feet]
anglaise, jardin *m* **à l'**   landscape(d) garden
anglaise, limon *m* **à l'**   cut or open string [of staircase]
anglaise, mèche *f*   centre bit
anglaise, parquet *m* **à l'**   strip flooring
anglaise, siège *m* **à l'**   toilet/lavatory bowl or pan
angle droit, à   square [at right angles]
angle(s) vif(s), à   square-edged
angle *m* **aigu**   acute angle
angle *m* **arrondi**   half-round fillet
angle *m* **azimut**   azimuth angle
angle *m* **d'affûtage**   sharpness angle [of a cutting tool]
angle *m* **d'attaque**   cutting angle [of a cutting tool]
angle *m* **d'éboulement**   angle of repose [of a slope]
angle *m* **d'enroulement**   angle of contact [of a driving belt]
angle *m* **d'entrée**   inlet angle
angle *m* **d'incidence**   angle of incidence
angle *m* **d'inclination**   angle of inclination
angle *m* **de 90 degrés**   90 degree angle
angle *m* **de chanfrein**   bevel angle
angle *m* **de cisaillement**   angle of shear
angle *m* **de coupe**   cut-off angle
angle *m* **de dépouille**   clearance angle [of a cutting tool]
angle *m* **de diffusion**   discharge angle
angle *m* **de friction sur paroi**   wall friction angle
angle *m* **de frottement**   friction angle
angle *m* **de frottement interne**   internal friction angle
angle *m* **de phase**   phase angle
angle *m* **de sortie**   angle of discharge
angle *m* **de talus**   angle of slope
angle *m* **de talus naturel**   angle of repose [of a slope]
angle *m* **droit**   right angle
angle *m* **du bec**   sharpness angle [of a cutting tool]
angle *m* **mort**   dead corner [in a rectilinear space]
angle *m* **oblique**   bevel square, sliding bevel
angle *m* **rentrant**   re-entrant angle, valley transition [road vertical alignment])
angle *m* **saillant**   salient angle, crest transition [road vertical alignment]
angledozer *m*   angledozer
angles droits, à   rectangular
anglet *m*   channelled joint [in building stone]
angulaire *adj*   angular
anhydre *adj*   anhydrous
anhydride *m* **carbonique**   carbon dioxide
anhydride *m* **sulfureux**   sulphur dioxide
anisotrope *adj*   anisotropic
anneau *m*   annulus, ring, ring connector [between two pieces of wood]
anneau *m* **à ressort**   spring washer
anneau *m* **brisé**   split ring

anneau *m* **d'étanchéité**   sealing ring
anneau *m* **de garniture**   packing ring
anneau *m* **de joint**   gasket ring
anneau *m* **de puits**   shaft ring
anneau *m* **de rideau**   curtain ring
anneau *m* **de support**   supporting ring
anneau *m* **elliptique**   elliptical ring
annexe *f*   annex or extension [building], outbuilding, appendix [to a document]
annonce *f*   advertisement
annuel *adj*   annual [e.g. accounts, charges]
annuellement *adv*   annually, yearly
annuité *m*   annuity
annulation *f*   cancellation [e.g. of a contract]
annuler *vb*   cancel [e.g a contract]
anode *f*   anode
anodisé *adj*   anodized
anormal *adj*   abnormal
anse *f*   bight or cove [in a shore]
anse *f* **de panier**   basket-handle curve [approximating to an ellipse]
antéfixe *f*   raised decorative element at ridge ends
antenne *f*   aerial
anthracite *m*   anthracite
antibélier *m*   water hammer damper
anticlinal *m*   anticline, saddle [geology]
anticorrosif *adj*   anti-corrosive
antidérapant *adj*   anti-skid, non-skid
antidérapant, revêtement *m*   anti-skid surfacing or coating
antiéblouissant *adj*   anti-glare
antiéblouissant, écran *m*   anti-dazzle or -glare screen [between or beside carriageways]
antigel *adj*   anti-frost
antigel *m*   antifreeze [solution], frost proofing (agent)
antigel, couche *f*   anti-frost layer, frost blanket
antigel, matériau *m*   frost protection material
antimousse, agent *m*   anti-foaming agent
antiphare, dispositif *m*   anti-dazzle or -glare screen [between or beside carriageways]
antirefouleur *m*   cowl [anti-backdraught, on flue or chimney]
antirouille *adj*   anti-corrosion or anti-rust, rustproof [of a treatment]
antirouille *f*   anti-rust, rustproofing
antitartre *adj*   anti-scale
antivandale *adj*   vandal resistant
antivibratile *adj*   anti-vibration
aplanissement *m*   levelling [flattening]
aplanisseuse *f*   planer [for road surfaces]
aplomb, (d') *adj*   plumb, vertical
aplomb, hors d' *adj*   out of plumb
appareil *m*   apparatus, appliance, equipment or device, bond [of brickwork or masonry], camera
appareil *m* **à adduction d'air**   fresh air hose breathing apparatus
appareil *m* **à assises réglées**   common/stretcher/stretching bond
appareil *m* **à volume variable**   variable volume unit

appareil *m* **alterné simple**   English bond
appareil *m* **anglais**   English bond
appareil *m* **croisé**   English cross bond
appareil *m* **d'appui**   bearing mechanism [e.g. for a bridge]
appareil *m* **d'appui à pot**   pot bearing [rubber]
appareil *m* **d'appui en élastomère fretté**   laminated (neoprene) rubber bearing
appareil *m* **d'appui fixe**   fixed bearing
appareil *m* **d'appui multidirectionnel**   free bearing
appareil *m* **d'appui néoprène fretté**   laminated (neoprene) rubber bearing
appareil *m* **d'appui unidirectionnel**   guided bearing
appareil *m* **d'essai**   test apparatus
appareil *m* **de battage**   piling rig
appareil *m* **de chasse d'eau**   flushing device
appareil *m* **de chauffage**   heating apparatus/appliance
appareil *m* **de chauffage à accumulation**   storage heater
appareil *m* **de chauffage instantané**   instantaneous heater
appareil *m* **de chauffage ménager**   domestic calorifier
appareil *m* **de chauffage par convection**   convection or convector heater
appareil *m* **de chauffage par rayonnement**   radiant heater
appareil *m* **de cheminée**   chimney bond
appareil *m* **de cisaillement circulaire**   ring shear apparatus
appareil *m* **de cisaillement directe**   direct shear apparatus
appareil *m* **de condensation refroidi par air**   air-cooled condensing unit
appareil *m* **de conditionnement d'air**   air-conditioning unit
appareil *m* **de construction**   masonry bond
appareil *m* **de contrôle**   control(ler) [device]
appareil *m* **de coupure HT**   high-voltage circuit-breaker
appareil *m* **de levage**   hoist, hoisting device or equipment, lifting gear
appareil *m* **de mesure**   measuring device, equipment or instrument
appareil *m* **de photogrammétrie**   photogrammetric camera
appareil *m* **de prise d'échantillon(s)**   sampler
appareil *m* **de séchage de l'air**   air drying apparatus
appareil *m* **de stéréométrie**   stereometric camera
appareil *m* **de stérilisation**   sterilizer
appareil *m* **de ventilation**   ventilating unit
appareil *m* **différentiel**   residual current device (RCD)
appareil *m* **électrique mobile/portatif**   portable electrical equipment
appareil *m* **élévateur**   lifting apparatus or equipment, vertical transport
appareil *m* **en besace**   toothing [of masonry]
appareil *m* **en boutisses**   heading bond
appareil *m* **en panneresses**   common, stretcher or stretching bond
appareil *m* **enregistreur**   recording apparatus or instrument
appareil *m* **éteignoir**   extinguisher

appareil *m* **filtrant**   respirator
appareil *m* **flamand**   Flemish bond
appareil *m* **isolant à prise d'air libre**   fresh air hose breathing apparatus
appareil *m* **ménager**   domestic appliance
appareil *m* **normal**   standard unit [equipment]
appareil *m* **périphérique**   peripheral equipment
appareil *m* **polyvalent**   multi-purpose appliance or unit
appareil *m* **radiographique**   X-ray equipment
appareil *m* **réfrigérant**   cooler
appareil *m* **respiratoire**   breathing apparatus
appareil *m* **sanitaire**   sanitary appliance
appareil *m* **totalement équipé**   packaged unit [HVAC]
appareils *m,pl* **d'éclairage**   lighting equipment
appareils *m,pl* **ménagers**   domestic equipment
appareillage *m*   accessories
appareillage *m*   bonding [of brickwork or masonry]
appareillage *m*   equipment
appareillage *m*   fittings [equipment, etc]
appareillage *m* **d'essai**   test equipment [for trials]
appareillage *m* **de contrôle**   control equipment [for testing and verification], test equipment [for verification]
appareillage *m* **électrique**   electrical equipment
appareillage *m* **pour tube**   pipe fitting
appareiller *vb*   choose and lay out timber and mark and cut joints
apparent *adj*   exposed, visible
apparition *f* **de fragilité**   embrittlement
appartement *m*   apartment, flat
appartement *m* **(en) duplex**   maisonette
appartement *m* **témoin**   show flat
appartenance *f*   appurtenance
appauvrir *vb*   weaken [a cement or mortar mix]
appel *m*   call, invitation
appel *m* **d'offres**   invitation to tender
appel *m* **d'offres, dossier** *m* **d'**   tender documents
appel *m* **d'offres public**   public invitation to tender
appel *m* **d'offres restreint**   restricted tender invitation
appel *m* **de candidature**   invitation to prequalify
appentis *m*   lean-to [roof, or building itself], outhouse or shed [lean-to], penthouse roof
application *f*   application
application *f* **d'une force**   application of a load
appliqué *adj*   applied
applique *f*   wall finishing in marble or mosaic, wall lamp
appliquée, contrainte *f*   applied stress
appontement *m*   jetty, landing stage, wharf
apport *m*   financial contribution or input
apport *m*, **bassin** *m* **d'**   catchment area
apport *m* **calorifique**   heat input
apport *m* **d'air**   supply of air
apport *m* **d'énergie**   energy input or supply of energy
apport *m* **de chaleur**   heat input
apports *m,pl* **de chaleur solaire**   solar heat gain
apports *m,pl* **solaires**   solar (heat) gain
appréciation *f*   assessment [e.g. of data]
apprenti *m*   apprentice
apprêt *m*   priming, size or sizing [for decorating]

apprêt *m*, **(papier** *m* **d')**   lining paper
apprêter *vb*   dress [metals], prime/size [a surface]
approbation *f*   (official) approval [of drawings, plans, etc]
approche *f*   approach, drawing near, slate cut at an angle adjacent to ridge or valley
approuvé *adj*   approved
approuver *vb*   approve
approvisionnement *m*   provision [reserve or supply], supply [of goods]
approximatif *adj*   approximate
appui libre, sur   simply supported
appui *m*   abutment, bearing or support, balustrade or handrail, window ledge or sill
appui, barre *f* **d'**   handrail [across an opening]
appui, mur *m* **d'**   low wall or parapet [on which one can lean]
appui *m* **à glissement**   sliding bearing
appui *m* **à rotule**   rocker bearing
appui *m* **à rouleau(x)**   roller bearing [support]
appui *m* **articulé**   pin-joint(ed) support
appui *m* **cylindrique**   roller bearing [support]
appui *m* **d'escalier**   banisters
appui *m* **de baie**   window ledge or sill
appui *m* **de fenêtre**   window ledge or sill
appui *m* **élastique**   elastic bearing
appui *m* **encastré**   built-in bearing or support
appui *m* **fixe**   fixed bearing
appui *m* **flottant**   floating bearing
appui *m* **glissant**   sliding bearing
appui *m* **intermédiaire**   intermediate bearing
appui *m* **libre**   simple support
appui *m* **médian**   central bearing
appui *m* **mobile**   expansion bearing
appuyé *adj* **(sur, contre)**   supported (on, against)
appuyé *adj* **sur quatre côtés**   supported on four sides
appuyer *vb*   prop (up), shore (up), stay [support], support
appuyer *vb* **(qch) (sur, contre)**   lean or rest (something) (on, against)
s'appuyer *vb* **(sur, contre)**   lean or rest (on, against)
aptitude *f* **à la déformation**   deformability
aquaplanage *m*   aquaplaning
aquastat *m*   aquastat, water thermostat [in a boiler]
aqueduc *m*   aqueduct
aqueux *adj*   aqueous
aquifère *adj*   water-bearing [ground]
aquifère *m*   aquifer
araignée   rainwater outlet grating [to prevent blockage]
arase *f*   horizontal bearing face [course at top of wall]
arase *f*   levelling course [of masonry or thin stone]
arase *f* **inférieure**   underside
arasement *m*   shoulder [of a tenon]
araser *vb*   cut off [the heads of piles], level, make level, set or trim to level [top of a wall, etc], square [timber]
arbalétrier *m*   principal rafter
arbitrage *m*   arbitration
arbitre *m*   arbitrator
arboré *adj*   wooded
arbre *m*   axle, shaft [rotating], spindle, tree

arbre *m* **à feuillage caduc**   deciduous tree
arbre *m* **de commande**   drive/driving shaft
arbre *m* **de transmission** *f*   transmission shaft
arbre *m* **manivelle**   crankshaft
arc *m*   arc [curve], arc [in welding], arch
arc, soudage *m* **à l'**   arc welding
arc *m* **à deux articulations**   two-hinged arch
arc *m* **à tirant**   bowstring girder, tied arch
arc *m* **à treillis**   trussed arch
arc *m* **à trois rotules**   three-pin arch
arc *m* **angulaire**   triangular arch
arc *m* **angulaire tronqué**   truncated triangular arch
arc *m* **anse de panier**   basket or three-centred arch
arc *m* **articulé**   hinged arch
arc *m* **bombé**   circular arch with centre lower than its springs, segmental arch
arc *m* **boutant**   abutment pier
arc *m* **brisé**   triangular arch
arc *m* **de cloître, voûte** *f* **en**   cloister(ed) vault
arc *m* **de décharge**   discharging or relieving arch
arc *m* **de radier**   breakwater wall [bridge pier protection]
arc *m* **déprimé**   flat arch
arc *m* **doubleau**   transverse arch
arc *m* **droit**   flat arch
arc *m* **électrique**   electric arc
arc *m* **en accolade**   keel or ogee arch
arc *m* **en carène**   tudor arch
arc *m* **en doucine**   convex arch
arc *m* **en lancette**   lancet arch
arc *m* **en ogive**   ogive or pointed arch
arc *m* **en plein cintre**   round arch
arc *m* **en tiers-point**   lancet arch
arc *m* **en treillis**   trussed arch
arc *m* **encastré**   built-in arch [rotation restrained at springs]
arc *m* **épaulé**   shouldered arch
arc *m* **équilibré**   balanced arch
arc *m* **exhaussé**   semi-circular arch [with centre above imposts]
arc *m* **gothique prolongé**   ogee arch
arc *m* **gothique surbaissé**   tudor arch
arc *m* **infléchi**   inflected or tented arch
arc *m* **lanciolé**   lance-head or lanciolate arch
arc *m* **outrepassé**   horseshoe arch
arc *m* **plein cintre**   semi-circular arch
arc *m* **plein cintre surhaussé**   semi-circular arch [with centre above springs]
arc *m* **polylobé**   multi-lobed arch
arc *m* **rampant**   asymmetrical arch
arc *m* **renversé**   reversed arch [with key beneath springs]
arc *m* **roman**   romanesque or semi-circular arch
arc *m* **surbaissé**   depressed arch
arc *m* **surhaussé**   stilted arch
arc *m* **trilobé**   trefoil arch
arc *m* **tudor**   tudor arch
arc-boutant *m*   flying buttress
arc-bouter *vb*   buttress
arcade *f*   archway
arcade *f* **composée**   archway consisting of multi-lobed arches

arcades *f,pl*   arcade
arcature *f*   blind or decorative arcade on the face of a wall
arcature *f* **en claire voie**   decorative arcade just separated from backing wall
arceau *m*   arch [of a vault]
arche *f*   arch, vault in the form of an arch
arche *f* **maîtresse**   main span arch of a bridge
archée *f*   span of an arch
architecte *m*   architect
Architecte DESA   architect with diploma from the *École Supérieure d'Architecture* [France]
Architecte DPLG (Diplômé par le Gouvernement)   architect with state architectural school diploma [France]
architecte *m* **agréé**   architect with higher qualifications and experience [France]
architecte *m* **d'intérieur**   interior designer
architecte *m* **diplômé**   architect with basic qualifications [France]
architecte-conseil *m*   consultant/consulting architect
architectonique *adj*   architectural
architectural *adj*   architectural
architecture *f*   architecture
architrave *f*   architrave
archives *f,pl*   archives, records
ardoise *f*   roofing slate, shale, slate
ardoisier *m*   roof slater and tiler
are *m*   100m^2
arête *f*   arris or edge [of a beam, etc], ridge [of a mountain], spine
arête *f* **chanfreinée**   bevelled or chamfered edge
arête *f* **de poisson**   herringbone pattern of paving setts
arête *f* **vive**   sharp or square edge
arêtes *f,pl*, **voûte** *f* **d'**   groin(ed) vault
arêtier *m*   hip [roof], hip rafter
argilacé *adj*   argillaceous, clayey
argile *f*   clay
argile, bouchon *m* **d'**   clay seal
argile, scellement *m* **d'**   clay seal
argile *f* **à blocaux**   boulder or glacier clay
argile *f* **compactée**   puddled clay
argile *f* **expansée**   expanded burnt clay
argile *f* **fissurée**   fissured clay
argile *f* **glacière**   glacier clay
argile *f* **sableuse**   sandy clay
argile *f* **schisteuse**   clay shale
argile *f* **sensible**   quick clay
argileux *adj*   argillaceous, clayey
argileux, schiste *m*   clay shale
argon *m*   argon
aride *adj*   arid, dry [country, land], barren [ground, etc]
arithmétique *adj*   arithmetic
armature *f*   armature [of d.c. electric motor], glazing rods, reinforcement (bar) or reinforcing steel
armature, crochet *m* **d'**   reinforcement hook
armature *f* **crénelée**   deformed or ribbed (steel) bar [reinforcement]
armature *f* **de compression**   compression reinforcement
armature *f* **de liaison**   connecting bars [reinforcement]

armature *f* **de précontrainte**   prestressing reinforcement

armature *f* **de répartition**   distribution bar [reinforcement]

armature *f* **de traction**   positive or tensile/tension reinforcement

armature *f* **en treillis soudé**   welded mesh reinforcement

armature *f* **inférieure**   bottom bars, positive or tensile/tension reinforcement

armature *f* **longitudinale**   longitudinal reinforcement

armature *f* **principale**   main reinforcement

armature *f* **supérieure**   compression or negative reinforcement, top bars

armature *f* **tendue**   positive or tensile/tension reinforcement

armature *f* **verticale**   vertical reinforcement

armatures écartement *m* **des**   reinforcement spacing

armatures enrobage *m* **des**   reinforcement cover

armatures entredistance *f* **des**   reinforcement spacing

armatures recouvrement *m* **des**   lap length [reinforcement]

armoire *f*   cupboard

armoire *f* **basse tension**   low-tension or low-voltage cabinet

armoire *f* **(de chambre à coucher)**   clothes cupboard, wardrobe

armoire *f* **de commande/contrôle**   control cabinet or cupboard

armoire *f* **de toilette**   bathroom cabinet or cupboard

armoire *f* **électrique**   electrical cabinet or cupboard

armoire *f* **frigorifique**   cold room

armoire *f* **ignifuge**   fireproof cabinet or cupboard

armoire *f* **murale**   built-in cupboard

aronde *f*   dovetail or fishtail tie

aronde, queue *f* **d'**   dovetail

arpent *m*   acre [approx]

arpentage *m*   chainage [surveying], land survey

arpenter *vb*   measure or survey [land]

arpenteur *m*   land surveyor

arrachage *m*   extraction [uprooting, tearing out]

arrachage *m*   pulling out [e.g. sheet piles]

arrache-clous *m*   nail puller

arrachement *m*   toothing [key between old & new masonry]

arracher *vb*   extract [e.g. sheet piles]

arracher *vb*   pull out [e.g. sheet piles]

arracher *vb* **le filet de**   strip the thread of

arracheur *m* **de palplanches**   sheet pile extractor

arrêt *m*   catch [for a shutter or other non-rebated closure], check [of a door], shut-down or stop switch, stop button, stop, stoppage, stopping [of machinery, work, etc]

arrêt *m*, **bande** *f* **d'**   shoulder [stopping lane]

arrêt *m*, **distance** *f* **d'**   stopping distance

arrêt *m*, **robinet** *m* **d'**   stop cock

arrêt *m*, **piste** *f* **d'**   bus stopping lane

arrêt *m*, **zone** *f* **d'**   bus layby (bus stop)

arrêt *m* **d'autobus**   bus stop

arrêt *m* **d'urgence**   emergency stop

arrêt *m* **d'urgence, bande** *f* **d'**   hard shoulder or emergency stop lane

arrêt *m* **de barre**   curtailment or cut-off (of a reinforcement bar)

arrêt *m* **de coulage**   limit of pour [concrete work]

arrêt *m* **de volet**   shutter catch

arrêt *m* **de paiement d'un chèque**   stopping a cheque

arrêté *m*   decree [administrative], official order

arrêté *m* **de compte**   settlement of account [closure]

arrêté *m* **de suspension**   injunction

arrêté *m* **municipal**   by-law

arrêter *vb*   stop

arrhes *f,pl*   deposit [on a purchase]

arrière *adj*   back *adj*, rear *adj*

arrière *m*   back [e.g. of a building], rear part

arrière-bras *m*   balance lever [of bascule bridge]

arrière-corps *m*   rear premises or set back part of a building

arrière-cour *f*   backyard

arrière-cuisine *f*   back kitchen, scullery

arrière-linteau *m*   secondary lintel [built into wall behind main lintel]

arrière-plan *m*   background

arrière-pays *m*   back country, hinterland

arrière-port *m*   inner harbour

arrivée *f* **d'eau**   water inlet

arrondi *adj*   round, rounded (off)

arrondir *vb*   round (off) [a corner or edge, a number]

arrondissement *m*   district

arrosage *m*   sprinkling [irrigation], watering

arroser *vb*   sprinkle [irrigate], water

arroseur *m*   sprinkler [irrigation or watering]

arroseur *m* **à flèche**   boom sprinkler

arroseur *m* **rotatif/tournant**   revolving or rotating sprinkler

arrosoir *m*   watering can

artère *f*   artery [traffic], electrical transmission line [as opposed to distribution line]

artère *f* **principale**   main thoroughfare

artésien *adj*   artesian

artichaut *m*   spiked ironwork on top of a wall or fence

article *m*   clause [in agreement, contract, etc], item

articulation *f*   hinge or link [structural], pin-joint [mechanism or structural]

articulation *f* **à la clé**   crown hinge [of a structure]

articulation *f* **à rotule**   ball or ball-and-socket joint

articulation *f* **aux naissances**   abutment hinge

articulation *f* **de culée**   abutment hinge

articulé *adj*   articulated, hinged, pin [of a joint]

articuler *vb*   hinge

artificier *m*   blaster

artisan *m*   craftsman, tradesman

asbeste *m*   asbestos

asbeste-ciment *m*   asbestos cement

ascendant *m* [soudage]   vertical [welding]

ascenseur *m*   elevator, lift

ascenseur *m* **à plan incliné**   inclined elevator

ascenseur *m* **de chantier**   site hoist

ascenseur *m* **des pompiers**   fire fighting or firemen's lift

ascenseur *m* **hydraulique**   hydraulic lift

ascension *f* **capillaire**   capillary rise

asperseur *m*   irrigator

asperseur *m*   sprinkler [irrigation or watering]

aspersion *f*   irrigation [by spraying], sprinkling [irrigation, watering]

asphalte *m*   asphalt

asphalte *m*, **mastic** *m* **d'**   mastic asphalt

asphalte *m* **coulé**   mastic asphalt

asphalte *m* **naturel**   natural asphalt

asphalteur *m*   asphalter

aspirateur *m*   exhaust or suction fan, extractor [dust]

aspirateur *m* **balai**   upright vacuum cleaner

aspirateur *m* **de copeaux**   chip extractor

aspirateur *m* **de poussière(s)**   dust extractor

aspirateur *m* **statique**   cowl [to improve draught of a chimney or flue]

aspirateur *m* **universel**   all purpose [wet and dry] vacuum cleaner

aspirateur-chariot *m*   cylinder vacuum cleaner

aspirateur-traîneau *m*   cylinder vacuum cleaner

aspiration *f*   exhaust [by suction], induction [of air], sucking up [of water, etc], suction, suction connection

aspiration *f*, **ventilateur** *m* **à**   exhaust fan

aspirer *vb*   breathe (in) or inhale, draw up [water, etc], exhaust [air, etc], suck in or up [water]

assainir *vb*   clean up, decontaminate, purify

assainissement *m*   cleaning, drainage, purification, sanitation, sewage disposal, sewerage

assainissement, canalisation *f* **d'**   drainage piping

assainissement, ouvrage *m* **d'**   drainage system

assainissement *m* **d'habitation**   domestic or house drainage

assainissement *m* **des eaux usées**   foul water sewerage or sewage disposal

assainissement *m* **pluvial**   rainwater or stormwater drainage

assainissement *m* **séparatif, système** *m* **d'**   separate drainage system

assainissement *m* **unitaire, système** *m* **d'**   combined drainage system

assainissement *m* **urbain**   town drainage

assèchement *m*   dewatering

assécher *vb*   dewater, drain, empty [drain], dry up

assemblage *m*   assembling, assembly, connection [structural], joint [timber, metal structure], jointing

assemblage *m* **à biellette**   link-rod assembly

assemblage *m* **à clef**   pegged joint

assemblage *m* **à clin**   lapped joint

assemblage *m* **à double épaulement**   double skew notch (joint)

assemblage *m* **à emboîtement**   spigot-and-socket joint

assemblage *m* **à embrèvement simple**   single skew notch joint

assemblage *m* **à enfourchement**   forked mortice-and-tenon joint

assemblage *m* **à entaille**   halved or halving joint

assemblage *m* **à fausse languette**   loose tongue-and-groove joint

assemblage *m* **à grain d'orge**   V-shaped edge or end joint [timber]

assemblage *m* **à l'atelier**   shop assembly

assemblage *m* **à mi-bois**   halved joint [in timber]

assemblage *m* **à mi-bois à queue d'aronde**   dovetail halving joint

assemblage *m* **à onglet**   mitre(d) joint

assemblage *m* **à queue d'aronde**   dovetail joint

assemblage *m* **à queues droites**   combed joint

assemblage *m* **à rainures et languettes**   tongue-and-groove joint(ing)

assemblage *m* **à rivets**   riveted assembly, connection or joint

assemblage *m* **à tenon et mortaise**   mortise-and-tenon joint

assemblage *m* **à trait de Jupiter droit**   straight scarf(ed) joint

assemblage *m* **d'about/par aboutage**   butt joint

assemblage *m* **en fausse coupe**   bevel joint

assemblage *m* **par boulons**   bolted assembly

assemblage *m* **par brides**   flange jointing

assemblage *m* **par couvre-joint**   butt-strap joint, splicing

assemblage *m* **rigide**   rigid assembly

assemblage *m* **rivé**   riveted assembly, connection or joint

assemblage *m* **soudé**   welded connection or joint

assemblage *m* **vissé**   screwed joint

assembler *vb*   assemble, erect [assemble, install], joint [join together]

assembler *vb* **bout à bout**   butt joint

asservissement *m*   control of a mechanism, sequence operation

asseoir *vb*   bed [foundations]

assiette *f*   lower face of a paving block or stone, stable position [of a body such as a beam]

assignation *f*   summons, writ

assise *f*   base [of a dam], course [of bricks or stone], layer [of bricks, cement, stone, road material], seating [of the end of a prop or strut]

assise *f* **de boutisses**   heading course

assise *f* **de briques**   brick course

assise *f* **de champ**   brick-on-edge course

assise *f* **de chaussée**   base (course) (of a road)

assise *f* **de panneresses**   stretching course

assise *f* **debout**   brick-on-end course

assise *f* **imperméable (à l'humidité)**   damp-proof course (DPC)

assise *f* **rocheuse**   bedrock

assistance *f* **à l'exploitation**   assistance with implementation

assistance *f* **aux réceptions**   assistance with commissioning

assistance *f* **au dépouillement des offres**   assistance in the tender analysis

assisté *adj* **par ordinateur**   computer-aided

Association *f* **Française de Normalisation (AFNOR)**   French national standards institution

association *f* **professionnelle**   trade association

associé *m*   partner

assurance *f*   insurance

assurance *f* **(contre l') incendie**   fire insurance

assurance *f* **(contre le) vol**   theft insurance

assurance *f* **décennale**   decennial insurance
assurance *f* **multirisques**   comprehensive insurance
assurance *f* **qualité (AQ)**   quality assurance (QA)
assurance *f* **responsabilité civile**   civil liability insurance
assurance *f* **(sur la) vie**   life assurance
assurance *f* **tous risques**   all-risk/comprehensive insurance
assuré *m*   insured [person]
assureur *m*   insurer, underwriter
astragale *m*   astragal [of a column, facade, etc], nosing [of a stair tread]
asymétrie *f*   asymmetry
asymétrique *adj*   asymmetric(al)
asynchrone *adj*   asynchronous [e.g. motor]
atelier *m*   workshop
atelier *m* **de constructions mécaniques**   machine shop
atelier *m* **de constructions métalliques**   structural steel workshop
atelier *m* **de décapage**   pickling plant [metal]
atelier *m* **de montage**   assembly shop
atelier *m* **de réparation(s)**   repair shop
atelier *m* **de soudage**   welding shop
atmosphère *f*   atmosphere
atmosphère *f* **normale**   standard atmosphere
atmosphérique *adj*   atmospheric
atmosphérique, pression *f*   atmospheric pressure
atomiser *vb*   atomize, spray [a liquid]
atomiseur *m*   atomizer, sprayer
âtre *m*   fireplace hearth, hearth(stone) or raised hearth
atrium *m*   atrium
attache *f*   cable clip, fastener, tie [to hold reinforcement]
attache *f* **de fixation**   fixing clip
attachement *m*   as-built record of work subsequently covered up
attacher *vb*   attach, clip, fasten, fix, tie (up)
attaque *f* attach [by corrosion, etc]
atteindre *vb*   spread to [e.g. fire], reach [arrive at]
attenant *adj*   adjoining, contiguous
attente *f*   starter bar, waiting
attente, obligation *f* **d'**   obligation to give way [road traffic]
attente, pierre *f* **d'**   toothing stone [key for subsequent stonework]
attentes *f,pl* **de ferraillage**   starter bars
atténuation *f*   attenuation
atténuation *f* **de la lumière**   dimming [light]
atténuation *f* **des sons/du son**   silencing, sound attenuation
attestation *f*   certificate
attique *m*   attic [top storey, generally set back]
attraction *f*   attraction, pull [of a magnet]
attraction *f* **capillaire**   capillary action
attribuer *vb*   allocate
attrition *f*   abrasion, attrition, wearing away
au bord de (la) mer   seaside *adj*
au bord de (la) rivière   riverside *adj*
au bourg   in the town
au fil de l'eau   with the stream
au point mort   out of gear or in neutral

au sommet de   at the top of [a hill, mountain]
aubage *m*   blading
aubage *m* **directeur**   splitter [in ductwork]
aubages *m,pl* **d'entrée à inclinaison variable**   variable inlet vanes
aube *f*   blade [of a turbine], vane [of a fan]
aube *f* **directrice**   directional, guide or turning vane
auberge *f*   inn
aubette *f*   bus/passenger/queue shelter
aubier *m*   sapwood
au-dessus du niveau de la mer   above sea level
audibilité *f*   audibility
audible *adj*   audible
audit *m*   audit
audit *m* **énergétique**   energy audit
auge *f*   galvanized iron sink, plasterer's trough
auge *f* **à mortier**   mortar pan, trough or tub
auget *m*   bedding of parquet backing strips, plaster infill [concave] between ceiling joists, pocket [for building a fixing into a wall, etc]
augmentation *f*   increase, increment, rise [prices, temperature, wages]
augmentation *f* **constante de la charge**   constant rate of loading
augmentation *f* **de pression**   pressure rise
augmenter *vb*   increase, raise/rise [wages, prices, taxes, etc], scale up [an activity]
augmenter *vb* **proportionnellement**   scale up [a drawing]
au(l)ne *m*   alder
austénite *f*   austenite
authentifier *vb*   authenticate
autobétonnière *f*   truckmixer
autoblocage *m*   self-locking
autocollant *adj*   self-adhesive
autocommutateur *m*   PABX (private automatic branch exchange)
autocurage *m*   self-cleansing [of drains or sewers]
autofinancement *m*   self-financing
automatique *adj*   automatic, self-acting
automatiquement *adv*   automatically
automatisation *f* **des systèmes des bâtiments**   building automation system (BAS)
automontable *adj*   self-erecting [e.g. tower crane]
automoteur *adj*   motorized, self-propelled
autonettoyant *adj*   self-cleaning
autonivelant *adj*   self-levelling
autonome *adj*   free-standing, independent, packaged [of a unit, equipment], self-contained
autopont *m*   flyover, overpass
autoportant *adj*   self-supporting
autoporteur *adj*   self-supporting
autorisation *f* **(de)**   authorization, permission or permit (for or to do)
autoriser *vb* **(qn à)**   authorize (someone to), give or grant permission (to someone to)
autorité *f*   authority
autorité *f* **de surveillance**   supervising authority
autoroute *f*   motorway

autoroute, branche d'   motorway fork
autoroute, intersection *f* **d'**   motorway interchange
autoroute *f* **à péage**   toll motorway
autoroute *f* **de rase campagne**   rural motorway
autoroute *f* **urbaine**   urban motorway
autorupteur *m*   make-and-break device [electrical]
autostable *adj*   self-supporting
autosurveillé *adj*   auto-checked
autotransformateur *m*   auto-transformer
autovaporisation *f*   flash steam
auvent *m*   canopy [over a porch], open shed, penthouse, porch roof [without posts], weather-board [on a door, window]
aval, en   downstream
aval *m*   downstream part
avalanche *f*   avalanche
avale, tête *f*   outlet [of a culvert]
avaloir *f*   gully [drainage outlet], throat [of a fireplace]
avance *f*   advance or loan
avancement *m*   progress
avancement *m* **des travaux**   progress of the works
avant-bec *m*   cutwater [of a bridge]
avant-corps *m*   part of building projecting in front of facade
avant-foyer *m*   fore-hearth [of a fireplace]
avant-métré *m*   preliminary measurement of building [before works]
avant-pieu *m*   dolly [piling]
avant-projet *m*   pilot, feasibility, preliminary or preparatory study, pre-design, sketch design
avant-projet *m* **détaillé (APD)**   detailed pre-project or preliminary design, final design study
avant-projet *m*, **étude** *f* **d'**   pilot, feasibility, pre-design, preliminary, preparatory or sketch design study
avant-projet *m* **sommaire (APS)**   preliminary design sketch or study, scheme stage design report, summary pre-project
avant-propos *m*   introduction [to document]
avant-solier *m*   corbel beam
avant-toit *m*   eaves or verge [roof]
avant-train tracteur *m*   tractor-scraper
avant-trou *m*   pilot hole
avec éclips   shuttered [of a socket outlet]
avenant *m*   amendment or rider [to a contract]
averse *f*   shower [of rain]

avertir *vb*   warn
avertissement *m*   warning
avertisseur *m*   alarm, buzzer
avertisseur *m* **d'incendie**   fire alarm
aveugle *adj*   blind [of a non-existent opening]
aveugler *vb*   tank [i.e. to make waterproof]
avis *m* **d'appel public à la concurrence (AAPC)**   public invitation to tender
avis *m* **de réception (AR)**   advice of delivery [of a letter, parcel]
avis *m* **préalable**   preliminary or prior notice [of start of works]
avis *m* **technique**   technical advisory note [e.g. as issued by CSTB in France]
aviver *vb*   break down or mill [a log to square-edged timber], (re)saw [a log to square-edged timber], repair damaged edge(s), sharpen [put an edge on]
avivés *m,pl*   square-edged timber
avocat *m*   advocate [lawyer]
avoir *m*   credit note
avoir *vb* **du balourd**   run out of true
avoir *vb* **l'intention de**   plan [intend to]
avoyer *vb*   set [saw teeth]
axe *m*   arbour [grindstone, etc], axle, spindle [grindstone, pump, etc], axis, centre line or line
axe *m*, **grand/petit**   major/minor axis
axe *m* **central**   centre-line
axe *m* **d'implantation**   centre-line, setting out line
axe *m* **de balancement**   axis of symmetry [architectural]
axe *m* **de chaussée**   road centre-line
axe *m* **de circulation**   traffic route [major]
axe *m* **de référence**   reference axis
axe *m* **de rotation**   axis of rotation
axe *m* **de symétrie**   axis of symmetry
axe *m* **longitudinal**   longitudinal axis
axe *m* **neutre**   neutral axis
axe *m* **repère de piquetage**   intersection of setting out
axe *m* **vertical**   vertical axis
axial *adj*   axial
axial, pression *f*   axial pressure
axonométrique *adj*   axonometric *adj*
axonométrique, perspective *f*   axonometric perspective
ayant droit *m*   interested party
azimut *m*   azimuth
azote *m*   nitrogen

bac *m*   bin, tank or vat, ferry(boat)
bac *m* **à eau**   water tank
bac *m* **à mortier**   mortar pan, trough or tub
bac *m* **à ordures**   dustbin
bac *m* **à saumure**   brine tank
bac *m* **de décantation**   sedimentation tank

bac *m* **de douche**   shower base or tray
bac *m* **de récupération**   drain or drip pan
bâche *f*   cistern, tank, tarpaulin, waterproof sheet
bâche *f* **alimentaire/d'alimentation**   feed tank
bâche *f* **d'aspiration**   suction tank
bâcle *f*   bar [door reinforcement]

bactéricide *adj*   bactericidal

bactéricide *m*   bactericide *n*

bactérie *f*   bacterium

bactérien *adj*   bacterial

bactérien, lit *m*   bacteria bed [in sewage works]

badigeon *m*   colour wash, distemper [paint]

badigeon *m* **(blanc)**   whitewash

badigeonnage *m*   colour washing, distempering, whitewashing

badigeonner *vb*   whitewash, colour wash, distemper

baffle *m*   'splitter [in ductwork]

baffle *m* **acoustique**   acoustic baffle

bague *f*   collar [on metallic downpipe]

bague *f*   ring or washer

bague *f* **d'étanchéité**   joint or sealing ring

bague *f* **de mise au point**   focusing ring [e.g. of a surveying telescope]

bague *f* **de paumelle**   hinge washer [to raise door or shutter]

bague *f* **isolante**   insulating collar

baguette *f*   beading or moulding [quarter or half round]

baguette *f*   came [of a leaded light], cover [strip wiring], strip [of wood]

baguette *f*, **demi-**   flat moulding [used as a cover strip]

baguette *f* **à souder/de soudure**   welding rod

baguette *f* **d'angle**   angle or corner bead [wood], corner moulding, fillet

baguette *f* **d'apport**   welding filler rod

baguette *f* **de soudure**   welding rod

baguette *f* **électrique**   flat casing or conduit for surface electric wiring

baguette *f* **pour soudage**   welding rod

bahut, mur *m*   dwarf or low wall [with coping to carry railings, etc]

baie *f*   opening or bay [architectural]

baie *f* **vitrée**   bay or picture window

baie, fenêtre *f* **en**   bay window

baignoire *f*   bath(tub)

baignoire *f* **à brassage**   whirlpool bath

baignoire-douche *f*   shower bath

baignoire-sabot *f*   hip bath

bail *m*   lease

bail *m* **à construction**   building lease

bail *m* **emphytéotique**   long term lease

bailleur *m*   lessor

bain *m*   bath

bain *m* **de décapage**   pickling bath

bain *m* **de mortier**   mortar bed [to receive cut stone, etc]

baïonnette, douille *f* **à**   bayonet fitting [lamp cap]

baisse *f*   decline [e.g. in prices, etc], drop or fall [of water level, pressure, price, etc]

baisser *vb*   decrease, drop or lower [water level, pressure, price, etc]

bajoyer, mur *m*   dock or quay wall, side wall of a canal lock

baladeuse *f*   hand or portable inspection lamp

balai *m*   broom

balai, enduit *m* **au**   Tyrolean or alpine finish rendering

balai *m* **à encoller**   paperhanger's brush

balance *f*   balance, pair of scales

balance *f* **à bascule**   weighbridge

balance *f* **à fléau**   beam scales

balance *f* **de précision**   analytical balance/scales

balance *f* **romaine**   steelyard

balancelle *f*   flying scaffolding

balancer *vb*   balance [make symmetrical], square [books]

balancier *m*   bearing [metal, of an arch bridge]

balayage *m*   air change rate, scavenging, sweeping (out or up)

balayer *vb*   scavenge, sweep (across) [of light], sweep [ground, etc, *not* chimney or flue]

balayeuse *f*   road sweeper [machine]

balcon *m*   balcony

balcon *m*, **deuxième**   upper circle [of a theatre]

balcon *m*, **premier**   dress circle [of a theatre]

balèvre *f*   lip or projecting irregularity in masonry or concrete

balisage *m*   beaconing, setting up of survey poles

balise *f*   marker stake or pole [civil works], survey pole

baliser *vb*   mark out [with survey poles]

ballast *m*   ballast [of road, railway track], gravel [ballast], lamp ballast

ballaster *vb*   ballast

ballastière *f*   gravel pit

ballon *m*   cylindrical reservoir or tank

ballon *m* **à chauffage mixte**   storage calorifier

ballon *m* **d'eau (chaude)**   (hot) water tank

balourd *m*   lack of balance [of rotating element]

balourd, avoir *vb* **du**   run out of true

balustrade *f*   balustrade, handrail, railing

balustre *m*   baluster, banister post

balustres *m,pl*   banisters

banc *m*   bed or layer [geological], workbench

banc *m* **d'essai**   test bench or stand

bancaire *adj*   banking, bank related *adj*

banchage *m*   pouring [of concrete using formwork or shuttering]

banche *f*   formwork or shuttering for casting a wall [i.e. two faced]

banché *adj*   cast [of concrete, in a mould or form], formed or shuttered [of concrete]

bancher *vb*   pour concrete [using formwork or shuttering]

bandage *m*   voussoirs and keystone of an arch

bande *f*   band [of material], belt [of land], joint tape, lane [of road], stretch [of ground], strip [of a material or slab]

bande *f* **à joint**   joint tape, scrim

bande *f* **commune**   busbar

bande *f* **d'accélération**   acceleration lane

bande *f* **d'arrêt**   shoulder [stopping lane]

bande *f* **d'arrêt d'urgence (BAU)**   emergency stopping lane, hard shoulder

bande *f* **de calfeutrement**   draught strip, weather strip(ping)

bande *f* **de chant**   edging strip

bande *f* **de circulation**   traffic lane

bande *f* **de décélération**   deceleration lane

bande *f* **de guidage**   guide markings or strips [on road surfaces]

bande *f* **de poussière**   dust band or strip
bande *f* **de recouvrement**   cover strip
bande *f* **de seuil**   threshold strip
bande *f* **de solin**   flashing (strip)
bande *f* **engazonnée**   grass verge
bande *f* **latérale**   shoulder [verge of a roadway]
bande *f* **non stabilisée**   soft verge
bande *f* **transporteuse**   conveyor belt
bandes *f*, **route** *f* **à deux/trois/quatre**   two-/three-/four-lane highway or road
bande-support *f*   backing or backing bar [welding]
bandeau *m*   band or string course [in masonry], pelmet
bandelette *f*   metal strip joining railings on which handrail is fixed
bander *vb*   assemble the voussoirs and keystone of an arch
banlieue *f*   suburb
banne *f*   awning, shop blind
banque *f*   bank
banque *f* **de données**   data bank
banquette *f*   bank [of earth], bench [in cut or fill], berm, footway [of bridge, tunnel], walkway between sloping roof and gutter
banquette *f* **de fenêtre**   window ledge or sill
banquette *f* **de halage**   tow(ing) path
bar *m*   bar [unit of pressure]
baraque *f* **(de chantier)**   site hut
baraquement *m*   site hut
barbacane *f*   barbican [of a fortress], drain channel [of a bridge], weephole [of a retaining wall]
barbelé *m*   barbed wire
barbelé, fil *m* **de fer**   barbed wire
barbotine *f*   cement grout applied to protect reinforcing steel or improve adhesion of new masonry
bardage *m*   cladding [with boards or panels], moving, shifting or transport [of heavy materials, on site]
bardage *m* **à clin**   weatherboard cladding
bardage *m* **en planches**   weatherboard cladding, weatherboarding
bardé *adj* **(de)**   clad (with), covered (in)
hardeau *m*   hollow brick used to form ceiling between steel joists, cladding board [timber], weatherboard, roof sheathing board, shingle [roofing, cladding]
bardeau *m* **d'asphalte**   asphalt shingle
barder *vb*   load and transport materials, offer up [stone in masonry construction]
barème *m*   ready reckoner
barème *m*   scale or schedule [of prices, salaries, etc], table [list, figures, etc]
barème *m* **des tarifs**   scale of charges
barillet *m*   cylinder lock (or cylinder of a lock)
barium *m*   barium
baromètre *m*   barometer
baromètre *m* **anéroïde**   aneroid barometer
baromètre *m* **enregistreur**   barograph, recording barometer
barrage *m*   barrage, barrier, barring or closing [of a road], obstruction, dam, weir, damming [of a valley]
barrage *m* **à contreforts**   buttress dam

barrage *m* **à vannes**   sluice dam
barrage *m* **à voûtes multiples**   multi-arch dam
barrage *m* **de régulation**   control dam or weir
barrage *m* **de retenue**   control dam or weir, retaining dam
barrage *m* **déversoir**   spillway dam
barrage *m* **en arc(he)**   arched dam
barrage *m* **en argile**   clay dam
barrage *m* **en béton**   concrete dam
barrage *m* **en enrochement(s)**   rockfill dam
barrage *m* **en maçonnerie**   masonry dam
barrage *m* **en remblai**   earth(fill) dam
barrage *m* **en terre**   earth(fill) dam
barrage m poids   **gravity dam**
barrage-poids *m*   gravity dam
barrage-poids *m*   gravity dam
barré *adj*   closed [of a road]
barre *f*   bar or rod [metal, wood], member, rail [of boarded door or shutter]
barre *f* **à haute adhérence**   deformed or ribbed (steel) bar [reinforcement]
barre *f* **à mine**   jumper [bar with flattened ends]
barre *f* **à rideau**   curtain rod
barre *f* **à U**   channel iron/section
barre *f* **collectrice**   busbar
barre *f* **comprimée**   compression member, strut
barre *f* **d'appui**   handrail [across a window opening]
barre *f* **de seuil**   door threshold (strip)
barre *f* **de treillis**   lattice or truss member
barre *f* **droite**   straight bar [reinforcement]
barre *f* **en attente**   starter bar
barre *f* **omnibus de distribution**   distribution busbar
barre *f* **omnibus principale**   main busbar
barre *f* **relevée**   bent-up bar [reinforcement]
barre *f* **tendue**   stay, tension member
barreau *m*   baluster or banister post, small bar
barreau *m* **d'essai**   test piece [bar]
barreau *m* **entaillé**   notched test bar
barreaudage *m*   protective grille in front of an opening
barrer *vb*   bar [a door], block or close [a road], cross [a cheque]
barrette *f*   short bar, strip [of a material]
barrette *f* **de connexion**   terminal strip
barrette *f* **de raccordement**   connecting strip
barrière *f*   barrier [e.g. level crossing], division [separator], gate [field, road]
barrière *f* **à l'air**   air seal
barrière *f* **au vent**   wind barrier
barrière *f* **d'étanchéité**   damp-proof course (DPC)
barrière *f* **de vapeur**   vapour barrier
barrière *f* **naturelle**   shelter belt
barytine *f*   barytes
bas *adj*   low
bas *m*   bottom
basalte *m*   basalt
basaltine *f*   reconstituted stone based on crushed basalt
basculant *adj*   swing, swinging or swivelling [bridge, crane, etc]
basculante, fenêtre *f*   horizontally pivoted window

bascule, en *adj*   cantilevered
bascule *f* à agrégats/ciment   weigher for aggregates/cement
bascule *f* à wagon   weighbridge [railway]
bas-côté *m*   verge [road]
base *f*   base, basis, bottom
base, couche *f* de   base course [road construction]
base, de   basal
base, niveau *m* de   base level
base, prix *m* de   basic price
base *f* de calcul   basis of design
base *f* de colonne   column base
base *f* de données   database
base *f* de la soudure   weld root
baser *vb* (sur)   base (on) [e.g. theory]
bases *f,pl* d'exploitation   operating data
basse pression *f*   low pressure
basse tension *f*   low voltage
basses eaux *f,pl*   low tide
bassin *m*   basin or dock, masonry lined pit, ornamental lake, pond or pool
bassin *m* d'aération   aeration tank
bassin *m* d'amortissement   stilling basin [of a dam]
bassin *m* de coagulation   coagulation tank
bassin *m* de crue   flood plain
bassin *m* de décantation   drain pit [sewage], sedimentation basin or tank
bassin *m* de dissipation   stilling basin [of a dam]
bassin *m* de rétention   reservoir
bassin *m* de stabilisation   stabilization basin
bassin *m* de tranquilisation   stilling basin [of a dam]
bassin *m* versant   drainage basin or catchment area
bast(a)ing *m*   plank 65×170mm
bastide *f*   bastide [fortified town in SW France], country house [southern France]
bâtard, mortier *m*   mortar with equal proportions of cement and lime
batardeau *m*   cofferdam
batardeau *m*   sheet pile caisson
bateau *m*   lowered kerb [for passage of vehicles]
bâti *adj*   built
bâti *m*   frame(work)
bâti *m* de fenêtre   window frame
bâti *m* de porte   door frame
bâti *m* dormant   built-in framework [to accept door, window, etc]
bâtiment *m*   building, construction
bâtiment *m* à ossature métallique   steel frame building
bâtiment *m* à étages multiples   multi-storey building
bâtiment *m* à plusieurs étages   multi-storey building
bâtiment *m* commercial   commercial building
bâtiment *m* des chaudières   boiler house
bâtiment *m* de jonction/liaison   link building
bâtiment *m* et travaux publics *m,pl* (BTP)   building and civil engineering works
bâtiment *m* industriel   industrial building
bâtiment *m* malsain   sick building
bâtiment *m* résidentiel   residential building
bâtiment *m* sans étage   single storey building

bâtiment *m* scolaire   school building
bâtiment *m* très élevé   tower building
bâtir *vb*   build, construct, raise [a building, etc]
bâtisse *f*   bricks and mortar, masonry, building [house] or premises
bâtons rompus, travailler à *vb*   work in fits and starts
battage *m*   driving in [piles]
battage, appareil *m* de   piling rig
battage, sabot *m* de   driving shoe [piling]
battage, tête *f* de   driving head [piling]
battage *m* de palplanches   sheet pile driving
battage *m* de pieux   pile driving
battant *m*   leaf [of a door, window, etc]
battant *m* mouton   leaf of a window with rounded edge fitting into groove in the frame
battants, porte *f* à deux   double or folding door
batte *f*   hand rammer
batte *f*   punner
battellement *m*   eaves courses [of roof tiles or slates]
battement *m*   stop rod [for door, shutter, etc, in non-rebated opening], weather strip [on edge of outer leaf of a double door or window]
batterie *f*   battery [of electrical accumulators or condensers], set [of equipment, etc]
batterie *f* à/de détente directe   direct expansion coil unit
batterie *f* d'ascenseurs   lift bank
batterie *f* d'évaporation   evaporation coil
batterie *f* de chauffage   heater battery
batterie *f* de condensation   condenser coil
batterie *f* de plafond   ceiling coil
batterie *f* de préchauffage   preheat coil
batterie *f* de rechange   refill or replacement battery
batterie *f* de réchauffage   reheat coil
batterie *f* de refroidissement   cooler battery
batterie *f* électrique   electric battery
battre *vb*   drive [a pile]
battre *vb* la ligne   chalk a line
battre *vb* le beurre   form fixing pockets in fresh masonry with a crowbar
baudrier *m* de sécurité   safety harness [for working at height]
bauge *f*   mortar of clay and straw
bavette *f*   flashing [sheet metal]
bavette *f* garde-boue   mud-flap [vehicle]
bavure *f*   burr or fin [on a casting], defect [in mortar, plaster, paintwork, etc]
bec *m*   catch [of a latch], cutwater [of a bridge], jib [crane], nose [of a latch or tool], nozzle [of a pipe or tube]
bec *m* Bunsen   Bunsen burner
bec *m* d'âne   mortise chisel
bec *m* de cane   lever handle [cupboard or door]
bec *m* de corbin   crow's beak [any projection resembling one]
bec *m* de gaz   gas burner
bêche *f*   shovel, spade
becquet *m*   deflector [metal, for rain water on the slope of a roof]

bédane *m* **(à bois)**   mortise chisel
beffroi *m*   belfry
bel *m*   bel (10 decibels)
bélier *m*   heavy beam used as a ram in demolition work, pile driver, ram [hydraulic, etc]
bélier, coup *m* **de**   water hammer
bélier *m* **hydraulique**   hydraulic ram
belvédère *f*   belvedere, summerhouse, viewpoint [at a beauty spot]
bénarde, clef *f*   flat key [which can be used from both sides]
bénéfice *f*   profit
bénéfice *f* **avant impôt**   pre-tax profit
bénéfice *f* **d'exploitation**   trading profit
bénéfice *f* **mise en réserve**   retained profit
bénéficiaire *m* **de l'adjudication**   successful bidder or tenderer
benne *f*   bucket [of dredger], skip
benne *f* **à bascule**   revolving skip
benne *f* **à béton**   concrete skip or bucket
benne *f* **à demi-coquilles**   clamshell grab
benne *f* **à fond ouvrant**   bottom opening bucket
benne *f* **d'alimentation**   feed(ing) hopper or skip
benne *f* **de bétonnage**   concrete or concreting skip
benne *f* **de chargement**   loading hopper or skip
benne *f* **preneuse**   grab(bing) crane
benne *f* **racleuse**   dragline bucket
bentonite *f*   bentonite
bentonitique *adj*   bentonitic
béquette *f*   steelfixer's flat tongs [for bending fine reinforcement]
béquille *f*   lever handle [cupboard or door], support leg [for trailer, mobile crane, etc]
béquille *f* **de portique**   frame stanchion
béquille *f* **hydraulique**   hydraulic jack [for steadying a crane, etc]
berceau *m*   plunger [router], cradle vault or vault springing from two parallel walls
berge *f*   bank [steep] [of a river], edge of an excavation
bergerie *f*   sheepfold or sheep pen
berme *f*   bench or berm
berme *f* **centrale**   central reservation [roads]
besace *f*   saddle (piece) [flashing above chimney, etc]
besace *f*, **appareil en**   header and stretcher bond [of stonework quoins], quoin bond [at corners of stonework]
besace *f* **de dilatation**   expansion joint cover [metallic roofing]
besoin *m*   demand or requirement
besoin *m* **d'énergie**   power requirement
besoin *m* **en eau**   water demand or requirement(s)
bétoire *f*   sink-hole or swallow-hole [geology], soakaway for rainwater [pebble-filled]
béton *m*   concrete
béton *m*, **gros**   concrete fill, unreinforced concrete
béton, pavé *m* **de**   concrete block paving
béton, socle *m* **de**   concrete base course
béton *m* **à air occlus**   aerated or air-entrained concrete
béton *m* **à entraînement d'air**   aerated or air-entrained concrete

béton *m* **à granulats apparents**   exposed aggregate concrete
béton *m* **à occlusion d'air**   aerated or air-entrained concrete
béton *m* **à remplissage**   concrete made by injection of mortar into ready placed aggregate
béton *m* **aéré**   aerated or air-entrained concrete
béton *m* **alumineux**   high alumina cement (HAC) concrete
béton *m* **alvéolaire**   cellular concrete
béton *m* **apparent**   exposed or visible concrete
béton *m* **armé**   reinforced concrete
béton *m* **asphaltique**   asphalt cement
béton *m* **(asphaltique) bitumineux**   bituminous concrete
béton *m* **au bitume-goudron**   asphaltic concrete with a tar binder
béton *m* **au goudron**   tar-concrete
béton *m* **au vide**   vacuum dewatered concrete [accelerated setting]
béton *m* **aux cendres volcaniques**   lightweight concrete (pumice aggregate)
béton *m* **aux résines**   resin modified concrete
béton *m* **avec entraîneur d'air**   aerated or air-entrained concrete
béton *m* **avec occlusion d'air**   aerated or air-entrained concrete
béton *m* **banché**   formed concrete
béton *m* **bitumineux**   asphaltic concrete
béton *m* **bitumineux coulé**   guss asphalt
béton *m* **bouchardé**   bush hammered concrete
béton *m* **brut (de décoffrage)**   as-struck or untreated concrete
béton *m* **caverneux**   no-fines concrete
béton *m* **cellulaire**   cellular concrete
béton *m* **centrifugé**   centrifugally cast concrete
béton *m* **chaud**   concrete made with pre-heated constituents
béton *m* **chauffé**   concrete heated in situ by heated formwork
béton *m* **compact**   heavy concrete
béton *m* **coulé en deuxième phase**   added concrete
béton *m* **coulé sur place**   cast in situ concrete
béton *m* **cyclopéen**   cyclopean concrete
béton *m* **d'agrégats légers**   lightweight aggregate concrete
béton *m* **d'argile expansé**   lightweight concrete (expanded clay aggregate)
béton *m* **d'asphalte**   asphalt concrete with oven-dried gravel aggregate
béton *m* **damé**   rammed or tamped concrete
béton *m* **de blocage**   concrete fill
béton *m* **de bois**   woodwool concrete
béton *m* **de briquaillons**   brick aggregate concrete
béton *m* **de cendres volantes**   fly ash or pulverized fuel ash (PFA) concrete
béton *m* **(de ciment), chaussée** *f* **en**   concrete carriageway
béton *m* **de diatomée**   diatomaceous (lightweight) concrete
béton *m* **de granulats légers**   lightweight concrete [general]

béton *m* **de gravats**   demolition waste aggregate concrete
béton *m* **de graviers**   sand and gravel aggregate concrete
béton *m* **de gravillons lavés**   concrete with carefully graded exposed aggregate
béton *m* **de kieselguhr**   diatomaceous (lightweight) concrete
béton *m* **de mâchefer**   ground granulated blastfurnace slag (ggbfs) concrete
béton *m* **de masse**   mass concrete
béton *m* **de (mousse de) laitier**   lightweight concrete (ggbfs)
béton *m* **de parement**   facing concrete
béton *m* **de perlite**   ultra-lightweight concrete (expanded perlite aggregate)
béton *m* **de pierres**   concrete using aggregate containing stone (>30mm)
béton *m* **de ponce**   pumice aggregate concrete
béton *m* **de pouzzolane**   pozzolanic concrete
béton *m* **de propreté**   bedding or blinding concrete, mat, mattress or sealing coat [blinding]
béton *m* **de résine**   resin concrete [i.e. using resin binder]
béton *m* **de riblons**   metallic waste aggregate concrete
béton *m* **de sable**   sand only aggregate concrete
béton *m* **de sciure de bois**   wood sawdust aggregate concrete
béton *m* **de scorie**   pulverized fuel ash (PFA) concrete
béton *m* **de terre stabilisé**   stabilized earth concrete
béton *m* **de vermiculite**   vermiculite concrete
béton *m* **de verre multicellulaire/vésiculaire**   foamed glass aggregate concrete
béton *m* **drainant**   no-fines concrete
béton *m* **dur**   hard concrete [metallic or carborundum aggregate]
béton *m* **durci**   hardened concrete
béton *m* **essoré sous vide**   vacuum dewatered concrete [accelerated setting]
béton *m* **fibré**   fibre-reinforced concrete
béton *m* **fondu**   high alumina cement (HAC) concrete
béton *m* **frais**   fresh, green or wet concrete
béton *m* **gommé**   lightly sandblasted concrete
béton *m* **gras**   rich concrete
béton *m* **hydraulique**   hydraulic concrete
béton *m* **hydrocarboné**   bitumen or tar binder concrete
béton *m* **jeune**   green concrete
béton *m* **lavé**   exposed aggregate concrete [obtained by hosing]
béton *m* **léger**   aerated, air-entrained or lightweight concrete
béton *m* **lisse de parement**   fair-faced concrete
béton *m* **lourd**   heavy aggregate concrete [e.g. for radiation shielding]
béton *m* **maigre**   lean or weak concrete
béton *m* **malaxé à sec**   dry-mixed concrete
béton *m* **manufacturé**   precast concrete
béton *m* **moulé**   cast concrete
béton *m* **mousse**   foamed concrete
béton *m* **non armé**   mass, plain or unreinforced concrete
béton *m* **plastique**   easily workable concrete
béton *m* **pompé**   pumped concrete [generally with plastifier]

béton *m* **poreux**   porous aggregate concrete
béton *m* **précontraint**   prestressed concrete
béton *m* **préfabriqué**   precast concrete
béton *m* **préparé**   ready-mix(ed) concrete
béton *m* **prépack/prépakt**   prepack concrete
béton *m* **prêt à l'emploi**   ready-mix(ed) concrete
béton *m* **projeté**   gunite, shotcrete
béton *m* **réfractaire**   refractory concrete
béton *m* **sablé**   sandblasted concrete
béton *m* **sans éléments fins**   no-fines concrete
béton *m* **sans sable**   no-fines concrete
béton *m* **Siporex**   lightweight concrete (Siporex)
béton *m* **translucide**   pavement or vault light [of glass blocks cast in concrete]
béton *m* **vibré**   vibrated concrete
béton *m* **volcanique**   pozzolanic aggregate concrete
béton-mousse *m*   aerated or foam concrete
bétonnage *m*   concreting
bétonnage *m* **par temps chaud/froid**   hot/cold weather concreting
bétonné, revêtement *m*   concrete pavement
bétonner *vb*   concrete
bétonneur *m*   concreter (worker)
bétonnière *f*   concrete mixer
bétonnière *f*, **grande**   concrete batcher
bétonnière *f* **à cuve basculante**   tilting (concrete) mixer
biais, en *adj*   bevel *adj*, oblique *adj*, skewed, on the skew, slantwise, on the slant
biais *m*   skew, slant
biaxial *adj*   bi-axial
bibliothèque *f*   library
bicolore *adj*   bi-coloured
bidet *m*   bidet, small joiner's bench
bidon *m*   drum [for oil, etc]
bief *m*   canal length between two locks
bief *m* **d'amont**   headrace [of a water mill]
bief *m* **d'aval**   tailrace [of a water mill]
bief *m* **(du moulin)**   millstream
bielle *f*   brace or strut [steel roof], strut supporting a purlin or principal rafter
bien affilé/aiguisé *adj*   sharp [blade, tool, etc]
bien calibré *adj*   well-graded/-sorted
biennale, garantie *f*   two-year guarantee [on minor work in France]
biens *m,pl*   assets, possessions, property, goods [as in goods and chattels]
biens *m,pl* **bâtis**   premises
biens *m,pl* **immeubles/immobiliers**   real-estate
bifurcation *f*   three-leg or -way junction [highways]
bigéminé *adj*   divided into four [of an opening, by mullions]
bigot *m*   navvy pick [two-pointed]
bigues *f,pl*   shearlegs or shears
bilame *f*   bimetallic strip
bilan *m*   assessment, balance sheet, consequences, final result [accounts, consequences]
bilan, déposer *vb* **son**   file one's petition in bankruptcy
bilan *m* **énergétique**   energy balance
bilan *m* **financier**   financial appraisal

bilan *m* **thermique**   heat balance
billage *m*   ball hardness test
bille *f*   billet [of timber or metal]
billes *f,pl*, **roulement** *m* **à**   ball bearing
bimétallique *adj*   bimetallic
binaire *adj*   binary
binder *m*   bitumen or tar-based binder
bipale *adj*   twin-bladed
bipolaire *adj*   double- or two-pole
birloir *m*   sash bolt
biseau *m*   bevel or chamfer, taper [highways]
biseauté *adj*   bevelled, chamfered
biseauter *vb*   bevel, chamfer
bistre *m*   tar-like deposit in a flue or chimney
bitord *m*   packing rope for pump glands
bitte *f* **d'amarrage**   bollard
bitumage *m*   asphalting [the process]
bitume *m*   asphalt, bitumen
bitume, émulsion *f* **de**   bitumen emulsion
bitume *m* **armé**   bitumenized sheeting
bitume *m* **asphaltique**   asphaltic bitumen [bitumen plus fine clay or sand]
bitume *m* **fluidifié**   cut-back bitumen
bitume-goudron *m* **(BG)**   bitumen-tar asphaltic concrete
bitume-goudron *m*, **béton au**   asphaltic concrete or hot-rolled asphalt with a tar binder
bitumer *vb*   asphalt
bitumineux *adj*   bituminous
bitumineux, coulis *m*   slurry seal [road construction]
bitumineux, enrobé *m*   bitumen binder
bitumineux, revêtement *m*   bituminous surfacing or wearing course [roads]
blanc, chèque *m* **en**   blank cheque
blanc, fer *m*   tinned steel sheet
blanc *m* **de céruse**   white lead
blanc *m* **de chaux**   whitewash
blanc *m* **de plomb**   white lead
blanc *m* **de titane**   titanium dioxide [white pigment]
blanchiment *m*   painting white [and its results]
blanchir *vb*   dress [joinery, timber]
blanchir *vb*   paint white, pickle [metal], smooth [joinery, timber]
blanchisserie *f*   laundry (room)
bleu *adj*   blue
bleu[+] *m*   blueprint
bleu *m* **de cobalt**   cobalt blue
bleu *m* **de Prusse**   Prussian blue
bleuissement *m*   bluestain [timber defect]
blindage *m*   earthwork support, planking and strutting, lining, sheeting or shoring [of an excavation], timbering [of excavations, action and result], screening or shrouding [electrical]
blindage *m* **de câble**   cable screening
blindage *m* **de puits**   well casing or lining
blindage *m* **extérieur**   armouring [dam]
blindé *adj*   armoured [cable, etc], reinforced [of a door, etc], shored, strutted or timbered [of an excavation], screened or shrouded [electrically]
blindée, fouille *f*   lined excavation

blindée, fouille *f* **non**   unlined excavation
blinder *vb*   case or line [a structure], sheet, shore or timber [an excavation], screen or shroud [electrically]
bliner *vb*   ram [paving stones, etc], tamp (down)
bloc *m*   block, building block
bloc *m* **à poncer**   sanding/sandpaper block
bloc *m* **autonome lumineux**   self-powered light fitting
bloc *m* **creux (de béton de ciment)**   hollow concrete block
bloc *m* **creux (de terre cuite)**   hollow clay brick or block
bloc *m* **de fondation**   foundation block [e.g. for a pole or post]
bloc *m* **de jonction**   terminal block
bloc *m* **multiprise (à cordon)**   trailing sockets [electrical]
bloc *m* **perforé**   hollow block
bloc *m* **refroidisseur**   cooling unit
bloc, serrer *vb* **à**   screw or tighten home
bloc, visser *vb* **à**   screw home
blocage *m*   random rubble construction, clamping or locking [in position]
blocage *m* **de profondeur**   depth stop
blocage *m* **d'inclinaison**   vertical clamp [of a theodolite]
blocaille *f*   ballast, hardcore, rubble
bloc-croisée *m*   prefabricated window unit
bloc-eau *m*   prefabricated pipework unit for kitchen and bathroom
bloc-porte *m*   prefabricated door unit
blondin *m*   aerial ropeway, cableway
bloqué *adj*   clamped or locked [wheel, machinery, etc], jammed, overtightened, seized [bearing, machinery, etc], filled up with rubble [of a wall]
bloquer *vb*   block, clamp [a piece of machinery], jam [a door, machinery, etc], jam on [a brake], lock [wheels], overtighten [a bolt], fill up [a wall] with rubble, point [a wall], lump together
se bloquer *vb*   jam, seize [e.g. a bearing]
bobinage *m*   winding [electrical]
bobinage *m* **de moteur**   motor winding
bobinage *m* **de transformateur**   transformer winding
bobine *f*   coil or reel, induction or resistance coil
bobine *f* **de câble**   cable drum
bobine *f* **de réaction/réactance**   choke (coil)
bocage *m*   coppice, copse
bois, à/de/en   wood *adj*, wooden *adj*
bois *m*   timber, wood
bois, coeur *m* **de**   heartwood
bois *m*, **petit**   glazing bar [wood]
bois *m* **aggloméré**   (wood) chipboard
bois *m* **amélioré**   improved wood
bois *m* **apparent**   exposed timber
bois *m* **avivé**   sawn timber
bois *m* **blanc**   deal (timber)
bois *m* **contreplaqué**   plywood
bois *m* **courbe**   bent(wood)
bois *m* **d'entrée**   semi-dry timber or wood
bois *m* **d'équarrissage**   square(d) timber
bois *m* **d'œuvre**   building or square-edged timber
bois *m* **de bout**   end grain timber
bois *m* **de brin**   roundwood

bois *m* de charpente   building or framing timber
bois *m* de cœur   heartwood
bois *m* de coffrage   shutter boards or boarding
bois *m* de construction   long or structural timber, constructional
bois *m* de fil   side grain timber
bois *m* (de) grume   roundwood
bois *m* de long   long timber
bois *m* de sciage   sawn timber
bois *m* dur   hardwood
bois *m* équarri   square-edged timber
bois *m* feuillard   lath
bois *m* flache   waney-edged timber
bois *m* gauche   timber which has twisted after squaring
bois *m* imparfait   sapwood
bois *m* imprégné   treated timber or wood [impregnated]
bois *m* lamellé   laminated timber or wood
bois *m* lamellé-collé   glued-laminated wood
bois *m* parfait   heartwood
bois *m* raboté   planed timber
bois *m* refait   adzed timber
bois *m* rond   round timber
bois *m* synthétique   plastic wood
bois *m* tendre   softwood [timber]
boisage *m*   earthwork support, timbering [of excavations, action and result]
boisé *adj*   wooded
boiser *vb*   panel [in wood] or wainscot, shore or timber [an excavation]
boiserie *f*   wainscot(ing), wooden panelling, woodwork
boiseur *m*   timberer [specializing in timbering excavations]
boisseau *m*   chimney or flue block or brick
boissellerie *m*   lath(s) used in wooden trellis-work
boîte *f*   case
boîte *f* à coupe(s)   mitre box
boîte *f* à fusibles   fuse board
boîte *f* à graisse   grease trap
boîte *f* à onglets   mitre box
boîte *f* à outils   tool-box
boîte *f* aux lettres   letter box
boîte *f* d'onglet   mitre box
boîte *f* de branchement   distribution box
boîte *f* de cisaillement   shear box
boîte *f* de commande   switch box
boîte *f* de connexion   connection box
boîte *f* de crémone   casement (cremona) bolt mechanism
boîte *f* de dérivation   junction box
boîte *f* de détente à volume constant   constant volume unit
boîte *f* de détente à volume variable   variable volume unit
boîte *f* de jonction   cable connection or connecting box, conduit or junction box
boîte *f* de mélange   mixing box
boîte *f* de parquet   floor socket box or fitting
boîte *f* de parquet à couvercle pivotant   floor socket with hinged lid [electrical]
boîte *f* de parquet pivotante   hinged floor socket [electrical]

boîte *f* de raccordement   cable distribution box, connection box
boîte *f* de sable   sand box
boîte *f* de tirage   cable drawing-in box
boîte *f* de vitesses   gearbox
boiteuse *f*   trimmed joist [built-in one end, other joist supported]
boîtier *m*   housing or case
boîtier *m* brise-vitre   break-glass fire alarm
boîtier *m* de diffusion d'air   air distribution box
boîtier *m* de prises de courant   socket outlet box
bollard *m*   bollard
bolomètre *m*   bolometer
bombe *f* fumigène   smoke bomb
bombé *adj*   domed, raised, rounded
bombement *m*   bulge, bulging, camber(ing), convexity [of a road surface]
bonde *f*   plug (hole) [of a bath, etc], sluice-gate
bondériser *vb*   bonderize
bondieu *m*   wedge jammed into saw cut to prevent saw binding
bord à bord   edge to edge
bord *m*   bank [of river, etc], shore [of a lake, sea], board, border, edge, kerb, rim, verge [road]
bord *m* à souder   fusion face [welding]
bord *m* chanfreiné   bevelled edge
bord *m* de coupe   cutting edge [of a tool]
bord *m* débité sur dosse   flat-sawn board
bord *m* intérieur de la chaussée   inner swept path [roads]
bord de (la) mer, au   seaside *adj*
bord de rivière, au   riverside *adj*
bordage *m*   plugging and filling prior to wallpapering
bordé de   bordered by
border *vb*   border
bordereau *m*   note or slip [summary]
bordereau *m* d'armatures   reinforcement schedule
bordereau *m* de livraison   delivery note
bordereau *m* de prix   price list or schedule
bordereau *m* de salaire   salary slip
bordereau *m* des travaux   schedule of works
bords arrondis, à   round-edged
bords de, sur les   on the banks of
bordure de, en   at the edge of
bordure *f*   edge or edging, kerb(stone)
bordure *f* arasée/basse   flush kerb
bordure *f* caniveau   kerb channel
bordure *f* creuse   hollow kerb channel
bordure *f* droite   vertical faced kerb
bordure *f* du pignon   barge board, gable board
bordure *f* du terrain   plot or site border or edge
bordure *f* du trottoir   pavement edge
bordure *f* en saillie   raised kerb
bordure *f* enterrée   flush kerb
bordure *f* haute   raised kerb
bordure *f* inclinée   splayed kerb
borgne *adj*   blank [without openings, as in blank wall]
bornage *m*   marking out or pegging out [boundaries of land]

bornage *m*, **pierre** *f* **de**   boundary stone
borne *f*   boundary marker [may be stone, plastic, a post, etc], distance marker, cable or electrical terminal
borne *f* **d'incendie**   fire hydrant [pillar type]
borne *f* **de câble**   cable socket
borne *f* **de réserve**   spare terminal
borne *f* **kilométrique**   milestone [strictly = kilometre stone]
borne-fontaine *f*   drinking fountain
borner *vb*   install boundary markers, mark out [land boundaries]
bornes plaque *f* **de**   terminal board
bornier *m* **de terre**   earthing terminal strip
bornoyer *vb*   mark out [with pegs or stakes], peg (out)
bosse *f*   bulge, bump, lump, swelling
bosseler *vb*   bruise [sth], dent
bossette *f*   round-headed tack
botte *f*   coil [of wire], boot [wellington]
botte *f* **de sécurité**   safety boot
bouchardage *m*   bush hammering
bouchardé *adj*   bush hammered
boucharde *f*   bush hammer or tool for bush hammering
boucharde *f* **d'enduiseur**   roller covered with pyramidal points for surface finishing
boucharder *vb*   bush hammer [stone]
bouché *adj*   obstructed
bouche *f*   grille [for ventilation], opening, outlet, mouth [of a river, or vent or inlet]
bouche *f* **à ailettes**   vaned outlet
bouche *f* **à clé**   key-operated outlet or valve
bouche *f* **à grille**   drainage inlet [road]
bouche *f* **à lames fixes**   fixed-direction grille
bouche *f* **à section variable**   differential grille
bouche *f* **d'aération**   air brick [foundation ventilator], ventilator
bouche *f* **d'aération à lames**   louvred grille
bouche *f* **d'air**   air nozzle
bouche *f* **d'alimentation**   supply air diffuser
bouche *f* **d'égout (BE)**   drain [bottom outlet], (street) gully [drainage outlet]
bouche *f* **d'évacuation**   exhaust opening or vent
bouche *f* **d'évacuation du bain**   bath drain
bouche *f* **d'évacuation ouvrable**   openable vent
bouche *f* **d'extraction**   extract grille
bouche *f* **d'incendie**   fire hydrant [ground level]
bouche *f* **d'incendie à clé sous trottoir**   underground fire hydrant
bouche *f* **de lavage**   cleansing hydrant [ground level]
bouche *f* **de plafond**   ceiling ventilator
bouche *f* **de reprise**   recirculation outlet
bouche *f* **de reprise d'air**   return air inlet
bouche *f* **de soufflage**   supply outlet
bouche *f* **de soufflage d'air**   supply air opening or outlet
bouche *f* **de soufflage orientable**   directional supply air terminal
bouche *f* **de ventilation**   air vent
bouche *f* **linéaire**   slotted outlet
bouché *adj*   blocked or plugged [openings, holes], stopped up

boucher *vb*   block [a hole], choke (up) [block], fill (up) [a crack, gap], obstruct [a pipe, a road], plug or stop up [a hole, a pipe, etc], seal off [close or plug]
bouchon *m*   bung or stopper, plug [for blocking or stopping up], cap [e.g. vehicle radiator], filler or make-up tile
bouchon *m* **à vis**   screw (on) cap, screwed plug
bouchon *m* **cadenassable**   lockable plug
bouchon *m* **d'air**   airlock [in a pipe]
bouchon *m* **de contact**   wall plug [socket outlet]
bouchon *m* **de dégorgement**   access cover or cleaning eye [plumbing]
bouchon *m* **de prise**   wall plug [socket outlet]
bouchon *m* **de raccord**   adaptor plug
bouchon *m* **de remplissage d'huile**   oil filler cap
bouchon *m* **de vidange**   drain plug
bouchon *m* **fusible**   fusible or safety plug
bouchon *m* **mâle**   spigot plug
bouchons *m,pl* **d'oreilles**   ear plugs
boucle *f*   loop
boucle, rue *f* **en**   loop road
boucle *f* **d'asservissement**   control loop
boucle *f* **d'hystérésis**   hysteresis loop
boucle *f* **de contrôle**   control loop
boucle *f* **de dilatation**   compensating loop
boucle *f* **de rideau**   curtain ring
boucle *f* **équilibrée**   balanced loop
boucler *vb*   buckle [metal]
bouclier *m*   bricklayer's float, shield [tunnelling]
bouclier *m* **d'électrode**   disintegration shield [in a lamp]
bouclier *m* **en béton**   concrete shield [nuclear power station]
boudin *m*   round moulding
boue *f*   deposit [sludge, etc], mud, sediment
boue *f* **de forage**   drilling mud or slurry
boues *f,pl*   sludge [sewage]
boues *f,pl* **activées**   activated sludge
boues *f,pl* **activées, procédé** *m* **des**   activated sludge process
boueux *adj*   muddy
bougie *f*   spark plug
bouillant *adj*   boiling
bouillir *vb*   boil
bouillon *m*   air bubble in glass
boulants, sables *m,pl*   quicksand
bouldozeur *m*   bulldozer
bouleau *m*   birch [timber or tree]
boules *f,pl* **d'oreilles**   ear plugs
boulevard *m* **périphérique**   ring road
boulin *m*   hole [in masonry for end of putlog or scaffolding tie], putlog or scaffolding tie
boulon *m*   bolt, pin, stud [generally threaded]
boulon *m* **à écrou**   screw bolt
boulon *m* **à haute résistance**   high-strength bolt
boulon *m* **à tête hexagonale**   hexagon head bolt
boulon *m* **à tête ronde**   round head bolt
boulon *m* **barbelé**   rag bolt
boulon *m* **brut**   black bolt
boulon *m* **d'ancrage**   anchor, holding down or rock bolt

boulon *m* **d'éclisse**   fishbolt [railway line]

boulon *m* **d'œil**   eye-bolt

boulon *m* **de charnière**   hinge pin

boulon *m* **de fixation**   fixing, holding down or securing bolt

boulon *m* **de montage**   erection bolt

boulon *m* **de retenue**   retaining bolt

boulon *m* **de scellement**   expansion or rag bolt

boulon *m* **fileté**   screw bolt

boulonnage *m*   bolting

boulonné *adj*   bolted

boulonner *vb*   bolt (down)

boulonnière *f*   bolt-hole borer [for timber]

bourg *m*   small town

bourg, au   in the town

bourg *m* **commerces**   shopping town

bourrage *m*   backfilling [of underground quarries, mines], filling [material], ramming down, seal [to prevent escape of air, water, etc], stemming [blasting holes], stuffing [of a gland, etc]

bourrelet *m*   draught excluder or strip, weather strip, flange or rim [of a wheel]

bourrer *vb*   ram down, stuff (full)

boursouflure *f*   blister [on paint]

bousillage *m*   cob

bousiller *vb*   construct in cob

boussole *f*   compass

bout *m*   butt, end, tip [extremity]

bout *m* **de terrain**   patch of land

bout *m* **de tube**   pipe end

bout *m* **en attente/escalier**   racking (raking) back

bout à bout   end-to-end [butted]

boutefeu *m*   blaster, shot-firer

bouteille *f*   bottle

bouteille *f* **à gaz/de gaz**   gas bottle or cylinder

bouteille *f* **de purge**   steam trap

bouterolle *f*   rivet-snap, riveting die

bouteroller *vb*   rivet over

bouteroue *f*   guard post or stone at corner of a building

bouteur *m*   bulldozer

bouteur *m* **à pneus**   turnadozer, wheeled dozer

bouteur *m* **biais**   angledozer

bouteur *m* **inclinable**   tiltdozer

bouteur *m* **sur chenilles**   crawler dozer

boutisse *f*   header [masonry], long paving slab used as kerb

bouton *m*   door knob

bouton *m* **d'appel**   call button

bouton *m* **de carillon (de porte)**   bell-push

bouton *m* **de commande/démarrage**   starter button

bouton *m* **de manivelle**   crank-pin

bouton *m* **(de porte)**   doorknob

bouton *m* **de poussoir**   press button, pushbutton

bouton *m* **de réarmement**   reset button

bouton *m* **de sonnette**   bell-push

bouvement *m*   quarter-round moulding

bouvet *m*   plough plane, tongue-and-groove

bouveté *adj*   tongued-and-grooved

bracon *m*   brace or strut [supporting a wall], stay [structural]

brai *m*   pitch (tar)

branchage *m*, **tapis** *m* **de**   brushwood raft [road foundation on peat]

branche *f* **d'autoroute**   motorway fork

branche *f* **de carrefour** *m*   arm, branch or leg of an intersection

branchement *m*   branch circuit, branching, connection, connecting up [to main supplies], lead [electrical], consumer connection, junction [in pipework], tapping [hydraulic, electrical]

branchement *m* **d'abonné/de l'abonné**   consumer connection

branchement *m* **d'air**   air connection

branchement *m* **d'appareil**   appliance connection [water]

branchement *m* **d'eau général**   water main connection

branchement *m* **d'égout**   sewer connection

branchement *m* **d'électricité**   mains (electricity) connection

branchement *m* **d'immeuble**   building or house connection

branchement *m* **de distributeur**   supplier connection

branchement *m* **de gaz**   gas service pipe

branchement *m* **de prise d'air**   air inlet connection

branchement *m* **de sortie**   outlet branch

branchement *m* **en attente**   dead leg

branchement *m* **fermé**   dead leg

branchement *m* **oblique simple**   single oblique junction [pipework]

branchement *m* **particulier de gaz**   gas main connection

branchement *m* **pluvial**   rainwater connection

branchement *m* **particulier**   house or individual connection

brancher *vb*   connect (up), plug in or switch on [electrically], tap off

brandille *f*   hole [in a rafter for peg jointing on to purlins]

brandir *vb*   joint notched timbers with a peg

bras *m* **de levier**   lever arm

bras *m* **mort**   ox-bow lake

brasage *m*   brazing, hard soldering

brasage *m* **à la louche**   wiped solder joint

brasé *adj*   brazed

braser *vb*   braze, solder

braser *vb* **à l'étain**   soft solder

brasserie *f*   brewery

brasure *f*   brazing

brasure *f* **à plomb**   lead soldering

brasure *f* **forte**   hard solder

brayage *m*   slinging [for lifting]

brayer *m*   sling [for lifting]

brayer *vb*   sling [for lifting]

bréchage *m*   marbling [paintwork]

brèche *f*   breccia

brèche *f*   breach [in a wall]

brécher *vb*   marble [paintwork]

bretèche *f*   bay or bow window

bretelle *f*   crossover [railway], link road

bretelle *f* **d'accès**   entry ramp or slip road

bretelle *f* **de raccordement**   link road

bretelle *f* **de sortie**   exit ramp or slip road

brette *f*   French or toothed drag

bretteler *vb*   drag [finish, with French/toothed drag]

brettelure *f*   dragged finish, teeth of a toothed drag

brevet *m*   patent

breveté *adj*   patented

bricolage *m*   do-it-yourself

bricole *f*   strap [leather, for carrying glass]

bricoler *vb*   bodge up[†], do odd jobs oneself

bricoleur *m*   do-it-yourself enthusiast, handyman

bride *f*   collar or flange [pipe, tube], thimble [in pipe joints]

bride *f* **à capote**   G-cramp

bride *f* **à emboîtement**   facing flange

bride *f* **à emboîtement double**   tongue-and-groove facing flange

bride *f* **à emboîtement simple**   male–female facing flange

bride *f* **d'accouplement**   coupling flange

bride *f* **de raccordement**   connecting/connection flange

bride *f* **de serrage**   clamp or cramp

bride *f* **de tube**   pipe flange

bride *f* **filetée**   screwed flange

bride *f* **folle**   slip flange

bride *f* **mobile**   loose flange

bride *f* **non percée**   blank flange

bride *f* **pleine**   blind flange

bride *f* **support**   support flange

brider *vb*   connect up [pipes], flange

brifier *m*   ridge capping [lead, of a slate roof]

brillant *adj*   gloss [paint]

brin *m*   strand [of a cable], roundwood

brin, bois *m* **de**   timber in log form

brique *f*   brick [normally hollow in France], building block made of burnt clay

brique *f* **à air**   air brick

brique *f* **alvéolée/à alvéoles**   air/perforated brick

brique *f* **ajourée**   perforated brick

brique *f* **armée**   reinforced brick

brique *f* **creuse**   hollow clay brick or block

brique *f* **crue**   air-dried brick

brique *f* **cuite**   burnt brick

brique *f* **de laitier**   compressed brick made with clay and ggbfs

brique *f* **de parement**   facing brick

brique *f* **de pavage**   paving brick

brique *f* **non ajourée**   brick without perforations

brique *f* **perforée**   air, perforated or ventilation brick

brique *f* **plâtrière**   hollow clay brick [30–50mm thick]

brique *f* **pleine**   solid brick

brique *f* **pleine calibrée**   standard brick

brique *f* **réfractaire**   refractory brick

brique *f* **silico-calcaire**   calcium silicate brick

brique *f* **tubulaire**   hollow clay brick or block

brique *f* **vernissée**   glazed brick

brique *f* **vitrifiée**   vitrified brick

briquetage *m*   brick masonry, brickwork

briqueterie *f*   brickworks

briqueteur *m*   bricklayer

briqueteuse *f*   bricklayer's trowel

bris *m*   break, breakage, breaking

brisé *adj*   folding [of a door leaf or a shutter]

brisé, arc *m*   triangular arch

brise *f*   breeze

brise-béton *m*   concrete breaker, pneumatic drill [for breaking concrete]

brise-circuit *m*   circuit-breaker, cut-out [electrical], make-and-break key

brise-glace *m*   ice apron or breaker on a bridge pier

brise-jet *m*   anti-splash tap nozzle

brise-lames *m*   breakwater, groyne

brise-roc(he) *m*   rock breaker

brise-soleil *m*   sunshade

brise-vent *m*   windbreak, wind shield

briser *vb*   break (up), shatter, smash

briser *vb* **un joint**   break a joint

brisis *m*   angle between the two planes of a curb or mansard roof, lower slope of a curb or mansard roof

broche *f*   drift [tool], curved steel rod for bending or bell-mouthing pipe, peg, pin [of electrical plug, fluorescent tube, etc], spike [long, headless nail], pin [of a hinge]

broche *f* **de charnière**   hinge pin

broche *f* **de tour**   lathe spindle

brocher *vb*   spike together

brochure *f*   booklet, brochure, pamphlet

brodequin *m*   anklet boot

brome *m*   bromine

bronze *m*   bronze

broquette *f*   tack [for wood, carpets, etc]

brosse *f*   brush

brosse *f* **à encoller**   paste brush

brosse *f* **à étaler**   dry brush [paperhanger's]

brosse *f* **circulaire**   circular wire brush

brosse *f* **dure**   scrubbing brush

brosse *f* **métallique**   wire brush

brosser *vb*   brush [a surface, finish], scrub

brouette *f*   wheelbarrow

brouette *f* **à bascule**   hand barrow for carrying wet concrete

brouillard *m*   fog, mist

brouillard *m* **de tour de refroidissement**   cooling tower fogging

brouillon *m*   rough draft

broussailles *f,pl*   brushwood, scrub [bushes], undergrowth

broyage *m*   crushing

broyer *vb*   crush

broyeur *m*   crusher, macerator [plumbing]

broyeur *m* **à barres**   bar mill

broyeur *m* **à boulets**   ball mill

broyeur *m* **à cylindres**   roll crusher

broyeur *m* **à marteau**   hammer mill

broyeur *m* **à ordures**   waste crusher

bruine *f*   drizzle

bruit *m*   noise, sound [noise]

bruit *m* **aérien**   airborne noise or sound

bruit *m* **d'air**   air noise

bruit *m* **d'ambiance**   ambient noise

bruit *m* **d'impact**   impact noise

bruit *m* **de la circulation**   traffic noise

bruit *m* **émis**   break-out noise

bruit *m* **émis par les conduits**   duct break-out noise

bruits *m,pl* **à dominantes aiguës**   predominantly high frequency noise
brûlage *m*   burning off [of paint]
brûler *vb*   be alight or on fire, burn (up, down, away, out)
brûleur *m*   arc tube [lighting], burner [gas, oil]
brûleur *m* **à atomisation**   atomizing burner
brûleur *m* **à coupelle**   pot burner
brûleur *m* **à courants parallèles**   parallel flow burner
brûleur *m* **à évaporation**   atomizing burner
brûleur *m* **à haute pression**   high pressure burner
brûleur *m* **à injection**   pressure jet burner
brûleur *m* **à jet de gaz**   gas jet burner
brûleur *m* **à jets croisés**   nozzle mixing burner
brûleur *m* **à plusieurs becs**   multi-jet burner
brûleur *m* **à pot**   pot burner
brûleur *m* **à vaporisation**   vaporizing burner
brûlure *f*   burn
brume *f*   fog, haze, mist
brun *adj*   brown
brut *adj*   crude or rough [unworked, without finishes], raw [untreated, e.g. water], untreated [without finishes], unworked [rough]
brut *adj* **de décoffrage**   as-struck [of concrete]
brute, eau *f*   raw/untreated water
bruyant *adj*   noisy
buanderie *f*   laundry (room), utility room, wash-house, washroom
bûcher *m*   woodshed
bûcher *vb*   rough hew [wood, stone]
budget *m*   budget
budget *m* **global**   total budget
buée *f*   condensation, mist [on glass, etc]
buée, trou *m* **de**   condensation drain [window joinery]
buffet *m*   sideboard
bulbe *m*   bulb [horticultural]
bulbe *m* **de thermomètre**   thermometer bulb
bulbe *m* **sec**   dry bulb
bull(dozer) *m*   bulldozer
bullage *m*   blowholes on the formed surface of concrete
bulle *f*   air hole [in casting]
bulle *f*   blister [on casting]
bulle *f* **d'air**   air bubble
bulletin *m* **météo(rologique)**   weather report
bureau *m*   office [building or room], room, study
bureau *m* **d'études**   consulting engineer's office, design agency or office, drawing (design) office, multi-disciplinary design and consulting firm

bureau *m* **d'études techniques (BET)**   design agency or technical consultancy office [generally contractor owned]
bureau *m* **de contrôle**   building design quality control office, checking agency [design and construction] [in France]
bureau *m* **de contrôle**   quality control office for building design
bureau *m* **de dessin**   drawing office
bureau *m* **de planification**   planning department/office [sequencing work]
Bureau *m* **de Recherches Géologiques et Minières Français (BRGM)**   French mining and geological research institution
bureau *m* **des hypothèques**   land registry
bureau *m* **en espace fonctionnel**   open-plan office
bureau *m* **paysagé**   open-plan office
burette *f*   oil-can
burin *m*   cold chisel
burin *m* **de tourneur**   parting tool [wood turning]
buriner *vb*   chisel [work with a]
bus *m* **(électrique)**   electrical bus(bar)
buse *f*   channel [pipe], concrete pipe, nozzle, pipe, tube, flue adaptor pipe [joining different size pipes], sleeving
buse *f* **d'éjecteur**   ejector nozzle
buse *f* **d'injection**   spray nozzle
buse *f* **de soufflage**   outlet nozzle
but *m*   aim [e.g. of lamp], purpose
butane *m*   butane
butée *f*   abutment, buttress, fence [of a power saw], prop, thrust [of the ground], travel limiting stop [of a lift/elevator]
butée *f* **de coude**   anchor block [pipe bend]
butée *f* **de profondeur**   depth stop
butée *f* **des terres**   passive earth pressure
buter *vb*   prop up
butoir *m*   buffer-stop, buffer [fixed railway], stop
butoir *m* **de porte/d'une porte**   doorstop
buton *m*   prop [to side of excavation]
butonnage *m*   propping [side of excavation]
butonner *vb*   prop [side of excavation]
butte, pelle *f* **équipée en**   face shovel [excavator]
butter *vb*   ridge [ground for drainage purposes]
by-pass *m*   bypass (pipe)

cabane *f*   cabin, hut
cabanon *m*   cabin, small hut
cabestan *m*   winch, windlass
cabine *f*   cabin
cabine *f* **d'aiguillage**   signal box
cabine *f* **d'ascenseur**   lift car

cabine *f* **de commande**   control cabin
cabine *f* **de douche**   shower cabinet
cabine *f* **de peinture**   paint spraying booth
cabine *f* **du conducteur**   driver's cabin
cabine *f* **du grutier**   crane driver's cabin
cabinet *m*   closet, office [of lawyer, etc], small room

cabinet *m* **d'aisances**   cloakroom, toilet (WC)

cabinet *m* **d'étude**   study [room]

cabinet *m* **de toilette**   cloakroom, dressing room

cabinet *m* **sec**   dry or earth closet

cabinets *m,pl*   toilet(s) (WCs)

câblage *m*   cabling, wiring

câblage *m* **de circuit**   circuit wiring

câblage *m* **de télécommande**   remote control wiring

câblage *m* **électrique intérieur**   house wiring

câblage *m* **permanent**   permanent wiring

câblage *m* **voix données image**   voice and data cabling

câble *m*   (electrical) cable, (wire) rope, tendon [stressing]

câble *m* **aérien**   overhead cable

câble *m* **armé**   armoured cable

câble *m* **bouclé**   looped cable

câble *m* **blindé**   screened cable

câble *m* **coaxial/concentrique**   coaxial cable

câble *m* **d'acier**   steel cable, steel-wire rope

câble *m* **d'alimentation**   feeder [electrical]

câble *m* **d'amenée de courant**   supply cable

câble *m* **de commande**   control cable

câble *m* **de compensation**   balancing wire

câble *m* **de dérivation**   bypass cable

câble *m* **de distribution**   mains cable

câble *m* **de masse**   earth(ing) cable

câble *m* **de (mise à la) terre**   earthing cable

câble *m* **de soudage**   welding cable

câble *m* **de suspension**   suspension cable

câble *m* **de transmission**   power cable

câble *m* **du secteur**   supply cable

câble *m* **du type pyrocâble**   MICC cable

câble *m* **en quatre âmes**   four-core cable

câble *m* **enterré**   buried cable

câble *m* **flexible**   electrical flex

câble *m* **métallique**   wire rope

câble *m* **multifilaire**   multi-core cable

câble *m* **multiple**   multi-conductor cable

câble *m* **pendentif**   trailing cable [lift/elevator]

câble *m* **pilote**   pilot cable

câble *m* **porteur**   suspension cable

câble *m* **porteur principal**   main (suspension) cable [bridge]

câble *m* **principal**   feeder mains, main cable

câble *m* **tendeur**   tension cable

câble *m* **toronné**   stranded cable

câble *m* **tracteur**   haulage cable [cableway]

câble *m* **transporteur**   aerial ropeway

câbles *m,pl* **apparents**   surface wiring

câble *m,* **poser** *vb* **un**   lay a cable

câble-ruban *m*   cable tape

cache *f*   escutcheon, mask [used in painting]

cache-borne *m*   terminal shield

cache-entrée *m*   keyhole cover

cachet *m*   seal [on a document]

cachetage *m*   encasement, sealing [of documents]

cache-vis *m*   screw-head cover, soft metal seal over a screw in a meter casing

cadastre *m*   cadastral register, land register/registry, plan [of property parcels in a commune]

cadastrer *vb*   register [property, in the cadastral record], survey [for land registration purposes]

cadenas *m*   padlock

cadenassable *adj*   lockable [with a padlock]

cadence *f*   rate [pace, rhythm]

cadmiage *m*   cadmium plating

cadmium *m*   cadmium

cadran *m*   dial [of a gauge or meter]

cadran *m* **solaire**   sundial

cadranure *f*   heart shake [timber defect]

cadre *m*   border [of a map, plan], frame(work)/framing, picture frame, crate or packing case, bounds, compass or limit, setting or surroundings, cage, link or stirrup [reinforcement], case [of a shaft, well], executive or management staff [member or one of]

cadre *m* **à étages**   multi-storey frame

cadre *m* **à trois articulations**   three-hinge frame

cadre *m* **d'armature**   hoop reinforcement [rectangular] or stirrup

cadre *m* **de serrage**   clamping frame

cadre *m* **fermé**   closed link or tie, or stirrup [reinforcement]

cadre *m* **incliné**   inclined link or stirrup [reinforcement]

cadre *m* **portique**   portal frame

caduc *adj*   crumbling or decrepit [e.g. building, wall], deciduous [bushes, trees]

cage *f*   carcass/carcase [of a house], shell [of a building], contained (vertical) space, shaft or well [for stairs, lift]

cage *f* **(d'armature)**   reinforcement cage

cage *f* **d'ascenseur**   elevator or lift shaft or well

cage *f* **d'escalier**   stairwell

cage *f* **technique**   services shaft

cahier *m* **des charges**   project contract conditions, specification

cahier *m* **des charges détaillé**   detailed specification

cahier *m* **des charges techniques**   technical specification

cahier *m* **des clauses administratives générales (CCAG)**   general contract conditions

cahier *m* **des clauses administratives particulières (CCAP)**   specific contract conditions

cahier *m* **des clauses générales**   general contract conditions

cahier *m* **des clauses particulières**   specific contract conditions

cahier *m* **des clauses spécifiques**   specific contract conditions

cahier *m* **des clauses techniques**   technical specification

caillasse *f*   road metal

caillebotis *m*   grating

caillou *m*   pebble, stone [40–60mm]

caillou *m,* **gros**   cobble

cailloutage *m*   pebbledash

caillouter *vb*   ballast or metal [a roadway], pave a road with pebbles

cailloutis *m*   crushed stone, gravel

caisse *f*   case, crate, packing case, cash desk

caisse *f* **à outils**   tool-box

caisse *f* **de chargement**   feed or loading skip

caissier *m*   cashier

caisson *m*   caisson, carcass [of built-in furniture], box, case or casing, coffer [e.g. in coffered slab]

caisson *m* **à air comprimé**   compressed air or pneumatic caisson

caisson *m* **cellulaire**   cofferdam

caisson *m* **de filtre à l'air**   air filter unit/section [in an air handling unit]

caisson *m* **de mélange**   mixing box [in an air handling unit]

caisson *m* **ventilateur**   fan unit [in an air handling unit]

caisson, poutre *f* **à**   box girder

caissons, plafond *m* **en**   coffered ceiling

calage *m*   keying [fixing with a key], wedging

calamine *f*   mill scale

calcaire *adj*   calcareous

calcaire *m*   limestone

calcaire *m* **compact**   hard limestone

calcaire *m* **oolithique**   oolitic limestone

calcifier *vb*   calcify

calcin *m*   hardpan [formed on the surface of limestone]

calcite *m*   calcite

calcium *m*   calcium

calcul *m*   calculation, computation

calcul *m* **approché**   approximate calculation

calcul *m* **de frais**   costing

calcul *m* **de la consommation de courant**   calculation of current consumption

calcul *m* **de probabilités**   probability calculation

calcul *m* **de structures**   analysis of structures

calcul *m* **des coûts**   cost calculation

calcul *m* **des déperditions calorifiques**   heat loss calculation

calcul *m* **des pertes de chaleur**   heat loss calculation

calcul *m* **élastique**   elastic analysis

calculateur *m* **de devis**   estimator

calculatrice *f* **(de poche)**   (pocket) calculator

calculer *vb*   analyse, calculate

cale *f*   choc, wedge, prop, strut, sanding/sandpaper block

cale *f* **à béton**   reinforcement spacer

cale *f* **sèche**   dry dock

calendrier *m*   calendar, programme [calendar, timetable]

calendrier *m* **du chantier**   site programme

calepin *m*   joint layout drawing [stonework, formwork, tiling, etc]

calepinage *m*   joint layout [stonework, formwork, tiling, etc]

calepiner *vb*   draw layout of joints [stonework, formwork, tiling, etc]

caler *vb*   blind [in road construction], chock (up), wedge (up)

calfatage *m*   caulking

calfatage *m* **de plomb**   lead caulking

calfater *vb*   caulk

calfeutrage *m*   caulking, draught-proofing

calfeutré *adj*   draught-proofed

calfeutrement *m*   caulking, draught-proofing, cover moulding or strip [e.g. between wall and door frame]

calfeutrer *vb*   fill or stop up [cracks, joints, etc], make good [around an element fixed in masonry, etc]

calibrage *m*   calibration

calibré *adj*   graded, sorted

calibre *m*   gauge [measuring tool], template

calibrer *vb*   calibrate, sort [grade]

caloduc *m*   heating pipe

caloporteur *m*   heat transfer medium

calorie *f*   calorie

calorifère *adj*   heat conveying

calorifère *m*   heating apparatus, appliance or installation

calorifère *m* **à air chaud**   air heater

calorifère *m* **d'eau chaude**   hot water heater

calorifique *adj*   calorific

calorifuge *adj*   heat-insulating, lagged [insulated]

calorifuge *m*   thermal insulator or insulating material

calorifugeage *m*   (thermal) insulation or lagging [the process]

calorifuger *vb*   insulate, lag

calorifugeur *m*   insulator [tradesman]

calorimètre *m*   calorimeter

calotte *f*   small spherical vault

calotte *f* **argentée**   silvered crown lamp bulb

calque *m*   tracing paper

calque *m*, **papier-**   tracing paper

calquer *vb*   trace

calqueur *m*   tracer [in a drawing office]

camarteau *m*   sleeper tower [temporary support]

cambium *m*   cambium

cambrer *vb*   make curved [wood, metal]

cambrure *f*   camber [of an arch, a piece of wood]

came *f*   cam

caméra *f*   camera

caméra *f* **thermique de surveillance**   thermal imaging camera

camion *m*   lorry, truck, paint kettle

camion *m* **benne**   dumper truck, tipper (truck)

camion *m* **malaxeur**   ready-mix concrete lorry, transit mixer or truckmixer

camion *m* **tribenne**   three-way tipper (lorry)

camion-citerne *m*   tanker lorry

camion-grue *m*   truck crane

camionnette *f*   light truck, van

camionner *vb*   carry, cart or convey [goods]

campagnard *adj*   country *adj*, rustic *adj*

campagne, maison *f* **de**   country cottage

campagne *f*   campaign [marketing, etc], country [as opposed to town]

campagne *f* **de reconnaissance**   site investigation

campanile *m*   bell tower, campanile

camping *m*   camp(ing) site

canal *m*   canal, channel, conduit, culvert, duct, flume, pipe

canal *m* **à ciel ouvert**   open culvert

canal *m* **d'adduction**   aqueduct

canal *m* **d'alimentation**   feeder or supply channel

canal *m* **d'amenée**   water intake channel

canal *m* **d'écoulement principal**   main drain

canal *m* **de chasse**   flushing channel

canal *m* **de dérivation**   diversion canal

canal *m* **de fuite**   tailrace

canal *m* **de rejet**   water discharge channel
canal *m* **de sortie**   discharge duct
canal *m* **navigable**   navigable channel
canal *m* **souterrain**   underfloor or underground duct
canalisation *f*   cable duct or its contents, duct(work), pipe(line), pipe(work) or piping
canalisation *f* **d'air de retour**   return air duct
canalisation *f* **d'air évacué**   exhaust air duct
canalisation *f* **d'air frais**   fresh air duct
canalisation *f* **d'alimentation**   flow (feed) pipework
canalisation *f* **d'assainissement**   drainage piping or pipework
canalisation *f* **de gaz**   gas pipeline
canalisation *f* **de prise d'air**   air inlet duct
canalisation *f* **de refoulement**   discharge channel
canalisation *f* **de retour**   return pipework
canalisation *f* **de trop-plein**   overflow pipe
canalisation *f* **de ville**   town mains
canalisation *f* **des câbles de contrôle**   control cable duct
canalisation *f* **des eaux de pluie**   stormwater drainage
canalisation *f* **électrique**   electrical conduit or mains
canalisation *f* **pour tuyaux**   pipe duct
canalisation *f* **sous pression**   pressure pipeline
canaliser *vb*   lay [pipes, mains]
candélabre *m*   (road) lighting column
candidats *m,pl* **sélectionnés**   invited tenderers
candidature *f*   prequalification
canevas *m*   skeleton map or triangulation
canevas *m* **d'ensemble**   general triangulation
caniveau *m*   cable duct, duct [for services], (drainage) channel [road], road or street drain or gutter, undercut [in welding]
caniveau *m* **à fentes**   slotted channel [road]
caniveau *m* **couvert (en U)**   box gutter [road or street]
caniveau *m* **technique**   services channel
canne *f* **d'injection**   injection rod [in a water treatment unit]
canne *f* **thermoélectrique**   pyrometer, thermocouple thermometer
cannelure *f*   channel [architectural], groove [fluting, striation], slot [groove, channel], fluting [of a column]
canon *m*   slotted tube to guide key into a lock
canon *m* **à ciment**   cement gun
cantilever *m*   cantilever
cantine *f*   site canteen
canton *m*   district [second layer of local government in France]
cantonal *adj*   cantonal
cantonal *adj*   district *adj*
cantonnier *m*   roadman
cantonnière, maison *f*   roadman's hut
canyon *m*   canyon, gorge
caoutchouc *m*   rubber
caoutchouc, joint *m*   rubber joint
caoutchouc *m* **toilé**   rubberized cloth
capable de fonctionner   in working order, workable [of a machine, etc]
capacitance *f*   capacitance
capacité *f*   capacity or volume

capacité *f* **calorifique**   heat capacity
capacité *f* **d'absorption de la poussière**   dust absorption capacity
capacité *f* **d'un filtre**   filter capacity
capacité *f* **de charge**   bearing load
capacité *f* **de déformation**   deformability
capacité *f* **de filtrage**   filtering capacity
capacité *f* **de production**   production capacity
capacité *f* **de puits**   well capacity
capacité *f* **de rétention des poussières**   dust holding/retaining capacity
capacité *f* **de stockage**   holding or storage capacity
capacité *f* **de stockage thermique**   heat storage capacity
capacité *f* **de surcharge**   overload capacity
capacité *f* **de ventilateur**   fan capacity
capacité *f* **installée**   installed capacity
capacité *f* **journalière**   daily capacity
capacité *f* **normale**   standard capacity
capacité *f* **officielle**   official capacity
capacité *f* **portante**   bearing capacity, load-carrying capacity
capacité *f* **thermique**   thermal capacity
capacité *f* **totale**   total capacity
capillaire *adj*   capillary *adj*
capillaire *m*   capillary
capillaire, ascension *f*   capillary rise
capillaire, eau *f*   capillary water
capillarité *f*   capillarity, capillary action
capital *m*   capital [share], funds
capital *m* **social**   share capital
capot *m*   bonnet [chimney] cowl, casing, cover, hood
capot *m* **de protection**   cover
capotage *m*   capping [with insulation] or insulating [end of an electrical conductor]
capote *f*   cowl [of a chimney]
captage *m*   catchment, collection [of water, or current from overhead wire]
captage *m* **de source**   spring catchment
capter *vb*   pick up [a signal], tap [pick up electrical current]
capter *vb* **la lumière**   capture the light
capte-suie(s) *m*   dust eliminator, grit or soot arrestor
capteur *m* **à distance**   remote sensor
capteur *m* **à gaz**   gas detector
capteur *m* **acoustique**   acoustic pickup
capteur *m* **(de mesure)**   sensor
capteur *m* **de niveau**   level detector
capteur *m* **de pression**   pressure transducer
capteur *m* **de rayonnement solaire**   solar sensor
capteur *m* **plan**   flat plate collector [solar]
capteur *m* **solaire**   solar collector
capuchon *m*   cover [e.g. of a lamp], cowl [e.g. on a chimney]
capuchon *m* **de valve**   tyre valve cap
capucine *f*   dripstone, fireplace surround [rectilinear and undecorated]
caractère *f*   nature [character]
caractéristique *adj*   characteristic

caractéristique *f*   attribute, characteristic, feature, property

caractéristique *f* **du local**   room characteristic

caractéristiques *f,pl* **d'une pompe**   pump characteristics

caractéristiques *f,pl* **de fonctionnement**   operating characteristics

caractéristiques *f,pl* **du projet**   design criteria

carbonatation *f*   carbonation

carbonate *m*   carbonate

carbone *m*   carbon

carbonisation *f*   carbonization

carborundum *m*   carborundum, silicon carbide

carburant *m*   fuel

carbure *m*   carbide

carbure *m* **de calcium**   calcium carbide

carbure *m* **de silicium**   silicon carbide

carbure *m* **de tungstène**   tungsten carbide

carcasse *f*   carcass/carcase [of electric motor], frame(work) or shell [of building], grounds [carpentry]

carderonner *vb*   round off the corner of [e.g. of exposed timbers]

carie *f*   rot [timber]

carie *f* **humide**   wet rot

carie *f* **sèche**   dry rot

carillon *m*   carillon, chime, peal of bells

carillon *m* **de porte**   doorbell

carne *f*   corner or edge [of a stone, mantlepiece]

carneau *m*   flue [between boiler and vertical flue]

carneau *m* **de chauffage**   heating flue

carneau *m* **principal**   main flue

carnet *m* **de battage**   pile (driving) record

carnet *m* **de chèques**   cheque book

carottage *m*   core sample extraction or removal

carotté, échantillon *m*   core sample

carotte *f*   borehole or core sample

carotte *f* **de roche**   rock core

carottier *m*   coring bit, sampler [for deep cores]

carottier, tube *m*   core barrel

carottin *m*   small pieces of building stone

carrare *m*   Carrara marble

carré *adj*   square

carré *m*   square [figure, second power]

carreau *m*   cut stone or stonework stretcher, pane [of glass] or window pane, wall or floor tile

carreau *m* **de liège**   cork tile

carreau *m* **de pierre**   flagstone

carreau *m* **de plâtre**   plaster building block [for partitions, etc]

carreau *m* **de revêtement**   wall tile

carreau *m* **de vitre**   window pane

carrée, racine *f*   square root

carrefour *m*   crossroads, intersection or junction [road], interchange [highway]

carrefour *m* **à feux**   signal controlled road junction

carrefour *m* **à niveau**   at grade or single level intersection

carrefour *m* **dénivelé**   grade-separated intersection

carrefour *m* **en trèfle**   cloverleaf intersection or junction [highways]

carrefour *m* **giratoire**   roundabout

carrelage *m*   tiled floor, tiling

carrelé *adj*   tiled

carreler *vb*   tile [a floor, a wall]

carrelet *m*   ceramic tile that is twice as long as it is wide

carreleur *m*   tiler [floor or wall]

carrer *vb*   square [a number, or make square]

carrier *m*   quarryman

carrière *f*   gravel pit, quarry

carrière, déchets *m,pl* **(de)**   quarry waste

carrière *f* **de sable**   sand pit or quarry

carte *f*   map

carte *f* **avec courbes de niveau**   contour map

carte *f* **de coloris**   colour or shade card or chart

carte *f* **perforée**   punched card

cartelette *f*   small sized slate

carter *m*   casing, housing [case], crankcase

carter *m* **de pompe**   pump casing

carter *m* **en tôle**   sheet metal casing

carton *m*   cardboard (box)

carton *m* **bitumé**   roofing paper

carton *m* **feutre bitumé**   roofing felt

cartouche *m*   cartouche [architectural], cartridge [for blasting], key map [inset on a drawing] or title block [of drawing]

cas *m*   case [instance]

cas *m* **d'urgence**   emergency

cas *m* **d'utilisation**   application

cas *m* **de charge**   load(ing) case

cas *m* **limite**   extreme case

cascade *f*   cascade, waterfall

caserne *f*   barracks

caserne *f* **des pompiers**   fire station

cash-flow *m*   cash flow

casque *m*   (safety) helmet

casque *m* **de battage**   pile helmet

casque *m* **de pieu**   pile cap

casque *m* **(de protection)**   hard hat, protective or safety helmet

casque *m* **de soudage**   welding helmet

casque *m* **enveloppant/intégral**   full helmet [protective]

casser *vb*   break, fracture, rupture

casser *vb* **un ton**   decrease purity of a (paint) colour by adding another tint

cassure *f*   break, fracture, rupture

cassure *f* **à grain fin**   fine-grained fracture

cassure *f* **de fatigue**   fatigue fracture

cassure *f* **ductile**   deformation or ductile fracture

catalogue *m*   catalogue

catalyseur *m*   catalyst

cataracte *f*   waterfall

catégorie *f*   category, class

caténaire *adj*   catenary *adj*

caténaire *f*   catenary *n*

cathédrale *f*   cathedral

cathédrale, toit *m*   high-vaulted ceiling

cathédrale, verre *m*   semi-opaque glass [type of]

cathode *f*   cathode

cation *m*   cation

caution *f*   bail (bond), deposit, guarantee [security], guarantor

caution *f* de soumission   bid/tender bond

cautionnement *m*   bond [in a contract], performance bond or guarantee, guarantee [security]

cavalier *m*   spoil bank [roadworks], waste bank [excavations], staple [e.g. for fixing electrical wiring]

cavalier *m* à deux scellements   heavy staple for building into masonry [e.g. step iron]

cavalier, plan *m*   bird's eye view [plan/drawing]

cave *f*   basement, cellar

caveau *m*   crypt, small cellar, staircase cupboard

caverne *f*   cave(rn)

caverneux *adj*   cavernous

cavitation *f*   cavitation

cavité *f*   cavity

cédant *m*   assignor

céder *vb*   give way [under load, pressure], part, [rope, cable, etc, under load], yield [collapse], convey [property], dispose of, make over or transfer [a lease], grant [rights, etc]

cèdre *m*   cedar

ceindre *vb* (de)   enclose [surround] (with)

ceint *adj* (de)   enclosed (by) or surrounded (by) [e.g. walls]

ceinture *f*   belt [of hills], metal ring holding pipe against a wall, ring beam, ring of strip metal covering a joint

ceinture *f* d'appartement   horizontal distribution pipe from rising main

ceinture *f* d'étage   horizontal distribution pipe from rising main

ceinture *f* de protection d'arbre   tree guard

ceinture *f* de sécurité   safety belt

ceinture *f* principale   horizontal distribution pipe between meter and appliance connections

cellier *m*   larder, storeroom [for food and wine]

cellulaire *adj*   cellular

cellule *f*   cell

cellule *f* de contrainte   load cell

cellule *f* de fluage   creep cell

cellule *f* photo-électrique   photo-electric cell

cellule *f* solaire   solar cell

cémentation *f*   case-hardening

cendre *f*   ash

cendres *f,pl*   cinders

cendres *f,pl* volantes   fly ash, pulverized fuel ash (PFA)

cendrier *m*   ash bin/box/pan/pit

central *adj*   central

central *m* automatique   automatic telephone exchange

central *m* automatique urbain   local automatic telephone exchange

centrale *f* d'enrobage   asphaltic concrete batching plant

central *m* téléphonique   telephone exchange

centrale *f*   plant [fixed], powerhouse, power station

centrale *f* à béton   concrete batching or mixing plant

centrale *f* à bitume   bitumen storage system

centrale *f* atomique   atomic/nuclear power plant or station

centrale *f* combinée/de chaleur force   combined heat and power (CHP) station

centrale *f* d'enrobage   asphalt plant

centrale *f* de chauffage   central boiler plant

centrale *f* de chauffage urbain   district heating plant or installation

centrale *f* de climatisation   air-conditioning or air handling plant

centrale *f* de malaxage à sec   dry batching plant

centrale *f* de production d'énergie   power plant

centrale *f* électrique   electric power or generating station, powerhouse [electric]

centrale *f* hydraulique/hydro-électrique   hydro-electric power station

centrale *f* modulaire de climatisation   modular air-conditioning plant/system

centrale *f* nucléaire   nuclear power plant/station

centrale *f* thermique   steam generating or thermal power station

centré *adj*   axial, central

centre *m*   centre, middle

centre *m* commercial   shopping centre

centre *m* communautaire   community centre

centre *m* d'entretien autoroutier   motorway maintenance area

centre *m* d'entretien routier   highway or road maintenance area

centre *m* de cisaillement   shear centre

centre *m* de gravité   centre of gravity or mass, centroid

centre *m* de recherches   research centre

centre *m* de rotation   centre of rotation

Centre Scientifique et Technique du Bâtiment (CSTB)   French national building research institution

centrer *vb*   centre

centrifuge *adj*   centrifugal

centrifuger *vb*   centrifuge

centrifugeur *m*   centrifuge

céramique *adj*   ceramic

céramique *f*   ceramic(s)

cerce *f*, *see also* cerceau   hoop

cerceau *m*   curved template [for archways, etc]

cercle *m*   circle, dial, ring [figure or shape]

cercle *m* à calculer   circular slide rule

cercle *m* de frottement   friction circle

cercle *m* de Mohr   Mohr's circle [of stress]

cercle *m* de perçage   bolt hole circle

cerne *m*   annual/growth ring [of a tree]

certificat *m*   certificate

certificat *m* d'épreuve   test certificate

certificat *m* d'essai   test certificate

certificat *m* d'homologation   certificate of approval [compliance with standard(s)]

certificat *m* d'urbanisme   town or urban planning certificate

certificat *m* d'usine   workshop certificate

certificat *m* de réception   acceptance certificate

céruse *f*   white lead

cession *f*   assignment

cessionnaire *f*   assignee

chaînage *m* chainage [surveying], ring beam [on top of or within a wall], ties [for walls]

chaînage *m* **périphérique** peripheral tie

chaîne *f* chain, pier in a wall [of cut stone, etc]

chaîne *f* **à mortaiser** mortising chain

chaîne *f* **à mortaiser sans fin** endless mortising chain

chaîne *f* **d'acquisition** data logger

chaîne *f* **d'angle** corner pier [to stiffen junction of two walls]

chaîne *f* **d'entraînement** drive chain [e.g. in an escalator]

chaîne *f* **de commande** driving chain

chaîne *f* **de fabrication** assembly line

chaîne *f* **de la scie** saw chain

chaîne *f* **transporteuse** chain conveyor

chaîner *vb* chain [measure], tie together [with a ring beam, etc]

chaînette, arc *m* **en** catenary curve

chaise *f* level board [two pegs + a horizontal], seating [of a bearing], sole piece or plate [of wood, under a prop, etc], support frame [fixed into masonry, to support first course of chimney, etc]

chalet *m* chalet, country cottage

chaleur *f* heat

chaleur *f* **d'évaporation** heat of vaporization

chaleur *f* **d'hydratation** heat of hydration

chaleur *f* **de combustion** heat of combustion

chaleur *f* **de compression** heat of compression

chaleur *f* **de condensation** heat of condensation

chaleur *f* **de convection** convection heat

chaleur *f* **de mélange** heat of mixing

chaleur *f* **de prise** setting heat [of concrete]

chaleur *f* **de réaction** heat of reaction

chaleur *f* **indirecte** indirect heating

chaleur *f* **latente** latent heat

chaleur *f* **latente d'évaporation** latent heat of evaporation/vaporization

chaleur *f* **latente de fusion** latent heat of fusion/melting

chaleur *f* **massique** specific heat

chaleur *f* **perdue** waste heat

chaleur *f* **radiante/rayonnante** radiant heat

chaleur *f* **résiduelle** afterheat

chaleur *f* **sensible** sensible heat

chaleur *f* **spécifique** specific heat

chaleur *f* **volumique** volumetric heat

chalumeau *m* blowlamp, blowpipe, torch [for heating, brazing, etc]

chalumeau *m* **à gaz** gas cutting torch

chalumeau *m* **oxyacétylénique** oxyacetylene torch

chalumeau *m* **(oxycoupeur)** cutting torch

chalumeau *m* **(soudeur)** welding torch

chalumeau-braseur *m* brazing torch

chalumeau-coupeur *m* cutting torch

chalumeau-soudeur *m* brazing torch, welding blowpipe

chambranle *m* architrave, casing [of a door, window], frame [of door]

chambre *f* room [bedroom or hotel], chamber

chambre *f* **à fumée** throat [of a fireplace]

chambre *f* **climatique** environmental or climatic chamber

chambre *f* **climatisée** air-conditioned chamber

chambre *f* **collectrice** collecting chamber

chambre *f* **d'amortissement** stilling basin [of a dam]

chambre *f* **d'air** inner tube [of a tyre]

chambre *f* **d'amis** guest room

chambre *f* **d'essai** test chamber

chambre *f* **de bonne** garret

chambre *f* **de chauffage** heating chamber

chambre *f* **de combustion** combustion chamber, firebox, furnace

chambre *f* **de mélange** mixing chamber

chambre *f* **de mise en charge** plenum chamber

chambre *f* **de mise en pression** plenum chamber

chambre *f* **de pulvérisation** spray chamber

chambre *f* **de réfrigération** cooling chamber

chambre *f* **de vannes** valve chamber

chambre *f* **de visite** inspection chamber, manhole

Chambre *f* **des Ingénieurs Conseils de France (CICF)** association of consulting engineers [France] [near equivalent of UK ACE]

chambre *f* **forte** strongroom

chambre *f* **frigorifique/froid** cold room

chambre *f* **propre** clean room

chambre *f* **réverbérante** reverberation chamber

chambre *f* **sourde** anechoic chamber

Chambre *f* **Syndicale des Sociétés d'Études et de Conseils Techniques (SYNTEC)** association of designing and consulting engineers [France] [generally those owned by contractors]

chamotte *f* fireclay

champ *m* edge, field, range [field]

champ, angle *m* **de** angle of view

champ, posé *adj* **sur** set on edge

champ *m* **de mesure** measuring range

champ *m* **électrique** electric field

champ *m* **magnétique** magnetic field

champ *m* **saillant** unmoulded architrave

champ *m* **tournant** rotating (electrical) field

champ *m* **visuel** field of view

champêtre *adj* rural, rustic

champignon *m* column head [in flat-slab construction], rail head [of rail], fungus

champlever *vb* chase out [for an inlay, or to form bas relief]

champs *m,pl* **d'épandage** sewage farm

chancelant *adj* unsteady [wall, structure, etc]

chancir *vb* go mouldy [of wood]

chandelle *f* prop, shore or stay [generally small, temporary], shuttering prop or strut

chandelle, distribution *f* **en** branched parallel rising main system of water supply

chandelle *f* **à crémaillère** adjustable telescopic prop [Acrow type]

chanfrein *m* bevel, chamfer

chanfreiné *adj* bevelled, chamfered

chanfreiner *vb* bevel, chamfer

changement *m* **(de)** change (in)

changement *m* **adiabatique d'état**   adiabatic change of state

changement *m* **d'échelle**   change of/in scale

changement *m* **de niveau d'eau**   change in water level

changement *m* **de pente**   gradient transition [highway]

changement *m* **de potentiel**   change in potential

changement *m* **de pression**   pressure change

changement *m* **de section**   change in section

changement *m* **(de température) été-hiver**   seasonal changeover

chanlatte *f*   chamfered batten under metallic roofing, eaves batten [double thickness, of a tiled roof]

chanlattée, lambourde *f*   chamfered batten on side of floor beam to support ends of abutting joists

chant, de   on edge

chant *m*   edge [of a board, a door leaf]

chantepleure *f*   weephole [vertical slot in a wall]

chanterelle *f*   bevel square, sliding bevel

chantier *m*   building or construction site, site [where works are taking or to take place], load platform consisting of thick boards

chantier *m* **de construction**   building or construction site

chantier *m* **de triage**   marshalling yard

chantier *m* **mobile**   mobile or temporary work site

chantignole *f*   batten fixed to principal rafter to prevent purlins slipping, brick 220×110×30mm used for fireplaces

chantourner, scie *f* **à**   jigsaw, scroll saw

chantourner *vb*   cut to a required profile or shape

chanvre *f*   hemp

chape *f*   coating [e.g. of mortar, cement], coping [of a bridge], screed

chape *f* **d'étanchéité**   damp-proof membrane, waterproof membrane [bridge deck], concrete compression screed

chape *f* **de bitume armé**   reinforced bitumen waterproof membrane

chape *f* **en ciment**   cement screed

chape *f* **flottante**   floating screed

chape *f* **incorporée**   monolithic screed

chapeau *m*   bolster [on top of posts or props], capping [of row of piles], cross beam [linking piles], cap, hood, upper part [cap, of a bearing, a pile]

chapeau *m* **de borne-fontaine**   hydrant cap

chapeau *m* **(de cheminée)**   cowl [on a chimney]

chapeau *m* **de palier**   bearing cap

chapeau *m* **de presse-étoupe**   packing or stuffing gland

chapeau *m* **de scie**   blade guard [circular saw]

chapeau *m* **pare-pluie**   weatherproof cowl

chapelle *f*   chapel

chapelle, toit *m*   high-vaulted ceiling

chaperon *m*   coping [of a wall]

chaperonner *vb*   provide with a coping [a wall]

chapiteau *m*   capital [of a column], cap(ping) [of a bridge pier]

chapoter *vb*   dress or trim timber with a drawing knife

charbon *m*   charcoal, coal

charbon *m*, **filtre à**   carbon filter

charbon *m* **à filtrer**   filter(ing) charcoal

charbon *m* **à vapeur**   boiler or steam coal

charbon *m* **actif**   activated carbon

charbon *m* **de bois**   charcoal

charbon *m* **flambant**   medium volatile coal

charbon *m* **maigre**   non-bituminous coal

charbon *m* **pulvérisé**   pulverized coal

charbon *m* **sans fumée**   smokeless coal

chardon *m*   spiked ironwork on top of a wall or fence

chardonnet *m*   hanging post [of a carriage gateway, etc]

charge *f*   bulking material [non-binding] in concrete, charge, expense, electrical power generated or absorbed, hydraulic pressure [total], load(ing) [electrical, structural, etc], measured constituents of concrete ready for mixing

charge, premier/etc cas *m* **de**   first/etc load(ing) case

charge *f* **à la rupture**   ultimate load

charge *f* **admissible**   allowable or safe load

charge *f* **alternée**   alternating load

charge *f* **au sol**   floor load

charge *f* **axiale**   axial load, thrust

charge *f* **calorifique**   fire load, heat load

charge *f* **concentrée**   concentrated or point load

charge *f* **connectée**   connected load

charge *f* **constante**   dead or steady load

charge *f* **critique d'Euler**   Euler buckling load

charge *f* **d'affaissement**   buckling or collapse load

charge *f* **d'épreuve/d'essai**   test or trial load

charge *f* **d'exploitation**   operating, service or working load

charge *f* **d'incendie**   fire load

charge *f* **d'un pieu**   pile bearing load

charge *f* **d'utilisation**   live load

charge *f* **de calcul**   design load

charge *f* **de connexion**   connected load

charge *f* **de flambement**   buckling load

charge *f* **de grille**   grate loading

charge *f* **de longue durée**   sustained load

charge *f* **de main d'œuvre**   labour cost

charge *f* **de pointe**   maximum or peak load

charge *f* **de rafraîchissement**   ventilation heat loss

charge *f* **de refroidissement**   cooling load

charge *f* **de rupture**   breaking load

charge *f* **de service**   allowable/operating/safe/service/ working load

charge *f* **de vent**   wind load

charge *f* **décentrée**   eccentric load

charge *f* **électrique**   electrical load

charge *f* **équilibrée**   balanced load

charge *f* **excentrée/excentrique**   eccentric load

charge *f* **exceptionnelle**   exceptional load

charge *f* **frigorifique**   cooling load

charge *f* **hydraulique**   head of water

charge *f* **limite**   ultimate load, lifting capacity [maximum load]

charge *f* **linéaire**   line load

charge *f* **maximale**   maximum/peak load

charge *f* **militaire**   military load

charge *f* **minimum**   base/minimum load

charge *f* **mobile**   moving load

charge *f* **morte**   dead load

charge *f* **ondulée**   pulsating load
charge *f* **par unité de surface**   load per unit area
charge *f* **permanente**   dead/permanent/steady load
charge *f* **ponctuelle**   point load
charge *f* **répartie**   distributed load (DL)
charge *f* **statique**   static head/load
charge *f* **sur le plancher**   floor load
charge *f* **thermique**   heat load
charge *f* **thermique de ventilation**   ventilation heat gain
charge *f* **totale**   total load
charge *f* **uniformément répartie**   uniformly distributed load (UDL)
charge *f* **unitaire**   unit load
charge *f* **utile**   live load, load capacity [e.g. of a lorry]
charge *f* **variable**   imposed/live/variable load
charge *f* **vive**   live load
charges *f,pl* **à payer**   accrued expenses
charges *f,pl* **comprises**   service charges included
charges *f,pl* **constatées d'avance**   charges fixed in advance
charges *f,pl* **d'exploitation**   operating expenses
charges *f,pl* **différées**   deferred charges
charges *f,pl* **directes**   direct charges
charges *f,pl* **fixes**   fixed charges
chargement *m*   charging [of a furnace, etc], loading
charger *vb*   charge, load
chargeur *m*   battery charger, loader [machine or person], feed or loading skip
chargeur *m* **d'huile**   oil filler pipe
chargeur *m* **de foyer**   stoker
chargeuse *f*   front-end loader
chargeuse *f* **à bande**   belt loader
chargeuse *f* **à direction par ripage**   skid-steer loader
chargeuse *f* **à godets**   bucket loader
chargeuse *f* **à/sur chenilles**   crawler loader
chargeuse-pelle(teuse) *f*   backhoe loader
chariot *m*   cart, trolley, truck, waggon
chariot *m* **à bras**   handcart
chariot *m* **de manutention**   handling cart or trolley, industrial truck
chariot *m* **de scie**   saw carriage
chariot *m* **élévateur (à fourche)**   forklift truck
charnier, montant *m*   hanging jamb or stile [carrying the hinge], upright carrying the hinge [of a door or window frame]
charnière *f*   (butt) hinge
charnière *f* **à broche/à fiche**   pin hinge
charnière *f* **universelle**   universal joint
charnon *m*   knuckle [of a hinge]
charpente *f*   structural frame(work)
charpente *f* **apparente**   exposed timber
charpente *f* **de comble**   roof truss, timber roof structure
charpente *f* **en bois**   timber frame
charpente *f* **métallique**   steel construction
charpenterie *f*   structural timber or steel work
charpentier *m*   carpenter
charpentier-coffreur *m*   formwork carpenter
chartreuse *f*   isolated country house
chas *m*   perforated plate [of a plumb line]

chasse *f*   drift [tool], flush [of water down a pipe or drain], set [of saw teeth]
chasse *f* **d'eau**   toilet flush cistern
chasse-clou *m*   nail punch
chasse-goupille *m*   nail punch
chasse-pointe *m*   nail punch
châssis *m*   chassis, casement, frame [door, window], mounting frame
châssis *m* **(à guillotine)**   sash window
châssis *m* **d'aération**   louvred window
châssis *m* **de fenêtre**   window frame
châssis *m* **de fondation**   foundation raft
châssis *m* **de porte**   door frame
châssis *m* **de translation**   bogie [of a crane on rails]
châssis *m* **tabatière**   skylight [hinged]
châssis *m* **vitré**   skylight
châtaignier *m*   chestnut [wood, and sweet chestnut tree]
château *m*   castle, château, country house, manor
château *m* **d'eau**   water tower
chatière *f*   cat hole, eyebrow dormer, roof ventilator, ventilation tile
chaud *adj*   hot, warm
chaud, mettre *vb* **en œuvre à**   hot lay [e.g. asphalt]
chaudière *f*   boiler
chaudière *f* **(à) basse pression**   low-pressure boiler
chaudière *f* **à bouilleurs croisés**   cross-tube boiler
chaudière *f* **à circulation d'eau forcée**   forced circulation boiler
chaudière *f* **à coke**   coke boiler
chaudière *f* **à eau chaude**   hot water (heating) boiler
chaudière *f* **à foyer surpressé**   pressure-fired boiler
chaudière *f* **à fuel**   oil-fired boiler
chaudière *f* **à gaz**   gas-fired boiler
chaudière *f* **à grande capacité**   high capacity boiler
chaudière *f* **(à) haute pression**   high pressure boiler
chaudière *f* **à plusieurs parcours**   multi-pass boiler
chaudière *f* **à récupération**   waste heat boiler
chaudière *f* **à tube de fumée**   fire tube boiler
chaudière *f* **à tubes de fumée**   Scotch boiler
chaudière *f* **à tubes d'eau**   water tube boiler
chaudière *f* **à tubes inclinés**   inclined tube boiler
chaudière *f* **à tubes-foyers verticaux**   vertical tube boiler
chaudière *f* **à vapeur**   steam boiler
chaudière *f* **à vapeur à basse pression**   low-pressure steam boiler
chaudière *f* **auxiliaire**   auxiliary boiler
chaudière *f* **convertible**   convertible boiler
chaudière *f* **de récupération**   waste heat boiler
chaudière *f* **de réserve**   standby boiler
chaudière *f* **équipée**   packaged boiler
chaudière *f* **mixte**   alternative fuel or combination boiler
chaudière *f* **multitubulaire**   fire tube or multi-tubular boiler
chaudière *f* **sectionnée**   sectional boiler
chaudronnerie *f*   boilermaker or boilermaking plant, boilers
chauffage *m*   heating [system or action of]
chauffage *m* **à accumulation**   storage heating
chauffage *m* **à air chaud**   hot or warm air heating

chauffage *m* **à la vapeur**   steam heating
chauffage *m* **à rayonnement**   radiant heating
chauffage *m* **à surface rayonnante**   panel heating
chauffage *m* **à vapeur**   steam heating
chauffage *m* **à vapeur (à) basse pression**   low-pressure steam heating
chauffage *m* **à vapeur (à) haute pression**   high-pressure steam heating
chauffage *m* **à vapeur d'échappement**   exhaust steam heating
chauffage *m* **au charbon**   coal fired heating
chauffage *m* **au fuel**   oil-fired heating
chauffage *m* **au gaz**   gas-fired heating
chauffage *m* **au mazout**   oil-fired heating
chauffage *m* **au plafond**   (radiant) ceiling panel heating
chauffage *m* **au rayons infrarouges**   infra-red heating
chauffage *m* **central**   central heating
chauffage *m* **central au gaz**   gas central heating
chauffage *m* **central général**   full central heating
chauffage *m* **collectif**   district heating
chauffage *m* **d'air**   air heating
chauffage *m* **d'air à gaz**   gas-fired air heating
chauffage *m* **d'air frais**   fresh air heating
chauffage *m* **d'appoint**   supplementary heating
chauffage *m* **de quartier**   district heating
chauffage *m* **de(s) locaux**   room heating
chauffage *m* **électrique**   electric heating
chauffage *m* **électrique intégré (CEI)**   integrated electric heating [insulation + mechanical ventilation + controlled electric heating]
chauffage *m* **en allège**   window sill heating
chauffage *m* **intégré par le sol**   underfloor heating
chauffage *m* **intermittent**   intermittent heating
chauffage *m* **mural**   wall panel heating
chauffage *m* **par accumulation d'énergie hors pointe**   off-peak storage heating
chauffage *m* **par éjecto-convecteur**   induction heating
chauffage *m* **par induction**   induction heating
chauffage *m* **par le plafond**   ceiling heating
chauffage *m* **par le sol**   floor heating
chauffage *m* **par les murs**   wall heating
chauffage *m* **par panneau**   panel heating
chauffage *m* **par rayonnement**   radiant heating
chauffage *m* **par tuyauteries de petit diamètre**   small bore heating
chauffage *m* **par tuyauteries de très faibles diamètres**   microbore heating
chauffage *m* **solaire**   solar heating
chauffage *m* **sous-sol**   underfloor heating
chauffage *m* **thermoélectrique**   thermoelectric heating
chauffage *m* **urbain**   district heating
chauffage *m***, ventilation et air climatisé**   heating, ventilation and air-conditioning (HVAC)
chauffagiste *m*   heating engineer
chauffé *adj*   heated
chauffé *adj* **à la vapeur**   steam heated
chauffe *f*   firing or stoking [of a boiler], heating
chauffe *f* **à/au charbon/coke/fuel/gaz**   coal/coke/oil/gas burning/firing

chauffe *f* **avec alimentation par trémie**   hopper-fed firing
chauffe-bain *m*   bath water heater or geyser [instantaneous]
chauffe-bain *m* **à gaz**   gas geyser [instantaneous bath water heater]
chauffe-eau *m*   immersion or water heater
chauffe-eau *m* **à accumulation**   water storage heater
chauffe-eau *m* **à gaz**   gas water heater
chauffe-eau *m* **à gaz à plusieurs points**   multi-point gas water heater
chauffe-eau *m* **instantané**   instantaneous water heater
chauffe-serviette(s) *m*   heated towel rail, towel warming rail
chauffer *vb*   become or get warm/hot, run hot, overheat, [of a bearing, etc], heat, warm, fire [a boiler]
chaufferie *f*   boiler house, plant, room
chaufferie *f* **à gaz**   gas firing [heating]
chaufferie *f* **centrale**   central boiler or heating plant
chauffeur *m*   driver [of a vehicle]
chaume *m*   stubble [of crops], thatch
chaume, toit *m* **de**   thatched roof
chaumière *f*   thatched cottage
chaussée *f*   carriageway, pavement [of a road], road(way)
chaussée, corps *m* **de**   pavement [of a road or runway]
chaussée *f* **à sens unique**   one-way road
chaussée *f* **en rase campagne**   rural road
chaussée *f* **rigide**   rigid pavement
chaussée *f* **souple**   flexible pavement
chaussure *f* **à semelle anti-chaleur**   shoe with heat resistant sole
chaussure *f* **(avec coquille) de sécurité**   safety boot
chaux *f*   lime [for mortar, etc]
chaux, lait *m* **de**   lime suspended in water
chaux, pierre *f* **à**   limestone, natural calcium carbonate
chaux *f* **anhydre**   quicklime
chaux *f* **éteinte**   hydrated or slaked lime
chaux *f* **grasse**   fat lime
chaux *f* **hydratée**   hydrated or slaked lime
chaux *f* **hydraulique**   hydraulic lime
chaux *f* **vive**   quicklime
chef *m*   edge of a slate, head or leader
chef, robinet *m*   gas stop tap [for a whole building]
chef *m* **d'atelier**   shop foreman
chef *m* **d'équipe**   chargehand, construction foreman, ganger
chef *m* **d'exploitation**   works manager
chef *m* **de chantier**   contract or site manager, general or site foreman
chef *m* **de projet**   project manager
chemin *m*   path, road, track
chemin *m* **d'accès**   access or approach road
chemin *m* **d'exploitation**   farm or farm road or track
chemin *m* **d'infiltration**   percolation path
chemin *m* **de câbles**   cable tray
chemin *m* **de drainage**   drainage path
chemin *m* **de fer**   French or toothed drag, mason's tool with lines of teeth for surfacing stone by scraping, railway

chemin *m* **de fer à crémaillère** rack railway

chemin *m* **de fer à simple/double voie** single/double track railway

chemin *m* **de fer aérien** overhead railway

chemin *m* **de fer souterrain** underground railway

chemin *m* **de fer surélevé** elevated/overhead railway

chemin *m* **de fer urbain** urban railway

chemin *m* **de halage** towpath

chemin *m* **de la grue** crane runway

chemin *m* **de roulement** roller track

chemin *m* **de service** access or service walkway

chemin *m* **départemental** county or secondary road

chemin *m* **empierré** metalled road

chemin *m* **en madriers** plank roadway or track

chemin *m* **ferré** metalled road

chemin *m* **forestier** forest road or track

chemin *m* **piéton** footpath

chemin *m* **rural** rural road or lane

chemin *m* **vicinal** by-road

cheminée *f* chimney, fireplace

cheminée *f* **à foyer ouvert** open hearth fireplace

cheminée *f* **auto-porteuse** self-supporting chimney

cheminée *f* **d'accès** access shaft, manhole

cheminée *f* **d'aération** air or ventilation/ventilating shaft

cheminée *f* **d'angle** corner fireplace

cheminée *f* **d'appel** emergency smoke vent

cheminée *f* **d'équilibre** surge tank [for pressure regulation]

cheminée *f* **de happement** waste gas chimney

cheminée *f* **de refoulement** discharge chimney [of fan coil unit]

cheminée *f* **de ventilation** ventilation shaft

cheminée *f* **verticale** shaft [of a manhole]

cheminement *m* **d'exfiltration** leakage path [HVAC]

chemisage *m* casing [jacketing, action and result], lining [of a conduit, duct, pipe, etc], sheathing [cladding or lagging – the action]

chemisé *adj* cased, jacketed, lagged [insulated], lined, sheathed, sleeved [e.g. of concrete piles]

chemise *f* case [around something], facing [interior], jacket or lagging [of a boiler], sheathing [cladding or lagging]

chemise, mur *m* **de** revetment wall

chemise d'air, à air-lagged

chemise *f* **d'eau** water jacket

chemise *f* **de refroidissement** cooling jacket

chemise *f* **de vapeur** steam jacket

chemise *f* **réfrigérante (d'eau)** (water) cooling jacket

chemiser *vb* case or jacket [equipment], sheath [clad or lag, e.g. a boiler], line [a conduit, duct, pipe, etc], channel

chenal *m* channel [port]

chêne *m* oak [tree, wood]

chéneau *m* eaves gutter, gutter [roof]

chéneau *m* **à l'anglaise** metal roof gutter [channel in wooden trough]

chéneau *m* **encaissé/à encaissement** box gutter [roof]

chenet *m* andiron, fire-dog

chenil *m* kennel

chenillard *m* crawler tractor

chenilles, à/sur tracked [e.g. of caterpillar vehicle]

chenilles *f,pl* tracks [of caterpillar vehicle]

chèque *m* cheque

chèque *m*, **paiement** *m* **par** payment by cheque

chèque *m* **bancaire** bank cheque

chèque *m* **de bois**[†] rubber cheque[†]

chèque *m* **postal** postal order

chèque *m* **sans provision** cheque without funds to meet it

chéquier *m* cheque book

cherche-fuite(s) *m* leak detector [generally for gas]

chevalement *m* props used during underpinning

chevaler *vb* prop or strut prior to underpinning

chevalet *m* trestle, frame or stand [support], bearing pedestal

chevalet *m* **pour tirer au cordeau** profile board [in excavations]

cheval-vapeur (CV) horsepower

chevauchement *m* lap [e.g. of tiles], overlap(ping)

chevaucher *vb* overlap

chevauchure *f* overlap(ping)

chevelu *m* small starter bars in reinforced concrete

chevet *m* bedside cabinet/table

chevêtre *m* joist trimmer beam

chevêtre *m* trimmer [round an opening, timber frame construction]

cheville *f* dowel, pin [used in jointing], wall plug [for fixing]

cheville *f* **à/de bois** wooden dowel/nail

chevillette *f* clamp [to hold a rule used in dressing a rendering flat]

chèvre *f* derrick, shearlegs or shears

chevron *m* common rafter, spar

chevron *m* **arbalétrier** principal rafter

chevron *m* **d'arête** hip rafter

chevron *m* **intermédiaire** common rafter

chevronner *vb* install rafters

chicane *f* baffle, deflector

chicane, joints *m,pl* **en** staggered joints

chien *m* heavy mason's hammer with hatched faces, tool used for grooving a substrate before applying rendering or stucco

chien *m* **taillé** heavily toothed sledgehammer

chien-assis *m* dormer with backward sloping roof

chiffonnage *m* rubbing down with a cloth or felt and a fine abrasive

chiffrage *m* assessing or reckoning [numerically], calculating, calculation or figuring [the process], numbering

chiffre *m* amount [total], total [in figures], digit, figure or number, profile of a flay key

chiffre *m* **d'affaires** business [volume of], turnover

chiffre *m* **de dureté Brinell** Brinell hardness number

chiffre *m* **rond** round number

chiffrer *vb* calculate, figure out or work out [amounts, numbers, etc]

chignole *f* pinion hand drill

chimère *f* gargoyle
chimique *adj* chemical
chœur *m* chancel or choir [of a cathedral, church]
chloration *f* chlorination
chlore *m* chlorine
chlorer *vb* chlorinate
chlorite *m* chlorite
chlorofluorocarbone *m* (CFC) chlorofluorocarbon (CFC)
choc *m* impact, shock
choc *m* de l'eau water hammer
choc *m* de foudre lightning surge
choc *m* électrique electric shock
choc *m* thermique thermal shock
chon *m* rough plank [for casing a well, timbering a tunnel]
chromage *m* chromium plating
chromaticité *f* chromaticity [lighting]
chromaticité *f*, coordonnées *f,pl* de chromaticity coordinates
chrome *m* chromium
chute *f* drop or fall, head of water, offcut, vertical waste pipe, below a WC
chute *f* d'eau waterfall
chute *f* d'objets fall of objects
chute *f* d'un toit pitch of a roof
chute *f* de flux lumineux lumen depreciation
chute *f* de hauteur fall from height
chute *f* de pierre(s) rockfall, stonefall
chute *f* de plain pied fall on the same level
chute *f* de potentiel potential drop
chute *f* de pression pressure drop/loss
chute *f* de pression dynamique dynamic head loss
chute *f* de température temperature drop
chute *f* de température adiabatique adiabatic lapse rate
chute *f* de température entrée-sortie cooling range
chute *f* de tension stress loss, voltage drop
chute *f* eaux pluviales (chute EP) rainwater stack
chute *f* eaux usées (chute EU) waste stack
chute *f* eaux vannes (chute EV) soil stack
cible *f* target [surveying]
ciel ouvert, à in open cut
cimaise *f* cyma
ciment *m* cement
ciment, coulis *m* de cement grout
ciment *m* à durcissement rapide rapid-hardening cement
ciment *m* à faible chaleur d'hydratation low heat cement
ciment *m* à haute résistance high-strength cement
ciment *m* à haute résistance initiale rapid-hardening cement
ciment *m* à la pouzzolane Portland pozzolana cement
ciment *m* à mater caulking compound
ciment *m* à prise lente slow-setting cement
ciment *m* à prise rapide quick-setting cement
ciment *m* alumineux aluminous cement, high alumina cement (HAC)
ciment *m* anglais type of alum plaster

ciment *m* blanc white cement
ciment *m* colle tile fixing adhesive cement
ciment *m* de fer (CF) blastfurnace cement (ggbfs) with lower lime sulphate content
ciment *m* de haut fourneau (CHF) blastfurnace cement (ggbfs) with higher lime sulphate content
ciment *m* de laitier slag cement [OPC + ggbfs]
ciment *m* de laitier à la chaux (CLX) lime slag cement [fat or hydraulic lime + ggbfs]
ciment *m* en vrac bulk cement
ciment *m* expansif expansive cement
ciment *m* fondu aluminous or high alumina (HAC) cement
ciment *m* hydraulique hydraulic cement
ciment *m* hydrofuge waterproof or water repellent cement
ciment *m* latex cement-rubber latex
ciment *m* magnésien magnesium cement [refractory magnesite + magnesium chloride]
ciment *m* métallique metallic cement [magnesium cement with zinc replacing magnesium]
ciment *m* métallurgique metallurgic cement [artificial cement + bfs]
ciment *m* métallurgique mixte (CMM) mixed metallurgic cement [equal parts OPC and ggbfs + lime sulphate]
ciment *m* naturel (CN) natural cement
ciment *m* pierre stone cement [for artificial stone facings]
ciment *m* Portland à haute résistance initiale (HRI) rapid-hardening Portland cement
ciment *m* Portland artificiel (CPA) artificial Portland cement
ciment *m* Portland de fer cement with 70% min OPC + 30% max basic slag
ciment *m* Portland de laitier cement with 20% min OPC + 80% max basic slag
ciment *m* Portland ordinaire ordinary Portland cement (OPC)
ciment *m* projeté shotcrete
ciment *m* prompt quick-setting cement
ciment *m* réfractaire refractory cement
ciment *m* résistant aux sulfates sulphate resisting cement
ciment *m* romain Roman cement
ciment *m* sursulfaté supersulphated cement
cimentation *f* cementation
cimenter *vb* cement
cimenterie *f* cement works
cimentier *m* cement finisher, renderer
cimentier *m* en béton armé reinforcement concreter
cimentier-ravaleur-applicateur *m* cement finisher, renderer
cimetière *m* cemetery, graveyard
cinématique *adj* kinematic
cinétique *adj* kinetic
cingler *vb* mark a level line along a wall
cintrage *m* bend(ing) [of pipes, etc], curvature, deflection [bending]

cintrage *m* **des aciers**   bar-bending

cintrage *m* **des tubes**   pipe-bending

cintré *adj*   arched, vaulted

cintre *m*   arch [of a tunnel], curve [in plate, surface, etc], falsework or formwork [curved, for arches, vaults, etc], rigging loft [above a stage], soffit [of an arch or tunnel], template [for an arch]

cintrer *vb*   arch [form into an arch], vault [make sth vault shaped], bend or curve [i.e. form sth into a curved profile]

cintres *m,pl*   flies [of a theatre]

cintreuse *f* **(à aciers)**   bar bender

cintreuse *f* **de tube**   pipe bender

circonférence *f*   circumference, perimeter

circonférentiel *adj*   circumferential

circuit, hors (de)   out of circuit

circuit, mise *f* **en**   connecting up

circuit *m*   circuit

circuit *m,* **en**   in circuit

circuit *m* **à haute tension**   high-voltage circuit

circuit *m* **à très basse tension (fonctionnelle)**   extra-low voltage circuit

circuit *m* **d'alimentation**   main circuit

circuit *m* **d'appel**   call circuit

circuit *m* **d'eau**   water system

circuit *m* **d'éclairage**   lighting circuit

circuit *m* **de contrôle**   control circuit

circuit *m* **de dérivation**   bypass circuit

circuit *m* **de fumées équilibré**   balanced flue

circuit *m* **de l'eau**   water cycle

circuit *m* **de mélange**   mixing circuit

circuit *m* **de régulation**   control loop

circuit *m* **de terre**   earth (circuit)

circuit *m* **dérivé**   shunt circuit

circuit *m* **en parallèle**   parallel circuit

circuit *m* **étanche à vase fermé**   sealed circuit [pipework]

circuit *m* **fermé**   closed circuit

circuit *m* **frigorifique**   refrigerant circuit

circuit *m* **imprimé**   printed circuit

circuit *m* **inductif**   inductive circuit

circuit *m* **local**   local circuit

circuit *m* **primaire**   primary circuit

circuit *m* **résistant**   resistive circuit

circuit *m* **secondaire**   secondary circuit

circuit *m* **TBTF**   extra-low voltage circuit

circuit, mettre *vb* **en**   connect up

circuit, mettre *vb* **en court**   short circuit

circulable *adj*   trafficable

circulaire *adj*   circular

circulateur *m*   circulating pump [heating system]

circulation *f*   circulation, traffic

circulation, à faible   lightly trafficked

circulation, eau *f* **souterraine en**   gravitational groundwater

circulation, espace *m* **interdit à la**   ghost island [road]

circulation, sens *m* **de**   traffic direction

circulation, surface *f* **de**   traffic area

circulation accéléré   forced circulation

circulation automobile   motor vehicle traffic

circulation *f* **d'air**   air circulation

circulation *f* **(d'air) forcée**   forced (air) circulation

circulation *f* **naturelle**   natural (air) circulation

circulation *f* **normale, voie** *f* **de**   general-purpose lane [of a highway]

circulation *f* **principale**   through traffic

circulation *f* **routière**   road traffic

circulation *f* **uni-directionnelle**   one-way traffic

circuler *vb*   circulate, flow [traffic, electricity, air, etc]

cire *f*   wax

cirer *vb*   wax

cisaillage *m*   cutting, shearing [metal, etc]

cisaille *f*   shearing machine, shears [for cutting]

cisaille *f* **à main**   hand shears

cisaille *f* **à tôle**   plate shearing machine

cisaille *f* **américaine/coupe-boulons**   bolt-cutting shears

cisailles *f,pl*   shears [for cutting]

cisaillement, de *adj*   shear(ing)

cisaillement *m*   shear(ing)

cisaillement, contrainte *f* **de**   shear stress

cisaillement, déformation *f* **de**   shear strain

cisaillement, essai *m* **de**   shear test

cisaillement, fissure *f* **de**   shear crack

cisaillement, module *m* **de**   shear modulus

cisaillement, plan *m* **de**   shear plane

cisaillement, résistance *f* **au**   shear strength

cisaillement, rupture *f* **par**   shear failure

cisaillement, surface *f* **de**   shear surface

cisaillement, zone *f* **de rupture par**   shear zone

cisailler *vb*   shear

cisailleuse *f* **(à aciers)**   bar cropper

ciseau *m*   chisel

ciseau *m* **à bois**   firmer or wood chisel

ciseau *m* **à bords biseautés**   framing chisel

ciseau *m* **à mater**   caulking tool

ciseau *m* **biseauté**   bevel edged chisel

ciseau *m* **biseauté à brides**   framing chisel

ciseau *m* **de briqueteur**   bolster [chisel]

ciseau *m* **de menuisier**   joiner's chisel

ciseau *m* **de tourneur**   parting tool [wood turning]

ciseau *m* **ordinaire**   firmer or wood chisel

ciseaux *m,pl* **de tapissier**   paperhanger's scissors

ciseler *vb*   chisel

ciselure *f*   chiselled decorations [and art of making these], smoothed edge of cut stone around the face

cité *m*   city, housing estate, nucleus of town

citerne *f*   cistern, (rain)water tank

citerne *m,* **camion-**   tanker (lorry)

citerne *f* **à bitume**   bitumen tank [truck mounted]

citerne *f* **à bitume avec rampe de répandage**   bitumen tank with spraying equipment

citerne *f* **de stockage**   storage tank

citerne *f* **de transport**   tanker lorry

citerne *f* **thermique répandeuse de liants bitumineux**   bitumen tank with spraying equipment

claie *f*   riddle or screen

clair *adj*   bright, clear [sky], light [colour]

clair, en *adj*   in clear or uncoded

clair *m*   light area, joist spacing

claire-voie *f*   clerestory, skylight, openwork fence

clairière *f*   clearing [in woodland]

clameau *m* (à deux points)   timber dog, cramp iron [timber]

clapet *m*   check valve, valve [generally flap type], rectifier [electrical], tap washer

clapet *m* à bille   ball valve

clapet *m* à guillotine   damper plate [in ductwork]

clapet *m* à membrane   disc valve

clapet *m* anti-retour   check, non-return or reflux valve

clapet *m* coupe-feu   fire damper

clapet *m* crépine   strainer check valve

clapet *m* d'aspiration   foot or suction valve

clapet *m* d'échappement   exhaust flap

clapet *m* d'entrée d'air   air inlet valve

clapet *m* d'étranglement d'air   exhaust air damper

clapet *m* de non-retour   non-return or reflux valve, check

clapet *m* de prise d'air   air inlet damper

clapet *m* de refoulement   flow valve

clapet *m* de réglage   damper [in ductwork]

clapet *m* de réglage de retour   return air damper

clapet *m* de retenue   check valve

clapet *m* de ventilation   air damper

clapet *m* fléchissant   flapper valve

clapet-pointeau *m*   needle valve

claquage *m*   burning out [electrical]

claquer *vb*   burn out [electric motor, portable tool, lamp]

clarificateur *m*   sand-bed filter

clarté *f*   brightness, clarity, light [clarity]

classé *adj*   listed [as of architectural or historic interest]

classe *f*   class, category

classe *f* d'abri   shielding class [HVAC]

classe *f* de forme   shape category or class

classe *f* de forme des grains   particle shape category

classe *f* de protection   shielding class [HVAC]

classe *f* de résistance   strength class

classe *f* granulaire   grain size bracket, class, fraction or range

classement *m*   classification, gradation, grading, sorting

classement *m* granulométrique   grading [by grain or particle size]

classer *vb*   classify, file [documents]

classification *f*   classification

classification *f* des eaux   water classification

classification *f* des terrains   terrain classes [HVAC]

classique *adj*   conventional

clastique *adj*   clastic

clause *f*   clause [in agreement, contract, etc]

clause *f* de contrat   condition of contract, contract clause

clause *f* pénale   penalty clause

clausoir *m*   last stone placed in an arch, vault or stonework course

clavage *m*   joint grouting, keying up [an arch], stone elements of flat or curved arch

claveau *m*   arch-stone, segment [of an arch], voussoir

clavetage *m*   keying [e.g. impeller on to shaft]

claveter *vb*   key [on to a shaft]

clavette *f*   cotter pin, key [on a shaft], wedge

clavette *f* à talon   gib(-headed key)

clavette *f* fendue   split pin

clavette *f* normale/ordinaire   sunk key

clavier *m*   keyboard [computer]

clef en main   turnkey [projects]

clef *f*   key [of a lock or arch, or in a joint], spanner or wrench

clef *f* à chaîne   chain wrench [plumbing]

clef *f* à choc   torque wrench

clef *f* à crémaillère   adjustable or shifting spanner or wrench

clef *f* à fourches   open-ended spanner or wrench

clef *f* à griffe   bar bender

clef *f* à molette   adjustable or shifting spanner or wrench

clef *f* à œil   ring spanner

clef *f* à sangle   strap wrench [plumbing]

clef *f* à tube   socket spanner

clef *f* à tubes   pipe wrench

clef *f* anglaise   screw or monkey wrench

clef *f* bénarde   flat key [used from both sides]

clef *f* de cisaillement   shear key

clef *f* de mandrin   chuck key

clef *f* de robinet   stopcock key

clef *f* de tuyau   pipe or stillson wrench

clef *f* de vanne   valve key

clef *f* de voûte   keystone

clef *f* plate   open spanner

clef *f* polygonale   ring spanner

clenche *f*   latch

cliché *m*   negative [photographic]

client/e *m/f*   client, customer

clignotant *adj*   blinking, flashing or winking [light]

clignotant, feu *m*   flashing light

clignotant *m*   warning light [flashing or winking]

clignoter *vb*   flash or wink [of a signal light], flicker [of a light]

climat *m*   climate

climat *m* intérieur   indoor climate

climat *m* tempéré   temperate climate

climatisation *f*   air-conditioning

climatisé *adj*   air-conditioned

climatiser *vb*   air-condition

climatiseur *m*   air-conditioner

climatiseur *m* autonome   packaged air-conditioner

climatiseur *m* individuel   room air-conditioner

climatologie *f*   climatology

climatologique *adj*   climatological

clin *m*   board used in horizontal cladding, siding, weatherboard

clinique *f*   clinic, nursing home, private hospital

clinomètre *m*   clinometer, inclinometer

cliquet *m*   pawl, ratchet

clivage *m*   cleavage

clivage, plan *m* de   cleavage plane

clivage, résistance *f* au   cleavage strength

cliver *vb*   cleave [e.g. slate]

cloche *f*   bell

cloche *f* d'effondrement   pothole [in the ground]

clocher *m*   belfry, bell or church tower, steeple

cloison *f*   interior or internal wall, partition (wall), division or separator [in equipment]

cloison *f*, contre-   wall lining

cloison, mur *m* de   dividing wall

cloison *f* coupe-feu   fire (division) wall

cloison *f* de doublage   cavity wall

cloison *f* en charpente *f*   timber frame partition or wall

cloison *f* en palanches   sheet piling cut-off

cloison *f* en treillis   stud partition

cloison *f* étanche   watertight bulkhead

cloison *f* lattée et plâtrée   stud wall or partition [lath and plaster]

cloison *f* maçonnée   masonry partition

cloison *f* mobile   sliding partition

cloison *f* pare-feu   fire (division)wall

cloison *f* porteuse   loadbearing partition or wall

cloison *f* sèche   curtain wall

cloisonnage *m*   partitioning

cloisonnement *m*   partitioning

cloisonnement *m* de sécurité   fire compartmentation

cloisonner *vb*   divide up [into compartments, rooms, etc], partition

cloître *m*   cloister

cloquage *m*   blisters/blistering [of paint]

clos *adj*   enclosed

clos *m*   close [enclosed space], enclosure

closerie *f*   small estate [enclosed]

clothoïde *f*   clothoid, spiral transition curve

clôturé *adj*   enclosed [fenced, walled], fenced

clôture *f*   enclosure [fence or wall], fence

clôture, mur *m* de   enclosing wall

clôture *f* à siphon   siphon trap

clôture *f* à eau   water seal

clôture *f* de chantier *m*   site fence

clôture *f* électrifiée   electric fence

clôturer *vb*   enclose [with a fence, wall, etc], fence [land]

clou *m*   nail or stud

clou *m*, petit   tack

clou *m* à ardoise   slate nail

clou *m* à tête d'homme   lost-head nail

clou *m* à tête de diamant   clout nail

clou *m* à tête large   felt nail

clou *m* à tête perdue   lost-head nail

clou *m* à tête plate   round-head nail

clou *m* calot(t)in   clout nail

clou *m* cavalier   staple [for fixing electrical wiring]

clou *m* de tapissier   upholsterer's tack

clou *m* découpé   cut nail

clou *m* en U   staple [U-shaped]

clou *m* étampé   cut nail

clou *m* sans tête   brad

clouage *m*   nailing

clouage *m* invisible   secret nailing

clouer *vb*   nail

cloueuse *f*   nail gun

clozot *m*   brick 60×60×220mm [sometimes used for chimneys in France]

coaguler *vb*   coagulate

coaltar *m*   coal tar [used as protective or waterproof coating]

coaxial *adj*   coaxial

cochère, porte *f*   carriage gateway, or gateway large enough to admit a vehicle

code *m*   code, regulations

Code *m* d'Urbanisme   town planning regulations [in France]

code *m* de couleurs normalisées   standard colour code

Code *m* de l'Environnement   development regulations [in France]

Code *m* de la Construction   building or construction regulations [in France]

Code *m* des Marchés Publics   public sector procurement regulations [in France]

code *m* du bâtiment   building code or regulations

Code *m,pl* pour Structures Élevées   high rise building regulations [in France]

coefficient *m*   coefficient, factor

coefficient *m* brassage   air change coefficient

coefficient *m* charge   load factor

coefficient *m* conductibilité   coefficient of thermal conductivity

coefficient *m* d'effet frigorifique   cooling efficiency ratio

coefficient *m* d'élasticité   coefficient or modulus of elasticity

coefficient *m* d'exfiltration   leakage coefficient [HVAC]

coefficient *m* d'occupation du sol (COS)   plot ratio

coefficient *m* d'uniformité   coefficient of uniformity

coefficient *m* d'usure   wear index

coefficient *m* d'utilisation   diversity factor, load factor, utilization factor [lighting]

coefficient *m* de correction   correction factor

coefficient *m* de décharge   discharge coefficient

coefficient *m* de dilatation   coefficient of expansion

coefficient *m* de dilatation thermique   coefficient of thermal expansion

coefficient *m* de dureté   temper [of steel]

coefficient *m* de forme   shape coefficient, form factor

coefficient *m* de frottement   coefficient of friction

coefficient *m* de performance   coefficient of performance (COP)

coefficient *m* de perméabilité   permeability coefficient

coefficient *m* de Poisson   Poisson's ratio

coefficient *m* de protection du soleil   shading coefficient

coefficient *m* de régression   regression coefficient

coefficient *m* de résistance   coefficient of resistance

coefficient *m* de rugosité   roughness coefficient

coefficient *m* de ruissellement   run-off coefficient

coefficient *m* de sécurité   safety factor

coefficient *m* de transfert thermique   heat transfer coefficient

coefficient *m* de transmission de chaleur   heat transfer coefficient

coefficient *m* de transmission du son   sound transmission coefficient

coefficient *m* de transmission thermique   thermal conductivity (k-value)

coefficient-G *m* **d'un logement**   heat loss coefficient of a habitation (watts/m^3/°C)

coefficient-k *m*   k-value (thermal conductivity)

coefficient-U *m*   U-value (thermal transmittance)

cœur *m*   core

cœur *m* **d'aiguille**   frog [of railway points]

cœur *m* **(de bois)**   heartwood

cœur *m* **du réacteur**   reactor core

coffine, tuile *f*   plain tile [convex]

coffrage *m*   casing or formwork, lining or shuttering

coffrage, fond *m* **de**   bottom shuttering

coffrage, joue *f* **de**   side shutter [of an edge beam]

coffrage *m* **assisté par ordinateur (CAO)**   computer-aided design of formwork

coffrage *m* **bois**   timber formwork

coffrage *m* **d'arrêt**   stop end [in formwork]

coffrage *m* **de la poutre maîtresse**   main joist or beam shuttering

coffrage *m* **du poteau**   column shuttering

coffrage *m* **en contreplaqué** *m*   plywood formwork

coffrage *m* **glissant**   sliding formwork, slipform

coffrage *m* **grimpant**   climbing formwork

coffrage *m* **métallique**   steel formwork

coffrage *m* **mobile**   travelling formwork

coffrage *m* **perdu**   lost or sacrificial formwork

coffrage *m* **provisoire**   temporary formwork

coffrage *m* **roulant**   travelling formwork

coffrage *m* **tunnel**   tunnel formwork

coffre *m*   ballast bed [of railway track], cofferdam

coffre *m* **congélateur**   chest freezer

coffre *m* **de serrure**   lock box or case

coffret *m*   box, cabinet, cupboard

coffret *m* **de compteur**   meter cupboard

coffret *m* **de coupe-circuit**   cut-out box [electrical]

coffret *m* **de distribution**   distribution box [electrical]

coffret *m* **électrique**   cable box

coffreur *m* **(pour béton armé)**   formwork carpenter

coffreur-boiseur *m*   formwork carpenter

cogénération *f*   combined heat and power (CHP)

cognée *f*   axe

cohérent *adj*   coherent, cohesive

cohérent, matériau *m*   cohesive material

cohérent, sol *m*   cohesive soil

cohérents, éléments *m,pl*   cohesive matter

cohésion *f*   cohesion

coiffé *adj*   capped [of a pile]

coiffe *f*   protective hood over a circular saw blade

coiffe *f* **de pieu**   pile helmet

coiffer *vb*   cap [a pile]

coiffeuse *f*   dressing table

coin *m*   corner, chock, wedge, quoin

coin *m* **à manger**   kitchen-dining area

coin *m* **cuisine**   kitchen corner

coin *m* **de bois dur**   hardwood wedge

coin *m* **de serrage**   wedge used to tighten props or shoring

coin *m* **du feu**   inglenook

coin *m* **repas**   dining area

coinçage *m*   cottering, keying [fixing with a key], wedging

coincé *adj*   jammed or stuck

se coincer *vb*   bind, jam or stick [a door, machinery, etc]

coke *m*   coke

col *m*   pass [between mountains]

col de cygne, à   swan-necked

col *m* **de diffuseur**   diffuser neck

col *m* **de cygne**   swan-neck pipe or tap

colimaçon *m*, **(escalier** *m* **en)**   spiral staircase

colis *m*   parcel [package]

collage *m*   adhesion, gluing

collé *adj*   glued

collé à la résine synthétique   resin-bonded

colle *f*   adhesive, glue, size [glue]

colle *f* **à dispersion**   emulsion paste [for wallpaper]

colle *f* **à tapisser**   wallpaper glue

colle *f* **blanche**   wood(working) glue

colle *f* **de pâte**   wallpaper paste

colle *f* **de pré-encollage**   glue size, or size [for wallpapering]

colle *f* **néoprène**   neoprene glue

collecte, voie *f* **de**   local distributor [collector] road

collecteur *m*   collector [drain, road, sewer], main drain or sewer, header [pipe or duct], manifold, receiver [tank], sliding pickup [electrical]

collecteur, grand   main collector drain or sewer, trunk sewer

collecteur *m* **à vide**   vacuum receiver

collecteur *m* **d'appareils**   collector drain pipe from several appliances

collecteur *m* **d'échappement**   exhaust manifold

collecteur *m* **de distribution**   supply manifold

collecteur *m* **de terre**   earth collector, main earth lead

collecteur *m* **(eaux) pluvial(es)**   stormwater drain or sewer

collecteur *m* **plan**   flat plate collector

collecteur *m* **principal**   main collector drain or sewer

collecteur *m* **secondaire**   sub-main sewer

collectif *adj*   collective, common, communal, joint, shared

collectivité *f*   community

collègue *m/f*   business associate, colleague

coller *vb*   glue

collerette *f*   flange [on pipe]

collet *m*   flange

collier *m*   cable tie, pipe clip or collar

collier *m* **à lunette**   device for installing a branch pipe, requiring supply interruption

collier *m* **à ressort**   spring clip

collier *m* **à vis**   Jubilee clip

collier *m* **d'attache**   pipe clip [collar]

collier *m* **d'écartement**   spacing clamp

collier *m* **de fixation**   clamp

collier *m* **de fixation de câble**   cable clip

collier *m* **de mise à la terre**   earth(ing) clamp

collier *m* **de prise en charge**   device for installing a branch pipe without interruption of supply

collier *m* **de serrage**   clamping ring

collier *m* **de tuyauterie**   pipe hanger

colline *f*   hill

colloïde *m*   colloid

colloïdal *adj*   colloidal

colmatage *m* clogging (up) [filters, etc], choking (up) [drains, etc], draught strip, filling in [potholes in a road], plugging [of leaks, holes]

colmater *vb* clog (up) [filter, etc], choke [drain, etc], fill in [potholes in a road], plug [a leak, a hole], seal (off) [a leak, leaking joint, etc], stop up [cracks, joints, etc]

colocataire *m/f* co- or joint tenant

colombage *m* half-timbering, studwork [timber]

colombage, maison *f* **en** half-timbered house

colombe *f* stud [timber structure]

colombier *m* dovecote

colonie *f* **de vacances** holiday camp

colonnade *f* colonnade

colonne, en forme de columnar

colonne *f* column, pillar, riser, vertical pipe

colonne *f* **de drainage** drainage pipe [in a dam]

colonne *f* **de refoulement** rising main [in a borehole]

colonne *f* **de ventilation** vent pipe/stack

colonne *f* **des ventilations secondaires** vent stack

colonne *f* **descendante** dry falling main

colonne *f* **en charge** pressurized or wet riser

colonne *f* **en fonte** cast iron column

colonne *f* **humide** wet riser

colonne *f* **montante** riser [vertical duct or pipe]

colonne *f* **montante d'alimentation** supply riser

colonne *f* **pilote** auxiliary wiring duct [communications wiring, etc]

colonne *f* **sèche** dry riser

colorant *m* colouring, dye

colorante, matière *f* colouring matter

coloration *f* colour

coloré *adj* coloured

colorimétrique *adj* colorimetric

coloris *m* colour [of paint]

combinaison *f* arrangement, combination

combinaison *f* **de cas de charges** design load arrangement or case

combiné *m* **(téléphonique)** telephone receiver

comble *m* garret, loft, roof space

comble *m*, **faux** roof space too low to be occupied

comble *m* **à deux pentes/versants** span roof

comble *m* **à l'impériale** roof (space) with convex slopes

comble *m* **à plancher suspendu** queen truss

comble *m* **mansardé** mansard roof [structure, roof space]

comble *m* **sur pignon** gable roof

comblement *m* filling (up, in) [of a ditch, a well, etc]

combler *vb* fill [a ditch, a well]

combles *m,pl* attic, roofspace [loft], roof timbers

combles *m,pl* **aménageables** convertible roof space

comburant *m* combustion medium, oxidizing substance

combustible *adj* combustible

combustible *f* fuel

combustible *f* **de chauffe** boiler fuel

combustible *f* **fossile** fossil fuel

combustible *f* **liquide** liquid fuel

combustible *f* **pauvre** low grade fuel

combustible *f* **solide** solid fuel

combustion *f* combustion

combustion *f* **par lit fluidisé** fluid bed combustion

commande *f* control, drive, order [for goods or services]

commande *f* **à distance** remote control

commande *f* **d'alimentation** feed control

commande *f* **de changement de température été-hiver** changeover control

commande *f* **hors marché** order not included in the contract

commande *f* **manuelle** manual control

commande *f* **par bouton poussoir** press-button control

commande *f* **par courroie** belt drive

commande *f* **par horloge** time clock control

commande *f* **ultérieure** repeat order

commande mécanique, à mechanically controlled

commande pneumatique, à pneumatically controlled

commander *vb* provide access to [e.g. one room to another], control, order [goods or services]

commandes *f,pl* **électriques** electric(al) controls

commandes *f,pl* **temporisées** timed controls

commencer *vb* initiate [a programme, a project]

commerçant *adj* business [e.g. business area], commercial

commerçant *m* retailer, shopkeeper

commerçant *m* **en matériaux de construction** builders' merchant

commerçant, quartier *m* shopping district

commerçante, rue *f* shopping street

commerce *m* business

commerces *m,pl* shopping facilities, shops

commercial *adj* commercial

commercial, centre *m* shopping centre

Commissaire *m* **aux Comptes** statutory auditor

Commissariat *m* **Général du Plan (CGP)** national planning commission [of France]

commission *f* commission [percentage]

commission *f* **d'agence** agency commission

commission *f* **de réception** commissioning authority [for acceptance of works]

commode *adj* convenient

commode *f* chest of drawers

commodité *m* comfort, convenience [for living, etc]

commodités *m,pl* convenience (WC), toilet(s) (WCs)

commun *adj* common, ordinary

communal *adj* communal, public

commune *f* commune [lowest tier of local government in France]

communiquer *vb* communicate [e.g. rooms with each other], communicate [ideas, information, etc], spread [ideas, news, plans, etc]

commutateur *m* changeover, commutator or selector switch

commutateur *m* **de démarrage** starter

commutateur *m* **de va-et-vient** two-way switch

commutateur *m* **principal** main switch

compacité *f* compactness or consolidation [of ground, road], relative compaction

compact *adj* compact, dense

compactage *m* compaction

compactage, contrôle *m* **du** compaction testing

compactage, degré *m* **de**   degree of compaction
compactage, énergie *f* **de**   compaction [amount of]
compactage, engin *m* **de**   compactor
compactage, engins *m,pl* **de**   compacting equipment or plant
compactage *m* **de sol**   soil compaction
compactage *m* **par arrosage**   compaction by watering
compactage *m* **par cylindrage**   compaction by rolling
compactage *m* **par vibration**   compaction by vibration
compacté *adj*   compacted [e.g. of fill], consolidated
compacter *vb*   compact
compacter *vb* **au rouleau**   roll [to compact]
compacteur *m*   compactor, roller [compactor]
compacteur *m* **à jante lisse**   smooth-wheeled roller
compacteur *m* **à plaque vibrante**   vibrating plate compactor
compacteur *m* **à pneu(matique)s**   rubber-tyred compactor or roller
compacteur *m* **d'ordures ménagères**   rubbish compactor
compacteur *m* **de berges**   embankment or slope compactor
compacteur *m* **de déchets**   refuse compactor
compacteur *m* **pour enfouissement**   landfill compactor
compacteur *m* **vibrant à plaque**   vibrating plate compactor
compacteur *m* **vibrant à rouleau(x)**   vibrating roller compactor
compactibilité *f*   compactibility
compagnie *f*   management company
compagnon *m*   qualified craftsman
comparaison *f*   comparison
comparateur *m* **à cadran**   dial gauge [measuring tool]
compartiment *m*   compartment
compartimentage *m*   compartmentation
compas *m*   drawing or pair of compasses
compas *m* **à pointes sèches**   dividers
compas *m* **à verge**   beam compass or trammel
compas *m* **d'alésage**   inside or internal callipers
compas *f* **d'épaisseur**   external or outside callipers
compas *f* **de mesure**   dividers
compas *f* **de sécurité**   window retaining strap, limiting opening
compassement *m*   measurement [with compasses/dividers]
compatible *adj*   compatible
compensateur *adj*   balancing, compensating, equalizing
compensateur *m*   compensator
compensateur *m* **à joints glissants**   sliding expansion joint
compensateur *m* **de dilatation**   expansion bellows
compensateur *m* **de pression**   pressure compensator
compensation *f*   balancing or equalization [of forces, etc], compensation [mechanical, electrical]
compensation, tube *m* **de**   expansion pipe
compensation *f* **de facteur de puissance**   power factor correction
comportement *m*   behaviour
composant *m*   component [part], part [constituent]
composant *m* **d'ouvrage**   building component or element

composante *f*   component [of voltage, force, etc]
composé *adj*   compound
composition *f*   composition [e.g. of sewage]
composition *f* **du béton**   concrete mix
composition *f* **granulométrique**   grain or particle size distribution
compostage *m*   composting
comprenant *adj*   including
compresseur *m*   (air) blower, compressor
compresseur *m* **à ailettes**   sliding vane compressor
compresseur *m* **à deux étages**   two-stage compressor
compresseur *m* **(à) double/simple effet**   double/single-acting compressor
compresseur *m* **à flux inversé**   return flow compressor
compresseur *m* **à piston**   reciprocating compressor
compresseur *m* **à vis**   screw compressor
compresseur *m* **alternatif**   reciprocating compressor
compresseur *m* **d'air**   air compressor
compresseur *m* **frigorifique**   refrigeration compressor
compresseur *m* **jumelé**   twin compressor
compresseur *m* **multi-étage**   multi-stage compressor
compresseur *m* **rotatif multicellulaire**   multi-vane rotary compressor
compressibilité *f*   compressibility
compressibilité, essai *m* **de**   compression test [for soils]
compressibilité, module *m* **de**   modulus of compressibility
compressible *adj*   compressible
compression *f*   compression, pressure
compression, résistance *f* **à la**   compressive strength
compression *f* **adiabatique**   adiabatic compression
compression *f* **simple**   unconfined compression
compression triaxial, essai *m* **de**   triaxial compression test
comprimé *adj*   compressed
comprimer *vb*   compress
compromis *m*   compromise
compromis, solution *f* **de**   compromise solution
compromis *m* **de vente**   preliminary sale agreement [to buy and sell]
comptabilité *f*   accounting, accountancy
comptabilité *f* **analytique**   cost accounting
comptable *m*   accountant, book-keeper
comptage *m*   counting
comptage *m* **d'impulsions**   pulse counting
comptant, acheter *vb*   buy for cash
comptant, argent *m*   ready money, spot cash
comptant, payer *vb* **(au)**   pay cash
compte *m*   account
compte *m* **de prorata**   pro rata account/calculation
compte *m* **rendu**   minutes [of a meeting]
compte-rendu *m*   report
compte-rendu *m* **d'essai**   test report
compteur *m*   meter [electricity, water]
compteur *m* **à Venturi**   Venturi meter
compteur *m* **d'eau**   water meter
compteur *m* **d'eau à turbine à aube**   vane water meter
compteur *m* **d'eau domestique**   domestic water meter
compteur *m* **d'électricité**   electricity meter

compteur *m* **d'énergie thermique** heat meter
compteur *m* **d'heures** hourly meter, hours run indicator
compteur *m* **d'huile** oil meter
compteur *m* **de calories/de chaleur** heat meter
compteur *m* **de gaz** gas meter
compteur *m* **divisionnaire** local or sub-meter
compteur *m* **électrique** electricity meter
compteur *m* **général** main or principal meter
concassage, sable *m* **de** crushed rock or stone fines
concassé *adj* crushed [of stone or rock]
concassé, gravier *m* gravel or crushed rock chippings
concassé, gravillon *m* fine gravel chippings
concassée, grave *f* crushed aggregate
concassée, roche *f* **naturelle** crushed rock
concasser *vb* crush [rock, etc], comminute
concasseur *m* crusher [for asphalt, rock, etc]
concasseur *m* **à mâchoires** jaw crusher
concasseur *m* **à percussion** impact crusher
concentration *f* concentration
concentration *f* **d'équilibre** equilibrium concentration
concentration *f* **de données** data logging
concentration *f* **de vapeur** absolute humidity
concentration *f* **des contraintes** stress concentration
concentré *adj* concentrated
concentrique *adj* concentric
concept *m* concept
concepteur *m* designer [architect or engineer]
conception *f* conception, design
conception *f* **assistée par ordinateur (CAO)** computer-aided design (CAD)
conception *f* **d'éclairage** lighting design
conception-réalisation *f* design-build
concessionnaire *m* statutory undertaker, utility services company
concevoir *vb* conceive or design, plan [a house, an estate, etc]
concluant *adj* decisive
conclure *vb* tie up [a deal]
concours *m* competition [e.g. architectural]
concrétion *f* concretion
concurrence *f* competition [commercial]
condamnation *f* fixing or locking [of a door, window, etc]
condamnation, pointe *f* **de** locking device [on a door, window, etc] [insurance terminology]
condensat *m* condensate
condensateur *m* capacitor [electrical], condenser
condensateur *m* **à double faisceau** double bundle condenser
condensateur *m* **à plaques** plate-type condenser
condensation *f* condensation
condensation *f* **en surface** surface condensation
condensation *f* **interstitielle** interstitial condensation
condenser *vb* condense
condenseur *m* condenser
condenseur *m* **à aspersion** sprayed coil condenser
condenseur *m* **à calandre et serpentin** shell-and-coil condenser

condenseur *m* **à tubes concentriques/jumelés** double pipe condenser
condenseur *m* **évaporatif** evaporative condenser
condenseur *m* **refroidi par l'air** air-cooled condenser
condition *f* condition
condition *f* **d'équilibre** balance condition
condition *f* **de sécurité** safety requirement
condition *f* **requise** requirement
condition *f* **suspensive** precedent condition
conditionnement *m* **d'air** air-conditioning
conditionnement *m* **d'air de pièce** room air-conditioning
conditionner *vb* condition, package, prepack
conditionneur *m* **d'air** air conditioner
conditionneur *m* **d'air autonome** packaged air-conditioner
conditionneur *m* **d'air de fenêtre** window air-conditioner
conditionneur *m* **d'air de paroi** through-the-wall air-conditioner
conditionneur *m* **d'air en toiture** roof-mounted/top air-conditioner
conditionneur *m* **d'air indépendant** unitary air-conditioner
conditionneur *m* **d'air monobloc** packaged air-conditioner
conditions *f,pl* terms [of a contract]
conditions *f,pl* **à la rupture** failure condition(s)
conditions *f,pl* **ambiantes** ambient conditions
conditions *f,pl* **aux limites** boundary conditions
conditions *f,pl* **climatiques** weather conditions
conditions *f,pl* **contrôlées** controlled conditions
conditions *f,pl* **d'ambiance** ambient conditions
conditions *f,pl* **de fonctionnement** operating conditions
conditions *f,pl* **de livraison** terms of delivery or supply
conditions *f,pl* **de référence** reference conditions
conditions *f,pl* **de vente** sale conditions
conditions *f,pl* **du contrat** contract conditions
conditions *f,pl* **locales** local conditions
conditions *f,pl* **standards** reference conditions
conductance *f* conductance
conductance *f* **thermique** thermal conductance
conducteur *m* conductor [electrical], driver [of a vehicle]
conducteur *m* **à la masse** earth(ing) conductor or lead
conducteur *m* **actif** current-carrying or live conductor
conducteur *m* **d'engin** machine or plant operator
conducteur *m* **de phase** phase conductor
conducteur *m* **de protection** protective conductor
conducteur *m* **de terre** earth(ing) conductor or lead
conducteur *m* **de travaux** general foreman, section engineer, site agent
conducteur *m* **neutre** neutral conductor
conducteur *m* **principal** main supply cable
conductibilité *f* conductibility
conductibilité *f* conductivity
conductibilité *f* **spécifique** conductance
conductibilité *f* **thermique** thermal conductivity
conduction *f* conduction
conduction *f* **thermique** heat or thermal conduction

conductivité *f*   conductivity
conductivité *f* spécifique   conductance
conductivité *f* thermique   heat or thermal conductivity
conduire *vb*   drive [a vehicle]
conduire *vb* à   lead to [e.g. water to a drain]
conduit *m*   (cable) duct, channel [duct], conduit, flue, passage
conduit *m* accolé à un mur   duct or flue fixed against a wall
conduit *m* adossé à un mur   duct or flue fixed to but standing free of a wall
conduit *m* coudé   elbow connection
conduit *m* d'aération   ventilation duct or shaft
conduit *m* d'air   air duct
conduit *m* d'air flexible   flexible duct
conduit *m* d'air mélangé   mixing duct
conduit *m* d'alimentation   supply duct or pipe
conduit *m* d'évacuation   discharge drain pipe, exhaust or extract duct or flue
conduit *m* d'extraction   exhaust or extract duct or flue
conduit *m* de cheminée   chimney flue
conduit *m* de crémone   Cremona bolt guide
conduit *m* de dérivation   bypass duct
conduit *m* de fuite   leakage duct
conduit *m* de fumée   boiler or smoke flue
conduit *m* de fumée unitaire   common flue [collecting from several installations]
conduit *m* de refroidissement   cooling duct
conduit *m* de ventilation   ventilation duct or shaft
conduit *m* de ventilation de la cuisine   kitchen extract duct
conduit *m* en grès   earthenware drain
conduit *m* en tôle   sheet metal duct
conduit *m* incorporé   built-in duct or flue
conduit *m* principal   main or trunk duct
conduit *m* principal de condensation   main condensate pipe
conduit *m* secondaire   branch duct
conduit *m* souterrain   underground culvert or drain
conduit *m* unitaire   common flue [for several hearths]
conduite *f*   channel [pipe], conduit, pipe, mains, direction [management], running [e.g. of an installation], driving [of a vehicle]
conduite *f* à air   air-duct, air-passage
conduite *f* collective   common duct
conduite *f* d'alimentation   delivery or supply pipe, supply conduit, duct or pipe, feed line [pipe]
conduite *f* d'aspiration   suction line or pipe
conduite *f* d'eau   water supply pipe
conduite *f* d'eau de refroidissement   cooling water pipe
conduite *f* de câbles   cable channel
conduite *f* de cheminée   chimney flue [pipe]
conduite *f* de condenseur   condenser main
conduite *f* de distribution   service pipe
conduite *f* de gaz   gas pipe
conduite *f* de raccordement   connecting conduit, service pipe
conduite *f* de refoulement   delivery main [from a pump], rising main [in a borehole]

conduite *f* de retour   return pipe
conduite *f* de vapeur   steam main
conduite *f* des eaux pluviales   stormwater drain or sewer
conduite *f* enterrée   buried pipe, culvert
conduite *f* forcée   penstock [pressure pipe of a dam]
conduite *f* montante   riser pipe, rising main
conduite *f* principale   trunk main
conduite *f* principale d'eau   water main
conduite *f* principale de chauffage à distance   district heating main
conduite *f* principale de distribution   distribution main
conduite *f* principale du gaz   gas main
conduite *f* secondaire   branch line [main or pipe]
conduite *f* verticale   dropper [dropping conduit]
cône *m*   cone, fan [as in alluvial fan], taper [pipework]
cône *m* d'Abrams   slump test cone
cône *m* d'ancrage   anchoring cone
cône *m* d'éboulis   talus cone
cône *m* de déjection   alluvial fan
cône *m* de diffusion   diffusion cone
cône *m* tronqué   truncated cone
configuration *f*   configuration
configuration *f* d'écoulement   flow pattern
confirmer *vb*   confirm
conforme à l'échantillon   up to sample
conformément à   in accordance with
conformément aux normes   in accordance with (the) standards
conformément aux règlements   according to (the) regulations
confort *m*   comfort [as in air-conditioning]
confort *m* thermique   thermal comfort
conforter *vb* (un bâtiment, etc)   strengthen (a building, etc)
congé *m*   bead or fillet [concave profile]
congélateur *m*   deep-freeze, freezer
congélation *f*   freezing
congélation *f* du sol   ground freezing
congeler *vb*   freeze
conglomérat *m*   conglomerate
conifère *m*   conifer
conjoncteur *m*   automatically closing switch [electrical]
conjoint *adj*   linked together e.g. of partners in a joint venture]
connecter *vb*   connect
connecteur *m*   connector [electrical]
connecteur *m* (de câble)   cable connector
connecteur *m* de circuit   circuit connector
connexion *f*   connection or lead [electrical]
connexion *f* au réseau de distribution   mains connection
connexion *f* en parallèle   parallel connection
connexion *f* en série   series connection
connexion *f* flexible   flexible connection
connexion *f* rigide   rigid connection
conseil *m*   advice or counsel
conseil *m* d'administration   board of directors
conseil *m* des architectes   architects' council [in France]
conseil *m* fiscal   tax adviser

conseil *m* juridique   lawyer [legal adviser, roughly equivalent to solicitor in England and Wales]

Conseil *m* National des Ingénieurs de France (CNIF)   national engineering council [in France]

conseiller *m* technique   technical adviser

conseiller *vb*   recommend

conseiller *vb* (à qn de faire qch)   advise (someone to do sth)

conservateur *m*   keeper, registrar, warden

conservateur *m* des hypothèques   registrar of mortgages and land charges

conservation *f*   preservation

conservation, produit *m* de   preservative

conservation *f* d'énergie/de l'énergie   energy conservation

Conservation *f* des Eaux et Forêts   national forest authority in France [similar to British Forestry Commission]

conservation *f* par le froid   cold storage

conserver *vb*   store [documents]

consignes *f,pl* d'utilisation   operating instructions

consistance *f*   consistency [e.g. of concrete before setting]

console *f*   cantilever, corbel, (wall) bracket

console *f* de couvreur   roofing bracket

console *f* de visualisation   visual display unit (VDU)

console *f* murale   wall bracket

consolidation *f*   consolidation, strengthening [of foundations]

consolidation, tassement *m* de   consolidation settlement

consolidation *f* initiale/instantanée   initial consolidation

consolidé *adj*   consolidated

consolider *vb*   consolidate, fund [debts], strengthen [foundations]

consommation *f*   consumption, usage

consommation *f* d'air/etc   air/etc consumption

consommation *f* par tête   per capita consumption

consortium *m*   consortium

constant *adj*   constant, invariable, stable, steady

constante *f*   constant [mathematical, physical]

constante *f* de Boltzmann   Boltzmann constant

constante *f* de réverbération   room constant [reverberation]

constante *f* diélectrique   dielectric constant

constat *m*   description of work to be subsequently covered up

constituant *m*   component [element], constituent

constructeur *m*   builder, constructor, erector, maker

constructible *adj*   buildable, developable [can be built upon]

construction, en   under construction

construction *f*   building, construction, erection, design(ing) [of machines, etc], structure

construction *f* à panneaux   construction, panel

construction *f* agricole   farm building

construction *f* d'habitation   house building

construction *f* en béton armé   reinforced concrete construction

construction *f* en bois   timber frame construction

construction *f* en charpente   frame construction

construction *f* en dur   permanent structure

construction *f* en éléments préfabriqués   prefabricated construction

construction *f* en fond de parcelle   back-land building development

construction *f* en fouille (ouverte)   cut-and-cover construction

construction *f* en régime accéléré   fast track (construction)

construction *f* en tranchée (ouverte)   cut-and-cover construction

construction *f* industrielle   industrial building

construction *f* légère   light gauge design

construction *f* mécanique   mechanical engineering

construction *f* métallique   steel construction

construction *f* mixte   composite construction/design

construction *f* porteuse   loadbearing structure

construction *f* provisoire   temporary construction or structure

construction *f* soudée   welded construction

construction *f* tubulaire   tubular construction

construire *vb*   assemble, build, construct, erect, lay out [a road, etc]

construit *adj*   built, constructed

consultant *m*   consultant

consultation *f* des travailleurs   worker consultation

contact *m*   contact

contact *m* d'étanchéité   seal contact [in a valve]

contact *m* de porte   door contact

contact *m* de repos   normally closed contact

contact *m* de terre   earth contact

contact *m* de travail   normally open contact

contact *m* direct   direct contact [with a bare conductor]

contact *m* électrique   electric contact

contact *m* indirect   indirect contact [through faulty insulation]

contact *m* normalement fermé   normally closed contact

contact *m* normalement ouvert   normally open contact

contact *m* ouvert   opening contact

contacteur *m*   contactor [electrical]

contacteur *m* bipolaire   double-pole switch

contacteur *m* de porte   door switch

contacteur *m* principal   main contactor

contacteur-disjoncteur *m*   circuit-breaker

contaminant *m*   contaminant

contamination *f*   contamination, pollution

contaminer *vb*   contaminate

contemporain *adj*   contemporary

contenance *f*   capacity [volumetric], contents, volume [contents]

contenance *f* thermique   heat content

conteneur *m*   container [transport]

contenir *vb*   contain

contenu *m*   content(s), scope

contenu *m* de chaleur   heat content

contigu *adj*   adjacent, adjoining, contiguous

contigu *adj* à   next to

continu *adj* continuous
continu *adj* en deux directions [dalle] spanning in two directions or two-way [of a slab]
continuité *f* continuity
contour *m* bend or twist [of a road], circumference, contour, outline [drawn]
contournement *m* bypass(ing), skirting round, shaping, outlining [drawing a line around]
contournement *m* d'une ville town bypass
contourner *vb* bypass [regulations, a place], contort, twist, warp, skirt, wind or work round, trace the outline of
contraction *f* shrinkage
contractuel *adj* contractual, design [i.e. contractual]
contractuelles, pièces *f,pl* contract documents [specification]
contracture *f* tapering [of a column, upwards]
contrainte *f* stress
contrainte *f* à la limite d'écoulement yield stress
contrainte *f* admissible allowable stress
contrainte *f* alternée alternating stress
contrainte *f* appliquée applied stress
contrainte *f* au bord extreme fibre stress
contrainte *f* d'adhérence bond stress
contrainte *f* d'affaissement buckling/collapse stress
contrainte *f* dans la roche rock pressure
contrainte *f* de cisaillement shear stress
contrainte *f* de compression compression/compressive stress
contrainte *f* de flexion bending/flexural stress
contrainte *f* de fluage creep stress
contrainte *f* de retrait shrinkage stress
contrainte *f* de rupture breaking/failure stress
contrainte *f* de service working stress
contrainte *f* de traction tensile stress
contrainte *f* effective effective stress
contrainte *f* horizontale horizontal stress
contrainte *f* initiale dans le massif rock pressure
contrainte *f* limite ultimate stress
contrainte *f* normale axial/normal stress
contrainte *f* permissible permissible/working stress
contrainte *f* principale principal stress
contrainte *f* propre internal stress
contrainte *f* résiduelle residual stress
contrainte *f* secondaire secondary stress
contrainte *f* thermique temperature/thermal stress
contrainte *f* triaxiale triaxial stress
contraintes, distribution *f* des stress distribution
contra-rotation *f* contra-rotation
contraste *m* contrast
contraste avec, faire *vb* contrast with
contraste, mettre *vb* en contrast
contrat *m* agreement [written], contract
contrat *m* clef en main turnkey contract
contrat *m* d'architecte architect–client agreement
contrat *m* de vente sale contract
contrat *m* passé à une entreprise contract awarded to a contractor
contrebalancé *adj* balanced [of a wooden panel]

contre-barrage *f* isolating valve [in an emergency water pipe system]
contre-bordure *f* strip of road surface next to the kerb forming gutter
contre-bride *f* mating flange
contre-buter *vb* resist pressure/thrust
contre-cadre *f* counter frame, sub-frame
contre-calque *f* transparency [drawing]
contre-chambranle *f* moulding applied to emphasize an architrave
contre-champ *m* flat portion of a moulding
contre-châssis *m* double sash [of a window]
contre-clef *f* voussoir on each side of keystone
contre-cloison *f* wall lining
contre-cœur *m* back-plate or rear wall of a chimney hearth
contre-courant *m* back current, counterflow
contre-courant, à counterflow *adj*
contre-courbe *f* counter curve
contre-dosse *f* second piece sawn from a log, after the slab
contre-écrou *m* counter- or lock-nut
contre-épreuve *f* duplicate or repeat test
contre-éprouvette *f* duplicate test piece
contre-essai *m* check or duplicate test, retest
contre-expertise *f* reappraisal, re-survey [of condition, causes, etc]
contre-fenêtre *f* inner sash [of a double window], secondary window placed against another, storm window
contre-fer *m* top-iron [of a plane]
contre-fiche *f* brace [roof], strut
contre-ficher *vb* strut
contre-fil, à against the grain [of wood], against the current [of water]
contre-fil de l'eau, à upstream [direction]
contre-flèche *f* camber
contrefort *m* buttress
contreforts, mettre *vb* des buttress
contre-foulement *m* surging [in a pipe]
contre-fruit *m* overhang [of a wall]
contre-garde *f* heavy stone revetment around a bridge pier
contre-hacher *vb* cross-hatch
contre-hachure *f* cross-hatching
contre-latte *f* counter lath
contre-limon *m* wall string(er)
contremaître *m* foreman
contremarche *f* riser [of stair]
contre-mastiquage *m* bedding with putty [of glazing]
contre-mur *m* lining wall, wall built against another
contre-noix *f* weather rebate in a jamb mating with a corresponding moulding in the hanging stile
contre-pente *f* counterfall, reverse slope
contre-placage *m* plywood construction
contreplaqué *adj* laminated [built up in layers]
contreplaqué *m* plywood
contreplaqué *m* latté laminboard
contreplaqué *m* moulé moulded plywood

contreplaqué *m* **sandwich**   plywood with lightweight core

contrepoids *m*   balance weight, counterbalance, counterweight

contre-porte *f*   screen or storm door

contre-pression *f*   back pressure

contre-profil *m*   matching profile [of a moulding], reversed moulding

contre-rail *m*   check or guard [at points] rail

contre-réaction *f*   feedback

contre-sens, à   in the wrong direction

contre-tirage *m*   back-draught [in a furnace]

contrevent *m*   back-draught [in a furnace], shutter [exterior, solid, of a window or door], windbrace

contreventement *m*   windbracing

contrôle *m*   check(ing)

contrôle *m*   control [of a machine, etc]

contrôle *m* [surveillance]   supervision

contrôle *m* [vérification]   control [check], inspection

contrôle *m* **à distance**   remote control

contrôle *m* **acoustique**   noise/sound control

contrôle *m* **d'accès**   access control

contrôle *m* **de dégivrage**   defrost control

contrôle *m* **de potabilité**   potability check [for drinking water]

contrôle *m* **de pression différentielle**   differential pressure control

contrôle *m* **(de) qualité**   quality control (QC)

contrôle *m* **de température de chaudière**   boiler water temperature control

contrôle *m* **des changes**   exchange control

contrôle *m* **des déformations**   strain control

contrôle *m* **des frais**   cost control

contrôle *m* **des mesures**   recheck or verification [of measurements, readings, etc]

contrôle *m* **du compactage**   compaction testing

contrôle *m* **en boucle fermée**   closed loop control

contrôle *m* **général des travaux (CGT)**   general work supervision

contrôle *m* **par pressostat de sécurité d'huile**   oil pressure cut-out control

contrôle *m* **par thermostat de sécurité d'huile**   oil temperature cut-out control

contrôle *m* **pas à pas**   step control

contrôle *m* **pneumatique**   pneumatic control

contrôler *vb*   check, inspect, control, monitor, supervise, test [check, verify]

contrôleur *m*   auditor, controller [person], regulator [control device]

contrôleur *m* **de chantier**   site agent

contrôleur *m* **de niveau (de liquide)**   (liquid) level controller

contrôleur *m* **de pression**   pressure regulating valve

contrôleur *m* **de température ambiante**   room temperature controller

contrôleur *m* **de travail**   labour supervisor

contrôleur *m* **de volume**   volume controller

contrôleur *m* **des travaux**   statutory representative

contrôleur *m* **électro-pneumatique**   electro-pneumatic controller

contrôleur *m* **pas à pas**   step controller

contrôleur *m* **proportionnel**   proportional action controller

contrôleur *m* **technique**   building inspector employed by client, checking engineer [for structural soundness, construction and safety]

convecteur *m*   convector

convecteur *m* **mural**   wall mounted unit heater

convecteur *m* **naturel**   natural convector

convection *f*   convection

convection *f* **de chaleur**   heat convection

convection *f* **forcée**   forced convection

convection *f* **naturelle**   natural convection

convenant *adj*   suitable

convenir *vb* **à**   be suitable for, suit

convention *f*   agreement [written], covenant

convention *f* **de signe**   sign convention

conversion *f*   conversion

convertible *adj*   convertible

convertisseur *m*   convertor

convertisseur *m* **c.c. en c.a.**   d.c./a.c. convertor

convertisseur *m* **de données**   data converter

convertisseur *m* **de fréquence et de tension**   frequency and voltage convertor

convertisseur *m* **rotatif**   rotary transformer

convoyeur *m* **(à bande)**   conveyor (belt)

convoyeur *m* **en continu**   continuous (belt) conveyor

coordinateur *m*   coordinator [e.g. of construction team]

coordinateur *m* **'projet de l'ouvrage'**   coordinator 'project design'

coordinateur *m* **'réalisation de l'ouvrage'**   coordinator 'project execution'

coordination *f*   coordination [of plans, works, etc]

coordination *f* **de la sécurité**   safety coordination

coordination *f* **modulaire**   modular coordination [dimensional, in design]

coordonnée *f*   coordinate

copeau *m*   chip [wood, metal], shaving

copeaux *m,pl*   woodwool

copie *f*   copy

copier *vb*   copy

copropriétaire *m*   co-owner, co-proprietor

copropriété *f*   co-ownership, joint ownership

coque *f*   shell [e.g. surround to column]

coque *f* **en béton**   concrete shell roof

coque *f* **précontrainte**   prestressed concrete shell roof

coquillart *m*   shell limestone

coquille *f*   body [case] of a centrifugal pump, casing [of a motor], dome of a wall niche [architectural], preformed half-round pipe insulation, shell-like decoration, sloping vault beneath treads of a spiral staircase

coquilles *f,pl* **de protection d'oreille**   ear cups, defenders or protectors

corbeau *m*   corbel, pipe hook

corde *f*   rope

cordeau *m*   bricklayer's or builder's line, chalk-line, cord, string [used for lining]

cordeau *m* **Bickford**   fuse [for igniting a detonator]

cordeau *m* **détonant**   detonating fuse

cordoir *m*   tool for ramming packing into a spigot-and-socket joint

cordon *m*   band or string course, circular moulding, cord, line or row [e.g. of posts, trees], seam [welded], strand [of rope]

cordon *m* **chauffant**   radiant strip heater

cordon *m* **(de soudure)**   weld

cordon *m* **d'amiante**   asbestos cord

cordon *m* **d'angle**   fillet weld

cordon *m* **d'étanchéité**   sealing run [in welding]

cordon *m* **(de) prolongateur**   extension cable or lead

cordon *m* **de soudure**   welded seam

cordon *m* **de soudure bout à bout**   butt-welded joint

cordon *m* **latéral**   side weld

cordon *m* **longitudinal**   longitudinal seam

corindon *m*   corundum

cormier *m*   service tree [and its hard wood]

corne *f* **d'un rabot**   plane handle

cornette *f*   angle-iron reinforcement of corner of a wall

corniche *f*   cornice

corniche *f* **cintrée**   cornice consisting of successive arc(he)s

corniche *f* **coupée**   cornice interrupted by other elements

corniche *f* **en chanfrein**   rectangular cornice without mouldings

cornier *adj*   angle [i.e. at meeting of two planes], corner [i.e. at a corner]

cornière *f*   angle iron (section), valley [between roofs]

cornière *f* **à ailes égales**   equal-leg angle (section)

cornière *f* **à ailes inégales**   unequal-leg angle (section)

cornière *f* **à angles arrondis**   round-edged angle (section)

cornière *f* **à angles vifs**   square-edged angle (section)

cornière *f* **à boudin**   bulb angle (section)

cornière *f* **de membrure**   flange angle (section)

corps *m*   body [e.g. of a dam, a lamp]

corps *m* **creux**   preformed coffer [between beams/joists], hollow concrete block

corps *m* **d'état**   building trade [masonry, carpentry, plumbing, etc]

corps *m* **de chauffe**   heating apparatus/appliance, radiator

corps *m* **de chaussée**   pavement [of a road or runway], road construction [the complete pavement]

corps-de-garde *m*   guardhouse/guardroom

corps *m* **de pompe**   pump body

corps *m* **de serrure**   lock housing

corps *m* **de vanne**   valve body

corps *m* **du barrage**   dam body

corps *m* **noir**   black body

correctement *adv*   right [correctly]

correction *f*   correction, rectification

correction *f* **de température**   temperature correction

corridor *m*   corridor, passage(way)

corriger *vb*   correct

corrodé *adj*   corroded, eaten away

corroder *vb*   corrode, rust [sth]

se corroder *vb*   become corroded, rust

corroi *m* **d'argile**   puddle clay

corrosif *adj*   corrosive

corrosion *f*   corrosion

corrosion *f* **atmosphérique**   atmospheric corrosion

corrosion *f* **aux intempéries**   atmospheric corrosion

corrosion *f* **des métaux**   metallic corrosion

corrosion *f* **ponctuelle**   pitting

corrosion *f* **sous contrainte/tension**   stress corrosion

corroyage *m*   dressing [of sawn timber]

corroyer *vb*   dress [metal or timber], puddle [line with, a pond, basin, etc]

corten, acier *m*   corten (steel)

corvée *f*   minor conversion or maintenance work

cosse *f*   terminal tag on an electrical lead

cotation *f*   indication of figures on a drawing [levels, dimensions]

coté *adj*   dimensioned

cote *f*   dimension [indication of], level [height], spot height [on drawing]

cote *f* **brute**   dimension without finishes

cote *f* **de la crête**   crest level [of a dam]

cote *f* **de niveau**   altitude above a reference height

cote *f* **extérieure**   outside dimension

cote *f* **nominale**   nominal dimension

cotes *f,pl* **d'encombrement**   space requirements [dimensions, for equipment]

côte *f*   coast, shore [of an ocean, sea], gradient or slope [hillside]

côté *m*   side

côté *m* **basse pression**   low-pressure side

côté *m* **d'aspiration**   suction side [of a fan, etc]

côté *m* **de refoulement**   delivery or discharge side [of a fan, etc]

côté *m* **haute pression**   high-pressure side

côté, point *m*   landmark, reference point

coteau *m*   slope or hillside

coter *vb*   dimension

cotisation *f*   premium [insurance]

cotisation *f* **de la sécurité sociale**   social security charge or contribution

côtoyer *vb*   border on

cou *m* **de cygne**   swan-neck

couche *f*   coat [layer of paint, tar, etc], layer [ground, etc], stratum, foot plate, ground sill, sole piece or plate [of wood, under a prop, etc]

couche *f* **antigel**   frost-proof layer [road construction]

couche *f* **antirouille**   corrosion protection layer

couche *f* **bitumineuse**   bituminous layer

couche *f* **compressible**   compressible layer

couche *f* **d'air**   air cushion, air gap [in a cavity wall]

couche *f* **d'enrochements**   stone base [of foundations]

couche *f* **d'étanchéité**   damp-proof course (DPC)

couche *f* **d'habillage**   decorative finish(ing) or top coat

couche *f* **d'impression**   primer [first coat on absorbent substrate], undercoat

couche *f* **d'isolation thermique**   thermal insulation layer

couche *f* **d'ozone**   ozone layer

couche *f* **d'usure**   wearing course [road construction]

couche *f* **de base**   base, substructure [road construction]

couche *f* **de base, sous-**   sub-base (course) [road construction]

couche *f* **de base rigide**   rigid roadbase

couche *f* **de base souple** flexible roadbase

couche *f* **de briques** brick course

couche *f* **de finition** finish(ing) or top coat

couche *f* **de fond** primary coat, primer [paint], undercoat

couche *f* **de fondation** foundation course or sub-base (course) [road construction]

couche *f* **de fondation, sous-** foundation sub-base [road construction]

couche *f* **de forme** sub-base (course) [road construction]

couche *f* **de gravillons** layer of chippings

couche *f* **de liant** grout course [in penetration surfacing of roads]

couche *f* **de liaison** binder course

couche *f* **de poussière** dust layer

couche *f* **de propreté** blinding (layer), sealing coat [blinding]

couche *f* **de protection** protective coat or layer, surface protection [road construction]

couche *f* **de remblai** layer of backfill

couche *f* **de renforcement** undercoat [reinforcing a primer or first undercoat]

couche *f* **de reprise** cement grout applied to cured concrete before pouring fresh

couche *f* **de rouille** rust layer

couche *f* **de roulement** road surface layer, running surface, wearing course [road construction]

couche *f* **de roulement hydrocarbonée** bituminous wearing course

couche *f* **de scellement** sealing coat [road construction]

couche *f* **de sol** soil horizon

couche *f* **de surface** surface course or layer

couche *f* **de surface bitumineuse** bituminous bound surfacing

couche *f* **de surface hydrocarbonée** bituminous surfacing or wearing course

couche *f* **de teinte** final coat

couche *f* **drainante** drainage layer

couche *f* **en matière plastique** plastic coating

couche *f* **filtrante** filter layer [road construction]

couche *f* **isolante** damp-proof course (DPC), insulating layer

couche *f* **limite** boundary layer

couche *f* **perméable** permeable layer

couche *f* **primaire** undercoat [on non-absorbent substrate]

couche *f* **protectrice** brush coat

couche *f* **protectrice en béton** protective concrete layer

couche *f* **superficielle** surface layer

couches *f,pl* **de feutre** felting

couches *f,pl* **successives** successive layers

coucher *vb* coat [apply one or more coats]

couchis *m* bedding of sand, etc, under paving, etc

couchis *m* **de voûte** planking between templates [arch/vault construction]

coudage *m* cranking or offset

coudé *adj* bent, cranked, kneed

coude *m* angle [iron bar, etc], bend [road, iron bar, pipe- or ductwork, etc], elbow [pipe- or ductwork, etc], knee [pipework, etc]

coude *m* **à angle droit** square elbow

coude *m* **à emboîtement et cordon** spigot-and-socket bend (>90°)

coude *m* **au quart à emboîtement et cordon** spigot-and-socket bend (90°)

coude *m* **de dilatation** expansion bend

coude *m* **de réduction** reducing elbow

coude *m* **de renvoi** S-bend, offset elbow

coude *m* **de tube** pipe bend

coude *m* **en U** return bend or U-bend

coude *m* **grand rayon** large-radius elbow

coude *m* **lisse** smooth bend

coude *m* **soudé** welded bend

couder *vb* bend [pipe, reinforcement, etc], crank [a pipe, shaft, etc]

se couder *vb* bend [take on a bent shape]

coulage *m* casting [the action], pouring [of concrete, asphalt road material]

coulé *adj* cast *adj*

coulé en/sur place cast in situ/in place

coulée *f* cast(ing) [the action], melt [of metal]

couler *vb* flow [river, stream], run [paint], cast or pour [concrete]

couler *vb* **sous pression** die-cast

couler *vb* **sur place** cast in situ/in place

couleur *f* colour, hue, shade

couleuvre *f* crack in a ceiling or arch [construction defect]

coulis *m* jointing mortar, slurry

coulis *m* **bitumineux** slurry seal [road construction]

coulis *m* **d'argile/de ciment/etc** clay/cement/etc grout

coulis *m* **d'injection** grout

coulis *m* **goudronneux** bitumen slurry [road construction]

coulissante, fenêtre *f*/**porte** *f* sliding window/door

coulisse *f* chute [of wood, for building materials or rubbish], runner [of a drawer], sleeve [cast or two-piece, for pipe repair], slide(way)

coulisse, équerre *f* **à** sliding calipers

coulisse, pied *m* **à** particle shape gauge [for road material]

coulisse *f* **de ramonage** opening with sliding cover for flue/chimney-sweep's brush

coulisse *f* **de tiroir** drawer slide

coulisse *f* **de vasistas** slotted strap holding a fanlight in open position

coulisseau *m* runner [of a drawer], slide(way)

coulisseau *m* **de cabine** cage slide [lift/elevator]

couloir *m* corridor, passage(way), gully [natural feature]

coulotte *f* chute [for building materials or rubbish]

coulure *f* run [of excess paint]

coup *m* blow, impact, knock, shock, inclination [of a wall about to fall]

coup *m* **de bélier** water hammer

coup *m* **de froid** cold snap or wave

coup *m* **de sabre** crack, long and thin, in a piece of stone

coup *m* **de vent** gust or squall [of wind]

coup, prendre *vb* get out of plumb, settle or sink in

coupage *m*   cutting
coupage *m* **à l'arc**   arc cutting
coupe *f*   profile or section [drawing]
coupe *f* **biaise**   mitre cut
coupe *f* **d'équerre**   straight cut [at 90°]
coupe *f* **de sondage**   soil profile
coupe *f* **droite**   straight cut [at 90°]
coupe *f* **embrevée**   tongued-and-grooved edges
coupe *f* **en long**   long(itudinal) section
coupe *f* **en travers**   cross-section, transverse section
coupe *f* **fausse**   bevel cut
coupe *f* **longitudinale**   long(itudinal) section
coupe *f* **micrographique**   micro-section
coupe *f* **mince**   thin section
coupe *f* **polie**   polished section
coupe *f* **(suivant) X–X**   section X–X [on a drawing]
coupe *f* **transversale**   cross-section
coupé aux dimensions nécessaires   cut to size
coupe-ardoises *m*   slate cutter, slater's iron
coupe-bordure *m*   edge trimmer [for lawns]
coupe-boulons *m*   bolt cropper or cutter
coupe-câble *m*   cable cutter
coupe-circuit *m*   circuit-breaker
coupe-circuit *m* **à cartouche**   cartridge fuse circuit-breaker, switched cartridge fuse
coupe-circuit *m* **à fusibles**   fuse circuit-breaker
coupe-circuit *m* **sectionnable**   modular fuse carriers
coupe-feu (CF)   fire rated
coupe-feu *m*   fire barrier, break or stop
coupe-fil *m*   wire cutters
coupe-froid *m*   weather strip [on a window]
coupe-larme *m*   drip [under a soffit], window ledge
coupelle *f*   cup burner
coupement *m*   cutting
couper *vb*   cut (out), shut off [electricity, gas], switch off
couper *vb* **à angle droit**   square [a corner]
couper *vb* **de longueur**   cut to length
couper *vb* **l'eau**   cut/turn off the water
couper *vb* **la lumière**   turn off/out the light(ing)
couper *vb* **la vapeur**   cut off or shut down (or off) the steam
couper *vb* **le courant**   turn off the current
couper *vb* **le gaz**   cut off the gas
couper *vb* **un circuit**   break or open a circuit
coupe-tirage *m*   air inlet to a flue to reduce excessive draught, spoil draught
coupe-tubes *m*   pipe cutter
coupe-tuiles *m*   tile cutter
coupe-tuyaux *m*   pipe cutter
coupe-type *f*   typical section
coupe-vapeur *m*   vapour barrier
coupe-vapeur *m* **dans un mur creux**   cavity barrier
coupe-vent *m*   wind barrier, windbreak
coupe-verre *m*   glass cutter
couplage *m*   (conduit) coupling
couplage *m* **de sécurité**   security interlock
couplage *m* **en étoile**   Y-connection
couplage *m* **en parallèle**   parallel connection
couplage *m* **en triangle**   delta connection
couplage *m* **en série**   series connection
couplage *m* **fermé**   close coupling [electrical]

couple *m*   couple or moment [of forces], torque
couple *m* **de démarrage**   starting torque
couple *m* **de renversement**   overturning moment
couple *m* **de rotation**   torque, turning moment
couple *m* **de torsion**   torque, twisting moment
couplement *m*   connecting or coupling together
coupler *vb*   connect
coupler *vb* **en parallèle**   connect in parallel
coupler *vb* **en série**   connect in series
coupler *vb*, **mal**   connect wrongly, misconnect
coupleur *m*   coupler [reinforcement]
coupleur *m* **acoustique**   acoustic coupler [computer]
coupleur *m* **d'armatures**   reinforcement coupler
coupole *f*   cupola, dome [architectural]
coupole *f* **vitrée**   clerestory, skylight [raised]
coupure *f*   cut [wound], cut(-off) or disconnection
coupure *f* **en charge**   full-load break [electrical]
cour *f*   court(yard), yard
cour *f* **anglaise**   area below street level giving light to basement windows
cour *f* **close**   enclosed courtyard
courant *adj*   current [present, running], flowing, running
courant *m*   current or flow [of air, fluids, electricity], drift [of a fluid], electric current or power, stream [current or of traffic]
courant *m* **alternatif (c.a.)**   alternating current (a.c.)
courant *m* **basse tension**   low-tension or low-voltage current
courant *m* **continu (c.c.)**   direct current (d.c.)
courant *m* **convergent**   converging [traffic] stream
courant *m* **d'air**   air current or stream, draught
courant *m* **d'air descendant**   downdraught
courant *m* **d'allumage**   inrush current [of a lamp]
courant *m* **d'appel**   surge current
courant *m* **d'eau**   current or stream [of water]
courant *m* **de convection**   convection current
courant *m* **de crête**   peak current
courant *m* **de démarrage**   starting current
courant *m* **de fonctionnement**   operating current
courant *m* **de pointe**   peak current
courant *m* **de retour**   return or reverse flow
courant *m* **direct (c.d.)**   direct current (d.c.)
courant *m* **divergent**   diverging [traffic] stream
courant *m* **électrique**   electric current
courant *m* **faible**   light current
courant *m* **induit**   induced current
courant *m* **polyphasé**   polyphase current
courant *m* **primaire**   primary current
courant *m* **principal (de trafic)**   through-traffic stream
courant *m* **triphasé**   three-phase current
courant *m* **vagabond**   stray current
courants *m,pl* **croisés**   cross currents, cross-flow
courants *m,pl* **de Foucault**   eddy currents
courbé *adj*   bent, curved
courbe *f*   bend, curve [of a road, etc], diagram or graph, curve or line [on a graph]
courbe *f* **caractéristique**   characteristic curve
courbe *f* **caractéristique du ventilateur**   fan (performance) curve

courbe *f* **charge-tassement**   load-settlement curve
courbe *f* **composée**   compound curve
courbe *f* **contrainte-allongement**   stress-strain curve
courbe *f* **contrainte-déformation**   stress-strain curve
courbe *f* **d'efficacité**   operating curve
courbe *f* **d'espacement**   spacing curve
courbe *f* **d'hystérésis**   hysteresis curve
courbe *f* **d'intensité lumineuse**   light distribution curve
courbe *f* **de chargement**   loading curve, load-settlement curve
courbe *f* **de compression**   compression diagram
courbe *f* **de consommation**   consumption curve
courbe *f* **de déchargement**   rebound curve
courbe *f* **de fonctionnement**   performance characteristic curve
courbe *f* **de fréquence**   frequency curve
courbe *f* **de liquidité Atterberg**   Atterberg flow curve
courbe *f* **de mortalité**   mortality curve [lighting]
courbe *f* **de niveau**   contour line
courbe *f* **de raccordement**   transition curve [roads]
courbe *f* **de refroidissement**   cooling curve
courbe *f* **de résistance à la pénétration**   penetration resistance curve
courbe *f* **de tassement**   load-settlement curve
courbe *f* **de tassement en fonction du temps**   settlement-time curve
courbe *f* **de teneur en eau-densité**   moisture-density curve
courbe *f* **efforts-déformations**   load-deflection/deformation curve
courbe *f* **granulométrique**   grading curve, grain or particle size distribution curve
courbe *f* **pression-indice des vides**   pressure–void ratio curve
courbe *f* **Proctor**   Proctor curve
courbe *f* **rampante**   string or stringer of a wooden spiral staircase
courber *vb*   bend [a road, etc], bend or curve [a beam, pipe, etc]
courbure *f*   camber or sagging [of a beam], curvature, curve [of a road, etc]
courçon *m*   piece [e.g. of wood] of the wrong length
courette *f*   air well, small enclosed courtyard
courge *f*   corbel supporting a mantlepiece
couronne *f*   drilling bit, driving band [round a pile head], pile head
couronne *f* **à diamants**   diamond bit
couronne *f* **au métal dur**   hard metal bit
couronne *f* **de forage**   boring bit
couronne *f* **mobile**   pump impeller
couronnement *m*   capping [of a chimney, pier, dam, etc], coping or crowning [of a wall], cover or grating [of a road gulley], crest [of a dam]
couronnement *m* **libre**   free crest [of a dam]
couronnement *m* **massif**   solid cap
couronner *vb*   cap [chimney, wall, etc]
couronner *vb* **de**   top by [surmount with]
courroie *f*   belt [drive belt]
courroie *f* **courante**   running belt

courroie *f* **de ventilateur**   fan belt
courroie *f* **transporteuse**   conveyor belt
courroie *f* **trapézoïdale**   V-belt
cours *m*   course [of a river, lectures, etc], flow [of a river], price [of base metals, securities], exchange rate [currency]
cours, avoir *vb*   be current, be legal tender
cours *m*, **dernier**   closing price
cours *m* **d'assises**   line of stones of equal height all along a facade
cours *m* **d'eau**   river or stream, watercourse or waterway
cours *m* **d'eau navigable**   navigable waterway
cours *m* **d'ouverture/de clôture**   opening/closing price
cours *m* **de change**   exchange rate [currency]
cours *m* **du marché**   market prices or rates
course *f*   stroke [of a jack, a piston, etc], travel [of a tool, etc]
course *f* **de compression**   compression stroke
course *f* **de détente**   expansion stroke
course *f* **de refoulement**   discharge stroke
coursive *f*   access balcony, walkway
courson *m*   flooring strip less than 1 metre long
court terme, à   short-term, in the short term
court-circuit *m*   short circuit
court-circuit, mettre *vb* **en**   short circuit
court-circuit, mise *f* **en**   short circuiting
court-circuiter *vb*   short circuit
coussin *m* **d'air**   air cushion
coussinet *m*   bearing bush(ing), sleeve bearing, springer or springing stone [of an arch]
coussinet *m* **à billes**   ball-bearing
coussinet *m* **à deux pièces**   split bush (bearing)
coussinet *m* **de glissement**   slide chair [railway points]
coussinet *m* **palier**   thrust bearing
coût *m*   cost, price
coût *m* **d'achat**   purchase cost
coût *m* **d'exploitation**   operating cost(s) or expenses
coût *m* **de centrale**   plant cost
coût *m* **de conversion**   conversion cost
coût *m* **de l'installation**   installation cost
coût *m* **de l'opération**   operating cost(s)
coût *m* **de la construction**   construction or erection cost
coût *m* **de remplacement**   replacement cost
coût *m* **de reproduction**   reproduction cost
coût *m* **de revient**   cost price
coût *m* **direct**   direct or prime cost
coût *m* **du cycle de vie**   life-cycle cost
coût *m* **estimé/estimatif**   estimated value [cost]
coût *m* **final**   final cost
coût *m* **indirect**   indirect cost [e.g. manufacturing overhead]
coût *m* **marginal**   marginal cost
coût *m* **moyen**   average cost
coût *m* **moyen pondéré**   weighted average cost
coût *m* **supplémentaire**   extra cost
coûts *m,pl* **de viabilisation**   cost of works for (site) services
couteau *m*   blade [of electrical switch], knife [e.g. painter's]
couteau, interrupteur *m* **à**   knife switch
couteau *m* **à araser**   trimming knife
couteau *m* **à deux manches**   draw knife or drawshave
couteau *m* **à émarger**   trimming knife

couteau *m* **à mastiquer**   putty knife
couteau *m* **décolleur**   stripping knife [for wallpaper]
couteau *m* **diviseur**   riving knife [of circular saw]
couteau *m* **universel**   Stanley or universal cutter or knife
coûter *vb*   cost
couvercle *m*   cap, cover [lid]
couvercle *m* **à bride**   flanged cover
couvercle *m* **d'inspection**   inspection cover
couvercle *m* **de trou d'homme**   manhole cover
couvercle *m* **démontable**   removable cover
couvert *adj*   covered
couverture *f*   covering, roof covering or cladding, roofing, sheath
couverture, terre *f* **de**   topsoil
couverture *f* **de chaume**   thatched roofing
couverture *f* **de sol**   floor covering
couverture *f* **en tuiles**   tiled roofing
couverture *f* **industrielle**   space-frame roof
couvrant, pouvoir *m*   covering capacity or hiding power [of paint]
couvre-bornes *m*   terminal cover
couvre-câble *m*   cable cover
couvre-chant *m*   edging strip [on a door]
couvre-joint *m*   cover plate or strip, cover(ing) bead, batten or strip, flashing, trim [cover strip]
couvreur *m*   roofer, tiler [roofs]
couvreur *m* **en ardoise**   slater
couvreur *m* **en chaume**   thatcher
couvrir *vb*   cover [generally, and one colour by another], roof
coyau *m*   arris or tilting fillet [roof], firring [at the edge of a timber roof], eaves board [roof]
crachat *m*   air-bubble in glass
craie *f*   chalk
crampon *m*   clamp, dog or cramp [iron], staple [U-shaped]
crapaud *m* **de rail**   rail or sleeper clip
crapaudine *f*   gravel/leaf guard in a rainwater outlet
craquelage *m*   surface cracking [of tiles, etc]
craquelure *f*   cracking [array of fine cracks], surface crack [in paintwork]
cratère *m*   crater [welding]
cratère *m* **d'effondrement**   sink-hole [geology]
crayon *m*   pencil
crayon *m* **de maçon** *m*   bricklayer's pencil
créance *f*   debt
créancier *m*   creditor
création *f*   creation
crédit *m*   credit, loan, mortgage
crédit *m* **à courte terme**   short-term credit
crédit bail *m*   leasing [with terminal purchase option, in France]
crédit *m* **par acceptation renouvelable**   revolving credit
crémaillère *f*   notched or drilled stay for holding rooflight open, notched or toothed lath for cupboard shelf supports, rack [toothed] of mountain railway
crémaillère *f,* **fausse**   false string(er) applied to edge of a staircase
crémaillère, limon *m* **à**   cut or open string(er)
crème *f* **barrière/de protection**   barrier cream

crémone *f*   casement or Cremona bolt
créneau *m*   battlement
créosote *f*   creosote
crépi *m*   rough or roughcast rendering, stucco
crépi *m* **moucheté**   pebbledash, roughcast [rendering]
crépi *m* **tyrolien**   Tyrolean or alpine finish rendering, roughcast [rendering]
crépine *f*   strainer [at foot of a suction pipe], suction strainer
crépine *f* **d'aspiration**   suction screen
crépine *f* **d'entrée**   inlet strainer
crépine *f* **eau potable**   drinking water strainer
crépine *f* **incendie**   emergency water strainer
crépissage *m*   rendering [of a wall, etc]
crépir *vb*   render [a wall, etc]
crête *f*   coping or capstone [of a wall], crest [of a dam], decorative ridge tile, ridge [of a mountain]
crête *f* **de barrage**   dam crest
crête *f* **de la digue**   embankment crest [dike]
crête *f* **de remblai**   embankment or fill crest
crête *f* **déversante**   spillway crest
creuser *vb*   bore or dig [tunnel, pile], dig [a hole, a trench], sink [dig, e.g. a well]
creuser *vb* **une tranchée**   trench
creux *adj*   hollow *adj*
creux *m*   hollow *n*
crevasse *f*   crevasse, crack, crevice
criblage *m*   screening [grading], riddling, sieving, sifting
criblé *adj*   riddled
crible *m*   riddle, sieve, screen [for gravel, etc]
crible *m* **à gravier**   gravel screen
crible *m* **à secousses**   shaking screen
crible *m* **classeur**   sizing screen
crible *m* **de classement**   classifying screen
crible *m* **rotatif**   rotary screen
crible *m* **scalpeur**   scalping screen, vibrating grizzly
crible *m* **vibrant à balourd**   vibrating screen
cribler *vb*   pierce with many holes, screen [grade], riddle, sieve, sift
criblures *f,pl*   riddled or sieved material
cric *m*   rack, ratchet-and-pawl or tooth-and-pinion jack
cric *m* **à vis**   screw jack
cric *m* **hydraulique**   hydraulic jack
cric, soulever *vb* **qch au**   jack something up
crinoline *f*   safety hoop [on a fixed ladder]
crique *f*   creek, crack, fissure
crique *f* **de retrait**   shrinkage crack
crique *f* **de tension**   stress crack
crique *f* **due aux tensions thermiques**   thermal stress crack
crique *f*   crack [in metal]
criquer *vb*   crack [of metals]
criqûre *f*   fissure
critère *m*   criterion
critique *adj*   critical, crucial, decisive, dangerous [situation]
critique *f*   criticism
crochet *m*   hook
crochet *m* **d'ardoise**   slate clip
crochet *m* **d'échelle**   ladder hook [to hold top of a ladder]

crochet *m* **de contrevent/volet**   shutter retaining clip
crochet *m* **de gouttière**   gutter bracket or clip
crochet *m* **de levage**   lifting hook
crochet *m* **de sécurité**   safety hook [of a roof]
crocheter *vb*   pick [a lock]
crocodilage *m*   paint defect due to painting on top of paint not yet dry
croisée *f*   casement window
croisement *m*   crossing, intersection
croisement *m* **à niveaux différents**   flyover
croisement *m* **de voie**   cross-over [railway]
croisement *m* **en double huit**   cloverleaf intersection or junction
croisement *m* **oblique**   diamond crossing [railway lines]
se croiser *vb*   intersect
croisillon *m*   cross brace, cross-piece, crossed glazing bars, tile spacer
croissance *f*   growth
croissance *f* **composée**   compound growth
croix *f*   cross [plumbing fitting], crossed windbracing [timber frame structure], cross-piece
croix *f* **de Saint-André**   herringbone strutting
croquis *m*   (rough) sketch
crosse *f* **(d'armature)**   projecting reinforcement, starter bar
croupe *f*   hip(ped) end [of a roof]
croupe *f* **faîtière**   half-hip [roof]
croûte *f* **de mortier**   outer mortar skin
cruciforme *adj*   cruciform
crue *f*   flood [water], rise [of water level in a river]
crue *f* **nominale**   design flood
crue, en   in spate
cryogénique *adj*   cryogenic
crypte *f*   crypt
cubature *f*   volume measurement [of solids]
cube *m*   cube
cubique, granulat *m*   cubic particle aggregate
cueillie *f*   re-entrant angle between two rendered surfaces
cuisine *f*   kitchen
cuisine-salle à manger *f*   kitchen-dining room
cuisinette *f*   kitchenette
cuisinière *f*   cooker, cooking stove, kitchen range [cooker]
cuisinière *f* **à gaz**   gas cooker
cuivre *m*   copper
cuivrer *vb*   copper plate
cul-de-four *m*   vault in form of quarter sphere
culée *f*   abutment [of bridge]
culée *f* **poids**   abutment [of a dam]
culière *f*   gutter stone [between down pipe and drain]
culot *m*   base or cap of a bulb/lamp
culot *m* **à baïonnette**   bayonet fitting [bulb/lamp]
culot *m* **à broches**   pin base or cap [fluorescent tube]
culot *m* **à vis**   screw base or fitting [bulb/lamp]
culotte *f*   breech or pipe junction fitting
culotte *f* **à regard**   breech fitting with cleaning eye
culotte *f* **double**   three-way junction [plumbing]
culotte *f* **(simple)**   Y-junction [plumbing, ductwork]
culture *f*   cultivation
cultures *f,pl*   cultivated land, land under cultivation

cumulé *adj*   cumulative
cumulus *m*   diaphragm tank [central heating system], expansion tank or vessel
cumulus *m* **électrique**   hot water tank with immersion heater
cunette *f*   drain [of a sewer], gutter [road or street]
curage *m*   cleaning out, clearing or flushing [of a drain, etc]
curage *m* **des égouts**   sewer cleaning
cure *f*   curing
cure, produit *m* **de**   curing agent/compound [e.g. for concrete]
cuve *f*   collecting chamber, cistern, tank, tub, vat, vessel [receptacle], deep excavation with vertical sides
cuve *f* **à immersion**   dip tank
cuve *f* **à mazout**   oil tank [domestic]
cuve *f* **du réacteur**   reactor vessel
cuvelage *m*   foundation slab and walls of a basement, lining [of a well, shaft]
cuveler *vb*   case or line [a borehole, shaft, well]
cuvette, en   dished [cover, etc]
cuvette *f*   basin [geographical], bowl, pan [of a WC, or to collect condensate, etc], rainwater head(er) or hopper, wash-basin, wash-hand basin
cuvette *f* **d'ascenseur**   lift pit
cuvette *f* **de cabinet/WC**   toilet bowl or pan
cuvette *f* **de chéneau**   rainwater head(er) or hopper
cuvette *f* **de WC**   lavatory basin
cycle *m*   bicycle, cycle
cycle *m* **de Carnot/Rankine/Stirling/Otto**   Carnot/Rankine/Stirling/Otto cycle
cycle *m* **de chargement/déchargement**   loading/unloading cycle
cycle *m* **de commutation**   switching cycle
cycle *m* **de dégivrage**   defrosting cycle
cycle *m* **de l'eau**   water cycle
cycle *m* **de transfert de chaleur**   cooling or heat transfer cycle
cycle *m* **de vie**   life cycle
cycle *m* **frigorifique**   refrigeration cycle
cycle *m* **inversé**   reverse cycle
cycle *m* **opératoire**   operating cycle
cycliste *m/f*   cyclist
cyclistes, passage *m* **pour**   crossing for cyclists
cyclone *m*   cyclone
cyclopéen, béton *m*   cyclopean concrete
cylindrage *m*   rolling [to compact]
cylindrage *m* **définitif**   final compacting or rolling
cylindré *adj*   rolled [compacted]
cylindre *m*   cylinder, roller
cylindre, méthode *f* **du**   cylindrical mould sampling method
cylindre *m* **à gaz**   gas cylinder
cylindre *m* **articulée**   hinged cylinder
cylindre *m* **d'orientation**   slewing cylinder [crane, etc]
cylindre *m* **de pompage**   pumping cylinder
cylindre *m* **vibrant**   vibrating/vibratory roller
cylindrée *f*   cylinder displacement
cylindrer *vb*   roll [to compact]
cymaise *f*   cyma, dado [architecture]

dallage *m*   ground slab, slab on grade, flagstones, paving or pavement [and the action of laying it]

dallé *adj*   paved

dalle *f*   paving slab or stone, plate or slab [reinforced concrete, stone, etc]

dalle *f* **avec prédalle**   composite slab

dalle *f* **caissonnée**   waffle slab

dalle *f* **de béton**   concrete slab

dalle *f* **de fond**   ground slab

dalle *f* **de fondation**   base or foundation slab

dalle *f* **de plancher**   floor slab

dalle *f* **de transition**   transition slab

dalle *f* **en béton**   concrete slab

dalle *f* **en corps creux**   hollow block, coffer or pot floor slab

dalle *f* **fléchie**   hollow slab

dalle *f* **flottante**   floating slab

dalle *f* **nervurée**   ribbed slab

dalle *f* **pleine**   solid slab

dalle *f* **thermoplastique**   thermoplastic or vinyl asbestos tile

dalle *f* **translucide**   glass block or brick (paving), pavement light

dalle *f* **travaillant dans un sens**   one-way slab

daller *vb*   flag, pave or slab [a pavement, with stones or slabs], tile [a floor]

dalleur *m*   pavior

dallot *m*   small paving slab

dalot *m*   opening in e.g. dock or retaining wall for drainage or escape of water, larger than a weephole

damage *m*   tamping (down)

damage, engin *m* **de**   rammer or ramming appliance

dame *f*   rammer, tamper

dame *f* **à béton**   concrete vibrator

dame *f* **vibrante**   vibratory rammer

damer *vb*   ram [earth, etc], tamp (down)

dameur *m*   rammer

dameuse *f* **mécanique**   power rammer

dans œuvre (D/O), mesure *f*   measurement between interior faces

dansante, marche *f*   winder [stair tread of varying width]

darse *f*   harbour basin [tidal]

date *f* **d'achèvement**   completion date

date *f* **limite**   deadline

dauphin *m*   rainwater shoe

de chant   on edge

de dérivation   bypass *adj*

de droit fil   with or along the grain

de fonctionnement   operating *adj*

de fond en comble   from top to bottom

de rase campagne   in the open country

de réserve   spare *adj*, standby *adj*

de secours   standby *adj*

de son domaine/ressort   within his/ her province

dé *m*   dado or die [of a pedestal], pane [individual] in a glazed panel

débarder *vb*   unload [timber, from a barge, ship]

débardeur *m* **forestier**   skidder

débarras *m*   boxroom, lumber room

débarrer *vb*   re-open [a road], unbar

débattre *vb*   discuss

débattre, à   to be discussed

débit *m*   debit [a person, an account], cutting up [of logs, etc], discharge or yield [output], output [of a motor, pump, fan, etc], (mass) flow rate, throughput

débit *m* **caractéristique moyen**   characteristic average discharge flow

débit *m* **d'air**   air flow [quantity/rate]

débit *m* **d'air de reprise**   recirculated air flow [quantity/rate]

débit *m* **d'air soufflé**   supply air flow

débit *m* **d'aspiration**   suction capacity

débit *m* **d'écoulement d'air**   air flow [quantity/rate]

débit *m* **d'évacuation**   discharge rate

débit *m* **de dose**   dose rate

débit *m* **de l'air**   air flow or throughput [volumetric]

débit *m* **de l'écoulement**   flow rate [volumetric]

débit *m* **de volume de sortie**   discharge or outlet flow rate [volumetric]

débit *m* **du ventilateur**   fan capacity

débit *m* **équilibré**   balanced flow

débit *m* **journalier**   daily flow

débit *m* **maximum**   peak load [output]

débit *m* **permanent, source** *f* **à**   constantly flowing spring

débit *m* **sur maille**   quarter-sawing [of timber from the log]

débit *m* **volumique**   flow rate [volumetric]

débiter *vb*   cut up [logs, etc], debit

débiteur *m*   debtor

débitmètre *m*   flow meter

déblai *m*   cutting [excavation, not the process]

déblai, matériau *m* **de**   excavated material

déblais *m,pl*   excavated material, muck, spoil

déblais, en   in cutting

déblais et remblais   cut and fill

déblayer *vb*   cut [earthworks]

déboisage *m*   deforestation

débordement *m*   overflow [of liquid]

déborder *vb*   overflow, run over

débordoir *m*   edge trimmer [plumber's tool]

débouché *m*   outfall, outlet [commercial, or opening]

déboucher *vb*   unblock [a hole, a pipe, etc]

déboucher *vb* **(sur)**   emerge, issue or open (on to)

débourrage *m*   removal of fill [from a hole, an excavation, etc]

débours(é) *m*   cost excluding profit and overheads, out-of-pocket expenses

debout *adj*   across the grain [of wood], upright

débranchement *m*   disconnection [electrical]

débrancher *vb*   disconnect [electrically]

débrayer *vb*   disconnect [mechanically], unsling

débrider *vb*   unhook or unfix [a lifting cable]

débris *m*   debris

débris *m* **de forage**   drilling spoil

décalage *m*   displacement backwards/forwards [in position, time], drift [from zero, of an instrument], phase difference or shift, stagger [positional], time lag [separation]

décalage *m* **de phase**   phase difference or shift

décalaminage *m*   decarbonization

décalaminer *vb*   decarbonize, descale [a casting, etc]

décalciner *vb*   descale

décaler *vb*   displace backwards/forwards [in position, time], shift the zero [of an instrument], offset or stagger [in position], unkey

décalquer *vb*   trace [a drawing, etc]

décamètre *m*   decametre [10 metres], 10-metre measuring tape

décantation *f*   sedimentation [of sewage]

décanteur *m*   sedimentation basin

décapage *m*   pickling [of metal], removal of surface layer, stripping [paint, topsoil]

décapage *m* **au chalumeau**   flame cleaning

décapage *m* **au (jet de) sable**   sandblasting

décapage *m* **de terre végétale**   topsoil stripping

décapant *m*   paint stripper

décaper *vb*   pickle [metal], remove the surface layer, strip [paint], sandblast

décapeur *m* **thermique**   heat gun [for paint stripping]

décapeuse *f*   scraper, tractor-scraper

décapeuse *f* **automotrice**   motor scraper

décapeuse *f* **avec avant-train moteur**   tractor-scraper

décapeuse *f* **élévatrice**   elevating scraper

décapeuse *f* **niveleuse**   trimmer [earthworks]

déceler *vb*   detect [leaks]

décélération *f*   deceleration

décélération, bande *f*/**voie** *f* **de**   deceleration lane

décélération, zone *f* **de**   deceleration section or zone

déceleur *m* **de fuites**   leak detector

décennal *adj*   10-year

décennale *f*   10-year guarantee [on major work, in France]

décharge *f*   discharge or discharging [electrical], dump or tip, outfall [of sewer], unloading

décharge, arc de *m*   arch built into wall above a lintel to distribute loads

décharge, égout *m* **de**   outfall sewer

décharge, lampe *f* **à**   discharge lamp

décharge, tuyau *m* **de**   discharge, exhaust or waste pipe

décharge, voûte *f* **de**   relieving arch [of a bridge]

décharge *m* **d'une poutre**   brace or strut of a beam

décharge *m* **de roche**   rock waste

décharge *m* **des eaux usées**   sewage outfall

décharge *m* **électrique**   electric shock

décharger *vb*   discharge, tip, unload

se décharger *vb*   run down [accumulator or battery]

déchausser *vb*   remove ground around foundations

déchets *m,pl*   refuse

déchets *m,pl* **(de) carrière**   quarry waste

déchets *m,pl* **recyclables**   recyclable waste(s)

décheviller *vb*   unpeg [remove pegs]

déchirer *vb*   tear [a fabric, etc]

déchirure *f*   tear [in fabric, etc]

décibel *m*   decibel

décimal *adj*   decimal

décintrage *m*   striking of arch or vault formwork

décintrer *vb*   strike [arch or vault formwork]

décisif *adj*   decisive [of an experiment, test]

décision *f* **de principe**   policy decision

déclaration *f* **de sincérité**   declaration that no other payments are involved

déclassement *m*   derating [of electrical equipment]

déclenchement *m*   break [of an electrical switch], disengagement [of gears], release, setting in motion [releasing], starting [activating, releasing]

déclenchement rapide, à   quick-release

déclencher *vb*   activate [release, set in motion], release [set off, start], disconnect, disengage, switch off, trip [disengage, switch off], unlatch [door]

déclencheur *m*   pressure switch, disconnecter [of an electrical breaker]

déclinaison *f*   declination

décoffrage *m*   formwork removal or striking

décoffrer *vb*   strike [concrete formwork]

décoincer *vb*   loosen [sth jammed]

décollement *m*   loss of adhesion [of mortar or rendering]

décollement *m* **de flamme**   flame separation

décolleuse *f*   stripper [for glued-on surfacings, carpets, etc]

décolleuse *f* **(à vapeur)**   steam stripper [for wallpaper]

décoloration *f*   discoloration

décolorer *vb*   discolour

décombres *m,pl*   debris, rubbish/rubble [from buildings]

décompactage *m*   bulking [of soil]

décomposer *vb*   decay, decompose or rot

se décomposer *vb*   decay, decompose or rot [by itself]

décomposition *f*   decay, decomposition, disintegration, rotting

décompresseur *m*   relief valve

décompression *f*   decompression

décompte *m*   deduction [from sum to be paid]

décompte *m* **final**   final account

décongélation *f*   thawing

déconnecter *vb*   disconnect [electrically, e.g. a lead]

déconnexion *f*   disconnection

déconseillé *adj*   not recommended

décontaminer *vb*   decontaminate

décor *m*   decoration [of a room, house, etc]

décorateur *m*   decorator

décorateur *m* **d'intérieur**   interior decorator

décoratif *adj*   decorative

décoré *adj*   decorated

décorer *vb*   decorate

découpage *m*   cutting out [patterns, etc]

découpage *m* **à l'oxygène**   oxygen cutting

découpage *m* **au chalumeau**   flame or gas cutting

découpage *m* **autogène**   oxyacetylene cutting

découpe *f*   cut-out [for fitting or passage of services], cutting up [wood, etc]

découpe *f* **au jet d'eau**   water-jet cutting

découper *vb*   clip, cut off, cut out [design, pattern, etc], cut up [subdivide land, a site, etc]

découpeuse *f*   shearing machine

découvert *adj*   overdrawn [bank account], uncovered

découverte *f*   removal of overburden in open-cast mining

décrassage *m*   clean(s)ing, scouring [cleaning], scraping

décrasser *vb*   clean(se), scour [clean], scale [a boiler, etc]

décrasser *vb* **(un fourneau)**   draw (a furnace)

décrépir *vb*   remove rendering

décrépitude *f*   disrepair

décret *m*   (government) decree or order

décrochement *m*   change in plane [of a facade or architectural feature], set back portion of a wall

décrochement *m* **anguleux**   angular break [e.g. in a kerb]

décroissance *f*   decrease, decrement

décroissement *m*   decrease, decrement

décrottage *m*   cleaning off mortar, etc, from used masonry materials

décrotter *vb*   clean off/down [used materials]

décrottoir *m*   boot, foot or shoe scraper [metal]

dedans *adj*   inside *adj*

dedans *m*   inside

dédosser *vb*   dress or edge timber to produce square edges

défaillance *f*   failure [e.g. of insulation, lighting]

défaillance *f* **de l'isolement**   insulation breakdown

défaillance *f* **opérationnelle**   operational fault

défaire *vb*   demolish, knock or pull down, tear down [dismantle], pull to pieces

défaut *m*   defect, fault

défaut *m* **apparent**   visible defect

défaut *m* **caché**   concealed/hidden/latent damage or defect

défaut *m* **d'isolement**   insulation defect or fault

défaut *m* **de construction**   structural defect

défaut *m* **de fabrication**   manufacturing defect or fault

défaut *m* **de laminage**   rolling defect

défaut *m* **de mise à la terre**   earth(ing) fault

défaut *m* **de surface**   surface defect or imperfection

défaut *m* **électrique**   electrical fault

défaut *m* **fugitif**   intermittent defect or fault

défaut *m* **latent/occulte**   concealed/hidden/latent damage or defect

défectueux *adj*   defective, faulty, unsound [work, etc]

défectuosité *f*   defect(iveness), fault(iness), unsoundness, flaw

défense *f*   fender [on a bridge pier]

défense *f* **mobile**   loose fender

défilé *m*   gorge, pass [between mountains]

défilement *m*   cut-off [lighting]

définitif *adj*   definitive, final or permanent

définition *f*   definition

déflecteur *m*   baffle(r) or baffle plate, deflector, guide vane

déflecteur *m* **de fumée**   smoke deflector

déflexion *f*   deflection [structural movement]

défoncement *m*   rebate [in a panel or surface]

défoncer *vb*   break or cut up [ground or road]

défonceuse *f*   router, trenching plough

défonceuse *f* **portée**   ripper [rock breaker]

défonceuse *f* **tractée**   rooter [earthworks], towed scarifier

deformabilité *f*   deformability

deformabilité *f* **d'un sol**   deformation properties of a soil

déformation *f*   buckling, deflection, deformation, distortion, strain [deformation], warping

déformation, jauge *f* **de**   strain gauge

déformations, ellipse *f* **des**   strain ellipse

déformation *f* **à chaud/froid**   hot/cold forming/shaping

déformation *f* **à la rupture**   strain at failure

déformation *f* **de cisaillement**   shear strain

déformation *f* **de retrait**   shrinkage distortion

déformation *f* **élastique**   elastic deformation

déformation *f* **permanente**   permanent deformation or distortion

déformation *f* **plane**   plane strain

déformation *f* **plastique**   plastic deformation

déformé *adj*   deformed, distorted

déformé *adj* **au delà de la limite élastique**   strained beyond the elastic limit

déformer *vb*   buckle, deform, distort, twist, warp

déformer *vb* **de manière plastique**   yield plastically

se déformer *vb*   buckle, twist or warp [by itelf]

dégagé *adj*   open [of a view]

dégagement *m*   clearance, headroom, emission, escape or release [of gas, heat, steam, etc], landing [large, upstairs], lobby, passage(way), paying off [mortgage]

dégagement *m* **de chaleur/etc**   heat/etc emission

dégagement, tuyau *m* **de**   bypass (pipe), waste pipe

dégager *vb*   clear [to free or open], unblock [a road, traffic], free [an object], emit or give off [heat, etc], release [heat, steam, etc]

dégarnir *vb*   pick or rake out [joints in masonry]

dégâts *m,pl*   damage [physical]

dégats *m,pl* **matériels**   property damage

dégauchir *vb*   dress [a wall, stone, etc], rough plane or surface, straighten [a piece of wood, metal], trim [stone, etc], true up [a piece of wood, stone]

dégauchissage *m*   dressing or trimming [of stone], straightening [of a piece of wood, metal, etc], surfacing [of wood], truing up [of a piece of wood, stone]

dégauchisseuse *f*   surface planer or planing machine [for straightening wood]

dégazer *vb*   de-gas

dégeler *vb*   defrost, thaw, unfreeze

dégivrage *m*   defrosting

dégivrer *vb*   defrost

déglacer *vb*   de-ice, defrost, melt the ice on, thaw

dégonder *vb*   take off its hinges or unhinge [a door, shutter, etc]

dégonflé *adj*   flat [of a tyre]

dégorger *vb*   press excess paint out of a brush, unblock [a pipe, sink, etc]

dégradation *f*   degradation, dilapidation, wear [of a road, etc], wear and tear [of a house, rented property, etc], weathering

dégradation *f* **(des joints)**   cleaning out (of joints) [prior to pointing]

dégradé *adj*   damaged, dilapidated, in disrepair or ruins

dégrader *vb*   damage, decay, deface, degrade, dilapidate

se dégrader *vb*   decay, dilapidate [become degraded], fall into disrepair

dégraisser *vb*   degrease [prior to painting], rough out to size or trim edges [of joinery]

dégraisseur *m*   grease remover

degré *m*   degree

degré *m* **d'efficacité**   degree of efficiency

degré *m* **d'épuration**   degree of purity

degré *m* **d'humidité**   moisture content

degré *m* **d'occupation**   occupancy rate

degré *m* **de compactage**   degree of compaction

degré *m* **de coupe-feu**   fire resistance

degré *m* **de décharge**   discharge rate

degré *m* **de dureté**   degree of hardness

degré *m* **de perméabilité**   degree of permeability

degré *m* **de pollution**   degree of pollution

degré *m* **de protection**   degree of protection

degré *m* **de résistance au feu**   fire rating

degré *m* **de saturation**   degree of saturation

degré *m* **de surchauffe**   degree of superheat

degré *m* **hygrométrique**   moisture (content) of air, relative humidity

degré-jour *m*   degree day

degrés *m,pl* **de liberté**   degrees of freedom

degrés, par   gradually

dégressivité *f* **des charges, règle** *f* **de**   loading reduction rule for multi-storey buildings

dégrillage *m*   screening [sewage]

dégrilleur *m*   screen or strainer [for sewage]

dégrossir *vb*   rough out [trim to size, of woodwork]

dégrossissage *m*   roughing out

dégueulement *m*   notching in a roof post for ends of braces, etc

déguisage *m*   thin coat of paint applied to a surface prior to filling

déguiser *vb*   apply a thin coat of paint prior to filling a surface

dehors *adj*   outside *adj*

dehors *m*   outside

dehors, en   exterior [outside]

déioniser *vb*   deionize

déjeté *adj*   buckled [metal], distorted or warped [wood]

déjeter *vb*   buckle [metal], warp [wood]

déjoutement *m*   narrowing of ends of timbers abutting same roof post

délabré *adj*   dilapidated, ruinous [building, etc]

délabrement *m*   dilapidation, disrepair

délai *m*   delay, extension of time, period of time, respite or time allowed [for completion of task, payment, etc]

délai *m* **d'exécution**   contract period, deadline

délai *m* **de livraison**   delivery period [time for delivery]

délai *m* **de réponse**   lag [in response]

délai *m* **de temporisation**   time delay [e.g. of a time lock]

délaissé *adj*   abandoned [e.g. redundant industrial building]

délardement *m*   adjustment/reduction of size to make sth fit, notching [of edge of a piece of timber]

délarder *vb*   adjust, pare down or reduce size [to make sth fit], notch [edge of a piece of timber]

délavage *m*   leaching, washing out [of colours]

délaver *vb*   leach, wash out [colours]

délayer *vb*   add water to [e.g. mixing concrete, plaster, etc]

Délégation *f* **à l'Aménagement du Territoire et à l'Action Régionale (DATAR)**   regional development agency [official national body in France]

délégué *m* **du personnel**   staff representative

délégué *m* **syndical**   shop steward, trade union representative

délestage *m*   load shedding

délesteur *m*   power switch [disconnector]

déligner *vb*   edge [a slab of wood], trim [with a saw]

déligneuse *f*   edger [saw]

délimitation *f*   delimitation, limits

délimité *adj*   demarcated

délit *m*   grain of stone at right angles to natural bedding

délit *m***, poser** *vb* **en**   lay [stone] at 90° to natural bedding plane

délitage *m*   slaking [lime]

déliter *vb*   cut or split [stone in its bedding plane], slake [lime]

delta *m*   delta [of a river]

démaigrir *vb*   remove a thin layer from or thin down [timber or stone]

démaigrissement *m*   thinning down [of timber, stone]

demandé *adj*   required

demandé, prix *m*   asking price

demande *f* **biochimique en oxygène (DBO)**   biological oxygen demand (BOD)

demande *f* **(de)**   application (for), claim (for), demand (for), request (for)

demande *f* **de chaleur**   heat requirement

demande *f* **de dérogation**   waiver request

demande *f* **de dommages-intérêts**   claim for damages

demande *f* **de froid**   refrigeration demand

demande *f* **de permis de construire**   planning application

demande *f* **de pointe**   peak demand

demander *vb* **renseignements**   make inquiries

démanteler *vb*   dismantle

démarche *f*   step [action, measure]

démarches *f,pl* **administratives**   administrative steps

démarrage *m*   starting (up) [of machines, equipment]

démarrage, bouton *m* **de**   starter button or switch

démarrage *m* **à vide**   no-load starting

démarrage *m* **delta-étoile**   star-delta starting

démarrer *vb*   start

démarreur *m*   starter

démarreur *m* **à opération directe**   direct-on-line (DOL) starter

démarreur *m* **automatique**   automatic starter

démarreur *m* **monophasé**   single-phase starter

démarreur *m* **sur fraction d'enroulement**   part-winding starter

déménagement *m*   house removal [furniture, etc], moving house

déménager *vb*   remove [moving house]

déménageur *m*   (household) remover

demeure *f*   dwelling, home [abode], house, residence

demi-botte *f*   ankle or half boot

demi-ceint *m*   half column [applied to face of a wall], semi-column

demi-double, verre *m*   thick glass

demi-ferme *f*   half truss

déminéralisation *f*   demineralization

déminéralisé *adj*   demineralized

déminéraliser *vb*   demineralize

demi-ombre *f*   half-shade

demi-opaque *adj*   semi-opaque

demi-palier *m*   quarter-space landing, square half-landing

demi-pilier *m*   half pillar [applied to face of a wall]

demi-rond *m*   half-round [moulding, section]

demi-ronde *f*   half-round file

demi-varlope *f*   jack plane

démodulation *f*   demodulation

demoiselle *f*   pavior's rammer or punner

démolir *vb*   demolish

démolisseur *m*   demolition contractor

démolition *f*   demolition

démontable *adj*   demountable, removable

démontage *m*   disassembling, dismantling, removal [taking out or down]

démonter *vb*   dismantle, strip down [a machine etc]

démoulage, produit *m* **de**   release agent [for concrete formwork]

démoulant *m*   release agent [for concrete formwork]

dénivellation *f*   contouring or determination of level [surveying], departure from level or gradients [of a roadway, etc], drop, change/difference/variation of level, difference or variation in level [of supports, etc], unequal settlement, sinking or subsidence [of supports, etc]

dense *adj*   dense

densification *f*   densification [of wood]

densimètre *m*   densimeter

densité *f*   density

densité *f* **apparente humide**   wet density

densité *f* **apparente sèche**   dry density

densité *f* **d'arrosage**   water discharge density [irrigation]

densité *f* **d'un gaz de fumée**   density of a gas/smoke

densité *f* **de flux**   flux density

densité *f* **de l'air**   air density

densité *f* **de la population**   population density

densité *f* **de saturation**   saturation vapour density [HVAC]

densité *f* **de vapeur**   vapour density [HVAC]

densité *f* **des logements**   housing density

densité *f* **des particules**   particle density

densité *f* **en vrac**   bulk density

densité *f* **Proctor**   Proctor density

densité *f* **sèche**   dry density, dry unit weight

dent *f* **de scie**   sawtooth

dénude-fil *m*   wire stripper(s)

dénuder *vb*   bare or strip [a cable or wire of insulation]

déodorant *m*   deodorant

dépanneuse *f*   breakdown [rescue] vehicle

départ *m* **d'eau**   water outlet

département *m*   county [third layer of local government, above canton in France]

départementale, route *f*   secondary road

dépassement *m* **de contrainte**   overstress

dépasser *vb*   exceed, overshoot

dépendance *f*   outbuilding

dépense *f*   expenditure, expense

dépense *f* **de chaleur**   heat consumption

dépenses *f,pl* **contrôlées**   cost-reimbursement [type of contract]

dépenses *f,pl* **d'investissement**   capital, first or investment cost

déperdition *f*   loss [of energy, heat, etc]

déperdition *f* **calorifique/thermique**   heat loss

déperdition *f* **de gaz**   escape of gas

déperdition *f* **par dispersion**   leakage [electrical]

déphasage *m*   phase difference or shift

déplacement *m*   deflection, displacement

déplacement *m* **de phase**   phase displacement

déplacement *m* **des marais à l'aide d'explosifs**   bog blasting

déplacer *vb*   displace, move, relocate, shift

dépliable *adj*   straightenable

dépoli *adj*   frosted [glass]

dépolir *vb*   frost [glass]

dépolluer *vb*   clean (up), depollute [e.g. a watercourse]

déposer *vb*   deposit

se déposer *vb*   settle [dust, etc], silt (up)

dépôt *m*   deposit [e.g. of dust], sediment, depot, warehouse

dépôt, mettre *vb* **en**   stockpile

dépôt *m* **bancaire**   bank deposit

dépôt *m* **d'explosifs**   explosive store

dépôt *m* **d'une chaudière**   boiler scale

dépôt *m* **de garantie**   guarantee deposit

dépôt *m* **de givre**   frost deposit

dépôt *m* **de poussière(s)**   dust deposit

dépouille *f*   formwork rake or taper [to facilitate striking]

dépouillement *m*   analysis [e.g. of accounts, tenders]

dépouiller *vb*   analyse [e.g. tenders, an account], sort through [e.g. mail, drawings, etc], strip [a cable, rendering from a wall, etc]

dépoussiérage *m*   dust extraction or removal

dépoussiéreur *m*   dust collector or extractor

dépoussiéreur *m* **à multicyclones**   multicyclone dry dust collector

dépoussiéreur *m* **à sec**   dry dust collector

dépoussiéreur *m* **à voie humide**   wet dust collector

dépoussiéreur *m* **électrostatique**   electrostatic precipitator

dépréciation *f*   depreciation

dépression *f*   depression

dépressurisation *f* depressurization
dérangé *adj* out of order [of a machine]
dérangement *m* disturbance
dérasement *m* lowering [of a wall, stonework, etc, to required level]
déraser *vb* cut down the top or lower the level of [a wall, a piece of stonework]
dérivation *f* bypass [pipework, or valve controlling it], diversion [of a watercourse], tapping or shunting of electrical current
dérivation, conduite *f* **de** bypass (pipe)
dérivation, excité en shunt-wound
dérivée *f* derivative, differential coefficient
dériver *vb* branch or shunt [current], derive [a mathematical function], divert [the course of]
dernier *adj* last
dernier étage *m* top floor
dernière marche *f* top step
dérobement *m* line of cut marked on face of a stone
dérochage *m* pickling [of metal], removal of rocks
dérocher *vb* pickle [metal], clear of rocks, strip paint from timber
dérocheuse *f* rock breaker
dérogation *f* special dispensation
dérouiller *vb* derust
dérouler *vb* peel [a log to produce veneer, e.g. for plywood], unwind
derrick *m* guy or pole derrick, drilling rig of timber
désaération *f* de-aeration
désagréger *vb* disintegrate
désamiantage *m* asbestos removal
désamorcer *vb* unprime [a pump, siphon]
se désamorcer *vb* run dry [of pump]
désaxer *vb* offset
descellement *m* loosening [of sth fixed into masonry, etc]
desceller *vb* loosen or remove [sth fixed into masonry, etc], unseal or remove seal from
se desceller *vb* come loose
descente *f* downpipe
descente *f* **des charges** calculation of vertical loads on a structure
descente *f* **eaux pluviales** rainwater downpipe
descriptif *m* description [of works], specification
description *f* description
description *f* **de travaux** description of works
description *f* **en détail** detailed description
désémulsionner *vb* demulsify
désenclavement *m* improvement of communications [e.g. with an isolated community]
désencombrer *vb* unblock [clear a way]
désenfumage *m* smoke clearance, fire ventilation
déshabiller *vb* remove cover strips, beadings, decorations, etc
désherbant *m* herbicide
déshuileur *m* oil collector, separator
déshumidifier *vb* dehumidify
déshumidificateur *m* dehumidifier
déshumidification *f* dehumidification
déshydratant *m* dehydratant

déshydratation *f* dehydration
déshydrater *vb* dehydrate
déshydrateur *m* dehydrator, dryer
design *m* aesthetic harmonization in architecture, furniture, etc
désignation *f* designation, name
désincrustation *f* descaling, scale removal
désincruster *vb* descale
désinfection *f* disinfection
désintégration *f* breaking up, disintegration, weathering [of rock]
désodorisant *m* deodorant
désorption *f* desorption
dessablage *m* sand trapping [sewage treatment]
dessableur *m* sand trap [sewage plant], sediment basin
desséchant *m* desiccant
desséché *adj* dried up
desséché à l'air air-dried
dessécher *vb* dry up
se dessécher *vb* run dry [spring, well, river, stream]
desserrage *m* loosening, releasing, slackening, release [undoing, of a clamp, nut, handbrake, etc]
desserré *adj* loose, slack [of a nut]
desserrer *vb* loosen, release [a clamp, nut, handbrake, etc], slacken [a nut, etc]
desserte *f* provision of transport services
desserte, voie *f* **de** access/service road
desserte, voie *f* **secondaire de** minor access/service road
desservir *vb* lead into [e.g. one space to another], provide access to, provide regular transport services to or serve with regular transport
dessiccateur *m* desiccator, dryer
dessiccation *f* desiccation, drying out
dessin *m* design or pattern, drawing or plan
dessin *m* **à l'échelle** scale drawing
dessin *m* **assisté par ordinateur (DAO)`** computer-aided draughting (CAD)
dessin *m* **au crayon** pencil drawing
dessin *m* **axonométrique** axonometric (drawing)
dessin *m* **coté** dimensioned drawing
dessin *m* **d'atelier** (work)shop drawing
dessin *m* **d'exécution** working or (work)shop drawing
dessin *m* **de détail** detail drawing
dessin *m* **de recollement** as-built drawing
dessin *m* **des plans** plan drawing
dessin *m* **en coupe** section (drawing)
dessin *m* **en élévation** elevation (drawing)
dessin *m* **en perspective** perspective (drawing)
dessin *m* **industriel** manufacturing drawing
dessin *m* **isométrique** isometric (drawing)
dessin *m* **préliminaire** preliminary design
dessin *m* **schématique** diagrammatic drawing
dessinateur *m* draughtsman
dessiner *vb* design, draw, sketch
dessous *adv* below
dessous *m* bottom, underside
dessous *m* **de poutre** beam soffit
dessus *adv* above

dessus *m*   top [side or surface]

déstabiliser *vb*   destabilize

destruction *f*   destruction

désuet *adj*   obsolete

détachable *adj*   detachable

détail *m*   detail

détail *m* **estimatif**   detail description and estimate

détail *m* **quantitatif estimatif (DQE)**   bill of quantities (BOQ or BQ), detailed quantitative estimate

détaillé *adj*   detailed

détailler *vb*   detail, enumerate, itemize [in accounts, etc]

détamponner *vb*   unplug [a duct, pipe, etc]

détartrage *m*   descaling

détartrer *vb*   descale, scale [a boiler, etc]

détartreur *m*   descaler

détecteur *m*   detector

détecteur *m* **d'incendie**   fire detector

détecteur *m* **d'intrus**   intruder detector

détecteur *m* **de contact à la terre**   earth (continuity) tester

détecteur *m* **de fuite(s)**   leak detector

détecteur *m* **de fuites d'haloïdes**   halide leak detector

détecteur *m* **de fumée**   smoke detector

détecteur *m* **de lumière**   light sensor

détecteur *m* **de mouvement**   movement detector

détecteur *m* **de présence**   occupancy detector

détecteur *m* **de pression**   pressure sensing device

détecteur *m* **de tension**   continuity tester [electrical]

détecteur *m* **immergé**   immersion detector

détection *f* **incendie**   fire detection system

détendeur *m*   expansion valve, pressure reducing valve or reducer

détendeur *m* **à double détente**   two-stage pressure reducing valve

détendeur *m* **à ressort**   spring-loaded pressure reducing valve

détendeur *m* **de pression**   pressure reducing valve

détendeur *m* **de vapeur**   steam pressure reducing valve

détendeur *m* **thermostatique**   thermostatic expansion valve

détendre *vb*   loosen, slacken, relax [stress, tension, etc]

détente *f*   pressure reduction, wedge under foot of a timber strut or stay

détergent *m*   cleaning agent, detergent

détérioration *f*   deterioration

détériorer *vb*   damage, deteriorate, spoil

détermination *f* **des efforts**   stress analysis

déterminer *vb*   determine, evaluate

déterrer *vb*   dig up or unearth [opposite of bury]

détonateur *m*   detonator [blasting]

détonateur *m* **électrique**   electric blasting cap [detonator]

détrempé *adj*   annealed

détremper *vb*   anneal, lose temper or soften [steel]

détritus *m*   debris, refuse, waste [rubbish]

détritus *m* **récupérable**   reclaimable waste

détruire *vb*   destroy

dette *f*   accounts payable, debt

devant *adv*   in front of

devant *m*   front

devanture *f*   frontage

devanture *f* **de magasin**   shop front or window

développement *m*   development [building, construction]

développer *vb*   develop

déverrouillage *m*   unbolting or unlocking

déverrouillage *m* **manuel**   manual override

déverrouiller *vb*   unbolt [a door, etc]

dévers *adj*   leaning, out of plumb, sloping

dévers *m*   banking, crossfall or super-elevation [of a road], inclination [slant, banking], slope [banking, and of a wall, etc], out of plumb, twist or warp(ing) [of wood]

déversé *adj*   sloping, banked

déversement *m*   discharge or overflow [of liquid], lateral buckling

déverser *vb*   incline, lean [at an angle], slant, bank [road at a bend], slope [bank a road, slant a wall], warp [wood]

déversoir *m*   overflow [pipe, chute], spillway, weir [of a dam]

déversoir *m* **d'eau de pluie**   stormwater overflow

dévêtissement *m*   clearance for placing a structural fixing

déviation *f*   diversion, deviation, leakage [of electric current]

déviation *f* **permanente**   permanent set

déviation *f* **standard**   standard deviation

dévidoir *m*   hose reel

devis, hors   not included in estimate

devis *m*   estimate, quote

devis *m* **descriptif**   estimate of quantities and costs

devis *m* **descriptif des travaux**   estimate of the cost of the works

devis *m* **détaillé**   detailed estimate

devis *m* **estimatif**   cost estimate, preliminary estimate

devis *m* **quantitatif**   bill of quantities

dévissage *m*   release, unscrewing

se dévisser *vb*   come unscrewed

dévisser *vb*   loosen, unscrew

dévoiement *m*   diversion or offset [of a pipe, flue, etc]

diabase *f*   diabase

diagonale *f*   cross-brace, strut, diagonal

diagramme *m*   diagram

diagramme *m* **contrainte–déformation**   stress–strain diagram

diagramme *m* **de Carrier**   psychrometric chart

diagramme *m* **de circulation**   flow diagram [pipework]

diagramme *m* **de confort**   comfort chart

diagramme *m* **de l'effort tranchant**   shear diagram

diagramme *m* **de Mollier**   Mollier diagram

diagramme *m* **de plasticité**   plasticity chart

diagramme *m* **de pression de vapeur**   vapour pressure diagram

diagramme *m* **des contraintes**   stress diagram

diagramme *m* **effort–déformation**   load–extension diagram, stress–strain diagram

diagramme *m* **en coordonnée polaire**   polar diagram

diagramme *m* **enthalpique**   enthalpy diagram

diagramme *m* **entropique**   entropy diagram

diagramme *m* **psychrométrique**   psychrometric chart

diagramme *m* **x–y de CIE**   CIE colour triangle

diamant *m* diamond, glass cutter
diamétral *adj* diametrical
diamètre *m* diameter
diamètre *m* **à fond de filet** core diameter [of a bolt]
diamètre *m* **absolu** clear diameter
diamètre *m* **des grains** grain size
diamètre *m* **efficace** effective diameter
diamètre *m* **extérieur/intérieur/nominal**
  exterior/interior/nominal diameter
diamètre *m* **primitif** pitch diameter [of a gear wheel]
diaphragme *m* diaphragm, orifice plate
diaphragme *m* **en béton** concrete diaphragm, grout
  curtain
diapositive *f* transparency [photographic slide]
diatomées *f,pl* diatomaceous earth or diatomite
diatomite *f* diatomite
dichroïque *adj* dichroic
dichroïque, lampe *f* **à réflecteur** dichroic reflector lamp
diélectrique *adj* dielectric
diesel *m* diesel [engine, or fuel]
différence *f* difference
différence *f* **de pression, etc** difference of pressure, etc
différence *f* **moyenne de température, etc** mean
  temperature, etc difference
différence *f* **psychrométrique** wet bulb depression
différencier *vb* differentiate
différend *m* **(à)** disagreement or dispute (about)
différentiel *m* differential
différer *vb* differ, defer, postpone
diffraction *f* diffraction
diffusant *adj* diffusing [lighting]
diffuser *vb* distribute [e.g. document], diffuse or spread
  [heat, light, news, etc]
diffuseur *m* diffuser [light or ventilation]
diffuseur *m* **à fente** slot diffuser
diffuseur *m* **(d'extincteur automatique d'incendie)**
  sprinkler head
diffuseur *m* **d'air** air diffuser
diffuseur *m* **de plafond/plafonnier** ceiling diffuser
diffusion *f* diffusion or spreading, distribution [e.g. of
  air]
diffusivité *f* **thermique** thermal diffusivity
digesteur *m* digester [sewage], sludge digestion plant
digestion *f* **(des boues)** (sludge) digestion
digital *adj* digital
digitaliseur *m* digitizer
digue *f* dike, embankment, flood bank
digue *f* **de rivage** riverside dike
dilapidation *f* misappropriation
dilatation *f* dilatation, expansion, extension [expansion]
dilatation, joint *m* **de** expansion joint
dilatation *f* **linéaire** linear expansion
dilatation *f* **thermique** thermal expansion
dilater *vb* cause to expand
se dilater *vb* expand [of itself]
diluant *m* dilutant, thinner
diluer *vb* dilute

dilution *f* dilution
dimension *f* dimension or size, measurement
dimension *f* **brute** raw [unsawn] size
dimension *f* **brute des particules** particle size
dimension *f* **de(s) grain(s)** grain size [in soil samples]
dimension *f* **de conduit** duct size
dimension *f* **du grain** grain size [in steel]
dimension *f* **hors tout** overall dimension or size
dimension *f* **inférieure à la normale** undersize
dimension *f* **intérieure** internal dimension
dimension *f* **variable** variable dimension
dimension, mettre *vb* **à** size [e.g. a hole]
dimensionnement *m* dimensioning or sizing [of ducts,
  pipes, structural members, etc], proportioning [size]
dimensions (de qch), prendre *vb* **les** measure (sth)
déminéralisation *f* demineralization
déminéraliser *vb* demineralize
diminution *f* decrease, lessening, reduction, lowering [of
  price], tapering [decreasing, reduction]
diplôme *m* **d'architecte** architect's diploma [in France]
diplôme *m* **d'ingénieur** engineering diploma [in France]
diplômé par le gouvernement (DPLG) with a state
  diploma [official designation, may apply to architects,
  surveyors]
directeur *m* director, manager
directeur *m* **administratif** administrative director
directeur *m* **commercial** business manager
directeur *m* **de travaux** project manager or director
directeur *m* **technique** technical manager or director
direction *f* direction, management, range [surveying],
  strike [geological]
direction, signal *m* **de** direction sign [road]
direction *f* **de l'écoulement** flow direction
direction *f* **de projet** project management
direction *f* **des travaux** supervision of the works
Direction *f* **Départementale de l'Équipement (DDE)**
  county planning, housing and highways department [in
  France
Direction *f* **Générale des Impôts (DGI)** Inland Revenue
  [French equivalent]
directionnel *adj* directional [e.g. lighting]
directrice *f* director [female], manageress
diriger *vb* direct, manage
discontacteur *m* breaker-contactor
discontinu *adj* discontinuous
discordance *f* discordance or dissonance [acoustics],
  unconform(abil)ity [geology]
disjoncteur *m* circuit- or contact breaker
disjoncteur *m* **à bilame** bimetallic contact breaker
disjoncteur *m* **à maximum** overload [cut-out, trip]
disjoncteur *m* **bipolaire** two-pole circuit-breaker
disjoncteur *m* **d'écoulement** flow switch [cut-out]
disjoncteur *m* **de branchement** consumer outlet circuit-
  breaker
disjoncteur *m* **de fuite de terre** earth leakage circuit-
  breaker
disjoncteur *m* **de perte à la terre** earth leakage circuit-
  breaker
disjoncteur *m* **de sécurité** safety cut-out

disjoncteur *m* différentiel   earth leakage circuit-breaker
disjoncteur *m* général   main circuit-breaker
disjoncteur *m* miniature   miniature circuit-breaker
disjoncteur *m* principal   main circuit-breaker
dislocation *f*   breaking up or dismantling, dislocation, dislodgement [of parts of a structure], fault [geological]
dispersant *m*   dispersant
disperser *vb*   disperse, scatter [light]
dispersion *f*   dispersion, liquid containing fine particles in suspension, scatter(ing) [of light]
dispersion *f* des contraintes   stress dispersion or dissipation
dispersion *f* des résultats   scatter of results
disponibilité *f*   availability
disponibilité *f* technique   serviceability [of equipment]
disponible *adj*   available
dispositif *m*   device, equipment, mechanism
dispositif *m* antiphare   anti-dazzle/-glare screen [roads]
dispositif *m* antistatique   anti-static device
dispositif *m* antivibratoire   anti-vibration device
dispositif *m* automatique de sécurité   automatic safety device
dispositif *m* d'alarme   alarm (device)
dispositif *m* d'alerte   warning device
dispositif *m* d'allumage   igniter
dispositif *m* d'arrêt (automatique)   shut-off device
dispositif *m* d'avertissement   warning device
dispositif *m* d'écartement   offset bracket [for safety barriers], spacer
dispositif *m* d'essai   testing equipment
dispositif *m* d'étanchéité   seal
dispositif *m* d'insonorisation   sound attenuator
dispositif *m* d'obturation coupe-feu   fire damper
dispositif *m* de commande   control gear
dispositif *m* de compensation de dilatation   expansion bellows
dispositif *m* de couplage   switchgear
dispositif *m* de déconnexion   cut-out device
dispositif *m* de dégivrage   defrosting device
dispositif *m* de délestage   unloading device or mechanism
dispositif *m* de démarrage   starter device or gear
dispositif *m* de démarrage automatique   automatic starter
dispositif *m* de dosage   batching equipment
dispositif *m* de fixation   clamping device
dispositif *m* de mesure   measuring device or equipment
dispositif *m* de pare-soleil   shading device
dispositif *m* de pesage *m*   weigh batcher [concrete]
dispositif *m* de projection de béton   shotcreting unit
dispositif *m* de protection   guard [safety], safety device
dispositif *m* de sectionnement   isolating device
dispositif *m* de sécurité   safety device
dispositif *m* de sécurité de manque d'eau   low water level cut-out
dispositif *m* de sécurité par coupure   safety cut-out
dispositif *m* de verrouillage   locking device
dispositif *m* de vidange   draining device

dispositifs *m,pl* d'étanchéité   waterproofing [of a road bridge]
disposition *f*   arrangement, disposition, layout, pattern
disposition *f* des lieux   site arrangements [for installation]
dispositions *f,pl*   measures [arrangements]
disque *m*   disc/disk, plate [circular]
disque *m* d'embrayage   clutch plate or disc
disque *m* de papier abrasif   paper sanding disc
disque *m* de refroidissement   cooling disc
disque *m* de sécurité   bursting disc
disque *m* dur   hard disk [computer]
disquette *f*   diskette, floppy disk [computer]
dissemblable *adj*   dissimilar
dissipation *f*   dissipation [e.g. of floodwater]
dissipation *f* des contraintes   stress dissipation or dispersion
dissolution *f*   decomposition, dissolving [of sth in a liquid], solution [result of dissolving]
dissolvant *adj*   solvent
dissolvant *m*   solvent
dissoudre *vb*   dissolve
dissous *adj*   dissolved
dissuader *vb* les intrus   deter intruders
distance, commande *f* à   remote control
distance *f*   distance
distance *f* au bord   edge distance
distance *f* cumulée   cumulative distance
distance *f* d'arrêt   stopping distance
distance *f* d'espacement   pitch or spacing distance
distance *f* de freinage   braking distance
distance *f* de visibilité   sight or visibility distance
distance *f* entre les supports   support spacing distance
distancier *m*   spacer [reinforcement]
distillation *f*   distillation
distillé *adj*   distilled
distiller *vb*   distil
distorsion *f*   distortion
distribuer *vb*   distribute
distributeur *m*   distribution board [electrical], header [pipe or duct]
distributeur *m* d'essence   petrol pump
distributeur *m* (de savon, vapeur, etc)   distributor (soap, steam, etc)
distribution *f*   division [sharing] or distribution
distribution, voie *f* de   primary distributor [road]
distribution *f* d'air/de l'air   air distribution
distribution *f* d'eau   water supply
distribution *f* d'eau potable   drinking water supply
distribution *f* d'eau rurale   rural water supply
distribution *f* de chaleur   heat distribution
distribution *f* de chaleur à distance   district heating
distribution *f* de la température   temperature distribution
distribution *f* de pression   pressure distribution
distribution *f* des contraintes   stress distribution
distribution *f* en chandelle   branched parallel rising main system of water distribution
distribution *f* en parapluie   drop system of water distribution, overhead distribution [of services]

distribution *f* **inférieure**   distribution from below
distribution *f* **par le haut**   distribution from above, overhead distribution [of services]
distribution *f* **urbaine de froid**   district cooling
distribution *f*, **voie** *f* **de**   primary distributor [road]
diurne *adj*   diurnal
divers *adj*   miscellaneous
diversité *f*   diversity
dividende *m*   dividend
divisible *adj*   divisible [e.g. a plot]
division *f*   division [mathematical]
division *f* **(en)**   division (into)
dock *m*   dockside warehouse, unloading dock
document *m*   document
document *m* **quantitatif estimatif (DQE)**   bill of quantities (BOQ or BQ), detailed quantitative estimate
document *m* **technique unifié (DTU)**   code of practice for building works [in France]
documentation *f*   documentation, documents
documentation *f* **étrangère**   foreign literature [documentation]
documentation *f* **technique**   technical literature
doler *vb*   pare or shave [a piece of wood to required thickness], square [a piece of wood by adzing]
doline *f*   sink-hole [geology]
doloire *f*   adze
domaine *m*   estate, property [land], field [of action, of a science, etc], department [area of activity]
domaine, de son   within his/her province
domaine *m* **critique**   critical range
domaine *m* **d'application/d'emploi**   area or field of application
domaine *m* **de dispersion**   range of scatter
domaine *m* **de régime transitoire**   transition region or zone
domaine *m* **de températures/etc**   range of temperature/etc
domaine *m* **plastique**   plastic range
domaine *m* **public**   public domain [i.e. not copyright]
dôme *m*   dome
dôme *m* **à bulbe**   imperial dome [roof]
domestique *adj*   domestic
domicile *m*   abode, dwelling place, domicile [legal], residence [place of]
dommage *m*   damage or injury
dommage *m* **ouvrage**   insurance taken out by client for building works [in France]
dommages *m,pl* **corporels**   physical injury
dommages *m,pl* **matériels**   property damage
dommages-intérêts *m,pl*   damages [legal]
domotique *f*   house automation [intelligent control systems]
donjon *m*   keep [of a fortress], turret [architecture]
données *f,pl*   data
données *f,pl* **analogiques**   analogue data
données *f,pl* **caractéristiques**   characteristic data
données *f,pl* **critiques**   critical data
données *f,pl* **du projet**   design conditions
données *f,pl* **en place**   field data

données *f,pl* **météorologiques**   weather data
donne-jour *m*   small opening to admit light
donner *vb* **suite à une commande**   process an order
dormant *m*   casing [of a door or window], frame [door, window], dead bolt, dead light, fixed fanlight
dormant *m* **de fenêtre**   window jamb [pier or post]
dortoir *m*   dormitory
dorure *f*   gilding
dos *m*   back
dos *m* **d'âne**   sleeping policeman
dosage *m*   dosage, gauging [of cement], proportioning [in mixtures], rate of spread [of tars, bitumen, etc]
dosage *m* **de constituants**   batching of constituents [of concrete]
dosage *m* **de mélange**   mixing rate
dosage *m* **en ciment**   cement content [of concrete]
dosage *m* **en eau**   water content [of concrete]
dose *f*   amount, content, rate of application
doser *vb*   batch [concrete], measure out [a dosage], mix [in correct proportions], proportion correctly [mixtures]
doseur *m*   measure [for liquids, powders], metering system [in mixing plant], weigh batcher [concrete]
dosse *f*/**dosseau** *m*   slab [cut from the outside of a log]
dosseret *m*   back of a tenon saw, buttress, pier in a wall [reinforcing], splashback [above kitchen worktop, lab bench]
dossier *m*   documents or file
dossier *m* **adapté**   project health and safety archive
dossier *m* **d'appel d'offres**   tender documents
dossier *m* **de consultation (des entreprises) (DCE)**   tender documents
dossier *m* **de la solution d'ensemble préconisée**   scheme stage design report
dossier *m* **de permis de construire**   planning application documents
dossier *m* **des ouvrages exécutés (DOE)**   as-built records, record of finished work
dossier *m* **technique**   detail design report
doublage *m*   lining [of a wall, to improve strength, insulation, etc]
doublage *m* **sec**   dry lining
double *adj*   double
double effet, à   double-acting *adj*
double fenêtre *f*   double glazed window
double huit, croisement *m* **en**   cloverleaf intersection or junction
double manchon *m* **mâle/mâle**   double male reducing coupler/coupling
double porte *f*   double door
double vitrage *m*   double glazing
doubleau *m*   ceiling beam, double floor joist
doubleau, arc *m* **en**   transverse arch
doubler *vb*   double
doublette *f*   oak board 5.5 × 32.5 mm [for stair treads]
douce, eau *f*   soft water
douche *f*   shower
douchette *f*   shower head [hand held]
douci *m*   polishing of mirrors
doucine *f*   cyma or ogee moulding [concave above, convex below], ogee or moulding plane

doucine *f* **renversée**   ogee moulding [convex above, concave below]

doucir *vb*   polish [a mirror]

doucissage *m*   polishing [of plate glass, to produce parallel faces]

douelle *f*   soffit [of an arch or vault]

douelle *f* **extérieure**   extrados

douelle *f* **intérieure**   intrados

douille *f*   bush [bearing, collar], bulb socket, lamp holder

douille *f* **à baïonnette**   bayonet fitting [lamp cap/socket]

douille *f* **à plafond**   ceiling lamp holder or socket fitting

douille *f* **pour lampe fluorescente**   fluorescent lamp socket fitting

douve *f*   moat [of a fortress]

dragage *m*   dredging

drag(ue)line *f*   dragline

dragline-marcheuse *f*   walking dragline

drague *f*   dredge(r)

drague *f* **à câble**   dragline

drague *f* **à godets**   bucket dredger

drague *f* **aspirante/suceuse**   suction dredger

draguer *vb*   dredge

drain *m*   drain

drain *m* **d'interception**   catch or collector drain, interceptor (drain)

drain *m* **de la fondation**   subsoil drain

drain *m* **de pierraille**   French drain

drain *m* **de sable**   sand drain

drain *m* **en grès**   earthenware drain

drain *m* **en sable**   sand drain

drain *m* **français**   French drain

drain *m* **pleureur**   weeping drain

drain *m* **profond**   subsoil drain

drain *m* **souterrain**   buried drain

drainage *m*   drainage [of ground]

drainage, bassin *m* **de**   drainage basin

drainage, chemin *m* **de**   drainage path

drainage, galerie *f* **de**   drainage gallery

drainage, tuyau *m* **de**   drainage pipe

drainage *m* **de surface**   surface drainage

drainage *m* **souterrain**   subsurface drainage

drainage *m* **taupe**   mole drainage

drainant *adj*   drainage *adj*

drainé *adj*   drained

drainer *vb*   drain

drain-taupe *m*   mole-drain

dressage *m*   adjustment [of size, position, shape], alignment, drafting or drawing up [a plan, a report], dressing [of a piece of stone, timber, etc], erection [of scaffolding, a monument, etc], straightening [of a component, a piece of material], truing up

dresser *vb*   adjust, align, line or true up [a machine], square [stone]

se dresser *vb*   stand or tower up [building, monument, peak, etc]

dresser *vb* **les plans de**   plan [a house, an estate, etc]

dressing *m*   dressing room

droit *adj*   straight, true, upright

droit fil, de   along or with the grain

droit *m*   authority [to do sth], duty, tax, law, plumb [of a wall, etc], right

droit *m* **d'enregistrement**   registration duty

droit *m* **de bail**   letting duty [in France]

droit *m* **de préemption**   pre-emptive right [generally to acquire, in France]

droit *m* **de renouvellement**   security of tenure

droits *m,pl*   charges [administrative], dues

droit *m* **de timbre**   stamp duty

droite *f*   straight line

droite, ligne *f*   straight line

droite, marche *f*   straight step [of a staircase]

du bas   from below [on pipework drawing]

du haut   from above [on pipework drawing]

ductile *adj*   ductile

ductilité *f*   ductility

dumper *m*   dump-truck, dumper

dune *f*   dune, sand dune

duplex *m*   (duplex) apartment

dur *adj*   hard, tough

durabilité *f*   durability

durable *adj*   durable

duralumin *m*   duralumin(ium)

durci *adj*   hardened

durcir *vb*   harden, toughen

durcissement *m*   hardening

durcissement *m* **du béton**   hardening of concrete

durcissement *m* **par écrouissage**   hardening by cold working

durcisseur *m*   hardener [for cement, glues]

duramen *m*   heartwood

durée *f*   duration

durée *f* **coupe-feu réglementaire**   required fire resistance (period)

durée *f* **de contrat**   contract period

durée *f* **de marche**   operating time

durée *f* **de mise en température**   heating period

durée *f* **de prise**   curing period or time

durée *f* **de vie**   life [length of]

durée *f* **de vie effective/réelle**   effective life

durée *f* **de réparation**   mean time to repair

durée *f* **effective de service**   actual working period

dureté *f*   hardness [metal, water, etc]

dureté *f* **de l'eau**   water hardness

dureté *f* **permanente**   permanent hardness

dureté *f* **résiduelle**   residual hardness

dureté *f* **temporaire**   temporary hardness

dureté *f* **totale**   total hardness

duromètre *m*   rebound hardness measurer

dynamique *adj*   dynamic

dynamomètre *m*   dynamometer

E

eau *f* water
eau, jet *m* d' fountain
eau, prise *f* d' tap or valve [for drawing off water], water intake
eau, teneur *f* en water content
eau, venue *f* d' inrush of water
eau *f* aval tailwater
eau *f* brute raw or untreated water
eau *f* capillaire capillary water
eau *f* condensée condensate
eau *f* courante running water
eau *f* d'alimentation de chaudière boiler feed water
eau *f* d'appoint make-up water
eau *f* d'égout sewage water
eau *f* d'équilibre, teneur *f* en equilibrium moisture content
eau *f* d'hydratation water of hydration
eau *f* d'infiltration infiltration water
eau *f* de chauffage heating water
eau *f* de condensation condensation water
eau *f* de condenseur condenser water
eau *f* de dilution make-up water [in treatment unit]
eau *f* de distribution mains or tap water
eau *f* de gâchage mixing water [for mortar, concrete, etc]
eau *f* de gravité free water
eau *f* de Javel bleach
eau *f* de percolation seepage water
eau *f* de puits well water
eau *f* de refroidissement cooling water
eau *f* de rétention absorbed or retained water
eau *f* de retour recirculated water
eau *f* de ruissellement surface water
eau *f* de surface surface water
eau *f* de traitement industriel process water
eau *f* de ville mains or tap water
eau *f* disponible available moisture
eau *f* distillée distilled water
eau *f* douce fresh or soft water
eau *f* dure hard water
eau *f* filtrée filtered water
eau *f* fixée bound water
eau *f* frache fresh water
eau *f* glacée chilled/iced water
eau *f* interstitielle pore water
eau *f* interstitielle, pression *f* d' pore water pressure
eau *f* libre free water
eau *f* liée bound water
eau *f* morte dead storage [water]
eau *f* phréatique ground water
eau *f* potable drinking water
eau *f* pour incendie fire-fighting water
eau *f* pour la lutte contre l'incendie fire-fighting water
eau *f* pulvérisée spray water

eau *f* salée salt water
eau *f* saumâtre brackish water
eau *f* séléniteuse sulphate bearing groundwater
eau *f* souterraine groundwater
eau *f* souterraine, abaissement de l' groundwater lowering
eau *f* souterraine en circulation gravitational groundwater
eau *f* souterraine stagnante stagnant groundwater
eau *f* stagnante stagnant water
eau *f* superficielle surface water
eau *f* traitée/non traitée treated/untreated water
eau *f* utilisable effective water storage capacity, usable water
eaux *f,pl* d'égout sewage
eaux *f,pl* d'égout brutes raw sewage
eaux *f* de pluie rainwater, stormwater
eaux *f,pl* de ruissellement surface water run-off
eaux *f,pl* de retour return water
eaux *f,pl* ménagères domestic waste water
eaux *f,pl* mortes stagnant water
eaux *f,pl* pluviales (EP) rainwater, stormwater
eaux *f,pl* usées (EU) waste water, sewage
eaux *f,pl* usées domestiques domestic sewage
eaux *f,pl* usées industrielles industrial sewage
eaux *f,pl* vannes (EV) effluent [from a cesspit, septic tank, factory], foul or toilet/lavatory waste water
ébarber *vb* deburr or fettle [castings], remove part of a moulding to make a 90° joint
ébarbeur *m* trimming shears [for metal]
ébardoir *m* triangular shave hook
ébauche *f* outline [written], rough outline or sketch, roughing out or sketching out
ébaucher *vb* rough out [a plan, etc], sketch
ébéniste *m* cabinet maker
ébénisterie *f* cabinet making
éblouir *vb* dazzle
éblouissant *adj* dazzling
éblouissement *m* dazzle [from headlights], glare
éblouissement, indice *m* d' glare index
ébouillanter *vb* scald
éboulement *m* caving in, collapsing, rockfall
éboulement *m* (de terre) landslide or landslip
éboulis *m* debris or rubbish [earth, etc], talus
ébousiner *vb* remove soft parts of dimension stone
ébrasement *m* door or window jamb [splay]
ébrasure *f* door or window jamb [splay]
ébullition *f* boiling, ebullition
écaillage *m* flaking (off) [of paint or other surface], spalling
écailler *vb* spall, scale [a boiler. etc]

s'écailler *vb*   flake or scale off [by itself], peel off [paint, etc, by itself]

écale *f*   paving stone 60–80mm thick

écales *f,pl*   road metal

écart, à l'   in the background, out of the way, to one side

écart *m*   deviation, difference [between numbers, temperatures, etc], dimensional variation [in relation to size specified], distance [apart], gap, spacing, margin [amount], variation [difference]

écart *m* **admissible**   allowable or permissible deviation or variation

écart *m* **au feu**   distance between inside of a flue and nearest wooden elements

écart *m* **de réglage**   control deviation

écart *m* **de température**   temperature difference

écart *m* **du point de rosée**   dew point depression

écart *m* **moyen**   average deviation

écart *m* **quadratique moyen**   mean square deviation

écart *m* **type**   standard deviation

écartement *m*   clearance

écartement *m* **(de, entre)**   distance, gap, space or spacing (between), separation (between) [distance]

écartement, dispositif *m* **d'**   offset bracket [for highway safety barriers], spacer

écartement *m* **des ailettes**   fin spacing

écartement *m* **des électrodes**   plug gap

écartement *m* **des rivets**   rivet pitch or spacing

écarteur *m*   spacer, separator [reinforcement], distance piece

échafaudage *m*   falsework, scaffolding [and erection of scaffolding]

échafaudage *m* **à chaise**   truss-out scaffolding

échafaudage *m* **à/sur échelles**   ladder scaffolding

échafaudage *m* **cantilever**   cantilever or outrigger scaffolding

échafaudage *m* **de/sur pied**   ground scaffolding

échafaudage *m* **en encorbellement/porte-à-faux**   cantilever or outrigger scaffolding

échafaudage *m* **en tubes d'acier**   steel tube or tubular steel scaffolding

échafaudage *m* **métallique tubulaire**   steel tube or tubular steel scaffolding

échafaudage *m* **par échelles et taquets**   ladder jack scaffolding

échafaudage *m* **roulant**   mobile or tower [wheeled] scaffolding

échafaudage *m* **sur cadre**   frame or square scaffolding

échafaudage *m* **sur consoles**   truss-out scaffolding

échafaudage *m* **sur taquets d'échelles**   ladder jack scaffolding

échafaudage *m* **sur tréteaux**   trestle scaffolding

échafaudage *m* **suspendu**   hanging stage [scaffolding], projecting or suspended scaffolding

échafaudage *m* **volant**   flying scaffolding, travelling cradle

échancrure *f*   indentation [e.g. in a coast line], notch

échange *m*   exchange

échange *m* **d'énergie**   energy exchange

échange *m* **d'ions**   ion exchange

échange *m* **de chaleur**   heat exchange

échanger *vb*   exchange

échangeur *m*   exchanger [heat], interchange [highway]

échangeur *m* **à condensation**   flue condenser

échangeur *m* **à courants croisés**   cross-flow heat exchanger

échangeur *m* **à flux croisés**   cross-flow heat exchanger

échangeur *m* **autoroutier**   motorway interchange

échangeur *m* **de chaleur**   heat exchanger

échangeur *m* **de chaleur à contre-courants**   counterflow heat exchanger

échangeur *m* **de chaleur air–gaz**   gas–air heat exchanger

échangeur *m* **de chaleur de gaz d'échappement**   waste gas heat exchanger

échangeur *m* **en trèfle**   cloverleaf intersection or junction [highway]

échangeur *m* **rotatif**   rotary heat exchanger

échangeur *m* **thermique**   heat exchanger

échangeur *m* **thermique à contact direct**   direct contact heat exchanger

échantignole *f*   block or cleat [fixed to a principal rafter to prevent a purlin slipping], brick 220×110×30mm used for fireplaces

échantillon *m*   test piece, sample or specimen

échantillon *m* **carotté**   drill core sample

échantillon *m* **d'eau**   water sample

échantillon *m* **d'essai/pour essai**   test piece

échantillon *m* **de sol**   soil sample

échantillon *m* **non remanié**   undisturbed sample

échantillon *m* **prélevé par hasard**   control sample, random test piece

échantillon *m* **remanié**   disturbed sample

échantillons, pavés *m,pl*   large sett paving

échantillonnage *m*   sampling

échantillonneur *m*   sampler

échappée *f*   headroom (in a staircase or passage), stair clearance

échappement *m*   exhaust [escape]

échappement *m*   headroom (in a staircase or passage), stair clearance

échappement *m* **d'air**   air outlet

écharde *f*   splinter [of wood]

écharpe *f*   brace [on a boarded shutter or door], diagonal tie [roof]

échasse *f* **(d'échafaudage)**   scaffolding pole [vertical]

échauffement *m*   overheating, temperature rise

échauffement *m* **limite**   maximum temperature rise

échauffer *vb*   fire [a boiler], overheat

échauffure *f*   breakdown of wood attacked by fungi

échéance *f*   (date of) maturity, (due) date of payment, settlement date, expiry [of a time limit], term [lapse of time], redemption date [of a loan], target date

échéance, payable à l'   payable when due

échelage *m*   easement to erect ladder(s) for repairs on another's land

échelle, à grande   large-scale

échelle *f*   ladder, range [scale, e.g. of prices, salaries], scale [of a map, ruler, thermometer, etc]

échelle *f* **à coulisse**   extending or extension ladder

échelle *f* à incendie/d'incendie   fire escape (ladder)
échelle *f* coulissante   extending or extension ladder
échelle *f* de Beaufort   Beaufort scale
échelle *f* de couvreur   roof(er's) ladder
échelle *f* de dureté   hardness scale
échelle *f* de Fahrenheit   Fahrenheit scale
échelle *f* de meunier   narrow, steep flight of stairs, staircase without risers
échelle *f* de réduction   scale ruler
échelle *f* de sauvetage   fire escape ladder
échelle *f* de visite   inspection ladder
échelle *f* double   stepladder
échelle *f* escamotable   extending or extension ladder
échelle *f* fixe   fixed ladder
échelle *f* isolante   insulated ladder
échelle *f* mécanique   mechanical ladder
échelle *f* plate   roof(er's) ladder
échelle *f* pliante   folding ladder, stepladder
échelle *f* roulante   tower ladder or ladder tower [wheeled]
échelon *m*   rung or step iron [of a ladder]
échelonné *adj*   stepped [spaced out, staggered]
échelonnement *m*   gradation, spacing (out) or staggering [spacing]
échelonner *vb*   spread out [e.g. payments, over time]
échenal *m*   wooden gutter
échiffre *f*   string wall
échine *f*   ovolo moulding [on a capital]
écimer *vb*   demolish upper part of a structure
éclair *m*   lightning
éclairage *m*   illumination, lighting
éclairage *m* autoroutier   motorway lighting
éclairage *m* d'ambiance   ambient lighting
éclairage *m* d'appoint   task lighting
éclairage *m* d'effet   effect lighting
éclairage *m* de fond   background illumination
éclairage *m* de scène   stage lighting
éclairage *m* de secours   emergency lighting
éclairage *m* de sécurité   safety or security lighting
éclairage *m* des routes   road lighting
éclairage *m* du lieu de travail   workplace lighting
éclairage *m* moyen   average illumination
éclairage *m* routier   road lighting
éclairage *m* tamisé   soft lighting
éclairage *m* urbain   street lighting
éclairage *m* vertical   overhead lighting
éclairagiste *m/f*   lighting engineer
éclairement *m*   illuminance
éclairer *vb*   illuminate, light
éclat *m*   chip or splinter [of wood, stone]
éclatement *m*   bursting [boiler, pipe, tyre, etc], explosion [e.g. of a boiler]
éclater *vb*   burst [e.g. tyre], explode, shatter [glass], split [wood]
éclimètre *m*   clinometer
éclissage *m*   connection with fishplates
éclisse *f*   butt strap, fishplate, splice plate
écluse *f*   lock [e.g. on canal]
écluse *f* d'air   air lock
écoinçon *m*   corner piece [wood or stone]

école *f*   school [buildings]
écologie *f*   ecology
écologique   ecological
écologiste *m/f*   ecologist
économie *f*   economics, economy, saving
économie *f* d'énergie   energy economics, economy or saving
économie *f* de combustible   fuel economy or saving
économie *f* de chauffage   heating economics/economy
économie *f* thermique   heat or thermal economy
économique *adj*   economic(al) [i.e. cheap or good value for money]
économique, science *f*   economics
économique *f*   economics
économiser *vb*   cut down or economize on [e.g. heating]
économiseur *m*   economizer
économiseur *m* d'eau alimentaire   feed water economizer
économiste *m/f*   economist, quantity surveyor [near French equivalent]
écoperche *f*   gin pole, scaffolding pole [wood or metal, may have pulley at upper end]
écorce *f*   bark
écornure *f*   chipped corner or edge [of a stone, etc]
écouenne *f*   plumber's rasp
écoulement *m*   current or discharge [gas, liquid], drainage [flow], (out)flow, run-off, waste pipe [of bath, shower, etc]
écoulement *m*   yielding
écoulement, limite *f* d'   yield point
écoulement *m* annulaire   annular flow
écoulement *m* d'air   air current or flow
écoulement *m* d'air stratifié   stratified air flow
écoulement *m* de chaleur   heat flow
écoulement *m* de retour   flow reversal, reverse flow
écoulement *m* de transition   transition flow
écoulement *m* des eaux   (water) run-off
écoulement *m* du trafic   traffic flow
écoulement *m* en spirales   spiral flow
écoulement *m* en vortex   vortex flow
écoulement *m* laminaire   laminar or streamline flow
écoulement *m* non perturbé   streamline flow
écoulement *m* plastique   plastic flow
écoulement *m* selon la ligne de courant   streamline flow
écoulement *m* turbulent   turbulent flow
écoulement *m* uniforme   uniform flow
écoulement *m* visqueux   viscous flow
s'écouler *vb*   flow out or stream [liquids], run off [rainwater, etc], yield [give way]
écouvillon *m*   brush for smoothing inside of joints in concrete pipe, flue brush
écran *m*   screen [computer, wall, etc]
écran *m* à main   hand-held (welding) shield
écran *m* anti-éblouissant   anti-dazzle/-glare screen [roads]
écran *m* anti-projection   kick-back guard
écran *m* d'injection   grout curtain [dam]
écran *m* de cheminée   fire screen
écran *m* de palplanches   sheet pile screen or wall

écran *m* de soudeur   welding screen
écran *m* étanche   waterproof barrier or membrane
écran *m* facial   face shield
écran *m* (pare-)vapeur   vapour barrier
écrasé *adj*   crushed
écrasement *m*   crushing
écrémage *m*   screening off floating matter from surface of
    sewage
écrêtage *m* des pointes   maximum demand limiting or
    peak lopping [electrical supply]
écrêtement *m* des crues   flood peak attenuation
écrêtement *m* des pointes   peak lopping or smoothing
écrêter *vb*   lop or smooth off [a peak]
écrou *m*   nut [for bolt]
écrou *m* à collet *m*   collar nut
écrou *m* à oreilles   wing nut
écrou *m* à six pans   hexagonal nut
écrou *m* borgne   blind nut
écrou *m* cage   cage nut
écrou *m* carré   square nut
écrou *m* crénelé   castellated nut
écrou *m* de blocage   lock-nut
écrou *m* de fixation   fixing nut
écrou *m* du presse-étoupe   gland nut
écrou *m* hexagonal   hexagonal nut
écrou *m* indesserrable   lock-nut
écrou *m* papillon   wing nut
écroui *adj*   cold worked, hardened [of steel]
écrouir *vb*   work harden [cold]
écrouissage *m*   cold working, work hardening
écroulement *m*   collapse, crumbling or giving away [of
    ground, a structure], fall [of earth, etc], falling down/to
    pieces [e.g. of a building]
écroulement *m* d'un système   breakdown of a system
s'écrouler *vb*   collapse or fail, give way, crumble away
    [ground, a structure], fall (down) [earth, etc], fall in/fall
    to pieces or tumble down [a structure]
écumage *m*   degreasing of sewage
écume *f*   foam [on the surface of a fluid], froth, scum
écumer *vb*   foam, froth
écurie *f*   stable
écuyer *m*   handrail bracket [built into a wall]
édifice *f*   building [large or important]
édifier *vb*   build, construct or erect
édition *f* d'ordinateur   computer print-out
effacer *vb*   efface
effaceuse *f* (de marquage routier)   road marking eraser
    or remover
effectif *adj*   effective [real]
effectif *m*   complement [number of people]
effectifs *m,pl*   manpower
effectuer *vb*   carry out [e.g. works]
effet *m*   effect
effet *m* acoustique   acoustic performance
effet *m* biochimique   biochemical action
effet *m* calorifique   heating effect
effet *m* d'échelle *f*   scale effect
effet *m* d'entaille *f*   notch effect
effet *m* de cheminée   chimney or stack effect

effet *m* de diffusion   diffusion effect
effet *m* de filtre   filtering effect
effet *m* de pointe   resistance to the penetration of the
    point of a pile
effet *m* de refroidissement   chilling/cooling/refrigerating
    effect
effet *m* de serre   greenhouse effect
effet *m* de tirage   chimney effect
effet *m* de trempe   quenching effect [on metals]
effet *m* de voûte   arching
effet *m* du vent   wind effect
effet *m* refroidissant   cooling effect
effets *m,pl* du terrain   terrain effects
efficace *adj*   effective, efficient
efficacité *f*   effectiveness, efficacy, efficiency
efficacité *f* lumineuse   luminous efficacy
effilé *adj*   tapered
effluent *m*   effluent [from a cesspit, septic tank, factory]
effluent *m* urbain   urban sewage [combined soil, waste
    and rainwater]
effleurissement *m*   powdery weathering or deterioration
    of stone surface
efflorescence *f*   efflorescence
effondrement *m*   collapse, subsidence
effondrement *m* en cascade   progressive collapse
s'effondrer *vb*   collapse or fail
effort *m*   effort, force, stress
effort *m* admissible   permissible stress
effort *m* alterné   alternating stress
effort *m* axial   axial stress
effort *m* centré   axial stress
effort *m* de cisaillement   shear force
effort *m* de compactage   compactive effort [of a roller or
    vibratory compactor]
effort *m* (de) compression   compression/compressive
    stress
effort *m* de fatigue   repeated (fatigue) stress
effort *m* de flexion   bending force
effort *m* de freinage   braking force
effort *m* de serrage   clamping force
effort *m* de torsion   torsional stress
effort *m* de traction   tensile stress
effort *m* latéral   lateral force
effort *m* limite de traction   ultimate tensile stress
effort *m* longitudinal   longitudinal force
effort *m* maximum   maximum stress
effort *m* normal   axial load, normal force
effort *m* pulsatif/pulsatoire   alternating or pulsating
    stress
effort *m* résiduel   residual stress
effort *m* secondaire   secondary stress
effort *m* statique   static stress
effort *m* tangentiel   tangential force
effort *m* tranchant   transverse shear force
effritement *m*   crumbling or disintegration [of solid
    materials], weathering [of stone]
s'effriter *vb*   weather [crumble]
égal *adj*   equal
égalisateur *m*   flow equalizer

égaliser *vb*   equalize
égaliseur *m*   equalizer
église *f*   church
égoïne *f*   handsaw
égout *m*   drain(age) [sewerage system], sewer, sewer pipe [large]
égout, bouche *f* **d'**   gulley
égout *m*, **tout à l'**   main drainage
égout *m* **à canalisation unique**   combined sewer
égout *m* **collecteur**   main sewer, sewer main
égout *m* **mixte**   combined sewer
égout *m* **pluvial**   stormwater drain or sewer
égout *m* **principal**   main drain or sewer
égout *m* **séparatif**   separate sewer
égouts *m,pl*   sewerage system
égoutter *vb*   drip
égouttoir *m*   draining board [of a sink], drip tray
égrenage *m*   smoothing of plaster prior to decorating
égrener *vb*   smooth a plaster surface prior to painting or papering
égruger *vb*   grind [into small fine gravel or powder]
éjecteur *m*   ejector
éjecto-convecteur *m*   induction unit
élaborer *vb*   plan [work out, a project, etc]
élagage *m*   lopping [of trees], tree pruning
élancement *m*   slenderness (ratio)
élargi *adj*   enlarged, widened
élargir *vb*   enlarge, widen
s'élargir *vb*   spread [a river, a stain], stretch [become wider]
élargissement *m*   enlargement, stretching, widening
élasticité *f*   elasticity, flexibility
élasticité, module *m* **d'**   modulus of elasticity
élastique *adj*   elastic
élastomère *f*   elastomer
électricien *m*   electrician
électricité *f*   electricity
électricité *f* **statique**   static electricity
électrique *adj*   electric(al)
électro-aimant *m*   electro-magnet
électro-béton *m*   electrically heated concrete [poured in cold weather]
électrode *f*   electrode
électrode, filament *m*   coil [lamp electrode]
électrode *f* **d'allumage**   ignition electrode
électrode *f* **composée**   composite electrode
électrode *f* **de niveau**   water level electrode
électrode *f* **enrobée**   coated electrode
électrolyse *f*   electrolysis
électrolyte *m*   battery solution, electrolyte
électrolytique *adj*   electrolytic
électronique *adj*   electronic
électronique *f*   electronics
électro-osmose *f*   electro-osmosis
électrotechnique *f*   electrical engineering
électrovanne *f*   solenoid control valve
élément *m*   cell [of a battery], component, element, member, unit [e.g. of built-in furniture]
élément *m* **bas**   base unit [e.g. in a kitchen]

élément *m* **chauffant**   heating element
élément *m* **comprimé**   compression member
élément *m* **de chauffe**   heating element
élément *m* **de construction**   structural member
élément *m* **de façade**   cladding unit
élément *m* **de glissement**   beam [of a highway safety barrier]
élément *m* **de machine**   machine part
élément *m* **de rangement**   storage unit
élément *m* **de refroidissement**   cooling element
élément *m* **de structure**   structural member
élément *m* **détecteur**   detecting element
élément *m* **filtrant**   filter element
élément *m* **filtrant non régénérable**   disposable filter element
élément *m* **mural transversal**   crosswall unit
élément *m* **préfabriqué**   prefabricated member
élément *m* **préfabriqué de cuisine**   prefabricated kitchen unit
élément *m* **sensible**   sensor
élément *m* **sensible à l'humidité**   humidity sensor
élément *m* **sensible à la température**   temperature sensor
élément *m* **sensible au point de rosée**   dew point sensor
élément *m* **suspendu**   wall unit
élément *m* **tendu**   member in tension, tension member
éléments *m,pl*   basic principles
éléments *m,pl* **compris dans le contrat**   items covered by the contract
éléments *m,pl* **d'insonorisation**   soundproofing [materials, devices]
éléments *m,pl* **finis, méthode** *f* **des**   finite element method
élévateur *m*   elevator
élévateur *m* **à godets**   bucket elevator or hoist
élévateur *m* **à nacelle**   aerial work or access platform
élévateur *m* **de chantier**   builder's hoist or lift, site hoist
élévateur *m* **(monté) sur camion**   lorry mounted access platform, cage lift, cherry picker
élévateur *m* **télescopique**   cherry picker, telescopic access platform
élévateur-chargeur *m*   elevating loader
élévateur-transporteur *m*   mobile or portable belt conveyor/elevator
élévation *f*   elevation [drawing, or of the sun], height [of the sun]
élévation, parking *m* **en**   multi-storey carpark
élévation *f* **de**   rise in
élévation *f* **de côté**   side elevation or view
élévation *f* **de face**   front view or elevation
élévation *f* **de la température**   temperature rise
élévation *f* **du point de rosée**   dew point rise
élevé *adj*   high
élever *vb*   construct, raise [a monument]
éliminateur *m*   eliminator
éliminateur *m* **de gouttelettes**   moisture eliminator
élimination *f*   elimination or removal
élimination *f* **des déchets**   waste disposal or removal
élimination *f* **des décombres**   rubble or building rubbish removal
éliminer *vb*   remove [pollutants, waste]

élingue *f*   sling [for lifting]
ellipse *f*   ellipse
éloigné *adj*   distant, far (away), remote
élongation *f*   elongation
émail *m*   enamel
emballage *m*   packaging
emballeur *m*   packer
embarrure *f*   pointing of a ridge [mortar or plaster]
embase *f*   base [of an instrument, switch, chimney, etc],
     baseplate or bedplate [of a machine], pedestal, plinth
embasement *m*   base [structural], footing
embaucher *vb*   hire [labour], take on [staff]
emboîtement *m*   fitting or jointing together, interlocking,
     spigot joint [in pipework], spigot-and-socket jointing
emboîtement, assemblage *m* à   mortise joint
emboîture *f*   rail [of a door or shutter]
emboîture *f* à épaulement   spigot with a shoulder formed
     in a pipe end
emboîture *f* conique   conical spigot formed in a pipe end
emboîture *f* physiquée   round shoulder formed in a pipe
     end
embouchure *f*   river mouth
embout *m*   gland [on electrical equipment], nozzle [of a
     pipe or tube]
embout *m* à collerette isolante   ferrule with insulated
     flange
embout *m* de câblage   ferrule {electrical]
embout *m* de réduction   reducing nipple
embout *m* femelle   socket end
embouti *adj*   embossed or pressed [generally of metal
     sheet]
emboutir *vb*   press, stamp or swage [metals]
embranchement *m*   branching [e.g. of ductwork], three-
     way junction [highway]
embrasure *f*   reveal [of a door, window]
embrasse *f*   curtain loop
embrèvement *m*   tongue-and-groove joint(ing)
embrever *vb*   join by tongue-and-groove
embu *m*   dullness of paint due to absorption by substrate
émeri *m*   emery (powder)
émeri, papier *m* d'   emery paper
émeri, toile *f*   emery cloth
émetteur *m*   transmitter
émetteur-récepteur *m*   transceiver
émietteuse *f*   cultivator [earthworks]
émissaire *m*   outlet [of a lake, reservoir, etc]
émission *f*   emission
émission *f* de chaleur   heat emission
émissivité *f*   emissivity
émittance *f*   emittance
emmagasinement *m*   accumulation [of electrical energy],
     storage [of water]
emmagasinement *m* annuel   annual storage
emmagasinement *m* d'eau   water storage
emmagasinement *m* saisonnier   seasonal storage
emmagasiner *vb*   store [heat, electricity, etc]
emmarchement *m*   cut-out in stringer to receive the end
     of a stair tread, tread width of a staircase
emmétrer *vb*   pile regularly to work out volume [e.g. sand]

emmortaiser *vb*   press a mortise into a tenon
émoluments *m,pl*   emoluments, fees [architect's, etc]
empannon *m*   hip rafter
empannon *m* à noulet   valley rafter
empannon *m* de croupe   jack rafter
empannon *m* de long pan   jack rafter
empattement *m*   footing [of a wall], foundation [footing],
     joining [of timbers], side connection [between two ducts
     or pipes], tenoning
empatter *vb*   joint [timbers], support [equipment], tenon
empênage *m*   mortise in a piece of metal
empierré *adj*   metalled [road]
empierré, chemin *m*   metalled road
empierrement *m*   hardcore, stone layer
empilage *m*   pile [of sth], stack [of wood, etc]
empirique *adj*   empirical
emplacement *f*   location, placing, position, site, situation
emploi *m*   use, situation [job]
emploi, mode *f* d'   directions for use
emploi, prêt d'   ready for use
employeur *m*   employer
emprise *f*   acquisition [of land for public purposes],
     expropriation, area occupied [e.g. by building, highway]
emprise *f* au sol   area of ground occupied, required or
     requisitioned
emprise *f* au sol d'un engin   machine working area
emprise *f* d'un bâtiment   footprint of a building
emprise *f* du chantier   work site area
emprunt *m*   loan [borrowed]
emprunt *m* (de matériau)   borrow [of fill material]
emprunt-logement *m*   mortgage [house]
emprunter *vb*   borrow
emprunteur *m*   borrower
émulsifiant *m*   emulsifier
émulsion *m*   emulsion [paint]
émulsion *m* d'accrochage   bitumen(-based) emulsion
     [used as a tack coat], tack coat [road construction]
émulsion *m* de bitume   bitumen emulsion
émulsion *m* de goudron   tar emulsion
en acier   steel *adj*
en amont   upstream, upwind
en arc   arched *adj*
en attente   pending *adj*, in abeyance
en aval   downstream *adj*
en bascule   cantilevered *adj*
en biais   bevel *adj*, oblique *adj*, skewed, on the skew,
     slantwise, on the slant
en bordure de   at the edge of
en caoutchouc   rubber *adj*
en crue   in spate
en cuivre   copper *adj*
en cuvette   dished *adj* [cover, etc]
en dehors   exterior *adj*, outside *adj*
en encorbellement   cantilevered *adj*
en état de marche   in working order
en fonction de   as a function of
en fonte   cast iron *adj*
en forme de   shaped *adj* [e.g. U-shaped]
en forme de losange   diamond *adj* [motorway interchange]

en grandeur nature   full-scale *adj*
en gros   wholesale *adj*
en haut   upstairs *adj*
en haut de   at the top of [a building, a ladder, stairs]
en jeu   at work [e.g. forces]
en longueur   lengthwise *adv*
en masse   in a body
en monophasé   single-phased
en ordre de marche   in working order
en paires   paired, in pairs
en phase   in phase
en place   in situ, in place
en plafond   (at) high-level [on pipework drawing]
en pleins champs   in the open fields
en pointe   tapering [to a point]
en porte-à-faux   balanced *adj* [as in balanced cantilever], overhanging *adj*, cantilevered *adj*
en prise   in gear
en projet   at planning stage
en quinconce   staggered
en quinconces   staggered *adj*, in staggered rows
en redan   stepped [e.g. a gable, an embankment]
en relief   raised *adj*
en retard   late, overdue
en retrait (de)   set back (from)
en saillie   jutting out, projecting
en sifflet   slantwise, on the slant
en stock   in stock
en surplomb   overhanging *adj*
en surpression   pressurized *adj*
en vrac   bulk *adj*, in bulk
encadrement *m*   framing, training and supervision
encaissement *m*   collection [action of, of rents]
encastré *adj*   flush mounted, restrained [built-in]
encastré *adj* **(dans)**   built-in, embedded (into)
encastré *adj* **dans le plafond**   flush mounted in the ceiling
encastré, luminaire *m*   recessed lighting fixture
encastrement *m*   building-in, embedding, fixing [building in], end restraint, restraining [by building in]
encastrer *vb* **(dans)**   build in, embed (into), mount flush (with)
enceinte *f*   close [of a cathedral, etc], enclosure [fence or wall, or the ground enclosed]
enceinte *f* **du chantier, dans l'**   within the site
enceinte *f* **du réacteur**   containment vessel [of a reactor]
enchère *f*   bid [e.g. at auction]
enchères, vente *f* **aux**   sale by auction
enchevalement *m*   shoring/propping to facilitate underpinning
enchevauchure *f*   overlap(ping) [of tiles, boarding, etc]
enchevêtrure *f*   trimming [structure round opening in floor]
enclave *f*   enclave [of land], recess for sluice gates
enclavement *m*   isolation [of a community], wedging in
enclaver *vb*   build end of a joist in or let it into a main beam, fit in [into a recess or housing], house [timber, in a housing], wedge in
enclenchement *m*   engaging [of drive, gear], interlock, make [of an electrical switch], engage [drive, gear]

enclencher *vb*   throw into gear, switch on
enclipsable *adj*   clip-on
enclos *m*   close or enclosure [enclosed space]
enclume *f*   anvil
encoche *f*   notch or slot [e.g. in a flat key]
encollage *m*   sizing with glue
encoller *vb*   size [with glue]
encombrant *adj*   encumbering [in the way]
encombré *adj*   obstructed
encombrement *m*   obstruction [congestion]
encorbellement *m*   cantilever, corbelling [out], outstand, overhang
encorbellement, en   cantilevered
encrassé *adj*   choked (up), fouled, sooted up [of a spark plug]
encrassement *m*   choking, clogging, fouling
encrasser *vb*   choke (up) [clog], foul
encroûtement *m*   scale [deposit]
endentement *m*   toothed joint [between two pieces of timber]
endettement *m*   indebtedness
s'endetter *vb*   run into debt
endommager *vb*   damage
endosser *vb* **un chèque**   endorse a cheque
endothermique *adj*   endothermic
endroit *m*   locality, place, position or spot [place]
enduire *vb* **(de)**   coat (with), render (with)
enduisage *m*   coating, rendering [application of]
enduit *m*   coating, rendering [plaster, mortar], filler, size/sizing [for decorating]
enduit *m* **à la truelle**   rendering trowelled to produce a rustic effect
enduit *m* **armé**   reinforced rendering
enduit *m* **au balai** *m*   alpine, brushed or Tyrolean finish rendering
enduit *m* **bâtard**   bastard rendering [with a half-and-half lime/cement binder]
enduit *m* **bitumeux**   bituminous coat or paint
enduit *m* **brossé**   rendering brushed to remove roughness before complete setting
enduit *m* **bouchardé**   rendering bush-hammered three weeks after setting
enduit *m* **colle**   adhesive coating
enduit *m* **de ciment**   cement rendering
enduit *m* **de lissage**   smooth filler
enduit *m* **de parement plastique**   polymer-based rendering with mineral filler
enduit *m* **de protection**   protective coating
enduit *m* **de ragréage**   surfacing [filler]
enduit *m* **de rebouchage**   coarse filler
enduit *m* **de scellement**   sealing coat [road construction]
enduit *m* **de surfaçage**   filler or coating providing base for surface treatment
enduit *m* **dressé**   rendering levelled with a rule
enduit *m* **étanche**   waterproof coating
enduit *m* **extérieur**   external rendering
enduit *m* **feutré**   rendering finished with a felt covered trowel
enduit *m* **fin**   plaster skim coat(ing)

enduit *m* **frotté**   rendering trowelled with a float to draw laitance to surface

enduit *m* **garnissant**   rendering applied in a thick coat to masonry facings

enduit *m* **gras**   painter's filler for plaster [oil mastic + zinc white]

enduit *m* **gratté**   rendering finished with a devil float two to four hours after application

enduit *m* **grésé**   rendering polished after two to six days with a brick or rotating sander

enduit *m* **hourdé**   rendering

enduit *m* **jeté**   rendering thrown at a surface to produce a rustic effect

enduit *m* **lavé**   rendering sprayed with water and sponged some hours after application

enduit *m* **lisse/lissé**   rendering with a smooth finish

enduit *m* **lisse de plâtre**   fair-faced plaster

enduit *m* **maigre**   painter's filler for wood or old base

enduit *m* **moucheté**   roughcast or Tyrolean finish rendering

enduit *m* **non repassé**   painter's filler applied in a single coat

enduit *m* **peigné**   rendering raked with a metal comb or toothed trowel two to six hours after application

enduit *m* **poli**   rendering polished with a fine abrasive disc

enduit *m* **poncé** [=enduit *m* grésé]

enduit *m* **protecteur**   protective coating

enduit *m* **raclé** [=enduit *m* peigné]

enduit *m* **ravalé**   rendering treated mechanically after hardening

enduit *m* **repassé**   painter's filler applied in several layers to produce smooth surface

enduit *m* **ribbé/strié**   rendering with a ribbed or scored finish

enduit *m* **superficiel**   surface treatment [filler, coating]

enduit *m* **taloché**   rendering with floated or trowelled finish

enduit *m* **tramé**   rendering with a special finish applied with a roller

enduit *m* **tyrolien**   alpine or Tyrolean finish rendering

énergie *f*   energy

énergie, maîtrise *f* **d'**   energy saving

énergie *f* **accumulée**   stored energy

énergie *f* **acoustique**   sound energy

énergie *f* **cinétique**   kinetic energy

énergie *f* **consommée**   power consumption

énergie *f* **de choc**   impact energy

énergie *f* **de compactage**   compactive effort [amount of compaction]

énergie *f* **de déformation**   deformation work

énergie *f* **de rayonnement**   radiant energy

énergie *f* **électrique requise**   electrical demand

énergie *f* **emmagasinée**   stored energy

énergié *f* **éolienne**   wind power

énergie *f* **marémotrice**   tidal power

énergie *f* **nécessaire**   energy or power requirement

énergie *f* **nucléaire**   nuclear energy

énergie *f* **potentielle**   potential energy

énergie *f* **récupérée**   recovered energy

énergie *f* **solaire**   solar energy

énergie *f* **sonore**   sound energy

énergie *f* **spectrale**   spectral energy

énergie *f* **spectrale, répartition** *f* **d'**   spectral energy distribution

énergie *f* **thermique**   thermal energy [heat]

énergie *f* **thermique fournie**   heat input

enfaîteau *m*   semi-circular ridge tile

enfaîtement *m*   ridge capping [tile or metal]

enfaîter *vb*   ridge [i.e. cover ridge of a roof]

enfilade *f*   series or succession [of doors, columns, etc]

enflure *f*   swelling of plasterwork due to saltpetre

enfoncé *adj*   driven in, embedded, recessed, sunken

enfoncée, vallée *f*   drowned valley

enfoncement *m*   driving in [e.g. piles], settlement or sinking [of foundations, etc], recess [alcove]

enfoncer *vb*   drive in, embed [hammer or push in]

enfouissement *m*   undergrounding [e.g. of power lines]

enfourchement *m*   forked/open mortise-and-tenon joint(ing), jointing of rafters into a ridge

enfouir *vb*   bury

engagé *adj*   built-in [e.g. chimney into a wall]

engagement *m*   agreement [to do sth], commitment, liability, promise, contract

engagement *m* **hors bilan**   contingent liability

engager *vb*   commit [someone to sth], initiate [negotiations], take on [staff], mortgage

s'engager *vb*   agree, bind or commit oneself, covenant or promise [to do sth]

engazonné *adj*   grassed

engazonnée, bande *f*   grassed strip

engazonnée, surface *f*   grassed surface

engin *m*   machine plant, piece of equipment or plant

engin *m* **automoteur**   self-propelled machine or equipment

engin *m* **d'épandage**   spreader [machine]

engin *m* **de battage**   piling rig

engin *m* **de compactage**   compacting machine, compactor

engin *m* **de damage**   rammer

engin *m* **de déblaiement**   excavator

engin *m* **de forage**   drill rig

engin *m* **de levage**   lifting equipment

engin *m* **de mise à feu**   blasting equipment

engin *m* **de reprise à roue-pelle**   bucket-wheel reclaimer

engin *m* **de traitement des sols**   soil stabilizing machine

engin *m* **élévateur à nacelle**   aerial platform, cherry picker, sky lift

engin *m* **pour poser les poteaux**   post driver

engin *m* **stabilisateur**   soil stabilizing machine

engin *m* **tracté**   towed appliance or equipment

engins *m,pl* **de terrassement**   earth-moving equipment or plant

englué *adj*   thick [of paint or varnish]

engorgement *m*   blocking, stoppage, clogging, obstruction [clogging, of a pipe]

engraver *vb*   flash [fit flashing]

engravure *f*   flashing or slot to accept its edge

engrenage *m*   gear

engrenage *m* conique   bevel gear

engueulement *m*   notch in a principal rafter to accept end of roof post

enhachement *m*   part of a structure built into a neighbouring property

enjamber *vb*   span (over)

enlaçure *f*   dowelled joint, hole through several joining timbers to receive a peg

enlever *vb*   carry or take away, remove

enluminer *vb*   colour with bright tints

enquête *f*   inquiry [official], investigation

enrayure *f*   horizontal framing of roof structure, radiating from central point

enregistrement *m*   record [reading], recording

enregistrer *vb*   record [a reading]

enregistreur *adj*   recording *adj*

enregistreur *m*   recorder, recording apparatus/instrument

enregistreur *m* de débit   flow recorder

enregistreur *m* de données   data recorder

enregistreur *m* de température   thermograph

enregistreur *m* de volume   volume recorder

enrobage *m*   bedding [of a pipe, etc], coating [e.g. road material], cover [of reinforcement]

enrobage *m* bétonné   concrete bedding

enrobage *m* des aciers   reinforcement cover

enrobé *m*   binder [road material]

enrobé *m* au bitume-goudron   tar-bitumen binder

enrobé *m* bitumineux   bitumen binder

enrobé *m* dense au bitume   dense bitumen binder

enrobé *m* fin   sand asphalt

enrober *vb* (de)   coat (with)

enrober *vb* de liant   coat with binder

enrochement *m*   riprap, stone pitching, rockfill

enrocher *vb*   apply stone pitching, riprap, etc

enroulement *m*   winding [electrical]

enroulement *m* de moteur   motor winding

enrouler *vb*   coil or wind (up), roll up [e.g. carpet, cable]

enrouleur *m*   drum [for winding up a hose, extension cable, etc]

ensabler *vb*   blind [with sand]

enseigne *f* au néon   neon sign

enseigne *f* lumineuse   illuminated sign

ensemble *adv*   at the same time

ensemble *m*   collection [set or group, e.g. of houses], assembly [the whole]

ensemble† *m*   housing project

ensemble *m* d'aubes directrices   guide vane assembly

ensemble *m* de bâtiments   block of buildings

ensemble *m* de régulation   regulating unit

ensemblier *m*   interior designer, package dealer

enserrer *vb*   enclose [hem in]

enseuillement *m*   height of a door sill or window ledge above the ground surface

ensoleillé *adj*   sunny

ensoleillement *m*   insolation

entablement *m*   coping, entablature

entablure *f*   halved joint [in timber]

entaille *f*   kerf [width of cut made by cutting torch in steel], groove, notch, slot

entaille *f* à mi-bois   halved joint [in timber]

entaille *f* à mi-fer   halved joint [in iron/steel]

entaille *f* à queue d'aronde   dovetail cut-out or slot

entailler *vb*   groove, notch

entamer *vb*   break ground, cut into, incise or score

entartrage *m*   scale formation

enter *vb*   extend the length of [piece of wood, ladder, etc], join two pieces of wood

enterré *adj*   buried

enterrée, bordure *f*   flush kerb

enterrer *vb*   bury

enthalpie *f*   enthalpy, heat content

enthalpie *f* totale   total heat

entièrement *adv*   totally

entonnoir *m*   funnel or tundish, tremie pipe, sink hole

entourer *vb* (de)   enclose [surround] (with)

entraînement *m*   drive

entraînement *m* à courroie trapézoïdale   V-belt drive

entraînement *m* d'air   air entrainment

entraînement *m* direct   direct drive

entraîner *vb*   drive, entrain [air, not railway]

entraîneur *m* d'air, (adjuvant *m*)   air entraining agent [for concrete]

entrait *m*   binder, collar beam or tie-beam [timber roof], ceiling joist, roof beam [horizontal], tie-rod [steel roof]

entrait *m* retroussé   collar beam [timber roof]

entraver *vb*   obstruct [action, plans, progress]

entraxe *f*   spacing between centre lines [e.g. of beams]

entrebâilleur *m*   stay [door or window]

entrecolonnement *m*   distance between columns

entrecroisement *m*   weaving [of traffic]

entrecroisement, section *f* d'   weaving section [of road]

entrecroisement, voie *f* d'   weaving lane [of road]

entredistance *f* (de)   spacing (between) [e.g. reinforcement bars]

entrée *f*   entrance, entry, way in, influx, inlet, insertion

entrée, accès *m* spécial d'   limited access (point)

entrée, point *m* d'   entry point

entrée *f* analogique   analogue input

entrée *f* d'air   air inlet

entrée *f* d'eau   water inlet

entrée *f* d'ordinateur   computer input

entrée *f* de câble   cable entry

entrée *f* de serrure   escutcheon, keyhole plate

entrée *f* de ventilateur   fan inlet

entrée *f* latérale   side entrance

entrées *f,pl* reparties   distributed inlets

entrefer *m*   air-gap [between metal surfaces, e.g. rotor and stator]

entreposage *m*   storage, storing

entreposage *m* thermique   thermal storage

entreposer *vb*   put into store or storage

entrepôt *m*   store [depot], warehouse

entrepôt *m* frigorifique   cold store

entrepreneur *m*   builder, contractor

entrepreneur *m* de bâtiment/de construction   building contractor

entrepreneur *m* général   general, main or prime contractor

entrepreneur *m* **principal**　main or prime contractor

entreprise *f*　business, company, enterprise or firm, contracting firm, contractor, project, undertaking [task]

entreprise *f* **adjudicataire**　successful bidder or tenderer

entreprise *f* **contractante**　direct contractor

entreprise *f* **de constructions métalliques**　steel fabricator

entreprise *f* **de service public**　utility (company)

entreprise *f* **de travaux publics**　civil engineering firm

entreprise *f* **générale**　general contractor

entreprise *f* **intervenante**　labour-only (sub-)contractor

entreprise *f* **pilote**　(management) contractor responsible for coordination of the whole works

entreprise *f* **principale**　main or prime contractor

entreprise *f* **sous-traitante**　subcontractor

entresol *m*　mezzanine (floor)

entretenir *vb*　maintain or service

entretenu *adj*　maintained or serviced

entretien *m*　conversation or interview, maintenance, servicing or upkeep

entretien, centre *m* **d'**　maintenance area

entretien *m* **autoroutier, centre** *m* **d'**　motorway maintenance area

entretien *m* **routier, centre** *m* **d'**　highway maintenance area

entretoise *f*　angle or diagonal brace, corner brace [timber structure], cross-beam or girder, cross-bar, tie (rod), spacer [in equipment, or reinforcement], spacing piece

entretoise *f* **d'appui**　sill rail [timber structure]

entretoise *f* **de contreventement**　diagonal tie or windbrace

entretoisement *m*　bracing, bridging [in a structural frame]

entretoiser *vb*　brace, cross-brace, stay or strut

entrevous *m*　ceiling infill between joists or girders

entropie *f*　entropy

enture *f*　end or end-to-end jointing [of timber], splice

envasement *m*　sedimentation [silting], siltation, silting (up)

enveloppe *f*　envelope, casing [jacket, of a boiler, etc], jacket or lagging [boiler, tank], housing [protective], sheathing [cladding or lagging]

enveloppe *f* **de chaudière**　boiler casing

enveloppe *f* **de vapeur**　steam jacket

enveloppe *f* **de ventilateur**　fan casing

enveloppe *f* **du bâtiment**　building envelope

envelopper *vb*　case [equipment], lag [insulate], sheath [clad or lag, e.g. a boiler]

environ *adv*　approximately

environnant *adj*　surrounding *adj*

environnement *m*　environment, surroundings

environs *m,pl*　surroundings, vicinity

éolien *adj*　aeolian

éolienne *f*　aerogenerator, windmill, wind generator or pump

épaisseur *f*　depth [e.g. of slab], thickness

épaisseur *f* **à la base**　base width

épaisseur *f* **d'aile**　flange thickness [of a beam]

épaisseur *f* **de couche**　coating or layer thickness [paint]

épaisseur *f* **de couche limite**　boundary layer thickness

épaisseur *f* **de l'âme**　web thickness [of a beam]

épaisseur *f* **de mur**　wall thickness

épaisseur *f* **de paroi**　wall thickness

épaisseur *f* **de tôle**　sheet metal gauge, plate thickness

épaisseur *f* **du cordon**　throat thickness

épaisseur *f* **en crête**　crest width [of a dam]

épaisseur *f* **isolante**　insulation thickness

épandage *m*　spreading [of sewage on land]

épandage, champs *m,pl* **d'**　sewage farm

épandeuse *f*　paver [highways], spreader [for fertilizer, sewage, etc]

épandoir *m*　distributor [in a sewage treatment filter bed]

épandre *vb*　spread [sewage, fertilizer]

s'épandre *vb*　spread [fire, water]

épaufrure *f*　breakage or spalling of stone edge or surface

épaulement *m*　transverse notch to receive shoulders of a tenon, side of a mortise, or material removed to form a tenon

épaulement *m* **de remblai**　edge of an embankment

éperon *m*　buttress, cutwater [of a bridge]

épi *m*　finial [on a roof], groyne or jetty, spiked hook on top of a wall or fence

épi, assise *f* **en**　diagonal brick course

épi *m* **de faîtage**　finial [on roof ridge]

épicéa *m*　spruce [tree, wood]

épicentre *m*　epicentre

épinçoir *m*　large pavior's hammer

épingle *f*　hooked tie bar, link [shear reinforcement]

épissure *f*　splice [in a rope, cable, conductor, etc]

épistyle *m*　architrave

éponge *f*　sponge

époque, d'　period *adj*

épreuve *f*　test

épreuve *f* **d'étanchéité**　leak or leakage test

épreuve *f* **de pression**　pressure test

éprouver *vb*　test

éprouvette *f*　test tube, sample, test piece or specimen

éprouvette *f* **cubique**　test cube [concrete]

éprouvette *f* **de traction**　tensile test piece

éprouvette *f* **normalisée**　standard test piece

épuisé *adj*　exhausted or out of stock

épuisement *m*　dewatering [of excavations], exhaustion [of stocks]

épuiser *vb*　exhaust or use up

s'épuiser *vb*　run out [of supplies, stocks]

épurateur *m*　purifier [air, water]

épurateur *m* **d'air**　air cleaner, purifier or scrubber

épurateur *m* **de vapeur**　steam separator

épuration *f*　cleansing, filtering [liquids], purification, purifying, scrubbing [of a gas]

épuration *f* **de l'air**　air cleaning or purification

épuration *f* **de l'eau**　water purification or treatment

épuration *f* **de l'eau d'alimentation**　feed water treatment

épuration *f* **des eaux usées**　sewage treatment

épuration *f* **naturelle**　natural purification

épure *f*　diagram, finished design, layout or working drawing

épurer *vb*   clean or scrub [a gas], filter [gas, water], purify

équarri *adj*   squared

équarrir *vb*   square [timber, stone]

équarrissage *m*   dimensions [of a squared piece of timber], scantling [timber size], squaring [of timber, stone]

équation *f*   equation

équation *f* **approximative**   approximate equation

équation *f* **d'état**   equation of state

équation *f* **de régression**   regression equation

d'équerre *adj*   square [at right angles]

équerre *f*   right angle bracket, set square or square [for marking right angles]

équerre *f*, **fausse**   bevel square, sliding bevel

équerre *f* **à coulisse**   sliding callipers

équerre *f* **à dessin**   set square

équerre *f* **d'arpenteur**   surveyor's square

équerre *f* **d'assemblage**   L-shaped clip or metal cleat

équerre *f* **d'onglet** *m*   mitre square

équidistance *f*   contour interval

équilibrage *m*   balancing

équilibré *adj*   balanced

équilibre, teneur *f* **en eau d'**   equilibrium moisture content

équilibre *m*   balance, equilibrium

équilibre *m* **d'humidité**   moisture balance

équilibre *m* **des températures**   temperature balance

équilibre *m* **thermique**   thermal balance or equilibrium

équilibrer *vb*   (counter)balance

équipage *m*   apparatus [general term], equipment

équipé *adj*   equipped

équipe *f*   gang or shift [work team], team

équipe *f* **de travail**   shift [work team]

équipement *m*   equipment

équipement *m* **à frein absorbeur d'énergie cinétique**   braking equipment to absorb kinetic energy

équipement *m* **d'usine**   factory plant

équipement *m* **de commande**   control device or equipment

équipement *m* **de diffusion d'air**   air supply fixtures

équipement *m* **de protection collective**   collective protective device or equipment

équipement *m* **de protection individuelle (EPI)**   personal protective equipment

équipement *m* **de réserve**   standby equipment

équipement *m* **de secours**   emergency equipment

équipement *m* **extérieur**   outdoor equipment

l'Équipement[†] *m*   planning, housing and highways department [in France]

équipement *m* **ménager**   household equipment

équipements *m,pl*   facilities, plant [equipment]

équipements *m,pl* **portuaires**   port/harbour facilities

équiper *vb*   equip

équipotentiel *adj*   equipotential

équivalent *adj*   equivalent *adj*

équivalent *m*   equivalent *n*

équivalent *m* **mécanique de la chaleur**   mechanical equivalent of heat

érable *m*   maple

ergot *m*   cam, lug, stop

ériger *vb*   raise [a monument]

éroder *vb*   abrade, eat away, erode, wear away

érosion *f*   erosion

erratique *adj*   erratic

erreur *f*   error, fault [mistake]

erreur *f* **accidentelle**   accidental error

erreur *f* **aléatoire**   random error

erreur *f* **de lecture**   reading error

erreur *f* **de mesure**   measuring error

erreur *f* **moyenne**   mean error

escabeau *m*   pair of steps, stepladder

escalator *m*   escalator

escalier *m*   flight of stairs, stair, staircase, stairway, steps

escalier *m* **à crémaillère**   open or cut string staircase

escalier *m* **à échelle de meunier**   shallow, open-tread staircase

escalier *m* **à l'anglaise**   open or cut string staircase

escalier *m* **à la française**   close or housed string (straight) staircase

escalier *m* **à limon (droit)**   close or housed string (straight) staircase

escalier *m* **à marches mobiles**   escalator

escalier *m* **à quartier tournant**   quarter-turn staircase with winders

escalier *m* **à vis**   spiral staircase

escalier *m* **à vis à noyau**   spiral staircase with solid newel

escalier *m* **à vis sans noyau**   spiral staircase with open newel

escalier *m* **d'évacuation**   escape staircase

escalier *m* **de secours**   escape staircase

escalier *m* **de service**   service staircase

escalier *m* **dérobé**   hidden or secret staircase

escalier *m* **en colimaçon**   spiral staircase

escalier *m* **en échelle**   skeleton or open tread staircase

escalier *m* **en escargot/en hélice/en limaçon**   spiral staircase

escalier *m* **encloisonné**   staircase between walls

escalier *m* **hors d'œuvre**   external staircase

escalier *m* **mécanique/roulant**   escalator

escalier *m* **tournant**   spiral or winding staircase

escalier *m* **tournant à noyau** *m* **creux**   open newel spiral or winding staircase

escalier *m* **tournant à noyau** *m* **plein**   solid newel spiral or winding staircase

escamotable *adj*   retractable

escargot *m*   spiral staircase

escarpe *f*   scarp [of a fortress]

escarpement *m*   escarpment, scarp

escompte *m*   discount, rebate [financial]

espacé *adj* **de**   spaced at

espace *m*   gap, space

espace *m* **d'air**   air gap or space

espace *m* **découvert**   open area

espace *m* **fonctionnel, bureau** *m* **en**   open plan office

espace *m* **interdit à la circulation**   ghost island [road]

espace *m* **libre**   clearance, open space

espace *m* **ouvert**   open area

espace *m* **vert**   green space [in town or city]

espacement *m*   pitch or spacing

espaceur *m*   spacer

espagnolette *f*   espagnolette or shutter bolt

esquisse *f*   outline [drawn or written], (preliminary) sketch

essai *m*   attempt, test, trial

essai *m* **à la plaque**   plate bearing test

essai *m* **à vitesse de chargement constante**   constant loading rate test

essai *m* **à vitesse de déformation constante**   constant rate of increase of strain test

essai *m* **au bleu de méthylène**   methylene blue test

essai *m* **au choc**   impact test

essai *m* **au gaz**   gas test

essai *m* **au pénétromètre**   penetration test

essai *m* **aux ultrasons**   ultrasonic test

essai *m* **billage**   ball hardness test

essai *m* **comparatif**   comparative test

essai *m* **compression**   compression test

essai *m* **courant**   routine test

essai *m* **d'Abrams**   slump test

essai *m* **d'adhérence**   bond (strength) test

essai *m* **d'écrasement**   crushing/flattening test

essai *m* **d'élargissement**   bulging test

essai *m* **d'endurance**   endurance test

essai *m* **d'étanchéité**   leak or leakage test

essai *m* **d'étanchéité à la fumée**   smoke test [tightness]

essai *m* **d'opacité de fumée**   smoke test [opacity]

essai *m* **de cassure**   breaking test

essai *m* **de chargement d'un pieu**   pile loading test

essai *m* **de chargement à la plaque**   plate bearing test

essai *m* **de chargement statique**   static loading test

essai *m* **de cisaillement**   shear test

essai *m* **de choc**   impact test

essai *m* **de compactage**   moisture-density test

essai *m* **de compactage Proctor**   Proctor compaction test

essai *m* **de compressibilité**   compression test [for soils]

essai *m* **de compression**   compression test

essai *m* **de compression triaxial**   triaxial compression test

essai *m* **de consolidation**   consolidation test

essai *m* **de contrôle**   (control) test

essai *m* **de ductilité**   ductility test

essai *m* **de dureté**   hardness test

essai *m* **de dureté Brinell/Rockwell**   Brinell/Rockwell hardness test

essai *m* **de fatigue**   fatigue test

essai *m* **de fatigue par choc sur le barreau entaillé**   repeated notched bar impact bending test

essai *m* **de flambement**   buckling test

essai *m* **de fluage**   creep test

essai *m* **de fluidité**   slump test

essai *m* **de forme des grains**   particle shape test

essai *m* **de gélivité**   freezing or frost sensitivity test

essai *m* **de laboratoire**   laboratory test

essai *m* **de pénétration**   penetration test

essai *m* **de perméabilité**   permeability test

essai *m* **de pliage**   bend(ing) test

essai *m* **de pliage en travers**   transverse bending test

essai *m* **de pliage sur éprouvette entaillée**   notch bend test

essai *m* **de poinçonnage**   punching test

essai *m* **de pompage**   pumping test

essai *m* **de pression**   pressure test

essai *m* **de recette**   acceptance test

essai *m* **de rendement de chaudière**   boiler (output) test or trial

essai *m* **de résilience**   notch impact bending test

essai *m* **de résistance à la compression**   crushing test [e.g. concrete]

essai *m* **de traction**   tensile test

essai *m* **de vérification**   control test

essai *m* **des matériaux**   materials test

essai *m* **en charge**   load test

essai *m* **en laboratoire**   laboratory test

essai *m* **en place/in situ**   field test or trial, in situ test

essai *m* **en vraie grandeur**   full-scale test/trial

essai *m* **flambement**   buckling test

essai *m* **grandeur nature**   full-scale test/trial

essai *m* **in situ**   in situ test

essai *m* **isolé**   single test

essai *m* **non destructif**   non-destructive test

essai *m* **préliminaire**   preliminary test

essai *m* **sur modèle reduit**   model test

essais *m,pl* **de sol**   soil investigation

essai, mettre *vb* **à l'**   put to the test, try out

essayer *vb*   test, try out [a method, experiment with]

essayer *vb* **de faire qch**   try to do sth

essence *f*   petrol

essences *f,pl* **conifères**   softwood [trees]

essente *f*   shingle [roofing, cladding]

essenter *vb*   shingle [a roof, a wall]

essieu *m*   axle

essieu *m* **arrière/avant**   rear/front axle

essorage *m*   dewatering [e.g. of concrete]

est *adj*   east, eastern, easterly, eastward

est *m*   east

estacade *f*   pier [on piles]

estampille *f*   identification mark

estimatif *adj*   estimated

estimation *f*   appraisal, (approximate) estimate, quotation/ quote [price], valuation

estimer *vb*   estimate

estrade *f* **de travail**   work platform

estuaire *m*   estuary

étable *f*   cowshed

établi *m*   (work)bench

établi *m* **de menuisier**   woodworker's bench

établi *m* **étau**   collapsible bench [e.g. Workmate]

établir *vb*   draft or draw up [documents, etc], put up [a building]

établir *vb* **un prix**   fix, quote or set a price

établir *vb* **une facture**   make out an invoice

s'établir *vb*   settle [in a country]

établissement *m*   drafting or drawing up [documents, etc], erecting, laying out of structural timbers on a full-scale plan, establishment [building, enterprise]

établissement *m* **des dossiers d'appel d'offres** preparation of tender documents

étagé *adj* **en section** stepped in section

étage *m* first or upper floor [two-storey building], floor [storey], level [of a building], storey, stage [of a pump, a compressor]

étage, hauteur *f* **d'** storey height

étage *m* **carré** storey of uniform height

étage *m* **courant** typical floor

étage *m* **d'attique** attic storey

étage *m* **en sous-sol** basement storey

étage *m* **hors-toit** penthouse

étage *m* **lambrissé** roof space storey

étager *vb* set or lay out in tiers or tiered rows

étagère *f* shelf or shelf unit

étai *m* prop or shore

étai *m* **stabilisateur** brace [prop]

étai *m* **tirant–poussant** push–pull prop

étaiement *m* bracing, propping, shoring, falsework

étain *m* tin

étain *m* **à braser** solder

étaler *vb* spread out [e.g. gravel, objects]

étalon *m* full-scale layout [of axes of a structural framework], standard [e.g. example, trial]

étalonnage *m* calibration or standardization

étalonné *adj* standardized

étalonnement *m* calibration or standardization

étalonner *vb* calibrate or standardize [an instrument, etc]

étamage *m* tinning

étamer *vb* tin

étamperche *f* scaffolding pole

étanche *adj* airtight, impermeable, impervious, sealed [water- or airtight], watertight or waterproof, weathertight or weatherproof

étanche *adj* **à l'air** airtight

étanche *adj* **à l'eau** waterproof, watertight

étanche *adj* **à l'huile** oil-tight

étanche *adj* **à l'immersion** watertight when immersed

étanche *adj* **à la fumée** smoke-tight

étanche *adj* **à la lance** watertight against water jet at any angle

étanche *adj* **à la poussière/aux poussières** dust-tight

étanche *adj* **à la pression** pressure tight

étanche *adj* **à la vapeur** steamproof, -tight

étanche *adj* **au gaz** gas-tight

étanche *adj* **aux vapeurs** vapour-tight

étanchéisation *f* waterproofing

étanchéité *f* airtightness, watertightness

étanchéité, chape *f* **d'** waterproofing [of a road bridge]

étanchéité, système *m* **d'** waterproofing system

étanchéité *f* **à l'air** airtightness

étanchéité *f* **multi-couches** multi-layered waterproofing

étanchement *m* proofing [against water ingress, etc], sealing off, stopping of flow of a liquid

étanchement *m* **bitumineux** bituminous seal

étancher *vb* stop the flow [of a liquid], waterproof

étançon *m* pit-prop, inclined prop or shore

étançonnement *m* propping, shoring

étançonner *vb* prop, shore (up), support with props or shoring

étang *m* mere, pond, pool, reservoir [small]

étape *f* phase [stage], stage or step [of a process, work, etc]

état *m* condition or state

état, à remettre *vb* **en** needing renovation

état *m* **brut de laminage** as rolled condition or mill state [steel]

état *m* **d'air/de l'air** air condition

état *m* **d'équilibre** (state of) equilibrium

état *m* **de contrainte** state of stress

état *m* **de fonctionnement** working condition [of equipment, etc]

état *m* **de régime permanent** steady state (condition)

état *m* **de surface** surface condition

état *m* **de tension** state of stress

état *m* **des lieux** existing conditions or state [of a building, site, etc]

état *m* **hypothécaire** land charges [as found by search]

état *m* **limite** limit state

état *m* **limite d'utilisation** serviceability limit state

état *m* **limite de service (ELS)** serviceability limit state

état *m* **limite ultime (ELU)** ultimate limit state (ULS)

état *m* **normal** normal state

état *m* **permanent** steady state (condition)

état *m* **satisfaisant (de fonctionnement)** serviceability [of equipment]

état *m* **stationnaire** steady state (condition)

étau *m* vice

étau *m* **à agrafe** small vice

étayage *m* propping, shoring, timbering [propping or shoring]

étayer *vb* brace, prop or shore (up), strut, support

éteignoir *m* extinguisher

éteindre *vb* extinguish, put out

ételon *m* full-scale layout [of axes of a structural framework]

étendre *vb* expand or extend [spread, spread out], widen, increase the scope of, spread [lay out, a map, a carpet, etc], spread [paint]

s'étendre *vb* expand or extend [e.g. forest, town, by growing], reach [spread to]

étendue *f* extent, range [area, duration], sprawl [of a town, etc], stretch [of ground]

étiage *m* lowest level [of watercourse]

étincelle *f* spark

étiquette *f* label

étirage *m* drawing [of metals]

étirage *m* **à froid** cold drawing

étiré à froid cold drawn

étirer *vb* draw out [stretch], stretch [elastically or plastically]

étirer *vb* **à froid** cold draw

étoile *f* star connection [electrical]

étoile-triangle *adj* star-delta

étonner *vb* crack or fissure [an arch], loosen or shake [a structure]

étonner *vb* **à la boucharde** bush hammer [stone]

étouffer *vb*   damp [deaden, noise]
étranglement *m*   constriction, restriction
étrangler *vb*   throttle
étrangleur *m*   throttle
être *vb* électrisé   get an electric shock
étrésillon *m*   angle or diagonal brace, strut between joists
étrésillon *m* à vérin   adjustable telescopic (Acrow type) prop
étrésillonnage *m*   diagonal bracing, strutting between joists
étrésillonnement *m*   diagonal bracing or strutting [the action]
étrésillonner *vb*   brace diagonally or strut
étrier *m*   binder or stirrup [reinforcement], connector [of galvanized metal, for timber structures], joist hanger
étrier *m* de suspension   hanger
étrier *m* de suspension des tubes   pipe hanger
étroit *adj*   narrow
étude *f*   design [in the sense of study], study
étude *f* (notariale)   notarial firm/office
étude *f* d'avant-projet   concept design
étude *f* d'exécution   working design
étude *f* d'impact   environmental (impact) study
étude *f* de base (du projet)   basic (project) design
étude *f* de bassins hydrologiques   hydrological study
étude *f* de béton armé   reinforced concrete design
étude *f* de cas   case study
étude *f* de faisabilité   feasibility study
étude *f* de marché   market study
étude *f* de pré-faisibilité   pre-feasibility study
étude *f* de pré-investissement   pre-investment study
étude *f* de/du projet   design study, project design
étude *f* de reconnaissance du site   site reconnaissance or survey
étude *f* de stabilité   stability analysis
étude *f* de tassement   settlement analysis
étude *f* de terrain   field investigation or survey
étude *f* des travaux   design of the works
étude *f* du site   site survey
étude *f* du sol   soil investigation or survey
étude *f* et supervision des travaux   design and supervision of the works
étude *f* foncière   land tenure/ownership investigation/survey
étude *f* géophysique   investigation, geophysical
étude *f* hydrologique   hydrological study
étude *f* notariale   notarial practice/office
étude *f* par ordinateur   computer design
étude *f* pédologique   soil investigation or survey
étude *f* préliminaire   preliminary study
étude *f* topographique   topographic(al) survey
études *f,pl*   design (work), studies
études *f,pl*, bureau d'   consulting engineers' office [roughly speaking!], design agency
études *f,pl* de conception   design studies/work
étudier *vb*   study
étuvage *m*   kiln drying [of timber], steam curing [of concrete]
étuve, séché *adj* à l'   oven dried

étuve *f*   heating chamber
étuve *f* de désinfection   sterilizer
étuve *f* de séchage   drying cupboard/chamber/oven/kiln
étuver *vb*   dry [in a kiln, oven, etc], kiln dry
évacuateur *m*   extract fan, outfall drain
évacuateur *m* de crue   spillway
évacuateur *m* de secours   emergency spillway
évacuation *f*   discharge, evacuation or removal, drain [outlet], drainage [of water], draining off, extraction [air, dust, liquid, etc], eviction [of a tenant]
évacuation *f* d'urgence   emergency evacuation
évacuation *f* de(s) boues   dirt pocket
évacuation *f* des cendres   ash removal
évacuation *f* des copeaux   chip extraction
évacuation *f* des décombres   rubble or building rubbish removal
évacuation *f* des déchets   waste removal
évacuation *f* des eaux d'égouts   sewage disposal
évacuation *f* des eaux du toit   roof drainage
évacuation *f* des eaux pluviales   rain water drainage
évacuation *f* des eaux usées   waste water drainage
évacuation *f* des eaux vannes   foul water drainage
évacuation *f* des fumées/gaz   flue gas removal
évacuation *f* des gravats   rubble or building rubbish removal
évacuer *vb*   blow out/down [water from a boiler], discharge, drain (off), evacuate, evict [tenants], exhaust [steam]
évaluation *f*   appraisal/assessment [e.g. of a project], evaluation/valuation, estimate [e.g. of costs]
évaluer *vb*   evaluate
évaporateur *m*   evaporator
évaporateur *m* à ailettes   finned evaporator
évaporateur *m* à calandre et serpentin   shell-and-coil evaporator
évaporateur *m* à détente directe   direct expansion evaporator
évaporateur *m* à faisceau tubulaire   shell-and-tube evaporator
évaporateur *m* à plaques   plate-type evaporator
évaporateur *m* à tubes verticaux   vertical-type evaporator
évaporateur *m* de plafond   ceiling-type evaporator
évaporateur *m* type vitrine   showcase-type evaporator
évaporation *f*   evaporation
évaporation *f* à détente directe   direct expansion evaporation
évaporer *vb*   evaporate
évapo-transpiration *f*   evapo-transpiration
évaseur *m*   socket forming tool [plumbing]
événement *m*   event
évent *m*   air vent [tap or cock], vent
évent *m* de décharge   discharge vent
éventail *m*   access scaffolding with angled support
éventail, imposte *m* en   semi-circular fanlight
éventail, ouvrage *m* en   work fanning out from one centre
éventail, parquet *m* en   wood strip flooring in a fan-shaped pattern

éventailler *vb*   assemble two ladders in an inverted V
éventuel *adj*   possible
évidement *m*   cavity, hollow [cut into something], recess [cavity, groove]
évider *vb*   channel, groove, hollow out
évier *m*   kitchen sink, sink [stone, porcelain, etc]
évier *m* **agent de ménage**   cleaner's sink
évier *m* **d'hôpital**   hospital sink
évolution *f*   development [of prices, a market]
exact *adj*   right [accurate], true [measure, report]
exactitude *f*   accuracy
examen *m*   examination
examen *m* **à rayons X**   X-ray examination
examen *m* **infrarouge**   infra-red survey
examen *m* **médical d'embauche/avant embauche**   compulsory medical examination before hiring
examen *m* **radiographique**   radiographic examination
examen *m* **visuel**   visual examination
examiner *vb*   examine, inspect, survey [a building]
excavateur *m*   mechanical digger, excavator or shovel, power shovel
excavateur *m* **à godets**   bucket digger, ditcher or trencher
excavateur *m* **à roue-pelle**   bucket-wheel excavator
excavation *f*   cut, excavation, undercutting
excavatrice *f*, *voir aussi* excavateur   general purpose excavator
excaver *vb*   dig (out), excavate, drive [a tunnel]
excédent *m* (**sur**)   excess or surplus (over)
excédent *m* **de consommation**   excess consumption
excédentaire *adj*   excess *adj*
excentré *adj*   eccentric *adj*
excentricité *f*   eccentricity
excentrique *adj*   eccentric *adj*
excès *m* (**de**)   excess/surplus (of)
excitation *f*   excitation [of a generator]
exciter *vb*   excite [a generator]
exclusivité *f*   exclusivity [e.g. of selling agency]
exécuté à la main   handmade or done by hand
exécuter *vb*   carry out [works, etc], run [a computer program]
exécution *f*   execution, workmanship
exécution *f* **d'un programme informatique**   computer run
exemplaire *m*   copy [e.g. of document or drawing]
exemple *m*   example
exempt d'huile   oil-free
exercer *vb*   exercise or practise [a trade, skill or profession]
exfiltration *f*   exfiltration
exhaussé *adj*   raised [heightened]
exhaussement   height increase [of building, wall, etc]
exhausser *vb*   heighten [a building, wall, etc], raise [heighten, increase the height of]
exigence *f*   requirement
exigences *f,pl* **techniques**   technical requirements
exiger *vb*   demand, insist upon, require
exigible *adj*   due [payment, etc]
expansé *adj*   expanded
expansibilité *f*   expansibility

expansion *f*   expansion
expansion *f* **directe**   direct expansion
expansion *f* **linéaire**   linear expansion
expédier *vb*   dispatch, forward, send off
expéditeur *m*   despatcher, sender
expédition *f*   despatch, shipping [sending off]
expérience *f*   experience, experiment, test
expérience *f* **en service**   operating experience
expérience *f* **pratique**   practical experience
expérimental *adj*   experimental
expérimenter *vb*   experiment [make an], try [test]
expert *m*   appraiser, independent expert, surveyor, valuer
expert *m* **comptable**   chartered accountant [near French equivalent]
expert *m* **immobilier**   property valuer
expert *m* **métreur**   quantity surveyor [near French equivalent]
expert-conseil *m*   consultant
expert-géomètre *m*   land surveyor
expertise *f*   (technical) appraisal, survey or valuation [independent, of work, materials, etc], expert opinion
expertiser *vb*   appraise, survey or value [work, materials, etc, independently]
exploitation *f*   exploitation, operation [e.g. of a railway], running [an enterprise, railway, etc], working [of a mine, railway, etc]
exploitation, chemin d'   farm or forest road or track
exploitation, mettre *vb* **en**   develop, exploit
exploitation, sentier *m* **d'**   farm track
exploitation *f* **à ciel ouvert**   opencast mining/working
exploitation *f* **agricole**   farm
exploitation *f* **commerciale**   business concern
exploitation *f* **forestière**   forestry development
exploitation *f* **industrielle**   industrial concern
exploitation *f* **minière**   mining development
exploiter *vb*   operate or run [a quarry, a railway, etc]
exploser *vb*   blow up, explode
explosif *adj*   explosive *adj*
explosif *m*   explosive
explosion *f*   explosion
exponentiel *adj*   exponential *adj*
exponentielle *f*   exponential
exposé *adj* **à**   exposed to
exposé *adj* **sud, etc**   south, etc facing
exposé *m*   account, report or statement [of facts, etc]
exposé *m* **d'un cas concret**   case history
exposer *vb*   exhibit
exposition *f*   exhibition, exposition or statement [of facts, etc], exposure [of a building, facade, etc]
exposition, temps *m* **d'**   exposure period or time
exposition *f* **abritée**   sheltered exposure
exposition *f* **aux agents chimiques**   exposure to chemical agents
expropriation *f*   compulsory purchase, expropriation
exproprier *vb*   purchase compulsorily, expropriate
expulsion *f*   eviction [of a tenant], extrusion
exsudation *f*   bleeding [water to surface of concrete, one colour through another]
extension *f*   extension [stretching or expansion]

extensomètre *m*   extensometer, strain gauge, tensiometer
extérieur *adj*   outer *adj*, outside *adj*
extérieur *m*   exterior
externe *adj*   external
extincteur *m*   fire extinguisher
extincteur *m* **à main**   portable fire extinguisher
extincteur *m* **automatique**   automatic sprinkler
extincteur *m* **d'incendie**   fire extinguisher
extincteur *m* **d'incendie à main**   hand operated fire extinguisher
extinction *f*   switching off [of lighting]
extracteur *m*   extractor

extraction *f*   extraction
extraction *f* **d'air**   air extraction
extrados *m*   external surface of an arch or its segments, extrados [architectural]
extradossé *adj*   of an arch whose voussoirs are cut on the extrados so as not to interfere with the surrounding work
extraire *vb*   extract
extraire *vb* **les boues**   desludge
extrapoler *vb* **(de)**   extrapolate (from)
extrémité *f*   end [of a cable, etc], extremity or tip
extrudé d'étanchéité, profilé *m*   extruded sealing profile
extrusion *f*   extrusion
exutoire *m*   outlet [drainage], smoke vent

fabricant *m*   maker, manufacturer
fabrication *f*   fabrication, manufacture, production
fabriqué *adj*   manufactured
fabriqué *adj* **en série**   mass-produced
fabrique *f*   factory [building], manufacture
fabriquer *vb*   fabricate, manufacture
façade *f*   elevation or face [of a building], facade
façade *f* **en ligne droite**   straight line facade
façade *f* **légère**   lightweight facade [e.g. metal, glass or light panel]
façade *f* **panneau**   lightweight facade installed between floors
façade *f* **respirante**   facade with vapour permeable exterior skin
façade *f* **rideau**   lightweight facade installed wholly outside the floors
façade *f* **ventilée**   ventilated facade [double skin with exterior air circulation]
façadier *m*   renderer
face à   facing [e.g. facing the station]
face *f*   face [side]
face *f* **avant**   front
facilité *f* **de crédit**   credit facility
facilités *f,pl* **de transport**   transport facilities
façon *f*   work only [excluding materials]
façon, travail *m* **à**   work with materials supplied free
façonnage *m*   forming, shaping or working [of stone, timber, etc]
façonnage *m* **à chaud**   hot forming
façonnage *m* **à froid**   cold forming
façonné à froid   cold shaped
façonner *vb*   form, shape, work
facteur *m*   factor
facteur *m* **d'adhérence**   bonding strength
facteur *m* **d'amortissement**   damping factor
facteur *m* **d'écoulement**   flow factor
facteur *m* **d'encrassement**   fouling factor
facteur *m* **d'uniformité**   uniformity ratio [lighting]

facteur *m* **d'utilisation**   utilization factor [lighting]
facteur *m* **de calcul**   design factor
facteur *m* **de charge**   load factor
facteur *m* **de conductance**   conductance factor
facteur *m* **de dépréciation**   lumen maintenance factor
facteur *m* **de diversité**   diversity factor
facteur *m* **de fonctionnement**   coefficient of performance
facteur *m* **de forme**   form factor
facteur *m* **de lumière du jour**   daylight factor
facteur *m* **de puissance**   power factor
facteur *m* **de rayonnement**   radiation factor
facteur *m* **de réflexion**   reflection factor [lighting]
facteur *m* **de sécurité**   design factor, factor of safety
facteur *m* **de simultanéité**   diversity factor
facteur *m* **de sûreté**   factor of safety
facteur *m* **de transmission**   transmission factor [lighting]
facteur *m* **décisif**   decisive factor
facteur *m* **dimensionnel**   size factor
facturation *f*   billing, invoicing
facture *f*   account, bill or invoice
facturer *vb*   bill, invoice
factures *f,pl*, **vérification** *f* **de**   checking certificates [for payment]
facturier *m*   invoice clerk, invoice register
facultatif *adj*   optional
faible *adj*   low [as in low density, etc], weak
faïençage *m*   surface cracking
faïence *f*   earthenware, stoneware, wall tiles
faille *f*   fault(ing) [geological]
faille, ligne *f* **de**   fault line [geology]
faille *f* **de chevauchement**   thrust fault [geology]
failli *adj*   bankrupt
faillite *f*   bankruptcy
faire *vb*   make [in a calculation]
faire *vb* **circuler**   circulate [sth]
faire *vb* **des essais**   carry out tests/trials, test
faire *vb* **des études**   study
faire *vb* **des réparations (à)**   do a repair (on), make repairs (to)

faire *vb* **éclater** burst [sth]

faire *vb* **fonctionner** operate [sth], run [a machine]

faire *vb* **l'installation électrique de** wire up [a building]

faire *vb* **le plein** fill up [a tank]

faire *vb* **le levé (topographique) de** survey

faire *vb* **marche arrière** back, reverse [of a vehicle]

faire *vb* **marcher** run [a machine]

faire *vb* **pivoter** swivel [sth]

faire *vb* **sauter** blast [rock], blow up [e.g. a bridge], fuse [blow the fuse of]

faire *vb* **sauter à l'explosif** blast

faire *vb* **tourner** swivel [sth]

faire *vb* **un essai** test

faire *vb* **une soumission** submit a bid or tender

faisabilité *f* feasibility

faisable *adj* feasible

faisceau *m* beam [of light], bundle [e.g. of pipes, wires], pile group

fait *adj* **à l'aplomb de** plumbed with

fait *adj* **sur commande** custom-made, made to order

faîtage *m* ridge [roof], ridge capping [tile or metal]

faîte *f* height [summit], ridge [roof], rooftop, summit [of a roof, a hill]

faîteau *m* decorative ridge end tile

faîtière *adj* ridge [as in ridge tile, etc]

faîtière *f* ridge tile

faîtière *f* **à bourrelet** spigot-and-socket ridge tile

faîtière *f* **simple** plain ridge tile

falaise *f* cliff (face), steep rock face

fascine *f* fascine mattress

fanton *m* square steel rod used for reinforcing plaster work

farinage *m* lightening of paint colour with loss of gloss

farine *f* extremely fine grain aggregate [flour-like]

fatigue *f* fatigue [due to alternating stress]

fatigue, résistance *f* **à la** fatigue resistance

fatigue *f* **du métal** metal fatigue

faubourg *m* suburb

faussé *adj* buckled

fausse crémaillère *f* false string(er) applied to edge of a staircase

fausse équerre *f* bevel square, sliding bevel, T-bevel

fausse fenêtre *f* deadlight

fausse porte *f* non-functional [blind] door

fausse-aire *f* filling between joists [on top of which floor is placed]

fausse-alette *f* set back pier supporting an archway or lintel course

fausser *vb* break [a lock], buckle [metal], distort [a calculation or statistic]

fausser *vb* **une vis** cross a thread

faute *f* fault [blame]

faute *f* **de construction** building defect

faute *f* **dolosive** deliberately concealed defect

fauteuil *m* armchair

fauteuil *m* **de bureau** desk chair

faux *adj* distorted [report, etc], erroneous, untrue, false [ceiling, etc], inaccurate [instrument]

faux aplomb untrue [apparently vertical but in fact out of plumb]

faux entrait *m* collar beam [timber roof]

faux onglet *m* mitre cut [at other than 45°]

faux parement *m* brick facing [of a wall]

faux plafond *f* false or suspended ceiling

faux plancher *m* false or raised floor

faux, porter *vb* **à** be out of plumb

faux, tourner *vb* **à** run out of true

faux-attique *m* attic [small] above and behind an entablature

faux-bois *m* painted imitation wood or wood-graining

faux-comble *m* upper part of a mansard roof

faux-jour *m* small window providing very little light

faux-limon *m* wall string(er)

faux-marbre *m* marbling [paintwork]

faux-niveau *m* unlevel part

faux-pieu *m* extension of an insufficiently long pile

Fédération Nationale des Travaux Publics (FNTP) national federation of civil engineering contractors [France]

Fédération Nationale du Bâtiment (FNB) national federation of building contractors [France]

femelle *adj* female [connection, thread, etc]

femme *f* **d'affaires** businesswoman

fendre *vb* cleave, rive [wood, slate, etc], split [lengthwise], crack, fissure or part [ground, etc]

se fendre *vb* crack [of itself]

fendu *adj* cracked, split, slotted

fenestration *f* fenestration

fenêtre *f*, window

fenêtre, banquette *f* **de** window ledge or sill

fenêtre, ouverture *f* **de** window opening

fenêtre *f* **à axe horizontal** horizontally pivoted window

fenêtre *f* **à battants** casement window

fenêtre *f* **à charnières** hinged window

fenêtre *f* **à coulisse** sash window

fenêtre *f* **à guillotine** sash window

fenêtre *f* **à jalousies** window with openable glass louvres

fenêtre *f* **à l'anglaise** outward opening casement window

fenêtre *f* **à l'australienne** window with linked leaves, each pivoting about vertically sliding axis at one edge

fenêtre *f* **à l'italienne** single leaf window opening outwards with a vertically sliding axis near its upper edge

fenêtre *f* **à la canadienne** single leaf window opening inwards with a vertically sliding axis at its lower edge

fenêtre *f* **à la française** inward opening casement window

fenêtre *f* **à simple vitrage** single-glazed window

fenêtre *f* **à soufflet** fanlight [bottom hung]

fenêtre *f* **à un vantail/deux vantaux** single/double casement window

fenêtre *f* **basculante** horizontally pivoted window

fenêtre *f* **coulissante** sliding window

fenêtre *f* **de toit** rooflight

fenêtre *f* **en accordéon** folding window

fenêtre *f* **en baie** bay window

fenêtre *f* **en mansarde**   mansard window [in the lower slope of the roof]

fenêtre *f* **en saillie**   bay or bow window

fenêtre *f* **fixe**   deadlight

fenêtre *f* **mansardée**   mansard dormer window

fenêtre *f* **pivotante**   vertically pivoted window

fenêtre *f* **pli(ss)ante**   folding window

fenêtre *f* **simple**   single-glazed window

fenêtre *f* **treillagée**   lattice window

fente *f*   cleft, crack, cranny, slit, split, slot [e.g. in a screw head]

fente *f* **d'aération**   ventilation slot

fente *f* **de cœur**   heart shake [timber defect]

fente *f* **de prise d'air**   air inlet slot

fente *f* **de retrait**   drying or seasoning check, crack or split [timber defect], shrinkage crack

fente *f* **de rive**   edge shake [timber defect]

fenton *m*   dowel [for timber structure], tre(e)nail, metal tie

fer *m*   iron, plane iron or blade

fer *m* **à béton**   reinforcement (bar), reinforcing steel

fer *m* **à braser**   soldering iron

fer *m* **à joints**   jointing tool [for types of recessed pointing]

fer *m* **à souder**   soldering iron

fer *m* **à souder instantané**   high-speed soldering iron

fer *m* **à vitrage**   glazing bar, sash bar

fer *m* **battu**   wrought iron

fer *m* **blanc**   tinned sheet steel

fer *m* **carré**   square iron bar

fer *m* **coulé**   cast iron

fer *m* **d'armature**   reinforcing steel/reinforcement

fer *m* **de fonte**   cast iron

fer *m* **de liaison**   connecting rod [not locomotive]

fer *m* **de masse**   scrap iron

fer *m* **de rabot**   plane iron [blade]

fer *m* **de répartition**   distribution bars [reinforcement]

fer *m* **de reprise**   starter bars

fer *m* **(en) T/té**   T-bar, T-iron or section

fer *m* **en T/té à vitrage**   glazing T-bar

fer *m* **en tôle**   sheet iron

fer *m* **(en) U**   channel iron

fer *m* **forgé**   wrought iron

fer *m* **marchand**   bar steel, merchant bar

fer *m* **rond**   round reinforcement (bar)

fer *m* **soudé**   wrought iron

fers *m,pl* **plats**   steel strip up to 160mm wide

fers *m,pl* **larges**   steel strip wider than 160mm

fer-blanc *m*   tin-plate

ferblanterie *f* **en plomb**   lead flashing

ferme *adj*   firm, fixed, shut, switched off

fermé, à joint   close-jointed [rock, etc]

fermé *adj* **hermétiquement**   hermetically sealed

fermée, à texture   close-grained or -textured [rock, etc]

ferme *f*   farm (building), farmhouse, (roof) truss

ferme *f* **à âme pleine**   solid web truss

ferme *f* **à entrait brisé**   scissors truss

ferme *f* **à la mansarde**   mansard truss

ferme *f* **anglaise**   double pitch truss

ferme *f* **en acier**   steel roof truss

ferme *f* **en écharpes**   scissors truss

ferme *f* **en shed**   north light or sawtooth truss

ferme-imposte *m*   fanlight opener [remote]

ferme-porte *m*   door closer

fermer *vb*   close [a door, a circuit], lock, shut, shut off [electricity, gas]

fermer *vb* **à clef**   lock up [premises]

fermer *vb* **hermétiquement**   seal [hermetically]

fermette *f*   cottage, farmhouse [small]

fermeture *f*   catch, fastening, latch, closing (down, off, up), covering [e.g. of a manhole], shutting (down, off, up), switching off or out

fermeture *f* **à crémone**   casement bolt or espagnolette closure

fermeture *f* **à eaux**   water seal

fermeture *f* **annuelle**   annual closure

fermeture *f* **hydraulique**   water seal [in a siphon]

fermier *m*   farmer

fermoir *m*   heavy chisel [for cramping floorboards for nailing]

ferrage *m*   fitting door or window hardware

ferraillage *m*   reinforcement [for concrete: design, material and installation of]

ferraille *f*   scrap (metal)

ferrailler   reinforce [i.e. fit metal reinforcement]

ferrailleur *m*   scrap metal merchant, steelfixer, steel bender

ferritique *adj*   ferritic

ferrer *vb*   fit door or window hardware

ferrociment *m*   ferrocement

ferronnerie *f*   hardware, ironmongery, ironwork, wrought iron work

ferroviaire *adj*   rail [or railways]

ferrure *f*   door or window hardware, iron fitting, ironwork [piece of]

feu *m*   burner [on a cooking stove], fire

feu *m* **clignotant**   flashing signal light

feu *m* **ouvert**   open fireplace

feux *m,pl*   traffic lights

feux *m,pl* **de circulation**   traffic lights

feux *m,pl* **de signalisation**   traffic lights

feuil *m*   film [deposited on and adhering to something]

feuillard *m*   thin straight metal strip or sheet

feuille *f*   foil, sheet

feuille *f* **d'aluminium**   aluminium foil

feuille *f* **d'étanchéité**   flashing [lead], waterproof membrane [dam]

feuille *f* **de caoutchouc**   rubber sheet

feuille *f* **de goudron**   asphaltic layer

feuille *f* **de placage**   veneer

feuille *f* **de plomb**   sheet lead

feuille *f* **de polyéthylène**   polyethylene sheeting

feuille *f* **de recouvrement**   flashing [lead]

feuillet *m*   board 10–13mm thick [used in door panels, etc], leaf or layer [of a partition or wall, etc], thin sheet

feuilleté *adj*   laminated [in thin sheets]

feuilleter *vb*   divide or split into sheets

feuillure *f*   groove, rebate or recess [in woodwork, etc]

feutré *adj* felted, muffled, padded
feutre *m* felt [roofing, carpet, etc]
feutre *m* **à toit** roofing felt
feutre *m* **bitumé/bitumeux** bitumen felt
feutre *m* **bituminé chargé d'amiante** bitumen asbestos sheeting
feutrer *vb* felt
feux *m,pl* **nus** naked flame
fiabilité *f* dependability, reliability
fiable *adj* reliable
fibragglo *m* woodwool concrete
fibre *f* fibre
fibre *f* **comprimée** compression fibre
fibre *f* **de verre** fibreglass, glass fibre
fibre *f* **extrême** extreme fibre
fibre *f* **inférieure** bottom fibre
fibre *f* **moyenne** middle fibre
fibre *f* **neutre** neutral fibre
fibre *f* **optique** fibre optic(s), optical fibre
fibre *f* **tendue** tension fibre
fibre *f* **torse** spiral grain [timber defect]
fibreux *adj* fibrous
fibrociment *m* asbestos cement
ficeler *vb* tie up [a parcel]
fiche *f* electric plug [not socket], form or sheet [to be filled in], index card, peg or pin [used in jointing], pin [of a hinge]
fiche, interrupteur *m* **à** switched plug
fiche *f* **à nœud** loose-pin hinge
fiche *f* **à vase** loose-butt hinge
fiche *f* **de connexion** connector plug
fiche *f* **de travail** work sheet
fiche *f* **(femelle) de prolongateur** extension socket
fiche *f* **(mâle) à trois broches** three-pin plug
fiche *f* **(mâle) de prolongateur** extension plug
fiche *f* **(mâle) de sécurité** earthed plug
fiche *f* **(mâle) pour triphasé** three-phase plug
fiche *f* **technique** data sheet
ficher *vb* drive in [nail, post, etc], fix [sth, e.g. a nail, into sth], ram sand down into joints between slabs or stones
fichier *m* card index box, file [computer]
figure *f* figure [drawing]
fil *m* cable, wire, crack or fissure [in stone], grain [of wood]
fil *m* **à plomb** plumb bob or line, plummet
fil *m* **à souder** welding rod
fil *m* **d'acier** steel wire
fil *m* **d'amenée** feed wire
fil *m* **d'appel** call wire
fil *m* **d'étain** tin-based solder wire
fil *m* **de fer** steel wire
fil *m* **de fer barbelé** barbed wire
fil *m* **de l'eau** current [of water, direction of]
fil *m* **de liaison** connecting or connection wire
fil *m* **de masse** earth wire
fil *m* **de retour** return wire
fil *m* **de soudure** lead tin solder wire
fil *m* **de sous tension** live wire

fil *m* **de terre** earth wire
fil *m* **de treillis soudé** welded wire mesh
fil *m* **du bois** fibre direction [in wood]
fil *m* **fusible** fuse wire
fil *m* **métallique** wire
fil *m* **neutre** neutral wire
fil *m* **recuit** annealed wire
fil *m* **tors** spiral grain [timber defect]
fil *m* **tréfilé** drawn wire
filament *m* filament [of a lamp]
filament *m* **bispirale** coiled coil filament
filament *m* **électrode** coil [filament]
file *f* line [row]
file *f* **de rivets** line of rivets
filet *m* fillet [moulding separating two elements], fillet or water check [of mortar, between roof slope and wall], fillet [thin painted line], net, thread [of a screw], trickle [i.e. small flow]
filet *m* **à droite/gauche** right-/left-handed thread
filet *m* **allongé** long thread
filet *m* **de recueil/de sécurité** safety net
filet *m* **extérieur/mâle** male thread
filet *m* **intérieur** female/internal thread
filet *m* **renversé** left-handed thread
filet *m* **sec** very fine painted line
filet *m* **Whitworth** Whitworth thread
filetage *m* threaded connection, thread(ing) [of a screw]
filetage *m* **conique** taper thread
filetage *m* **de raccordement** connecting thread
filetage *m* **de tubes** pipe thread
filetage *m* **de vis** screw thread
filetage *m* **extérieur/intérieur** external/internal thread
filetage *m* **femelle/mâle** female/male thread
filetage *m* **pour bois** wood screw thread
fileter *vb* thread [cut or form a thread]
fil(et)eur *m* painter specializing in fillets
filiale *f* branch [of a company or organization], subsidiary (company)
filière *f* die [for threading], purlin
filler *m* filler [added to certain types of concrete, to plastics]
fillérisation *f* filling [i.e. adding a filler]
fillériser *vb* add a filler
film *m* film [thin, deposited on something], layer
film *m* **bitumineux** asphaltic layer
film *m* **protecteur** protective film
filon *m* seam or vein [geological]
fils *m,pl* **adhérents** bonded wires
filtrage *m* filtering or filtration
filtration *f* filtering or filtration, percolation
filtration *f* **de l'eau** water filtration
filtration *f* **naturelle** natural filtration
filtration *f* **par le sol** soil filtration
filtre *m* filter
filtre *m* **à air** air cleaner or filter
filtre *m* **à air à bain d'huile** oil bath air filter
filtre *m* **à air mouillé** wetted air filter
filtre *m* **à boues** sludge filter
filtre *m* **à cartouche** cartridge filter

filtre *m* **à charbon** charcoal filter
filtre *m* **à charpie** lint filter [dryer or washing machine]
filtre *m* **à choc** impact filter
filtre *m* **à combustible** fuel filter
filtre *m* **à déroulement** roll filter
filtre *m* **à eau** water filter
filtre *m* **à éléments non régénérables** disposable (element) filter
filtre *m* **à graisse(s)** grease filter
filtre *m* **à gravier** sand filter
filtre *m* **à haute efficacité** high-efficiency filter
filtre *m* **à huile** oil filter
filtre *m* **à manche** bag filter
filtre *m* **à permutite** permutite or zeolite filter
filtre *m* **à poussières** dust filter
filtre *m* **à sac** bag filter
filtre *m* **à tamis** strainer [for fluids]
filtre *m* **à vide** vacuum filter
filtre *m* **automatique** autoroll filter
filtre *m* **bactérien** bacterial filter
filtre *m* **cellulaire** cellular filter
filtre *m* **centrifuge** rotary filter
filtre *m* **d'air évacué** exhaust air filter
filtre *m* **d'huile** oil filter
filtre *m* **de graviers** gravel filter
filtre *m* **de la canalisation** pipeline strainer
filtre *m* **de tube** pipe filter
filtre *m* **désodorisant** odour filter
filtre *m* **duplex** duo filter
filtre *m* **électrostatique** electrostatic filter
filtre *m* **en charbon actif** activated charcoal filter
filtre *m* **en fibre sec** dry fabric filter
filtre *m* **en tissu** cloth filter
filtre *m* **épurateur** aerobic filter in septic tank or sewage system
filtre *m* **grossier** coarse filter
filtre *m* **inversé** inverted filter
filtre *m* **métallique** metal filter
filtre *m* **multicellulaire aspirateur** air filter cell
filtre *m* **primaire** primary filter
filtre *m* **rotatif** rotary filter
filtre *m* **sec** fabric filter
filtre *m* **secondaire** secondary filter
filtre *m* **ultra-fin** micro-filter
filtre-presse *m* sludge pressing plant
filtrer *vb* filter
fin *f* end
fin *f* **d'élévation** end elevation
fin *f* **de course** end of travel [e.g. of a lift]
finance *f* finance
financement *m* financing
financement, agence *f* **de** lending agency
financier *adj* financial
financier *m* financier, lender [lending company]
fines *f,pl* filler [in road material], fines [aggregate, etc]
fines *f,pl* **de charbon** slack [fine coal]
fini *m* finish
fini *m* **à la lisseuse** trowelled finish
fini *m* **à la taloche** float finish

fini *m* **au profileur** screeded finish
finir *vb* end [work, a task], finish
finissage *m* finishing
finisseur *m* **à asphalte** asphalt finisher or paver
finisseur *m* **à béton** concrete finisher [in a paving train]
finisseur *m* **routier** road finisher or paver
finition *f* finish, surface treatment
finitions *f,pl* second fixings
fioule *m* fuel/heating oil, (domestic) heating oil
firme *f* firm [company]
fisc[†] *m,* **le** Inland Revenue [French equivalent]
fiscal *adj* fiscal
fiscalité *f* financial system [national], tax situation
fissuration *f* cracking, fissuring
fissuration *f* **dans la soudure** weld cracking
fissuré *adj* cracked, fissured
fissure *f* crack, crevice, fissure, slit
fissure *f* **à la base** root crack
fissure *f* **capillaire** hair(line) crack
fissure *f* **de retrait** shrinkage crack
fissure *f* **fine** hair(line) crack
fissure *f* **superficielle** surface crack
fissure *f* **transversale** transverse crack
(se) fissurer *vb* crack, fissure
fixation *f* fastening, fixing [accessory for]
fixation *f* **de câble** cable fitting
fixe *adj* fast or permanennt [colour], fixed, determined, set [predetermined]
fixé *adj* **(à, sur)** fastened (to) or secured (to)
fixer *vb* arrange [e.g. a date, a meeting], fasten, determine, lay down, set
fixer *vb* **(à, sur)** fasten (to), fix (to), secure (to)
fixe-tapis *m* carpet tack
flache *f* depression or pothole [in a road, pathway, etc], dishing of a horizontal surface, wane [on the edge of sawn timber]
flacheux *adj* waney-edged [of timber]
flambage *m* buckling, lateral flexion
flambement *m* buckling, lateral flexion
flambement *m* **avec flexion** flexural buckling
flambement *m* **avec torsion** torsional buckling
flamber *vb* buckle
flamme *f* flame
flamme *f* **de coupe** cutting flame
flamme *f* **explosive** explosive flame
flamme *f* **neutre** neutral flame
flamme *f* **oxydante/réductrice** oxidizing/reducing flame
flamme *f* **soudante** welding flame
flanc *m* side [sloping, e.g. of an embankment, hill, etc]
flanc *m* **de coteau** hillside
flatslab *m* flat-slab
fléau *m* latch [straight bar, for gate, shutter, etc]
flèche *f* arrow, boom or jib [crane], conical roof space of a tower or spire, spire [conical or pyramidal], deflection [deflected distance], sag [distance]
flèche *f* **(d'un arc)** rise (of an arch)
flèche *f* **de cote** dimension arrow [on a drawing]
flèche *f* **de distribution de béton** concrete placing boom
flèche *f* **en treillis** lattice boom or jib

flèche *f* **permanente**   permanent deflection
fléchir *vb*   bend, deflect, sag, fail [beam]
fléchissement *m*   bending, bowing, deflection or sagging [of a beam, etc]
fleuret *m*   bit of a pneumatic hammer
fleuve *m*   river [with direct outflow to the sea]
flexibilité *f*   flexibility
flexible *adj*   flexible
flexible *m*   flexible connector or lead
fleximètre *m*   deflection indicator
flexion *f*   bending, flexing or flexion [under axial load], sagging [of a beam, etc]
flexion *f* **composée**   compound bending
flexion *f* **déviée**   bi-axial bending
flexion *f* **simple**   simple bending
flipeau/flipot *m*   piece or strip of wood used to disguise a crack, etc
flocage *m*   spraying with textile + glue or binder, for acoustic damping
flocon *m*   flake [snow, vermiculite, etc]
floculation *f*   flocculation
floculent *adj*   flocculant *adj*
floculer *vb*   flocculate
flottabilité *f*   buoyancy
flottage *m*   soaking of timber in water
flottant *adj*   floating, suspended [in a liquid]
flottant, pieu *m*   friction pile
flotter *vb*   float
flotteur *m*   ballcock, float [in a cistern], float valve
flotteur *m* **de commande**   float control
fluage *m*   creep, plastic flow
fluage *m*, **contrainte** *f* **de**   creep stress
fluatation *f*   hardening the surface of soft limestone with fluorosilicates
fluctuation *f*   fluctuation
fluctuation *f* **de charge**   load fluctuation
fluide *adj*   fluid *adj*
fluide *m*   fluid
fluide *m* **caloporteur**   heat transfer fluid
fluide *m* **de forage/de perforation**   drilling fluid
fluide *m* **frigoporteur**   coolant
fluide *m* **frigorigène**   refrigerant
fluidifiant *m*   plasticizer [for asphalt, bitumen, concrete]
fluidifié, bitumé *m*   cut-back bitumen
fluidique *f*   fluidics
fluidité *f*   fluidity, plasticity [of concrete]
fluor *m*   fluorine (gas)
fluorescence, lampe *f* **à**   fluorescent lamp
fluorescent, tube *m*   fluorescent tube
fluorine *m*   calcium fluoride
fluorocarbure *m*   fluorocarbon
fluvial *adj*   river *adj*
flux *m*   flow, flux [for brazing, etc]
flux *m* **à souder**   welding flux
flux *m* **de brasage**   brazing flux
flux *m* **de chaleur**   heat flux
flux *m* **de cisaillement**   shear flow
flux *m* **de trésorerie**   cash flow
flux *m* **laminaire**   laminar flow

flux *m* **lumineux**   luminous flux
flux *m* **massique**   mass flow
foisonnée, surface *f*   loose surface [of ground]
foisonnement *m*   bulk(ing) [increase in volume]
foisonner *vb*   bulk, expand, increase in volume, swell
fonçage *m*   boring or sinking [a well or shaft], driving (in) [piles]
fonçage *m* **horizontal**   thrust boring [e.g. for pipes under roads]
foncé *adj*   dark [colour]
foncer *vb*   darken a colour, drive [a pile], sink [a pile, shaft, well]
foncier *adj*   of or pertaining to land
fonction de, en   as a function of
fonctionnaire *m*   civil or public servant
fonctionnaire *m* **délégué**   statutory representative
fonctionnel *adj*   easy to run, functional, practical
fonctionnement *m*   functioning, operation or running [of equipment, machines], working [of apparatus, a company, etc]
fonctionnement *m* **continu**   continuous operation or running
fonctionnement *m* **du chauffage**   heater operation
fonctionnement *m* **silencieux**   quiet or silent operation or running
fonctionner *vb*   run [machine]
fond *m*   background, base, bottom [e.g. valley, swimming pool, etc], floor [of a valley], fireplace back, preliminary coat of paint
fond *m* **de coffrage**   bottom shuttering
fond *m* **de fouille**   trench bottom
fondant *adj*   melting *adj*, thawing *adj*
fondant *m*   flux, plasticizer [for asphalt, bitumen, concrete]
fondation *f*   base, foundation
fondation, couche *f* **de**   sub-base [road construction]
fondation, dalle *f* **de**   base or foundation slab
fondation, niveau *m* **du terrain de**   sub-grade level
fondation, sol *m* **de**   sub-grade, subsoil
fondation *f* **de la chaussée**   road foundation
fondation *f* **en redans**   stepped foundation
fondation *f* **isolée**   pad footing
fondation *f* **parasismique**   earthquake resistant foundation
fondation *f* **profonde**   deep foundation
fondation *f* **superficielle**   shallow foundation
fondation *f* **sur pieux/pilotis**   piled foundation
fondation *f* **sur puits**   pier foundation
fondation *f* **sur radier**   mat or raft foundation
fondation *f* **sur semelle**   strip foundation
fondation *f* **sur semelle continue**   long strip foundation
fondation *f* **traitée**   improved foundation [ground]
fondement *m*   basis, foundation
fonder *vb* **(sur)**   base (on), found (on)
fonderie *f*   foundry
fondre *vb*   melt (down, away), mix [several colours of paint together], thaw
fondre *vb* **un fusible**   blow a fuse
fonds *m*   capital, funds

fonds *m* **de roulement**   working capital
fonds *m* **de roulement net**   net current assets
fondu *m*   melting or fading of one colour into another
fongicide *adj*   fungicidal
fongicide *m*   fungicide
fontaine *f*   fountain, spring [of water]
fonte *f*   cast iron, casting [the action]
fonte *f* **blanche**   white iron
fonte *f* **d'acier**   cast steel
fonte *f* **ductile**   ductile cast iron
fonte *f* **émaillée**   glazed cast iron
fonte *f* **grise**   grey cast iron
fonte *f* **malléable**   malleable cast iron
fonte *f* **salubre**   spun iron pipes with spigot-and-socket joints
forage *m*   boring or drilling [a tunnel, pile, hole], sinking [a borehole, a well], bore-hole, drill-hole, hole
forage *m*, **fluide** *m* **de**   drilling fluid
forage *m* **au tube carottier**   tube sample boring
forage *m* **d'essai**   trial boring
forage *m* **d'injection**   grout hole
forage *m* **par délavage**   wash-out boring
forage *m* **tubé**   cased or lined borehole
forcé *adj*   forced
forcée, vibration *f*   forced vibration
force *f*   force, strength
force *f* **centrifuge**   centrifugal force
force *f* **de cisaillement**   shear force
force *f* **de collage**   adhesive strength
force *f* **de compression**   compressive force
force *f* **de freinage**   braking force
force *f* **de frottement**   friction force
force *f* **de liaison**   bond strength
force *f* **de rappel**   restoring force
force *f* **de traction**   drawbar pull, tensile force
force *f* **électromotrice**   electromotive force
force *f* **frictive**   drag, friction force
force *f* **longitudinale**   longitudinal force
force *f* **motrice**   motive force
force *f* **portante**   bearing capacity [e.g. of a pile]
force *f* **transversale**   transverse force
forée, clef *f*   key with a hollow shank
forer *vb*   bore or drill [tunnel, pile, hole], sink [a borehole, a well]
forestier *adj*   forest *adj*
foret *m*   twist drill (bit)
foret *m* **de maçonnerie**   masonry drill
foret *m* **hélicoïdal**   twist drill (bit)
forêt *f*   forest
forêt *f* **domaniale**   communal/public forest
forêt *f* **feuillue**   deciduous forest
foreur *m*   driller
foreuse *f*   boring machine, drilling machine or unit
forfait *m*   fixed price, lump sum
forfait, prix *m* **à**   contract price, lump sum (price)
forfait, vente *f* **à**   outright sale
forfaitaire *adj*   all-in or inclusive [e.g. of fee, payment]
forge *f*   forge, smithy
forger *vb*   forge [metal]

forgeron *m*   smith [ironworker]
formaldéhyde *m*   formaldehyde
formation *f*   bed [e.g. under paving stones], formation or forming [process of], training
formation *f* **d'incrustation**   scale formation
formation *f* **de sédiments**   build-up or deposit of sediment
formation *f* **des travailleurs**   worker training
formation *f* **professionnelle**   professional training
forme *f*   bed(ding) [for paving stones, tiling, etc], form or mould, formation [undisturbed ground], shape
forme, classe *f* **de**   particle shape category
forme *f* **de radoub**   graving dock
forme *f* **des barres**   shape of bars [reinforcement]
forme *f* **des grains**   particle shape
forme *f* **sèche**   dry dock
formé *adj*   trained
formé à froid   cold formed
former *vb*   build up or form [a collection, deposit, etc], train [in a skill]
formeret *m*   wall rib [of an arch]
formulaire *m*   form [document]
formule *f*   formula
formule *f* **chimique**   chemical formula
formule *f* **de battage**   pile driving formula
formule *f* **de calcul/de dessin**   design formula
formule *f* **de révision des prix**   fluctuations formula [costing, pricing], price revision formula
fort *adj*   strong
fort *m*   fort(ress)
fosse *f*   hole, sump or pit [in the ground], inspection pit
fosse *f* **à cendres**   ash pit
fosse *f* **à compteur**   meter pit
fosse *f* **à graisse**   grease trap
fosse *f* **à purin**   midden pit, sump for liquid manure
fosse *f* **d'aisance**   cesspit or cesspool
fosse *f* **d'orchestre**   orchestra pit
fosse *f* **de vidange**   blow down-pit, drainage pit [e.g. for draining oil from a vehicle]
fosse *f* **étanche**   sealed sewage tank
fosse *f* **septique**   septic tank
fossé *m*   open ditch or drain, surface drain, trench, moat [of a fortress]
fossé *m* **collecteur**   collector drain or drainage ditch
fossé *m* **couvert**   French drain
fossé *m* **d'irrigation**   irrigation channel
fossé *m* **de crête**   intercepting ditch or channel [road construction]
fossé *m* **de drainage**   drainage channel or ditch, open drain
fossé *m* **herbacé**   grassed ditch
fossile *adj*   fossil *adj*
fossile *m*   fossil
fouille *f*   cavity, pit, digging, excavation, trenching
fouille, fond *m* **de**   trench bottom
fouille, réalisation *f* **d'une**   excavation [the process]
fouille *f* **blindée**   lined or timbered excavation
fouille *f* **de fondation**   foundation digging or excavation
fouille *f* **en rigole**   trenching, step-by-step excavation during underpinning

fouille *f* **en tranchée**   trenching
fouille *f* **non blindée**   unlined excavation
fouille *f* **ouverte**   open excavation
fouiller *vb*   dig, excavate, trench
foulée *f*   flight [of stairs]
four *m*   furnace, oven
four *m* **à chaux**   lime kiln
four *m* **à coke**   coke oven
four *m* **à gaz**   gas-fired furnace, gas oven
four *m* **à micro-ondes**   microwave oven
four *m* **à sécher**   drying oven
four *m* **d'incinération**   incinerator
four *m* **de chaudière**   heating furnace
four *m* **de normalisation**   normalizing furnace
four *m* **de recuit**   annealing furnace
fourche *f*   fork
fourchette *f*   bracket or range [of prices]
fourgon *m*   lorry, van [large], security van
fourgonnette *f*   light van
fourneau *m*   furnace, hole [into rock for blasting charge]
fourneau *m* **à gaz**   gas stove
fourneau *m* **à gaz industriel**   industrial gas furnace
fourneau *m* **de construction**   heating or cooking stove enclosed in masonry
fournir *vb*   furnish, provide, supply
fournisseur *m*   supplier
fourniture *f*   materials [supply of, as opposed to labour], supply or provision [of goods, supplies, etc], furnishing, providing or supplying
fourreau *m*   pipe chase or sleeve [in masonry, concrete work], protective sleeve or duct [for services]
fourrer *vb*   fill up a space [e.g. between ridge and sloping tiles]
fourrure *f*   firring, packing [of a joint]
foyer *m*   firebox, grate, fireplace, hearth(stone), focus [optical], home [house], household
foyer *m* **à grille (mécanique)**   chain grate stoker
foyer *m* **à tirage mécanique**   forced draught combustion chamber
foyer *m* **d'antifriction**   anti-friction bush or lining [bearing]
foyer *m* **de chaleur**   heat source
foyer *m* **de l'incendie**   seat of the fire
foyer *m* **soufflé**   forced draught combustion chamber
fraction *f*   fraction or percentage
fragile *adj*   brittle
fragile, rupture *f*   brittle failure
fragilité *f*   brittleness, fragility
fragment *m*   fragment
fragmentaire *adj*   fragmental, fragmentary
fraîcheur *f*   coolness, freshness [of colours]
frais *adj*   fresh [air, concrete, etc]
frais *m,pl*   costs, expenditure, expenses, charge [fee, etc]
frais *m,pl* **d'acquisition**   costs of buying or purchase
frais *m,pl* **d'agence**   agency fees
frais *m,pl* **d'entretien**   maintenance costs
frais *m,pl* **d'envoi**   postage (costs)
frais *m,pl* **d'exploitation**   operating costs or expenses, running costs

frais *m,pl* **de chauffage**   heating costs
frais *m,pl* **de déplacement**   travelling expenses
frais *m,pl* **de maintenance**   maintenance costs
frais *m,pl* **de notaire**   legal costs/expenses [buying/selling property]
frais *m,pl* **fixes**   fixed costs
frais *m,pl* **généraux**   business expenses, overheads
frais de qch, faire *vb* **les**   bear the costs of something
frais *adj*   cool, fresh [e.g. of concrete]
fraisage *m*   countersinking, cutting, milling
fraise *f*   countersink (bit), cutter, drilling tool, reamer
fraise *f* **à dénoder**   knot-hole cutter
fraiser *vb*   countersink, mill, open out [e.g. the end of a pipe]
fraiseuse *f*   milling machine
fraiseuse *f* **à chaîne**   chain mortiser
fraiseuse *f* **à froid**   cold planer [roadworks]
fraiseuse *f* **à queue**   tenoner
fraiseuse *f* **routière**   pavement profiler
France Télécom   national telephone service [in France]
franchir *vb*   span (over)
franchise *f*   excess [on insurance claim], exemption
franchise *f* **de loyer**   rent-free period
franchissable *adj*   crossable [e.g. of a flush kerb]
frein *m*   brake
frein *m* **de secours**   emergency brake
freins *m,pl* **à main**   handbrake
freinage *m*   braking
frêne *m*   ash [timber]
fréon *m*   freon
fréquence *f*   frequency [electrical, statistical]
fréquence *f* **de mise en circuit**   switching cycle
fréquence *f* **de résonance**   resonant frequency
fréquence *f* **de vibration**   vibration frequency
fréquence *f* **naturelle/propre**   natural frequency
fréquence *f* **porteuse**   carrier frequency
fret *m*   freight, cargo
frettage *m*   binding [around something]
frette *f*   binding in ring or spiral form to restrain splitting or swelling, hoop [reinforcement]
fretter *vb*   bind [to prevent splitting or swelling]
friable *adj*   friable
friction *f*   friction
frigidaire *m*   refrigerator
frigo *m*   fridge, refrigerator
frigorie *f*   French unit of cold [= calorie]
frigorifier *vb*   refrigerate
frigorigène *m*   refrigerant
frise *f*   frieze [architectural], floorboard or strip 60–80mm wide
frisette *f*   planed and grooved floorboard, 65–80mm wide
frisette *f*, **double**   planed and grooved floorboard, 45–65mm wide
frittage *m*   sintering
friture *f*   crackle, crackling or intereference [radio/PA/phone]
froid *adj*   cold, cool
froid *m*   cold *n*, refrigeration
froid, démarrer *vb* **à**   cold start

froid, laminer *vb* **à**   cold roll
froid, mettre *vb* **en œuvre à**   lay cold [road materials]
froid, souder *vb* **à**   cold weld
front *m*   front [meteorological]
front *m* **d'attaque**   face [in underground excavation]
front *m* **de mer**   seashore
fronteau *m*   fronton [small, above a door]
fronton *m*   fronton, pediment
frottement *m*   friction, rubbing
frottement, angle *m* **de**   friction angle
frottement, résistance *f* **au**   frictional resistance
frottement *m* **interne**   internal friction
frottement *m* **interne, angle** *m* **de**   natural angle of
   repose
frottement *m* **latéral/superficiel**   skin friction
frottement *m* **négatif**   negative skin friction
frotter *vb*   rub
frottis *m*   thinnest possible coat of paint
frotture *f*   damage to wood surface resulting from an
   impact
fruit *m*   batter, or inwards inclination or leaning [opposite
   of overhang]
fuel *m* **(de chauffage)**   fuel or heating oil
fuel *m* **domestique**   domestic heating oil
fugitif *adj*   intermittent [of a defect]
fuir *vb*   give way [ground], leak
fuite *f*   escape, seepage, leak or leakage
fuite *f* **d'air**   air exfiltration or leak

fuite *f* **d'eau**   water leak
fuite *f* **de conduit/tuyau**   duct/pipe leak or leakage
fuites *f,pl*   leakage
fumée *f*   fumes, smoke
fumée *f*, **trappe de**   control damper or throat restrictor
   [fireplace chimney]
fumisterie *f*   flue, technology of chimneys, flues,
   fireplaces, stoves, etc
fumivore *adj*   retaining smoke particles
fumoir *m*   smoking room
funiculaire *m*   cable or funicular railway, catenary
   curve
furet *m*   flexible pipe cleaner
fusibilité *f*   fusibility
fusible *adj*   fusible
fusible *m*   (safety) fuse
fusible *m* **à cartouche**   cartridge fuse
fusible *m* **de sûreté**   cut-out or safety fuse
fusible-cartouche *m*   cartridge fuse
fusion *f*   fusion or melting [of metals, etc], merger
   [companies, etc]
fusion *f* **de glace**   de-icing
fusionné *adj*   merged
fût *m*   barrel or cask, bole [of a tree], brace [of brace and
   bit], handle [of a saw], stock [of a plane], central part of
   pier [bridge], shaft [of a pile], shank [of a rivet], trees
   [cluster, plantation, etc]
futaie *f*, **haute**   mature (standing) timber

gabarit, hors   out of gauge or oversize [e.g. of load]
gabarit *m*   gauge [template or size gauge], jig, template,
   outline [of a building, gauge, template]
gabarit *m* **d'écartement**   spacing gauge or template
gabarit *m* **de chargement**   loading gauge
gabbro *m*   gabbro
gabion *m*   gabion
gable *m*   gable [old fashioned word in French]
gâchage *m*   mixing [of mortar, etc]
gâche *f*   pipe clip [collar], keeper or strike plate of a
   lock
gachée *f*   batch [of concrete]
gâcher *vb*   add water [to plaster, etc, for mixing], mix
   [adding water to]
gâcheur *m*   mason's labourer, setter out [of timber
   structural frameworks], site foreman [structural timber]
gâcheux *adj*   muddy
gâchis *m*   mud, wet mortar
gain *m*   gain
gain *m* **de chaleur**   heat gain
gain *m* **de chaleur occasionnel**   casual heat gain
gain *m* **par conduction**   conduction (heat) gain

gains *m,pl* **de chaleur variable**   incidental heat gain
gaine *f*   conduit, duct or shaft, sheath [of a cable], sleeve
   [sheath]
gaine *f* **d'aération**   ventilation duct or shaft
gaine *f* **d'air**   air duct
gaine *f* **d'alimentation**   supply air duct
gaine *f* **d'ascenseur**   elevator or lift shaft or well
gaine *f* **d'entrée**   inlet duct
gaine *f* **d'évacuation**   exhaust duct
gaine *f* **d'extraction**   extract duct
gaine *f* **de câbles**   cable duct or protection sleeve
gaine *f* **de distribution d'air**   air distribution duct
gaine *f* **de recirculation**   recirculation duct
gaine *f* **de reprise**   return air duct
gaine *f* **de retour d'air**   return air duct
gaine *f* **de service**   service duct
gaine *f* **de ventilation**   ventilation duct or shaft
gaine *f* **haute vitesse**   high-velocity duct
gaine *f* **informatique**   communications duct, riser or shaft
gaine *f* **souple**   flexible duct
gaine *f* **technique**   service(s) duct or riser
galandage *m*   partition [of uprights and plaster, brick or
   similar infill]

galbe *m*   contour or profile [of an architectural element]

galerie *f*   balcony [of a theatre], culvert [dam], duct [underground, for services], gallery [long room, mining, theatre, etc], landing [large, serving several rooms], tunnel [gallery, in a dam, a mine]

galerie *f* **d'accès**   access gallery or tunnel, adit [dam]

galerie *f* **d'amenée**   headrace (tunnel) or inlet tunnel [dam]

galerie *f* **d'exploration**   adit [mining]

galerie *f* **d'inspection**   inspection gallery

galerie *f* **de dérivation**   diversion tunnel

galerie *f* **de fuite**   outlet tunnel [dam], tailrace

galerie *f* **de prise**   intake tunnel [dam]

galerie *f* **de reconnaissance**   heading [geology]

galerie *f* **de vidange**   outlet tunnel [dam]

galerie *f* **de visite**   inspection gallery

galerie *f* **des câbles**   cable tunnel

galerie *f* **en charge**   pressure tunnel [dam]

galerie *f* **technique**   service(s) gallery

galet *m*   cobble, pebble, roller [metal]

galet *m*, **petit**   pebble

galets *m,pl*   shingle [small stones]

galetas *m*   garret

galipot *m*   tar-based wood protective coating

galvanisation *f*   galvanizing, zinc coating

galvanisé *adj*   galvanized

galvanisé *adj* **à chaud/au bain**   hot-dip galvanized

galvaniser *vb*   galvanize

gamme *f*   range [of colours, products, etc]

gamme *f* **d'accessoires**   range of accessories

gant *m*   glove

gant *m* **de protection**   protective glove

gants *m,pl* **de travail**   work/working gloves

gants *m,pl* **en caoutchouc**   rubber gloves

garage *m*   docking [of boats], garage [*not* petrol station], storage [of wheeled vehicles]

garage, voie *f* **de**   siding [railway]

garant *m*   guarantor

garantie *f*   guarantee, security [for loan]

garantie *f* **biennale**   two-year guarantee [on minor work, in France]

garantie *f* **de bonne fin**   completion bond, performance bond or guarantee

garantie *f* **décennale**   10-year guarantee [on major work, in France]

garantir *vb*   guarantee, warrant

garantir *vb* **contre**   insure against

garçon *m* **de relais**   general site labourer

garçon-maçon *m*   mason's labourer [French regional expression]

garçon-plombier *m*   plumber's mate [French regional expression]

garde *f* **au gel**   depth of base of foundations below ground to avoid freezing

garde *f* **d'eau**   seal or effective height of water seal in a siphon

garde *f* **de serrure**   metal plate preventing wrong-shaped key entering a lock

garde-corps *m*   balcony [guard] rail, guard or safety rail, handrail, parapet

garde-feu *m*   fire guard or screen

garde-fou *m*   balcony [guard] rail, guard or safety rail, handrail, parapet

garde-gravois *m*   guard board to retain building rubbish from falling

garde-jambe *m*   leg guard

garde-main *m*   hand guard

garde-manger *m*   larder, pantry

garde-neige *m*   snow guard, rail or trap [on a roof]

garde-pieds *m*   footboard, guard board, toeboard

garde-robe *f*   closet, clothes cupboard, wardrobe

garde-verre *m*   protective grille over glazing

gardien *m*   caretaker, janitor, watchman

gardiennage *m*   security measures [e.g on site]

gare *f*   (railway) station

gare *f* **de marchandises**   goods station

gare *f* **de triage**   marshalling yard

gare *f* **de voyageurs**   passenger station

gare *f* **routière**   bus station, lorry or truck depot

gargouille *f*   gargoyle, shoe [at foot of downpipe]

garni *m*   random rubble fill

garnir *vb*   fill [a road or bridge joint], pack or stuff [a gland], render [heavily to bring surface to desired plane]

garnissage *m*   lagging or lining [of a boiler], packing or stuffing [of a gland, etc], rendering used to bring a surface to desired plane

garnissage, produits *m,pl* **de**   joint filling materials [road or bridge joints]

garniture *f*   packing [gland, tap, valve, etc]

garniture *f* **d'amiante**   asbestos packing

garniture *f* **d'étanchéité**   sealing strip

garniture *f* **de frein**   brake lining

garniture *f* **de joint**   joint sealing compound

garniture *f* **de presse-étoupe**   packing [for a gland or stuffing-box]

garrigue *f*   heath and moorland

garrot *m*   piece of wood used for tightening a bow saw

gas-oil *m*   diesel fuel/oil, gas oil

gaspiller *vb*   waste

gauche *adj*   deformed, twisted

gauchi *adj*   out of true [bent, distorted]

gauchir *vb*   buckle [metal]

se gauchir *vb*   distort or warp

gauchis *m*   curvature or twist [out of required plane]

gauchissement *m*   buckling, deviation or departure from planarity, twist [of timber], warping

gaz *m*   gas

gaz *m* **à l'eau**   water gas

gaz *m* **brûlé**   burnt or waste gas

gaz *m* **carbonique**   carbon dioxide

gaz *m* **combustible**   fuel gas

gaz *m* **comprimé**   compressed or high pressure gas

gaz *m* **d'échappement**   exhaust gas

gaz *m* **d'huile**   oil gas

gaz *m* **de chauffage**   heating gas

gaz *m* **de combustion**   combustion gas

gaz *m* **de curage**   sewage or sewer gas

gaz *m* **de fumée**   fumes, flue or waste gas

gaz *m* **de gazogène**   producer gas

gaz *m* **de pétrole liquéfié**   liquefied petroleum gas (LPG)

gaz *m* de remplissage   refill gas [of a lamp]
gaz *m* de ville   town gas
gaz *m* extrait   exhaust gas
gaz *m* naturel   natural gas
gaz *m* rare   inert or rare gas
gaz *m* rejeté   exhaust gas
gaz *m* traceur   tracer gas
gazéification *f*   gasification
gazéifier *vb*   gasify
gazier *m*   gas fitter
gazeux *adj*   gaseous
gazinière *f*   gas cooker
gazoduc *m*   gas pipeline
gazogène *adj*   gas producing
gazogène *m*   gas producer (plant)
gazole *m*   diesel fuel/oil
gazomètre *m*   gasometer or gas holder
gazon *m*   grass [short], lawn, sward, turf [in situ]
gazon, plaque *f* de   turf [piece for laying]
gazonnement *m*   turfing
gazonner *vb*   turf
gel *m*   freezing, frost
gel, gonflement dû au   frost heave
gel, insensible au   frost resistant
gel, résistance *f* au   frost resistance
gel, résistant au   frost resistant
gel, zone *f* de pénétration du   freezing zone
gel *m* de silice   silica gel
gel *m* des sols   ground freezing
gelé *adj*   frozen
gelée *f*   frost
gelée *f* blanche   hoar-frost
geler *vb*   freeze [sth]
se geler *vb*   solidify [by itself by freezing]
gélif *adj*   frost-riven [e.g. stone], frost-sensitive/
   -susceptible [of stone, etc]
gélivité *f*   frost sensitivity or susceptibility [of stone, etc]
gélivité *f*, essai *m* de   freezing or frost sensitivity test
gélivure *f*   frost crack [e.g. in stonework, masonry,
   timber]
géminé *adj*   paired or twin [e.g. of columns, arches]
gendarmerie *f*   police station
générateur *m*   generator
générateur *m* d'acétylène   acetylene generator
générateur *m* d'air chaud   warm air heater/heating unit
générateur *m* de chaleur   heat generator
générateur *m* de gaz   gas generator
générateur *m* de vapeur   steam boiler or generator
génératrice *f*   generator [electrical]
génératrice *f* à courant continu   direct current
   generator
génératrice *f* pour soudage   welding generator
génie *m*   art of construction
génie *m* civil   civil or construction engineering
génie *m* climatique   air-conditioning or HVAC
   engineering
génie *m* industriel   industrial engineering
génie *m* maritime   maritime engineering
génie *m* mécanique   mechanical engineering

génie *m* sanitaire   sanitary or public health (PH)
   engineering
genouillère *f*   articulation [between two rigid members],
   jointed lamp support [one to three sections], kneepad
gentilhommière *f*   country seat, manor house
géodésie *f*   geodesy, surveying
géologie *f*   geology
géologie *f* de l'ingénieur   engineering geology
géologique *adj*   geological
géologue *m*   geologist
géométral *m*   drawing showing overall dimensions of a
   building
géomètre *m*   land surveyor
géomètre-expert *m*   land surveyor [dealing with local
   matters in France]
géomètre-topographe *m*   land surveyor [dealing with
   larger areas, major developments, etc in France]
géométrie *f*   geometry
géomorphologie *f*   geomorphology
géomorphologique *adj*   geomorphological
géophysique *adj*   geophysical
géophysique *f*   geophysics
géotechnique *adj*   geotechnical
géotechnique *f*   geotechnics
géothermie *f*   study of geothermal phenomena
géothermique *adj*   geothermic
gérant *m*   manager
gerce *f*   drying/seasoning check, crack or split [timber
   defect]
gercement *m*   cracking [of timber]
gercer *vb*   crack [ground, timber]
gerçure *f*   hair(line) crack, shake [timber defect]
gérer *vb*   administer, manage
germe *f*   germ
germicide *adj*   germicidal
gestion *f*   administration, management
gestion *f* d'accès   access control/management
gestion *f* d'immeuble   building management
gestion *f* de l'énergie   energy management
gestion *f* du patrimoine   property management
gestion *f* du projet   project management
gestion *f* technique centralisée (GTC)   building
   management system (BMS)
gestion *f* technique de bâtiment   building management
   system (BMS)
gestionnaire *m* de chantier   site agent [manager]
gestionnaire *m* de projet   project manager
giboulée *f*   sudden shower [of rain]
gicleur *m*   atomizer [fuel jet in oil-fired boiler]
gicleur *m* de brûleur à fuel   oil burner nozzle or jet
gilet *m* de sauvetage   life jacket
girafe *f*   side-tipping wagon
giron *m*   tread [of a stair]
gironné *adj*   trapezoidal [of plain tiles for pyramidal
   roofs]
girouette *f*   weathercock, weather vane, wind vane
gîtage *m*   total structure supporting a floor
gîte *m*   floor support, lodging
gîte *m* rural   country lodging to let

givrage *m*   frosting/frost formation

givre *m*   frost [surface deposit, not weather], hoar-frost

givrer *vb*   frost

givrure *f*   frost crack, heart shake [timber defect]

glace *f*   plate or sheet of glass, ice, mirror

glace, lentille *f* **de**   ice lens [in ground]

glacial *adj*   glacial

glacier *m*   glacier

glacis *m*   bank or ramp [regular and with even slope], slope [regular], semi-transparent oil paint

glaise *f*   clay

glaiseux *adj*   clayey

glaisière *f*   clay pit

glisse-main *m*   banister rail, handrail (of balustrade)

glissement *m*   sliding, slip(ping)

glissement, élément de   safety fence [road]

glissement, plan *m* **de**   slip plane

glissement, surface *f* **de**   slip surface

glissement *m* **circulaire**   rotational slip [geotechnics]

glissement *m* **de pente**   slope failure

glissement *m* **de remblai**   embankment failure

glissement *m* **(de terrain)**   landslide or landslip

glisser *vb*   slide, slip

glissière *f*   slide or slide rail, sliding channel

glissière, panneau *m* **à**   sliding panel

glissière, porte *f* **à**   sliding door

glissière, système *m* **à**   sliding device or system

glissière *f* **de sécurité**   crash or safety [highways] barrier

glissière *f* **des lumières**   lighting track

glissière *f* **double de sécurité**   double-sided crash or safety barrier [highways]

glissière *f* **métallique de sécurité**   steel beam crash or safety barrier [highways]

glissière *f* **simple de sécurité**   single-sided crash or safety barrier [highways]

gneiss *m*   gneiss

gnomon *m*   blade of a sundial

gobelet *m*   ring to which upper ends of rafters in a conical roof are fixed

gobetage *m*   roughing-in [plastering or rendering]

gobeter *vb*   rough in [before final coat of plaster or rendering]

gobetis *m*   roughing-in [plaster or render]

godet *m*   bucket [of a dredger, excavator, etc]

godet *m* **de pelle** *f*   excavator bucket

godet *m* **graisseur**   lubricating cup

golf *m*   golf course

golfe *m*   bay, gulf

gommage *m*   light sandblasting

gommé *adj*   lightly sandblasted

gommer *vb*   lightly sandblast

gond *m*   hinge bracket and pin [of a strap hinge]

gonflage *m*   inflation [physical rather than monetary], swelling [inflating]

gonflement *m*   heaving [of ground, road surface], inflation [physical rather than monetary], swelling [inflation]

gonflement *m* **dû au gel/par le gel**   frost heave

gonfler *vb*   blow up [inflate, *not* explode], inflate, pump up, swell, expand [inflate, swell], heave [ground, etc]

gonne *f*   barrel/drum [for tar]

gorgé d'eau   saturated [ground]

gorge *f*   concave circular moulding, gorge, narrowest part of throat [of a chimney]

gorge *f* **d'écoulement**   condensation gutter

goudron *m*   tar

goudron, béton *m* **au**   tar-concrete

goudron, émulsion de   tar emulsion

goudron *m* **fluide**   cold tar

goudron-bitume *m* **(GB)**   dense tar

goudronné *adj*   tarred

goudronneuse *f*   hot-mix (tar) spreader, tar-sprayer/tar spraying machine

goudronneux, coulis *m*   tar slurry

gouge *f*   gouge, hollow chisel

goujon *m*   dowel, pin, shear connector [composite construction], stud bolt or screw

goulet *m*   gully [natural feature]

goulotte *f*   cable ducting, dado/skirting trunking, mini-trunking [electrical], chute [of wood, for building materials or rubbish], rainwater channel on top of cornice, shallow trench to carry away rainwater

goupille *f*   pin [e.g. to retain a doorknob]

goupille *f* **(fendue)**   split pin

gousset *m*   bracket [supporting a shelf], gusset (plate), haunch [in concrete structure]

goutte *f*   drip [drop]

goutte *f* **d'eau**   drip [under a soffit, window ledge, etc], dripstone

gouttelette *f*   droplet

gouttière *f*   gully [gutter], gutter [on building], rain gutter, rainwater downpipe

gouttière *f* **à l'anglaise**   metal roof gutter [rounded, metal feet]

gouttière *f* **collectrice**   collecting channel

gouttière *f* **havraise/nantaise**   triangular section gutter fixed under the eaves course

gouttière *f* **pendante**   suspended gutter

gradateur *m*   dimmer (switch)

gradation *f*   gradation, gradual process

grader *m*   grader

gradient *m*   gradient

gradient *m* **de hydraulique**   hydraulic gradient

gradient *m* **de pression**   pressure gradient

gradient *m* **de température**   temperature gradient

gradient *m* **thermique**   thermal gradient

gradin *m*   step [of terracing, or control system]

gradine *f*   mason's toothed chisel

gradué *adj*   graded [in terms of particle size]

grain *m*   grain or particle, grain [texture], grain structure [of material]

grain, forme *f* **de**   particle shape

grain d'orge, assemblage *m* **à**   V-shaped edge or end joint [timber]

grain *m* **de la cassure**   fracture grain

grain *m* **de poussière**   dust particle

grain fin, à   fine-grained

grains, dimension *f* **des**   particle size

grains, essai *m* **de forme des**   particle shape test

graisse *f*   grease

graisser *vb*   grease, lubricate, oil

graisseur *m*   grease nipple, lubricator

grammage *m*   weight [grams/m^2]

grand collecteur *m*   main collector drain or sewer

grand magasin *m*   department store

grandeur *f*   size [general, and height, of a tree, etc]

grandeur nature, en   full-scale *adj*

grandeur *f* **nature**   full scale

grange *m*   barn

granit(e) *m*   granite

granito *m*   terrazzo

granulaire *adj*   granular

granulaire, classe *f*   grain size class or range

granularité *f*   particle size distribution

granularité, de grosse   coarse-textured [sand, aggregate]

granularité fine, de   fine-textured [sand, aggregate]

granulats *m,pl*   (particles of) aggregate

granulats *m,pl* **anguleux**   angular particles of aggregate

granulats *m,pl* **bien gradués**   well-graded particles of aggregate

granulats *m,pl* **concassés**   broken particles of aggregate

granulats *m,pl* **cubiques**   cubic particles of aggregate

granulats *m,pl* **fins**   fine aggregate

granulats *m,pl* **gros**   coarse aggregate

granulats *m,pl* **minéraux**   mineral aggregate

granulite *m*   fine-grained granite

granulométrie *f*   grain or particle size grading or distribution

granulométrique *adj*   granulometric

granulométrique, analyse *f*   analysis of grain or particle size, sieve analysis

granulométrique, composition *f*   grain or particle size grading or distribution

graphique *adj*   graphic

graphique *m*   graph(ic), graphical representation, chart [diagram], plot [graph]

graphite *m*   graphite

grappin *m*   grab [bucket]

gras *adj*   fat or rich [of a plaster or cement rich in binder]

grattage *m*   stripping of paint or wallpaper by scraping

grattage *m* **à vif**   scraping down to bare plaster before repapering

gratter *vb*   scrape, scratch

grattoir *m*   scraper

gratton *m*   devil float

gravats *m,pl*   demolition rubbish

grave *f*   aggregate [for road construction]

grave *f* **bien graduée**   well-graded aggregate

grave *f* **ciment**   lean concrete with pebble, gravel or sand aggregate

grave *f* **concassée**   broken aggregate

graver *vb*   cut or engrave [a design on a material]

gravier *m*   gravel

gravier, petit   pebble

gravier *m* **concassé**   crushed gravel

gravier *m* **d'alluvion**   alluvial gravel

gravier *m* **de filtrage**   filter gravel

gravier *m* **roulé**   rolled gravel

gravière *f*   gravel pit

gravillon *m*   fine or pea gravel

gravillonnage *m*   surfacing with chippings or grit spread on tar

gravillonner *vb*   spread gravel or chippings on tar

gravillonneuse *f*   grit or chippings spreader, gritter, spreader

gravillons *m,pl*   (loose) chippings [stone]

gravillons, teneur *f* **en**   stone (chippings) content

gravillons *m,pl* **concassés**   crushed rock or gravel chippings

gravillons *m,pl* **enrobés**   coated chippings

gravillons *m,pl* **préenrobés/prétraités**   pre-coated chippings

gravillons *m,pl* **gros**   coarse chippings

gravimétrie *f*   gravimetry

gravité *f*   gravity

gravois *m*   demolition rubbish

gravure *f*   engraving

gravure *f* **à l'eau forte**   etching

gré à gré   negotiated [e.g. of contracts]

gré à gré, de   by (mutual) agreement

greffier *m*   clerk or registrar of the court

grégeoir *m*   tool for smoothing cut edges of glass

grêle *f*   hail

grelet *m*   brick axe, bricklayer's hammer

grelichonne   trapezoidal trowel

grenaillage *m*   (steel) gritblasting

grenaillé *adj*   grit- or shotblasted

grenaille *f*   steel grit

grenouille *f*   frog rammer

grenailler *vb*   shotblast

grenier *m*   attic or loft

grenu, sol *m*   coarse-grained soil

grès *m*   salt-glazed ware, stoneware, sandstone

grès *m* **à bâtir**   freestone

grès *m* **cérame**   vitrified clay

grès *m* **de construction**   freestone

grès *m* **fin**   siltstone

grès *m* **flambé**   glazed stoneware

grésage *m*   smoothing [a surface by rubbing, or the cut edge of glass]

gréser *vb*   smooth by rubbing [a surface], smooth cut edge of glass

grésillement *m*   crackle, crackling or interference [radio/PA/phone]

grève *f*   coarse sand, sandy beach or sea shore, strike or walk-out

griffe *f*   clamp, claw

griffe, marteau *m* **à**   claw hammer

griffe *f* **à couder**   bending tool [reinforcement]

griffe *f* **d'établi**   bench stop

grignoteur *m*   nibbler

gril *m*   grid, grill [generally within oven in France]

grillage *m*   grating, grillage [foundations], wire fencing, mesh or netting

grillage *m* **soudé**   welded mesh reinforcement

grille *f* bars [window], grating, grid, grille or screen, railing
grille *f* **à gradins** step grate
grille *f* **à secousse** shaking grate
grille *f* **anti-volatile** bird guard [grille]
grille *f* **d'aération** air grille
grille *f* **d'entrée** intake screen
grille *f* **d'entrée d'air** air inlet grille
grille *f* **de circulation** walkway grid
grille *f* **de protection** guard grill or netting
grille *f* **de radiateur** radiator grille
grille *f* **de reprise** recirculation grille
grille *f* **de retour** return (air) grille
grille *f* **de soufflage** discharge or supply air grille
grille *f* **de soufflage murale** wall (mounted) supply grille
grille *f* **de structure** structural grid
grille *f* **de transfert** transfer grille
grille *f* **en tablette de fenêtre** sill-mounted grille
grille *f* **essoreuse** dewatering screen
grille *f* **mécanique** travelling grate
grille *f* **oscillante** rocking/shaking/vibrating grate
grille *f* **perforée** perforated grille
grille *f* **roulante** travelling grate
griller *vb* blow [a fuse, a circuit], burn out [electric motor, portable tool, lamp bulb]
gripper *vb* seize (up) [bearing, machine]
se gripper *vb* seize [itself, e.g. a bearing]
gros agrégats *m,pl* coarse aggregate
gros béton *m* concrete fill
gros granulats *m,pl* coarse aggregate
gros mur *m* loadbearing wall
gros œuvre *m* building structure and envelope, civil or (main) structural works
gros ouvrage *m* building structure and envelope, fabric [of a building]
grosseur *f* size [weight, thickness, volume], bulk [size]
grosseur *f* **de grain** grain size
grossier *adj* coarse
grossiste *m* wholesaler
grotte *f* cave [natural], grotto
groupe *m* cluster or group [e.g. of houses, piles, trees]
groupe[†] *m* generator [portable]
groupe *m* **classeur** screening unit [for granular materials]
groupe *m* **compresseur-condenseur hermétique** sealed condensing unit
groupe *m* **condenseur** condensing unit
groupe *m* **convertisseur** rotary converter
groupe *m* **d'alimentation** feeding unit
groupe *m* **d'eau glacée** water chiller
groupe *m* **d'eau glacée refroidi par l'air** air-cooled water chiller
groupe *m* **d'essorage** dewatering unit
groupe *m* **de broyage** breaking unit [for aggregates]
groupe *m* **de criblage** screening unit [for granular materials]
groupe *m* **de dosage** proportioning unit [for mixing]
groupe *m* **de pieux** pile cluster
groupe *m* **de réfrigération** cooling unit
groupe *m* **de travail** working party

groupe *m* **doseur** weighing unit [for mixing materials]
groupe *m* **électro-compresseur** air compressor set
groupe *m* **électrogène** generating set, generator
groupe *m* **électrogène de secours** emergency or standby generating set
groupe *m* **électro-pompe** electric pump set
groupe *m* **malaxeur** mixing unit
groupe *m* **sécheur** drying unit
groupe *m* **serpentin refroidisseur** cooling coil
groupe *m* **ventilo-convecteur** fan coil unit
groupement *m* **conjoint** joint contracting with separate, individually responsible trade contractors
groupement *m* **d'entreprises** consortium, joint venture
groupement *m* **solidaire** joint contracting with joint responsibility of all contractors
grue *f* crane
grue *f* **à flèche horizontale à chariot** crane with trolley jib
grue *f* **à flèche relevable** luffing jib crane
grue *f* **à flèche télescopique** telescopic boom crane
grue *f* **à flèche treillis** lattice boom crane
grue *f* **à mât pivotant** slewing tower crane
grue *f* **à portique** gantry crane
grue *f* **à tour** tower crane
grue *f* **à tour à flèche distributrice** slewing tower crane
grue *f* **à tour pivotante** rotary tower crane
grue *f* **automobile/automotrice** self-propelled crane
grue *f* **automotrice tout terrain** rough-terrain (RT) crane
grue *f* **auxiliaire hydraulique** lorry/truck-mounted crane, lorry loader
grue *f* **de chantier** site crane
grue *f* **de chargement** loading crane [vehicle mounted]
grue *f* **de manutention** handling crane
grue *f* **de montage** erection crane
grue *f* **flottante** floating crane
grue *f* **mobile** mobile or lorry/truck-mounted crane
grue *f* **routière tout-terrain** all-terrain (AT) crane
grue *f* **sapine** fixed jib crane
grue *f* **sur camion** lorry/truck mounted crane, lorry loader
grue *f* **sur chenilles** crawler crane
grue *f* **sur porteur** vehicle mounted crane
grue *f* **sur porteur à flèche télescopique** vehicle-mounted telescopic boom crane
grue *f* **sur rails** rail-mounted crane
gruger *vb* notch, smooth cut edge of glass, break with a bush or pointed hammer a material which cannot be worked with a cutting tool
grugeoir *m* tool for smoothing cut edges of glass
grume *f* log [large, suitable for sawmilling], round timber
grume, bois *m* **en** rough or undressed timber
grutier *m* crane driver or operator
gué *m* ford
gueulard *m* gutter overflow pipe
gueule-de-loup *f* weather rebate [semi-circular, in the edge of a casement]
guêtre *f* **de protection** legging or protective gaiter
guette *f* windbracing [in a metal or timber wall panel]

guichet *m*   pedestrian opening in large gate, small door, spyhole or grille in a larger door, small opening in a partition or wall

guidage *m*   guide (rails) [e.g of a lift or site hoist]

guidage, bande *f* **de**   guide markings or strips [on road surfaces]

guide *m*   guide

guide *m* **à onglets**   fence, angling or mitre

guide *m* **(de battage)**   leader [of piling rig]

guide *m* **d'air**   air guide

guide *m* **de courroie**   belt guide

guide *m* **de la cabine**   cage guide [lift/elevator]

guide *m* **de refend**   fence [of a power saw]

guillaume *m*   French or toothed drag, rabbet or rebate plane

guillotine *f*, **(fenêtre à)**   sash window

guimbarde *f*   grooving or router plane

guirlande *f* **lumineuse**   roadworks lighting [to guide traffic]

guitare *f*   curved structural timber member

gunitage *m*   gunite or guniting, shotcrete or shotcreting, sprayed concrete

gunite *m*   gunite

gypse *m*   gypsum

ha[+] **(= hectare** *m***)**   hectare [10,000m^2 = 100 ares = 2.47 acres]

habillage *m*   casing [of equipment], decorative cladding, coating, finish or joinery feature, trim, metallic grill or mesh reinforcement of a wall for rendering

habiller *vb*   affix light wooden mouldings to decorate or cover joints, affix metal grill or mesh to a surface to reinforce its rendering, encase, wrap up

habitable *adj*   fit for habitation, inhabitable [in theory!]

habitant *m*   inhabitant

habitat *m*   habitat

habitat *m* **collectif**   collective habitation [e.g. block of flats]

habitation *f*   dwelling

habitation *f* **à loyer modéré (HLM)**   public housing [in France]

habité *adj*   inhabited

habiter *vb*   dwell or live in, inhabit

hache *f*   axe

hache *f* **à main**   hatchet

hache *f* **de bûcheron**   felling axe

hachement *m*   picking off of cement or plaster rendering or coating, preparation of a surface for rendering by light picking

hacheteau *m*   plasterer's tool for hacking plaster

hachette *f*   brick axe, bricklayer's hammer, hatchet

hachette *f* **à marteau**   hammer-head(ed) hatchet

hachotte *f*   lath hatchet [roofer's]

hachuré *adj*   hatched

haie *f*   hedge

halage, banquette *f* **de**   towpath

haler *vb*   tow [a barge]

hall *m*   arrival or departure hall, concourse [e.g. of station], (entrance) hall, lobby, bay [of a building]

hall *m* **d'entrée**   entrance hall

hall *m* **de gare**   station hall or concourse

hall *m* **des passagers**   passenger concourse

halle *f*   covered market

halle *f* **à marchandises**   freight hall, goods shed

halogène *adj*   halogenous

halogène *m*   halogen

hameau *m*   hamlet, village [small]

hangar *m*   barn or shed [generally open-sided], lean-to, hangar, warehouse

hangar *m* **d'avions**   aircraft hangar

happe *f*   cramp iron or dog [holding stones or timbers together]

harnais *m* **de sécurité**   body or safety harness [for working at height]

harpe *f*   toothing brick or stone

harpe *f* **d'attente**   toothing [for subsequent brick- or stonework]

harpon *m*   L-shaped cleat [for holding two walls or timbers together]

hauban *m*   cable stay [bridge], guy [line], span wire

hauban(n)age *m*   guying or staying with guys, staying/stay ropes

hauban(n)er *vb*   brace or stay [with guys], guy

hausse *f*   block or piece added to raise something

hausse *f* **de**   increase or rise in [prices, temperature, wages]

hausser *vb*   heighten, raise

haut *adj*   high

haut *m*   top [of stairs, a wall, a ladder]

haut rendement   high-efficiency *adj*

haute pression   high-pressure

haute tension (HT)   high-tension, high-voltage *adj*

haute tension *f*   high-voltage

hauteur *f*   depth [e.g. of an arch], height [e.g. of a building], elevation [height], high ground

hauteur *f* **d'ascension capillaire**   capillary rise height

hauteur *f* **d'aspiration**   suction head or lift

hauteur *f* **d'assise**   course depth/height [of bricks, stone]

hauteur *f* **d'eau**   depth of water

hauteur *f* **d'encombrement**   overall height

hauteur *f* **d'étage**   floor/storey height

hauteur *f* **dalle à dalle**   slab-to-slab height
hauteur *f* **de charge**   head [pressure]
hauteur *f* **de chute d'eau**   head of water, pressure head
hauteur *f* **de coupe**   depth of cut
hauteur *f* **de déblai**   depth of excavation
hauteur *f* **de l'âme**   web depth/height
hauteur *f* **de l'ouvrage**   structure height
hauteur *f* **de la pression statique**   static head
hauteur *f* **de levage**   lift(ing) height
hauteur *f* **de montée**   rise [small hill]
hauteur *f* **de pièce (sous plafond)**   floor-to-ceiling height
hauteur *f* **de poutre**   beam depth or height
hauteur *f* **de pression**   pressure head
hauteur *f* **de refoulement**   pressure head
hauteur *f* **du profil**   depth of section
hauteur *f* **dynamique**   velocity head
hauteur *f* **échappée**   clearance (height)
hauteur *f* **hors tout**   overall height
hauteur *f* **libre**   clear height, headroom
hauteur *f* **limite**   maximum height
hauteur *f* **manométrique**   pressure head
hauteur *f* **sous plafond**   ceiling height
hauteur *f* **sous poutres**   clear height beneath beams
hauteur *f* **statique**   static head
hauteur *f* **totale**   overall depth/height
hauteur *f* **utile**   effective depth [e.g. of beam]
haut-fourneau *m*   blastfurnace
haut-parleur *m*   loudspeaker
havage *m*   under-cutting used for sinking a caisson
havre *m*   harbour
hayon *m* **élévateur**   tailboard lift
héberge *f*   height to which lower building rises on a party wall
hectare *m* [† = ha]   hectare [10,000m² = 100 ares = 2.47 acres]
hélice *f*   helix, propeller
hélice *f* **bipale**   twin-bladed propeller/rotor [wind generator]
hélicoïdal *adj*   helical, spiral, twist [as in twist drill]
hélicoptère *f*   power float
héliostat *m*   heliostat
herbacé *adj*   herbaceous
herbe *f*   grass, herb, plant
herbe *f*, **mauvaise**   weed
herbeux *adj*   grassy
herbicide *m*   herbicide
hérisson *m*   chimney-sweep's or flue brush, spiked ironwork on top of a wall, stone bedding or pitching
hérissonnage *m*   laying of stone bedding or pitching
hermétique *adj*   airtight, hermetic, sealed, watertight
hermétique *f*   cast iron access/inspection cover
herminette *f*   adze, small hammer with one wedge shaped peen parallel to its handle
herse *f*   batten [for stage lights], metal division across a balcony between two flats
herses *f,pl*   stage lights
hétérogène *adj*   heterogeneous
hêtre *m*   beech

heures *f,pl* **creuses**   off-peak periods [for electrical tariffs], slack periods
heures *f,pl* **d'obscurité**   hours of darkness
heures *f,pl* **de travail**   work(ing) hours
heures *f,pl* **pointes**   peak hours/periods
heures *f,pl* **supplémentaires**   overtime
heurtoir *m*   buffer(s)
hie *f*   pavior's rammer or punner, pile driver
hisser *vb*   haul or heave up, hoist, pull up
homme *m* **d'affaires**   businessman
homogène *adj*   homogeneous
homologation   official approval [in accordance with standards, etc]
honoraires *m,pl*   fees [architect's, etc]
hôpital *m*   hospital
horizon *m*   horizon [visual and geological]
horizontal *adj*   horizontal
horloge *f*   clock
horloge *f* **à contacts**   time clock [time switch]
horloge *f* **de réglage**   control clock
horloge *f* **hebdomadaire**   seven-day clock/timer
hors circuit   disconnected [electrically], out of circuit
hors d'aplomb   out of true [not vertical]
hors d'eau   sheltered from water
hors de   away from, outside
hors exploitation *adj*   non-trade
hors normes *adj*   non-standard [not in accordance with standards]
hors série *adj*   made to order, non-standard
hors-œuvre *adj* **(H/O)**   external face to external face, overall [of measurement]
hors-œuvre *m*   external dimension
hors taxe *adj* **(HT)**   tax excluded, tax-free [excluding VAT, etc]
hors-tension *adj*   de-energized
hors-tout, largeur *f*/**longueur** *f*   overall width/length
hôtel *m*   hotel, mansion, town house
hôtel *m* **de ville**   town hall
hôtel *m* **des impôts**   tax office
hôtel *m* **des ventes**   salerooms [generally for auctions]
hôtel *m* **particulier**   private mansion
hôtellerie *f*   hostelry, inn
hotte *f*   cooker or fume hood
hotte *f* **aspirante**   extract hood
hotte *f* **d'évacuation/d'extraction**   exhaust or extract hood
hotte *f* **de cuisine**   kitchen hood
hotte *f* **de laboratoire**   fume cupboard
houille *f*   coal
houlice *f*   mitred mortise-and-tenon joint [tenon normally vertical, mortised piece angled]
hourdage *m*   infill [of a framework, with bricks, plaster, etc], pugging, rough flooring of rubble, broken bricks, etc
hourder *vb*   infill [a framework, with bricks, plaster, etc], pug, roughcast [render]
hourdis *m*   pugging [of a timber frame panel], roughcast [rendering], in situ concrete topping
hourdis *m* **(creux)**   hollow clay brickwork or blockwork

hublot *m* bulkhead light
huilage *m* priming coat [oil-based, on a porous surface]
huile *f* oil
huile *f* **à frigorigène** refrigerant oil
huile *f* **combustible** burning oil
huile *f* **de décoffrage** mould oil, release agent [oil]
huile *f* **de lavage** flushing oil
huile *f* **de lin** linseed oil
huile *f* **légère** light oil
huisserie *f* door or window frame
huisserie *f* **à chapeau** framing with verticals stopped under the top member
huissier *m* bailiff, court process server
humecter *vb* dampen, moisten
humide *adj* damp, humid, moist, wet
humidificateur *m* (air) humidifier, moistening apparatus
humidificateur *m* **à évaporation** evaporative humidifier
humidificateur *m* **à vapeur** steam humidifier
humidification *f* humidification, moistening
humidifier *vb* damp(en), moisten
humidimètre *m* moisture meter
humidité *f* damp(ness), humidity, moisture, wetness
humidité *f* **de l'air** air humidity, atmospheric moisture
humidité *f* **du sol** soil moisture
humidité *f* **relative** relative humidity
humus *m* humus, mould [soil]
hydrater *vb* hydrate
hydratation *f* hydration
hydraulicien *m* hydraulic engineer
hydraulique *adj* hydraulic *adj*
hydraulique, chaux *f* hydraulic lime
hydraulique, liant *m* hydraulic binder
hydraulique, remblai *m* hydraulic fill
hydraulique *f* hydraulics
hydraulique *f* **urbaine** public health engineering
hydrocâblage *m* plastic pipework [for pressurized plumbing and heating systems]
hydrocarboné, liant *m* bitumen binder
hydrocarboné, revêtement *m* bituminous surfacing
hydrocarbonée, couche *f* **de roulement** bituminous wearing course
hydrocarbonée, couche *f* **de surface** bituminous surface course or surfacing

hydrocarbure *m* hydrocarbon
hydrodynamique *adj* hydrodynamic
hydro-éjecteur *m* hydraulic or silt ejector
Hydrofraise *m* Solétanche rotary diaphragm wall rig [trade name]
hydrofuge *adj* damp-proof, waterproof, water-repellent
hydrofuge *m* **(de masse)** mass waterproofing compound [for mortar, concrete]
hydrofuge *m* **(de surface)** surface waterproofing compound
hydrogène *m* hydrogen
hydrologie *f* hydrology
hydrologue *m* hydrologist
hydrolyse *f* hydrolysis
hydromètre *m* hydrometer
hydromètre *m* **à ailettes** vane water meter
hydromorphe *adj* waterlogged [earth, ground]
hydromorphe, terre *f* waterlogged land
hydromorphie *f* waterlogging [of earth, ground]
hydrophile *adj* absorbent, hydrophilic
hydrophobe *adj* hydrophobic
hydrostatique *adj* hydrostatic
hygiène *f* **du travail** industrial hygiene, occupational health
hygiène *f* **industrielle/professionnelle** industrial hygiene
hygromètre *m* hygrometer
hygromètre *m* **à aspiration** aspirated hygrometer
hygromètre *m* **à cheveu** hair hygrometer
hygromètre *m* **à point de rosée** dew point hygrometer
hygromètre *m* **enregistreur** hygrograph, recording hygrometer
hygrométrie *f* hygrometry
hygroscopique *adj* hygroscopic
hygrostat *m* hygrostat
hyperbolique *adj* hyperbolic
hyperstatique *adj* statically indeterminate
hypothèque *f* mortgage
hypothèque, privilège *m* **d'** mortgage charge
hypothèse *f* assumption, hypothesis
hypothèses *f,pl* **de calcul** design criteria
hystérèse *f* hysteresis
hystérésis *f* hysteresis
hystérésis, boucle *f* **d'** hysteresis loop

idée *f* **de départ** basic concept or idea
if *m* yew
ignifuge *adj* fireproof
ignifugeage *m* fireproofing
ignifugeant *m* fire retardant or fireproofing compound
ignifuger *vb* fireproof
illumination *f* floodlighting, illumination, lighting

illuminé *adj* floodlit, illuminated
illuminer *vb* illuminate, light up
illuminer *vb* **au moyen de projecteurs** floodlight
îlot *m* island [small, in water], (traffic) island
îlot *m* **directionnel** traffic island with direction signs
îlot *m* **séparateur** channelizing traffic island
image *f* **(de)** image, picture (of)

image *f* **symétrique**   mirror image
image-miroir *f*   mirror image
imbibé *adj* **de**   soaked with
imbibé *adj* **d'eau**   waterlogged
imbrûlable *adj*   fireproof
imbrûlé *adj*   unburnt
imbue *f*   priming coat [largely absorbed, on wood or masonry]
immerger *vb*   immerse
immeuble *m*   building
immeuble *m* **collectif**   collective habitation [e.g. block of flats]
immeuble *m* **d'appartements**   block of flats
immeuble *m* **d'habitation**   residential block
immeuble *m* **de bureaux**   office block
immeuble *m* **de grande hauteur**   high-rise building
immobilier *adj*   property [as in property sale, etc]
immobilier, agent *m*   house or real estate agent
immobilier *m*   property or real estate
immobilisation *f*   capitalization of assets, fixed asset
immobilisation *f* **amortissable**   depreciable asset
immobilisation *f* **corporelle**   tangible asset
immobilisation *f* **incorporelle**   intangible asset
immobiliser *vb*   lock up or tie up [capital]
impair   odd [of numbers]
impasse *m*   cul-de-sac
impasse *m* **avec boucle**   cul-de-sac with closed loop
impédance *f*   impedance [electrical]
impériale, comble *m* **à l'**   roof (space) with convex slopes
imperméabilisant *m*   waterproofing agent
imperméabilisation *f*   waterproofing
imperméabiliser *vb*   waterproof
imperméable *adj*   impermeable, impervious, waterproof, watertight
imperméable à l'air/au gaz   impermeable to air/to gas, airtight/gas-tight
imperméable, couche *f*   impermeable layer [e.g. of soil]
implantation *f*   layout, marking or setting out [layout of buildings, roads, etc]
implantation, axe *m* **d'**   centre-line or setting out line
implantation, plan *m* **d'**   setting out drawing or plan
implantation, point *m* **d'**   setting out point
implanter *vb*   mark, peg or set out, site
imposable *adj*   subject to tax(ation), taxable
imposer *vb*   tax [somebody]
imposition *f*   taxation
imposte *f*   fanlight, impost [of an arch]
imposte *f*, **traverse** *f* **d'**   transom
imposte *f* **en éventail**   semi-circular fanlight
impôt *m*   tax
impôts, centre *m* **des**   tax office
impôts, hôtel *m* **des**   tax office
impôt *m* **foncier**   land/real estate tax
impôt *m* **sur le chiffre d'affaires**   tax on turnover
impôts *m,pl* **locaux**   rates [local taxes]
impôts *m,pl* **sur le revenu**   income tax
imprégnation *f*   impregnation
imprégné *adj* **d'eau**   waterlogged
imprégner *vb* **(de)**   impregnate, permeate or soak (with)

impression *f*   priming [painting], pressure
impression, couche *f* **d'**   priming coat or primer [painting]
imprimante *f*   printer [computer]
imprimante *f* **à jet d'encre**   inkjet printer
imprimer *vb*   prime [for painting], print
impropre *adj* **à sa destination**   unfit for its purpose
impropre, sol *m*   unsuitable soil
impureté *f*   impurity
imputation *f* **des coûts**   cost allocation
imputrescible *adj*   rot-proof
inafouillable *adj*   water-resisting [e.g. of a canal lining]
inapplicable *adj*   unenforceable [of regulation, etc]
incandescence *f*   glow, incandescence
incandescent *adj*   incandescent
incassable *adj*   unbreakable
incendie *m*   fire [e.g. burning building]
incendier *vb*   set fire to
incident *adj*   incident
incinérateur *m*   incinerator
incinération *f*   incineration
incinérer *vb*   incinerate
inclinaison *f*   gradient, inclination, slope
inclinaison *f* **des ailes**   flange slope
incliné *adj*   inclined
inclus *adj*   enclosed [with a letter, etc]
inclusion *f*   inclusion
inclusion *f* **de laitier**   slag inclusion
incombustible *adj*   fireproof, incombustible
incongelable *adj*   freeze-resistant
incorporation *f*   incorporation
incorporé *adj* **(dans)**   built in(to), incorporated (in, into)
incrément *m*   increment
incrustation *f*   scale [in a boiler or pipes]
incrustation *f* **de chaudière**   boiler scale
incurvé *adj*   concave
indemnisation *f*   compensation [for loss]
indemniser *vb*   compensate [for loss]
indentation *f*   indentation
indépendant *adj*   free-standing, independent
indépendant, travailleur *m*   freelance or self-employed worker
indépendante, maison *f*   detached house
index *m*   index [of a catalogue, etc], indicator, pointer
indexation *f*   indexation/indexing [of costs, prices, etc]
indicateur *m*   gauge [indicator], indicator [as in wind speed indicator, etc]
indicateur *m* **à distance**   remote indicator
indicateur *m* **d'air**   air gauge
indicateur *m* **d'appel**   call indicator
indicateur *m* **de niveau**   level indicator or gauge
indicateur *m* **de niveau à flotteur**   float level indicator
indicateur *m* **de niveau d'eau**   water level indicator
indicateur *m* **de pression**   pressure gauge
indicateur *m* **de répétition**   repeater gauge or indicator
indicateur *m* **de tirage**   draught indicator
indicateur *m* **de vitesse**   flow or speed indicator, velometer
indicateur *m* **immobilier**   property gazette

indicatif *m*   call sign
indicatif *m* **téléphonique**   dialling code
indication *f*   information [piece of]
indication *f* **analogique**   analogue indication
indication *f* **de coupe**   section line [drawing]
indication *f* **routière**   road/traffic sign
indication *f* **topographique**   survey mark
indications *f,pl* **topographiques**   survey data
indice *m*   clue, indication, factor, sign, symptom, rating or
   index [cost, etc], suffix [mathematical]
indice *m* **CBR**   California bearing ratio (CBR)
indice *m* **d'amortissement**   decay factor
indice *m* **d'éblouissement**   glare factor or index
indice *m* **de confort**   comfort index
indice *m* **de confort thermique**   thermal comfort index
indice *m* **de liquidité**   liquidity index
indice *m* **de plasticité**   plasticity index
indice *f* **de protection (IP)**   index of protection (IP)
indice *f* **de rendu des couleurs (IRC)**   colour rendering
   index (CRI)
indice *m* **de retrait**   shrinkage ratio
indice *m* **des vides**   void ratio
Indice du Coût de la Construction (ICC)   construction
   costs index
indice *m* **du local**   room index
indice *m* **portant**   bearing capacity index or value
indice *m* **portant californien (CBR)**   California bearing
   ratio (CBR)
indiquer *vb*   indicate, register
indium *m*   indium
individuel *adj*   independent, separate
indivision *f*   joint ownership
inducteur *m*   induction unit
inducteur *m* **à basse pression**   low-pressure induction
   unit
induction *f*   induction
induire *vb*   induce
induit *adj*   induced
induit *m*   armature [of electric motor, dynamo]
industriel *adj*   industrial
industrie *f* **du bâtiment**   building industry
inefficace *adj*   inefficient
inégal *adj*   uneven
inégalité *f*   inequality, unevenness [of a surface, road, etc]
inerte *adj*   inert
inertie *f*   inertia [e.g. of geometric section]
inertie *f* **thermique**   thermal inertia
inférieur *adj*   lower
infection *f*   infection
infiltration *f*   infiltration [damp, air], percolation, seepage
infiltrer *vb*   infiltrate, filter or seep into, percolate
infirmerie *f*   first aid room, health centre, sick bay/room,
   surgery
inflammabilité *f*   inflammability
inflammable *adj*   inflammable
inflation *f*   inflation
influence *f*   influence
information *f* **des travailleurs**   worker information
informatique *adj*   computer *adj*

informatique *f*   computer science, computing
infranchissable, **ligne** *f*   continuous line [road marking]
infrarouge *adj*   infra-red
infrastructure *f*   bed [of a road or railway], substructure,
   infrastructure
ingénierie *f*   engineering [strictly, in French construction,
   through concept and design to supervision]
ingénieur *m/f*   engineer
ingénieur *m* **agréé**   chartered engineer [approximate
   French equivalent]
ingénieur *m* **chimiste**   chemical engineer
ingénieur *m* **conseil**   consulting engineer [approximate
   French equivalent]
ingénieur *m* **d'exécution**   site agent [engineer]
ingénieur *m* **de bâtiment**   building engineer
ingénieur *m* **de chantier**   field or site engineer
ingénieur *m* **de structure**   structural engineer
ingénieur *m* **des mines**   mining engineer [from l'École
   des Mines]
ingénieur *m* **des ponts et chaussées**   civil engineer [from
   l'École des Ponts et Chaussées]
ingénieur *m* **des travaux publics**   civil or construction
   engineer
ingénieur *m* **diplômé**   qualified engineer [title with legal
   protection in France]
ingénieur *m* **électricien**   electrical engineer
ingénieur *m* **en chef**   chief engineer
ingénieur *m* **(en) génie civil**   civil engineer
ingénieur *m* **(en) structure**   structural engineer
ingénieur *m* **permanent**   resident engineer
ingénieur *m* **sanitaire**   public health or sanitary engineer
ingénieur *m* **système**   systems engineer
ingénieur-conseil *m*   consulting engineer [approximate
   French equivalent]
ingénieur-géomètre *m*   engineer-surveyor [generally for
   civil works]
inhabité *adj*   uninhabited
inhibiteur *m*   inhibitor
inhibiteur *m* **de corrosion**   anti-corrosive agent,
   corrosion inhibitor
ininflammable *adj*   non-inflammable
initial *adj*   initial [load, etc]
injecter *vb*   grout, inject
injecter *vb* **de mortier**   grout up with (cement) mortar
injecteur *m*   injector
injection *f* **de ciment**   cement grouting, cementation
injection *f* **(de coulis)**   grouting
injection *f* **des joints**   joint grouting
inodore *adj*   odourless
inondable *adj*   liable to flooding
inondation *f*   flood(ing), inundation
inondé *adj*   flooded
inonder *vb*   drench, soak, flood [land]
inorganique *adj*   inorganic
inoxydable *adj*   rustproof, stainless [as of steel]
inscription *f*   registration [of mortgages, etc]
insecte *m*   insect
insensibilité *f* **(à)**   insensitivity (to)
insensible *adj* **à**   insensitive to

insensible *adj* **au gel**   frost-resistant
insertion *f*   weaving [of traffic]
insertion, voie *f* **d'**   weaving lane [of road]
insoluble *adj*   insoluble
insonore *adj*   soundproof
insonorisation *f*   soundproofing
insonoriser *vb*   soundproof
inspecteur *m*   inspector
inspecteur *m* **du travail**   health and safety inspector
[French equivalent]
inspecteur *m* **municipal des travaux**   local authority
building inspector
inspection *f*   inspection
inspection *f* **de détail**   detailed inspection
inspection *f* **périodique**   periodic inspection
installateur *m*   fitter, installer
installation *f*   installing, installation, plant [fixed], wiring
up [of a circuit]
installation *f* **à deux conduits**   dual duct installation or
plant
installation *f* **à faible vitesse**   low-velocity installation or
plant
installation *f* **à gaz à basse pression**   low-pressure gas
installation or plant
installation *f* **à haute vitesse**   high-velocity installation or
plant
installation *f* **à refroidissement par eau**   water cooling
installation
installation *f* **à très basse tension de sécurité/à TBTS**
extra-low voltage safety installation
installation *f* **à vide**   vacuum installation or plant
installation *f* **à volume variable**   variable volume
installation or plant
installation *f* **au gaz**   gas installation
installation *f* **d'adoucissement**   water softening
installation or plant
installation *f* **d'aération**   ventilating plant or system
installation *f* **d'alimentation en eau**   water supply
installation or plant
installation *f* **d'aspiration**   suction installation or plant
installation *f* **d'eau glacée**   chilled water plant
installation *f* **d'éclairage**   lighting installation
installation *f* **d'évaporation**   evaporating installation or
plant
installation *f* **d'extinction d'incendie**   fire extinguishing
equipment or installation
installation *f* **d'extinction par sprinkle(u)rs**   fire
sprinkler installation or system
installation *f* **d'humidification**   air humidifying
installation or plant
installation *f* **d'incinération**   incinerator [installation or
plant]
installation *f* **de chauffage**   heating installation or plant
installation *f* **de chauffage à air chaud**   hot air
installation or plant
installation *f* **de chauffage annexe**   separate heating
installation or plant
installation *f* **de chauffage central**   central heating
installation or plant

installation *f* **de chauffage urbain**   district heating plant
or installation
installation *f* **de chaufferie**   boiler plant
installation *f* **de climatisation**   air-conditioning
equipment, plant or installation
installation *f* **de climatisation à un conduit**   single-duct
air-conditioning plant
installation *f* **de concassage**   crushing plant [for
aggregates, rock, etc]
installation *f* **de conditionnement d'air**   air-conditioner,
air-conditioning plant or system
installation *f* **de congélation**   freezing equipment or plant
installation *f* **de contrôle**   control installation
installation *f* **de criblage**   screening plant [for granular
materials]
installation *f* **de dépoussiérage**   dust extraction/removal
installation or plant
installation *f* **de dosage et de mélange du béton**
concrete batching and mixing plant
installation *f* **de filtrage**   filter plant, filtration installation
or plant, screening installation or plant
installation *f* **de forage**   drilling rig
installation *f* **de lavage**   washing plant
installation *f* **de préparation du béton**   concrete
batching and mixing plant
installation *f* **de production de glace**   ice-making plant
installation *f* **de ramassage des ordures**   refuse or waste
collecting installation or plant
installation *f* **de récupération**   recovery installation or
plant
installation *f* **de recyclage**   recycling plant
installation *f* **de réfrigération**   cooling or refrigeration
installation or plant
installation *f* **de refroidissement**   refrigeration
installation or plant
installation *f* **de refroidissement à air**   air cooling
installation or system
installation *f* **de réserve**   duplicate plant, standby
installation or plant
installation *f* **de séchage**   dryer/drying installation or
plant
installation *f* **de séchage de l'air**   air drying equipment,
installation or plant
installation *f* **de triage**   separating installation or plant
installation *f* **de ventilation**   ventilating equipment or
plant
installation *f* **des chaudières**   boiler installation or plant
installation *f* **du transport des cendres**   ash handling
installation or plant
installation *f* **électrique**   electrical installation or services,
wiring installation
installation *f* **en plein air**   open air or outdoor installation
or plant
installation *f* **frigorifique**   refrigeration installation or
plant
installation *f* **sanitaire**   sanitary installation
installations *f,pl*   fittings [equipment, etc]
installations *f,pl* **collectives**   communal or shared
facilities

installations *f,pl* **de chantier**   site facilities, installations or plant

installations *f,pl* **portuaires**   port or harbour facilities

installations *f,pl* **sanitaires**   sanitary or toilet facilities

installer *vb*   install

s'installer *vb*   settle [take up residence]

instaurer *vb*   initiate [measures, a method, a policy]

Institut Géographique National (IGN)   Ordnance Survey (OS) [near French equivalent]

Institut *m* **International du Froid**   Air-conditioning and Refrigeration Institute (ARI)

Institut National de la Statistique des Études Économiques (INSEE)   French national statistical institute [responsible for official building cost indices, etc]

instructions *f,pl*   instructions [printed or written]

instructions *f,pl* **d'entretien**   maintenance instructions

instructions *f,pl* **d'exploitation**   operating instructions

instructions *f,pl* **d'installation**   installation instructions

instrument *m*   instrument

instrument *m* **à lecture directe**   direct reading instrument

instrument *m* **de contrôle de câble**   cable testing instrument

instrument *m* **de mesure**   measuring instrument

instrument *m* **étalonné**   calibrated instrument

insuffisant *adj*   inadequate, insufficient

intact *adj*   intact, undamaged, undisturbed or unweathered [e.g. sample, in geotechnics]

intégral *adj*   integral *adj*

intégral, paiement *m*   payment in full

intégral, texte *m*   full text [e.g. of a specification]

intégrale *f*   integral *n*

intégrateur *m*   integrator

intégration *f*   integration

intégré *adj*   integrated, integral

intégrer *vb*   integrate

intégrité *f*   integrity

intempérie *f*   bad or inclement weather

intensité *f*   current [electrical], intensity

intensité *f* **acoustique**   acoustic power

intensité *f* **de bruit**   noise intensity

intensité *f* **de bruit de fond**   background noise level

intensité *f* **de courant**   current intensity

intensité *f* **de la lumière**   light density

intensité *f* **de rayonnement**   radiation intensity

intensité *f* **des précipitations**   rainfall intensity

intensité *f* **lumineuse**   luminous intensity

intensité *f* **nominale**   nominal rating

intensité *f* **sonore**   loudness, sound intensity

intensité *f* **sonore, niveau** *m* **d'**   sound level

intensité *f* **thermique**   thermal rating [electrical equipment]

inter[+] *m*   switch [electrical]

interaction *f*   interaction

interchangeabilité *f*   interchangeability

interdit *adj* **aux poids lourds**   closed to heavy traffic/loads/lorries

intérêt *m*   interest, share or stake [in a business, etc]

intérêt *m* **composé/simple**   compound/simple interest

interface *f*   interface

interférence *f*   interference [of light rays, or electrical signals]

interférer *vb*   interfere [of light rays]

intérieur *adj*   inner, internal

intérieur *m*   inside *n*

intermédiaire *adj*   intermediate

intermittent *adj*   intermittent

interne *adj*   internal

interphone *m*   intercom (system), internal telephone

interprétation *f*   evaluation, interpretation

interprétation *f* **des essais**   evaluation or interpretation of test(s) (results)

interrompre *vb*   break or interrupt [a circuit]

interrupteur *m*   breaker, interrupter or switch [electrical], circuit- or contact-breaker, contactor

interrupteur *m* **à bascule**   rocker switch

interrupteur *m* **à deux pôles**   double-pole switch

interrupteur *m* **à distance**   remote switch

interrupteur *m* **à flotteur**   float switch

interrupteur *m* **à mercure**   mercury switch

interrupteur *m* **à minuterie**   time switch

interrupteur *m* **à poussoir**   pushbutton switch

interrupteur *m* **à tirette**   cord-operated or pull switch

interrupteur *m* **au pied**   foot switch

interrupteur *m* **automatique**   shut-down switch

interrupteur *m* **automatique de réglage**   automatic time switch

interrupteur *m* **crépusculaire**   daylight operated or light-sensitive switch

interrupteur *m* **de contrôle**   control switch

interrupteur *m* **de démarrage**   starting switch

interrupteur *m* **de fin de course**   end-of-travel switch

interrupteur *m* **de marche forcée**   manual override switch [on a time switch]

interrupteur *m* **de mise à niveau de liquide**   liquid level switch

interrupteur *m* **de sécurité**   safety switch

interrupteur *m* **de surcharge**   overload cut-out device/switch

interrupteur *m* **dérogation**   override switch

interrupteur *m* **différentiel**   residual current switch

interrupteur *m* **horaire**   time switch

interrupteur *m* **magnétique**   magnetic switch

interrupteur *m* **principal**   main switch or breaker

interrupteur *m* **rotatif**   rotating or rotary switch

interrupteur *m* **temporisé**   time-lag switch

interrupteur *m* **thermique**   thermal cut-out or switch

interrupteur *m* **unipolaire**   single-pole switch

interrupteur *m* **va-et-vient**   two-way switch

interrupteur-inverseur *m*   changeover switch

interruption *f*   breakage or interruption [of current]

interruption *f* **(de marche)**   stoppage [of machinery]

intersection *f*   intersection, junction [road]

interstitiel *adj*   interstitial

interstitielle, pression *f* **d'eau**   pore water pressure

intervalle *m*   clearance, distance or space [apart], gap, interval [of time or space]

intervalle *m* **de temps**   interval of time

intervenant *m*   contributor, party involved
intrados *m*   internal surface or soffit [of an arch or vault], intrados
inusable *adj*   hard-wearing
invalide *adj*   invalid
inventaire *m*   inventory, stock-list
inventaire *m* **des ressources**   resource inventory
inverser *vb*   reverse (current)
inversion *f*   inversion, reversal [electrical]
inversion *f* **de (la) polarité**   change, inversion or reversal of polarity
inversion *f* **thermique**   temperature inversion
investir *vb*   invest
investissement *m*   investment
iode *m*   iodine
ion *m*   ion
ionisation *f*   ionization
irradiation *f*   irradiation
irrégularité *f*   erratic performance, irregularity, unsteadiness
irrégulier *adj*   irregular, uneven, unsteady
irrigation *f*   irrigation
irrigation *f* **de surface**   surface irrigation
irrigation *f* **goutte à goutte**   drip irrigation
irrigation *f* **par aspersion**   sprinkler irrigation
irrigation *f* **souterraine**   subsurface irrigation
irrigation *f* **superficielle**   surface irrigation
irrigué *adj*   irrigated
irriguer *vb*   irrigate
isenthalpique *adj*   isenthalpic
isobare *adj*   isobaric
isobare *f*   isobar
Isobéton *m*   Isobéton [trade name, lightweight concrete of polyurethane foam binder and glass bead aggregate]
isolant *adj*   insulating, non-conducting [electrically], soundproofing *adj*
isolant, ruban *m*   tape, insulating
isolant *m*   insulating material, insulator
isolant *m* **calorifuge**   heat or thermal insulating material
isolant *m* **en coquille**   preformed pipe insulation
isolant *m* **en matelas**   blanket type (thermal) insulation
isolant *m* **en panneau rigide**   rigid panel/sheet insulation
isolant *m* **en vrac**   bulk or loose fill insulation
isolant *m* **moussé**   foam insulation [for cavity walls]

isolant *m* **thermique**   lagging, thermal insulation
isolateur *adj*   insulating
isolateur *m*   isolator [electrical]
isolation *f*   insulation [thermal, acoustic]
isolation *f***, double**   double insulation
isolation *f* **acoustique**   acoustic insulation
isolation *f* **contre le bruit**   noise or sound insulation
isolation *f* **expansée in situ**   foamed in situ (thermal) insulation
isolation *f* **par plaque de liège**   corkboard thermal insulation
isolation *f* **phonique/sonore**   acoustic insulation
isolation *f* **thermique**   thermal insulation
isolation *f* **thermique cellulaire**   cellular thermal insulation
isolation *f* **thermique en vrac**   bulk or loose fill (thermal) insulation
isolé *adj*   individual, isolated [e.g. house], insulated
isolément *adv*   individually, separately
isolement *m*   insulation [electrical], isolation, separation [fire or between buildings], space between adjacent buildings or parts of buildings
isolement *m* **acoustique normalisé**   standardized acoustic insulation
isolement *m* **coupe-feu**   fire separation
isolement *m* **recherché**   desired/required (level of) insulation
isoler *vb*   insulate [acoustically, electrically], soundproof
isonivelage *m*   levelling [of a lift/site hoist during loading/unloading]
Isorel *m*   hardboard [trade name]
isostatique *adj*   statically determinate
isotherme *adj*   isothermal
isotherme *f*   isotherm
isotope *m*   isotope
isotrop(iqu)e *adj*   isotropic
issue *f*   exit, outlet
issue *f* **de secours**   emergency exit
italienne, fenêtre *f* **à l'**   single leaf window opening outwards with a vertically sliding axis near its upper edge
itinéraire *m*   route [itinerary]

jalon *m*   surveyor's pole, rod or staff
jalousie *f*   adjustable louvres [of a window, blind, etc], sun blind
jambage *m*   small supporting wall [e.g. for a sink, a mantlepiece], supporting pier [between arches or parts of a wall]
jambage *m* **de porte**   door stud

jambe *f* **boutisse**   end of party wall in stone, visible from the street
jambe *f* **de force**   brace, prop, strut
jambe *f* **étrière**   pillar at end of party wall, visible from the street
jambette *f*   vertical member [small] in a roof truss
jante *f*   rim [of a wheel]

jaquette *f*   jacket cover [round a boiler]
jaquette *f* **calorifuge**   insulating jacket
jardin *m*   garden, park
jardin *m* **à l'anglaise**   landscape(d) garden
jardin *m* **à la française**   formal garden
jardin *m* **d'agrément**   decorative garden
jardin *m* **de rocaille**   rock garden
jardin *m* **en toiture**   roof garden
jardinet *m*   small garden
jardinier *m*   gardener
jarret *m*   bend of a straight beam, protuberance or break in continuity of an arch or vault profile
jauge *f*   dipstick, gauge
jauge *f*, **robinet** *m*   calibrated tap [used prior to advent of flow meters]
jauge *f* **d'allongement**   strain gauge
jauge *f* **d'épaisseur**   thickness gauge
jauge *f* **(de niveau) de liquide/de mazout/d'huile**   liquid/fuel oil/oil level gauge
jauge *f* **de pression**   pressure gauge
jauge *f* **de vapeur**   steam gauge
jauge *f* **de vide**   vacuum gauge
jauger *vb*   gauge, measure
Javel, eau *f* **de**   bleach
jet *m*   jet
jet *m* **d'eau**   drip moulding, weatherboard [on a door, window]
jet *m* **d'entrée**   inlet nozzle
jet *m* **de décharge**   discharge jet, outlet nozzle
jet *m* **de vapeur**   steam jet
jet *m* **premier**   first draft or sketch
jet *m* **sur banquette**   throwing excavated material up on to a bench in a trench wall
jet *m* **sur berge**   throwing excavated material directly to the surface from a shallow trench
jetée *f*   jetty, pier [marine, inland water or airport]
jeu, en   at work [e.g. forces]
jeu *m*   allowance, clearance, looseness, play, set [of drawings, tools, etc]
jeu *m* **de clés**   set of spanners
jeu *m* **libre**   running clearance
jeu, prendre *vb* **du**   work loose
jeux *m,pl* **d'eau**   fountains
joindre *vb*   close, shut [door, window], contact, get in touch with [a person], join, link [together]
joint *adj*   enclosed [with a letter, etc]
joint, à plat   butted together [boards or planks, side-to-side]
joint fermé, à   close-jointed [rock, etc]
joint *m*   connection, joint, splice, join [line of], gasket, pointing [masonry]
joint, faux   joint in a road pavement not extending to full depth
joint *m* **à brides/de bride**   flange connection or joint
joint *m* **à emboîtement**   rebated joint, spigot-and-socket joint
joint *m* **à emboîtement soudé**   welded socket joint
joint *m* **à haute pression**   high pressure joint or packing
joint *m* **à klingerite**   Klingerite jointing

joint *m* **à l'angle**   corner joint
joint *m* **à la pompe**   plastic joint sealing applied under pressure with a pump
joint *m* **à labyrinthe**   labyrinth seal
joint *m* **à recouvrement**   lap or splice joint, seam
joint *m* **à rotule**   ball/ball-and-socket joint, swivel joint
joint *m* **abouté**   butt joint
joint *m* **affleuré**   flush pointing
joint *m* **arrondi**   keyed pointing
joint *m* **articulé**   hinged joint
joint *m* **biais**   bevelled or skew joint
joint *m* **biseauté**   scarf joint
joint *m* **bout à bout**   butt joint
joint *m* **brasé**   brazed or soldered joint
joint *m* **caoutchouc**   rubber seal
joint *m* **carré**   butt joint
joint *m* **composé**   composite joint
joint *m* **conique**   cone joint
joint *m* **coulé à plomb**   lead-caulked or leaded joint
joint *m* **creux**   hollow or recessed pointing
joint *m* **creux chanfreiné**   weathered pointing
joint *m* **d'angle**   angle joint
joint *m* **d'assise**   bed joint
joint *m* **d'emboîture**   spigot-and-socket joint
joint *m* **d'étanchéité**   weatherproof joint or seal
joint *m* **de bord**   edge joint
joint *m* **de bride**   flange(d) joint
joint *m* **de chantier**   site connection, field/site joint
joint *m* **de chaussée**   expansion joint [highway/bridge]
joint *m* **de compression**   compression joint [plumbing]
joint *m* **de construction**   construction joint
joint *m* **de contraction**   contraction joint
joint *m* **de déplacement**   movement joint
joint *m* **de dilatation**   expansion or movement joint
joint *m* **de l'âme**   web joint or splice
joint *m* **de lit**   bed joint
joint *m* **de montage**   field or site joint
joint *m* **de plomb**   lead-caulked or leaded joint
joint *m* **de reprise**   construction joint [in concrete work]
joint *m* **de retrait**   contraction or shrinkage joint
joint *m* **de robinet**   tap washer
joint *m* **de rupture**   movement joint [to allow for unequal loading]
joint *m* **de tassement**   settlement joint
joint *m* **de terre glaise**   clay seal
joint *m* **debout**   standing seam joint [in sheet metal]
joint *m* **droit**   butt joint
joint *m* **en biseau**   chamfered joint
joint *m* **en canal**   keyed joint [mortar]
joint *m* **en caoutchouc**   rubber gasket
joint *m* **en caoutchouc mousse**   foam rubber gasket
joint *m* **en mortier**   mortar joint
joint *m* **en néoprène**   neoprene joint or seal
joint *m* **étanche**   leak-proof joint or seal
joint *m* **étanche à l'air**   air-tight joint or seal
joint *m* **fait en montant**   bedding and jointing in a single operation
joint *m* **glissant**   expansion joint, sliding or slip

joint *m* **horizontal à recouvrement**   horizontal tongue-and-groove joint [in cladding]

joint *m* **lisse**   flush pointing

joint *m* **mandriné**   flared joint

joint *m* **maté**   caulked joint

joint *m* **mousse**   foam seal

joint *m* **par recouvrement**   overlapping joint

joint *m* **plat**   butt joint, flush pointing

joint *m* **plein**   flush pointing

joint *m* **soudé**   welded joint

joint *m* **soudé par capillarité**   capillary or sweat(ed) joint

joint *m* **torique**   O-ring joint

joint *m* **vertical à pincement**   vertical loose tongue joint between facade panels

joint *m* **vissé**   screwed joint

joints *m,pl* **croisés**   alternate joints

joints *m,pl* **en chicane**   staggered joints

jointement *m*   assembling, bonding, fitting, jointing

jointoiement *m*   pointing [brickwork, tiles, the action and its result]

jointoyer *vb*   grout [tiles], joint or point [masonry]

jointoyeur *m*   grouter or jointer [tiles], jointer or pointer [masonry, tiles]

jointure *f*   join or joint

jointure *f* **de câbles**   cable joint

jonction *f*   junction

jonction *f* **de tuyaux**   pipe connection/joint

jonction *f* **par manchon**   socket connection [pipe]

joue *f*   cheek [of a bearing, mortise, plane, etc], side [of a plane], thickness of each jaw of the groove in tongue-and-groove floorboarding

jouée *f*   cheek [of a dormer window], wall thickness at a door or window opening

jouissance *f*   right of enjoyment/possessio/tenure

jour *m*   day, day(light), lighting, gap or chink, width of stair(well)

jour *m,* **mise** *f* **à**   revision, updating

jour *m* **d'atelier**   studio lighting

jour *m* **d'escalier**   stairwell

jour *m* **de congé**   day off, holiday

jour *m* **de coutume**   open window in a party wall

jour *m* **de ligne**   distance between a builder's line and the wall surface

jour *m* **de servitude**   opening or window in a wall in accord with a particular agreement or easement

jour *m* **de souffrance**   opening or window overlooking a neighbour to which the latter cannot object

jour *m* **de travail**   work(ing) day [of the week]

jour *m* **ouvrable**   business day, work(ing) day [of the week]

***Journal Officiel, le* (JO)**   *Official Gazette* [French]

journalier *adj*   daily or per day

jouxter *vb*   adjoin, be contiguous to

judas *m*   spyhole [in an entrance door]

jumelé *adj*   paired, twin(ned)

jumelée, maison *f*   semi-detached house

juridique *adj*   legal

jurisprudence *f*   precedents [of a legal case]

jus† *m*   juice† [electric current]

juste *adj*   right [accurate]

juste *adv*   right [correctly]

justesse *f*   accuracy, precision

justification *f*   supporting calculations

kaolin *m*   kaolin

kieselguhr *m*   diatomaceous earth or diatomite

kilobar *m*   kilobar

kilocalorie *f*   kilocalorie

kilogramme *m*   kilogram

kilométrage *m*   measurement in kilometres, marking with kilometre stones

kilomètre *m*   kilometre

kilométrer *vb*   mark with kilometre stones, measure off in kilometres

kilotonne *f*   kilotonne

kilowatt *m*   kilowatt

kilowatt–heures *m,pl*   kilowatt–hours

kiosque *m*   kiosk

kiosque *m* **de jardin**   summerhouse

krypton *m*   krypton

L

label *m*   seal or stamp [commercial]

label *m* **de circuit**   circuit label

labile *adj*   unstable

labo(ratoire) *m*   lab(oratory)

laboratoire *m* **d'essais**   test(ing) laboratory

laboratoire *m* **de recherches**   research laboratory or station

labourer *vb*   plough

lac *m*   lake

lac *m* **de barrage**   dam lake

laceret *m*   boring tool for peg holes in timber structures

lacet *m*   pin [of lift-off hinge]

lâche *adj*   loose, slack, slipshod, slovenly [work]

lâcher *vb*   fail [brakes], loosen, slacken, release [let sth go]

lagon *m*   lagoon

laie *f*   forest ride or track

laine *f* **d'acier/de fer**   steel wool

laine *f* **de bois**   woodwool

laine *f* **de laitier/scories**   slag wool

laine *f* **de minéral/minérale**   mineral wool (insulation)

laine *f* **de roche**   rock wool (insulation)

laine *f* **de verre**   fibreglass or glass wool (insulation)

laisser *vb* **tomber**   drop [let fall]

lait *m* **de chaux/ciment**   lime/cement whitewash

lait *m* **de ciment**   cement slurry

laitance *f*   laitance

laiterie *f*   dairy

laitier *m*   slag [for use as aggregate or in slag cement]

laitier *m* **expansé**   expanded blastfurnace slag

laitier *m* **d'aciérie**   steelworks slag

laitier *m* **de haut-fourneau**   blastfurnace slag

laitier *m* **de haut-fourneau concassé**   crushed blastfurnace slag

laiton *m*   brass

lambourde *f*   batten to support wood strip/parquet flooring, bearing joist [supporting ends of floor joists], bridging joist, soft stone from the Paris area

lambourde *f* **chanlattée**   chamfered batten on side of floor beam to support ends of abutting joists

lambrequin *m*   pelmet [solid]

lambris *m*   matchboarding, panelled joinery [e.g. doors, shutters, etc], panelling, wainscot(ing), wall panelling [thin stone or wood panel, etc]

lambris *m* **à table saillante**   panel proud of one side of its surrounding frame

lambris *m* **arasé**   panel flush with its frame

lambris *m* **d'appui**   dado

lambris *m* **de comble**   panelling on the sloping side of a curb or mansard roof

lambrissage *m*   panelling [of a room or space]

lambrissé *adj*   panelled

lambrisser *vb*   panel

lame *f*   blade [knife, saw], flooring strip 60–80mm wide, thin strip of wood or metal

lame *f* **d'air**   air-gap [between two sensibly parallel faces]

lame *f* **d'aiguille**   points blade

lame *f* **de cisaille**   shear blade

lame *f* **de parquet**   strip of wood to be machined for flooring

lame *f* **de persienne**   venetian blind leaf/blade

lame *f* **de ressort**   spring leaf [*not* botanic]

lame *f* **de scie**   saw blade

lame *f* **de scie à ruban**   bandsaw blade

lame *f* **de scie circulaire**   circular saw blade

lamellaire *adj*   lamellar

lamellation *f*   laminated assembly

lamellé *adj*   laminated [structure]

lamelle *f*   lamella, small strip of metal or plastic

lamellé-collé *m*   glued-laminated wood

lamellé-collé *m*, **(bois)**   laminated timber or wood

lamelliforme *adj*   lamellar

laminage *m*   lamination [the process], rolling [of steel]

laminage *m* **à chaud/froid**   hot/cold rolling

laminaire *adj*   laminar

laminé *adj*   laminated, rolled [e.g. steel]

laminé *adj* **à chaud**   hot rolled

laminé *adj* **à froid**   cold rolled

laminé *m* **marchand**   bar steel

laminer *vb*   roll [metal]

laminoir *m*   rolling mill

lampadaire *m*   lamp post, standard lamp

lampe *f*   bulb, lamp

lampe, durée *f* **de vie d'une**   lamp life

lampe *f* **à arc**   arc lamp

lampe *f* **à braser**   blowlamp

lampe *f* **à faisceau directionnel**   display lamp

lampe *f* **à fluorescence**   fluorescent lamp

lampe *f* **à halogène**   halogen lamp

lampe *f* **à incandescence**   filament or incandescent lamp/bulb

lampe *f* **à réflecteur dichroïque**   dichroic reflector lamp

lampe *f* **à vapeur de mercure à haute pression**   high pressure mercury (vapour) lamp

lampe *f* **à vapeur de sodium à haute pression**   high pressure sodium (vapour) lamp

lampe *f* **à vapeur métallique**   metal vapour lamp

lampe *f* **aux halogénures**   metal halide lamp

lampe *f* **de signalisation**   indicating light or lamp

lampe *f* **de table**   table lamp

lampe *f* **fluorescente**   fluorescent lamp or tube

lampe *f* **germicidale**   bactericidal or germicidal lamp

lampe *f* **témoin**   pilot lamp

lançage *m*   launching [e.g. of bridge girder], pile driving in sand with the aid of water injection

lancer *vb*   initiate [a programme, a project]

lance-rideau *m*   curtain push-rod
lanceur *m*   launching frame or truss [e.g. for bridge sections]
lancier *m*   gutter [of lead, from a roof or terrace]
lancis *m*   repair of masonry by replacement of individual brick or stones
lande *f*   heath, moor(land)
langue *f*   crack initiating breakage of a piece of glass, spit or tongue [of land]
langueter *vb*   tongue [the edge of a board]
languette *f*   small partition [e.g. between flues, or a division within a duct], weather moulding [tongue, on edge of a casement]
languette *f* **rapportée**   loose tongue [metal, forced into opposing grooves in wood]
lanterne *f*   chimney capping with closed top and side openings, lantern [open, above a dome, etc]
lanterne, tendeur *m* **à**   turnbuckle
lanterne *f* **de serrage**   turnbuckle
lanterneau *m*   lantern [skylight], rooflight or skylight [raised]
laque *f*   lacquer, shellac
lardée, serrure *f*   mortise lock
larder *vb*   stud timber with nails [as key for abutting masonry], insert nails into or fix laths to masonry joints to hold subsequent rendering
lardis *m* **de clous**   nails in timber face to be joined to masonry
large *adj*   wide
large plat *m*   flat [wide metal strip]
largeur *f*   breadth, width
largeur *f* **d'aile**   flange width
largeur *f* **de bande**   bandwidth
largeur *f* **de canal**   duct width
largeur *f* **hors tout**   overall width
largeur *f* **libre**   clear width
largeur *f* **nominale**   nominal width
largeur *f* **roulable**   width usable by vehicles
largeur *f* **totale**   overall width
largeur *f* **utile**   clear, net or usable width
larmier *m*   drip [under a soffit, window ledge,etc], drip edge or dripstone
laser *m*   laser
latent *adj*   latent
latéral *adj*   lateral
latitude *f*   latitude
latrines *f,pl* **de chantier**   site latrines
latte *f*   batten [slating, tiling], lath, slat
latté† *m*   blockboard
latte *f* **de garde**   scaffolding guard board
latte *f* **de recouvrement**   fillet
latte *f* **volige**   roof batten
latter *vb*   lath
lattis *m*   battens [of a roof], laths [of lath and plaster]
lause/lauze *f*   stone slates or paving slabs
lavable *adj*   washable
lavabo *m*   handbasin, wash-basin
lavabos *m,pl*   toilets [WCs]
lavage *m*   cleansing [washing]

lave-linge *m*   washing machine
lave-mains *m*   handbasin [small], wash stand
lave-vaisselle *m*   dishwasher, dishwashing machine
laver *vb*   wash
laverie *f*   laundry (room)
laverie *f* **automatique**   launderette
laveur *m*   scrubber [washer]
laveur *m* **d'air**   air washer
laveur *m* **de carreaux**   window cleaner
lavoir *m*   scullery, wash-house, washing room, wash tub
laye *f*   mason's toothed hammer for preliminary dressing of soft stone
layer *vb*   dress stone with a *laye*
lé *m*   width of wallpaper between edgings, width or strip [of sheet material, e.g. carpet, wallpaper]
lecteur *m* **de cartes**   card reader [computer]
lecture *f*   reading [of a meter, instrument]
légende *f*   key, legend [on a drawing, map, etc]
léger *adj*   light [not heavy]
(léger) tremblement *m* **de terre**   earth tremor
(légère) secousse *f* **s(é)ismique**   earth tremor
légers ouvrages *m,pl*   plasterwork
législation *f*   legislation
législation *f* **de la construction**   building code
lentille *f*   lens [light or geological]
lentille *f* **de glace**   ice lens
lessivage *m*   leaching, washing down [paintwork]
lessivé *adj*   leached
lessive *f* **St Marc**   sugar soap [French equivalent]
lessiver *vb*   wash down [paintwork, walls, etc]
lest *m*   ballast [load, e.g. for pile test], counterweight
lettre *f* **avec avis de réception (AR)**   recorded delivery advise letter
lettre *f* **de change**   bill of exchange
lettre *f* **recommandée**   registered letter
levage *m*   erection [of a structural frame assembled flat], hoisting, lifting
levé *m*   survey
levé *m* **aérien**   aerial survey
levé *m* **aéro-photogrammétrique**   photogrammetric aerial survey
levé *m* **cadastral**   cadastral survey
levé *m* **de terrain**   land survey
levé *m* **sur le terrain**   field survey
levé *m* **terrestre**   ground survey
levé *m* **topographique**   topographical survey
levée *f*   causeway, embankment, dike, flood bank, road along a dike/embankment
levée *f* **de terre**   earth bank
lever *vb*   lift (up), raise
lever *vb* **un plan**   draw a plan, carry out a (topological) survey
levier *m*   crowbar, lever
levier, bras *m* **de**   lever arm
levier *m* **à main**   hand lever
levier *m* **de changement de vitesse**   gear (change) lever
levier *m* **de chasse d'eau**   flushing lever [of a WC]
levier *m* **de commande**   control lever
levier *m* **de débrayage**   gear lever

levier *m* **de déclenchement/de dégagement**   release lever
levier *m* **de frein**   brake lever, handbrake
levier *m* **de manœuvre**   control or operating handle or lever
levier *m* **de réglage**   adjusting lever
levier *m* **de serrage**   clamping lever
levier *m* **de vanne**   valve handle
lézarde *f*   chink, crack, crevice or split [in wall, plaster, etc]
lias *m*   lias
liais *m*   hard limestone [frost resistant]
liaison *f*   bond(ing) [of masonry], connection, coupling, joining, link
liaison *f* **à sec**   dry joint [without mortar]
liaison *f* **aérienne**   air link
liaison *f* **de transmission**   data link [to or between computers]
liaison *f* **en pont**   bridge(d) link
liaison *f* **ferroviaire**   rail connection/link
liaison *f* **maritime**   sea link
liaison *f* **routière**   road connection/link
liant *m*   binder [in concrete or road materials], medium, vehicle [of paint], flexibility [of metal]
liant *m* **aérien**   air-setting [aerobic] binder
liant *m* **hydraulique**   hydraulic binder
liant *m* **hydrocarboné**   hydrocarbon-based binder
libération *f* **de tension**   stress relief
libre *adj*   clear, free, open
libre, sur appui   simply supported
libre *adj* **d'occupation**   vacant [of property]
libre service *m*   self-service
lice/lisse *f*   handrail, top rail [of a barrier, balustrade], rail [of post-and-rail fencing]
licence *f*   bachelor's degree, permit
licence, produit *m* **sous**   licensed product
licencié *adj*   redundant [out of work]
licencié *m*   graduate, licensee
liège *m*   cork
lien *m*   bond, diagonal brace, strap [connection], tie [structural], wall tie
lien *m* **d'angle**   angle brace, diagonal brace, corner tie
lier *vb*   join together [with a binder, a tie, a strap]
lierne *f*   diagonal brace, purlin in a pyramidal roof, strut [solid, between floor joists]
lierre *m*   ivy
lieu *m*   locality, spot [locality]
lieu *m* **de construction**   building or construction site
lieu *m* **de mise en place**   site [of installation of a piece of equipment]
lieux *m,pl*   premises
lieux *m,pl*, **état** *m* **des**   existing state [described in drawing, report, etc]
lieux *m,pl* **(d'aisances)**[†]   toilet(s) (WCs)
lifting *m*   refurbishment [e.g. of shop interior]
ligature *f*   splice [in reinforcing bar, a cable, wire, etc], tie [to hold reinforcement]
lignage *m*   marking [of line of cut with a chalk line], setting out of lines of slates or tiles

ligne *f*   contour, line, outline [drawn], cord marking face line of a wall under construction, wires [electrical transmission line]
lignes aérodynamiques, aux   streamlined
ligne *f* **aérienne**   overhead (power) line
ligne *f* **à haute tension**   high-tension/-voltage line
ligne *f* **à postes groupés**   party line
ligne *f* **blanche (continue/discontinue)**   (continuous/discontinuous) white line [road marking]
ligne *f* **continue/discontinue**   continuous/discontinuous line [road marking]
ligne *f* **d'alimentation**   supply line [electrical mains]
ligne *f* **d'écoulement**   flow line
ligne *f* **d'influence**   influence line
ligne *f* **de bris**   line of intersection of two planes of mansard roof
ligne *f* **de chemin de fer**   railway line
ligne *f* **de courant**   streamline
ligne *f* **de crête**   ridge [geology]
ligne *f* **de démarcation**   boundary
ligne *f* **de faille**   fault line
ligne *f* **de faîtage**   ridge [roof]
ligne *f* **de faîte**   catchment boundary, watershed
ligne *f* **de feu**   fire-break
ligne *f* **de force**   line of force
ligne *f* **de foulée**   line of flight [of stairs]
ligne *f* **de marée haute/de la marée**   high tide/water line/mark
ligne *f* **de mire**   line of sight
ligne *f* **de naissance**   springing line [of an arch]
ligne *f* **de partage des eaux**   catchment boundary, watershed
ligne *f* **de référence**   datum line
ligne *f* **de repère**   datum line
ligne *f* **de saturation**   saturation line
ligne *f* **de soudure bout à bout**   butt-welded seam
ligne *f* **des centres de gravité**   centroidal or gravity axis
ligne *f* **droite**   straight line
ligne *f* **hachurée**   dotted line
ligne *f* **infranchissable**   continuous line [road marking]
ligne *f* **principale (de chemin de fer)**   main (railway) line
ligne *f* **secondaire (de chemin de fer)**   branch (railway) line
ligner *vb*   mark with a chalk line
lignite *m*   brown coal, lignite
limaçon *m*   spiral staircase
limaçon, escalier *m* **en**   spiral staircase
limaille *f* **(d'acier, de fer)**   chips, filings, millings or turnings (steel, iron)
limande *f*   flat cover fillet or moulding, straight edge [of wood]
lime *f*   file [rasp]
lime *f* **à bois**   wood file
lime *f* **à dégrossir**   roughing file
limer *vb*   file [metal, wood, etc]
limite *adj*   limiting, maximum, ultimate [e.g. at failure]
limite *f*   boundary [of a garden, plot, site, etc], limit
limite, cas *m*   borderline case
limite, charge *f*   maximum load
limite, date *f*   closing date, deadline

limite, hauteur *f*   maximum height
limite, longueur *f*   maximum length
limite, valeur *f*   limiting value
limite, vitesse *f*   maximum speed
limite *f* d'écoulement   offset limit, yield point
limite d'écoulement, contrainte *f* à la   yield stress
limite d'écoulement, déformation *f* à la   yield strain
limite *f* d'écrasement   compressive yield point
limite *f* d'élasticité   elastic limit
limite *f* d'endurance   endurance or fatigue limit
limite *f* de chaussée   road edge
limite *f* de fatigue   fatigue limit
limite *f* de marée haute/de la marée   high water/tide
   line/mark
limite *f* de liquidité   liquid limit [soil property]
limite *f* de plasticité   plasticity limit [soil property]
limite *f* de proportionnalité   limit of proportionality,
   proportional limit
limite *f* de rupture   breaking point or strength, ultimate
   stress
limite *f* de température   temperature limit
limite *f* des arbres   tree line
limite *f* du terrain   plot or site boundary
limite *f* élastique   yield point, yield strength
limite *f* élastique à la compression   compressive yield
   point
limite *f* élastique apparente   offset limit
limite *f* inférieure   lower limit
limite *f* supérieure   upper limit
limiteur *m* de courant   current limiter
limiteur *m* de débit   flow limiting device
limiteur *m* de pression   pressure limiter/limiting device
limiteur *m* de tension   voltage limiter
limiteur *m* de tirage   draught limiter/limiting device
limon *m*   alluvium, silt, ooze [wet mud], string(er) [of a
   staircase]
limon *m*, faux   wall string(er)
limon *m* à crémaillère/à l'anglaise   cut or open string
limon *m* à la française   close string
limon *m* apparent   outer string(er)
limon *m* (d'escalier)   stair string(er)
limoneux *adj*   silty
limousin(ant) *m*   mason building rough or rubble masonry
limousinage *m*/limousinerie *f*   rough or rubble masonry
linceau *m*/linçoir *m*   lintel beam over an opening between
   two columns or posts
linçoir *m*   trimmer [supporting beams across a chimney,
   roof opening, etc]
linéaire *adj*   linear
linéairement *adv*   linearly
lingerie *f*   linen room
linoléum *m*   linoleum
linteau *m*   lintel (beam)
linteau *m*, arrière-   secondary lintel built into wall behind
   main lintel
linteau *m* de béton armé   reinforced concrete lintel
linteau *m* de décharge   reinforcing lintel [built into a
   wall above a weak point]
liquéfaction *f*   liquefaction

liquide *adj*   liquid *adj*
liquide *m*   liquid
liquide *m* de refroidissement   cooling liquid
liquidité *f*   liquidity
liquidité, limite *f* de   liquid limit [soil property]
liquidités *f,pl*   liquid assets
lisière *f*   edge [of a field or forest, or wallpaper]
lissage *m*   smoothing
lisse *adj*   glossy, smooth
lisse *f*   barrier rail [level crossing], hand-rail, rail [of post-
   and-rail fencing], smoothing tool, smoothness
lisse *f* d'appui   sill rail [timber structure]
lisse *f* d'assise   wall plate [timber]
lisseuse *f*   power float, machine trowel [for smoothing
   concrete]
lissoir *m*   smoothing tool, rule following a power float to
   ensure flatness of slab
liste *f*   list
liste *f* de prix   price list
listel *m*   fillet [in a moulding]
lit *m*   bed [furniture, and of mortar, river, etc], channel
   [river bed], layer of material, horizontal faces of cut
   stones in stone masonry, plane of cleavage of a stone bed
lit *m* bactérien   bacteria bed
lit *m* d'acier   layer of reinforcement
lit *m* d'argile   layer of clay
lit *m* de crue   flood plain
lit *m* de filtrage   filter-bed
lit *m* de fleuve/rivière   river bed
lit *m* de mâchefer   clinker bed
lit *m* de scories   slag layer
lit *m* de séchage (des boues)   sludge drying bed
lit *m* des hautes eaux   flood plain
lit *m* filtrant   filter-bed
lits *m,pl* superposés   bunk beds
liteau *m*   batten [slating or tiling]
litige *m*   lawsuit, legal action or dispute, litigation
litre *m*   litre
living *m*   living room
livraison *f*   delivery
livrer *vb*   deliver
local *adj*   local
local *m*   room [premises, and as in plantroom]
local *m* comptage   meter room
local *m* d'essai   test room
local *m* d'habitation   dwelling
local *m* de centrale   plant room
local *m* de restauration   canteen
local *m* de stockage   store room
local *m* occupé   occupied room or space
local *m* technique   equipment room, plantroom
local *m* technique des ascenseurs   lift machine/motor
   room
localisateur *m* d'incendie   addressable fire detection system
localisateur *m* de câbles   cable detector
localisation *f*   siting
locataire *m*   tenant
locatif *adj*   letting

location *f*   hire, rental, renting
location *f* **en meublé**   furnished let(ting)
location *f* **vente**   hire/instalment purchase
locaux *m,pl*   offices, premises
loess *m*   loess
loge *f*   loggia
logement *m*   accommodation, collective housing, dwelling, lodging, quarters
loggia *m*   balcony [inset], loggia
logiciel *m*   program [computer]
logiciel *m* **de base**   systems program
logique *adj*   logical
logique *f*   logic
loi *f*   law
loi *f* **d'Ohm/Gay-Lussac/Hooke/Marriotte**
    Ohm's/Charles'/Hooke's/Boyle's law
loi *f* **de la conservation d'énergie**   law of conservation of energy
lointain *adj*   far [distant]
lois *f,pl* **fondamentales des ventilateurs**   fan laws
loisirs, base *f* **de**   leisure or recreation centre
long *adj*   long
long terme, à   in the long term, long term [loan, etc]
longeron *m*   central girder [of a bridge], longitudinal girder or member, stringer [structural, not staircase]
longitude *f*   longitude
longitudinal *adj*   longitudinal
long-pan *m*   long face or side [of a building, a roof plane]
longrine *f*   beam on grade, cross beam [linking piles], foundation (tie) beam, ground beam, pile cap (beam)
longueur *f*   length [duration, space]
longueur *f* **d'ancrage**   anchorage length
longueur *f* **d'onde**   wavelength
longueur *f* **de fente**   crack length
longueur *f* **de flambement**   buckling length
longueur *f* **de l'arc**   arc length [in welding]
longueur *f* **de recouvrement**   lap length
longueur *f* **de serrage**   grip length
longueur *f* **hors-tout**   overall length
longueur *f* **limite**   maximum length
longueur *f* **totale**   overall length
longueur *f* **utile**   useful length
loquet *m*   latch [pivoted type, for a door, gate, etc]
loqueteau *m*   catch [door, shutter, cupboard, etc]
loqueteau *m* **magnétique**   magnetic catch
losange, en forme de   diamond *adj* [e.g. of a motorway interchange]
losange *m* **flamand**   Flemish bond
lot *m*   bid, contract or work package, building plot, batch [of materials, tools, etc], lot [at auction]
lot *m* **de terrain**   lot [plot of land]
lotir *vb*   parcel out [land]
lotissement *m*   housing estate or site, or its division into plots
lots *m,pl* **séparés**   separate trades [contracting]
louage *m*   labour-only (sub-)contract
loué *adj*   rented
louer *vb*   let, rent
loup *m*   bad workmanship [piece of]
lourd *adj*   heavy

louve *f*   rag bolt or expanding fitting used for lifting a piece of stone
loyer *m*   rent
lu et approuvé   read and approved [in contract documents]
lubrifiant *adj*   lubricating *adj*
lubrifiant *m*   lubricant
lubrification *f*   lubrication
lubrification *f* **forcée**   forced lubrication
lubrifier *vb*   lubricate, oil
lucarne *f*   attic or dormer window, skylight [dormer]
lucarne *f* **à croupe**   hip(ped) dormer
lucarne *f* **à demoiselle**   dormer with backward sloping roof
lucarne *f* **à deux pentes**   dormer with ridged roof
lucarne *f* **bombée**   eyebrow dormer
lucarne *f* **capucine**   hip(ped) dormer
lucarne *f* **en chien assis**   dormer with backward sloping roof
lucarne *f* **en défoncé**   internal dormer
lucarne *f* **rampante**   shed dormer
lucarne *f* **retroussée**   dormer with backward sloping roof
lueur *f*   glow
lumen *m*   lumen
lumière *f*   countersinking for nut and bolt head, light, mouth [of a plane]
lumière *f* **artificielle**   artificial light
lumière *f* **de contrôle**   pilot light
lumière *f* **du jour**   daylight
lumière *f* **électrique**   electric lighting
lumière *f* **naturelle**   natural light
lumière *f* **noire**   black light
luminaire *m*   light fitting or fixture, luminaire
luminaire *m* **encastré**   recessed fixture [light fitting]
luminaire *m* **extensif**   batwing [light fitting]
luminaire *m* **intensif**   downlighter
luminance *f*   brightness, luminance
luminescence *f*   luminescence
luminescent *adj*   luminescent
lumineuse, intensité *f*   luminous intensity
lumineux *adj*   illuminated [sign, etc], luminous, well-lit
lumineux, flux *m*   luminous flux
luminosité *f*   brightness, light [quality of], luminosity
lunette *f*   intersection of a minor and major road in a dip, secondary vault piercing a barrel vault, telescope [of a theodolite]
lunette, coup *m* **de**   observation or sighting [surveying]
lunette *f* **de visée**   sighting or surveyor's telescope
lunette *f* **viseur**   sighting or surveyor's telescope
lunettes *f,pl* **(à branches)**   spectacles
lunettes *f,pl* **de protection**   eye protectors, protective glasses or goggles
lunettes-loup *f,pl*   goggles [protective]
lunettes-masque *f,pl*   goggles [protective]
lustre *f*   chandelier [ornamented], glaze coat [paintwork], gloss, shine
lustré *adj*   glossy, shiny [paintwork]
lustrer *vb*   glaze [paintwork]
lut *m*   jointing mastic or paste
luter *vb*   seal with mastic or paste
lux *m*   lux
lyre *f* **de dilatation**   expansion loop

macabé *m*   painter's filling knife [French regional term]
macadam *m*   macadam, or aggregate for tarmacadam
macadam *m* **à l'eau**   waterbound macadam
macadam *m* **cylindré**   rolled macadam
macadam *m* **en pénétration**   grouted macadam
macadam *m* **enrobé**   coated macadam
macadam *m* **goudronné**   Tarmac(adam)
macadam *m* **traité en pénétration**   bitumen grouted or
  bitumen penetration macadam
macadam *m* **vibré**   dry-bound macadam
macadamiser *vb*   macadamize
macérateur *m*   macerator
mâchefer *m*   clinker or slag [for use as aggregate]
machine *f*   engine [railway], machine
machine *f* **à bois**   woodworking machine
machine *f* **à bois polyvalente**   universal woodworking
  machine
machine *f* **à calculer**   calculating machine, calculator
machine *f* **à cintrer**   bending machine
machine *f* **à coffrage glissant**   slipform paver or paving
  machine
machine *f* **à corroyer/à dresser**   surfacer or surface
  planer [for straightening wood], surfacing machine
machine *f* **à dessiner**   draughting machine
machine *f* **à eau glacée**   chiller
machine *f* **à écrire**   typewriter
machine *f* **à encoller**   pasting machine [for wallpaper]
machine *f* **à fileter**   screw cutting machine
machine *f* **à fileter des tubes**   pipe threading machine
machine *f* **à froid**   refrigerating machine
machine *f* **à laver**   washing machine
machine *f* **à laver la vaisselle**   dishwashing machine
machine *f* **à mortaiser**   mortising machine
machine *f* **à raboter**   planer
machine *f* **à souder**   welding machine
machine *f* **à/de traitement de texte**   word processor
machine *f* **à trancher le bois**   veneer slicing machine
machine *f* **d'oxycoupage**   gas cutting machine
machine *f* **de bureau**   business or office machine
machine *f* **décapeuse**   scraper [ground surfaces]
machine *f* **électrique mobile/portatif**   portable electrical
  equipment
machine *f* **frigorifique**   refrigerating machine
machine *f* **pour soudage à l'arc**   arc welding machine
machine *f* **tournante**   rotary machine
machine-outil *f*   machine tool
machine-outil *f* **à (travailler le) bois**   woodworking
  machine
mâchoire *f* **fixe/mobile**   fixed/movable jaw [of a vice, etc]
maçon *m*   bricklayer, jobbing builder, mason
maçonnage *m*   bricklaying or masonry [the work]
maçonné *adj*   masonry *adj*
maçonner *vb*   brick, or lay bricks or stone

maçonnerie *f*   brickwork, stonework, masonry [also
  plastering and rendering]
maçonnerie *f* **composite**   masonry consisting of two or
  more materials [e.g. brick and stone]
madrier *m*   beam [heavy timber, at least 60–80mm thick]
madrier *m* **de faîtage**   ridge beam
magasin *m*   shop, warehouse
magasinier *m*   storekeeper
magnésie *f*   magnesia
magnésium *m*   magnesium
maigre *adj*   lean [of cement or mortar], thin [of a column,
  of paint containing little oil]
maigreur *m*   thinness [of a rendering, paintwork, etc]
mail *m*   mall [avenue], sledgehammer, square [on a grid]
maillage *m*   grid [for analysis or drawing], grillage [e.g.
  for bridge deck analysis]
maille *f* **mesh**   [i.e. individual opening in a mesh]
maille, débit *m* **sur**   quarter sawing [from the log]
maille, scié *adj* **sur**   quarter-sawn
maille *f* **d'un tamis**   sieve opening size
maille *f* **du bois**   wood grain on a lengthwise cut face
maillé fer *m*   metal lattice, wire netting
maillet *m* **(à tête rectangulaire)**   wood mallet
maillon *m*   link [of a chain]
main *f* **courante**   balustrade, banister rail, handrail
main-d'œuvre *f*   labour, manpower, workforce
main-d'œuvre *f* **qualifiée**   skilled labour
mainlevée *f*   release [of a mortgage or charge]
maintenance *f*   maintenance, servicing
maintenance *f* **assistée par ordinateur (MAO)**
  computer-assisted planned maintenance
maintenance *f* **corrective**   corrective maintenance
maintenance *f* **préventive**   preventive maintenance
maintenir *vb*   hold up [support], maintain, support
maintien *m*   maintenance, preservation, upholding
maire *m*   mayor
mairie *f*   town hall
maison *f*   house
maison *f* **à étage**   two-storey house
maison *f* **à rez-de-chaussée**   single-storey house
maison *f* **cantonnière**   roadman's hut
maison *f* **d'amis**   weekend house
maison *f* **d'habitation**   dwelling house
maison *f* **d'hôtes**   guesthouse
maison *f* **de campagne**   country house
maison *f* **de garde/gardien**   caretaker's house
maison *f* **de maître**   country seat, gentleman's residence
maison *f* **de retraite**   retirement home
maison *f* **en bande/rangée**   terrace house
maison *f* **forte**   fortified house
maison *f* **individuelle/isolée**   detached house
maison *f* **jumelée**   semi-detached house
maison *f* **paysanne**   farmhouse**

maison *f* **sur terre-plein** single-storey house at ground-level

maître *m* courtesy title for lawyer

maître *m* **d'œuvre** architect in charge of the works, designer, lead consultant, project manager

maître *m* **d'ouvrage/de l'ouvrage** awarding authority, client, developer or owner [client for building work], employer [client for works]

maître *m* **de chantier** construction manager, site manager

maître ouvrier *m* chargehand, foreman

maître-chevron *m* principal rafter

maîtrise *f* workmanship [of a craftsman]

maîtrise *f* **d'œuvre** management of the works [by the *maître d'œuvre*]

maîtrise *f* **des coûts** cost control

maîtriser *vb* master

majeur *adj* major

mal *adv* badly

malandre *f* rot pocket or rotten knot hidden in sound wood

malaxer *vb* mix

malaxeur *m* mixer

malaxeur *m* **à coulis** grout mixer

malaxeur *m* **de béton** concrete mixer

malaxeur *m* **discontinu** batch mixer

malaxeur *m* **plat** pan mixer

malaxeur *m* **pompe à béton** truckmixer with concrete pump

mâle *adj* male [connection, thread, etc]

malfaçon *m* defect [in workmanship], faulty work(manship)

malléabilité *f* malleability

malléable *adj* malleable

malsain *adj* unhealthy

mamelon *m* hinge pin, nipple [plumbing], pipe joint between pipes of different diameter

manche *f* sleeve

manche *m* handle [of a tool or implement]

manche *f* **à air** windsock

manche *m* **à balai** broomstick, joystick [computer]

manche *m* **de scie** saw handle

manchette *f* oversleeve or sleeve [protective clothing]

manchette *f* **de protection** protective (over)sleeve [clothing]

manchette *f* **flexible** flexible sleeve

manchette *f* **souple** flexible sleeve

manchon *m* coupler [reinforcement], roller [of a painter's roller], (reducing) sleeve or union, casing

manchon *m* **à brides** flanged sleeve

manchon *m* **d'entrée** inlet or intake sleeve

manchon *m* **de raccordement** connecting sleeve or socket

manchon *m* **de tuyau** pipe socket

manchon *m* **élastique** flexible coupling

manchon *m* **en fonte** cast iron sleeve or socket

manchon *m* **fileté** threaded socket

manchon *m* **réducteur** reducing bushing

manchon *m* **taraudé** sleeve nut

manchon *m* **vissé** screwed socket

mandat *m* commission or mandate [to do sth]

mandat *m* **de vente** selling agreement [between owner and agent]

mandataire *m/f* holder of a commission or mandate [e.g. estate agent selling a property], representative [of somebody, e.g. client]

mandrin *m* chuck [of a drill, lathe], drift or punch [tool], expander [plumber's]

mandrin *m* **à deux mors** two-jaw chuck

mandrins *m,pl* **à serrage rapide** multi-head, quick-action chuck

mandrin *m* **creux** hollow chuck

manège *m* riding school

manette *f* hand lever

manette *f* **de chasse** flushing lever [WC cistern]

manganèse *m* manganese

maniabilité *f* workability [concrete]

manicle *f* protective glove

manier *vb* handle

manille *f* clevis or shackle, manilla rope

manipulation *f* handling

manipuler *vb* handle

manique *f* protective glove

manivelle *f* crank or crank-handle

manœuvré *adj* **à distance** remote(ly) controlled or operated

manœuvré *adj* **à pied d'œuvre** locally controlled or operated

manœuvre *m* (bricklayer's or builder's) labourer, manoeuvre or operation

manœuvre, travail *m* **de** unskilled labour or work

manœuvrer *vb* manoeuvre, operate or work [a machine]

manoir *m* country house, manor

manomètre *m* manometer, pressure gauge

manomètre *m* **(à tube) incliné** inclined tube manometer

manomètre *m* **différentiel** differential manometer or pressure gauge

manométrique *adj* manometric

manostat *m* pressure switch

mansarde *f* attic or garret [in a mansard roof], mansard (roof)

mansarde, fenêtre *f* **en** mansard window [in the lower slope of the roof]

manteau *m* **d'Arlequin** proscenium arch

manteau *m* **(de cheminée)** mantlepiece or mantelshelf

manuel *adj* hand-operated, manual

manuel *m* handbook, manual

manuel, réglage *m* manual adjustment

manuel *m* **d'exploitation** operating/operations manual

manuel *m* **de service** operating manual

manuellement *adv* by hand, manually

manutention *f* handling

manutention *f* **manuelle** manual handling

manutention *f* **mécanique** mechanical handling

maquette *f* mock-up, model [architectural]

maquettiste *m/f* model-maker

marais *m* fen, marsh, swamp

marbre *m* marble

marbre *m* **brèche** breccia marble

marbre *m* **composé**   marble of several different colours

marbre *m* **simple**   marble of a single colour

marchand *m* **de biens**   real-estate dealer

marchand *m* **de bois**   timber merchant

marche *f*   running or working [of equipment, machines], stair, step or tread [of a stair]

marche *f*, **dernière**   top step

marche, nez *m* **de**   nosing [of a stair tread]

marche *f* **à blanc/vide**   idling, running light, no-load operation/running [of a machine]

marche arrière, faire *vb*   back [of a vehicle]

marche *f* **biaise**   skewed step of varying width

marche *f* **carrée**   flier

marche *f* **continue**   continuous operation or running

marche *f* **d'essai**   test/trial run

marche *f* **dansante**   balanced or dancing step, French flier, winder [stair tread of varying width]

marche *f* **de départ**   bottom step

marche *f* **droite**   flier

marche *f* **du bas**   bottom step

marche *f* **gironnée**   winder [stair tread of varying width]

marche *f* **palière**   landing tread, top step

marche *f* **pleine**   block step

marche *f* **tournante**   winder [stair tread of varying width]

marché, à bon *adv*   cheaply

marché, bon *adj*   cheap

marché *m*   contract, deal or transaction, market

marché *m* **à forfait**   lump sum contract

marché *m* **clef en main**   turnkey contract

marché *m* **de gré à gré**   negotiated contract or tender

marché *m* **de travaux**   construction contract

marché *m* **du travail**   labour market

marché *m* **négocié**   negotiated contract or tender

marché *m* **par adjudication**   competitive contract or tendering

marché *m* **sur appel d'offres**   competitive contract or tendering

marchepied *m*   footpath [along a watercourse], towpath, running-board [vehicle]

marcher *vb*   operate, run [equipment or machinery], work [function]

marcher *vb* **à vide**   run on or at no-load

marcher *vb* **en parallèle**   operate or run in parallel

marcher *vb* **en série**   operate or run in series

mare *f*   pond

marécage *m*   marsh

marée *f*   tide

marée *f* **basse**   low tide

marée *f* **descendante**   ebb tide

marée *f* **haute**   high tide

marée *f* **montante**   flood tide

marge *f*   margin

marge *f* **bénéficiaire**   profit margin

marge *f* **brute**   gross margin

marge *f* **de sécurité**   safety margin

marge *f* **de tolérance**   tolerance [margin]

margelle *f*   coping [of a well or swimming pool]

marin *m*   muck from tunnelling

marinage *m*   mucking-out stage of tunnel works

mariner *vb*   muck out [a tunnel]

marmite *f* **d'érosion**   pothole [in the ground]

marnage *m*   tidal range

marne *f*   marl

marneux *adj*   marly

marouflage *m*   gluing a heavy fabric to a panel to hide joints and inhibit swelling

maroufler *vb*   reinforce by gluing a heavy fabric to a panel to hide joints and inhibit swelling

marquage *m*   etching [marking], marking [e.g. white lining of roads]

marquage *m* **latéral**   edge marking

marquage *m* **peint**   painted marking [on roads]

marquage *m* **routier**   road marking

marquage *m* **scellé**   embedded marking [on roads]

marqué *adj*   marked

marque *f*   brand or make

marque *f* **(de fabrique)**   trademark

marque *f* **déposée**   registered trademark

marqueur *m*   marker

marquer *vb*   mark, show [time, temperature]

marqueter *vb*   inlay

marqueterie *f*   marquetry

marquise *f*   canopy, glass roof over platform [station]

marronnage *m*   cracking in a road surface

marronnier *m*   chestnut [wood, and sweet chestnut tree]

marronier *m* **d'Inde**   horse chestnut

marteau *m*   hammer

marteau *m* **à bigorner**   chamfering hammer

marteau *m* **à décalaminer**   needle scaler

marteau *m* **à deux mains**   sledgehammer

marteau *m* **à frapper devant**   sledgehammer

marteau *m* **à main**   plumber's or hand hammer

marteau *m* **à piquer**   chipping hammer

marteau *m* **à planer**   flatter, or flat-face hammer

marteau *m* **à plomb**   lead hammer

marteau *m* **à poinçon**   punch hammer

marteau *m* **d'ardoisier**   slate hammer, slater's hammer

marteau *m* **dame**   drop hammer [pile-driver]

marteau *m* **de battage**   pile hammer

marteau *m* **de charpentier**   carpenter's hammer

marteau *m* **de forgeron**   blacksmith's (hand) hammer

marteau *m* **de maçon**   axe or bricklayer's hammer

marteau *m* **de vitrier**   glazier's hammer

marteau *m* **piqueur**   jackhammer, pneumatic drill or hammer

marteau *m* **pneumatique**   pneumatic drill or hammer, rock drill

marteau *m* **rivoir**   machinist's hammer

marteau-burineur *m*   chipping hammer

marteau-perforateur *m*   pneumatic hammer or rock drill

marteau-perforateur *m* **hors (du) trou**   down-(the-)hole hammer

martelage *m*   hammering or hammered finish

martelé *adj*   hammer finish [of glass]

marteler *vb*   hammer or peen

martelet *m*   small hammer, slater's or tiler's hammer [similar to bricklayer's hammer]

mas *m*   farmhouse [south of France]

mascaron *m*   cover plate for unused tap hole in a sanitary appliance

masquage *m*   masking [for painting]

masque *m*   cover, mask [shield], protection [mask]

masque *m* **amont/aval**   upstream/downstream facing or shell [of a dam]

masque *m* **de soudage à l'arc**   arc-welding mask

masque *m* **de soudeur**   welder's mask

masque *m* **en béton armé**   reinforced concrete facing or protection

masque *m* **en enrochement**   stone pitching [of an embankment, etc]

masque *m* **étanche**   damp-proof course (DPC), watertight curtain or surfacing

masque *m* **respiratoire**   breathing mask

masse, en   in a body

masse *f*   earth [electrical], mash or mason's hammer, sledgehammer, mass, bulk [volume], massing or overall [of buildings, etc]

masse, faire *vb*   act as an earth

masse, mettre *vb* **à la**   earth [electrically]

masse *f*, **plan** *m* **(de)**   overall layout plan

masse *f* **active**   assets

masse *f* **d'eau**   body or expanse of water

masse *f* **passive**   liabilities

masse *f* **volumique**   density, bulk density

massette *f*   club or lump hammer [small]

massif *adj*   bulky [of a building], massive, solid [e.g. solid timber, etc]

massif *m*   body or mass [of a structure], solid block or mass [of concrete, masonry, etc]

massif *m* **d'ancrage**   anchor block

mastic *m*   filler, mastic, putty

mastic *m* **acrylique**   resin based mastic

mastic *m* **d'asphalte**   mastic asphalt

mastic *m* **d'étanchéité**   sealing mastic

mastic *m* **de chaudière**   boiler mastic

mastic *m* **de vitrier**   glazing putty

mastic *m* **lin naturel**   linseed oil putty

mastic *m* **pâteux**   joint sealant, mastic paste

mastic *m* **pour canalisations**   duct sealant

mastic *m* **pour joints**   jointing compound

masticage *m*   filling or stopping up [of holes, cracks or joints]

mât *m*   mast, pole

mât *m* **d'éclairage**   lighting mast

mât *m* **de levage**   guy or pole derrick

mât *m* **en treillis**   trellis mast

mât *m* **haubané**   guyed or stayed mast

mât *m* **inclinable**   guy or pole derrick

mat *adj*   dull or matt [finish], unpolished

matage *m*   caulking, matting [of gilding]

matelas *m* **d'air**   air cushion or mattress

mater *vb*   caulk

matériau *m*   material

matériau *m* **antigel**   frost protection material

matériau *m* **cohérent**   cohesive material

matériau *m* **concassé**   broken-stone [road material]

matériau *m* **d'emprunt**   borrow material

matériau *m* **de construction**   building material

matériau *m* **de déblai**   excavated material

matériau *m* **de remblai**   fill(ing) material

matériau *m* **de remplissage**   fill(ing) material

matériau *m* **de revêtement**   lining material

matériau *m* **imperméable**   impermeable material

matériau *m* **isolant**   insulating material

matériau *m* **meuble**   loose material

matériau *m* **perméable**   permeable material

matériau *m* **rocheux compact**   solid rock [aggregate, road material]

matériel *m*   equipment or plant

matériel *m* **anti-déflagrant**   flameproof equipment

matériel *m* **assemblé en usine**   factory assembled plant or equipment

matériel *m* **d'injection**   grouting equipment

matériel *m* **de brasage**   soldering equipment

matériel *m* **de chantier**   site equipment/plant

matériel *m* **de communication**   communication equipment

matériel *m* **de construction**   building equipment

matériel *m* **de contrôle**   test(ing) equipment

matériel *m* **de creusement**   excavating equipment

matériel *m* **de dégel**   de-icing equipment

matériel *m* **de drainage**   drainage fitting

matériel *m* **de filtrage**   filter(ing) equipment

matériel *m* **de manutention**   handling equipment

matériel *m* **de perforation**   drilling or tunnelling equipment

matériel *m* **de relayage**   relay equipment

matériel *m* **de terrassement**   earth-moving equipment or plant

matériel *m* **de topographie**   (land) surveying equipment

matériel *m* **électrique mobile/portatif**   portable electrical equipment

matériel *m* **force**   power equipment

matériel *m* **informatique**   computer equipment or hardware

matériel *m* **roulant**   rolling stock [railway]

matière *f*   material, matter [substance]

matière *f* **absorbante**   absorbent or absorption material or medium

matière *f* **adhésive**   adhesive (material)

matière *f* **colorante**   dye

matière *f* **combustible**   combustible (material)

matière *f* **de filtrage**   filter medium

matière *f* **filtrante**   filter medium

matière *f* **inflammable**   inflammable material

matière *f* **isolante**   insulating material

matière *f* **première**   raw material

matière *f* **soluble**   soluble material

matière *f* **volatile**   volatile matter

matières *f,pl* **en solution**   dissolved matter

matoir *m*   marking punch, riveting hammer

matrice *f*   matrix

matrice *f* **carrée**   square matrix

matrice *f* **diagonale**   diagonal matrix

maturation *f*   curing [completion of, of concrete]

maximal *adj*   maximum

maximum *adj*   maximum

maximum *m*   maximum

maximum, débit *m*   peak load

maximum *m* **admissible/permissible**   maximum allowable or permissible

mazout *m*   fuel or heating oil

mazout *m* **chauffage**   heating oil

mazout *m* **diesel**   diesel fuel/oil

mazout *m* **domestique**   domestic heating oil

méandre *m*   bend or meander [of a watercourse]

mécanique *adj*   mechanical

mécanique, résistance *f*   load-carrying capacity [of road material]

mécanique *m* **des fluides**   fluid mechanics

mécanique *m* **des roches**   rock mechanics

mécaniser *vb*   mechanize

mécanisme *m*   mechanism

mécanisme *m* **d'alarme**   alarm mechanism

mécanisme *m* **de contrôle**   control mechanism

mécanisme *m* **de soupape**   valve linkage

mèche *f*   drill bit

mèche *f* **à bois à trois pointes**   live centre [for wood turning]

mèche *f* **à bois creuse**   rough turning gouge

mèche *f* **à cuiller**   spoon bit

mèche *f* **anglaise**   centre bit

médecin *m* **d'entreprise**   works doctor or medical officer

médecin *m* **du travail**   occupational health specialist, works doctor or medical officer

médecin *m* **inspecteur de travail**   medical (works) or occupational health inspector

médias-filtrants *m,pl*   filter media

mélange *m*   mixture

mélange *m* **air-gaz**   gas-air mixture

mélange *m* **bien gradué**   well-graded mixture [e.g. of road materials]

mélange *m* **de gaz explosive**   explosive gas mixture

mélange *m* **en centrale**   stationary plant mixing [soil stabilization method]

mélange *m* **en place**   mix-in-place [soil stabilization method]

mélange *m* **frigorifique**   freezing mixture

mélanger *vb*   mix

mélangeur *m*   agitator, mixer, combination or mixer tap [twin tap, single spout]

mélangeur *m* **à tambour tournant**   revolving drum mixer

mélèze *m*   larch

membrane *f*   diaphragm, membrane

membre *m*   member

membron *m*   metal-clad section at top of lower slope of curb or mansard roof

membrure *f*   chord or flange [of a girder]

membrure *f* **comprimée**   compression chord or flange

membrure *f* **inférieure**   lower chord or flange

membrure *f* **supérieure**   upper chord or flange

membrure *f* **tendue**   tension chord or flange

mémoire *f*   memory [computer, and generally]

mémoire *f* **à bande magnétique**   magnetic tape memory [computer]

mémoire *f* **à disques**   disk memory [computer]

mémoire *f* **(de réclamation)**   claim [by a contractor, with quantities + costs]

mémoire *f* **morte**   read only memory (ROM)

mémoire *f* **tampon**   buffer [computer]

mémoire *f* **vive**   random access memory (RAM)

mémoriser *vb*   store [in computer memory]

meneau *m*   mullion [both horizontals and verticals]

meneau, fenêtre *f* **à**   mullion window

meneau *m* **horizontal**   transom

méniane *f*   small veranda, on columns

méniane, colonne *f*   column supporting a small balcony or veranda

mensole *f*   keystone

mensuel *adj*   monthly

mensualité *f*   monthly payment or salary

mensuration *f*   measurement [the process]

mentonnet *m*   catch [of a pivoted/lifting type latch]

mentonnet *m* **de sûreté**   extremity of a fanlight stay with provision for locking

menuiserie *f*   joinery, woodwork

menuiserie *f* **métallique**   metal joinery [e.g. doors, windows, etc]

menuisier *m*   joiner

méplat *adj*   flat, flat-laid [of a joist or beam]

méplat *m*   thin, flat metallic strip

mercure *m*   mercury

merisier *m*   (wild) cherry [tree, wood]

merrain *m*   quarter-sawn board [normally 130–150×40–60mm]

mérule *m*   dry rot fungus, *Merulius lacrymans*

mesurage *m*   dimensioned plan [of a plot of land], measurement [the process] or measuring

mesure, sur   to measure

mesure *f*   extent, size [measure or measurement], measure(ment) [unit of, or dimension]

mesure *f* **à ruban**   tape measure

mesure *f* **d'humidité**   moisture measurement

mesure *f* **dans œuvre (D/O)**   internal dimension, measurement between interior faces

mesure *f* **de débit**   discharge or flow measurement

mesure *f* **de décharge**   discharge measurement

mesure *f* **de la consommation**   consumption measurement

mesure *f* **de la dureté**   hardness measurement

mesure *f* **de la lumière**   illumination measurement

mesure *f* **de la pression**   pressure measurement

mesure *f* **de la quantité d'air**   air quantity measurement

mesure *f* **de la teneur en suies**   soot test

mesure *f* **de la tension**   voltage measurement

mesure *f* **de la vitesse**   speed measurement

mesure *f* **de tassement**   settlement measurement

mesure *f* **des gaz de fumée**   flue gas test

mesure *f* **du bruit**   noise measurement

mesure *f* **du rendement**   efficiency measurement

mesure *f* **hors-œuvre**   external dimension

mesurer *vb* **(qch)**   measure (sth)
mesures *f,pl*   measures or steps
mesures *f,pl* **anti-corrosives**   corrosion prevention
mesures *f,pl* **automatiques**   monitoring
mesures *f,pl* **de protection**   protective measures
métairie *f*   tenanted farm [in France]
métal *m*   metal
métal *m* **d'apport de brasage**   soft solder
métal *m* **de base**   base or parent metal
métal *m* **déployé**   expanded metal
métal *m* **déposé**   weld metal
métal *m* **ferreux/non ferreux**   ferrous/non-ferrous metal
métal *m* **léger**   light metal
métal *m* **tendre**   soft metal
métallique *adj*   metal(lic), steel *adj*
métallisation *f*   metal spraying, metallization
métalline *f*   surfacing resembling stone of plaster, lime and sundry aggregates
métallographie *f*   metallography
métallurgie *f*   metallurgy
métamorphique *adj*   metamorphic [of rock]
métamorphisme *m*   metamorphism [of rock]
météo[†] *f*/**prévision** *f* **météo(rologique)**   weather forecast
météorologie *f*   meteorology
météo(rologique), bulletin *m*   weather report
méthane *m*   methane
méthode *f*   method
méthode *f* **à échange d'ions**   base exchange process
méthode *f* **brevetée**   patented or proprietary system
méthode *f* **d'enregistrement**   recording method
méthode *f* **d'équilibrage**   balancing method
méthode *f* **d'essai**   test(ing) method
méthode *f* **d'exécution**   execution procedure
méthode *f* **de calcul**   calculation method
méthode *f* **de chargement**   firing method [of a boiler], loading method
méthode *f* **de chauffe**   heating method
méthode *f* **de construction**   construction method
méthode *f* **de déchargement**   discharging or unloading method
méthode *f* **de dégivrage**   defrost(ing) method
méthode *f* **de nettoyage**   cleaning or scrubbing method
méthode *f* **de refroidissement**   cooling method
méthode *f* **de regain statique**   static regain or velocity reduction method [for duct sizing]
méthode *f* **de travail**   method of work(ing)
méthode *f* **du cylindre**   cylindrical mould sampling method [soil testing]
méthode *f* **graphique**   graphical method
métrage *m*   measure(ment) or measuring [of length, in metres], quantity surveying
métré *m*   estimate of quantities and costs, measurement or survey [of quantities and work done], quantity surveying
métré *m* **estimatif**   estimate of costs and quantities
métré *m* **quantitatif**   measurement of quantities or quantity survey
mètre *m*   metre, metre rule
mètre *m* **carré**   square metre
mètre *m* **cube**   cubic metre

mètre *m* **linéaire**   linear metre
mètre *m* **pliant**   folding rule or measure
mètre *m* **(à) ruban**   tape measure
métrer *vb*   measure [length], measure or survey [quantity survey]
métreur *m*   quantity surveyor [near French equivalent]
métreur-vérificateur *m*   quantity surveyor [near French equivalent, there are several grades and specialities]
métrique *adj*   metric
métro(politain) *m*   tube[†], underground railway [urban]
métro(politain) *m* **aérien**   elevated/overhead railway [urban]
mettre *vb* **à dimension**   size [adjust to size]
mettre *vb* **à l'épreuve**   prove [i.e. test]
mettre *vb* **à l'essai**   try out
mettre *vb* **à la masse/à (la) terre**   earth [electrically]
mettre *vb* **à la poste**   post [a letter, parcel]
mettre *vb* **en charge**   charge [a battery]
mettre *vb* **en circuit**   switch on [an electrical circuit]
mettre *vb* **en court-circuit**   short circuit
mettre *vb* **en dépôt**   run to spoil [excavated material]
mettre *vb* **en marche**   start (up) [a machine, etc]
mettre *vb* **en œuvre à chaud/froid**   lay hot/cold [road material]
mettre *vb* **en ordre**   put right
mettre *vb* **en place**   mount, position
mettre *vb* **en précontrainte**   prestress
mettre *vb* **en pression**   pressurize
mettre *vb* **en route**   start (up) [a machine, etc]
mettre *vb* **en service**   commission
mettre *vb* **hors circuit/tension**   de-energize or switch off [electrical circuit]
mettre *vb* **le chauffage**   turn on the heating
mettre *vb* **(le) feu à**   set fire to
mettre *vb* **qch en adjudication**   put sth out to tender, invite tenders for sth
mettre *vb* **qch en pratique**   put sth into practice
mettre *vb* **qch au rebut**   reject/scrap sth
mettre *vb* **sous forme graphique**   graph
mettre *vb* **sous tension**   energize or switch on [an electrical circuit]
meuble *adj*   loose, soft or uncompacted [of ground]
meublé *adj*   furnished
meublé, location *f* **en**   furnished let(ting)
meuble *m*   furniture [piece of]
meubler *vb*   furnish
meubler *vb* **de**   stock with
meule *f*   grinding wheel, millstone
meuler *vb*   grind
meuleuse *f*   (angle) grinder
meuleuse *f* **de sols**   floor grinder
meulier *adj*   grindstone *adj*, millstone *adj*
meulière *f*   grindstone, millstone
meurtrière *f*   loophole [in a battlement]
mezzanine *f*   mezzanine (floor), window [small at mezzanine level], window wider than it is tall
mi-bois *adj*   halved [of a joint in timber]
mica *m*   mica

microclimat *m*   microclimate
microcrique *f*   hair(line) crack
microfissure *f*   hair(line) crack, microcrack, microfissure
micrographie *f*   micrograph
micromètre *m*   micrometer
micron *m*   micron
micro-onde *f*   microwave
micro-pelle *f*   mini-digger, mini-excavator
micro-pieu *m*   micro-pile
microprocesseur *m*   microprocessor
microrupteur *m*   micro-switch
microstructure *f*   fine structure, microstructure
mi-dur *adj*   half-hard, medium hard
mieux-disant *adj*   best [of tenders]
mignonnette *f*   fine river gravel
mignonnette *f* **lavée**   washed fine gravel concrete finish
migrer *vb*   migrate [of water in a permeable medium]
milieu *m*   environment or surroundings, medium [e.g. air], mean, middle
milieu *m* **agressif**   aggressive environment
milieu *m* **humain**   human environment
milieu *m* **naturel**   natural environment or surroundings
milieu *m* **physique**   physical environment
millimètre *m*   millimetre
mince *adj*   thin
mine *f*   mine or pit, pencil or refill lead
mine, coup *m* **de**   blasting shot
mine *f* **à ciel ouvert**   opencast mine
mine *f* **anglaise**   red lead
minéral *adj*   mineral *adj*
minéral *m*   mineral
mineur *adj*   minor *adj*
minier *adj*   mining *adj*
minimal *adj*   minimum *adj*
minimum *adj*   minimum *adj*
minimum *m*   minimum *n*
ministère *m*   department [ministry]
minium *m*   red lead
minute *f*   minute [of time, or note of sth], original [of a drawing, estimate, etc]
minuter *vb*   draft [e.g. an agreement], minute, record [a minute, a judgement, etc]
minuterie *f*   automatic timing device, (electric) time switch, counting or recording dial(s) [e.g. of a watt–hour meter]
minuterie *f* **à commande**   control time switch
minuterie *f* **de contrôle**   control time switch
mire *f*   (surveyor's) levelling rod or staff
mire *f* **à glissière**   surveyor's rod with sliding target
miroir *m*   mirror
miroitier *m*   glazier
mis à jour   up-to-date
mis en œuvre à chaud   hot-laid [road materials]
mis sous terre   underground *adj* [e.g. services]
mise *f* **à (la) terre**   earth protection, protective earthing [electrical]
mise *f* **à exécution**   execution or implementation [e.g. of a project]
mise *f* **à jour**   revision, updating

mise *f* **au point**   adjustment, running in, tuning, focusing
mise *f* **de fonds**   capital outlay
mise *f* **en cavalier**   piling of earth
mise *f* **en circuit**   connecting up [sth electrically]
mise *f* **en dépôt**   stockpiling
mise *f* **en feu**   firing [of a boiler], or setting fire to
mise *f* **en jeu**   bringing to bear [of forces]
mise *f* **en ligne**   alignment
mise *f* **en marche**   starting (up) [of machines, equipment]
mise *f* **en marche à pleine charge**   full-load starting
mise *f* **en œuvre**   construction, implementation, workmanship
mise *f* **en place**   laying, placing [the action]
mise *f* **en route**   starting (up) [of machines, equipment]
mise *f* **en service**   commissioning, putting into service, setting to work [of equipment]
mise *f* **en température**   heating up
mise *f* **en tension**   tensioning, energizing electrically
mise *f* **en valeur**   development [of land], improvement [of buildings, etc]
mise *f* **en vigueur**   enforcement
mises *f,pl* **en chantier de logements**   housing starts
mise de fonds, faire *vb* **une**   lay out or put up some capital
mission *f*   assignment [of architect, engineer, etc], (professional) task
mitaine *f*   mitt [protective]
mitigeur *m*   mixing tap [monobloc, single lever controlled]
mitoyen *adj*   intermediate, party [as in party wall]
mitoyenneté *f*   joint ownership [of intervening wall, hedge, etc]
mi-travée *adj*   mid-span
mitre *f*   rectangular chimney pot, open-topped cowl [on chimney]
mitron *m*   round chimney pot
mixte *adj*   mixed [in the sense of combined or composite]
mnémonique *f*   mnemonic
mobile *adj*   movable
mobilier *m*   furniture
mobilier *m* **urbain**   street furniture
mobilisation *f*   mobilization
mobilisation *f* **de la ressource**   resource exploitation
mode *m*   method
mode *m* **d'emploi**   directions for use, operating, user or working instructions
mode *m* **de fonctionnement**   method of operation
mode *m* **de paiement**   method or mode of payment
mode *m* **opératoire**   operating conditions
modèle *adj*   model
modèle *m*   model
modèle *m* **de circulation**   circulation pattern
modèle *m* **réduit**   scale model
modem *m*   modem
modénature *f*   profile [of a moulding, cornice, etc]
modification *f*   alteration, amendment, change, modification
modifié *adj*   altered, amended, changed, modified
modifier *vb*   alter, amend, change, modify

modillon *m*   modillion [architectural]
modulaire *adj*   modular
modulateur-démodulateur *m*   modem
modulation *f*   modular coordination [the action, or its result]
modulation *f* **complète**   full modulation
module *m*   module [in design, or unit], standard unit, modulus
module *m* **d'affichage de données**   data display module
module *m* **d'élasticité**   modulus of elasticity or stiffness, Young's modulus
module *m* **d'élasticité transversale**   shear modulus
module *m* **de cisaillement**   shear modulus
module *m* **de contrôle**   control module
module *m* **de Poisson**   Poisson's ratio
module *m* **de résistance**   section modulus
module *m* **de rigidité**   modulus of rigidity
module *m* **de Young**   Young's modulus
module *m* **électronique de contrôle**   electronic control module
moduler *vb*   dimension in terms of a module
modulor *m*   proportion system of *Le Corbusier*
moellon *m*   block of stone, quarry stone
moellon *m* **bloqué**   rubble stone too irregular to be squared
moellon *m* **brut**   rubble stone
moellon *m* **d'appareil**   squared and rusticated quarry stone
moellon *m* **ébousiné**   roughly squared rubble stone
moellon *m* **en coupe**   rubble stone laid on edge
moellon *m* **grisant**   bedded rubble stone [needing only light tooling]
moellon *m* **piqué**   tooled rubble stone, with sharp edges
moellon *m* **plat**   rubble stone laid on its bedding plane
moellon *m* **préscié**   rubble stone sawn to dimensions at the quarry
moellon *m* **smillé**   rubble stone tooled with a mason's hammer
moie *f, voir* **moise**
moignon *m*   rain water outlet [from a roof gutter]
moins-disant *adj*   lowest [of tenders]
moins-value *f*   depreciation, drop in value
moins-value *f* **(de cession)**   capital loss (on disposal)
moise *f*   binding piece, corner brace, cross piece [sometimes double], (roof) tie or tie-beam [sometimes double] [all in timber structures], ledger tube [scaffolding]
moise *f* **en écharpe**   diagonal brace or tie
moisi *m*   mildew, mould
moisir *vb*   mould [mildew]
moisissure *f*   mould, mildew, mustiness
moite *adj*   damp, moist
môle *m*   breakwater, mole, pier
moleté *adj*   knurled, milled
moleter *vb*   knurl, mill
molets *m,pl*   boggy or swampy ground
molette, clef *f* **à**   adjustable spanner
molybdène *m*   molybdenum
moment *m*   moment, momentum

moment *m* **d'encastrement**   fixed-end or restraint moment
moment *m* **d'inertie**   moment of inertia, second moment of area
moment *m* **d'inertie polaire**   polar moment of inertia
moment *m* **de flexion**   bending moment
moment *m* **de renversement**   overturning moment
moment *m* **de torsion**   torsional or twisting moment
moment *m* **en travée**   moment at mid-span, panel or span moment
moment *m* **fléchissant**   bending moment
moment *m* **par rapport au grand/petit axe**   moment about the major/minor axis
moment *m* **résistant**   restoring or resisting moment
moment *m* **statique**   first moment of area, statical moment
moment *m* **sur appuis**   moment at supports
monnaie *f*   change [money]
mono-   single-
monobloc *adj*   one-piece, packaged [of a unit, equipment]
monocellulaire *adj*   single-cell [e.g. of bridge deck]
monocouche *adj*   single-layer [e.g. of paint system]
monocouche[†] *f*   single layer rendering
monodimensionnel *adj*   one-dimensional
mono-étagé *adj*   single-stage
monolithe *adj*   monolithic
monolithe *m*   monolith [made from a single piece of stone]
monophasé *adj*   single-phase, uniphase
monorail *m*   monorail
mont *m*   hill
montage *m*   assembling or assembly [the process], erecting or erection, fitting up, carrying up or hoisting, connecting or wiring up [of a piece of equipment]
montage, schéma *m* **de**   wiring diagram
montage *m* **de soudage**   welding jig
montage *m* **en atelier**   shop assembly
montage *m* **en étoile-triangle**   star-delta connection
montage *m* **en saillie**   surface mounting
montage *m* **en triangle**   delta connection
montage *m* **sur chantier**   site assembly
montagne *f*   mountain
montant *adj*   ascending, rising [ground, tide], vertical
montant *m*   (total) sum or amount [of money], post [of gate, roof], riser [of stair], stile [of door, window], stud [timber structure], upright or vertical [of ladder, scaffolding, window or door jamb, etc]
montant *m*, **petit**   muntin [of a door]
montant *m* **charnier**   hanging stile, upright carrying the hinge [of door or window]
montant *m* **de ferrage/de rive**   hanging jamb or stile
montant *m* **de serrure**   meeting or shutting stile
montant *m* **du devis**   budget estimate or figure
montant *m* **du marché**   contract price
monté en usine   factory assembled or set
monte-charge *m*   goods lift
monte-charge *m* **de chantier**   builder's or site hoist or lift
montée *f*   ascent, climb, rise, gradient [rise], slope [upwards], height [e.g. of an arch], height between staircase landings

monte-matériaux *m*   materials hoist

monte-plats *m*   dumb waiter, service lift [from a kitchen]

monter *vb*   assemble, erect [install], connect [electrically], wire up [a piece of equipment], climb, go up, rise

monter *vb* **à**   reach [rise to]

monter *vb* **l'escalier**   go upstairs

monter *vb* **sur une échelle**   climb up a ladder

monte-sacs *m*   sack hoist

monteur *m*   erector, fitter

monteur *m* **électricien**   electrical fitter

monticule *m*   hillock, knoll

montre *f*   shop window

monument *m*   monument

monument *m* **classé**   listed building

monument *m* **historique** [France]   historic or listed building [UK]

moquette *f*   fitted carpet

moraillon *m*   hasp

moraine *f*   moraine

morceau *m* **de gazon**   piece of turf, sod

mordre *vb*   pickle/strip off paint with acid

morfil *m*   wire edge on a cutting tool

mors *m*   jaw [of a chuck], tenon [large]

mors *m* **fixe/mobile**   fixed/movable jaw [of a vice, etc]

mort, corps *m*   kentledge

mort, poids *m*   dead weight

mortaisage *m*   morticing

mortaise *f*   mortise

mortaiser *vb*   mortise

mortaiseuse *f* **à chaîne**   chain mortising machine

mortier *m*   mortar

mortier, résistance *f* **du**   mortar strength

mortier *m* **à prise lente**   slow-setting mortar

mortier *m* **à prise rapide**   quick-setting mortar

mortier *m* **activé**   activated mortar [treated to improve workability]

mortier *m* **bâtard**   bastard mortar [for coating]

mortier *m* **colle**   tile fixing mortar [adhesive]

mortier *m* **d'enduit**   rendering mortar

mortier *m* **de chaux**   lime mortar

mortier *m* **de ciment**   cement mortar

mortier *m* **de pose**   bedding mortar

mortier *m* **de reprise**   bonding mortar

mortier *m* **fin blanc**   lime mortar

mortier *m* **fin gris**   cement mortar

mortier *m* **gras**   fat mix mortar

mortier *m* **joint**   grouting cement

mortier *m* **normal/ordinaire**   mortar [for brick/block laying]

mosaïque *f*   mosaic

mosaïque, dallage *m* **en**   mosaic (tile) floor

mosaïque, pavé *m*   mosaic sett paving

mot-code *m*   code word

motel *m*   motel

moteur *m*   engine, motor

moteur *m* **(à courant) triphasé**   three-phase motor

moteur *m* **à cage d'écureuil**   squirrel cage motor

moteur *m* **à courant alternatif**   alternating current (a.c.) motor

moteur *m* **à courant continu**   direct current (d.c.) motor

moteur *m* **de levage**   lifting motor

moteur *m* **de registre**   damper actuator or motor

moteur *m* **de réserve/secours**   standby motor

moteur *m* **diesel**   diesel engine

moteur *m* **électrique**   electric motor

moteur *m* **monophasé**   single-phase motor

moteur *m* **shunt**   shunt motor

moto-basculeur *m*   dumper [site]

moto-basculeur *m* **à quatre roues motrices**   four-wheel drive dumper

moto-compresseur *m*   compressor unit

moto-pompe *f*   motor pump

motorgrader *m*   motorgrader

motoriser *vb*   mechanize, motorize

motorscraper *m*   motorscraper

motosouffleur *m*   motor-blower [for road cleaning]

mouche *f*   cutting tip of an auger, tip of a drill

mouchetis *m*   Tyrolean or alpine finish rendering

mouchette *f*   demolition waste, outer edge/fillet of a dripstone, moulding plane

moufle *f*   block and tackle, mitten [protective]

mouillage *m*   dampening, moistening

mouillé *adj*   wet

mouiller *vb*   damp(en), moisten, wet

moulé *adj*   moulded

moule *f*   formwork or shuttering, mould [for casting]

mouler *vb*   mould [for casting]

mouleur *m*   moulder [mould maker]

moulin *m*   mill

moulin *m* **à eau**   watermill

moulin *m* **à vent**   windmill

moulin *m* **(à vent) hollandais**   smock (wind)mill

moulin *m* **(à vent) sur pile**   post (wind)mill

moulinage *m*   tooling stone surface with knurled wheel or hammer

mouliné, bois *m*   worm-eaten wood or timber

mouliner *vb*   tool stone surface with knurled wheel or hammer

moulinet *m*   current meter [paddle type]

mouluré *adj*   moulded

moulure *f*   moulding, mini-trunking [electrical]

moussant *m*   additive for producing foamed or aerated concrete

mousse *f*   foam

mousse *f* **à base d'urée formol**   thermo-setting resin foam thermal insulation

mousse *f* **de laitier**   expanded blastfurnace slag

mousse *f* **phénolique**   phenolic foam

mousseline *f*   glass with small, frosted patterns

mousser *vb*   foam

mouton *m*   monkey or pile hammer, weather moulding [tongue, on edge of a casement]

mouton, battant *m*   window casement with a tongue [*mouton*] on its mating edge

mouton *m* **à déclic**   drop hammer [pile-driver]

mouton *m* **à vapeur**   steam hammer [pile-driver]

mouton *m* **de battage**   pile-driver, piling rig
mouton *m* **sec**   gravity hammer [pile-driver]
mouvement *m*   movement
mouvement *m* **d'air**   air motion, movement or pattern
mouvement *m* **des terres/de terre général**   earth-moving, and earth-moving records
moye *f*   defect in face of cut stone [depression], soft layer on face of hard stone in its bedding plane
moyen *adj*   average, mean, middle, medium
moyen *m*   mean [of a progression or series, e.g temperature], means or way [of doing sth]
moyen *m* **d'assemblage**   fastener
moyen *m* **de pression**   means of applying pressure
moyen *m* **de protection contre la rouille**   anti-corrosion, anti-rust or rust protective agent
moyen *m* **de secours**   emergency measure
moyen *m* **de transport**   means of transport
moyenne *f*   average
moyenne[†] *f*   average or mean speed
moyenne *f* **arithmétique**   arithmetical mean
moyenne *f* **des températures**   average or mean temperature
moyenne *f* **géométrique**   geometrical mean
moyenne *f* **numérique**   average [numerical], numerical mean
moyeu *m*   boss or hub [e.g. of a fan, impeller]
mulot *m*   half-brick [special, 220×55×60mm]
multicellulaire *adj*   multi-cell [e.g. of bridge deck]
multicouche *adj*   multi-layer [e.g. of paint system]
multicouche[†] *f*   multi-layer waterproof surfacing
multimètre *m*   multimeter, universal test meter
multiniveau *adj*   multi-level
multiplicateur *m*   multiplier
multizone *adj*   multi-zone [e.g. air-conditioning system]
mur *m*   wall
mur *m*, **gros**   loadbearing wall
mur *m* **à double paroi**   hollow wall
mur *m* **bahut**   dwarf or low wall [with coping to carry railings, etc]
mur *m* **banché**   shuttered wall [cast in place]
mur *m* **coupe-feu**   fire (division) wall
mur *m* **creux**   cavity wall

mur *m* **cribwall**   cribwall
mur *m* **d'appui**   supporting wall
mur *m* **d'échiffre**   wall supporting first steps of a stair
mur *m* **d'écran**   baffle or screen wall
mur *m* **d'enveloppe**   wall surrounding a flue
mur *m* **de chaudière**   boiler shell
mur *m* **de clôture** *f*   enclosing wall
mur *m* **de pierres sèches**   drystone wall
mur *m* **de quai**   quay or wharf wall
mur *m* **de refend**   crosswall, loadbearing separating wall
mur *m* **de soutènement**   retaining wall
mur *m* **déversé**   leaning wall
mur *m* **dosseret**   wall supporting chimney pipe(s)
mur *m* **en agglomérés**   blockwork wall
mur *m* **en aile**   wing wall [at an angle to a bridge abutment]
mur *m* **en angle**   L-wall
mur *m* **en décharge**   wall supported by archways
mur *m* **en maçonnerie**   masonry wall
mur *m* **en parpaing**   blockwork wall
mur *m* **en retour**   side wall [at right angles to a bridge abutment]
mur *m* **extérieur**   external wall
mur *m* **mitoyen**   party wall
mur *m* **orbe**   wall with only simulated arches or openings
mur *m* **ossaturé**   framed or stud partition or wall
mur *m* **parafouille**   cut-off wall [dam]
mur *m* **pignon**   gable (end) wall
mur *m* **porteur**   loadbearing wall
mur *m* **rideau**   curtain wall
mur *m* **séparatif**   separating wall
murs *m,pl* **rideaux**   curtain walling
muraille *f*   defensive wall [e.g. round a town]
mural[†] *m*   wall units
murer *vb*   brick
muret *m*   low wall
murette *f*   low wall
mur-poids *m* **de soutènement**   gravity retaining wall
musoir *m*   head of a jetty or pier, tip or point of a traffic island
musoir *m* **d'une écluse**   wing wall of a lock

nacelle *f* **à bras/flèche télescopique**   cherry picker, telescopic boom access/work platform
nacelle *f* **à ciseaux**   scissor lift (work) platform
nacelle *f* **de nettoyage**   window cleaners' cradle
nacelle *f* **élévatrice articulée**   articulated aerial access/work platform
naissance *f*   spring [of an arch or vault]
nantissement *m*   collateral security

nappe *f* **aquifère/d'eau**   (ground)water table
nappe *f* **d'eau suspendue**   perched water table
nappe *f* **phréatique**   (phreatic) (ground)water table
nappe *f* **suspendue**   perched water
nappes *f,pl* **phréatiques/souterraines**   groundwater
naturel *adj*   natural
nauséabond *adj*   evil/foul smelling, nauseating
nébulisation *f*   fogging

nef *f* nave

négatif *adj* negative

négociateur *m* negotiator

négocier *vb* negotiate

neige *f* snow

neige, blanc *m* **de** zinc white

neige *f* **carbonique** dry ice

néoprène *m* synthetic rubber (neoprene)

néoprène, colle *f* **au** neoprene glue

nervuré *adj* finned, ribbed

nervure *f* rib, rib (arch) [of a vault], chase or rebate [e.g. on outside of a frame to provide key for walling]

nervurer *vb* fin, rib

net *adj* clean, net [price, weight, etc]

netteté *f* clearness [of water], sharpness [of reflection]

nettoiement *m* clearing [of ground, etc]

nettoyage *m* cleaning

nettoyer *vb* clean, wash

nettoyeur *m* cleaner

nettoyeuse *f* cleaning apparatus or machine

nettoyeuse *f* **à haute pression** high pressure cleaner

neutre *adj* neutral [e.g. axis, pressure, colour, conductor]

neutre, fil *m* neutral conductor

neutre *m* **commun** common neutral

nez *m* collar brazed on to metal downpipe to prevent it slipping down

nez *m* **d'un rabot** plane handle

nez *m* **de marche** *f* nosing [of a stair tread]

niche *f* alcove, niche, recess

nicher *vb* lodge itself

nickel *m* nickel

nickelage *m* nickel plating

nid *m* void [concrete defect]

nid *m* **d'abeilles** honeycomb [structure, etc]

nid *m* **de cailloux** honeycombing [e.g. in concrete surface]

nid *m* **de gravillons** honeycombing [e.g. in concrete surface]

nid *m* **de poule** pothole [in a road, pathway, etc]

nipple *m* nipple, washer between boiler or radiator sections

nipple *m* **double** barrel nipple

nitrate *m* nitrate

niveau *m* level [storey of a building], builder's or surveyor's level

niveau *m* **à bulle (d'air)** spirit level

niveau *m* **à bulle transversale** spirit level

niveau *m* **au plus bas** lowest or minimum level

niveau *m* **au plus haut** highest or maximum level

niveau *m* **aval** tailwater level [below a dam]

niveau *m* **bas de l'eau** low water level

niveau *m* **d'eau/de l'eau** water level

niveau *m* **d'eau supérieur** upper water level

niveau *m* **d'éclairage** lighting level

niveau *m* **d'huile** oil level

niveau *m* **d'intensité sonore** loudness or sound intensity level

niveau *m* **de base** base level

niveau *m* **de bruit** noise level

niveau *m* **de la fondation/du terrain de fondation** subgrade level

niveau *m* **de la mer** sea level

niveau *m* **de la nappe aquifère/phréatique** groundwater level

niveau *m* **de la retenue** highest/maximum water storage level [in a dam]

niveau *m* **de maçon** plumb level

niveau *m* **de poche** pocket level

niveau *m* **de pression acoustique** sound pressure level

niveau *m* **de pression acoustique normalisé** standardized sound pressure level

niveau *m* **de puissance sonore** sound power level

niveau *m* **de réglage** grade [finished level]

niveau *m* **du sol fini** finished floor level

niveau *m* **général de la France (NGF)** national level survey [France]

niveau *m* **haut de l'eau** high-water level

niveau *m* **phréatique** groundwater level

niveau *m* **piézométrique** piezometric surface level

niveau *m* **sonore** loudness/noise/sound level

niveau, ramener *vb* **à** top up [battery, tank, etc]

nivelage *m* **automatique** automatic levelling [of a lift cage]

nivelage *m* **de charge** load levelling

niveler *vb* level (out), line up [at same level], true up [make or bring to same level]

se niveler *vb* become level

niveleuse *f* **(à lame)** grader [earth- or roadworks]

niveleuse *f* **automotrice** motorgrader

nivelle *f* spirit level

nivellement *m* equalizing [of levels], evening out or grading [levelling], levelling [making level, and in surveying], surveying [levels]

nivellement *m*, **tige** *f* **de** levelling rod or staff

nivellement *m* **national** national level survey [France]

nocif *adj* harmful, noxious

nœud *m* connection, joint, junction, node [in a structural frame, a network], knot [in timber], knuckle [of a hinge]

nœud *m* **à embase** lead-caulked joint [spigot-and-socket pipes]

nœud *m* **bouchon/d'arête** loose/arris knot

nœud *m* **d'échafaudage** scaffolding joint

nœud *m* **d'empattement** brazed T-joint [in pipework]

nœud *m* **de face/rive** face/edge knot

nœud *m* **de soudure de jonction** brazed butt joint [in lead pipework]

nœud *m* **de tamponnage** brazed lead plug [blocking the end of a lead pipe]

nœud *m* **plat/tranchant** splay knot

nœud *m* **sain/vicieux** sound/unsound knot

nœuds *m,pl* **groupés** knot cluster

noisetier *m* hazel [tree or bush]

noix *f* weather rebate [semi-circular, in the edge of a casement]

nom *m* **chimique** chemical name

nom *m* **de marque** trademark

nombre *m* number [quantity]

nombre *m* **pair/impair/entier** even/odd/whole number

nomenclature *f*   catalogue, list, schedule
nomenclature *f* **d'/des aciers**   (bar) bending schedule
nominal *adj*   nominal
nomographie *f*   nomogram
non à l'échelle   not to scale (NTS)
non enterré *adj*   above ground
non potable *adj*   undrinkable
non remanié *adj*   undisturbed [e.g. sample, in geotechnics]
non visible *adj*   hidden
non-cohérent *adj*   non-cohesive
non-conducteur *m*   non-conducting material
non-corrosif *adj*   corrosion resistant
non-ferreux *adj*   non-ferrous
non-fixé *adj*   unfixed
non-goudronné *adj*   unmade [of a road]
non-métallique *adj*   non-metallic
non-meublé *adj*   unfurnished
non-valable *adj*   invalid [of a clause]
noquet *m*   valley gutter [folded metal strip]
nord *adj*   north, northerly, northern, northward
nord *m*   north
nord-est *m*   north-east
nord-ouest *m*   north-west
normal *adj*   normal, perpendicular, standard, usual
normaliser *vb*   normalize
norme *f*   code of practice, standard
norme *f* **anglaise**   British Standard (BS)
norme *f* **en vigueur**   current standard
norme *f* **française (NF)**   national standard [France]
norme *f* **nationale**   national standard
notaire *m* **(public)**   (public) notary [functions overlap those of solicitors in England and Wales]
notarial *adj*   notarial
notation *f*   notation
note *f*   account, note
note *f* **d'honoraires**   fee account
note *f* **de calcul**   calculations
note *f* **de frais**   claim form for expenses

note *f* **de service**   memo
notice *f*   account [e.g. report on sth]
notice *f* **d'assemblage**   assembly instructions
notice *f* **d'installation**   installation instructions
notice *f* **de documentation**   data sheet
notice *f* **de montage**   assembly instructions
noue *f*   valley channel or gutter, valley rafter or tile
noulet *m*   roof valley, valley rafter
nourrice *f*   header [multi-outlet for individual plumbing service pipes]
nourrir *vb*   prime or seal [a surface, with paint or varnish]
noyau *m*   core, newel [solid, of a spiral staircase]
noyau *m* **à corde**   newel of a spiral staircase with moulded-in handrail
noyau *m* **à jour**   newel of open tread spiral staircase
noyau *m* **central**   middle third
noyau *m* **d'argile**   clay core
noyau *m* **d'escalier**   newel of a spiral staircase
noyau *m* **de bois**   newel post
noyau *m* **de service**   service core
noyau *m* **étanche**   cut-off or impermeable core [of a dam]
noyau *m* **évidé**   open well [of a staircase]
noyau *m* **imperméable**   cut-off or impermeable core [of a dam]
noyau *m* **suspendu**   newel discontinuous between floors
noyau *m* **vide**   newel of open tread spiral staircase
noyé en   embedded in
noyer *m*   walnut [tree, wood]
nu *m*   plane [of a surface]
nuance *f* **d'acier**   grade of steel
nucléaire *adj*   nuclear
nue-propriété *f*   reversionary interest [in property]
nuisance *f* **sonore**   noise nuisance
nuisible *adj*   harmful [e.g. gas]
numérique *adj*   digital
numéro *m*   number [in series, of a telephone, etc]
numéro *m* **de série**   serial number
nu-propriétaire *m*   bare owner [legal term, not literal!]

obligation *f*   bond, debenture
obligation *f* **d'attente**   obligation to give way [road traffic]
obligation *f* **contractuelle**   contractual obligation
obligatoire *adj*   compulsory, mandatory
oblique *adj*   oblique
obliquité *f*   obliquity
objet *m*   subject [e.g of a report, study]
obscur *adj*   dark [room, etc], obscure [dark]
obscurité, heures *f,pl* **d'**   hours of darkness
observation *f*   comment, observation, remark
obstacle *m*   hindrance, obstruction [of or to progress]
obstruction *f*   blockage, obstruction, stoppage
obstrué *adj*   blocked, obstructed [e.g. a road, a pipe]

obstruer *vb*   block, choke [an outlet], obstruct
s'obstruer *vb*   get or become blocked up
obturateur *m*   stopper, shutter [photographic]
obturation *f*   blocking [of a duct, a pipe, an aperture], closing [of a cavity]
obturation *f* **par congélation**   freezing up [and hence blocking, of pipework, etc]
obturé *adj*   blocked [a duct, a pipe, an aperture], stopped up
obturer *vb* **une fissure**   fill/seal/stop up a crack
d'occasion *adj*   used [second-hand]
occulte, défaut *m*/**vice** *m*   concealed, hidden or latent defect
occupant *m*   occupant, occupier
occupation *f*   occupation [of land, premises]

occupation des sols (POS), plan *m* **d'**   local land use zoning plan, local plan [showing permissible land uses]
occupation *f* **précaire**   occupation under licence
occupé *adj*   busy, engaged
occuper *vb*   occupy [place, time, job, etc]
s'occuper de *vb*   take care or look after [i.e. deal with]
octer *vb*   fill very small holes [e.g. in plaster surface]
octogonal *adj*   octagonal
oculus *m*   circular window
odeur *f*   odour, smell
odomètre *m*   odometer
œdomètre *m*   consolidation meter, oedometer
œil *m*   circular window
œil *m* **roulé**   knuckle [folded, of a strap hinge]
œil-de-boeuf *m*   bull's eye window
œillet *m*   eyelet
œuvre, à pied d'   on site
œuvre, dans (D/O)   internal [of dimensions]
œuvre, en sous-   underpinning *adj*
œuvre, hors   external [overall, of dimensions or measurements]
œuvre, mesure *f* **dans**   internal dimension(s)
œuvre *f*   undertaking [task], work [book, drawing, painting, etc]
œuvre *m,* **gros**   structural works
œuvre *f,* **maître** *m* **d'**   architect in charge of the works
œuvre *m,* **second**   secondary works [e.g. finishes, fitting out]
officiel *adj*   official *adj*
offre *f*   bid, offer, proposal, tender
ogive *f*   gothic or pointed arch, ogive
ognette *f*   narrow mason's chisel
ohm *m*   ohm
ohmmètre *m*   ohmmeter
oignon *m*   bulb [horticultural]
oiseau *m*   hod
oléoduc *m*   pipeline [for oil products]
olive *f*   oblong rounded profile of handrail, ornament in the shape of an olive, olive [in plumbing joint]
olivier *m*   olive tree
oliveraie *f*   olive grove
ombragé *adj*   shaded [from light]
ombrager *vb*   shade [from light]
ombre *f*   shade, shadow
ombre *f* **portée**   shadow cast by one body on another
onctueux *adj*   greasy or oily [surface]
onde *f*   planer mark on wood surface, wave [physics, i.e. sound, light, etc]
onde *f* **sonore**   sound wave
ondulation *f*   ripple [electrical]
ondulé *adj*   corrugated
onglet *m*   mitre
onglet, boîte *f* **à**   mitre box
onglet, coup *m* **d'**   mitre cut [to form a right angle]
onglet *m,* **faux**   mitre cut at other than 45°
opacifier *vb*   render opaque
opacimètre *m*   opacity meter [for opacity or hiding power of paint]
opacité *f*   hiding power [of paint], opacity

opacité *f* **de fumée**   smoke density
opaline *f*   milk glass
opaque *adj*   opaque
opération *f*   operation, procedure, project [undertaking]
opération *f* **cyclique**   cycling/cyclic operation
opération *f* **de mise en marche**   starting procedure
opération *f* **en cascade**   series operation
opération *f* **périodique**   cycling/cyclic operation
optimal *adj*   optimum *adj*
optimum *m*   optimum
opus *m* **incertum**   random rubble [masonry]
orage *m*   storm, thunderstorm
orchestre *m*   front or orchestra stalls [of a theatre]
ordinateur *m*   computer
ordinogramme *m*   flow diagram [computer]
ordonnance *f*   edict, official order, statute, ruling, general arrangement, layout or plan [of a building]
ordonnance *f* **de paiement**   authorization of payment
ordre *m*   order [architectural, for goods, etc], sequence
ordre *m* **de grandeur**   order of size
ordre *m* **de marche, en**   in working order
ordre *m* **de montage**   assembly sequence
ordre *m* **de service**   architect's instruction (AI) to contractor
ordre *m* **de soudage**   welding sequence
Ordre *m* **des Architectes**   society of architects [statutory body in France]
Ordre *m* **des Géomètres-Experts**   society of land surveyors [in France]
ordre *m* **du jour**   agenda
ordre *m* **officiel**   official order
ordures *f,pl*   refuse, rubbish [waste]
ordures *f,pl* **ménagères**   garbage, domestic/household rubbish/refuse
ordures *m,* **vide-**   rubbish chute
oreille *f*   ear or fixing lug
oreillon *m*   end of window ledge built into a pier
organe *m*   instrument, medium [instrument]
organe *m* **de commande**   control block or unit
organe *m* **de contrôle**   control unit
organe *m* **de coupure**   isolating/isolation valve
organes *m,pl* **de commande**   controls
organes *m,pl* **de transmission**   transmission system
organigramme *m*   flow diagram [computer], flow chart, organization chart
organique *adj*   organic
organisme *m* **agréé de contrôle**   approved control authority, authorized control body
orge, grain *m* **d'**   decorative element shaped like head of barley
orgueil *m*   fulcrum in form of a large wedge of wood or stone
oriel *m*   oriel window
orientable *adj*   adjustable [in direction, angle]
orientation *f*   aspect [e.g. of a house], direction [e.g. of axis of a building], orientation [e.g. of a crane boom]
orienter *vb*   orientate
orifice *m*   aperture, opening, orifice, mouth [pipe, well, etc]

orifice *m* **d'admission**   intake port
orifice *m* **d'aération**   air vent
orifice *m* **d'aspiration**   suction port
orifice *m* **d'échappement**   blow out or exhaust port
orifice *m* **d'entrée**   inlet opening
orifice *m* **de chasse**   sluicing outlet [of a dam]
orifice *m* **de sortie**   exhaust opening or outlet
orifice *m* **de soufflage**   exhaust opening or outlet
orifice *m* **fendu**   slotted outlet
original *adj*   original *adj*
original *m*   original *n*
origine *f* **de propriété**   root of title
ormaie *f*   elm grove or plantation
orme *m*   elm [tree, wood]
orneaux *m,pl*   heavy slate paving slabs
ornement *m*   adornment, decoration, enrichment
   [decoration], ornament(ation)
ortie *f*   nettle
oscillation *f*   oscillation
osciller *vb*   oscillate
oscillogramme *m*   oscillogram
oscillographe *m*   oscillograph
oscilloscope *m*   oscilloscope
osmose *f*   osmosis
osmotique *adj*   osmotic
ossature *f*   frame(work), framing [timber or steel],
   skeleton, structure [framework]
ossature *f* **de construction**   structural frame(work)
ossature *f* **(en) béton armé**   reinforced concrete frame
   construction or structure
ossature *f* **en bois**   timber framing
ossature *f* **en portique**   portal frame
ossature *f* **métallique**   steel frame or structure
ôter *vb*   remove [edges, stains]
ouest *adj*   west, western, westerly, westward
ouest *m*   west
ouïe *f* **d'aération**   ventilation louvre
ouïe *f* **d'aspiration de ventilateur**   fan inlet
oulice *f*   mortise-and-tenon joint, with tenon normally
   vertical, mortised piece angled
ouragan *m*   hurricane
ourlet *m*   folded and rolled edge of metal sheeting
outil *m*   tool
outil *m* **à main**   hand tool
outil *m* **pneumatique**   air or pneumatic(ally operated)
   tool
outillage *m*   equipment [tools], implements, set of tools
outillage *m* **à façonner**   planing tools
outillage *m* **à main**   hand tools
outrepassé, arc *m*   horseshoe arch
ouvert *adj*   open [of a door, a valve, etc]
ouvert, à ciel   open to the sky [e.g. ditch, mine]
ouverture *f*   aperture, opening [the action], starting up [of
   work], switching on
ouverture *f* **d'entrée**   inlet opening
ouverture *f* **d'un arc**   span of an arch between springs
ouverture *f* **de fenêtre** *f*   window opening
ouverture *f* **de porte**   doorway

ouverture *f* **du plafond**   ceiling outlet
ouverture *f* **nominale (d'un tamis)**   nominal mesh
   opening size (of a sieve)
ouvrable *adj*   workable [of a material]
ouvrable, jour *m*   business day, work(ing) day [of the
   week]
ouvrabilité *f*   workability [of concrete, material]
ouvragé *adj*   treated aesthetically or architecturally
ouvrage *m*   structure, work [i.e. something built, done,
   made]
ouvrage, maître *m* **d'/ de l'**   client or employer [for
   building work], developer or owner [client for building
   work]
ouvrage *m* **accessoire**   auxiliary or secondary
   works/structure accessory
ouvrage *m* **d'art**   major civil engineering structure or
   work
ouvrage *m* **d'assainissement**   drainage system or works
ouvrage *m* **de drainage**   drainage works
ouvrage *m* **de soutènement**   supporting structure
ouvrage *m* **de terre**   earthwork
ouvrage *m* **en briques**   brickwork
ouvrage *m* **en éventail**   work fanning out from one centre
ouvrages *m,pl*, **gros**   structural masonry, main
   walls, etc
ouvrages, restes *m,pl* **d'**   existing structures [when
   starting construction of new]
ouvrages *m,pl* **annexes**   associated works
ouvrages *m,pl* **enterrés en tranchée ouverte**   cut-and-
   cover [the works]
ouvrages *m,pl* **légers/menus**   non-structural masonry and
   other minor works
ouvrant *adj*   opening
ouvrant *m*   leaf [of a door, window, etc], opening light
   [window]
ouvrier *m*   workman or worker
ouvrier *m* **de construction**   construction worker
ouvrier *m* **du bâtiment**   builder's labourer
ouvrier *m* **qualifié**   qualified or skilled craftsman, labour
   or worker
ouvrier *m* **spécialisé**   semi-skilled labour or worker
ouvrir *vb*   break [a pipe, circuit, contact, etc], initiate
   [negotiations], open [a window, valve, etc], switch or
   turn on [electricity, gas, etc], turn on [a tap]
ovale *adj*   oval [shaped]
ove *f*   egg-shaped decorative feature
oxyacétylène *m*   oxyacetylene gas
oxycoupage *m*   flame, gas or oxygen cutting
oxydation *f*   oxidation
oxydé *adj*   oxidized
oxyde *m* **de carbone**   carbon monoxide
oxyde *m* **de fer**   iron oxide
oxyde *m* **nitrique**   nitric oxide
oxyder *vb*   oxidize
oxygène *m*   oxygen
oxygène *m* **dissous**   dissolved oxygen
ozone *m*   ozone

paiement *m*   payment
paiement *m* **comptant**   payment in cash/full
paiement *m* **d'avance**   payment in advance
paiement *m* **échelonné**   payment by instalments or stages
paiement *m* **en liquide(s)**   cash payment
paiement *m* **par chèque**   payment by cheque
paiement *m* **partiel**   partial payment
paillasse *f*   masonry slab with waterproof surface [laboratory bench, kitchen slab, etc]
paillasson *m* **(d'entrée)**   doormat
paille *f*   straw
paille *f* **de fer**   steel or wire wool
pair *adj*   even [of numbers]
paire *f* **torsadée**   twisted pair
palace *f*   large luxury hotel
palan *m*   pulley block, tackle
palan *m* **à câble**   rope hoist
palan *m* **à chaîne**   chain hoist
palançon *m*   piece of wood reinforcing cob walling
palastre/palâtre *m*   base plate of a door lock
palastre/palâtre *m*, **serrure** *f* **à**   rim lock
palastre/palâtre (de serrure)   lock plate
pâle *adj*   pale [colour]
pale *f*   blade, sluice-gate
pale *f* **de guidage**   guide vane
palée *f*   bent [frame], sheet piling or sheet pile wall, trestle
palée *f* **de contreventement**   vertical framework with windbracing
palette *f*   pallet
palfeuille *f*   flat web sheet pile
palier *m*   bearing [rotating machine], landing [of stair, etc]
palier *m* **à billes**   ball bearing
palier *m* **à roulement**   roller bearing
palier *m* **circulaire**   landing in a spiral staircase
palier *m* **de butée**   thrust bearing
palier *m* **de communication**   landing serving several apartments or offices
palier *m* **de départ**   ground floor landing
palier *m* **de repos**   intermediate landing
palier *m* **en porte-à-faux**   outboard bearing
palier *m* **lisse**   plain bearing
palier *m* **support auxiliaire**   pedestal bearing
palière, porte *f*   door opening on to a landing
palifier *vb*   strengthen [ground with piles]
pâlir *vb*   go or grow dim, turn pale, fade
palis *m*   post [of a fence or sheet pile wall]
palissade *f*   hoarding [around a site], palisade, wooden fence
palissade *f* **de planches**   boarded fence
palissader *vb*   fence in [land, a site]
palissandre *m*   rosewood

palme *f*   palm tree
palonnier *m*   lifting beam
palplanche *f*   sheet pile
palplanche *f* **légère**   trench sheet pile
palplanches, rideau *m* **de**   sheet pile wall or sheet piling
pan *m*   face or side [of angular object, building, etc], panel [flat face of a building, a roof], section [length of wall, roof plane, etc]
pan *m* **coupé**   chamfered corner between two walls
pan *m* **de bois**   stud wall, timber frame panel, wood partition
pan *m* **de comble**   roof panel
pan *m* **de fer**   steel panel, bay or wall
pan *m* **de maçonnerie**   masonry bay or panel
pan *m* **de mur**   piece or length of wall [whole height]
panache *m*   plume [smoke, steam, etc]
panier *m*   basket filter [in a siphon], mud/grit bucket in a road gulley
panier *m* **siphon**   siphon basket
panne, en   broken down [machinery]
panne *f*   breakdown, pantile, peen [of a hammer], purlin
panne *f* **à lierne**   purlin jointed into rather than laid upon principal rafter
panne *f* **de bris**   curb [purlin between slopes of a mansard roof]
panne *f* **de communication**   communication breakdown or failure
panne *f* **de courant**   power failure or loss [electrical]
panne *f* **en treillis**   lattice purlin
panne *f* **en verre**   glass tile/slate
panne *f* **faîtière**   ridge board or purlin
panne *f* **fendue**   claw peen [of a hammer]
panne fendue, marteau *m* **à**   claw hammer
panneau *m*   panel [of a door, etc], sign
panneau *m* **à âme creuse**   cellular (hollow) core board
panneau *m* **à âme lamellée**   laminboard
panneau *m* **à âme lattée**   blockboard
panneau *m* **à glace**   plain panel, without a border
panneau *m* **à tubes enrobées**   buried pipe/tube heating panel
panneau *m* **arasé**   panel flush with its frame
panneau *m* **chauffant**   radiant (heating) panel
panneau *m* **d'accès**   access panel
panneau *m* **d'affichage**   advertisement hoarding
panneau *m* **de copeaux**   chipboard
panneau *m* **de dalle**   slab panel
panneau *m* **de fibre (de bois)**   fibreboard
panneau *m* **de fibres dures**   hardboard
panneau *m* **de particules**   chipboard, particle board
panneau *m* **de plâtre**   plasterboard
panneau *m* **de régulation**   control panel

panneau *m* **de signalisation**   road sign [board]
panneau *m* **de verrière**   large glazed panel
panneau *m* **dur perforé**   perforated hardboard
panneau *m* **incorporé**   built-in or embedded heating panel [floor, wall or ceiling]
panneau *m* **indicateur**   road or traffic sign
panneau *m* **isolant**   insulating board
panneau *m* **latté**   blockboard
panneau *m* **ouvrant**   opening panel
panneau *m* **perforé pour plafond**   perforated ceiling board
panneau *m* **préfabriqué**   precast panel
panneau *m* **publicitaire**   advertisement hoarding
panneau *m* **radiant au gaz**   gas radiant heater
panneau *m* **rayonnant**   radiant panel
panneau *m* **sandwich**   sandwich panel [generally with insulating core]
panneau *m* **stop**   stop sign
panneresse *f*   stretcher
panneton *m*   flat arm of espagnolette bolt holding shutter closed, flat part of a key
pannetonnage *m*   fixing from below of tiles against wind uplift
pans, à six/huit/etc   hexagonal/octagonal/etc
pans coupés, à   with corners cut off
panse *f*   swelling of a baluster or column
pantographe *m*   pantograph
papier *m* **anglais**   insulation paper used under sheet metal roofing
papier *m* **bitumé**   bitumen paper
papier *m* **cache**   masking tape
papier *m* **d'aluminium**   aluminium faced insulating paper
papier *m* **d'apprêt**   backing or lining paper
papier *m* **de construction**   building paper
papier *m* **de tenture**   wallpaper
papier *m* **de verre**   glass paper, sandpaper
papier *m* **filtrant**   filter paper
papier *m* **gaufré**   embossed wallpaper
papier *m* **hygiénique**   toilet paper
papier *m* **millimétré**   graph paper
papier *m* **peint**   wallpaper
papier-calque *m*   tracing paper
papillon *m*   key for operating a butterfly damper
papillotement *m*   flicker
paquet *m*   parcel [package]
paquet *m* **(de)**   bundle (of) [e.g. reinforcement bars]
par personne/tête   per capita [per head]
parabole *f*   parabola, satellite dish [aerial]
parabolique *adj*   parabolic
parachèvement *m*   finishing works [closure, services, partitions, finishes]
parachute *m*   emergency/safety arrestor [of lift or hoist]
parafoudre *m*   lightning protection, lightning surge diverter
parafouille *f*   cut-off drain or trench beneath a dam
parafouille *f* **en palplanches**   sheet pile cut-off
parage *m*   final dressing of joinery
parallèle *adj* **(à)**   parallel (to)

parallélogramme *m* **des forces/vitesses**   parallelogram of forces/velocities
paramètre *m*   parameter
parapet *m*   breastwork [of a fortress], guard rail, parapet
parapluie *m*   temporary covering/roof to shelter repairs/works
parapluie, distribution *f* **en**   drop system of water distribution, overhead distribution [of services]
parasismique *adj*   earthquake resistant [foundation]
parasites *m,pl*   static [interference]
paratonnerre *m*   lightning conductor or rod
paravent *m*   wind screen [*not* windscreen]
parc *m*   depot, storage yard, store [depot], grounds, park, stock [of equipment, etc]
parc *m* **à mazout**   oil fuel depot
parc *m* **de camions**   fleet of lorries
parc *m* **de stationnement**   car park, parking area or lot
parcellaire *adj*   divided into parcels or plots
parcellaire, plan *m*   plot layout plan, subdivision plan [showing layout of plots]
parcelle *f*   compartment [of a forest], lot [plot of land]
parcelle *f* **[de terrain]**   parcel or plot [of land]
parcelle *f* **riveraine**   bordering parcel or property
parclause *f*/**parclose** *f*   cover strip [over joint or crack], glazing bead or fillet
parcomètre *m*   parking meter
pare-brise *m*   windscreen [*not* wind screen]
pare-douche *m*   shower screen
pare-étincelles *m*   fire-guard
pare-feu *m*   fire-break
pare-flamme *m*   flame barrier or guard
pare-flammes *adj* **(PF)**   fire-resistant, flame-resistant
pare-flammes *m*   resistance to spread of flame
pare-vapeur *m*   vapour barrier
parement *m*   dressed face [of a stone, etc], (wall) facing, formed surface [of concrete], mortared edge of tiled roof, kerb-stone
parement *m* **assisé**   coursed face [of a masonry wall]
parement *m* **brut**   as-built face
parement *m* **débruti**   rough finished facing [of a stone wall]
parement *m* **layé**   tooled finish rubble facing
parement *m* **piqué**   picked wall finish [joints raked out of stonework]
parement *m* **smillé**   scabbled stone finish
paremente *vb*   face [a wall]
pare-pluie *m*   waterproof paper [used in roofs or facades]
pare-soleil *m*   sunshade
pare-vapeur *m*   vapour barrier
pare-vent *m*   wind barrier paper
parfaire *vb*   finish off [a piece of work], perfect
parking *m*   carpark, parking area or lot, space
parking *m* **en élévation**   multi-storey carpark
parking *m* **en sous-sol**   basement parking
parking *m* **ouvriers**   workmen's carpark
parking *m* **personnel**   staff carpark
parking *m* **souterrain**   underground carpark
parking *m* **visiteurs**   visitors' carpark
paroi mince, à   thin-walled

paroi *f*　face [of a hill, cliff, wall, etc], plane element of a building [wall, floor, roof, partition, etc]
paroi *f* **au coulis**　grout or slurry wall
paroi *f* **berlinoise**　Berlin wall [strutting system for deep excavations]
paroi *f* **de palplanches**　sheet pile wall
paroi *f* **en béton armé**　reinforced concrete wall
paroi *f* **étanche**　diaphragm wall
paroi *f* **extérieure**　external face or skin
paroi *f* **intérieure**　internal face or skin
paroi *f* **interne**　inner (wall) surface
paroi *f* **moulée (dans le sol)**　diaphragm wall (cast in situ)
paroi *f* **porteuse**　loadbearing partition or wall
paroi *f* **rocheuse**　rock face
paroisse *f*　parish
parpaing *m*　breeze or precast concrete (masonry) block, concrete building block, stone exposed on both faces of a wall
parpaing *m* **creux**　hollow block
parpaing *m* **d'appui**　breast stone under a window or door sill
parpaing *m* **d'échiffre**　stone string [part which supports treads]
parquet *m*　timber or wood (panel) floor(ing), parquet floor(ing), floorboards
parquet *m* **à bâtons rompus**　herringbone pattern parquet [square ends]
parquet *m* **à compartiments**　framed inlaid panel parquet
parquet *m* **à lames/à l'anglaise**　strip flooring [parallel]
parquet *m* **à l'anglaise à coupe de pierre**　alternate joint strip flooring
parquet *m* **à l'anglaise à coupe perdue**　random length strip flooring
parquet *m* **à la française**　framed inlaid panel parquet
parquet *m* **à/en point de Hongroie**　herringbone pattern parquet [mitred ends]
parquet *m* **contreplaqué**　plywood flooring
parquet *m* **ciré**　polished or waxed floor
parquet *m* **d'assemblage**　framed inlaid panel parquet
parquet *m* **de glace**　mirror backing panel
parquet *m* **en chevrons**　chevron pattern parquet
parquet *m* **en éventail**　fan pattern parquet
parquet *m* **en feuilles**　framed panel parquet
parquet *m* **en fougère**　herringbone pattern parquet [mitred ends]
parquet *m* **(en) mosaïque**　inlaid strip flooring, mosaic parquet
parquet *m* **latté**　plywood flooring
parqueteur *m*　parquet layer [craftsman]
partenariat *m*　joint venture, partnership
parterre *m*　flower bed, pit [of a theatre or concert hall]
parterre *m* **d'eau**　water garden
participation *f* **équilibrée des travailleurs**　equal worker participation
particule *f*　particle
particule *f* **en suspension**　suspended particle
particules *f,pl* **aériennes**　airborne particles
particules *f,pl* **d'aérosols**　aerosol particles
particules *f,pl* **de poussières**　dust particles

particulier *adj*　individual, particular
particulier *m*　person, private individual
partie *f*　part [fraction, portion]
partie *f* **en poids**　part by weight
partie *f* **en volume**　part by volume
partie *f* **inférieure/supérieure**　lower/upper part
partiel *adj*　partial [load, etc]
parties *f,pl* **communes**　common areas [of buildings]
parties *f,pl* **privatives**　private areas [of buildings]
parvis *m*　courtyard, square, etc, in front of public building
pas *m*　pitch [of gear teeth, a spiral, etc], spacing [pitch], thread [of a screw], step [of stair], threshold [of a door]
pas *m* **à droit/à gauche**　right-/left-handed thread
pas *m* **d'ailettes**　fin spacing
pas *m* **d'âne**　staircase with large going and relatively small rise
pas *m* **de porte**　goodwill [of a business]
pas *m* **des rivets**　rivet pitch or spacing
passable *adj*　passable
passage *m*　arcade with shops on each side, alley(way), passage(way), crossing, way (through), route [itinerary]
passage *m* **à demi-barrière**　half-barrier crossing [railway]
passage *m* **à niveau**　level crossing [railway]
passage *m* **clouté**　pedestrian or zebra crossing [UK]
passage *m* **inférieur (PI)**　subway, underpass
passage *m* **pour cyclistes**　cyclists' crossing
passage *m* **pour piétons**　pedestrian crossing
passage *m* **souterrain**　subway, underground passage(way)
passage *m* **supérieur (PS)**　overpass, viaduct
passager *m*　passenger
passation *f*　signing [e.g. of contract]
passe *f*　layer or pass [in welding], harbour/navigable channel, master key, saddle [between mountains], stone cut parallel its bedding plane
passe *f* **à poissons**　fish pass [dam]
passe-partout *m*　master key
passer *vb* **qch au rabot**　run a plane over sth
passer *vb* **un accord**　conclude an agreement
passer *vb* **un contrat**　enter into or sign a contract
passer *vb* **un pont**　cross a bridge
passer *vb* **une couche**　apply a coat [of paint]
passerelle *f*　catwalk, footbridge, gangway, walkway
passerelle *f* **de service**　service walkway
passerelle *f* **de travail**　work(ing) platform
passerelle *f* **de visite**　inspection walkway
passif *m*　liabilities
passif *m* **circulant**　current liabilities
passif *m* **éventuel**　contingent liabilities
passivation *f*　passivation
pastille *f*　bursting disc
pâte *f*　paste
pâte *f* **à bois**　wood filler
pâte *f* **de téflon**　Teflon paste
patenôtre *f*　paternoster
patente *f*　licence to trade or engage in a profession
patenter *vb*　licence

patère *f*   coat hook

patin *m*   baseplate [for scaffolding, a structural frame or wall], shoe [scaffolding base]

patin *m* **vibrant**   vibrating plate [for compacting]

patine *f*   patina

patio *m*   patio [enclosed courtyard]

patrimoine *f*   heritage

patte *f*   clamp, clip, fastening, fixing [accessory for]

patte *f* **à chambranle**   duckbill nail

patte *f* **à glace**   mirror fixing

patte *f* **à scellement**   flat fixing cleat [of which one end is grouted in]

patte *f* **d'agrafe**   fixing for metal sheet roofing

patte *f* **d'éléphant**   spread base [of a column] or footing [of a pile]

patte *f* **d'oie**   small plasterer's tool

patte *f* **de fixation**   mounting bracket

patte *f* **de gouttière**   gutter bracket or clip

patte *f* **de support**   support bracket

paume *f*   halved mortise-and-tenon joint

paumelle *f*   hinge [split]

paumelle *f* **à droite**   lift-off butt hinge

paumelle *f* **dégondable**   lift-off hinge

paumelle *f* **double**   lift-off hinge

pavage *m*   paving [the process]

pavage, brique *f* **de**   paving brick

pavage *m* **mosaïque**   radial-sett paving

pavé *adj*   paved

pavée, chaussée *f*   road paved with setts

pavé *m*   paving sett or stone

pavé *m* **à emboîtement**   interlocking paving block

pavé *m* **de béton**   concrete block paving

pavé *m* **de verre**   glass block or brick (paving)

pavé *m* **échantillon**   large paving sett

pavé *m* **mosaïque**   small paving sett

pavillon *m*   house or villa [generally detached, modern], housing into which a blind is raised, roof with more than four meeting slopes, structure rising above main body of a building, ward [of a hospital], wing [of a building]

pavillon *m* **d'aspiration**   inlet cone [on fan impeller]

pavillon *m* **d'entrée**   gate lodge

pavillon *m* **de chasse**   hunting lodge

pavillon *m* **de golf**   golf clubhouse

pavillon *m* **de jardin**   summerhouse

pavillon *m* **de musique**   music pavilion

payer *vb*   pay

payer *vb* **à terme**   buy on credit

payer *vb* **au comptant**   pay cash

pays *m* **boisé**   woodland

paysagé *adj*   open-plan [e.g. office]

paysage *m*   countryside, landscape, landscaping, scenery

paysager, jardin *m*   landscape garden

paysagiste *m/f*   landscape architect

péage *m*   toll

peau *f*   membrane, skin

peau *f* **d'orange**   orange peel [paint defect]

peau *f* **de coffrage**   formwork inner side

peau *f* **de crapaud/crocodile**   paint defect [second coat applied over still wet coat]

peau *f* **de laminage**   mill scale

pédibulateur *m*   foot valve

pédicule *m*   small column or pillar

pédiluve *f*   shower tray

pédologie *f*   soil science

pédologiste *m/f*   soil scientist

pédologue *m*   soil scientist

peigne *m*   comb [of an escalator, moving walkway]

peigne *m* **à fileter (le bois)**   thread chaser [wood turning]

peigner *vb*   comb or corrugate the surface of a rendering, grain [paint to resemble wood grain]

peindre *vb*   paint

peint *adj*   painted

peintre *m* **(en bâtiment)**   (house) painter

peinturage *m*   painting [the action, or its result]

peinture *f*   paint(ing)

peinture *f* **à dispersion**   emulsion paint

peinture *f* **à l'huile**   oil paint

peinture *f* **à la brosse**   brush painting

peinture *f* **antirouille**   anti-corrosion, anti-rust or rustproofing paint

peinture *f* **au latex**   latex (rubber) paint

peinture *f* **au plomb**   lead paint

peinture *f* **au pistolet**   paint spraying

peinture *f* **au pistolet sans air**   airless (paint) spraying

peinture *f* **au rouleau**   roller painting

peinture *f* **bitumineuse**   asphaltic or bitumastic paint

peinture *f* **brillante**   gloss paint

peinture *f* **émail**   enamel paint

peinture *f* **émulsion**   emulsion paint

peinture *f* **glycérophtalique**   oil-based paint

peinture *f* **intumescente**   intumescent paint

peinture *f* **laque**   lacquer

peinture *f* **par pulvérisation**   paint spraying

peinture *f* **protectrice**   protective coat [paint]

peinturer *vb*   paint

pelage *m*   peeling [paint defect]

pelle *f*   digger, excavator, shovel

pelle *f* **à benne traînante**   dragline [excavator]

pelle *f* **à câbles**   cable/dragline excavator

pelle *f* **à direction par ripage**   skid-steer digger/excavator

pelle *f* **à godets**   bucket excavator

pelle *f* **chargeuse**   loading or tractor shovel, (wheeled) loader [machine]

pelle *f* **en benne preneuse**   grab excavator

pelle *f* **en dragueline**   dragline [excavator]

pelle *f* **(équipée) en butte**   face or front shovel

pelle *f* **(équipée) en rétro**   backacter, backhoe

pelle *f* **hydraulique**   hydraulic excavator or shovel

pelle *f* **mécanique**   mechanical digger, excavator or shovel, power shovel

pelle *f* **mécanique marcheuse**   walking dragline

pelle-araignée *f*   walking excavator

pelle-grue *f* **(sur chenilles)**   crane-excavator (crawler mounted)

pelletée *f*   shovelful

pelleteuse *f*   mechanical digger/shovel, excavator, power shovel

pellicule *f*   film [thin layer or photographic]

pelouse *f*   lawn

pénalité *f* **de retard**   damages/penalty for delay

pendage *m*   angle of dip [geology]

penderie *f*   hanging cupboard or wardrobe

pêne *m*   bolt [of a lock or door catch]

pêne *m* **à demi-tour**   self-locking bolt [operating without a key]

pêne *m* **biseauté**   bevelled-end bolt

pêne *m* **dormant**   dead bolt

pénétration *f*   intersection of two vaults, penetration

pénétration, macadam *m* **en**   grouted macadam

pénétromètre *m*   penetrometer

penne *f*   heavy joist

pente *f*   fall, gradient, slope

pente, changement *m* **de**   gradient transition

pente *f* **ascendante**   up slope

pente *f* **de talus**   slope inclination

pente *f* **descendante**   down slope

pente *f* **longitudinale**   longitudinal gradient

pente *f* **maximale**   maximum gradient

pente *f* **naturelle**   natural fall

pente *f* **transversale**   crossfall

pente unique, profil *m* **en travers à**   straight crossfall

penture *f*   strap [of a strap hinge]

penture *f* **anglaise**   T-hinge

pénurie *f*   scarcity or shortage [of goods, etc]

pépin[†]**, avoir** *vb* **un**   hit a snag[†]

pépinière *f*   tree nursery

pépiniériste *m*   nurseryman

perçage *m*   boring [a hole], boring through [a material], drilling

percé *adj*   pierced

percée *f*   breach or gap [in a wall]

percement *m*   boring [with a drill], drilling, piercing, building or driving [a tunnel], making [i.e. piercing for, a window]

percepteur *m*   collector [of taxes, fines, etc]

perception *f*   collection [action of, of taxes]

percer *vb*   bore a hole through, drill, make a hole in, pierce, drive [a tunnel, etc], make an opening [for a window, etc]

perceuse *f*   drill

perceuse *f* **à dénoder**   knot hole moulding machine

perceuse *f* **à percussion**   hammer or percussion drill

perceuse *f* **électrique**   electric drill

perceuse-visseuse *f*   screwdriver-drill

percevoir *vb*   collect [taxes, etc]

perche *f*   rod [wood], scaffolding pole [may be of wood]

percolation *f*   seepage

percoler *vb*   seep

percussion *f*   percussion

percussion, sondage *m* **par**   percussion drilling [in the ground]

perdue, porte *f*   smooth door flush with a wall or partition

perfectionnement *m*   improvement

perfectionner *vb*   improve [make better]

perforateur *m* **de cartes**   card punch [computer]

perforateur *m* **hydraulique**   hydraulic drifter

perforation *f*   perforation

perforatrice *f* **hydraulique**   hydraulic drill rig

perforé *adj*   perforated

perforeuse *f*   hole punch

performance *f*   performance

performance *f* **pratique**   actual/practical performance

performance *f* **prévue**   design/predicted/target performance

performant *adj*   high-return [investment], high-performance

pergola *f*   pergola

périmé *adj*   expired, out-of-date, superseded

périmètre *m*   perimeter

périmètre *m* **mouillé**   wetted perimeter

période *f*   period or stretch [of time], phase [stage]

période *f* **d'arrêt**   shut-down period

période *f* **d'ensoleillement**   sunshine period

période *f* **d'instruction**   training period

période *f* **de brûlage**   burning period

période *f* **de chauffage**   heating period

période *f* **de fonctionnement**   on, operating or running period

période *f* **de mise en circuit**   switching period

période *f* **de pointe**   peak period

période *f* **de réchauffe**   heat-up time, reheat period

période *f* **de refroidissement**   cooling down time or period

période *f* **de repos**   rest period

période *f* **de rétention**   retention period

période *f* **de réverbération**   reverberation time

période *f* **des vacances**   holiday period

période *f* **hors pointe**   off-peak period

périodique *adj*   periodic

périodique *f* **professionnelle**   trade journal

périphérique *adj*   peripheral

périphérique[†] *m*   ring road

périphérique, boulevard *m*   ring road

périphérique *f* **d'entrée**   input peripheral [computer]

périphérique *f* **de sortie**   output peripheral [computer]

perlite *m*   perlite

permanent *adj*   permanent

perméabilité *f*   permeability

perméabilité, essai *m* **de**   permeability test

perméable *adj*   permeable, pervious

perméance *f* **à la vapeur**   vapour permeability

permis *m*   licence, permit

permis *m* **d'abattage**   tree felling permission or permit

permis *m* **de bâtir**   building permit

permis *m* **de conduire**   driving licence

permis *m* **de construire**   building permit, planning permission

permis *m* **de défrichage**   scrub clearance permit

permis *m* **de démolir**   demolition permission or permit

permis *m* **de travail**   work permit

permissible *adj*   allowable

perpendiculaire *adj* **(à)**   perpendicular (to)

perré *m*   drystone facing or stone pitching [of an embankment, etc], revetment [drystone], rip-rap

perré *m* **maçonné**   paved revetment

perron *m*   flight of steps [in front of a building], front steps [of a building]

perron *m* **de chargement**   loading platform [railway]

persienne *f*   louvred or slatted opening or shutter

personne *f* **compétente**   competent person

personne *f* **responsable du marché (PRM)**   person responsible for letting of the contract

perspectif *adj*   perspective *adj*

perspective *f*   perspective *n*

perspective *f* **cavalière**   bird's eye view perspective

perte *f*   loss

perte *f* **à la cheminée**   chimney or flue gas loss

perte *f* **d'eau**   loss of water

perte *f* **d'énergie**   energy loss(es)

perte *f* **de chaleur**   heat loss

perte *f* **de chaleur par unité de temps**   heat loss rate

perte *f* **de charge**   pressure loss

perte *f* **de charge à l'entrée**   entry loss

perte *f* **de charge dans un tube**   pipe friction loss

perte *f* **de charge de bouche d'air**   grille differential pressure

perte *f* **de courant à la terre**   earth leakage (current) loss

perte *f* **de froid**   cooling loss

perte *f* **de pression**   loss of head or pressure, pressure drop

perte *f* **de vitesse**   speed loss

perte *f* **de volume**   volume(tric) loss

perte *f* **en poids**   weight loss

perte *f* **par conduction**   loss by conduction

perte *f* **par évaporation**   loss by evaporation

perte *f* **par frottement**   friction(al) loss

perte *f* **par radiation**   loss by radiation

pertes *f,pl* **dans le réseau**   system losses

pertes *f,pl* **totales**   total loss(es)

pertuis *m*   sluicehole, opening, hole, narrows [of a river]

pertuis *m* **de vidange**   draining culvert or sluice

perturbation *f*   disturbance [atmospheric, magnetic]

pervibrateur *m*   immersion or poker vibrator

pervibration *f*   pervibration, internal vibration of concrete

pesage *m*   weighing

pesant *adj*   heavy, unwieldy, weighty

pesanteur *f*   gravity, weight [relative]

pesanteur *f* **spécifique**   specific gravity

peser *vb*   weigh

peser *vb* **sur**   press or weigh down on

peson *m*   spring-loaded balance

petit bois *m*   glazing bar [wood]

petit clou *m*   tack

petit fer *m*   glazing bar [metal]

pétrifier *vb*   calcify

pétrissage *m*   kneading [soil treatment]

pétrole *m, voir aussi* essence   petroleum [not petrol]

pétrole *m* **brut**   crude oil/petroleum

pétrole *m* **lampant**   kerosene, paraffin oil

peu profond *adj*   shallow

peuplier *m*   poplar [wood, tree]

peupleraie *f*   poplar plantation

phare *m*   headlamp/-light, lighthouse

phasage *m*   phasing, programme

phasage *m* **des études**   design programme

phasage *m* **des travaux**   (construction) works programme

phase, en   in phase

phase *f*   phase [electrical], phase or stage [of a process, work, etc]

phase, décalage *m* **de**   phase difference

phase, être *vb* **en avance de**   lead in phase

phase, être *vb* **en retard de**   lag in phase

pH-mètre *m*   pH indicator

phosphatisation *f*   phosphating

phosphore *m*   phosphorous

photo *f*   photo

photocalque *m* **bleu**   blueprint

photocopie *f*   photocopy

photocopieur *m*   photocopier

photogrammétrie *f*   photogrammetry

photographe *m*   photographer

photographie *f*   photograph(y)

photographie *f* **aérienne**   aerial photograph(y)

photographie *f* **aux rayons X**   X-ray photograph

photopile *f*   solar cell

photothéodolite *m*   photo-theodolite

phréatique, nappe *f*   phreatic/groundwater table

piazza *f*   pedestrian area, square

pic *m*   pick

pick-up *m*   pickup truck

picnomètre/pyknomètre *m*   pycnometer or pyknometer

picot *m*   mini-pile

pièce *f*   component, part, piece [fragment, object], paper [document], room

pièce *f* **comprimée**   compressed or compression member

pièce *f* **coulée**   casting [concrete, metal]

pièce *f* **d'appui**   window ledge or sill

pièce *f* **d'eau**   ornamental lake, pond [in garden or park], water feature

pièce *f* **d'origine**   guaranteed genuine or original spare part

pièce *f* **de jonction**   connection member or piece

pièce *f* **de rechange**   replacement or spare part

pièce *f* **de réduction**   transition piece [ductwork]

pièce *f* **de serrage**   clamp

pièce *f* **de transformation**   transformation piece

pièce *f* **de visite**   inspection chamber

pièce *f* **détachée**   spare (part)

pièce *f* **fondue**   casting [piece of cast metal]

pièce *f* **forgée**   forging

pièce *f* **jointe (PJ)**   enclosure [with letter]

pièce *f* **moulée**   casting

pièce *f* **noyée**   insert

pièce *f* **occupée**   occupied room

pièce *f* **préfabriquée**   precast element or unit

pièce *f* **scellée**   insert

pièce *f* **tendue**   tension member

pièces *f,pl* **contractuelles**   contract documents

pied, de plain   on one level

pied *m*   foot, heel [of a dam], lower part

pied *m* **à coulisse**   vernier callipers

pied *m* **à trois branches**   tripod

pied *m* **amont**   upstream toe [of a dam]

pied *m* **aval**   downstream toe [of a dam]

pied *m* **carré**   square foot

pied *m* **cube**   cubic foot

pied *m* **d'aile**   jointly-owned space on each side of a party wall

pied *m* **d'appui**   prop or stay

pied *m* **de poteau**   base of column or stanchion

pied *m* **de remblai**   base of the embankment

pied *m* **de talus**   toe (or foot) of a slope

pied-à-terre *m*   small flat

pied-de-biche *m*   claw bar, jemmy, nail extractor/puller, pinch bar, pry bar, wrecking bar

pied-droit/piédroit *m*   door or window jamb [pier or post], pier [of an arch, bridge, window], side wall [of doorway, arch, etc]

piédestal *m*   pedestal

piège *m* **à son**   noise absorber

pierraille *f*   broken stones, road metal, rubble, stone fill

pierraille *f* **compactée**   compacted hardcore

pierre *f*   stone

pierre *f* **à affûter**   sharpening stone

pierre *f* **à aiguiser**   whetstone

pierre *f* **à bâtir**   building stone

pierre *f* **à huile**   oilstone

pierre *f* **à macadam**   broken-stone [road material]

pierre *f* **à meuler**   grindstone

pierre *f* **à plâtre**   gypsum

pierre *f* **appareillée**   dressed stone

pierre *f* **apparente**   exposed stone(work)

pierre *f* **artificielle**   artificial stone

pierre *f* **calcaire**   limestone

pierre *f* **concassée**   crushed rock/stone

pierre *f* **d'angle**   quoin

pierre *f* **d'attente**   toothing stone

pierre *f* **d'évier**   stone sink

pierre *f* **de taille**   ashlar, freestone

pierre *f* **du hérisson**   pitching stone

pierre *f* **naturelle**   natural stone

pierre *f* **pelliculaire**   thin slab/sheet of stone

pierre *f* **ponce**   pumice stone

pierre *f* **reconstituée**   reconstituted stone

pierre *f* **tranchée**   riven stone

pierre, poser *vb* **la première**   lay the foundation stone

pierrée *f*   drain of loose stones

pierres sèches, mur *m* **de**   drystone wall

piéton *adj*   pedestrian *adj*

piéton *m*   pedestrian

piétons *m,pl*, **passage** *m* **pour**   pedestrian crossing

piétonné *adj*   pedestrianized

pieu *m*   pile [for foundations]

pieu *m* **à la tarière continue**   continuous flight auger pile

pieu *m* **à sable**   sand pile

pieu *m* **à vis**   screw pile

pieu *m* **battu/foncé**   driven pile

pieu *m* **d'accostage**   fender pile

pieu *m* **en béton armé**   reinforced concrete pile

pieu *m* **flottant**   friction pile

pieu *m* **foré**   bored pile

pieu *m* **in situ**   in situ pile

pieu *m* **incliné**   batter pile

pieu *m* **moulé battu**   driven cast in situ pile

pieu *m* **moulé dans le sol**   cast in situ pile

pieu *m* **moulé foré**   drilled cast in situ pile

pieu *m* **portant en pointe**   end-bearing pile

pieu *m* **résistant à la pointe**   end-bearing pile

pieux *m,pl* **berlinois**   king piles

pieux *m,pl* **contigus**   contiguous piles

pieux, faisceau *m* **de**   pile group

piézomètre *f*   piezometer

pige *f*   measure [rod of unit of length], measuring rod

pigeon *m*   lump [in lime], tongue [in a mitred joint]

pigeonnage *m*   moulded plaster work

pigeonnier *m*   dovecote

pigment *m*   pigment

pignon *m*   gable, gear wheel

pignon *m* **coupe-feu**   fire gable

pilastre *m*   newel (post), pilaster

pile *f*   battery, pier [of a bridge], pile or stack [of wood, etc]

pile *f* **au lithium**   lithium battery

pile *f* **sèche**   dry battery

pilier *m*   column [straight, not ornamented], pillar [also in underground workings], shaft [of a column]

pilon *m*   rammer or tamper

pilonnage *m*   ramming, tamping (down)

pilonner *vb*   beat, ram, tamp (down) [earth, etc]

pilonneuse *f*   plate compactor

pilonneuse *f* **vibrante**   vibrating/vibratory rammer, vibratory tamper

pilot *m*   (wooden) pile [for foundations]

pilotage *m*   pile driving

pilote *m*   construction coordinator, coordinator of works on site

pilote, conducteur *m*   communications conductor [for controls, alarms, etc]

pilote, entreprise *f*   coordinating contractor

pilotis *m*   foundation of wooden piles, pile group, piling [i.e. several piles]

pin *m*   pine [tree, timber]

pin *m* **d'Oregon**   Douglas fir [timber]

pin *m* **du nord**   redwood [Scandinavian]

pinacle *m*   pinnacle [on a building, buttress]

pince *f*   clamp, crowbar, pinch bar, pair of pliers

pince *f* **à avoyer**   saw setting tool

pince *f* **à becs ronds**   round nose pliers

pince *f* **à dénuder**   wire strippers

pince *f* **crocodile**   crocodile clip

pince *f* **de courant**   clamp type ammeter

pince *f* **de fixation**   (fixing) clip

pince *f* **de raccordement**   cable connecting clamp

pince *f* **de sertissage**   crimping tool

pince *f* **multiprise**   crocodile clip, multi-plier [adjustable wrench], pipe wrench

pince *f* **universelle**   all-purpose wrench, combination (cutting) pliers

pinceau *m*   (paint)brush

pinceau *m* **à rechampir**   cutting-in brush [painter's]

pinceau *m* **plat/rond**   flat/round paintbrush
pinceau *m* **pour radiateurs**   radiator brush
pince-étau *f*   monkey wrench
pincelier *m*   brush cleaning container
pinces *f,pl*   pair of pliers
pinces *f,pl* **à ligatures**   steelfixer's pliers
pinède *f*   pine wood [forest]
pioche *f*   mattock, pick, pickaxe
piochement *m*   axing off [of a lump or projection on stone, etc]
pipe *f*   pipe [generally lead] from a WC to its downpipe
pipe-line *m*   pipeline
pipette *f*   drip pipe
piquage *m*   chipping or scaling a boiler, dressing [of stone], pitting [of metal], tapping point, or action of tapping into a main pipe
piqué *adj*   dressed with sharp edges by a percussion tool [of stone, etc], pitted [of metal], spotted with holes [of coatings, rendering, varnish, etc]
pique *f*   spike on railings
pique-feu *m*   fire rake, poker [fire iron]
piquer *vb*   dress [masonry with a percussion tool], join/connect [a branch pipe or duct into a main], mark out [fixing position by outlining]
piquet *m*   peg, picket, post or stake [e.g. for setting out]
piquet *m* **de terre**   earthing rod
piquetage *m*   pegging out [for setting out, surveying]
piqueter *vb*   peg (out)
piqûre *f*   pitting [on metal surface]
piscine *f*   swimming bath or pool
pisé *m*   rammed earth construction
piser *vb*   beat earth, for rammed earth construction
piste *f*   airfield/airport runway or tarmac, earth road
piste *f* **cyclable**   cycle track
piste *f* **d'arrêt**   stopping lane (for buses)
piste *f* **d'atterrissage**   runway [airfield]
piste *f* **piétonne**   footway, pedestrian way
piste *f* **pour piétons**   footway, pedestrian way
pistolet *m*   drawing or French curve, gun for spraying paint, sandblasting, etc
pistolet *m* **à colle**   gluing gun
pistolet *m* **à mastiquer**   mastic gun
pistolet *m* **à peinture**   paint spray(ing) pistol
pistolet *m* **à souder (électrique)**   soldering gun
pistolet *m* **de scellement à cartouches explosives**   stud gun
piton *m*   banister support bracket on side of a stair string, ring with a tail for fixing to/into wood, masonry, etc, pivot
pivotant *adj*   hinged, pivoting
pivotement *m*   pivoting, swivelling, slewing [of a crane], turning [about a point]
pivoter *vb*   pivot, revolve [about a point], swivel [of itself]
placage *m*   finishing with a thin sheet of material, veneer(ing) [of wood, the action and the result]
placage *m* **en pierre**   stone facing
placage *m* **extérieur**   face veneer
placard *m*   cabinet, cupboard (built in), poster

placard *m* **suspendu**   wall cupboard
place *f*   place or square [as in a town], room [amount of space] or space
place *f* **de stationnement**   parking bay or space
place *f* **nécessaire**   required space, space requirement
placement *m*   investment [in securities]
placer *vb*   fit in, put in, place or put [an object or person], site, place [an investment]
placoplâtre *m* [marque commerciale]   gypsum board or plasterboard
plafond *m*   ceiling
plafond *m* **à caissons**   coffered ceiling
plafond *m* **chauffant**   heated ceiling
plafond *m* **diffuseur d'air**   ventilated ceiling
plafond *m* **en lambris**   boarded or panelled ceiling
plafond *m* **flottant**   floating ceiling [isolated from structure above]
plafond *m* **lambrissé**   boarded or panelled ceiling
plafond *m* **perforé**   perforated ceiling
plafond *m* **suspendu**   suspended ceiling
plafonné, grenier *m*   loft with a ceiling
plafonner *vb*   reach a ceiling/maximum, put in a ceiling
plafonneur *m*   plasterer
plafonnier *m*   ceiling fixture, lamp or light
plage *f*   beach, bracket or range [within a series or table], segment [of time]
plage *f* **d'erreur**   error band or range
plage *f* **de réglage**   range of adjustment
plage *f* **de régulation**   control range
plage *f* **de volume**   volume range
plain-pied *adj*   one-level, single-storey [house]
plain-pied, de   at ground/street level, on one floor, on the same level
plaine *f*   plain
plan *adj*   even, flat, level, plane [surface, etc], true [plane]
plan *m*   drawing, layout or plan (drawing), outline drawing, level or plane [surface]
plan *m* **approuvé**   approved drawing
plan *m* **cavalier**   bird's eye view [plan/drawing]
plan *m* **coté**   dimensioned drawing
plan *m* **d'aménagement de zone (PAZ)**   plan of a ZAC [zone d'aménagement concerté]
plan *m* **d'assurance (de la) qualité (PAQ)**   quality assurance plan
plan *m* **d'atelier**   (work)shop drawing
plan *m* **d'eau**   lake, pond
plan *m* **d'ensemble**   location or site plan
plan *m* **d'études**   study plan or programme
plan *m* **d'évacuation des eaux**   drainage plan
plan *m* **d'exécution**   construction or working drawing
plan *m* **d'hygiène et de sécurité (PHS)**   health and safety plan
plan *m* **d'implantation**   layout plan [of a site, etc]
plan *m* **d'implantation de chantier**   site layout plan
plan *m* **d'installation**   installation plan
plan *m* **d'installation du chantier**   site facilities or layout plan
plan *m* **d'intervention en cas d'urgence**   emergency (response) plan

plan *m* d'occupation des sols (POS)   land use zoning plan, local plan [showing permissible land uses]
plan *m* d'un rabot   sole of a plane
plan *m* de câblage   cable layout or plan
plan *m* de circuit   circuit diagram, drawing, plan or layout
plan *m* de cisaillement   shear plane
plan *m* de coffrage   formwork drawing
plan *m* de comparaison   datum/reference level
plan *m* de construction   building plan, construction or workshop drawing
plan *m* de détails   detail drawing
plan *m* de faille   fault plane
plan *m* de ferraillage   reinforcement drawing
plan *m* de financement   financing scheme
plan *m* de glissement   plane of sliding
plan *m* de l'âme   web plane [of as beam]
plan *m* de masse   location or site plan
plan *m* de mise en route   commissioning plan
plan *m* de modernisation   modernization plan
plan *m* de repérage   location or site plan
plan *m* de santé et sécurité   health and safety plan
plan *m* de secteur   area development plan
plan *m* de situation   location or site plan
plan *m* de stratification   bedding plane
plan *m* de travail   work plan, programme or schedule, work surface or worktop
plan *m* des canalisations   ductwork drawing
plan *m* directeur   master plan
plan *m* général d'implantation   overall layout plan
plan *m* général des transports *m,pl*   general transport or traffic plan
plan *m* guide   indicative drawing
plan *m* incliné   inclined plane
plan *m* moyen   mid-plane
plan *m* parcellaire   plot layout plan, subdivision plan [showing layout of plots]
plan *m* particulier de santé et sécurité   detailed health and safety plan
plan *m* provisoire   preliminary drawing
plans *m,pl* d'exécution des ouvrages (PEO)   working drawings
plans *m,pl* de recollement   as-built or as-constructed drawings
planage *m*   flattening [of metal sheet by hammering], planing or smoothing [making even of a wood or metal surface]
planche *f*   board or plank [less than 54mm thick], floorboard, page or sheet [e.g. in drawing numbers]
planche, prix *m*   floor price
planche *f* à arêtes vives   square-edged or trimmed board or plank
planche *f* à dessin(er)   drawing board
planche *f* avivée   square-edged or trimmed board or plank
planche *f* d'échafaudage   scaffolding board
planche *f* de cœur   heartwood plank
planche *f* de coffrage   shutter board [formwork]
planche *f* de garde   scaffolding guard board
planche *f* de topographe   plane table

planche *f* en grume/non équarrie   untrimmed or waney-edged board or plank
planche *f* équarrie   square edged/trimmed board/plank
planche *f* profilée   matchboard
planches sol *m* en   wooden floor [boarded]
planches *f,pl* à recouvrement   weatherboarding
planchéiage *m*   boarding, flooring [boarding], planking
planchéier *vb*   floor [with boards]
plancher *m*   floor or floor structure
plancher *m* à corps creux   hollow block, coffer or pot floor
plancher *m* champignon   flat slab on headed columns
plancher *m* dalle   flat slab floor
plancher *m* de service   horizontal scaffolding, walkway [for moving materials on site]
plancher *m* de travail   work(ing) platform
plancher *m* en hourdis creux   hollow block, coffer or pot floor
plancher *m* haut   ceiling structure
plancher *m* massif   solid floor
plancher *m* préfabriqué   prefabricated floor [of beams and panels or slabs]
plancher *m* saillant   projecting floor [visible between cladding panels]
plancher *m* suspendu   suspended ceiling
plane *f, voir* rabot   draw knife, drawshave
planer *vb*   plane or smooth [make even, a wood or metal surface]
planificateur *m*   planner [of work, rather than urban, town, etc]
planification *f*   planning [sequencing]
planification, bureau *m* de/service *m* de   planning department/office [sequencing work]
planifier *vb*   plan [programme]
planimétrie *f*   planimetry
planning *m*   programme
planning *m* à barres   bar chart programme
planning *m* d'exécution des travaux   construction/works programme
planning *m* directeur de réalisation   master construction programme
planning *m* général   construction schedule
planning *m* pert   PERT programming and network analysis
plantation *f* d'arbres   tree planting
plante *f*   plant [botanical]
planter *vb*   plant [grass, flowers, etc, or a stake]
plaque *f*   plate, sheet [of metal], slab, plaque [on a wall], street (name) sign
plaque[†] *f*   hotplate [cooking stove]
plaque *f*, essai *m* à la   plate bearing test
plaque *f* à bornes   terminal board
plaque *f* chauffante   hotplate
plaque *f* d'ancrage   anchor plate
plaque *f* d'appui   bearing plate
plaque *f* d'arrêt   check plate
plaque *f* d'assise/de base   baseplate
plaque *f* d'âtre   hearth plate
plaque *f* d'égout   manhole cover

plaque *f* **d'entrée**   keyhole surround
plaque *f* **d'entreprise**   contractor's name board
plaque *f* **d'obturation**   orifice plate
plaque *f* **(de cheminée)**   fireback [ornamental cast iron]
plaque *f* **(de cuisson)**   hotplate [on a cooking stove]
plaque *f* **de fibres-ciment**   fibre-cement board
plaque *f* **de fond**   hearth back plate
plaque *f* **de gazon**   sod or piece of turf
plaque *f* **de liège**   cork slab
plaque *f* **de plâtre**   plasterboard [sheet of]
plaque *f* **de propreté**   finger-plate [on a door]
plaque *f* **de protection**   cover plate
plaque *f* **de recouvrement**   fascia plate
plaque *f* **de remplissage**   packing [of a plate cooling tower]
plaque *f* **de serrage**   rail clip [railway track]
plaque *f* **du constructeur**   maker's nameplate
plaque *f* **ondulée**   corrugated sheeting
plaque *f* **perforée**   perforated sheet or plate
plaque *f* **signalétique**   identification plate or name plate [on a tool, machine, etc]
plaque *f* **vibrante**   vibrating plate compactor
plaque-peigne *f*   comb-plate [escalator]
plaquer *vb*   veneer
plaquette *f*   landing board with nosing, at top of stairs
plaquette[†] *f*   brake pad [disc brake]
plasticité *f*   plasticity
plasticité, limite *f* **de**   plastic limit [soil property]
plastifiant *m*   plasticizer [e.g. for concrete]
plastifier *vb*   plasticize [i.e. to add a plasticizer]
plastique *m*   plastic
plastique *m* **armé de verre**   rigid fibreglass
plastique *m* **expansé (rigide)**   plastic foam (rigid)
plat *adj*   flat *adj*
platane *m*   plane [tree]
plateau *m*   cradle [of a flying scaffolding], flat trailer, flitch [large timber not yet resawn into boards], plateau
plateau *m* **absorbant**   absorption bed [for purifying septic tank effluent]
plateau *m* **d'échafaudage**   work(ing) platform [on scaffolding]
plateau *m* **d'établi**   bench top
plateau *m* **de polissage**   sanding/polishing disk [backing disk]
plateau *m* **de sciage**   saw bench
plateau *m* **tellurien**   absorption bed [for purifying septic tank effluent]
plate-bande *f*   flat arch, flower bed, foundation [solid, of concrete], lintel course [of stone], panel edge 30–50mm wide [joinery], splice plate
plate-forme *f*   level area, platform, roof terrace, pile cap [footing for masonry], road area including footways and verges, wall plate [carrying feet of rafters]
plate-forme *f* **à ciseaux/à compas**   scissor lift (work) (platform)
plate-forme *f* **d'observation**   observation platform
plate-forme *f* **de commande**   control platform
plate-forme *f* **de forage**   drilling platform
plate-forme *f* **de levage**   lifting platform

plate-forme *f* **de madriers**   scaffolding board platform
plate-forme *f* **de travail**   work(ing) platform
plate-forme *f* **des terrassements**   formation level [of excavations]
plate-forme *f* **élévatrice**   elevating (work)platform
plate-forme *f* **hollandaise**   raft foundation [on soft ground]
platelage *m*   decking or flooring [of a bridge], board or plank platform, planking, scaffolding boards or scaffolding board platform
platelage *m* **en bois**   boarded floor
platelage *m* **léger**   lightweight decking or flooring [of a bridge]
platine *f*   metal plate, paving slab of thin sandstone
plâtrage *m*   plastering
plâtras *m*   plaster debris from demolition
plâtre *m*   (gypsum) plaster
plâtre *m* **à enduit**   coating plaster
plâtre *m* **aluné**   alum plaster [used for stucco work]
plâtre *m* **anhydre**   anhydrous plaster
plâtre *m* **au sas**   finely sieved plaster [for fine mouldings, etc]
plâtre *m* **boratée**   borax plaster [used for stucco work]
plâtre *m* **crue**   ground gypsum
plâtre *m* **d'albâtre**   alabaster plaster
plâtre *m* **de finition**   finish(ing) plaster
plâtre *m* **lissé**   trowelled plaster
plâtrer *vb*   plaster
plâtrier *m*   plasterer
plein *adj*   full, solid [without openings]
plein *m* **de camion**   full load [of a lorry]
plein, faire *vb* **le**   fill up [a tank]
pleine campagne, en   out in the country
pleine charge *f*   full load
pleine mer *m*   high tide
pleine ville, en   right in the town
pleins champs, en   in the open fields
plénum *m*   plenum
plénum *m* **de mélange**   mixing chamber
pleuvoir *vb*   rain
pli *m*   envelope [e.g. containing a letter], fold, ply or veneer [of plywood]
pli *m* **de terrain**   fold in the ground
pliage *m*   folding
pliant *adj*   folding *adj*
plier *vb*   bend (over), fold (up, down, back), sag
plinthe *f*   plinth [of a column] or plinth course [of a stone wall], skirting (board)
plinthe *f* **à convecteur**   skirting board convector
plinthe *f* **chauffante**   skirting board heating
plinthe *f* **technique**   perimeter trunking
plissé *adj*   folded
plisser *vb*   fold
plomb, à   plumb, vertical
plomb *m*   lead [metal]
plomb *m* **en feuille**   sheet lead
plombage *m*   filling, sealing or stopping [with lead]
plombagine *f*   black lead [for darkening lead or zinc work]

plomber *vb*   seal [with a lead seal, e.g. an electricity meter]

plomberie *f*   plumbing

plombier *m*   gas fitter, plumber

plonger *vb*   immerse, dive

plongeur *m*   diver

plot *m*   contact [electrical], log sawn through-and-through

plot *m* **(rétro-)réfléchissant**   cat's-eye, reflecting or reflective stud

plot *m* **réflectorisé**   cat's-eye, reflecting or reflective stud

plots *m,pl* **anti-vibratiles**   anti-vibration mounting

pluie *f*   rain

plumée *f*   edge tooling [round edges of the face of a stone to be dressed]

plumer *vb*   strip [rendering from a wall]

pluridisciplinaire *adj*   multi-disciplinary

plus *m*   plus [sign]

plus-value *f*   appreciation or gain [in value], capital gain, cost increase [e.g. during course of a contract]

plus-value *f* **(de cession)**   capital gain (on disposal)

pluviomètre *m*   rain gauge

pluviométrie *f*   pluviometry, rainfall [quantity]

pluviométrique, courbe *f*   rainfall chart

pneu *m*   tyre

pneu *m* **compacteur**   compacting tyre

pneumatique *adj*   air operated, pneumatic

pneumatique *f*   pneumatics

poche *f*   pothole [in the ground]

poche *f* **d'air**   air pocket

poche *f* **d'argile**   clay pocket

poche *f* **de résine**   resin pocket [timber defect]

pocher *vb*   grain [paint or a coating to imitate stone]

pochoir *m*   stencil (plate)

pochoir, peinture *f* **au**   stencilling

pochon *m*   stencil brush

poêle *f*   frying pan

poêle *m*   stove

poêle *m* **à bois**   wood(-burning) stove

poêle *m* **à feu continu**   slow-burning stove

poêle *m* **à gaz**   gas stove

poêle *m* **à mazout**   oil stove

poêle *m* **de chauffage**   heating stove

poêle *m* **de cuisine**   cooker, cooking stove

poids *m*   weight

poids *m* **atomique**   atomic weight

poids *m* **brut**   gross weight

poids *m* **de l'air normal**   weight of standard air

poids *m* **de matériau sec**   dry weight [of material]

poids *m* **lourd**   heavy (goods) vehicle

poids *m* **moléculaire**   molecular weight

poids *m* **mort**   dead load/weight

poids *m* **net**   net weight

poids *m* **propre**   dead load, self weight

poids *m* **spécifique**   (bulk) density, specific gravity, unit weight

poids *m* **volumique sec**   dry unit weight

poids *m* **utile**   live/net weight

poignée *f*   handle

poignée *f* **de porte**   doorhandle

poignée *f* **ouverte**   open handle [e.g. of a saw]

poignée *f* **latérale**   side grip handle

poignée *f* **revolver**   pistol grip handle

poignet *m* **de force**   wrist band [protection for heavy work]

poil *m*   bristle [of a brush], crack or fissure [in stone]

poinçon *m*   bradawl, drift, punch [tools], crown/king-post, roof/truss post [timber roof]

poinçon *m* **d'échafaudage**   scaffolding pole

poinçon *m* **rampant**   diagonal tie [roof]

poinçonnement *m*   punching, punching shear

poinçonner *vb*   punch

poinçonneuse *f*   punching machine, metal punch

point *m*   point or stage [in a process, programme, etc], point [of an idea]

point *m* **bas**   dip [in a road], low point

point *m* **culminant**   height, summit [highest point]

point *m* **d'ancrage**   anchorage point

point *m* **d'application**   application point

point *m* **d'appui**   fulcrum, point of bearing/support, support point

point *m* **d'ébullition**   boiling point

point *m* **d'éclair**   flash point

point *m* **d'enclenchement**   cut-in point

point *m* **d'entrée**   entry point [for traffic]

point *m* **d'équilibre**   balance point

point *m* **d'implantation**   setting out point

point *m* **d'inflammation**   burning/flash point

point *m* **d'inflexion**   point of contraflexure, change of gradient or inflexion

point *m* **d'intersection**   crossover point, point of intersection

point *m* **de condensation**   dew point

point *m* **de congélation**   freezing point

point *m* **de consigne**   set point

point *m* **de consigne de température**   temperature set point

point *m* **de contact**   point of contact

point *m* **de contrôle**   control point

point *m* **de coupure**   cut-off point

point *m* **de déclenchement**   cut-out point

point *m* **de départ**   departure point, starting point

point *m* **de dérivation**   branch-off point

point *m* **de détente**   pressure reducing point

point *m* **de fixation**   fixing point

point *m* **de fonctionnement**   operating point

point *m* **de fusion**   melting point

point *m* **de mesure**   measuring point

point *m* **de puisage**   draw-off point [water]

point *m* **de référence**   reference point

point *m* **de repère**   datum point [for setting out, etc], setting out peg

point *m* **de rosée**   dew point

point *m* **de rupture**   fracture [point of]

point *m* **de saturation**   saturation point

point *m* **de sortie**   exit point [for traffic]

point *m* **de tirage**   draw-off point

point *m* **de vue**   viewpoint

point *m* **fixe**   bench mark, fixed point

point *m* **géodésique**   triangulation point or station
point *m* **haut**   high point, crest [of a road]
point *m* **mort**   dead centre
point *m* **neutre**   neutral point
points *m,pl* **de suspension**   suspension points
points, soudure *f* **par**   spot welding
pointage *m*   check(ing) [on or from a list], levelling [on a map or plan], stitch or tack welding
pointe, en   tapering [to a point]
pointe *f*   nail with small head, panel pin, tack, point [sharp end], tip, head [of an arrow], headland, peak [as in peak hours, etc]
pointe, heures *f,pl* **de**   peak hours
pointe, jour *m* **de**   peak day
pointe, période *f* **de**   peak period
pointe *f* **à damas**   upholstery tack
pointe *f* **à papier bitumé**   roofing felt nail
pointe *f* **à tête** *f* **conique**   wire nail
pointe *f* **à tête** *f* **large**   clout nail
pointe *f* **capteur d'énergie**   lightning protection finial
pointe *f* **cruciforme**   screwdriver bit [for Phillips/Pozidrive/Supadrive recessed head screws]
pointe *f* **de cœur**   traffic island [generally triangular]
pointe *f* **de condamnation** *f*   locking device [insurance terminology]
pointe *f* **de jet**   nozzle tip
pointe *f* **de tension**   transient [electrical]
pointe *f* **de vitrier**   glazing sprig
pointe *f* **de Paris**   wire nail
pointe *f* **du pieu**   pile tip
pointe *f* **longue**   spike
pointe *f* **plate**   flat/plain slot bit [for a screwdriver]
pointe *f* **vive**   centre [in lathe tailstock]
pointeau *m*   centre punch, timekeeper
pointer *vb*   check (off), tick (off) [on/from a list], clock in/out [of employee], level, point, or train [a surveying instrument], mark
pointer *vb* **en arrivant**   sign in/on [on arrival at work]
pointer *vb* **en partant**   sign off/out [on leaving work]
pointeur *m*   timekeeper
pointier *m*   scaffolding pole
pointillé *m*   dotted line
pointu *adj*   sharp [pointed]
poirier *m*   pear tree
poitrail *m*   lintel beam [single or composite]
poivrière *f*   pepper-pot turret
polariser *vb*   polarize
polarité *f*   polarity
pôle *m*   pole [electrical, magnetic], terminal [of an accumulator/battery]
pôle *m* **négatif/positif**   negative/positive pole [electrical]
poli *adj*   polished
police *f*   policy [of insurance]
police *f* **d'assurance**   insurance policy
police-secours *f*   emergency services
police *f* **unique du chantier**   joint insurance policy covering all parties to the contract
polir *vb*   polish, surface [polish]
politique *f*   policy [company]

polluant *adj*   pollutant *adj*
polluant *m*   pollutant
polluer *vb*   pollute
pollution *f*   pollution
pollution *f* **d'air/de l'air**   air or atmospheric pollution
polychlorure *m* **de vinyl (PVC ou PCV)**   polyvinyl chloride (PVC)
polyester *m*   polyester
polyéthylène *m*   polyethylene
polygone *m* **des forces**   polygon of forces
polymère *m*   polymer
polymérisation *f*   polymerization
polyphasé *adj*   polyphase
polystyrène *m*   polystyrene
polystyrène *m* **expansé**   expanded polystyrene
polystyrène *m* **extrudé**   extruded polystyrene
polyvalent *adj*   multi-purpose
pomme *f*   knob
pommeau *m*   knob [of a hand plane]
pommeau *m* **d'un rabot**   plane handle
pomme(au) *f* **de douche**   shower head or nozzle
pommelle *f*   grating or grille [to trap rubbish at a pipe entry]
pommier *m*   apple tree
pompage *m*   pumping
pompage *m* **en régulation**   hunting
pompe *f*   pump
pompe *f* **à amorçage automatique**   self-priming pump
pompe *f* **à béton**   concrete pump
pompe *f* **à boue**   slurry pump
pompe *f* **à chaleur**   heat pump
pompe *f* **à diaphragme**   diaphragm pump
pompe *f* **à engrenage**   gear pump
pompe *f* **à haute pression**   high pressure pump
pompe *f* **à huile**   oil pump
pompe *f* **à incendie**   fire engine/pump
pompe *f* **à jet**   steam injector
pompe *f* **à main**   hand pump
pompe *f* **à matières épaisses**   sludge pump
pompe *f* **à mazout**   oil feed pump
pompe *f* **à piston**   piston/reciprocating pump
pompe *f* **à plusieurs étages**   multi-stage pump
pompe *f* **à simple effet**   single stage centrifugal pump
pompe *f* **à vide**   vacuum pump
pompe *f* **aspirante**   suction pump
pompe *f* **automobile à béton**   mobile concrete pump
pompe *f* **centrifuge**   centrifugal pump
pompe *f* **centrifuge à haute pression**   high pressure centrifugal pump
pompe *f* **centrifuge multicellulaire**   multi-stage centrifugal pump
pompe *f* **d'alimentation**   feed pump
pompe *f* **d'alimentation de chaudière**   boiler feed pump
pompe *f* **d'amenée d'huile**   oil feed pump
pompe *f* **d'épreuve**   pressure test pump
pompe *f* **d'injection**   grout pump
pompe *f* **de circulation**   circulating pump
pompe *f* **de circulation d'eau de refroidissement**   cooling water circulating pump

pompe *f* **de dosage**   proportioning pump
pompe *f* **de drainage**   drainage pump
pompe *f* **de relevage**   drainage pump
pompe *f* **de réserve**   standby pump
pompe *f* **de secours**   auxiliary or emergency pump
pompe *f* **de service**   run pump
pompe *f* **de suralimentation/de surpression**   booster
   pump
pompe *f* **des eaux d'égout**   sewage pump
pompe *f* **du chauffage à l'eau chaude**   hot water heating
   pump
pompe *f* **hélice**   axial-flow pump
pompe *f* **horizontale**   horizontal pump
pompe *f* **hydraulique**   hydraulic pump
pompe *f* **immergée**   submersible pump
pompe *f* **immersible**   submersible pump
pompe *f* **jumelle**   twin pump
pompe *f* **manuelle**   hand pump
pompe *f* **mélangeuse**   mixer/mixing pump
pompe *f* **rotative**   rotary pump
pompe *f* **submersible**   submersible pump
pomper *vb*   pump (up, out, etc)
pompiers *m,pl*   fire brigade
ponçage *m*   sanding
ponce *f*   pumice
ponceau *m*   bridge [small, over a culvert], culvert
poncer *vb*   rub down, sand
ponceuse *f*   sander
ponceuse *f* **à bande**   belt sander
ponceuse *f* **à sol**   pavement grinder
ponceuse *f* **vibrante**   orbital sander
pont *m*   bridge
pont *m* **à béquille**   portal (frame) bridge
pont *m* **à deux étages**   double-deck bridge
pont *m* **à haubans**   cable-stayed bridge
pont *m* **à péage**   toll bridge
pont *m* **à poutres**   girder bridge
pont *m* **à tablier inférieur**   bowstring truss bridge
pont *m* **autoroutier**   overpass [motorway]
pont *m* **basculant**   bascule or lifting bridge
pont *m* **biais**   skew bridge
pont *m* **cantilever**   cantilever bridge
pont *m* **dalle**   concrete slab deck bridge
pont *m* **de service**   temporary bridge [e.g. over an
   excavation]
pont *m* **en arc(s)**   arch bridge
pont *m* **en arcs en treillis**   trussed arch bridge
pont *m* **en treillis métallique**   steel lattice bridge
pont *m* **levant**   drawbridge, lift bridge
pont *m* **métallique**   steel bridge
pont *m* **mobile**   movable bridge
pont *m* **ouvrant**   double bascule bridge
pont *m* **phonique**   acoustic bridge
pont *m* **portique**   gantry/portal [jib] crane
pont *m* **poussé**   launched bridge
pont *m* **rail**   rail bridge
pont *m* **roulant**   overhead or travelling (gantry) crane
pont *m* **route**   road bridge
pont *m* **suspendu**   suspension bridge

pont *m* **thermique**   cold or thermal bridge
pont *m* **tournant**   swing bridge
pont *m* **transbordeur**   transporter bridge
pont *m* **voûté**   arch(ed) bridge
pontage *m*   bridge building
pont-bascule *m*   weighbridge
pontet *m*   clip used to fix down support of a *gouttière*
   anglaise
pont-levis *m*   drawbridge
ponton *m*   pontoon
ponton-grue *m*   floating crane
pont-poutres *m*   flat beam bridge
pont-rail *m*   railway bridge
pont-route *m*   road bridge
ponts et chaussées, ingénieur *m*   civil engineer
porche *m*   porch
porcherie *f*   pigsty
pore *m*   pore [e.g. of stone]
poreux *adj*   porous
porosité *f*   porosity
port *m*   harbour, port, (opening) port
port, franc de   carriage or postage paid
port payé   carriage or postage paid
portail *m*   gate(way), portal [e.g. of tunnel]
portance *f*   bearing/load-carrying capacity [per unit area]
portant *adj*   bearing, carrying
portant *m*   stay [support], strut
portatif *adj*   portable
porte *f*   door(way), entrance, gate(way)
porte, bloc- *m*   door set [complete with frame, hinges,
   etc]
porte *f*, **fausse**   false/imitation door
porte *f* **à barres et écharpes**   ledged and braced door
porte *f* **à charnière**   hinged door
porte *f* **à claire-voie**   barred gate, glazed door
porte *f* **à coulisse**   sliding door
porte *f* **à deux battants**   double door/gate, folding door
   with two leaves
porte *f* **accordéon**   multi-folding door
porte *f* **anti-incendie**   fire door
porte *f* **anti-panique**   panic bolt door
porte *f* **basculante**   up-and-over/swing-up door
porte *f* **bâtarde**   two-leaf gate too narrow for vehicles
porte *f* **battante**   swing door
porte *f* **blindée**   reinforced door
porte *f* **charretière**   carriage gate/entrance
porte *f* **cochère**   carriage gate/entrance, gateway large
   enough to admit a vehicle
porte *f* **coulissante**   sliding door
porte *f* **coup de poing**   panic bolt door
porte *f* **coupe feu**   fire (resisting) door
porte *f* **coupée**   stable door
porte *f* **croisée**   glazed door
porte *f* **d'accès**   access door
porte *f* **d'entrée**   front door
porte *f* **de cendrier**   ash box door
porte *f* **(de) coupe feu**   fire resisting door
porte *f* **de glace**   glazed door
porte *f* **de nettoyage**   cleaning door

porte *f* **de ramonage**   cleaning door [flue or chimney]
porte *f* **de secours**   emergency/escape door
porte *f* **de service**   tradesman's or service entrance,
   service door
porte *f* **de visite**   access/inspection door
porte *f* **dérobée**   flush door flush with wall surface
porte *f* **en planches**   batten/ledged and braced door
porte *f* **ignifugée**   fire resisting door
porte *f* **isoplane**   flush door
porte *f* **palière**   landing door
porte *f* **perdue**   flush door, flush with wall surface
porte *f* **pivotante**   door folding away to one side of an
   opening
porte *f* **plane**   flush door
porte *f* **pliante**   folding door
porte *f* **portefeuille**   folding door with two leaves
porte *f* **tournante**   hinged door
porte *f* **va-et-vient**   swing door [opening either way]
porte *f* **vitrée**   glazed door
porte-à-faux *m*   overhang
porte-à-faux, bras *m* **en**   cantilever arm
porte-à-faux, en   balanced [as in balanced cantilever],
   cantilevered, overhanging, out of plumb
porte-à-faux, palier *m* **en**   outboard bearing
porte-ampoule *m*   bulb or lamp holder
porte-auge *m*   hod carrier, mason's labourer
porte-bagages *m*   luggage/roof rack [on a vehicle]
porte-câble *m*   cable hanger
porte-chaîne *m*   chainman
porte-croisée *f*   French window
porte-glace *f*   glass door
portée, à   within reach
portée *f*   span [distance between supports], range
   [distance, scope], reach
portée *f* **d'air**   air throw
portée *f* **d'étanchéité en caoutchouc**   rubber gasket [in a
   valve]
portée *f* **d'un jet**   jet throw
portée *f* **de grue**   crane range or working radius
portée *f* **entre nus**   clear span
portée *f* **libre**   clear span
portée *f* **principale**   main span
porte-étiquette *m*   label holder
porte-fenêtre *f*   French window
porte-foyer *f*   fire door [of a wood stove]
porte-fusible *m*   fuse holder/terminal
porte-fusibles *m*   fuseboard
porte-lé *m*   ceiling paperhanger, paperhanger's ceiling
   prop
porte-mine *m*   clutch pencil
porte-mire *m*   rod man [surveying]
porte-outil *m*   tool rest
porter *vb* **au maximum**   maximize
porte-revolver *f*   revolving door
porte-rouleau *m*   toilet roll holder
porte-savon *m*   soap dish or holder
porte-serviette *m*   towel rail
porte-serviette *m* **chauffant**   heated towel rail
porte-verre *m*   glass holder [bathroom fitting]

porteur *adj*   loadbearing/-carrying
porteur *m*   bearer [e.g. of a letter, message, etc]
portier *m*   doorkeeper, janitor, porter
portier *m* **électrique**   electrically opened door
portière *f*   door curtain
portillon *m*   door within a carriage gate
portique *m*   gantry, portal frame, portico
portique, cadre *m*   portal frame
portique *m* **à étages**   multi-storey frame
portique *m* **à signaux**   signal gantry
portique *m* **de chargement**   loading frame
portique *m* **de manutention**   travelling (gantry) crane,
   overhead crane
portique *m* **de signalisation**   signal gantry
portique *m* **simple**   single portal frame
portiques *m,pl* **multiples**   multi-bay frame
portoir *m*   frame for transporting glass
posage *m*/**pose** *f*   fitting [positioning, fixing], fixing
   [fitting, installing], laying
posé bord à bord   butted [of wallpaper edges]
pose *f, voir*   posage
pose *f* **de papier peint**   paper hanging, wallpapering
pose-drains *m*   mole drainage pipe-layer, mole pipelayer
pose-tubes *m*   pipelayer [machine]
poser *vb*   drive [rivets], fit [install, fix], fix [install], lay,
   place
poseur *m*   layer [tradesman, layer of cables, pipes, etc]
poseur *m* **de carreaux**   tiler [floor or wall]
poseur *m* **de tuyaux**   pipe fitter/layer
positif *adj*   positive
position *f*   position or situation [location, or job], site or
   sitaution [of a house, town, etc]
position *f* **de coupure**   cut-out setting
position *f* **de fermeture**   closed/on position [of a switch]
position *f* **de fonctionnement**   burning position [of a
   lamp]
post-contrainte *f*   post-tension(ing)
poste *m*   element or item [e.g. in bill of quantities, a list],
   equipment
poste[†] *m*   telephone extension
poste *f,* **bureau** *m* **de**   post office
Poste *f,* **La**   French post office service
poste *m* **d'accostage**   berth [port]
poste *m* **d'aiguillage**   signal box
poste *m* **d'amarrage**   berth, mooring
poste *m* **d'enrobage**   asphalt plant, coating plant
poste *m* **d'enrobage à mélange discontinu**   batch type
   asphalt plant
poste *m* **d'équilibrage**   balancing station
poste *m* **de butte**   marshalling yard signal box
poste *m* **de chloration**   chlorination plant
poste *m* **de conditionnement**   treatment plant
poste *m* **de conditionnement d'eau**   water treatment
   plant
poste *m* **de contrôle**   checkpoint
poste *m* **de contrôle sécurité**   security checkpoint/control
   point
poste *m* **de jour**   day shift
poste *m* **de livraison**   delivery point, supply intake

poste *m* **de nuit** night shift

(poste *m* **de) péage** *m* toll booth [bridge, motorway]

poste *m* **de recyclage de béton résiduel** concrete recycling plant

poste *m* **de recyclage de chaussée bitumineuse** asphalt recycling plant

poste *m* **de secours** first aid point

poste *m* **de soudage** welding set

poste *m* **de transformation** transformer station

poste *m* **de travail** work station

poste *m* **groupé** party line telephone

poste *m* **secondaire** outstation

poste *m* **téléphonique** telephone (extension)

Postes, Télécommunications et Télédiffusion [PTT] Former French post office and telephone service [now separated as La Poste and France Télécom]

post-réchauffeur *m* reheater

post-refroidissement *m* re-cooling

post-tension *f* post-tension(ing)

pot *m* **d'échappement** silencer, exhaust pipe [vehicle]

pot *m* **de brûleur** burner pot

pot *m* **de cheminée** chimney pot

pot *m* **de décantation** sump [of a machine]

pot *m* **de vin**[†] bribe

potabilité *f*, **contrôle de** potability check [for drinking water]

potable *adj* drinkable

potager *adj* vegetable *adj*

potager *m* kitchen/vegetable garden, side oven [beside open fireplace]

poteau *m* column, stanchion, mast, pole, pier [of a bridge], post, stud [timber structure], vertical strut [in a roof truss]

poteau *m* **cornier** corner post/stud/stile

poteau *m* **d'angle** corner post/stud/stile

poteau *m* **d'huisserie/de baie** door/window jamb [pier or post]

poteau *m* **de décharge** supporting post/stud

poteau *m* **de fenêtre/de porte** window/door jamb [pier or post]

poteau *m* **de refend** principal post

poteau *m* **de remplissage** intermediate stud

poteau *m* **de réverbère** lamp post

poteau *m* **électrique** electricity supply pole

poteau *m* **en treillis** lattice stanchion

poteau *m* **métallique** steel column

poteau *m* **tenu en tête** braced column

potée *f* **d'émeri** emery powder [for polishing metal]

potelet *m* small vertical strut, post or stud

potence *f* triangulated bracket

potentiel *adj* potential *adj*

potentiel *m* potential [electrical, etc]

potentiel *m* **calorifique** heat load

potentiomètre *m* potentiometer

poterie *f* clay pipe [piece of], pottery or terracotta ware

poterne *f* postern [of a fortress], postern-gate

poubelle *f* dustbin, refuse container, waste bin

poucier *m* thumb-piece of a door/gate latch

poudingue *m* conglomerate [stone]

poudrage *m* coating [of a lamp]

poudre *f* dust, powder

poudre *f* **à souder** welding flux

poudre *f* **d'asphalte naturelle** natural rock asphalt, ground or powdered

poul *m* stone that breaks up when one tries to dress it

poulie *f* block, pulley

poulie *f* **d'entraînement** drive pulley [e.g. in electric lift/elevator]

poulie *f* **folle** loose pulley

poulie *f* **motrice** drive pulley

poupée *f* **mobile** tailstock

pour cent *m* per cent

pourcentage *m* percentage

pourcentage *m* **de pertes** percentage loss

pourcentage *m* **du poids** percentage by weight

pourriture *f* rot

pourriture *f* **sèche** dry rot

pourtour *m* circumference or periphery [e.g. surrounding a building], precincts [e.g. of a cathedral]

poussage *m* push-launching [e.g. of bridge girder]

pousse-tube *m* **(pneumatique)** pipe jacking machine, (pneumatic)

poussée *f* horizontal pressure, pushing, thrust

poussée *f* **ascensionnelle** uplift pressure/force

poussée *f* **axiale** axial thrust

poussée *f* **d'Archimède** buoyancy, uplift [hydraulic]

poussée *f* **de voûte** arch pressure

poussée *f* **des terres** active earth pressure

poussée *f* **du vent** wind pressure

pousser *vb* grow [of plants, trees], push, thrust

poussière *f* dust

poussiéreux *adj* dusty, dust covered

poussoir *m* push block [carpenter's or joiner's], pushbutton, press button

poutrage *m* framework of beams

poutraison *f* beam grillage/system, exposed beams

poutre *f* beam/joist [wooden, metal], girder

poutre *f* **à âme pleine** plate/solid web girder

poutre *f* **à treillis** truss [girder], trussed beam, lattice girder

poutre *f* **caisson** box girder

poutre *f* **ceinture** ring beam

poutre *f* **composée** built-up/composite beam

poutre *f* **continue** continuous beam

poutre *f* **cornière** angle beam

poutre *f* **creuse** hollow girder

poutre *f* **de bordure** edge beam

poutre *f* **de contreventement** windbracing beam

poutre *f* **de faîte** ridge beam

poutre *f* **de plancher** floor joist

poutre *f* **de rive** edge/spandrel beam

poutre *f* **de roulement** gantry girder

poutre *f* **de toiture** roof beam

poutre *f* **échelle** open frame/Vierendeel girder

poutre *f* **en H** H-beam

poutre *f* **en I** I-beam

poutre *f* **(en) T/té** T-beam

poutre *f* **en T/té (en béton armé)** T-beam (reinforced)

poutre *f* **(en) U** channel [metal section]

poutre *f* **encastrée** beam with fixed ends

poutre *f* **équerre** angle beam

poutre *f* **laminée** laminated beam

poutre *f* **maîtresse** main beam/girder

poutre *f* **palière** landing beam

poutre *f* **porte-cloison** wall joist [under internal wall in timber structure]

poutre *f* **préfabriquée** precast/prefabricated beam

poutre *f* **principale** main beam/girder

poutre *f* **renversée** upstand beam

poutre *f* **sur appuis élastiques** elastically supported beam

poutre *f* **sur appuis simples** simply supported beam

poutre *f* **Vierendeel** open frame/Vierendeel girder

poutre *f* **Warren** Warren girder

poutrelle *f* beam/girder [small], joist

poutrelle *f* **à ailes parallèles** parallel-flange beam

poutrelle *f* **à larges ailes** wide flange beam

poutrelle *f* **bois** wooden girder

poutrelle *f* **laminée (en acier)** rolled steel beam/joist (RSJ)

poutrelle *f* **métal** steel girder

poutres *f,pl* **apparentes** exposed beams

pouvoir *m* ability, capacity, power

pouvoir *m* **absorbant** absorbing/absorptive capacity

pouvoir *m* **auto-épurateur** self-purifying capacity

pouvoir *m* **calorifique** calorific value, heating power

pouvoir *m* **calorifique inférieur (PCI)** net calorific value

pouvoir *m* **calorifique supérieur (PCS)** gross calorific value

pouvoir *m* **couvrant** hiding power/opacity [of paint]

pouvoir *m* **de coupure** breaking/switching capacity, contact rating [electrical]

pouvoir *m* **réfléchissant** reflectivity

pouvoir *m* **réflecteur** reflectance

pouvoir *m* **réfrigérant** cooling capacity

pouzzolane *m* pozzolan(a)

prairie *f* grassland, meadow

pratique *adj* practical

pratique *f* practical experience, practice [actual, common]

pratique *f,* **dans la** in (actual) practice

pratique, mettre *vb* **qch en** put sth into practice

pratiquer *vb* exercise or practise [a trade, skill or profession]

pré *m* meadow

préalable, avis *m* preliminary or prior notice [of start of works

préassemblage *m* pre-assembly

préau *m* courtyard or covered area of a courtyard

préavis *m* advance notice

précadre *m* sub-frame [of an opening, prefabricated or cast in situ]

préchargement *m* preload(ing)

préchauffage *m* preheat(ing)

préchauffage *m* **d'air** air preheating

préchauffer *vb* preheat

préchauffeur *m* preheater

précipitation *f* deposit [e.g. of dust], precipitation, rainfall

précipitation *f* **de poussières** dust deposit

précipité *m* deposit [e.g. of dust]

préciser *vb* make clear, specify

précision *f* accuracy, precision

précision *f* **d'affichage** setting accuracy

précompactage *m* initial compaction, precompaction

précompression *f* precompression

préconiser *vb* advocate, recommend

précontraindre *vb* prestress

précontraint *adj* prestressed

précontrainte *f* prestress(ing)

précontrainte *f* **externe** external prestressing, post-tension

précontrainte *f* **interne** internal prestressing, pre-tension

précontrainte *f* **par câbles** prestressing by tendons

précontrainte *f* **par fils adhérents** prestressing by pre-tensioned bonded wires (Hoyer method)

précribleur répartiteur *m* placer spreader

précribleur *m* **vibrant** vibrating grizzly feeder

prédalle *f* permanent shuttering [precast concrete planks], lower layer of a concrete slab containing its main reinforcement

prédalle *f* **participante** structurally participating precast concrete plank

prédosé *adj* premixed

prédoseur *m* **des enrobés** cold aggregate feed unit

préenrobé *adj* pre-coated

préfabrication *f* precasting, prefabrication

préfabrication *f* **fermée** prefabrication of elements fitting together

préfabrication *f* **légère** prefabrication of lightweight elements

préfabrication *f* **lourde** prefabrication of major elements

préfabriqué *adj* precast, prefabricated

préfabriqué *adj* **en atelier/sur chantier** precast/prefabricated in the factory or works/on site

préfabriquer *vb* precast, prefabricate

préfabriquer *vb* **en atelier/sur chantier** precast/prefabricate in the factory or works/on site

préfet *m* administrative head of a **département** in France

préfiltre *m* coarse filter, prefilter

préforé *adj* prebored, predrilled

préforer *vb* prebore, predrill

préisolé *adj* pre-insulated

prélèvement *m* deduction in advance, levy, direct debit [on a bank account]

prélèvement *m* **d'échantillons** removal/selection of test pieces, sampling

prélever *vb* deduct in advance, levy, direct debit [from a bank account]

prélever *vb* **des carottes** take core samples

prélever *vb* **des échantillons** take samples

préliminaire *adj* preliminary

prélinteau *m* lintel [prefabricated lower part containing main reinforcement]

premier *adj* initial, first

premier étage *m*   first floor
première couche *f*   priming coat
premiers secours *m,pl*   first aid
premiers soins *m,pl*   first aid
prémonté *adj*   pre-assembled
prendre *vb*   set [concrete, glue]
prendre *vb* **coup**   settle [foundation]
prendre *vb* **de la vitesse**   pick up speed
prendre *vb* **le jus**[†]   get an electric shock
prendre *vb* **le mesure (de qch)**   measure (sth)
prendre *vb* **son assiette**   bed down [e.g. a foundation]
préparation *f*   preparation
préparation *f* **au poli**   final dressing of stone before polishing
préparer *vb*   plan [work out, a project, etc]
prérefroidissement *m*   pre-cooling
prérefroidisseur *m*   pre-cooler
préréglage *m*   pre-setting
préréglé *adj*   pre-set
prérégler *vb*   pre-set
presbytère *m*   parsonage, presbytery, rectory, vicarage
préscription *f*   order or instruction, regulation [rule]
préscriptions *f* **techniques**   technical regulations [e.g. building regulations]
présentoir *m*   display unit [shelf]
président *m*   chairman [e.g. of a meeting]
présignalisation, signal *m* **de**   advance direction sign
presqu'île *f*   peninsula, spit [of land]
presse *f*   press
presse *f* **à plaquer**   veneering press
presse *f* **arrière de l'établi**   tail vice
presse *f* **d'établi**   front vice
presse *f* **parisienne**   tail vice
presse-étoupe *m*   gland nut [of a tap, valve, etc], packing-box, stuffing-box
presse-étoupe, garniture *f* **de**   packing [for a gland or stuffing-box]
presse-étoupe *m* **de câble**   cable gland
pressiomètre *m*   pressure meter
pression *f*   compression, pressure
pression *f* **acoustique**   acoustic pressure
pression *f* **ambiante**   ambient pressure
pression *f* **atmosphérique**   atmospheric pressure
pression *f* **barométrique**   barometric pressure
pression *f* **d'alimentation**   distribution pressure
pression *f* **d'appui**   bearing stress
pression *f* **d'aspiration**   suction pressure
pression *f* **d'eau interstitielle**   pore water pressure
pression *f* **d'épreuve/d'essai**   test pressure
pression *f* **d'équilibre**   balance pressure
pression *f* **d'huile**   oil pressure
pression *f* **dans la chaudière**   boiler pressure
pression *f* **dans le brûleur**   burner pressure
pression *f* **de compensation**   equalizing pressure
pression *f* **de contrôle**   control pressure
pression *f* **de décharge**   discharge pressure
pression *f* **de gaz**   gas pressure
pression *f* **de l'eau**   water pressure
pression *f* **de marche**   operating or working pressure

pression *f* **de refoulement**   delivery pressure
pression *f* **de service**   operating or working pressure
pression *f* **de vapeur**   steam/vapour pressure
pression *f* **de vitesse**   velocity pressure
pression *f* **des terres**   earth pressure
pression *f* **différentielle**   differential pressure
pression *f* **du vent**   wind pressure
pression *f* **dynamique**   dynamic pressure
pression *f* **filtrante différentielle**   differential head of or pressure drop across a filter
pression *f* **hydraulique**   hydraulic pressure
pression *f* **limite de sécurité**   maximum safe pressure
pression *f* **nominale de travail**   design/nominal working pressure
pression *f* **normale de travail**   operating/working pressure
pression *f* **partielle**   partial pressure
pression *f* **réduite**   reduced pressure
pression *f* **statique**   static pressure
pression *f* **superficielle**   surface pressure
pression, être *vb* **sous**   be under pressure
pressostat *m*   pressure switch
pressostat *m* **basse pression**   low-pressure controller/cut-out
pressostat *m* **de manque d'air**   air pressostat
pressostat *m* **de sécurité**   safety cut-out [pressure switch]
pressostat *m* **de sécurité basse pression**   low-pressure safety cut-out
pressostat *m* **de sécurité haute pression**   high-pressure safety cut-out
pressostat *m* **différentiel**   pressure differential cut-out
pressostat *m* **haute pression**   high-pressure controller/cut-out
pressurisation *f*   pressurization
prestataire *m*   provider of services [e.g. of design or construction]
prestations *f,pl*   services [e.g. of design or construction]
prêt *adj* **à l'emploi**   ready for use
prêt *m*   loan
prêt *m* **à terme**   loan at notice
prêt *m* **bancaire**   bank loan
prêt *m* **garanti**   secured loan
prêt *m* **hypothécaire**   mortgage loan
prêt *m* **sur gage**   secured loan
prétaillage *m*   dressing [squaring of stone at the quarry]
prétension *f*   pretension(ing)
prêter *vb*   lend
prétraité *adj*   pretreated
prétraité, gravillon *m*   pre-coated gravel or crushed rock chippings
prévention *f* **des accidents**   accident prevention
prévention *f* **routière**   road safety
prévision *f*   estimate, forecast, prediction
prévision *f* **météo(rologique)**   weather forecast
prévoir *vb*   project [predict]
prieuré *m*   priory
primage *m*   priming [of a boiler, or painting]
prime *f*   grant, subsidy, premium [insurance]
primer *vb*   prime [a surface]

principal *adj* main	**prix** *m* **plafond** ceiling price
principe *m* principle	**prix** *m* **plancher** bottom/floor price
principe *m* **de la construction modulaire** modular construction principle	**prix** *m* **net** trade price
	prix *m* **unitaire** unit price
principe *m* **de superposition** principle of superposition	**probabilité** *f* probability
principes *m,pl* **comptables** accounting principles	**problème** *m* problem
prise, en engaged, in gear	**problème** *m* **de conception** design problem
prise *f* grip [hold], plug socket or socket outlet, tapping [for a sensor, a current take-off], setting [of concrete, glue, etc]	**problème** *m* **de dessin** design problem
	procédé *m* process [industrial, physical]
	procédé *m* **de conditionnement d'air** air-conditioning process
prise *f*, **chaleur** *f* **de** heat of hydration, setting heat [of concrete]	**procédé** *m* **de congélation** freezing process
prise *f* **d'air** air intake	**procédé** *m* **de dégivrage** defrosting process
prise *f* **d'air neuf** fresh air intake	**procédé** *m* **de fabrication** manufacturing method/ process
prise *f* **d'eau** cock/tap [for tapping off water], valve [for drawing off water], hydrant, water intake	**procédé** *m* **de séchage** drying process
	procédé *m* **de traitement** treatment process
prise *f* **d'échantillon** sampling	**procédé** *m* **de traitement chimique** chemical treatment process
prise *f* **d'incendie** dry riser outlet	
prise *f* **de courant** power point	**procédé** *m* **normalisé** standard method/process
prise *f* **de courant de sécurité** earthed socket outlet	**procédure** *f* procedure, process [administrative, legal]
prise *f* **(de courant) femelle** plug socket, socket outlet	**procès** *m* (legal) action or proceedings, process [legal action]
prise *f* **(de courant pour) rasoir** razor socket	
prise *f* **de débit-mètre** flowmeter tapping	**processeur** *m* **central** central processor (unit)
prise *f* **de masse** earthing clip [welding]	**processus** *m* operation, procedure, process
prise *f* **de soutirage** tap-off (point) [electrical]	**processus** *m* **de mélange** mixing process
prise *f* **de terre** earthing rod or system	**procès-verbal** *m* **(PV)** minutes/notes [of a meeting], record of evidence, written statement, report
prise *f* **électrique femelle** plug socket, socket outlet	
prise *f* **électrique mâle** plug [electrical]	**Proctor, densité** *f* Proctor density
prise *f* **en rivière** river intake	**Proctor, essai** *m* **(de compactage)** Proctor compaction test
prise *f* **multiple** adaptor plug [electrical]	
prisonnier *m* stud bolt	**se procurer** *vb* **les fonds** raise funds
privatif *adj* private	**production** *f* generation [e.g. of steam], output, product(ion)
privé *adj* private	
privilège *m* **d'hypothèque** mortgage charge	**production** *f* **de puissance électrique** electric(al) power generation
prix *m* cost, price, value, charge [fee, etc]	
prix *m* **à forfait** contract/firm/fixed/inclusive/lump sum price	**production** *f* **sur site** on-site generation
	productivité *f* productivity
prix *m* **au comptant** cash price	**produire** *vb* generate [electricity, steam], make, produce
prix *m* **calculé** estimated price	
prix *m* **courant** market price	**produit** *m* income [proceeds, yield], profit, product, yield
prix *m* **coûtant** cost price	
prix *m* **d'achat** purchase price	**produit** *m* **chimique** chemical
prix *m* **de catalogue** list price	**produit** *m* **d'addition** admixture
prix *m* **de départ** asking price	**produit** *m* **d'entretien** clean(s)ing product
prix *m* **de détail** retail price	**produit** *m* **de combustion** combustion product
prix *m* **de fabrique** factory price	**produit** *m* **de cure** curing compound
prix *m* **de gros** wholesale price	**produit** *m* **de garnissage** joint sealant or sealing compound
prix *m* **de l'énergie** energy cost	
prix *m* **de revient** basic/first cost, cost price	**produit** *m* **de substitution** alternative product, substitute
prix *m* **de revient plus honoraires** cost-plus-(fixed-)fee [type of contract]	
	produit *m* **moussant** foaming agent
prix *m* **de vente** selling price	**produit** *m* **stabilisant** stabilizer [compound or product]
prix *m* **départ usine** factory (gate) price	**produits** *m,pl* goods
prix *m* **en gros** wholesale price	**produits** *m,pl* **industriels** industrial goods or products
prix *m* **fixe** firm or fixed price	**produits** *m,pl* **laminés** rolling mill products
prix *m* **forfaitaire** [= prix à forfait]	**produits** *m,pl* **semi-finis** semi-finished products
prix *m* **initial** prime cost	**professionnel** *adj* professional *adj*, trade *adj*
prix *m* **marqué** marked price	**profil** *m* *voir aussi* profilé contour, profile, section, structural shape
prix *m* **mis à jour** updated price	
	profil *m* **à froid** cold rolled section
	profil *m* **circulaire** camber [of a road]
	profil *m* **composé** built-up or compound section

profil *m* **creux carré**   square hollow section (SHS)
profil *m* **creux rectangulaire**   rectangular hollow section (RHS)
profil *m* **creux rond**   circular hollow section (CHS)
profil *m* **de vitesse**   velocity contour
profil *m* **des vitesses**   velocity profile
profil *m* **des vitesses d'air**   air distribution envelope
profil *m* **en forme de toit**   cambered cross-section/profile [of a road]
profil *m* **en long**   long(itudinal) section
profil *m* **en travers**   cross-section
profil *m* **en travers à pente unique**   straight crossfall [road]
profil *m* **en travers en toit**   cambered cross-section/profile [of a road]
profil *m* **fermé**   closed section [of which some parts are inaccessible]
profil *m* **léger**   light section
profil *m* **normal**   standard section
profil *m* **ouvert**   open section [all faces are accessible]
profilé *adj*   shaped, streamlined
profilé *m voir aussi* profil   profile, metal section, rolled section, sectional steel
profilé *m* **de couverture**   profiled capping or coping
profilé *m* **en T/té**   T-iron or section
profilé *m* **laminé**   rolled section
profilé *m* **métallique**   steel section
profilé *m* **normalisé**   standard section
profit *m*   profit
profond *adj*   deep
profond, peu *adj*   shallow
profondeur *f*   deepness, depth
profondeur, peu *m* **de**   shallowness
profondeur *f* **d'enterrement**   burying depth [e.g. of services]
profondeur *f* **de la fondation**   foundation depth
profondeur *f* **de pénétration**   depth of penetration
profondeur *f* **de pose**   laying depth
profondeur *f* **hydraulique moyenne**   mean hydraulic depth
programmable *adj*   programmable
programmateur *m*   programmer [for electrical equipment, etc]
programme *m*   content of a project, programme
programme *m* **de construction**   building programme
programme *m* **de fabrication**   production schedule, manufacturing programme
programme *m* **numérique direct**   direct digital control
programme *m* **technique**   technical brief
progressif *adj*   phased
projecteur *m*   floodlight, spotlight
projecteur *m* **de scène**   spotlight [theatre]
projection *f*   projection [drawing], spraying [of concrete, mortar, rendering]
projection *f* **horizontale**   plan [drawing]
projection *f* **verticale**   elevation [drawing]
projet, en   at planning stage
projet *m*   design, plan [design or intention], project, scheme

projet *m***, direction** *f* **du**   project management
projet *m* **d'acte**   draft contract
projet *m* **d'exécution**   final/working design
projet *m* **de bail**   draft lease
projet *m* **de bâtiment**   building project
projet *m* **de construction**   building project
projet *m* **de construction soudée**   welded structural design
projet *m* **détaillé**   detailed design
projeter *vb*   design
projeter *vb* **de faire qch**   plan to do sth
projeteur *m*   designer, planner [of projects], project manager
prolongateur *m*   extension cable or lead
prolongation *f*   extension [generally of time], lengthening, prolongation
prolongation *f* **de temps**   extension of time
prolongement *m*   extension [of a building, road, etc]
prolonger *vb*   extend [make longer], lengthen, prolong [in space or time]
promesse *f* **de vente**   preliminary sale agreement [to sell]
promoteur *m*   developer, promoter [of development/construction schemes]
promoteur *m* **de construction**   property developer
promoteur *m* **immobilier**   property developer
promotion *f*   advertising campaign, promotion
prononcé *adj*   sharp [of contrast, difference]
propagation *f*   spread [of fire]
propane *m*   propane
proportion *f*   proportion, ratio
proportionnel   proportional
proposer *vb*   offer, propose, suggest
proposition *f*   proposal
propre *adj* **à sa destination**   fit for its purpose
propreté *f*   clean(li)ness
propriétaire *m*   landlord [owner], owner, proprietor
propriété *f*   property
propriété *f* **allégée**   unencumbered property
propriété *f* **foncière perpétuelle**   freehold property
propriété *f* **mécanique**   mechanical property, strength
propriété *f* **riveraine**   bordering property
propulseur *m*   impeller [air]
proscenium *m*   proscenium
prospect *m*   prospect, view, open space in front of a building [in regulations]
protecteur *adj*   protective
protection *f*   guard [against hazard, not a security guard]
protection *f* **(de, contre)**   protection (from, against)
protection, couche *f* **de**   surface protection (layer)
protection, enrochement *m* **de**   rip-rap
protection *f* **à la thermistance**   thermistor protection
protection *f* **anti-rouille**   anti-corrosion finish, rust prevention/protection
protection *f* **cathodique**   cathodic protection
protection *f* **contre les crues**   flood protection
protection *f* **contre la corrosion**   corrosion protection
protection *f* **contre les corps solides**   protection against solid bodies

protection *f* **contre les liquides**   protection against liquids
protection *f* **contre les surcharges**   overload protection
protection *f* **contre les surtensions**   over-voltage protection
protection *f* **contre l'incendie/les incendies**   fire protection
protection *f* **de machine**   machine guard, motor protection
protection *f* **de terre**   earth protection, protective earthing
protection *f* **du ventilateur**   fan shroud
protection *f* **incendie**   fire protection
protection *f* **pare-vapeur**   vapour barrier
protégé *adj*   protected
protège-genoux *m*   kneepad
protège-lame *m*   blade guard [circular saw]
protéger *vb* **(de, contre)**   protect (from, against)
prototype *m*   prototype
province *f*   province
provincial *adj*   provincial
provision *f*   provision, reserve or stock [supply], store [stock or reserve]
provisoire *adj*   interim, provisional, temporary
proximité de, à   in the neighbourhood of
proximité *f*   nearness, proximity
prunier *m*   plum tree
psychromètre *m*   psychrometer
psychromètre *m* **à aspiration**   aspiration psychrometer
psychromètre *m* **à rotation**   sling psychrometer
psychromètre *m* **fronde**   sling psychrometer
puisage *m*   water extraction [from the ground], catchpit, cesspool, sump, drain tank, floor drain
puisard *m* **de relevage**   drain(age) sump
puisard *m* **incendie**   emergency water supply well or sump, firemen's well, sump/water tank
puiser *vb*   draw water from
puissance *f*   capacity, power [of a machine, motor, etc], force, wattage
puissance, à haute   at high power
puissance, à toute   at full power
puissance *f* **absorbée**   power consumption
puissance *f* **acoustique**   sound power
puissance *f* **au frein**   brake power
puissance *f* **calorifique**   heat output (rating)
puissance *f* **calorifique nominale**   rated heat output
puissance *f* **de chaudière**   boiler output/rating
puissance *f* **de la lampe**   lamp power
puissance *f* **de levage**   lifting capacity [of a crane]
puissance *f* **de moteur**   motor rating
puissance *f* **de refroidissement**   cooling capacity
puissance *f* **de ventilateur**   fan power
puissance *f* **électrique**   electric power supply [amount]
puissance *f* **électrique absorbée**   electric power absorbed
puissance *f* **électrique appelée**   electrical demand
puissance *f* **en chevaux**   horsepower
puissance *f* **lumineuse**   luminous intensity
puissance *f* **minimale**   minimum output
puissance *f* **nominale**   rated duty or power
puissance *f* **nominale à l'heure**   one-hour rating
puissance *f* **nominale du moteur**   motor rating
puissance *f* **réactive**   reactive power

puissance *f* **thermique**   heat/thermal output
puissance *f* **thermique totale**   total heat output
puissance *f* **thermique utile**   useful heat output
puissance *f* **utile**   useful capacity
puissance *f* **vive**   kinetic energy
puits *m*   shaft [for foundations or underground access], well [water source]
puits *m* **à câbles**   cable pit/manhole
puits *m* **à ciel ouvert**   open well
puits *m* **à eau**   water well
puits *m* **artésien**   artesian well
puits *m* **d'accès**   manhole
puits *m* **d'aération**   ventilation shaft/well
puits *m* **d'exhaure**   pumping out shaft
puits *m* **d'emprunt**   borrow pit
puits *m* **d'infiltration**   seepage pit, soakaway
puits *m* **de décharge**   relief well
puits *m* **de lumière**   light well
puits *m* **de mine**   mine shaft, pit [mine]
puits *m* **de reconnaissance**   test/trial pit
puits *m* **de reconnaissance à la pelle (mécanique)**   machine dug test/trial pit
puits *m* **de sondage**   boring [into the ground]
puits *m* **de thermomètre**   thermometer well
puits *m* **de visite**   inspection chamber
puits *m* **drainant**   absorbing/drain well
puits *m* **en charge**   pressure shaft [dam]
puits *m* **filtrant**   absorbing/drain well
puits *m* **perdu**   cesspool, soakaway
puits, eau *f* **de**   well water
puits, sondage *m* **par**   trial boring
pulsatoire *adj*   pulsating
pulvérisateur *m*   atomizer, sprayer
pulvérisateur *m* **à main**   hand spray(er)
pulvérisateur-mélangeur *m*   pulverizing mixer
pulvériser *vb*   atomize or spray [a liquid, paint], grind to powder, pulverize
pulvérulent *adj*   cohesionless, non-cohesive [of soil]
pupitre *m* **de commande/de contrôle**   control console or desk [e.g. for stage lighting]
pureau *m*   gauge [of a tile or slate, i.e. visible area or height]
pureté *f*   purity
purge *f*   bleed(ing) [e.g. of brakes], draining, flushing out
purge *f* **d'air**   air purging or vent(ing)
purge *f* **de déconcentration**   blowdown
purger *vb*   blow off/out/through, cleanse [by purging through], purge, drain [a cylinder]
purgeur *m*   bleeder, blow-off/exhaust valve
purgeur, robinet *m*   air/drain/waste cock/valve
purgeur *m* **automatique**   automatic drain-cock
purgeur *m* **d'air**   air cock, purge/vent valve
purgeur *m* **d'air automatique**   automatic air valve (AAV)
purgeur *m* **de vapeur**   condensing/steam trap
purification *f*   purification, treatment
purifier *vb*   cleanse, purify
pylône *m*   mast, pole, pylon, support tower [for cableway]
pylône *m* **ligne haute tension**   high-tension/-voltage tower
pyromètre *m*   pyrometer

quadrillage, en with a grid layout
quadrillage *m* cross-ruling, grid layout
quadrillé *adj* crossed [e.g. lines], gridded
quadrillé, papier *m* squared paper
quadrillée, carte *f* grid map
quadriller *vb* rule in squares, lay out in a grid
quai *m* dock, pier, quay, wharf, platform [e.g. of station]
quai *m* **d'amarrage** berthing quay
quai *m* **de chargement** loading dock
quai *m* **de gare** station platform
qualification *f* qualification
qualifié *adj* qualified [as or to do sth]
qualifier *vb* qualify
qualité *f* quality
qualité *f* **d'acier** grade of steel
qualité *f* **d'air intérieur** indoor air quality
qualité *f* **d'eau/de l'eau** water quality
qualité *f* **de l'air** air quality
qualité *f* **de la vapeur** steam quality
qualité *f* **du gaz combustible** fuel gas quality
quantifier *vb* quantify
quantité *f* amount, quantity
quantité *f* **annuelle de pluie** annual rainfall
quantité *f* **approximative** approximate amount or quantity
quantité *f* **d'air** air quantity or volume
quantité *f* **d'air nécessaire** air requirement
quantité *f* **de chaleur produite** heat output
quantité *f* **de mouvement** momentum

quantité *f* **journalière de pluie** daily rainfall/precipitation
quantité *f* **par jour** quantity per day
quart *m* quarter [one-fourth]
quart *m* **de rond** quarter round moulding
quartier *m* area, district or quarter [of a town, e.g. commercial]
quartier *m* **résidentiel** residential area
quartier *m* **tournant** wreath [of a handrail]
quartier tournant, escalier *m* **à** quarter-turn staircase
quartier (à une pierre de taille), faire *vb* turn (a piece of cut stone) to lie on another face
quartz *m* quartz
queue *f* [d'une pierre de taille] header [of bond stone in a wall]
queue-d'aronde *f* dovetail halved joint
queue-de-carpe *f* dovetail joint, split end of a fishtail fixing lug
queue-de-moellon *f* stone depth perpendicular to wall face
queue-de-morue *f* flat paintbrush
queue-de-rat *f* rat-tail or round file
queue-de-vache *f* rafter end, metal reinforcement for plaster infill
quincaillerie *f* hardware, ironmongery [metal fittings]
quinconce, en staggered
quinconces, en staggered, in staggered rows
quitter *vb* discharge [a debt], vacate [premises, a property]

rabais *m* discount, reduction [in cost or price]
rabattement *m* **d'une nappe d'eau** lowering of a water table
rabattre *vb* close, shut down, deduct [from a price], reduce [price, etc], fold back
rabot *m* plane [for timber, etc]
rabot *m* **à coller/à dents/denté** tooth(ing) plane
rabot *m* **à dégrossir** jack plane
rabot *m* **à languette** grooving plane
rabot *m* **à main** hand plane
rabot *m* **à moulures** moulding plane
rabot *m* **(à semelle) cintré** compass plane
rabot *m* **long** trying plane
rabot *m* **plat** smoothing plane
rabotage *m* planing

raboté *adj* planed (off)
raboter *vb* plane
raboteuse *f* planing machine or planer, thicknesser
raboteuse *f* **(routière) à froid** cold pavement milling machine, cold planer [for road surfaces]
rabotin *f* small toothed drag [for dressing plastered surfaces]
rabouter *vb* butt, place or join end-to-end [two lengths of sth]
raccord *m* adaptor, connecting piece, connection, connector, coupler/coupling [reinforced concrete, scaffolding, etc], joint [pipework], junction, link, join, repair [in or of paintwork, plaster, etc], repeat [of a pattern, e.g. wallpaper]
raccord *m* **à brides** flanged joint

raccord *m* à **compression**   compression joint [plumbing]
raccord *m* à **emboîtement**   spigot-and-socket joint
raccord *m* à **visser**   screw (on) joint
raccord *m* **bouchon**   stop end [for a pipe]
raccord *m* **coude**   elbow joint
raccord *m* **d'alimentation**   dry riser inlet, supply connection [pipework]
raccord *m* **de branchement**   branch piece
raccord *m* **de câble**   cable joint
raccord *m* **de dérivation**   bypass connection
raccord *m* **de graissage**   grease nipple
raccord *m* **de réduction**   pipe reducer, reducing coupler/coupling or joint
raccord *m* **en fonte**   cast iron fitting [coupling/connection]
raccord *m* **fileté**   threaded connector
raccord *m* **flexible**   flexible connection
raccord *m* **manchon**   sleeve joint, straight pipe (sleeve) joint
raccord *m* **montage bicône**   compression fitting [pipe jointing]
raccord *m* **montage soudé**   solder fitting [pipe jointing]
raccord *m* **pompier/rapide**   fireman's valve or hose connection
raccord *m* **T/té**   T-joint
raccord *m* **union**   union [pipe joint], plumber's union [joint]
raccord *m* **vissé**   screwed connection/joint
raccord *m* **vissé pour tuyau**   screwed pipe connection/joint
raccordé *adj*   connected
raccordement *m*   connection, connecting, joining (together), transition [roads]
raccordement, **courbe** *f* **de**   transition curve [roads]
raccordement, **rayon** *m* **de**   transition radius
raccordement *m* **aux réseaux publics**   connection to (the) public mains, mains connection
raccordement *m* **circulaire**   circular transition curve
raccordement *m* **de départ**   flow connection [on a boiler]
raccordement *m* **en série**   series connection
raccordement *m* **par brides**   flange(d) connection
raccordement *m* **par soudure bout à bout**   butt-welded joint
raccordement *m* **par tuyau flexible**   hose connection
raccordement *m* **parallèle**   parallel connection
raccordement *m* **vertical**   vertical transition
raccorder *vb* **(à)**   connect (to, with), join (to), joint [join together]
raccorderie *f* **de chaudière**   boiler fittings
raccorderie *f* **de chauffage central**   central heating fittings
raccourci *m*   short cut
raccourcissement *m*   shortening
raccourcissement *m* **élastique**   elastic shortening
racheter *vb*   conceal, hide, make good [a minor defect, a difference of level]
racheux *adj*   knotty [wood]
racine *f*   root [of a plant, tree, and mathematical]
racine *f* **carrée**   square root

racine *f* **cubique**   cube root
racler *vb*   scrape
raclette *f*   scraper, squeegee
racleur *m*   scraper [earthworks]
racloir *m*   scraper
radial *adj*   radial
radiant *adj*   radiant
radian(t) *m*   radian
radiateur *m*   radiator
radiateur *m* **à colonnes**   column radiator
radiateur *m* **à gaz**   gas radiator
radiateur *m* **à gaz à rayonnement**   gas radiant heater
radiateur *m* **à haute pression**   high pressure radiator
radiateur *m* **à panneaux**   panel radiator
radiateur *m* **à plaque**   radiant strip heater
radiateur *m* **à tubes à ailettes**   finned-tube radiator
radiateur *m* **d'appoint**   back-up heater or radiator
radiateur *m* **électrique**   electric fire or radiator
radiateur *m* **électrique à rayonnement**   radiant electric heater
radiateur *m* **en fonte**   cast radiator
radiateur *m* **encastré**   recessed radiator
radiateur *m* **mural**   wall heater
radiateur *m* **panneau (en acier)**   (steel) panel radiator
radiateur *m* **sur plinthe**   skirting board radiator
radiation *f* **ionisante**   ionizing radiation
radier *m*   base slab, mat foundation, raft [base], invert [of a sewer, etc]
radier *m* **de fondation**   foundation raft
radier *m* **en béton**   concrete base
radiographie *f*   radiography
radoucir *vb*   re-smooth [a surface]
rafale *f*   gust, squall [of wind]
raffiner *vb*   refine
rafraîchir *vb*   add water [to stiffened mortar], cool, refresh, refurbish [e.g. joinery, masonry]
rafraîchir *vb* **(les joints)**   re-point
rafraîchissement *m*   cooling, refreshing, restoration [of masonry, joinery]
rafraîchissement *m* **d'ambiance/de confort**   comfort cooling
rafraîchissement *m* **de locaux**   space cooling
rafraîchissement *m* **libre**   free cooling
rafraîchissement *m* **par pulvérisation**   spray cooling
rafraîchisseur *m*   cooler [of liquids]
ragréage *m*   cleaning down/off/up [e.g. brickwork, stonework, etc], final making good [of surfaces]
ragréer *vb*   clean down/off/up [e.g. brickwork, stonework, etc]
ragrément *m*   cleaning down/off/up [e.g. brickwork, stonework, etc], final making good [of surfaces]
raideur *f*   rigidity, stiffness
raidir *vb*   stiffen
raidisseur *m*   gusset, stiffener
raidisseur *m* **d'âme**   web stiffener
raie *f*   line [e.g. on paper]
rail *m*   rail [railway track]
rail *m* **à base plate**   flat-bottomed rail
rail *m* **conducteur**   live/third rail

rail *m* **de contact**   conductor rail
rail *m* **de guidage**   guide rail
rail *m* **de scellement**   channel/concrete insert, fixing channel
rail-frein *m*   rail-brake, retarder [railway]
rainé *adj*   grooved [of the edge of a board]
rainer *vb*   groove, slot [cut a slot in]
rainure *f*   channel, groove, slot, rabbet
rainuré et languetté *adj*   tongued-and-grooved
rainureuse *f*   groover [for road concrete, asphalt]
ralenti *m* **de nuit**   night set-back
ralentir *vb*   slow (down)
rallonge *f*   extension piece
ramassage *m* **d'ordures (ménagères)**   rubbish collection
ramasser *vb*   pick up [sth dropped]
rambarde *f*   guard rail
rame *f*   train [group of wagons]
rame *f* **de métro**   underground train
ramification *f*   branch(ing)
ramollir *vb*   soften
ramollissement *m*   softening
ramonage *m*   chimney or flue cleaning or sweeping
ramoner *vb*   sweep [a chimney or flue]
ramoneur *m*   chimney-sweep
rampant *adj*   inclined, sloping
rampant *m*   sloping part of a building [stairs, roof, vault, etc]
rampe *f*   (access ramp), balustrade of a stair, battery [of lighting], footlights [theatre]
rampe *f* **d'accès**   access ramp
rampe *f* **d'escalier**   balustrade [of a stair], banisters
ranche *f*   peg of a peg ladder
rancher *m*   peg ladder
rang *m*   course or row [of tiles or slates], line [of trees, columns, etc]
rang *m* **de faîtage**   ridge course [of roof tiles or slates]
rang *m* **de gouttière**   eaves course [of roof tiles or slates]
range *f* **de pavés**   row of paving slabs of equal size
rangée *f*   array, line, row
rangée *f* **de chiffres**   array of figures/numbers
rangée *f* **des brûleurs**   burner array
rangement *m*   cupboard space, storage space [cupboards, shelving], putting away
ranger *vb*   arrange, put in order, put/stow away, put back in place, tidy (away or up), set in rows
râpe *f*   rasp
râpe *f* **cintrée**   compass plane
râpe-scie *f*   saw rasp
râper *vb*   rasp
rapide *m*   fast train
rapides *m,pl*   rapids [in a watercourse]
rappointis *m*   nails or metal fixing in timber [to support heavy plaster or finishings]
rapport avec, en   in harmony with, in keeping with
rapport *m*   income, profit, return, revenue, yield [financial], ratio, report
rapport *m* **air–combustible**   fuel–air ratio
rapport *m* **d'engrenage**   gear ratio
rapport *m* **d'essais**   test report

rapport *m* **de compression**   compression ratio
rapport *m* **de mélange**   mixing ratio
rapport *m* **de terre**   made ground
rapport *m* **eau–ciment**   water–cement ratio
rapport *m* **hebdomadaire**   weekly report
rapport *m* **mensuel**   monthly report
rapport *m* **qualité–prix**   quality–price ratio, value-for-money
rapporteur *m*   minute-taker
rase campagne, de   in the open country
raser *vb*   raze
rassemblement *m*   collection [action of, of facts, etc]
rassembler *vb*   collect [documents, evidence, facts, information, etc]
râteau *m*   rake
ratissage *m*   rendering with a large, flat trowel
ratisser *vb*   render with a large, flat trowel
ravalement *m*   finishing, rendering, (re)surfacing or restoration [of a facade, stonework, etc], refacing work
ravaler *vb*   reface [masonry], render, restore, (re)surface [a facade, stonework, etc]
ravaleur *m*   wall finishing specialist
ravin *m*   gully [natural feature], ravine
rayer *vb*   scratch
rayère *f*   narrow wall opening to provide light [e.g. on a stair]
rayon *m*   beam [of light], radius, shelf
rayon *m* **d'action**   working area/radius
rayon *m* **d'influence**   radius of influence
rayon *m* **de courbure**   radius of bending/curvature
rayon *m* **de giration**   radius of gyration, turning radius
rayon *m* **de raccordement**   transition radius
rayon *m* **de soleil**   sunbeam
rayonnage *m*   set of shelves
rayonnant *adj*   radiant/radiating
rayonnement *m*   radiation [of heat]
rayonnement *m* **à basse température**   low-temperature radiation
rayonnement *m* **de base**   background radiation
rayonnement *m* **des conduits d'air**   duct radiation
rayonnement *m* **infrarouge**   infra-red radiation
rayonnement *m* **ionisante**   ionizing radiation
rayonnement *m* **solaire**   solar radiation
rayonnement *m* **solaire incident**   insolation, incident solar radiation
rayonnement *m* **thermique**   heat or thermal radiation
rayonnement *m* **visible**   visible radiation [light]
rayonner *vb*   radiate
rayure *f*   scratch [on wood or metal]
réa *m*   pulley wheel, sheave
réaccoupler *vb*   reconnect [mechanically, electrically]
réacheminer *vb*   re-route
réactance *f*   reactance
réacteur *m*   reactor [nuclear power station]
réacteur, cuve *f* **de**   reactor vessel [nuclear power station]
réactif *adj*   reactive
réactif *m*   reagent
réaction *f*   reaction

réaction *f* **au feu**   fire performance

réaction *f* **d'appui**   bearing stress, support reaction

réaction *f* **en chaîne**   chain reaction

réalisation *f*   achievement, carrying out, implementation

réalisation *f* **d'ouvrages enterrés en tranchée ouverte**   cut-and-cover [carrying out the works]

réalisation *f* **du projet**   project implementation

réaliser *vb*   carry out, implement

réallumage *m*   re-ignition

réaménagement *m*   renovation

réaménager *vb*   renovate

réarpentage *m*   re-survey [of land, etc]

rebondissement *m* **Schmidt**   Schmidt rebound

rebord, à   with a raised edge

rebord *m*   window ledge or sill

rebouchage *m*   filling [of cracks, etc, prior to painting or papering]

reboucher *vb*   fill [cracks, etc, prior to painting or papering]

rebours, à   against the grain, backwards

rebours *adj*   cross-grained [of wood]

rebours *m*   cross/raised grain [of wood]

rebrancher *vb*   reconnect [electrically]

rebut *m*   reject, throw-out

rebut, mettre *vb* **qch au**   reject or scrap sth, throw sth away

rebut, papier *m* **de**   waste paper

rebut, pièces *f,pl* **de**   rejects

rebuter *vb*   reject, throw aside [goods]

recaler *vb*   reset [equipment], smooth [a joint, a groove, in wood]

recépage *m*   trimming [e.g. a pile, a tree]

recéper *vb*   trim [a pile]

récépissé *m*   receipt

récepteur[†] *m*   receiver, telephone receiver

réception *f*   acceptance or approval [of work done], acceptance or receipt [action or fact of], reception [room]

réception, accusé *m* **de**   acknowledgement of receipt [of letter, parcel]

réception *f* **définitive**   final acceptance [of works]

réception *f* **des travaux (RDT)**   acceptance of the works and final account

réception *f* **provisoire**   provisional acceptance of the works [with defects still to be remedied]

recette *f*   acceptance tests on samples, receipt

receveur *m*   collector [of taxes]

receveur *m* **à douches/de douche**   shower tray

recevoir *vb* **le courant/un décharge électrique**   get an electric shock

rechampir *vb*   line [contours or outline of paintwork against another colour]

rechampissage *m*   lining [contours or outline of paintwork against another colour]

recharge *f*   recharging, refill

rechargeable *adj*   rechargeable

rechargement *m*   deposit welding

recharger *vb*   recharge [battery, groundwater, etc], reload

recharger *vb* **par soudure**   build up by welding

réchaud *m*   portable stove

réchauffage *m*   reheating, warming (up)

réchauffer *vb* **qch**   reheat, warm (up) sth

réchauffeur *m*   heater, reheater

réchauffeur *m* **à immersion**   immersion heater [for water]

réchauffeur *m* **aval**   after-heater

réchauffeur *m* **d'air**   air heater

réchauffeur *m* **d'air à gaz**   gas-fired air heater

réchauffeur *m* **d'eau**   water heater

réchauffeur *m* **d'eau d'alimentation de chaudière**   boiler feedwater heater

réchauffeur *m* **d'huile**   oil heater [to heat oil]

réchauffeuse-répandeuse *f*   heater-sprayer [for bitumen, tar, etc]

rechausser *vb*   consolidate [the footing of a wall]

recherche *f*   investigation or research, replacement [repair, of individual roof slates or tiles]

recherche *f* **de dérangements**   fault location [electrical]

recherche *f* **en laboratoire**   laboratory investigation

rechercher *vb*   enquire into/investigate [causes, etc], search for

rechercher *vb* **un dérangement**   look for/try to locate a fault [electrical]

recherches sur qch, faire *vb* **des**   enquire into/investigate sth, do/make research into sth

récipient *m*   container [box, case, etc], vessel [receptacle], receiver [of fluid, gas]

récipient *m* **à vide**   vacuum receiver

réciproque *adj*   mutual, reciprocal

recirculation *f*   recirculation

recirculer *vb*   recirculate

réclamation *f*   claim [by a contractor, with quantities + costs], complaint

réclamer *vb*   complain

recolement *m*   verification of compliance with facade projection and height restrictions

recoler *vb*   verify compliance with facade projection and height restrictions

recommandation *f*   recommendation, registration [of post]

recommander *vb*   recommend, register [a letter, parcel]

recompacter *vb*   recompact [geotechnics]

recompression *f*   recompression

reconnaissance *f*   investigation, preliminary study/survey

reconnaissance, puits *m* **de**   test/trial pit [geotechnics]

reconnaissance *f* **de site**   field/site investigation

reconnaissance *f* **de sol**   ground/soil investigation, soil survey

reconnaissance *f* **photographique**   photo reconnaissance

reconstitué *adj*   reconstituted [e.g. stone]

reconstitution *f*   reconstruction or restoration [e.g. of sth damaged or worn]

reconstruction *f*   rebuilding, reconstruction

reconstruire *vb*   rebuild, reconstruct

recoupe *f*   waste stone from dressing stone to dimension

recoupement *m*   batter, stepping [of an embankment], cross-checking, intersection [surveying], reduction in thickness due to dressing

recouper *vb*   intersect, recut, step [wall, embankment]

recourber *vb*   bend (again/back/down/round)

recouvrement, à   lapped

recouvrement, soudé *adj* **à**   lap-welded

recouvrement *m*   collection [action of, of taxes], lap [e.g. of tiles], lap (joint) [in reinforcement bar], (over)lapping, recovering, lining [covering]

recouvrement, joint *m* **à**   lap joint

recouvrement, planches *f,pl* **à**   weatherboarding

recouvrement, terrain *m* **de**   overburden

recouvrement *m* **d'armatures/des barres**   overlapping of (reinforcement) bars

recouvrement *m* **intérieur**   internal lining

recouvrer *vb*   collect, recover [debts, etc]

recouvrir *vb*   cover, lap, overlap, re-cover [a roof, etc]

recouvrir *vb* **de terre**   blanket [with earth]

recouvrir *vb* **d'un tapis/d'une moquette**   carpet

recrépir *vb*   repair rendering, re-render, resurface

rectangle *m*   rectangle

rectangulaire *adj*   rectangular

rectificateur *m*   rectifier [electrical]

rectificateur *m* **de courants**   current rectifier

rectificatif *m*   correction [to a document, drawing]

rectification *f*   adjustment [of an instrument], readjustment, rectification, straightening [e.g. of a road]

rectification, vis *f* **de**   adjusting screw

rectifier *vb*   adjust [an instrument], amend [an account], correct [a drawing, error, price, etc], put a mistake right, rectify, straighten [an alignment, etc], true (up) [a surface]

rectifier *vb* **une erreur**   put a mistake right

rectiligne *adj*   rectilinear, straight

reçu *m*   receipt

recueil *m* **des données**   data collection

recueillir *vb*   collect [surface water], collect [documents, facts, information]

recuire *vb*   anneal

recuit *m* **de détente**   stress relief by annealing

recuit *m* **de normalisation**   stress relief by annealing

recul *m*   backwards movement

recul *m* **à l'ancrage**   anchorage seating

reculement *m*   setting back [in space, e.g. behind a building line]

récupérable *adj*   recoverable

récupérateur *m* **de chaleur**   heat recovery device/ equipment

récupérateur *m* **de poussière**   dust collector, collecting unit

récupération *f*   recovery, recuperation

récupération, freinage *m* **par**   regenerative braking

récupération, matériaux *m,pl* **de**   salvaged materials

récupération *f* **de chaleur**   (waste) heat recovery

récupération *f* **des boues**   sludge recovery

récupéré *adj*   recovered or salvaged [of used materials]

récupérer *vb*   recoup [a loss], recover [debt], recover or salvage [waste, etc]

recyclage *m*   recycling

recycleur *m* **(d'asphalte)**   (asphalt) recycler

recycler *vb*   recycle

rédaction *f*   drawing up, editing [e.g. manuals]

redan, en   stepped [e.g. a gable, an embankment]

redan/redent *m*   step [in a gable, an embankment, etc]

redenté *adj*   stepped [e.g. a gable, an embankment]

redos *m*   wavey-edged board

redressement *m*   rectifying [electrical], righting [setting upright], straightening, truing [a surface]

redresser *vb*   rectify [an error], re-erect [something fallen or lying down], straighten out, true [a surface]

redresseur *m*   rectifier [electrical]

réducteur *m*   reducer, reducing gear

réducteur *m* **d'éclairage**   dimmer

réducteur *m* **de pression**   pressure reducer, pressure regulating valve

réduction *f*   decrease, reduction

réduction *f* **d'échelle**   reduction in scale [of a drawing, etc]

réduction *f* **de bruit**   noise reduction

réduction *f* **de charge/débit**   load shedding

réduction *f* **de pression**   pressure decrease or reduction

réduction *f* **de section**   reduction in (cross-)section

réduction *f* **des coûts**   cost-cutting

réduction *f* **mâle–femelle**   male–female reducer [plumbing]

réduire *vb* **au minimum**   minimize

réduire *vb* **l'échelle de**   scale down [a drawing]

réduire *vb* **la largeur de**   narrow [make narrower]

réel *adj*   actual

réemploi *m*   re-use

refaire *vb*   do again, rebuild, remake

refaire *vb* **la surface de**   resurface [a road]

refait *adj*   done up [renewed]

réfection *f*   rebuilding, repair, restoration

réfectoire *m*   canteen, dining hall, refectory

refend *m*   splitting

refend, bois *m* **de**   wood cut parallel to the grain

refend, mur *m* **de**   cross-wall, loadbearing separating wall

refend, pierre *f* **de**   corner stone

refendre *vb*   cut [a key], cleave, split [e.g. slate], rip [timber]

référence *f*   reference

référence *f* **de commande**   order reference

reficher *vb*   rejoint [stonework]

réfléchissant, plot *m*   reflecting/reflective stud

réflecteur *m*   reflector

réflectivité *f*   reflectivity

réflectorisé, plot *m*   reflecting/reflective stud

reflux *m*   ebb(ing)

refouillement *m*   cutting or deepening of cuts into [stone, to heighten a sculpture or relief on a face]

refouiller *vb*   cut into [stone, to heighten a sculpture or relief on a face

refoulement *m*   back or reverse flow, delivery, discharge, output [of a pipe], lift [of a pump], pumping [of drainage]

réfractaire *adj*   refractory
réfracteur *m*   refractor
réfrigérant *m*   coolant, cooling/refrigerating medium,
   refrigerant
réfrigérant *m* **à cheminée**   cooling tower
réfrigérant *m* **atmosphérique**   cooling tower
réfrigérateur *m*   refrigerator
réfrigérateur *m* **à gaz**   gas refrigerator
réfrigérateur *m* **ménager**   domestic refrigerator
réfrigération *f*   chilling, refrigeration
réfrigération *f* **par compression**   compression
   refrigeration
réfrigéré *adj*   cooled, refrigerated
réfrigérer *vb*   refrigerate
refroidi *adj*   cooled
refroidi *adj* **par air/eau**   air-/water-cooled
refroidir *vb*   cool (down), get cold
refroidissement par eau, à   water-cooled
refroidissement *m*   cooling, refrigeration
refroidissement *m* **à air**   air cooling
refroidissement *m* **à convection forcée**   forced draught
   cooling
refroidissement *m* **à gaz**   gas cooling
refroidissement *m* **aval**   aftercooling
refroidissement *m* **excessif**   overcooling
refroidissement *m* **local**   spot cooling
refroidissement *m* **par absorption**   absorption cooling
refroidissement *m* **par circulation**   closed circuit cooling
refroidissement *m* **par dilatation**   expansion cooling
refroidissement *m* **par évaporation**   cooling by
   evaporation, evaporative cooling
refroidissement *m* **par liquide**   liquid cooling
refroidissement *m* **par pression d'air**   pressure cooling
refroidissement *m* **par rayonnement**   radiant cooling
refroidisseur *adj*   cooling *adj*
refroidisseur *m*   chiller, cooler
refroidisseur *m* **à air/d'air**   air cooler
refroidisseur *m* **à condensation par air**   air-cooled
   (packaged) chiller
refroidisseur *m* **aval**   after-cooler
refroidisseur *m* **d'eau potable**   drinking water cooler
refroidisseur *m* **par évaporation**   evaporative cooler
refuge *m* **pour piétons/cyclistes**   waiting area for
   pedestrians/cyclists
refus, enfoncer *vb* **à**   drive in to refusal
refus *m* **d'un pieu**   refusal of a pile
refus *m* **(supérieur)**   oversize [particles not passing
   through a sieve]
régalage *m*   spreading a thin layer of earth, gravel, etc, to
   level ground
régaler *vb*   level [ground, with a thin layer of earth,
   gravel, etc]
regard *m*   manhole [for sewer, etc]
regard *m* **d'égout**   manhole [for drain or sewer]
regard *m* **de visite**   inspection chamber or manhole
   [small]
regarnir *vb*   touch up [paintwork, masonry, pointing, etc]
regarnissage *m*   touching up [paintwork, masonry,
   pointing, etc]

régénérateur *m* **de chaussée**   road reclaimer
régénération *f*   regeneration
régie *f*   control room
régie, travaux *m,pl* **en**   cost-plus-percentage contract,
   labour, materials and percentage overheads work
Régie *f* **Autonome des Transports Parisiens (RATP)**
   transport authority of Paris
régie *f* **(de)**   local government or state control (of)
régie *f* **(de l'état)**   state-owned company or organization
   [France]
régime, à plein   at full or top speed
régime *m*   administrative system, flow rate, speed [of a
   motor]
régime *m* **d'écoulement transitoire**   transition flow
régime *m* **d'équilibre**   balance(d) condition
régime *m* **permanent**   steady state (condition, flow,
   running, etc)
régime *m* **variable**   non steady state (condition, flow,
   running, etc)
région *f*   area, region
région *f* **de basse/haute pression**   low/high pressure area
registre *m*   damper, register
registre *m* **à lames multiples**   multi-leaf damper
registre *m* **à papillon**   single-leaf damper
registre *m* **à persiennes**   multi-leaf damper
registre *m* **à volet unique mobile**   single-leaf damper
registre *m* **à volets multiples**   multi-leaf damper
registre *m* **à volets opposés/parallèles**   opposed/parallel
   blade damper
registre *m* **coulissant**   sliding damper
registre *m* **d'exploitation**   operating record/log
registre *m* **de réglage**   control damper
registre *m* **de réglage de débit**   volume control damper
registre *m* **de tirage**   draught control damper [in a flue
   pipe]
registre *m* **de vapeur**   steam throttle valve
registre *m* **extérieur**   outside air damper
registre *m* **motorisé**   motorized damper
registre *m* **papillon**   butterfly damper
registre *m* **tournant**   butterfly damper
réglable *adj*   adjustable
réglage *m*   adjustment, control or regulation, setting [of a
   clock, equipment], tuning
réglage, bouton *m* **de**   control button
réglage *m* **à distance**   remote setting
réglage *m* **approximatif**   rough adjustment/setting
réglage *m* **de température**   temperature control
réglage *m* **de vitesse automatique**   automatic speed
   control
réglage *m* **manuel**   manual adjustment/control/regulation
règle *f*   rule
règle *f* **à calcul**   slide rule
règle *f* **d'urbanisme/de construction**   planning/building
   by-law/regulation
règle *f* **de niveau inférieur/supérieur**   lower/upper edge
   board [excavations]
règle *f* **empirique**   empirical rule
réglé *f* **graduée**   ruler
règle *f* **(surfaceuse) vibrante**   vibrating screed(er)

règlement *m*   establishment of the amount of the final account, payment or settelement [of an account], regulation(s), rule(s)

règlement *m* **de la construction**   building code

règlement *m* **des litiges**   settlement of (legal) disputes

réglementation *f*   regulations

réglementation *f* **de la circulation**   traffic regulations

réglementation *f* **pour structures élevées**   high rise building regulations

réglementé *adj*   limited [e.g. access to a motorway, etc]

réglementé, accès *m*   limited access

réglet *m*   plane moulding separating two curved mouldings

régler *vb*   adjust, control, regulate or set [a clock, equipment, a thermostat, a valve, etc], decide (on), fix up or settle on [a date, a programme, etc], grade [ground, fill, etc, to required level], pay or settle [an account], settle or sort out [an affair, an argument, a problem]

régles *f,pl* **de l'art, dans les**   in accordance with good practice

Règles *f,pl* **Nationales d'Urbanisme**   national planning regulations [in France]

réglet *m*   carpenter's rule

régleur *m*   adjuster, regulator

regonder *vb*   rehang [a door or shutter on pin hinges]

regratter *vb*   re-dress [weathered stonework to give it a new appearance]

régressif *adj*   regressive

régularisation *f*   regulation [of flow of a watercourse]

régulateur *m*   control(ler), regulating valve, regulator, throttle valve

régulateur *m* **à cascade**   cascade controller

régulateur *m* **à deux positions**   two-position control(ler)

régulateur *m* **à haute pression**   high pressure regulator

régulateur *m* **à programme**   programme controller

régulateur *m* **d'expansion**   expansion regulator

régulateur *m* **d'humidité**   humidity controller

régulateur *m* **de basse pression**   low-pressure regulator

régulateur *m* **de débit**   delivery/discharge/output regulator

régulateur *m* **de gaz**   gas governor

régulateur *m* **de jet**   nozzle regulator

régulateur *m* **de la pression d'air**   air pressure regulator

régulateur *m* **de pression**   pressure controller/regulator

régulateur *m* **de pression d'aspiration**   back pressure regulator

régulateur *m* **de tension**   voltage regulator

régulateur *m* **de tirage**   draught stabilizer

régulateur *m* **de vitesse**   speed regulator

régulateur *m* **esclave**   slave/sub-master controller

régulateur *m* **par tension variable**   variable voltage regulator

régulateur *m* **pas à pas**   step controller

régulation *f*   adjustment, control, regulation

régulation *f* **à pointe de consigne fixe**   fixed setting control

régulation *f* **à valeur de consigne fixe**   constant value control

régulation *f* **automatique**   automatic control/regulation

régulation *f* **de débit d'air variable**   variable air volume control

régulation *f* **de la circulation**   traffic control

régulation *f* **de local**   room control

régulation *f* **de puissance**   capacity/power control

régulation *f* **de(s) débit(s)**   flow regulation

régulation *f* **en cascade**   cascade control

régulation *f* **manuelle**   hand/manual control/regulation

régulation *f* **numérique directe**   direct digital control

régulation *f* **par tout ou rien**   on–off control

régulation *f* **pas à pas**   step control

régulation *f* **séquentielle**   sequence control

réguler *vb*   control, regulate

réhabilitation *f*   rehabilitation

réhabiliter *vb*   rehabilitate, renovate, restore

rehaussé *adj*   raised [heightened]

rehaussement *m*   heightening, raising

rehausser *vb*   pick out [a colour], raise [heighten, increase height of]

reins *m,pl*   haunches [of an arch], sides [of a vault]

rejet *m* **en mer**   sea outfall

rejointoiement *m*   re-pointing [the action, and its result]

relai(s) *m*   relay [electrical]

relai(s) *m* **d'appel**   call relay

relai(s) *m* **de contrôle**   alarm/control relay

relai(s) *m* **de positionnement**   positioning relay

relai(s) *m* **de protection**   protective/safety relay

relai(s) *m* **de protection primaire**   primary protection relay

relai(s) *m* **de sécurité**   safety relay

relai(s) *m* **inverseur**   changeover relay

relai(s) *m* **temporisé**   time delay relay

relai(s) *m* **thermique**   thermal relay

relais *m*   distance used as a basis for calculation of haulage costs, sandbank, flats, site for dumping earth, provisional

relais, travail *m* **par**   shift work

relancer *vb*   replace a degraded brick or stone in masonry

relancis *m*   replacing a degraded brick or stone in masonry

relatif *adj*   relative [density/position/size/value etc]

relatter *vb*   rebatten [a roof]

relayer *vb*   relieve [i.e. to take over from]

se relayer *vb*   work in shifts

relaxation *f*   relaxation

relevé *m*   abstract or summary

relevé *m* **à distance**   remote reading [e.g. of meters]

relevé *m* **d'étanchéité**   height (of waterproof flashing above edges of a terrace)

relevé *m* **de compte**   statement of account [balance in an account]

relevé *m* **de comptes**   statement of account [balance of accounts unpaid]

relevé *m* **de l'ouvrage**   dimensional survey of a structure

relevé *m* **de terrain**   ground survey

relèvement *m*   lifting up, raising [increasing], righting, setting upright

relèvement *m* **de**   rise in [wages, etc]

relever *vb*   bank [road at a bend], pick up [an error], raise [increase], **raise** [heighten, increase height of]

relever *vb* **un compteur**   read a meter

relief *m*   higher/raised ground [e.g. hills, mountains], relief [projection from a surface]

relief, carte *f* **en**   relief map

relief, en   raised *adj*

relief *m* **tabulaire**   plateau

relier *vb*   connect, link up, reconnect

reloger *vb*   rehouse

remaniage *m*   altering, changing, reshaping, rebattening [of a roof]

remanié *adj*   disturbed [of soil, a soil sample]

remanié, échantillon *m* **non**   undisturbed sample

remanié, sol *m* **non**   undisturbed soil

remaniement *m*   alteration, change, disturbance

remanier *vb*   alter, disturb [soil, a sample, etc], rehandle [materials], relay [paving, etc], retile [roof]

remarque *f*   comment

rembarrure *f*   pointing of a ridge [mortar or plaster]

remblai, en   on embankment

remblai *m*   (back)fill, hardcore, made ground, embankment

remblai, crête *f* **de**   embankment crest

remblai, épaulement *m* **de**   embankment edge

remblai, matériau *m* **de**   fill(ing) material

remblai, pied *m* **de**   embankment base or foot

remblai *m* **compacté**   compacted fill

remblai *m* **(en) tout venant**   random fill

remblai *m* **et déblai** *m*   cut and fill

remblai *m* **pierreux**   rockfill

remblaiement *m*   backfilling [action of]

remblayer *vb*   backfill, bank (up), fill, embank

remblayeuse *f*   backfiller

rembourser *vb*   reimburse, repay

rembourser *vb* **l'emprunt**   repay the loan

remède *m*   cure, remedy [e.g. for a defect]

remédier *vb* **(à qch)**   cure, put right or remedy (sth) [e.g. a defect]

remettre *vb*   hand over, postpone, put back [postpone], put back in place

remettre *vb* **à neuf**   refurbish

remettre *vb* **de**   top up with

remembrement *m*   reallocation of land holdings to improve use [France]

remise *f*   allowance, discount, reduction [in cost or price], delivery, handing over [of a letter, parcel, etc], putting back [of sth in its place], remittance [of money], shed

remontée *f* **d'eau**   percolation of water to the surface of setting concrete

remontée *f* **mécanique**   mechanical lift [ski lift]

remorque *f*   trailer

remorque *f* **à déversement par le fonds**   bottom-dump trailer

remorque *f* **basculante/benne**   tipping trailer

remorque *f* **plateau/plate-forme**   flat-bed trailer

remorque *f* **porte-engins**   plant trailer

remorque *f* **surbaissée**   low loader

remorquer *vb*   tow [a vehicle]

remous *m*   eddy, swirl

rempart *m*   rampart

rempiéter *vb*   consolidate [for the footing of a wall]

remplacer *vb*   renew, replace

remplacement *m*   replacement

remplacement *m* **de filtre**   filter replacement

remplacement *m* **en groupe**   group replacement [of lamps]

remplir *vb*   fill in [e.g. a form]

remplir *vb* **avec**   top up with

remplir *vb* **(de)**   fill (with)

remplissage *m*   filling (up)

remplissage, matériau *m* **de**   backfill(ing) material

remplissage *m* **de tranchée**   trench backfilling

remploi *m*   re-use

rémunération *f*   remuneration

renard *m*   leak [the hole itself, in a hydraulic structure, a boiler]

rencontre *f*   meeting [individuals]

rendement *m*   efficiency, output, performance [output, yield], productivity

rendement *m* **(de)**   return (on), yield (of) [an investment]

rendement *m* **d'un filtre**   filter (separation) efficiency

rendement *m* **de chaudière**   boiler efficiency

rendement *m* **de combustion**   combustion efficiency

rendement *m* **de compression**   compression efficiency

rendement *m* **de condenseur**   condenser output

rendement *m* **de conversion**   conversion efficiency

rendement *m* **du brûleur**   burner efficiency

rendement *m* **énergétique**   energy efficiency ratio

rendement *m* **global**   overall efficiency

rendement *m* **maximal**   maximum/peak efficiency/ output

rendement *m* **mécanique**   mechanical efficiency

rendement *m* **volumétrique**   volumetric efficiency

rendez-vous *m*   appointment [e.g. for a meeting]

rendre *vb* **impropre à sa destination**   make unfit for its purpose

renduire *vb*   refill [a surface for second coat of paint]

renfoncer *vb*   knock further in [e.g. a nail], set back [recess]

renforcement *m*   reinforcement, strengthening [general terms], trussing [of a beam]

renforcement, barre *f* **de**   reinforcement bar

renforcer *vb*   reinforce, strengthen, tighten [restrictions, etc], truss [a beam]

renformir *vb*   apply preliminary rendering

renformis *m*   preliminary rendering

renfort *m*   reinforcement, stiffener, strengthening piece [supplementary, around holes, etc]

renfort, plaque *f* **de**   reinforcing/ stiffening plate

renfort, tôle *f* **de**   reinforcing/stiffening plate

reniflard *m*   blow-valve [on a boiler], breather pipe

reniveler *vb*   re-level, top up [battery, tank, etc]

renouvellement *m*   renewal

renouvellements *m,pl* **d'air à l'heure**   air changes per hour, ventilation rate

renouveler *vb*   renew, replace

rénovation *f*   renewal, renovation

rénovation *f* **urbaine**   urban renewal/renovation

rénové *adj*   refurbished, rehabilitated, renovated, restored

rénover *vb*   rehabilitate, renovate, restore

renseignement *m*   information [piece of]

renseignements *m,pl*   enquiries, information, particulars

renseignements, demander *vb* **des**   make enquiries, seek information

renseignements, service *m* **des**   directory enquiries [telephone]

rentabilité *f*   cost-effectiveness, profitability

rentabilité *f* **économique**   economic viability

rentable *adj*   cost-effective

rente *f*   annuity, pension [income]

rentrant, angle *m*   re-entrant angle

renversement *m*   change of direction, reversal, reversing, overturning

renversement *m* **de charge**   load reversal

renversement *m* **de l'écoulement**   flow reversal

renvoi *m*   reference [of a matter to an authority, etc], return [of goods, etc]

renvoi, levier *m* **de**   reversing lever

renvoi, poulie *f* **de**   return pulley

renvoi *m* **d'angle**   angle gearbox, bevel gear

renvoi *m* **d'eau**   arrangement for throwing water off from a roof

répandage *m*   spreading [of chippings, tar, etc]

répandeuse *f*   spreader [road works], sprayer [for bitumen, road emulsion, etc]

répandre *vb*   spread [distribute: loads, sand, chippings, etc]

se répandre *vb*   spread [flood, river, water]

reparage *m*   fettling

réparation *f*   repair

réparer *vb*   fix, mend, patch, repair

réparti *adj*   distributed

réparti *adj* **en équerre**   L-shaped

répartir *vb*   distribute [e.g. loads]

répartiteur *m*   distributing frame [telephone connections], distribution block or board

répartiteur *m* **d'écoulement**   flow equalizer

répartition *f*   distribution [e.g. of loads, stresses], division [distribution], sharing (out)

répartition *f* **de la charge**   load distribution

répartition *f* **des contraintes**   stress distribution

répartition *f* **triangulaire de charge**   triangular load distribution

réparure *f*   primer, thick white [prior to painting]

repassage *m*   raking [of a gravel path or drive], re-cutting [of the edge of paving slabs/stones]

repasser *vb*   recoat [of renderings, finishes]

repérage *m*   marking

repère *m*   indicator, mark(er), landmark, reference point

repère, point *m* **de**   (setting out) datum point

repère *m* **de niveau**   bench mark

repère *m* **de nivellement**   (reference) level marker

repère *m* **fixe**   monument [fixed datum point]

repérer *vb*   mark out

répéter *vb*   do again, repeat

repiquage *m*   hacking [of a rendering, etc], relaying [of individual paving slabs], shallow excavation

repiquer *vb*   hack [a rendering or surface finish], relay [individual paving slabs]

replacer *vb*   put back in place, replace

replanir *vb*   plane [flooring]

replanissage *m*   flattening, planing [of flooring]

replanir *vb*   flatten

replâtrage *m*   replastering [the action and its result]

replâtrer *vb*   replaster

repliement *m*   closing down of site

réplique *f*   justification for further revision of a claim

répondeur *m*   telephone answering machine

réponse *f*   reply, response

réponse, délai *m*/**retard** *m* **de**   lag

réponse *f* **de contrôle**   control response

réponse *f* **en fréquence**   frequency response

reporter *vb*   postpone, put off [defer]

repos *m*   bearing [of a beam or column]

repose *f*   refixing, relaying

repoussoir *m*   long chisel [for cutting mouldings in stonework], tool for driving out tre(e)nails

reprendre *vb*   strengthen [cracked or unstable work]

reprendre *vb* **en sous-œuvre**   underpin

reprendre *vb* **une charge**   take up a load

représentant *m*   representative

représentant *m* **des travailleurs**   staff or workers' representative

représentant *m* **syndical**   shop steward, trade union representative

représentation *f*   representation

reprise *f*   partial rebuilding of damaged work

reprise *f* **de bétonnage**   restart of concrete pouring

reprise *f* **en sous-œuvre**   underpinning

reprise *f* **par encastrement/incrustation**   strengthening [by replacing blocks of material, e.g. stone]

reproduire *vb*   copy, duplicate, reproduce [a drawing, etc], repeat [an error]

reprofileur *m* **(à froid)**   pavement milling machine, planer or profiler

reprofileuse *f*   pavement milling machine, planer or profiler

réseau *m*   grid [electrical supply], network, system [pipes, cables, etc]

réseau *m* **alimenté en retour**   distribution system fed from both ends

réseau *m* **circulaire**   ring distribution system

réseau *m* **d'alimentation**   supply network or system

réseau *m* **d'alimentation en eau**   water supply system

réseau *m* **d'assainissement**   sewerage system

réseau *m* **de canalisations**   pipe network

réseau *m* **de communication**   communication network

réseau *m* **de conduits**   duct system

réseau *m* **de distribution**   delivery or distribution network

réseau *m* **de distribution d'air**   ductwork system

réseau *m* **de drains/d'évacuation**   drainage network or system

réseau *m* **de réserve**   back-up system

réseau *m* **enterré**   buried main
Réseau *m* **Express Régional (RER)**   express transport system in Paris region
réseau *m* **ferroviaire**   railway network
réseau *m* **local**   local area network (LAN)
réseau *m* **maillé**   branching system [no loops], distribution system fed from both ends
réseau *m* **primaire**   primary distributor [roads]
réseau *m* **primaire en boucle**   ring main
réseau *m* **ramifié**   distribution system fed from one end, network system [with loops]
réseau *m* **routier**   road network/system
réseau *m* **secondaire**   district distributor [roads]
réseau *m* **séparatif/unitaire**   combined/separate sewer(age) system
réseau *m* **smillé**   distribution system fed from both ends
réseau *m* **téléphonique**   telephone network
réservation *f*   box out, hole or pocket [precast in structure, for services], pipe or services sleeve [hole in concrete, masonry, etc], zone allowed or reserved for finishes, services, etc
réserve, de   spare, standby *adj*
réserve *f*   reserve [stock, or as in nature reserve], stock, store [stock or reserve], storehouse, storeroom
réserve *f* **d'eau**   water reserve
réserve *f* **de carburant(s)**   fuel tank
réserve *f* **de matériaux**   stockpile [of materials]
réserve *f* **naturelle**   nature reserve/sanctuary
réservoir *m*   container [tank, etc], tank, reservoir [structural rather than dammed]
réservoir *m***, barrage-**   reservoir [dammed]
réservoir *m* **à air**   air receiver
réservoir *m* **à air comprimé**   compressed air tank/vessel
réservoir *m* **à carburant**   fuel tank [of a vehicle]
réservoir *m* **à filtration**   filter tank
réservoir *m* **collecteur**   collecting tank/vessel
réservoir *m* **d'accumulation**   storage reservoir
réservoir *m* **d'air**   air chamber/container/vessel, etc
réservoir *m* **d'air anti-bélier**   pressure/surge tank [anti-water hammer]
réservoir *m* **d'air comprimé**   compressed air tank/vessel
réservoir *m* **d'alimentation**   supply tank
réservoir *m* **d'arrêt**   balancing reservoir
réservoir *m* **d'eau à chauffe directe**   direct cylinder
réservoir *m* **d'eau de refroidissement**   cooling water tank
réservoir *m* **d'égalisation de pression**   surge tank
réservoir *m* **d'épuration**   filter tank
réservoir *m* **d'équilibre**   balance tank
réservoir *m* **de carburant**   fuel tank
réservoir *m* **de chasse (d'eau)**   toilet flush cistern
réservoir *m* **de combustible**   fuel tank
réservoir *m* **de décharge**   discharge tank or vessel
réservoir *m* **de décompression**   blow-down tank
réservoir *m* **de dilatation**   expansion tank or vessel
réservoir *m* **de mazout**   oil tank
réservoir *m* **de mélange**   mixing tank
réservoir *m* **de secours**   reserve tank
réservoir *m* **de stockage**   storage tank

réservoir *m* **égalisateur**   equalizer tank
réservoir *m* **en acier**   steel tank
réservoir *m* **en charge**   header tank
réservoir *m* **hydraulique**   hydraulic reservoir
réservoir *m* **pour autovaporisation**   flash tank
réservoir *m* **sous pression**   pressure vessel
réservoir *m* **supplémentaire**   additional, back-up or reserve tank
réservoir *m* **tampon**   buffer tank
réservoir-nourrice *m*   feed tank
résidence *f*   dwelling, residence
résidence *f* **principale**   principal residence
résidence *f* **secondaire**   second home/residence
résidence *f* **troisième âge**   retirement home
résidentiel *adj*   residential
résidu *m*   remainder, residue
résidu *m* **de compte**   amount still owing, balance [of an account]
résidus *m,pl* **(de mine)**   tailings
résiduel *adj*   residual
résiliation *f*   annulment, cancellation [of an insurance policy]
résilience *f*   impact strength, resilience
résilient *adj*   resilient [of a material]
résilier *vb*   cancel [a contract of insurance]
résille *f*   came [of a leaded light], leading
résine *f*   resin
résineux *adj*   resinous [of timber]
résistance *f*   heating element [electrical], resistance, strength, resistor
résistance, à haute   high-strength, high tensile
résistance *f***, classe** *f* **de**   strength class
résistance *f* **à l'altération**   weathering resistance
résistance *f* **à l'éclatement**   bursting strength
résistance *f* **à l'écoulement**   flow resistance
résistance *f* **à l'écrasement**   crushing strength
résistance *f* **à la chaleur**   heat resistance
résistance *f* **à la compression**   compressive strength
résistance *f* **à la corrosion**   corrosion resistance
résistance *f* **à la fatigue**   fatigue resistance
résistance *f* **à la flexion**   bending strength
résistance *f* **à la pénétration**   penetration resistance
résistance *f* **à la rupture**   breaking strength
résistance *f* **à la traction**   tensile strength
résistance *f* **aéraulique**   air flow resistance
résistance *f* **au choc**   impact resistance or strength
résistance *f* **au cisaillement**   shear resistance or strength
résistance *f* **au feu**   fire resistance
résistance *f* **au flambement**   buckling strength
résistance *f* **au froid**   cold resistance
résistance *f* **au frottement**   frictional resistance [to friction]
résistance *f* **au gel**   frost resistance
résistance *f* **au glissement**   sliding resistance
résistance *f* **au passage de l'air**   air flow resistance
résistance *f* **au vieillissement**   durability
résistance *f* **aux efforts alternés**   fatigue limit or strength
résistance *f* **aux intempéries**   weather resistance
résistance *f* **de démarrage**   starting resistance

résistance *f* **d'isolement**   insulation resistance [electrical]
résistance *f* **d'un conducteur**   resistance of a conductor
résistance *f* **de frottement**   frictional resistance [arising from friction]
résistance *f* **de l'air**   air resistance
résistance *f* **du mortier**   mortar strength
résistance *f* **du pieu au battage**   driving resistance of the pile
résistance *f* **électrique**   electrical resistance
résistance *f* **en traction**   strength, tensile
résistance *f* **mécanique**   mechanical resistance
résistance *f* **par friction**   flow resistance [frictional]
résistance *f* **résiduelle**   residual strength
résistance *f* **sur cube**   cube strength
résistance *f* **thermique**   thermal resistance
résistance *f* **thermique superficielle**   film resistance
résistant *adj*   strong
résistant, moment *m*   resisting moment
résistant à l'écrasement   crushproof
résistant à la chaleur   heatproof, heat-resistant/-resisting
résistant à la corrosion   corrosion resistant
résistant au feu   fire-resistant
résistant au gel   frost-resistant
résistant aux acides   acid-resistant
résistant aux alcali(n)s   alkali-resistant
résistant aux intempéries   weather-resistant, weatherproof/weathertight
résistivité *f*   resistivity
résistivité *f* **thermique**   thermal resistivity
résonance *f*   resonance
respiration *f*   breathing
responsabilité *f*   responsibility
responsabilité *f* **civile décennale**   10-year civil liability [on major work in France]
responsable *adj*   responsible
ressaut *m*   hydraulic jump, projection or shelf [from/above a surface], step [in metallic roofing, a gutter, etc]
ressaut, faire *vb*   project [stick out]
resserrer *vb*   tighten (up) [a screw, a nut]
se resserrer *vb*   narrow [get or grow narrower]
ressort *m*   competence, scope, spring [mechanical]
ressort, de son   within his/her province/responsibility/ scope
ressort *m* **à lames**   leaf spring
ressort *m* **annulaire**   circular spring
ressort *m* **de rappel**   return spring
ressource *f*   resource
ressources *f,pl* **en eau**   water resources
ressources *f,pl* **en hommes et en matériel**   resources of men and materials
ressources *f,pl* **naturelles**   natural resources
ressuage *m*   bleed-through, emergence of laitance, sweating of fresh concrete
ressuer *vb*   sweat [develop surface droplets], sweat [a pipe joint]
restaurateur *m*   restorer
restauration *f*   restoration
restauré *adj*   rehabilitated, restored

restaurer *vb*   rehabilitate, restore
restaurer, à   for restoration
reste *m*   remainder
restes *m,pl*   remains
restes *m,pl* **d'ouvrage**   remains of existing buildings
restitution *f*   recreation of former state, restoration [of a building]
résultant *adj*   resultant or resulting
résultante *f*   resultant, vector sum of forces [resultant]
résultat *m*   outcome, result
résultat *m* **d'essai**   test result
résultats, dispersion *f* **des**   scatter of results
résultats *m,pl* **des marchés**   tender awards or results
résumé *m*   summary
résurgence *f*   spring [of water]
retailler *vb*   re-cut [stone], remake [dismantle and reassemble joinery]
retard *m*   delay, (time) lag [delay]
retard *m* **de commencement**   delay in starting [e.g. of work]
retard *m* **de réponse**   lag [in response]
retard *m* **thermique**   thermal lag
retardateur *m* **de prise**   retarding agent [for concrete]
retarder *vb*   put or set back [a clock, a programme], delay, hinder, hold up, make late
retarder *vb* **de**   put back by [a clock]
retarder *vb* **la croissance (de)**   inhibit the growth (of)
retardeur *m*   retarder
retardeur *m* **de vapeur**   vapour retarder
rétention *f*   retention [e.g. of water]
rétention, eau de *f*   absorbed or retained water
retenue *f*   deduction [from salary, wages], impounding dam
retenue *f* **d'eau**   storage of water [by damming]
retenue *f* **de garantie**   retention [until final acceptance of works]
réticulé *adj*   in the form of a network
retirable *adj*   withdrawable
retiré *adj*   secluded
retirer *vb*   remove, take off or out, withdraw
retombée *f*   spring [of an arch or vault]
retoucher *vb*   touch up [paintwork, photos]
retour *m*   return [back, in space or time]
retour *m* **à la masse**   earth (return)
retour *m* **à la terre**   earth (return)
retrait *m*   contraction, shrinkage
retrait, contrainte *f* **de**   shrinkage stress
retrait, façade *f* **en**   set back facade or frontage
retraite *f*   offset between faces of adjoining walls, retirement, retreat, space created by setting back of part of a building
retraite, maison *f* **de**   old peoples'/retirement home
rétrécir *vb*   narrow [make narrower], shrink
rétrécissement *m*   contracting, narrowing [e.g. between hearth and mantlepiece], shrinkage
rétro, pelle *f* **équipée en**   drag shovel
rétro-chargeuse *f*   front-end loader that can discharge at rear
rétro-réfléchissant, plot *m*   reflecting/reflective stud

retroussis *m* banking [of ground]
réunion *f* meeting [several people]
réussir *vb* succeed [business, project, etc]
réutilisation *f* re-use
réutilisation *f* **des déblais en remblais** cut and fill
réutiliser *vb* re-use
revanche *f* vertical difference in height between maximum groundwater level and the top of an impermeable casing, e.g. tanking
revendre *vb* resell
revente *f* resale
revenu *m* income, tempering [of steel]
revenus *m,pl* **locatifs** rental income
réverbération *f* reverberation
réverbère *m* reflector, street lamp/light
réversible *adj* reversible
revêtement *m* coating, covering [interior wall, floor], lining [covering], cladding, (sur)facing or rendering [exterior wall], surfacing [of roads]
revêtement *m* **acoustique** acoustic insulation
revêtement *m* **anti-dérapant** anti-skid surface treatment/surfacing
revêtement *m* **antifriction** anti-friction lining
revêtement *m* **bétonné** concrete/rigid pavement
revêtement *m* **bitumineux** asphaltic/bituminous surfacing
revêtement *m* **de bitume** bitumen coating
revêtement *m* **de chaussée** carriageway or road surfacing
revêtement *m* **de feutre** felt covering
revêtement *m* **de murs** wall covering
revêtement *m* **de pont** bridge decking or surfacing
revêtement *m* **de protection** protective coating
revêtement *m* **des tubes** pipe wrapping
revêtement *m* **du sol** flooring [surfacing, covering]
revêtement *m* **du trottoir** footpath or pavement surfacing
revêtement *m* **hydrocarboné** bituminous surfacing
revêtement *m* **mural** wall covering
revêtement *m* **réfractaire** refractory lining
revêtement *m* **thermoplastique** thermoplastic coating
revêtir *vb* surface [a road]
revêtu *adj* **de** clad or covered in/with [surfaced, coated]
reviser/réviser *vb* audit [accounts], overhaul or service [equipment, plant], reappraise, re-examine, revise [proofs, documents, drawings]
revision/révision *f* examination, revision [of accounts], inspection, testing, overhauling, proof-reading
revision/révision *f* **des prix** cost or price revision
revoir *vb* inspect, examine, revise [accounts], read [proofs], re-examine, revise [documents, drawings]
révolution *f* revolution [turn]
revue *f* **professionnelle** trade journal
rez-de-chaussée *m* ground floor, street level
rhéologie *f* rheology
rhéostat *m* rheostat
rhéostat *m* **de champ** field rheostat

rhéostat *m* **de démarrage** starting resistance [the apparatus], rheostat
ridage *m* wrinkling [paint defect], curtain
rideau *m* **coupe-feu** fire safety curtain
rideau *m* **d'air** air curtain
rideau *m* **d'air chaud** warm air curtain
rideau *m* **d'eau** water curtain [fire barrier]
rideau *m* **d'étanchéité** impermeable curtain [in groundworks]
rideau *m* **d'injection** grout curtain
rideau *m* **de** screen of [e.g. smoke, trees, etc]
rideau *m* **(de cheminée)** fireplace curtain [slides down to block off]
rideau *m* **de fer** safety curtain [in a theatre]
rideau *m* **de/en palplanches** sheet pile screen or wall
rideau *m* **de pulvérisation** spray curtain
rideau *m* **de scène** stage curtain
rideau *m* **de soutènement** bulkhead [for ground support]
riflard *m* jack plane, roughing file
rifler *vb* smooth [timber with a jack plane]
rigide *adj* non-yielding, rigid, stiff [rigid]
rigidité *f* rigidity, stiffness
rigidité *f* **à la torsion** torsional rigidity
rigidité *f* **diélectrique** dielectric strength
rigole *f* channel [trench or ditch], drain [open], trench
rigole, fouille *f* **en** trench excavation
rigole *f* **couverte** enclosed channel or ditch
rigole *f* **d'évacuation** drainage channel
rigole *f* **ouverte** open channel or ditch
rincer *vb* flush, rinse, wash [rinse]
ringard *m* poker [for a boiler]
ringard *m* **à crochet** rake [for a boiler]
ripage *m* small sideways movement of a part of a structure, scraping of stone to smooth or polish it
ripe *f* scraper [for stone]
riper *vb* displace sideways, polish or scrape [stone]
rippeur *m* ripper [groundworks]
risberme *f* scour protection
risque *f* risk
risque *f* **de contamination** danger/risk of contamination
risquer *vb* risk
rissolé *adj* heat-stained [brown/yellow]
rivé *adj* **à chaud** hot riveted
rivé *adj* **à froid** cold riveted
rive *f* bank [of a river], shore [of a lake, sea], edge [of a roof, a piece of wood, etc], verge [roof]
rive *f* **de chaussée** *f* carriageway edge
rive *f* **de tête** upper edge of a roof panel
rive *f* **latérale** side edge of a roof panel
river *vb* clench or clinch [a nail], rivet (together)
riverain *adj* bordering [e.g. on a road], riparian
riverain *m* resident [living along a road or river], riparian proprietor
riveraine, parcelle *f* bordering property or plot
rivet *m* rivet
rivet *m* **à tête fraisée** countersunk rivet
rivet *m* **à tête ronde** button or round head rivet
rivetage *m* riveting
riveteuse *f* riveting machine

rivets, assemblage *m* **à**   riveted assembly or joint

riveur *m*   riveter [worker]

riveuse *f*   riveting tool

rivière *f, voir aussi* fleuve   river

rivière, prise *f* **en**   river intake

rivoir *m*   riveting hammer

rivure *f*   rivet head as set during riveting process, riveted assembly, connection or joint, riveting [the result]

rivure *f* **en quinconce**   staggered riveting

rivure *f* **fraisée**   flat countersunk rivet head

rivure *f* **sphérique**   snap head as set during riveting

roanne/rouanne *f*   carpenter's auger, scriber for wood

robinet *m*   cock, tap, valve [generally screw down type]

robinet *m* **à bec orientable**   swivel tap

robinet *m* **à boisseau**   cock valve

robinet *m* **à boule**   plug valve

robinet *m* **à clé**   key operated valve

robinet *m* **à double siège**   double-seated valve

robinet *m* **à équilibrage**   balancing valve [in radiator system]

robinet *m* **à flotteur**   ballcock, float valve

robinet *m* **à membrane**   diaphragm valve

robinet *m* **à papillon**   butterfly valve

robinet *m* **à passage direct**   direct flow valve, through-way cock

robinet *m* **à pointeau**   needle valve

robinet *m* **à soupape**   globe valve

robinet *m* **à trois voies**   three-way valve

robinet *m* **chef (de gaz)**   gas stop tap [for a whole building]

robinet *m* **d'air**   air cock

robinet *m* **d'alimentation**   feed check valve

robinet *m* **d'arrêt**   stopcock, stop tap

robinet *m* **d'arrêt d'urgence**   emergency stop valve

robinet *m* **d'arrêt général**   main stop valve

robinet *m* **d'arrêt sphérique**   globe valve

robinet *m* **d'eau**   water tap

robinet *m* **d'échantillonnage**   sampling cock

robinet *m* **d'écoulement**   outlet valve/cock

robinet *m* **d'équilibrage**   balancing valve

robinet *m* **d'essai**   test cock

robinet *m* **d'incendie**   fire valve

robinet *m* **d'incendie armé (RIA)**   fire hose reel, fire point

robinet *m* **d'isolement**   isolating/isolation valve

robinet *m* **de branchement**   junction cock

robinet *m* **de chasse**   flush valve

robinet *m* **de commande d'appareil à gaz**   gas appliance isolating valve

robinet *m* **de contrôle**   test cock

robinet *m* **de décharge**   blowdown valve

robinet *m* **de dérivation**   diverter valve

robinet *m* **de fermeture**   stopcock

robinet *m* **de gaz**   gas cock/tap/valve

robinet *m* **de puisage**   draw-off tap/valve

robinet *m* **de purge**   bleed(er) cock/tap/valve, purge valve

robinet *m* **de réglage**   (flow) regulating cock/valve

robinet *m* **de vidange**   drain cock/tap/valve

robinet *m* **de vidange de boue**   sludge cock

robinet *m* **décompresseur**   air relief cock, pressure relief valve

robinet *m* **électro-magnétique**   electromagnetically operated valve

robinet *m* **flotteur**   ballcock, float valve

robinet *m* **jauge**   calibrated tap [used prior to advent of flow meters]

robinet *m* **mélangeur**   mixer tap [twin tap, single spout]

robinet *m* **modulateur**   control or modulating valve

robinet *m* **principal**   main cock/valve/tap

robinet *m* **principal de gaz**   main gas cock/valve/tap

robinet *m* **purgeur**   drain/purging cock

robinet *m* **solénoïde**   solenoid valve

robinet-vanne *f*   gate/sluice-valve

robinetterie *f*   fittings [taps]

robinetterie *f* **de douche**   shower fittings

rocade *f*   bypass [relief road], ring road

rocaille *f*   loose stones, rocaille [decoration in rococo architecture], rock garden, rockery

roche *f*   rock [in general]

roche *f* **éruptive**   eruptive rock

roche *f* **métamorphique**   metamorphic rock

roche *f* **naturelle**   natural rock

roche *f* **saine**   bedrock

roche *f* **sédimentaire**   sedimentary rock

roche *f* **stratifiée**   sedimentary rock

roches *f,pl* **compactes**   hard rock

roche-mère *f*   parent rock

rocher *m*   boulder, rock [block of stone]

rocher *m* **intact**   sound rock

roches, mécanique *f* **des**   rock mechanics

rochet *m*   ratchet

rocheux *adj*   rocky

roder *vb*   grind (in) [generally with abrasive powder]

rogner *vb*   clip [pare edge], pare (down), trim [edges]

rognon *m* **de silex**   flint nodule

rompre *vb*   break [a contract, etc], break off [meeting, negotiations, etc]

se rompre *vb*   break [fall apart], fail [by breaking up or into pieces]

ronce *f*   barbed wire, bramble

rond *adj*   round *adj*

rond *m*   circle, ring, round [e.g. moulding or steel bar], washer [*voir aussi* rondelle]

rond *m* **à béton**   round reinforcement (bar)

rond *m* **à béton crénelé**   deformed round reinforcement (bar)

rond *m* **(en acier)**   round (steel) bar

rond-point *m*   traffic roundabout

rondelle *f*   washer

rondelle *f* **à denture extérieure**   lock washer (external tooth)

rondelle *f* **à denture intérieure**   lock washer (internal tooth)

rondelle *f* **à ressort**   spring washer

rondelle *f* **Belleville**   bellows washer

rondelle *f* **d'étanchéité**   seal [washer], sealing washer

rondelle *f* **de butée**   thrust collar or washer

rondelle *f* **de garniture**   packing washer

rondelle *f* **de robinet**   tap washer

rondelle *f* **Grower** lock washer [square section]
rondin *m* roundwood prop
ronfleur *m* buzzer
rogner *vb* trim the edges of
rogner *vb* **les dépenses/frais** cut down/reduce the expenses/costs
rognure *f* offcut [from trimming the edges of something]
rosace *f* ceiling rose
rosée *f* condensation, dew
rosette *f* rose [of a door handle or bell push]
rossignol *m* wedge for jamming a tenon in an oversize mortise
rotatif *adj* rotary/rotating
rotation *f* turn, rotation
rotation *f* **à droite** clockwise rotation
rotation *f* **à gauche** anti-clockwise rotation
rotonde *f* rotunda
rotor *m* rotor
rotule *f* ball-and-socket joint
rotule *f* **plastique** plastic hinge
rouanne *f* carpenter's auger, scriber for wood
rouanner *vb* scribe [wood]
roue *f* impeller, wheel
roue *f* **de chaleur** heat wheel
roue *f* **de secours** spare wheel
roue *f* **de ventilateur** fan impeller
roue *f* **mobile** pump impeller
rougeurs *m,pl* stains indicating rot in timber
rouillé *adj* rusty
rouille *f* rust
rouiller *vb* go or make rusty, corrode or rust [sth]
se rouiller *vb* corrode or rust, get rusty
roulant *adj* moving, rolling, sliding, travelling
roulant, escalier *m* escalator
roulante, rampe *f* moving ramp
roulant, tapis *m* moving walkway
roulé *adj* rounded [of a sand or gravel particle]
roule *f* roller [wood, for moving heavy loads]
rouleau *m* roll [e.g. of wallpaper], roller [painter's, road, etc]
rouleau, compacter *vb* **au** roll or compact
rouleau *m* **à grille** grid roller
rouleau *m* **à guidage à main** pedestrian-controlled roller
rouleau *m* **à peindre** paint roller
rouleau *m* **à pieds coniques** tapered foot roller
rouleau *m* **à pieds dameurs** tamping foot roller
rouleau *m* **à pieds de mouton** sheepsfoot roller
rouleau *m* **à pochoir** stippling roller
rouleau *m* **à pneus** rubber-tyred roller
rouleau *m* **(compacteur) à pneu(s)** rubber-tyred compactor or roller
rouleau *m* **compresseur** scroll compressor, road roller, steamroller
rouleau *m* **cylindre** cylinder roller
rouleau *m* **dameur** tamping roller
rouleau *m* **de colleur** seam roller [for wallpaper]
rouleau *m* **entraîneur** feed roller
rouleau *m* **lisse** smooth-wheeled roller

rouleau *m* **monobille automoteur** self-propelled single drum roller
rouleau *m* **pied de mouton** sheepsfoot roller
rouleau *m* **transporteur** conveyor roller
rouleau *m* **vibrant** vibrating/vibratory roller
rouleau *m* **vibrant à double bille** double-drum vibratory roller
rouleau *m* **vibrant à guidage à main** pedestrian-controlled vibratory roller
rouleau *m* **vibrant tracté** towed vibrating/vibratory roller
roulement *m* rotation [e.g. in shift working]
roulement *m* **à aiguilles** needle bearing
roulement *m* **à billes** ball bearing
roulement *m* **à rouleaux** roller bearing
roulement *m* **conique** conical bearing
rouler *vb* roll [to compact]
roulette *f* wallpaper seam roller
roulure *f* ring-shake [timber defect]
route *f,pl* road, route [itinerary]
route, catégorie *f* **de** highway classification
route, type *m* **de** highway type
route *f* **à grande circulation** major road
route *f* **à sens unique** one-way road
route *f* **à deux/trois/quatre bandes/voies** two-/three-/four-lane road
route *f* **collectrice** district or local distributor road
route *f* **d'accès** access or approach road
route *f* **départementale** secondary road, county road
route *f* **en remblai/déblai** road on embankment/in cutting
route *f* **latérale** turning [side road]
route *f* **nationale** major/trunk road
route *f* **principale** main/major road
route *f* **secondaire** minor/secondary road
routier *adj* highway *adj*, road *adj*
routier, projet *m* road project
ruban *m* **adhésif** adhesive tape
ruban *m* **de fixation** fixing tape
ruban *m* **de téflon** Teflon tape
ruban *m* **isolant** insulating tape
rubigineux *adj* rust coloured
rude *adj* rough [of a surface]
rue *f* street
rue *f* **commerçante** shopping street
rue *f* **en boucle** loop road
rue *f* **piétonne** pedestrianized street
rues *f,pl* **en quadrillage** roads with a grid layout
ruelle *f* alley(way)
ruellée *f* flashing of mortar between a roof and wall, verge pointing [of a roof]
rugosité *f* roughness
rugosité *f*, **coefficient** *m* **de** roughness coefficient
rugueux *adj* rough [of a surface]
ruine *f* decay or downfall [e.g. of a building], ruin
ruiner *vb* destroy, ruin, spoil completely
ruines, tomber *vb* **en** fall into ruin(s)
ruineux *adj* dilapidated, falling into ruins, ruinous, ruinous [expensive]

ruisseau *m*   brook, burn (Scots), rivulet, stream
ruisseau *m* **à sec**   dry stream(bed)
ruisseler *vb*   run off [rainwater, etc]
ruisselet *m*   brooklet, rivulet
ruissellement *m*   run-off of water
rupteur *m*   contact-breaker, interrupter [electrical]
rupture *f*   breach(ing) or bursting [of a dam, etc],
   breaking or collapse [under load], parting [rope, etc],
   failure, fracture, rupture
rupture, charge *f* **de**   breaking load, load at failure
rupture circulaire, analyse *f* **de stabilité en**   slip-circle
   analysis

rupture *f* **d'endurance**   fatigue failure
rupture *f* **de câble**   cable failure
rupture *f* **de fatigue**   fatigue fracture
rupture *f* **du contrat**   breach/breaking of the contract
rupture *f* **fragile**   brittle fracture
rupture *f* **par cisaillement**   shear failure
rupture *f* **par fatigue**   fatigue failure
rusticage *m*   rough [rustic] dressing of stone
rustique *adj*   rustic
rustiquer *vb*   dress stone to give a rustic (rough) finish

sablage *m*   sandblasting
sablé *adj*   sandblasted
sable *m*   sand
sable *m* **alluvionnaire**   river sand
sable *m* **boulant**   quicksand
sable *m* **compacté**   compacted sand
sable *m* **coquiller**   shell sand
sable *m* **de concassage**   crushed stone fines
sable *m* **de laitier concassé**   ground granulated
   blastfurnace slag (ggbfs)
sable *m* **de rivière**   river sand
sable *m* **doux**   soft sand
sable *m* **enrobé**   sand asphalt
sable *m* **éolien**   blown sand
sable *m* **fin**   fine sand
sable *m* **grossier**   coarse sand
sable *m* **mouvant**   quicksand
sable *m* **naturel**   natural sand
sabler *vb*   sandblast
sableux *adj*   sandy
sablière *f*   bottom/top/wall plate [of a wall panel], ground/
   sole plate [of a timber frame panel], string piece [in
   staircase], sand pit/quarry
sablière *f* **de chambrée**   wall plate [supporting a timber
   frame panel]
sablière *f* **de forme**   top plate [of a timber frame panel]
sablière *f* **supérieure**   head rail or top plate [of a timber
   frame panel]
sablon *m*   sand [very fine grain]
sabot *m*   pile shoe, shoe or socket [on end of a structural
   member]
sabot *m* **de balai**   brush clamp [electrical]
sabot *m* **de frein**   brake block/shoe
sabot *m* **de prise de courant**   contact shoe
sac *m*   bag, sack
sac *m* **filtrant**   filter bag
saignée *f*   chase or groove, drainage ditch/trench, grip [in
   a verge], saw cut [in wood]

saillant *adj*   jutting out, projecting, salient
saillie *f*   ledge, projection
saillie, bordure *f* **en**   kerb [road]
saillie, en   jutting out, projecting
saillie *f* **de rive**   eaves overhang
saillie, faire *vb*   jut out, project
sain *adj*   sound [of timber]
saisie *f* **d'une hypothèque**   foreclosure of a mortgage
saisie *f* **de données**   data acquisition
saisie *f* **immobilière**   seizure of real property
saisir *vb* **(un tribunal) (d'une affaire)**   refer (to a court)
   (a matter)
saison *m* **de chauffage**   heating season
saisonnier *adj*   seasonal
salage *m*   salting [of roads]
salaire *m*   pay, salary, wage
salarié *m*   wage earner
saleté *f*   dirt, dirtiness, filth, junk [worthless], rubbish
   [junk]
salin *adj*   saline
salinité *f*   salinity
salissure *f*   soiling
salle *f*   auditorium, room [large or formal]
salle *f* **à manger**   dining room, dining room suite
salle *f* **blanche**   clean room
salle *f* **d'attente**   waiting room
salle *f* **d'eau/de douches**   shower room
salle *f* **de bain(s)**   bathroom
salle *f* **de commande/de contrôle**   control room
salle *f* **de conférence**   conference room, lecture hall
salle *f* **de jeux**   games room
salle *f* **de réception**   drawing room [formal]
salle *f* **de réunion(s)/de séances**   meeting room
salle *f* **de séjour**   living room
salle *f* **de récupération/de repos**   rest room
salle *f* **de traite**   milking parlour
salle *f* **des appareils**   equipment room
salle *f* **des ventes**   auction room, saleroom

salon *m*   drawing/sitting room

salon *m* **de réception**   drawing room [formal]

salpêtre *m*   saltpetre

sandwich, panneau *m*   sandwich panel [generally with insulating core]

sanitaire *adj*   sanitary

sanitaires *f,pl*   bathroom suite or plumbing, bathroom

sanctionner *vb*   approve [e.g. quality of workmanship], confirm [approval, a report, etc], ratify

sans cloisons   open-plan

sans fil   cordless [of portable electric tools]

sans goutte   non-drip

sans huile   oil-free

sans soudure   seamless

sape *f*   undermining tunnel

saper *vb*   undermine

sapeurs-pompiers *m,pl*   fire brigade

sapin *m*   fir [wood, tree]

sapin *m* **du nord**   whitewood

sapine *f*   ba(u)lk of fir/spruce timber

sarrasine *f*   portcullis

sas *m*   lobby

sas *m* **à air**   airlock [of pressurized caisson, lobby]

sas *m* **d'écluse**   lock chamber [canal]

satellite *m*   satellite [building]

satiné *adj*   satin [paint finish]

saturation *f*   saturation

saturé *adj*   saturated

saule *m*   willow

saumâtre *adj*   brackish, briny

saumure *f*   brine

saut-de-loup *m*   area below street level giving light to basement windows, ha-ha

saut-de-mouton *m*   cloverleaf road intersection

sauterelle *f*   apron conveyor, (mobile) belt elevator, mobile/portable belt conveyor, bucket conveyor [for sand, aggregates], bevel square, sliding bevel

sauterelle *f* **cribleuse**   screening conveyor

sauvegardé *adj*   protected

sauvegarder *vb*   conserve, protect or safeguard

sauver *vb*   save [a building, etc]

scaphandre *m*   diver's/diving suit

scaphandrier *m*   (helmet) diver

scarificateur *m*   scarifier

sceau *m*   seal [on a document]

scellé *adj*   anchored, embedded, fixed, grouted [in or into masonry, etc], sealed

scellement *m*   anchorage bond [in concrete, masonry, etc], fixing, grouting in, mortaring in [into concrete, masonry, etc], seal or sealing

scellement, couche *f* **de**   seal(ing) coat [road construction]

scellement, enduit *m* **de**   seal(ing) coat [road construction]

sceller *vb*   bed [e.g. a beam into a wall], embed, fix, grout/mortar in, seal

schéma *m*   diagram, schematic drawing, sketch, outline [written]

schéma *m* **d'écoulement d'air**   air flow pattern

schéma *m* **de câblage**   circuit diagram, drawing, plan or layout

schéma *m* **de principe**   flow diagram, rough sketch

schéma *m* **directeur**   development plan [long term]

schéma *m* **directeur d'aménagement et d'urbanisme**   development plan [regional, urban, etc, long term]

scène *f*   stage [of a theatre]

schiste *m*   schist

schiste *m* **argileux**   shale

schiste *m* **brûlé**   burnt colliery shale

schiste *m* **expansé**   expanded shale

sciage *m*   sawing

sciage, bois *m* **de**   resawn timber

sciage *m* **en plots**   through-and-through sawing

scié *adj* **sur maille**   quarter-sawn

scie *f*   saw

scie *f***, chaîne** *f* **de la**   saw chain

scie *f***, traverse** *f* **de la**   saw guide [of a chainsaw]

scie *f* **à archet**   hacksaw

scie *f* **à béton**   concrete saw

scie *f* **à bûches**   bucksaw

scie *f* **à chaîne(tte)**   chainsaw

scie *f* **à chantourner**   fretsaw, jigsaw, scroll saw

scie *f* **à découper**   fretsaw

scie *f* **à deux mains**   crosscut saw [double ended]

scie *f* **à (dos)seret**   tenon saw

scie *f* **à guichet**   compass or keyhole saw

scie *f* **à main**   handsaw

scie *f* **à métaux**   hacksaw

scie *f* **à panneaux**   panel saw

scie *f* **à refendre**   ripsaw

scie *f* **à rocher**   rock cutter or cutting saw

scie *f* **à ruban**   bandsaw

scie *f* **à sol**   pavement saw

scie *f* **à travers**   bow saw

scie *f* **à tronçonner**   crosscut saw

scie *f* **anglaise**   jigsaw

scie *f* **articulée,** *voir aussi* tronçonneuse   chainsaw

scie *f* **circulaire**   circular saw

scie *f* **d'encadreur**   tenon saw

scie *f* **de délignage**   edger (saw), trim saw

scie *f* **de maçon**   masonry saw

scie *f* **égoïne** *f*   handsaw

scie *f* **manuelle**   handsaw

scie *f* **multiple**   frame saw

scie *f* **passe-partout**   crosscut saw [double ended]

scie *f* **sauteuse**   jigsaw

scie *f* **sans fin**   bandsaw

scie *f* **trépan**   hole saw

science *f* **économique** *f*   economics

scier *vb*   saw

scierie *f*   sawmill

sciure *f*   sawdust

scléromètre *m*   rebound hammer, sclerometer

scorie *f*   clinker, slag

scorie *f* **de laminage**   mill scale

scotie *f*   scotia [architectural moulding]

scrape(u)r *m*   scraper [earthworks]

seau *m*   bucket

sec *adj*   dry
sec *adj*, **absolument**   bone dry
sec *adj* **comme de l'amadou**   dry as a bone
sec *m*   dry (place)
sec, tenir *vb* **qch au**   keep something dry/in a dry place
séchage *m*   drying
séchage *m* **à l'air**   air drying
séchant rapide   quick-drying
séché *adj* **à l'air**   air dried
séché *adj* **à l'étuve**   oven dried
sèche-linge *m*   drying cabinet [for clothes], tumble dryer
sèche-mains *m*   hand dryer
sécher *vb*   dry (up, out)
sécheresse *f*   drought, dryness [of weather]
sécheur *m*   dryer
sécheuse *f*   clothes dryer
second œuvre *m*   finishing works [closure, services, partitions, finishes]
secondaire *m*   secondary
secours, de   standby
secousse *f*   jolt, shaking, shock
secousse *f* **électrique**   electric shock
secousse *f* **s(é)ismique**   earthquake
secousse *f* **s(é)ismique, légère**   earth tremor
secousse *f* **tellurique**   earth/ground tremor
secrétaire *m*   desk
secteur *m*   district, sector
secteur *m* **d'obturation**   blanked sector
secteur *m* **privé**   private sector
secteur *m* **public**   public sector
section *f*   cross-section, profile [section], section [cross-section, or structural]
section *f* **brute**   concrete section, gross section(al) (area)
section *f* **composée**   built-up/compound section
section *f* **creuse**   hollow section
section *f* **d'acier comprimé**   area of compression reinforcement
section *f* **d'acier tendu**   area of tension reinforcement
section *f* **d'entrecroisement**   weaving section [of road]
section *f* **droite**   cross-section
section *f* **fissurée**   cracked section
section *f* **homogène**   homogeneous section
section *f* **homogénéisée**   transformed section
section *f* **nette**   net section
section *f* **pleine**   solid section
section *f* **transversale**   cross-section, cross-sectional area
section *f* **utile**   effective cross-section, usable or useful section
sectionnel *adj*   sectional
sectionnement *m*   isolation [electrical]
sectionneur *m*   isolating switch, isolator
sécurité *f*   safety, security, safety device
sécurité, couplage *m* **de**   safety interlock
sécurité, disjoncteur *m* **de**   safety cut-out
sécurité, soupape *f* **de**   safety valve
sécurité *f* **au voilement**   safety against local buckling
sécurité *f* **contre l'incendie**   fire safety
sécurité *f* **de manque d'eau**   low water safety device
sécurité *f* **du travail**   site safety

sécurité *f* **routière**   road safety
sédiment *m*   deposit, sediment
sédiment *m* **limoneux**   muddy/silt sediment
sédimentaire *adj*   sedimentary
sédimentation *f*   sedimentation
ségrégation *f*   segregation
séisme *m*   earthquake
s(é)ismique *adj*   seismic
s(é)ismographe *m*   seismograph
s(é)ismographie *f*   seismography
s(é)ismologie *f*   seismology
s(é)ismologique *adj*   seismological
séjour *m*   living room, stay [at/in a place]
sélecteur *m*   selector switch
sélecteur *m* **d'étage**   floor selector [lift/elevator]
sélecteur *m* **local groupé**   local group switching
sélectionné *adj*   selected
self *f*   choke [electrical]
selle *f* **de rail**   baseplate or chair [railway track]
sellette *f*   bos'n's chair, painter's cradle [chair]
semelle *f*   abutment piece, baseplate, foot plate, ground sill, sole piece/plate [of wood, under a prop, etc], flange [of girder], footing [foundation], sole [of a plane]
semelle *f* **armée**   reinforced concrete footing
semelle *f* **continue**   continuous footing
semelle *f* **en gros béton**   mass concrete footing
semelle *f* **filante**   strip footing
semelle *f* **inférieure**   lower chord or flange [of truss, girder]
semelle *f* **isolée**   individual or pad footing
semelle *f* **résiliente**   resilient strip [under a partition, to allow for structural movement]
semelle *f* **superficielle**   shallow footing
semelle *f* **supérieure**   upper chord or flange [of truss, girder]
semelle *f* **sur pieu**   piled footing
semence *f*   tack [short nail]
semi-automatique *adj*   semi-automatic
semi-benne *f*   tipping semi-trailer
semi-circulaire *adj*   semi-circular
semi-conducteur *m*   semiconductor
semi-remorque *f*   semi-trailer
semi-remorque *f* **à benne**   tipping semi-trailer
semi-remorque *f* **plateau/plate-forme**   flat bed semi-trailer
semi-remorque *f* **porte-engins**   plant semi-trailer
sens *m*   direction, meaning or sense
sens *m* **contraire, en**   in the opposite direction
sens *m* **de circulation**   direction of circulation or traffic
sens *m* **de déplacement**   direction of motion
sens *m* **de l'écoulement de l'air**   direction of air flow
sens *m* **de la flèche**   direction of the arrow
sens *m* **de rotation**   direction of rotation
sens dessus dessous   upside down, wrong way up
sens devant derrière   back to front, wrong way round
sens *m* **du laminage**   direction of rolling
sens unique, à   one-way [road, street]
sensibilité *f* **(à)**   sensitiveness/sensitivity (to), susceptibility (to)

sensibilité à l'effet d'entaille   notch sensitivity
sensible *adj*   sensitive
sentier *m*   footpath, path
sentier *m* d'exploitation *f*   farm track
séparateur *m*   central reservation [roads], traffic separator, separator [in general]
séparateur *m* à expansion   expansion separator
séparateur *m* centrifuge   centrifugal separator
séparateur *m* d'air   air separator
séparateur *m* d'air centrifuge   centrifugal air separator
séparateur *m* d'eau   water separator
séparateur *m* d'eau de condensation   steam separator
séparateur *m* d'huile   oil separator
séparateur *m* de gouttelettes [tour *f* de refroidissement]   drift eliminator [cooling tower]
séparateur *m* de poussières   dust separator
séparateur *m* décanteur   interceptor, petrol
séparateur *m* en saillie   raised central reservation [roads]
séparatif *adj*   separating
séparatif d'assainissement, système *m*   separate drainage system
séparation *f*   dividing, division, partition, separation
séparation, mur *m* de   dividing/partition/separating wall
séparation entre, établir *vb* une   set up a boundary between
séparé *adj*   separate
séparer *vb*   divide, separate
séparer *vb* par congélation   freeze out
septain *m*   seven-strand cord [for blinds, or as sash cord]
séquence *f*   sequence
séquence *f* de contrôle   control sequence
séquence *f* de mise en route   starting sequence
séquoia *m*   redwood [North American]
sergent *m*   carpenter's or gluing clamp
série *f*   set [e.g. of drawings], series [electrical, mathematical]
série *f* des prix du bâtiment   builder's price book
serpenter *vb*   wind [e.g. a road]
serpentin *m*   coil [tubing], pipe coil
serpentin *m* à détente directe   direct expansion coil
serpentin *m* à vapeur   steam coil
serpentin *m* de chauffage   heating coil
serpentin *m* de condenseur   condenser coil
serpentin *m* de dilatation   expansion coil
serpentin *m* de réfrigération   cooling or refrigeration coil
serpentin *m* de refroidissement   cooling coil
serpentin *m* évaporatoire   evaporator coil
serpentin *m* rechauffeur   calorifier (coil)
serpette *f*   painter's curved knife for cutting heavy wallpaper, pruning knife
serrage *m*   tightening
serrage *m* du béton   compaction of concrete
serré *adj*   tight
serre *f*   greenhouse
serre *f* chaude   hothouse
serre-câbles *m*   cable clamp
serre-fil *m*   connector [electrical]
serre-joint *m*   cramp

serre-joint *m* à coller   gluing clamp
serre-joint *m* à vis   screw clamp
serre-nuque *m*   ear cups/defenders/protectors
serre-tête *m* antibruit   ear cups/defenders/protectors
serre-tubes *m*   plumber's chain tongs/vice, pipe wrench
serrée, à texture   close-grained
serrer *vb*   clamp, close up, grip, hold tight, tighten, turn off tight [a tap or valve]
serrer *vb* à vis   screw down
serrure *f*   lock [door, gate, etc]
serrure *f* à barillet   cylinder lock
serrure *f* à bec-de-cane   drawback lock with a bevelled bolt
serrure *f* à combinaisons   combination lock
serrure *f* à mortaiser   mortise lock
serrure *f* à palastre/palâtre   rim lock
serrure *f* à pêne demi-tour   lock with a bevelled bolt
serrure *f* à pêne dormant   dead lock
serrure *f* à ressort   spring lock
serrure *f* à tour et demi   lock closed with one-and-a-half turns
serrure *f* à tubulaire   tubular latch [door]
serrure *f* batteuse   Bramah lock
serrure *f* camarde   drawback lock
serrure *f* chronométrique   time lock
serrure *f* de sûreté à gorges   tumbler safety lock
serrure *f* de sûreté à pompe   Bramah safety lock
serrure *f* en applique   rim lock
serrure *f* en large   vertical lock [i.e. long edge vertical]
serrure *f* en longue   horizontal lock [i.e. long edge horizontal]
serrure *f* encastrée   mortise lock
serrure *f* encloisonnée   rim lock
serrure *f* entaillée   lock mortised into the face of a door
serrure *f* lardée/mortaisée   mortise lock
serrure *f* temporisée   time lock
serrurerie *f*   hardware, ironwork, metalwork [the work, and its result], locksmithing
serrurerie *f* d'art   ornamental ironwork
serrurier *m*   ironsmith, locksmith
sertir *vb*   crimp [an electrical connection], set [glass in a frame]
sertisseur *m*   crimper [electrical]
service *m*   administrative department, service
Service *m* Cartographique de l'État   mapping agency of France [Ordnance Survey in UK]
service *m* d'eau chaude   hot water supply
service *m* d'entretien   maintenance service
service *m* de courte durée   short-term operation
service *m* de l'urbanisme   planning department/office [town planning]
service *m* de planification   planning department/office [sequencing work]
service *m* des eaux   water supply (company)
service *m* médical du travail   occupational health service
services *m,pl*   services [as provided by architect, etc]
services *m,pl* concédés   statutory services or utilities
services *m,pl* d'ingénierie   engineering services
services *m,pl* publics   public services

se servir *vb* **de** use

servitudes *f,pl* charges attached to a property, constraints on the design/use of a property, easements

servocommande *f* servo-control

servomoteur *m* servo-motor

seuil *m* door sill, doorstep, threshold

seuil *m* **d'audibilité** threshold of hearing

seuil *m* **déversoir** sill [of a dam]

seuil *m* **du son** noise threshold level

seuil *m* **en rivière** weir

sgraffite *f* design [decoration] cut into a rendering

shunt *m* smoke extract or ventilation duct for a whole building, shunt [electrical]

shuntage *m* shunting

shunter *vb* shunt [electrically]

siccatif *adj* drying, siccative *adj*

siccatif *m* drying agent, siccative [as used in paint]

siccité *f* dryness

sidérurgique *adj* steel *adj*, relating to steel industry

siège *m* head office, headquarters, seat, seating [of a valve or tap]

siège *m* **à l'anglaise** toilet/lavatory bowl or pan

siège *m* **à la turque** Asiatic/squatting closet (WC)

siège *m* **d'obturateur** valve seating

siège *m* **de bureau** office chair

siège *m* **de cabinet** toilet seat

siège *m* **de soupape** valve seating

siège *m* **de trapillon** trapdoor seating

siège *m* **de vanne** valve seating

siège *m* **social** registered office

signal *m* signal

signal *m* **d'alarme** alarm signal

signal *m* **d'avertissement** warning signal [railway]

signal *m* **d'incendie** fire alarm

signal *m* **de direction** direction sign [road]

signal *m* **de limitation de vitesse** speed restriction sign

signal *m* **de présignalisation** advance direction sign

signal *m* **de ralentissement** speed restriction sign

signal *m* **géodésique** geodetic mark, survey station

signal *m* **lumineux** illuminated (road) sign

signal *m* **routier** road or traffic sign

signalisation *f* indication, indicator, signage [putting up signs, and the signs]

signalisation *f*, **panneau** *m* **de** traffic sign [board]

signalisation *f* **des risques** safety warnings

signalisation *f* **lumineuse** illuminated signing [e.g. for emergency exit]

signature *f* signature

signe *m* indication, sign, symbol, symptom

signe *m* **conventionnel** conventional sign/symbol

silencieux *adj* noiseless, silent

silencieux *m* (sound) attenuator, muffler, silencer

silex *m* chert, flint

silhouette *f* profile of a flat key

silicagel *m* silica gel

silicate *m* silicate

silice *f* quartz glass, silica

silicium *m* silicon

silicone *f* silicone

silo *m* bunker, silo

silo *m* **à ciment** cement silo

silo *m* **de décantation de boues** sludge settling tank

silo *m* **de stockage** storage bin

simbleau *m* carpenter's/bricklayer's/builder's line for marking out circles

simple *adj* single

simulation *f* simulation

simultané *adj* simultaneous

simultanément *adv* simultaneously

singe *m* crab [portable winch], drop hammer, monkey [pile driver]

sinistre *m* accident [loss under insurance policy]

siphon *m* (drain) trap, siphon

siphon *m* **à culotte démontable** bottle trap

siphon *m* **de sol** floor trap

siphon *m* **d'huile** oil trap

siphon *m* **en S** S-trap

siphon *m* **renversé** dip-pipe

siphonner *vb* siphon

siporex *m* Siporex [lightweight concrete]

sis *adj* located, situated

sismique *adj* seismic

sismographe *m* seismograph

site *m* lie of the ground, archaeological site

site *m* **classé** protected site [in France, similar to Conservation Area in UK]

situation *f* location, position, situation, record of work completed and payments made and outstanding

situation, plan *m* **de** layout plan

situé *adj* situated

situer *vb* locate, site, situate

smillage *m* scabbling [with a scabbling hammer, the action or result]

smillé *adj* scabbled [of stone dressed with many short parallel cuts]

smille *f* mason's two-pointed (scabbling) hammer

smiller *vb* scabble [with a scabbling hammer]

société *f* company, firm

société *f* **concessionnaire** concessionaire, or company holding a concession [e.g. statutory undertaker]

société *f* **d'économie mixte (SEM)** joint public/private company

Société *f* **Française des Urbanistes (SFU)** French society of (town) planners

Société *f* **Nationale des Chemins de Fer Belges (SNCB)** Belgian national railways

Société *f* **Nationale des Chemins de Fer Français (SNCF)** French national railways

société *f* **à responsabilité limitée (SARL)** limited liability company

sociologie *f* sociology

sociologue *m* sociologist

socle *m* base, bottom, mounting base, pedestal, plinth, socket outlet

socle *m* **de béton** concrete base course

socle *m* **de chaudière** boiler mounting

socle *m* **de deux prises de courant** double (plug) socket outlet

socle *m* **de fondation**   plinth foundation
socle *m* **de moteur**   motor bedplate
socle *m* **de prise à encastrer**   flush mounted socket outlet
socle *m* **de prise (de courant)**   socket outlet
socle *m* **de prises de courant à trois contacts**   three-pin socket outlet
socle *m* **de prise de sol**   floor mounted socket outlet
socle *m* **de tableau**   panel mounting socket outlet
socle *m* **en béton**   concrete base
socle *m* **(en) saillie**   surface mounted socket outlet
socle *m* **isolant**   insulating base
socle *m* **receveur**   shower tray
soffite *m*   soffit [architectural]
soie *f*   bristle [of a paintbrush], fine starter bars
soins *m,pl* **d'urgence**   first aid
sol *m*   floor [solid], ground, soil
sol-ciment *m*   soil-cement
sol, essais *m,pl* **de**   soil survey
sol, stabilisation *f* **du**   soil stabilization
sol, structure *f* **du**   soil structure
sol *m* **brut**   unfinished floor
sol *m* **cohérent**   cohesive soil
sol *m* **cultivable**   topsoil
sol *m* **de fondation**   sub-grade [formation]
sol *m* **en béton**   concrete floor
sol *m* **en planches**   wooden floor [boarded]
sol *m* **fin**   fine-grained soil
sol *m* **fini**   finished floor
sol *m* **grenu**   coarse-grained soil
sol *m* **impropre**   unsuitable soil
sol *m* **perméable**   permeable ground
sol *m* **pulvérulent**   non-cohesive soil
sol *m* **tourbeux**   peat
sol *m* **vivant**   organic soil, topsoil
solaire *adj*   solar
solarium *m*   solarium, sun terrace
solde *m*   balance [in an account]
solde *m* **débiteur**   debit balance, overdraft
soleil *m*   sun, polygonal facing stone
solénoïde *m*   solenoid
solidaire *adj*   jointly liable [e.g. of partners in a joint venture]
solidarisé *adj*   interlocked
solide *adj*   reliable, robust, solid, sound, strong, sturdy, tough
solidification *f*   solidification
solidité *f*   reliability, robustness, solidity, soundness, sturdiness, toughness
solin *m*   flashing between wall and abutting roof
solivage *m*   secondary joisting
solive *f*   beam, (secondary) joist
solive *f* **boiteuse**   trimmer joist
solive *f* **courante**   joist [normal, as opposed to trimmer or trimming]
solive *f* **d'enchevêtrure**   trimming joist
solive *f* **de ferme**   wall plate [carrying feet of rafters]
solive *f* **de remplissage**   trimmed joist
solive *f* **de rive**   edge joist
solive *f* **enchevêtrée**   trimming joist

solive *f* **porteuse**   trimming joist
soliveau *m*   small trimmer joist
soliveau *m* **en empannon**   angled trimmer joist
sollicitation *f*   externally applied effect, load or stress [externally applied]
sollicitation *f* **alternée**   alternating stress
sollicitation *f* **du trafic**   traffic load(ing)
sollicitation *f* **en traction**   tensile stress
solliciter *vb*   attract [of magnets], load or stress
solubiliser *vb*   make soluble
solubilité *f*   solubility
soluble *adj*   soluble [problem, substance]
solution *f*   solution
solution *f* **de base**   contract design
solution *f* **de compromis**   compromise solution
solution *f* **incongelable**   anti-freeze solution
solvant *m*   solvent
sombre *adj*   dark [colour]
sommaire *m*   list or table of contents, summary
somme *f*   sum or total [sum, amount]
somme *f* **forfaitaire**   lump sum
sommet *m*   crest [of a road, embankment], crown [of an arch], highest point [of a hill, tree], peak [of a hill], summit or top [of a hill, ladder, tree], vertex [of an angle, a curve]
sommier *m*   bed base [divan, etc], girt/ledger/ribbon board [timber frame structure], head beam, horizontal member of railings/window bars, springer or springing stone [of an arch]
son *m*   sound
sondage *m*   hole [borehole, drill hole], bore/test/trial hole [ground investigation], sounding [to measure depth of water]
sondage, coupe *f* **de**   soil profile
sondage, tige *f* **de**   sounding rod/probe
sondage *m* **à la pelle**   machine dug test/trial pit
sondage *m* **à pénétration dynamique, appareil** *m* **de**   dynamic penetration test apparatus
sondage *m* **carotté**   test core [gound investigation]
sondage *m* **du sol**   borehole investigation, trial boring
sondage *m* **par percussion**   percussion drilling [in the ground]
sondage *m* **par puits**   borehole sounding
sondage *m* **par tranchée**   test or trial pit
sonde *f*   probe
sonde, trou *m* **de**   borehole
sonde *f* **de température**   temperature probe
sonder *vb*   make borings, investigate [by boring or drilling], probe, sound [to investigate at depth], probe [to test, etc]
sonnerie *f*   alarm (bell), chimes, ringing [of a bell]
sonnerie *f* **d'alarme**   alarm bell
sonnerie *f* **électrique**   electric bell
sonnette *f*   doorbell
sonnette *f* **à vapeur**   steam hammer [pile-driver]
sonnette *f* **d'alarme**   alarm bell
sonnette *f* **(de battage)**   pile-driver, piling rig
sonomètre *m*   sound level meter
sonore *adj*   sound [acoustic]

sonorisation *f* **(générale)**   public address (PA) system
   [provision with]
sortie *f*   exit
sortie *f*, **point** *m* **(de)**   exit point [for traffic]
sortie *f* **analogique**   analogue output
sortie *f* **d'air**   air outlet
sortie *f* **d'eau**   water outlet
sortie *f* **d'évacuation de fumée**   smoke exhaust outlet
sortie *f* **d'ordinateur**   computer output
sortie *f* **de câbles**   cable outlet
sortie *f* **de secours**   emergency exit
sortie *f* **de ventilateur**   fan outlet
soubassement *m*   base/basement course, plinth
   [basement], bedrock, fireplace facade immediately under
   mantlepiece
souche *f*   stump [of a tree]
souche *f* **(de cheminée)**   chimney stack above roof level
soudabilité *f*   weldability
soudable *adj*   weldable
soudage *m*, *voir aussi* soudure   brazing, soldering,
   welding
soudage *m* **à l'arc**   arc welding
soudage *m* **à l'arc au charbon**   carbon arc welding
soudage *m* **au chalumeau oxhydrique**   oxyhydrogen
   welding
soudage *m* **au gaz**   gas welding
soudage *m* **autogène**   acetylene, gas or oxyacetylene
   welding
soudage *m* **bout à bout**   butt welding
soudage *m* **d'angle**   fillet welding
soudage *m* **de goujons**   stud welding
soudage *m* **discontinu**   discontinuous or intermittent
   welding
soudage *m* **électrique**   electric welding
soudage *m* **en bout**   butt welding
soudage *m* **manuel**   hand welding
soudage *m* **oxyacétylénique**   acetylene, gas or acetylene
   welding
soudage *m* **par fusion**   fusion welding
soudage *m* **par points**   stitch or tack welding
soudage *m* **par résistance**   resistance welding
soudé *adj* **en bout**   butt-welded
soude-brasage *m*   bronze welding
souder *vb*   braze, solder, weld
soudeur *m*   welder
soudoir *m*   soldering iron
soudure *f*, *voir aussi* soudage   brazing, seam [welded or
   brazed], solder(ing), weld(ing)
soudure *f* **à l'arc**   arc weld
soudure *f* **à la main**   hand or manual weld
soudure *f* **à l'étain**   soft soldering
soudure *f* **à plat**   flat weld
soudure *f* **à recouvrement**   lap-weld
soudure *f* **au cuivre**   hard soldering
soudure *f* **au plomb**   lead soldering
soudure *f* **bout à bout**   butt weld
soudure *f* **continue**   continuous weld
soudure *f* **d'angle**   fillet weld
soudure *f* **d'étanchéité**   seal weld

soudure *f* **de chantier**   field or site weld
soudure *f* **de pointage**   tack weld
soudure *f* **électrique par résistance**   electric resistance
   welding
soudure *f* **en angle**   fillet weld
soudure *f* **en bouchon**   plug weld
soudure *f* **en plusieurs passes**   multi-layer weld
soudure *f* **oxyacétylénique**   oxyacetylene welding
soudure *f* **par points**   spot welding
soufflage *m*   air supply
soufflante *f*   blower
souffler *vb*   blow
soufflerie *f*   blower, ventilating fan
soufflerie *f* **à gaz**   gas blower
soufflet *m*   bellows
soufflet, fenêtre *f* **à**   fanlight [bottom hung]
soufflet *m* **de dilatation**   expansion bellows
soufflure *f*   blowhole [casting defect]
soufre *m*   sulphur
souillard *m*   hole [through structure for a rainwater
   downpipe]
soulager *vb* **un palan**   ease off a block and tackle
soulager *vb* **une poutre**   take the strain off a beam
soulèvement *m*   heave or heaving [of ground], upheaval,
   upthrust [geological], lifting, uplift
soulever *vb*   lift up [sth], raise [a question or subject,
   sth]
soulever *vb* **qch au cric**   jack sth up
soulever *vb* **une réclamation**   make or put in claim
se soulever *vb*   heave [geologically]
soumission *f*   bid, submission, tender, tendering [i.e.
   submitting]
soumission, faire *vb* **une**   submit a bid or tender
soumissionnaire *m*   tenderer
soumissionner *vb*   submit a bid or tender
soupape *f*   valve [generally spring-loaded or operated]
soupape *f* **à action rapide**   quick-action/-acting valve
soupape *f* **à boulet**   ball (check) valve
soupape *f* **à contrepoids**   weight-loaded valve
soupape *f* **à diaphragme**   diaphragm valve
soupape *f* **à double arrêt**   double shut-off valve
soupape *f* **à flotteur**   float valve
soupape *f* **à haute levée**   high-lift valve
soupape *f* **à languettes**   flapper valve
soupape *f* **à moteur**   motorized valve
soupape *f* **à ouverture rapide**   quick-release valve
soupape *f* **à plusieurs voies**   multi-way valve
soupape *f* **à pointeau**   needle valve
soupape *f* **à ressort**   automatic (spring-loaded) valve
soupape *f* **à trois voies**   three-way valve
soupape *f* **anti-siphon (casse-vide)**   anti-siphon valve
soupape *f* **casse-vide**   anti-vacuum valve
soupape *f* **d'admission**   inlet valve
soupape *f* **d'arrêt**   shut-off or stop valve
soupape *f* **d'arrêt principale**   main stop valve
soupape *f* **d'échappement**   exhaust valve
soupape *f* **d'échappement d'air**   air release valve
soupape *f* **d'étranglement**   throttle valve
soupape *f* **d'évacuation**   exhaust valve

soupape *f* **de contrôle** control valve

soupape *f* **de contrôle thermostatique** thermostatic control valve

soupape *f* **de décharge** discharge/relief valve

soupape *f* **de dérivation** bypass valve

soupape *f* **de détente** expansion valve

soupape *f* **de modulation** modulating valve

soupape *f* **de pression d'eau** water pressure valve

soupape *f* **de purge** bleeder/blow-off valve

soupape *f* **de réduction** reducing valve

soupape *f* **de réduction à ressort** spring-loaded pressure reducing valve

soupape *f* **de régulation** control valve

soupape *f* **de retenue** non-return/reflux valve

soupape *f* **de sécurité** pressure relief valve

soupape *f* **de sûreté** safety valve

soupape *f* **de sûreté à ressort** spring-loaded safety valve

soupape *f* **de surpression** pressure control valve

soupape *f* **de trop-plein** balancing (overflow) valve

soupape *f* **de vidange** drain valve

soupape *f* **directionnelle** changeover valve

soupape *f* **réductrice** reducing valve

soupape *f* **réductrice de pression** pressure reducing valve

soupirail *m* basement window [small, generally barred]

source *f* spring [of water]

source *f* **alternative d'énergie** alternative energy source

source *f* **d'électricité** (electrical) power source

source *f* **d'énergie** energy source

source *f* **d'énergie perdue** waste energy source

source *f* **d'erreur** source of error

source *f* **de chaleur** heat source

source *f* **de chaleur souterraine** hot well

source *f* **lumineuse** light source

source *f* **sonore** sound source

sourcier *m*/**sourcière** *f* water diviner

sourdière *f* internal window shutter

souricière *f* mousetrap

souris *f* mouse [computer]

sous *adv* beneath, under

sous gaine *adj* sheathed [e.g. cable]

sous pression *adj* pressurized, under pressure

sous-affermer *vb* sublease or sublet [land]

sous-bois *m* undergrowth

sous-centrale *f* substation

sous-couche *f* underlying layer, sub-base or sub-grade [of a roadway], substratum, undercoat [paint]

sous-couche *f* **anti-capillaire** anti-capillary layer [road construction]

sous-couche *f* **anti-gel** frost-proof layer [road construction]

sous-couche *f* **drainante** filter layer [road construction]

souscrire *vb* take out [an insurance policy]

sous-estimation *f* undervaluation

sous-évaluation *f* undervaluation

sous-face *f* soffit, underside

sous-jacent *adj* underlying

sous-joint *m* joint tape

sous-lisse *f* middle rail [of a barrier]

sous-locataire *m*/*f* subtenant

sous-location *f* subleasing or subletting

sous-plancher *m* sub-floor

sous-pression *f* uplift pressure

sous-produit *m* by-product

sous-répartiteur *m* cable distribution plug

sous-sol *m* basement, subsoil

sous-station *f* substation

sous-toiture *f* cladding of the underside of a roof

sous-traitance *f* subcontracting

sous-traitant *m* subcontractor

sous-louer *vb* sublease or sublet [property]

sous-traiter *vb* subcontract, sublet [a contract]

soute *f* bunker, fuel storeroom

soute *f* **à charbon** coal bunker

soute *f* **à combustible** fuel bunker

soutènement *m* retention or support, structure resisting ground pressure

soutènement, mur *m* **de** retaining wall

soutènement *m* **de bois** timbering [of an excavation]

soutenir *vb* retain, support

souterrain *adj* underground, subterranean

souterraine en circulation, eau *f* gravitational groundwater

souterrain *m* subway, tunnel, underground passage(way)

souterrain, parking *m* underground carpark

soutien *m* prop, stay, support

soutirage *m* tapping [electrical supply]

soutirer *vb* bleed, draw off, tap [electrical supply]

soyeux *adj* silky

spatule *f* filling or stripping knife, spatula

spatule *f* **de vitrier** putty knife

spécial *adj* special

spécialiste *m* specialist

spécification *f* specification

spécifications *f,pl* **techniques détaillées** detailed technical specifications

spécifier *vb* to specify

spécifique specific

spectre *m* spectrum

spectre *m* **sonore** sound spectrum

spectrographique *adj* spectrographic

sphère *f* sphere

sphérique *adj* spherical

spirale *f* spiral

spire *f* hoop [reinforcement]

spirotringle *f* curtain wire [plastic covered spring]

spiter *vb* shot-fire [a nail or wall plug]

sprinklage *m* sprinkling [fire protection]

sprinkle(u)r *m* sprinkler [fire protection]

stabilisant *adj* stabilizing

stabilisant *m*/**stabilisateur** *m* stabilizer [compound or product], stabilizing agent

stabilisateur *m* **de sols** soil stabilizer

stabilisation *f* annealing, balancing, stabilization

stabilisation *f* **au ciment** cement stabilization

stabilisation *f* **du/d'un sol** soil stabilization

stabilisation *f* **du sous-sol** subsoil stabilization

stabilisé *adj* stabilized

stabilisée, bande *f*   paved verge [road]
stabilisée, bande *f* **non**   grass verge
stabilisée, terre *f*   stabilized earth
stabiliser *vb*   anneal, stabilize
stabilité *f*   stability
stabilité *f* **au feu**   fire resistance
stabilité *f* **au renversement**   stability against overturning
stabilité *f* **de la flamme**   flame stability
stabilité *f* **dimensionnelle**   dimensional stability
stabilité en rupture circulaire, analyse *f* **de**   slip-circle analysis
stable *adj*   firm, stable, steady
stable *adj* **au feu (SF)**   fire stable
stade *m*   sports stadium
staff *m*   fibrous plaster
staffeur *m*   fibrous plaster specialist
standard *adj*   standard
standard *m*   telephone switchboard
starter *m*   starter [for a lamp]
station *f*   position, station [place, position], resort, spa
station *f* **d'épuration**   cleaning/purification plant
station *f* **d'épuration des eaux usées**   sewage disposal/ treatment plant
station *f* **de pompage**   pump house, pumping station, waterworks
station *f* **de pompage intermédiaire**   booster pumping station
station *f* **de relevage**   pumping station
station *f* **de tamisage**   screening plant [for granular material]
station *f* **de traitement**   treatment plant
station *f* **de traitement des eaux usées**   sewage disposal or treatment plant
station *f* **métro**   underground (railway) station
station *f* **service** *voir* station-service
stationnement *m*   stopping [parking of vehicles]
stationnement, parc *m* **de**   carpark, hardstanding [for vehicle parking]
stationnement, zone *f* **de**   parking area or lot
station-service *f*   filling or petrol station, service station [vehicular]
statique *adj*   static
statique *f*   statics
statistique *f*   statistics
stator *m*   stator
stéréogramme *m*   stereogram or stereograph
stéréographe *m*   stereograph [instrument]
stéréographique *adj*   stereographic
stéréomètre *m*   stereometer
stéréométrie *f*   stereometry
stéréophotographie *f*   stereograph or stereophotograph
stéréotomie *f*   stereotomy
stérile *adj*   germ-free, sterile
stockage *m*   stocking [holding], storage
stockage *m*, **voie** *f* **de**   storage or waiting lane
stockage *m*, **zone** *f* **de**   storage or waiting [road] space
stockage *m* **d'eau chaude**   hot water storage

stockage *m* **de chaleur**   heat storage
stockage *m* **de combustible**   fuel storage
stockage *m* **de déchets**   waste storage
stockage *m* **de décombres**   building rubbish/rubble storage
stockage *m* **de gaz**   gas storage
stockage *m* **des données**   data storage [computer]
stockage *m* **thermique**   heat storage, thermal storage
store *m*   awning, blind, shade [blind]
store *m* **à rouleaux**   roller blind
store *m* **capote**   awning mounted on hoops
store *m* **vénitien**   venetian blind
strate *f*   stratum
stratégie *f* **de contrôle**   control strategy
stratégie *f* **de protection contre l'incendie**   fire protection strategy
stratification *f*   stratification
stratification *f* **de l'air**   air stratification
stratifié *adj*   laminated, stratified
stratifié *m*   laminate
stratifier *vb*   stratify
stratigraphie *f*   stratigraphy
striction *f*   necking/reduction of area [of a test piece]
structure *f*   structure
structure *f* **de chaussée**   road structure
structure *f* **de transfert**   transfer structure
structure *f* **du sol**   soil structure
structure *f* **en nid d'abeilles**   honeycomb structure
structure *f* **interne**   internal structure
stuc *m*   stucco
studio *m*   apartment [one room], flatlet
stylobate *m*   skirting 220–250mm high, stylobate [architectural]
styropor *m*   expanded polystyrene (Styropor)
subjectile *m*   ground or substrate [for paint, rendering]
submerger *vb*   flood [land], submerge
subsidence *f*   subsidence
subsidence *f* **minière**   mining subsidence
substance *f* **absorbée**   adsorbate
substance *f* **adhésive**   adhesive
substance *f* **comburante**   oxidizing substance
substance *f* **diluante**   diluting medium
substance *f* **explosible**   explosible or explosive substance
substance *f* **inflammable**   (in) flammable or ignitable substance
substance *f* **irritante**   irritant (substance)
substance *f* **toxique**   toxic material or substance
substituer *vb* **(à)**   substitute (for)
substitution *f* **(de, à)**   substitution (of, for)
substratum *m* **imperméable**   impermeable substratum
substratum *m* **rocheux**   bedrock
substruction *f*   foundations of old buildings found under new, underpinning
substructure *f*   substructure
subterrané *adj*   subterranean, underground
subvention *f*   grant, subsidy
subventionner *vb*   fund, subsidize
succession *f*   inheritance
succion *f*   suction

sud *adj*   south, southern, southerly, southward
sud *m*   south
sud-est *m*   south-east
sud-ouest *m*   south-west
suie *f*   soot
suintement *m*   leaking [of a boiler], oozing, seepage, sweating or weeping [e.g. of a damp wall], weeping [e.g. of a joint]
suivant *adj*   in accordance with, according to
suivant contrat   according to contract
suivant inventaire   according to stock-list
sujet *m*   subject
sujétion *f*   constraint
sujétions   subject to [in contract specifications]
sulfate *m*   sulphate
sulfater *vb*   spray (with copper sulphate)
superciment *m*   super cement [OPC mixed to improve long term strength]
superficie *f*   area, surface, surface area
superficie *f* **de section**   cross-section(al) area
superficiel *adj*   surface
superficielle, eau *f*   surface water
superficielle, tension *f*   surface tension
supérieur *adj*   upper
superposer *vb* **sur**   put on top of
superstructure *f*   pavement [of a road, runway], superstructure
superstructure *f* **métallique**   steel superstructure
support *m*   bearer, bearing, bracket, prop, support
support *m* **à étrier**   stirrup bracket
support *m* **anti-vibration**   anti-vibration mounting
support *m* **de câble**   cable bracket
support *m* **de perceuse**   drill stand
support *m* **de soutien**   bearing support
support *m* **de tube/tuyau**   pipe hanger/ support
support *m* **de tuyauterie**   pipe hanger/support
support *m* **du brûleur**   burner mounting
supporter *vb*   bear, carry, hold up, support
supporter *vb* **une charge**   carry/support a load
sur champ   edgewise, on edge
sur chenilles   tracked [of caterpillar vehicle]
sur place   in situ
surabondant *adj*   redundant [of a system able to redistribute loads]
surcharge *f*   additional, excess, extra or variable load, overloading, surcharge (loading)
surcharge *f* **de neige**   snow load
surcharge *f* **de vent**   wind load
surcharge *f* **mobile**   live load, moving load
surcharger *vb*   overload
surchauffé *adj*   overheated, superheated
surchauffe *f*   overheating, superheat(ing)
surchauffer *vb*   overheat, superheat
surchauffeur *m*   superheater
surcompactage *m*   over-compaction
surconsolidé *adj*   over-consolidated
surcontrainte *f*   overstress
surdimensionnement *m*   overdesigning, oversizing
surélévation *f*   heightening or raising [e.g. of a wall]

surélevé *adj*   elevated [e.g. of a railway], heightened, raised
surélever *vb*   raise or heighten [increase height of]
surépaisseur *f*   oversize [in thickness]
sûreté *f*   safety
surfaçage *m*   surfacing or smoothing [surface of a slab]
surface, de   cosmetic [e.g. changes]
surface *f*   area, surface, floor or surface area
surface, couche *f* **de**   surface course/layer
surface, eaux *f,pl* **de**   surface water run-off
surface, finition *f* **de**   surfacing [roads, etc]
surface *f* **chauffante**   heating surface
surface *f* **couverte**   floor area/space
surface *f* **corrigée**   amended area [for rent calculation]
surface *f* **d'appui**   bearing area
surface *f* **de chauff(ag)e**   heating surface
surface *f* **de chauffe d'une chaudière**   boiler heating surface
surface *f* **de circulation**   traffic area
surface *f* **de cisaillement**   shear surface
surface *f* **de conduit**   duct area
surface *f* **de contact**   contact area/surface
surface *f* **de déchargement**   unloading bay
surface *f* **de glissement**   sliding surface
surface *f* **de grille**   grate area
surface *f* **de l'eau**   water surface
surface *f* **de plancher**   floor area
surface *f* **de rayonnement**   radiant area/surface, radiator surface
surface *f* **de référence**   datum surface
surface *f* **de refroidissement**   cooling surface
surface *f* **de reprise**   construction joint [in concrete]
surface *f* **des étages**   floor area or space
surface *f* **des sols**   floor areas
surface *f* **du filtre**   filter area
surface *f* **du sol**   floor finish
surface *f* **filtrante**   filter area
surface *f* **frontale**   face area
surface *f* **habitable**   habitable area/space
surface *f* **plane**   plane surface
surface *f* **portante**   bearing area/surface
surface *f* **rugueuse**   rough surface
surface *f* **usinée**   machined surface
surface *f* **utilisable**   usable space
surface *f* **vitrée**   glazed area
surface, faire *vb*   surface
surface, refaire *vb*   resurface
surfaceuse *f*   finisher/finishing machine [road construction]
surfaceuse *f* **à coffrage glissant**   slipform paver/paving machine
surfaire *vb*   overcharge [price too high]
surfusion *f*   supercooling
surmoduler *vb*   overmodulate
surmonté *adj* **(de)**   surmounted (by), topped (by)
surmonter *vb* **de**   surmount by, top by
surpeuplement *m*   over-occupation, overpopulation
surplomb, en   overhanging

surplomb *m, voir aussi* fruit   outwards inclination or leaning, overhang

surplomber *vb*   be out of plumb, overhang

surplus *m*   excess, surplus

surpresseur *m*   pressure booster

surpression *f*   excess pressure

sursaturation *f*   supersaturation

sursaturé *adj*   supersaturated

sursaturer *vb*   supersaturate

surtemps *m*   overtime

surtension *f*   over-stretching, over-voltage

surtension, onde *f* **de**   voltage surge

surveillance *f*   control, inspection, supervision

surveillance *f* **des chaudières**   boiler inspection

surveillance *f* **hygiénique**   sanitary inspection

surveillance *f* **permanente**   constant/continuous supervision

surveillance *f* **préliminaire**   preliminary inspection

surveillant *m* **de travaux**   clerk/inspector of works

surveiller *vb*   control, observe, supervise

survitrage *m*   secondary glazing

survolteur *m*   voltage booster

suspendre *vb*   hang up, suspend

suspendu *adj*   suspended, hanging

suspendu, luminaire *m*   suspended light fitting/fixture

suspendu, pont *m*   suspension bridge

suspendue, couverture *f*   suspended roof [hung from cables]

suspension *f*   hanging/suspended light fitting/fixture, liquid containing fine particles in suspension

suspente *f*   suspension, suspender [e.g. wire or rod for fibrous plaster work], suspension [equipment supporting a lift cage], hanger [suspension bridge]

suspente *f* **à collier**   clamp support

syénite *f*   syenite

symbole *m*   symbol

symbole *m* **de représentation**   conventional sign/symbol

symétrie *f*   symmetry

symétrique *adj*   symmetrical

synchronisation *f*   synchronization

synclinal *m*   syncline, trough [geological]

syndical *adj*   union

syndicat *m*   association, trade union

syndrome *m* **de bâtiment malsain**   sick building syndrome

système *m*   system

système *m* **à absorption–diffusion**   diffusion–absorption system

système *m* **à aspiration**   induction system

système *m* **à détente directe**   direct expansion system

système *m* **à deux conduits**   dual duct system

système *m* **à eau chaude (à haute pression)**   (high pressure) hot water system

système *m* **à eau chaude à basse/moyenne pression**   low-/medium-pressure hot water system

système *m* **à volume d'air variable (VAV)**   variable air volume (VAV) system

système *m* **d'aération**   ventilating system

système *m* **d'alarme**   alarm system

système *m* **d'assainissement séparatif/unitaire**   separate/combined drainage system

système *m* **d'énergie totale**   total energy system

système *m* **d'expansion directe**   direct expansion system

système *m* **d'extinction automatique**   fire sprinkler system

système *m* **d'extinction à sec**   dry sprinkler system

système *m* **d'extraction d'air**   air extract(ion)/exhaust system

système *m* **d'imperméabilisation**   waterproofing system

système *m* **de câblage**   wiring system

système *m* **de canalisation (pour câbles)**   (cable) duct system

système *m* **de chauffage (à eau chaude)**   (hot water) heating system

système *m* **de chauffage à vapeur avec retour**   dry return (steam) heating system

système *m* **de circulation d'air**   air recirculating plant or system

système *m* **de climatisation (d'air)**   air-conditioning system

système *m* **de climatisation à débit d'air variable**   variable air volume (VAV) air-conditioning system

système *m* **de climatisation à double conduit**   dual duct air-conditioning system

système *m* **de commande**   control system

système *m* **de communication**   communication system

système *m* **de commutation**   changeover system

système *m* **de conditionnement d'air (tout air)**   (all-air) air-conditioning system

système *m* **de construction**   construction system

système *m* **de contre réaction**   feedback system

système *m* **de contrôle (automatique)**   (automatic) control system

système *m* **de contrôle d'accès (SCA)**   access control system (ACS)

système *m* **de déclenchement**   release system

système *m* **de dégivrage**   defrosting system

système *m* **de détection**   detection system

système *m* **de diffusion d'air**   air distribution system

système *m* **de distribution**   distribution system

système *m* **de distribution (de câbles)**   (cable) distribution system

système *m* **de drainage**   drainage system

système *m* **de gestion énergétique**   energy management control system

système *m* **de gestion de l'énergie des bâtiments**   building energy management system

système *m* **de gestion technique d'un bâtiment**   building management system (BMS)

système *m* **de lubrification**   lubrication system

système *m* **de mise à l'atmosphère**   direct vent system

système *m* **de plafond intégré**   integrated ceiling system

système *m* **de rafraîchissement par rayonnement**   radiant cooling system

système *m* **de récupération de chaleur**   heat recovery system

système *m* **de réfrigération à compression**   compression refrigeration system

système *m* **de réfrigération à détente directe** direct expansion refrigeration system

système *m* **de refroidissement (direct)** (direct) cooling system

système *m* **de régulation (de température)** (temperature) regulation system

système *m* **de repérage pour filerie** marking system for cables/wiring

système *m* **de soufflage d'air** supply-air system

système *m* **de tuyauteries** pipe system

système *m* **drainant** subsoil drain [road construction]

système *m* **du tout à l'égout** sewerage system

système *m* **mécanique/naturel de désenfumage** mechanical/natural smoke extract system

système *m* **monobloc de conditionnement d'air** self-contained air-conditioning system

système *m* **monotube** one/single-pipe system

système *m* **par gravité** gravity system

système *m* **piloté par ordinateur** computer-based system

système *m* **porteur** supporting structure or system

système *m* **tout air** all-air system

système *m* **VAV (à volume d'air variable)** VAV (variable air volume) system

tabatière *f* skylight [hinged]

tabatière, châssis *m* dormer skylight

table *f* face [of a hammer], flange [of girder], plain decorative panel [in a wall or panel], slab [slate], tablet [of stone], telephone switchboard

table *f* **à abattants** drop-leaf table

table *f* **à couder** bending table [reinforcement]

table *f* **à dessin** drawing board

table *f* **à encoller** wallpaper pasting table

table *f* **à rallonges** extending table

table *f* **à repasser** ironing board

table *f* **à secousse** shaking table [for testing concrete spread]

table *f* **anglaise** gate-legged table

table *f* **basse** coffee/occasional table

table *f* **coffrante** re-usable formwork for floor slabs

table *f* **de compression** compression flange

table *f* **de cuisson** cooking top unit

table *f* **de matières** list/table of contents

table *f* **de nuit** bedside table

table *f* **de rabotage (mobile)** thicknessing table (moving)

table *f* **de répartition** fuseboard

table *f* **de toilette** dressing table, wash stand

table *f* **défoncée** recessed plain decorative panel

table *f* **élévatrice** elevating work platform

table *f* **saillante** raised plain decorative panel

tableau *m* board [for notices, etc], panel [board], chart, diagram, reveal [of a door, window], switchboard, table [list, figures, etc]

tableau *m* **d'annonces** notice board

tableau *m* **de commande** control board/panel, switchboard

tableau *m* **de commutation** changeover panel

tableau *m* **de contrôle** control board/panel

tableau *m* **de contrôle incendie** fire control panel

tableau *m* **de contraintes** table of technical requirements

tableau *m* **de dépannage** fault finding chart

tableau *m* **de distribution** distribution panel, switchboard

tableau *m* **de graissage** lubrication chart

tableau *m* **de puissances** load table [electrical]

tableau *m* **de régulation** control board or panel

tableau *m* **de signalisation** indicator board

tableau *m* **de télécommande** remote control panel

tableau *m* **des alarmes** alarm panel

tableau *m* **des charges** load table [structural]

tableau *m* **des quantités des matériaux** bill of quantities (BOQ)

tableau *m* **des tarifs** scale of charges

tableau *m* **électrique** electrical distribution board/panel

tableau *m* **général basse tension (TGBT)** low-voltage distribution board/panel

tableau *m* **général de commande** central control panel

tableau *m* **HT** high-voltage switchboard

tables *f,pl* **de vapeur** steam tables

tablette *f* bearing surface [of a joist or beam], coping stone, flat slab [e.g. a worktop], mantelpiece, mantelshelf

tablette *f* **d'appui** stone handrail

tablette *f* **de cheminée** mantelpiece or mantelshelf

tablette *f* **de fenêtre** window ledge or sill

tableur *m* spreadsheet program

tablier *m* deck [e.g. bridge, floor], lowest projecting panel of a *porte cochère*

tablier *m* **(de cheminée)** fireplace curtain [slides down to block off]

tablier *m* **métallique** steel deck or decking plate [of a bridge]

tache *f* spot [of dirt, mark], stain [mark]

tâche *f* assignment, job

tachéomètre *m* tacheometer, tachymeter

tacher *vb* stain [mark], tarnish

tâcher *vb* **de faire qch** attempt/try to do sth

tâcheron *m* minor/small works/labour-only (sub-)contractor, jobbing workman, piece worker

tachymètre *m* tachometer

taillage *m* dressing [of stone]

taillant *m*   bit [for ground/rock drilling], cutting edge [of a tool]

taillant *m* **au tungstène**   tungsten (tipped) bit

taillant *m* **en tarière**   auger-type bit

taille *f*   cutting, dressing or shaping [of stone, brick, etc], dimensions, size [of an object], pruning [trees, etc]

taille *f* **bouchardée**   bush hammering

taille *f* **brute**   rough dressing

taille *f* **d'ébauche**   preparatory dressing [for a moulding, etc]

taille *f* **d'un joint**   dressing of side of stone to form a joint

taille *f* **d'un lit**   dressing to bed a stone on another

taille *f* **de parement**   surface dressing

taille *f* **de particule**   particle size

taille *f* **de sécurité**   safety pruning

taille *f* **douce**   gentle pruning

taille *f* **éclatée**   split-face dressing

taille *f* **layée**   dressing with a toothed hammer

taille *f* **lisse/plane**   smooth dressing

taille *f* **mémoire d'un ordinateur**   computer storage size

taille *f* **rustiquée**   dressing to give a 'rustic' finish

taille *f* **sévère**   hard pruning

taille-haies *f*   hedge trimmer

tailler *vb*   dress, cut or shape [stone, brick, etc], prune, sharpen [a pencil]

tailler *vb* **en redans**   excavate in steps

tailleur *m* **de pierre(s)**   stone cutter or finisher, stonemason

taillis *m*   coppice, copse, thicket

tailloir *m*   abacus, glazier's knife for cutting lead

talochage *m*   floating [of concrete surface]

taloche *f*   float [tool], (plasterer's) hawk, laying on trowel

talocher *vb*   float [plaster, etc]

talon *m*   cheque stub/counterfoil, closure over the end of a seam in metallic roofing, ogee moulding, nib [of a beam, a plain tile], sole [of a plane]

talus *m*   bank, embankment, slope [embankment, excavation], batter [excavations]

talus, pente *f* **de**   slope inclination

talus *m* **à forte pente**   steep slope

talus *m* **à pente faible**   gentle slope

talus *m* **en déblai**   cut(ting) slope

talus *m* **en remblai**   embankment slope

talus *m* **herbacé**   grassed slope

taluter *vb*   batter [an embankment, etc], slope [side of earthworks]

taluteuse *f*   sloper [earthworks]

tambour *m*   revolving door, drum or tambour [of a column, under a cupola or dome, etc]

tambour *m* **de frein**   brake drum

tambour *m* **de malaxage**   mixing drum

tambour *m* **enregistreur**   drum chart recorder

tambour *m* **filtrant**   filter drum

tambour *m* **mélangeur**   mixing drum

tambour *m* **sécheur enrobeur**   asphalt drum mix(ing) plant

tamis *m*   riddle, screen, sieve

tamis *m* **à mailles**   mesh screen

tamis *m* **de contrôle**   test sieve

tamis *m* **en toile métallique**   wire mesh screen [fine]

tamis, passer *vb* **au**   sieve, sift

tamisage *m*   filtering [sieving], riddling, screening [grading], sieving, sifting

tamisage, station *f* **de**   screening plant

tamisage *m* **de contrôle**   sieving test

tamisat *m* **inférieur**   undersize [in a sieving test]

tamisé *adj*   riddled, sieved, sifted, subdued [light]

tamiser *vb*   filter [light], riddle, screen [grade], sieve, sift

tampon *m*   buffer, cover [plate or slab], end plate, stopper [of a pipe, etc], pad [e.g. for spreading varnish], plug [for blocking or stopping up sth], wall plug [for fixing], rubber stamp

tampon *m* **avec étanchéité**   sealed/watertight cover

tampon *m* **d'égout**   manhole cover

tampon *m* **de forage**   borehole cover

tampon *m* **de visite**   inspection cover

tampon *m* **hermétique**   access cover/eye, cleaning eye, sealed inspection cover/fitting

tampon-fonte *m*   cast-iron cover

tamponné *adj*   plugged [wall, etc], stamped [letter, document, etc]

tamponner *vb*   dab [a surface, with a pad, etc], mop [a surface, to dry it], plug [a wall], stamp [a document or letter]

tangentiel *adj*   tangential

taper *vb*   dab or lay on [paint, etc], plug up, stop up [a hole, etc], tap [strike lightly]

taper *vb* **à la machine**   type (out) [on a keyboard]

tapette *f*   paintbrush used for wood-graining

tapis *m*   carpet, thin layer [of road construction]

tapis *m* **aiguilleté**   needle-punch carpet

tapis *m* **alimenteur**   feed(ing) belt (conveyor)

tapis *m* **bitumineux**   flexible carpet or layer [road foundations]

tapis *m* **d'argile**   clay blanket

tapis *m* **de branchage**   fascine mattress

tapis *m* **de drainage**   drainage layer [groundworks]

tapis *m* **de récolte**   collecting conveyor [pavement planing machine]

tapis *m* **drainant**   drainage blanket [groundworks]

tapis *m* **élévateur**   loading conveyor [pavement planing machine]

tapis *m* **filtrant**   filter layer [groundworks]

tapis *m* **imperméable**   impermeable blanket [groundworks]

tapis *m* **peseur**   belt weigher

tapis *m* **roulant**   travolator

tapisser *vb*   cover [furniture, a wall], line [with paper/fabric, etc], upholster, wallpaper

tapisser *vb* **(de papier peint)**   paper (with wallpaper)

tapisserie *f*   upholstery, wallpaper, wall-covering, tapestry wallpaper

tapissier *m*   interior decorator, paperhanger [tradesman], upholsterer

tapure *f*   shrinkage crack [in welds]

taquet *m*   cleat [for a blind cord], dowel [loose, e.g. to support a shelf], insert, plug [of wood, in masonry, etc to receive fixings], peg [for marking a layout or alignment], wedge

taquet *m* **d'échelle**   tray fixable to ladder rungs
taquets *m,pl*   grounds or groundwork [carpentry]
tarabiscot *m*   fine groove between two mouldings
taraud *m*   tap [for threading]
taraudage *m*   tapping [a thread]
taraudé *adj*   threaded
tarauder *vb*   tap [a thread], thread [cut or form a thread]
targette *f*   small flat bolt [for gate, etc], sash bolt
tarière *f*   auger
tarière *f* **à vis**   screw auger
tarière *f* **américaine**   Scotch screw auger
tarière *f* **cuiller**   spoon bit
tarif *m*   list of charges, price list, rate, tariff
tarif *m* **binôme**   two-part rate [electricity tariff]
tarif *m* **de base**   basic rate
tarif *m* **forfaitaire**   flat rate, fixed charge
tarif *m* **uniforme**   flat rate [charge]
tarifer *vb*   fix a price or rate for
(se) tarir *vb*   dry up or run dry [a spring, well, stream, etc]
tarmacadam *m*   tarmacadam
tartre *m*   fur or scale [in a boiler or pipes]
tartre *m* **de chaudière**   boiler scale
tas *m*   building under construction, heap, pile [of sth]
tas, sur le   on site
tas *m* **de charge**   springer or springing stone
tasseau *m*   bracket, batten, strip [of wood]
tassement *m*   packing, ramming or tamping down, settlement [of ground, earth]
tassement, sensibilité *f* **au**   settlement tendency [of ground or fill]
tassement *m* **de consolidation**   consolidation settlement
tassement *m* **des appuis**   settlement of supports
tassement *m* **différentiel**   differential settlement
tasser *vb*   pack, ram or tamp (down)
se tasser *vb*   bed down [earth, ground, etc], settle [compact or sink down], sink (down) or subside [ground, a structure, etc]
taupe *f*   tunnel boring machine
taupette *f*   paintbrush used for wood-graining
taux *m*   rate
taux *m* **d'air frais**   fresh or outside air rate
taux *m* **d'entraînement**   entrainment ratio
taux *m* **d'escompte**   discount rate
taux *m* **d'évaporation**   evaporation rate
taux *m* **d'induction**   induction ratio
taux *m* **d'intérêt**   interest rate
taux *m* **de base bancaire**   base lending rate
taux *m* **de change**   exchange rate [currency]
taux *m* **de charge**   charging rate [electrical]
taux *m* **de compression**   compression ratio
taux *m* **de consommation**   consumption rate
taux *m* **de croissance**   growth rate
taux *m* **de décroissance**   rate of decay [i.e. rate of decrease, not rotting]
taux *m* **de pollution de l'air**   level of air pollution
taux *m* **de prêt**   lending rate
taux *m* **de renouvellement d'air/de l'air**   air change rate
taux *m* **de travail**   stress
taux *m* **des salaires**   wage rate

taux *m* **officiel d'escompte**   bank rate
tavaillon *m*   shingle [roofing, cladding]
taxe *f*   tax
taxe *f* **d'habitation**   residential tax [in France]
taxe *f* **professionnelle**   tax on professional activity [in France]
taxe *f* **sur la valeur ajoutée (TVA)**   value added tax (VAT)
taxes *f* **foncières**   land or real estate taxes [in France]
taxer *vb*   tax [sth]
té *m*   T/tee [pipe fitting], T-joint, T-square
té, fer *m* **en**   T-iron, T-section
té, raccord *m*   T-joint
té *m* **de purge**   T/tee with cleaning eye [pipe fitting]
té *m* **de réduction**   reducing T/tee
té *m* **de réglage**   regulating T/tee
technicien *m*   technician
technicien *m* **en étude de prix**   estimator
technico-commercial, agent *m*   technical salesman
technique *adj*   technical
technique *f*   technique
technique *f* **d'assainissement**   public health engineering
technique *f* **de mesure**   measuring technique
technique *f* **du froid**   refrigeration engineering
technique *f* **sanitaire**   sanitary engineering
technologie *f* **du froid**   refrigeration engineering
teck *m*   teak
téflon *m* **(en ruban)**   Teflon tape
teint *m*   colour, dye
teintage *m*   tinting
teinte *f*   hue, shade [of colour], tint, stain
teinter *vb*   stain, tint
teinture *f*   dye(stuff)
télécabine *f*   cabin cableway/ropeway
télécommande *f*   remote control
télécommunications *f,pl*   telecommunications
télécontrôle *m*   remote control
télécopie *f*   facsimile/fax, telefax
télécopieur *m*   facsimile/fax[†] machine
téléphérique *f*   blondin, cableway, ropeway
téléphone *m*   telephone
téléphonique *adj*   telephone *adj*
téléphonique, central *m*   telephone exchange
télérupteur *m*   remote switch
télésiège *m*   chairlift
téléski *m*   drag lift
téléthermomètre *m*   remote reading thermometer
téléviseur *m*   television set
télévision *f* **par câble**   cable television
tellurien *adj*   earth [of current]
tellurique *adj*   from/of the ground/soil
tellurique, secousse *f*   ground tremor
tellurohmètre *m*   earth resistance meter
témoin *adj*   control, telltale, warning [of a lamp]
témoin *m*   boundary/height marker, telltale, witness
témoin, appartement *m*   show flat
tempérament, achat *m* **à**   hire/instalment purchase
tempérament, vente *f* **à**   sale on deferred payment terms
température *f*   temperature

température *f* **ambiante**   air or ambient temperature
température *f* **apparente**   apparent temperature
température *f* **au thermomètre**   dry bulb temperature
température *f* **au thermomètre globe**   globe temperature
température *f* **continue de service**   continuous working temperature
température *f* **d'alimentation**   supply temperature
température *f* **d'allumage**   ignition temperature
température *f* **d'ébullition**   boiling temperature
température *f* **d'équilibre**   balance point or equilibrium temperature
température *f* **d'évaporation**   evaporation temperature
température *f* **d'ignition**   ignition temperature
température *f* **d'injection**   injection temperature
température *f* **d'utilisation**   working temperature
température *f* **de brillance**   black body temperature
température *f* **de combustion**   combustion temperature
température *f* **de condensation**   condensing temperature
température *f* **de congélation**   freezing temperature
température *f* **de décharge**   discharge temperature
température *f* **de départ**   flow temperature
température *f* **de fonctionnement**   operating temperature
température *f* **de fumée**   flue gas temperature
température *f* **de l'air extérieur**   outside air temperature
température *f* **de la couche limite**   boundary layer temperature
température *f* **de la flamme**   flame temperature
température *f* **de la fumée**   smoke temperature
température *f* **de mélange**   mixing temperature
température *f* **de paroi**   wall temperature
température *f* **de refoulement**   discharge temperature
température *f* **de retour**   return temperature
température *f* **de service**   operating temperature
température *f* **de sortie**   outlet temperature
température *f* **de surface**   surface temperature
température *f* **de travail**   working temperature
température *f* **délivrée**   delivery temperature
température *f* **des gaz brûlés**   flue/waste gas temperature
température *f* **équivalente de corps noir**   black body equivalent temperature
température *f* **extérieure**   outside temperature
température *f* **fictive extérieure**   sol air temperature
température *f* **humide**   wet bulb temperature
température *f* **imposée**   design temperature
température *f* **initiale**   initial temperature
température *f* **intérieure du local**   room temperature
température *f* **radiante**   radiant temperature
température *f* **sèche**   dry bulb temperature
température *f* **superficielle**   skin temperature
tempête *f*   storm
temporaire *adj*   temporary
temporisation *f*   time delay
temporisé *adj*   slow-acting, time delay(ed)
temps *m*   time [length of], season [of the year], weather
temps *m* **d'arrêt**   idle time
temps *m* **de mise en marche**   running-up or starting time
temps *m* **de pause**   idle time
temps *m* **de prise**   setting time [concrete, glue]
temps *m* **de réponse**   response time

temps *m* **de retard**   time lag [delay]
temps *m* **de réverbération**   reverberation time
temps *m* **mort**   dead time
temps *m* **moyen entre pannes**   mean time between failures
temps *m* **réel**   real time
temps *m* **solaire**   solar time
ténacité *f*   toughness [of metals]
tenaille *f*   tongs
tenaille *f* **à vis**   hand vice
tenaille *f* **d'essai**   tong tester
tenaille *f* **de menuisier**   joiner's pincers
tenailles *f,pl*   pincers, pliers, tongs
tenant, d'un seul   continuous [of a block or plot of land]
tendeur *m*   strainer, tensioner [e.g. for wire fencing]
tendeur *m* **à la lanterne**   turnbuckle
tendre *vb*   tighten [tension]
tendu *adj*   very smooth [of paint or varnish]
teneur *f*   content [relative, e.g. in a solution, a gas]
teneur *f* **en**   content [relative, percentage], degree of, percentage of [proportion]
teneur *f* **en carbone**   carbon content
teneur *f* **en cendres**   ash content
teneur *f* **en eau**   degree of humidity, water content
teneur *f* **en eau d'équilibre**   equilibrium moisture content
teneur *f* **en fines**   filler or fines content [of road material]
teneur *f* **en humidité**   moisture content
teneur *f* **en poussière(s)**   dust content [of air]
teneur *f* **en vides**   voids content
tenon *m*   tenon
tenon et mortaise, assemblage *m* **à**   mortise-and-tenon joint
tenon *m* **simple**   mortise-and-tenon joint
tenonneuse *f*   tenoner
tenseur *m*   tensor
tensiomètre *m*   tensiometer
tension *f*   tensile stress, tension, voltage
tension *f***, haute/moyenne/basse**   high/medium/low voltage
tension *f* **à vide**   no-load voltage
tension *f* **d'allumage**   starting voltage [lighting]
tension *f* **d'isolement**   insulation voltage
tension *f* **d'origine thermique**   thermal stress
tension *f* **de claquage**   breakdown voltage [insulation]
tension *f* **de fonctionnement**   operating voltage
tension *f* **de grille**   grid voltage
tension *f* **du secteur**   mains/supply voltage
tension *f* **en charge**   load voltage
tension *f* **étoilée**   star voltage
tension *f* **interne**   internal stress
tension *f* **résiduelle**   residual stress
tension *f* **superficielle**   surface tension
tenu *adj*   fine [of particles, wire]
tenue *f*   behaviour [e.g. in fire], dress [e.g. working dress], running [of a house, shop, etc], upkeep
tenue *f* **à la flamme**   flame resistance
tenue *f* **au feu**   behaviour in fire
tenue *f* **de travail**   working dress
tenue *f* **des livres**   book-keeping

tenue *f* **limitée**   limited resistance
terme *m*   conclusion, end, term
terme, à   full-term, in due time
terme, à court   in the short term
terme, à long   in the long term, long term [loan, etc]
terme, arriver *vb* **à**   expire, fall due [e.g. payment]
terme *m* **de métier**   technical term
terminal *m*   terminal [airport]
terminer *vb*   complete, finish [make an end], round off [a job, etc], terminate
terminer *vb* **le gros œuvre**   top out [a building]
terminologie *f*   terminology
termite *m*   termite
ternir *vb*   dull, tarnish
ternissement *m*   clouding [glass], fading [colours], tarnishing [metal surfaces]
terrain *m*   ground, land, terrain, site [area or plot of land]
terrain *m* **à bâtir**   building plot, housing site
terrain *m* **agricole**   agricultural land
terrain *m* **constructible**   building land
terrain *m* **couvert**   covered/roofed-over area
terrain *m* **d'atterrissage**   landing strip
terrain *m* **d'aviation**   airfield
terrain *m* **de camping**   campsite
terrain *m* **de couverture**   overburden
terrain *m* **de fondation**   formation, sub-grade or subsoil
terrain *m* **de sport(s)**   sports field or ground
terrain *m* **de transport**   alluvial deposit
terrain *m* **meuble**   soft ground
terrain *m* **naturel**   formation, natural or undisturbed ground
terrain *m* **nu**   open [uncovered] area
terrain *m* **vague**   waste ground
terrasse *f*   terrace, nearly flat roof
terrasse, jardin *m* **en**   terraced garden
terrasse *f*, **toiture-**   roof terrace
terrasse-jardin *f*   roof garden
terrassement *m*   earthwork(s), embankment, excavation(s), groundworks
terrassement, engins *m,pl* **de**   earth-moving equipment/plant
terrassement, travaux *m,pl* **de**   earthworks
terrasser *vb*   dig out, excavate
terrassier *m*   earthworks labourer, navvy
terrasson *m*   nearly flat roof, upper part of a mansard roof
terrasson *m* **de sécurité**   overflow tray [beneath a head tank]
terrazzo *m*   terrazzo
terre *f*   earth [ground, or electrical], soil
terre *f* **armée**   reinforced earth
Terre *f* **Armée**   Reinforced Earth Company [trade name]
terre *f* **cuite**   earthenware, fired clay
terre *f* **d'infusoires**   diatomaceous earth or diatomite
terre *f* **de couverture**   topsoil
terre *f* **électrique**   electrical ground
terre *f* **réfractaire**   refractory earth
terre *f* **végétale**   organic soil, topsoil
terre-plein *m*   central reserve/reservation [roads], solid floor on formation, terrace [horizontal strip of ground]

terre-plein *m* **central**   central reservation [roads]
terre-plein *m* **spécial**   delimited railway right-of-way on highway [not UK]
terreau *m*   compost, loam
territoire *f*   area [e.g. part of a country], territory
tête *f*   head [of a nail, hammer], top [of a list, table]
tête *f* **amont**   inlet [of a culvert]
tête *f* **aval**   outlet [of a culvert]
tête *f* **bombée**   raised countersunk head [of a screw]
tête *f* **creuse**   recessed head [Robertson type screw]
tête *f* **cruciforme**   cross-head or recessed head [Phillips/Pozidrive/Supadrive type screw]
tête *f* **cylindrique**   cheese head [of a screw or bolt]
tête *f* **de boulon**   bolt head
tête *f* **de brûleur**   burner nozzle or tip
tête *f* **de câble**   cable end or tail
tête *f* **de chat**   small round piece of rubble
tête *f* **de chevalement**   underpinning beam [temporary support]
tête *f* **de chevron**   rafter end or head
tête *f* **de frappe**   striking face [of a hammer]
tête *f* **de mur**   end face of a wall
tête *f* **de pieu**   pile cap
tête *f* **de poteau**   column/stanchion head
tête *f* **fraisée**   countersunk head
tête *f* **hexagonale**   hexagonal head
tête *f* **noyée**   countersunk head
tête *f* **plate**   flat countersunk head [of a screw]
tête *f* **ronde**   button head [of a screw]
têtière *f*   side of a lock through which the bolt passes
tétrapode *m*   tetrapod
tétrapolaire *adj*   triple-pole
têtu *m*   mason's hammer
texture *f*   texture
théodolite *m*   theodolite
théorème *m*   theorem
théorie *f*   theory
théorique *adj*   theoretic(al)
théoriquement *adv*   theoretically
thermique *adj*   heat [conduction, engine, etc], temperature [stress, etc], thermal [insulation, stress, etc]
thermique, carte *f*   temperature/thermal map
thermique, centrale *f*   power station
thermique, moteur *m*   heat engine
thermistance *f*   thermistor
thermocouple *m*   thermocouple
thermodurcissable *adj*   thermo-setting
thermodynamique *f*   thermodynamics
thermoélectrique *adj*   thermoelectric
thermo-hygromètre *m* **enregistreur**   recording thermo-hygrometer, thermo-hygrograph
thermolaqué *adj*   powder coated
thermomètre *m*   thermometer
thermomètre *m* **à alcool**   alcohol thermometer
thermomètre *m* **à bulbe mouillé-sec**   wet and dry bulb thermometer
thermomètre *m* **à cadran**   dial thermometer
thermomètre *m* **à lecture à distance**   remote reading thermometer

thermomètre *m* à mercure   mercury thermometer
thermomètre *m* à plongeur   plunger thermometer
thermomètre *m* à résistance   resistance thermometer
thermomètre *m* bilame/bimétallique   bimetallic thermometer
thermomètre *m* enregistreur   recording thermometer, thermograph
thermomètre *m* globe   globe thermometer
thermomètre *m* indicateur à distance   remote indicating thermometer
thermomètre *m* mouillé/à bulbe mouillé   wet bulb thermometer
thermomètre *m* sec   dry bulb thermometer
thermoplastique *adj*   thermoplastic *adj*
thermoplastique *m*   thermoplastic
thermorégulateur *m*   temperature controller or regulator
thermostat *m*   thermostat
thermostat *m* à gel   frost thermostat
thermostat *m* à immersion   immersion thermostat
thermostat *m* d'ambiance   room thermostat
thermostat *m* d'arrêt eau glacée   chilled water thermostat
thermostat *m* d'huile (sécurité basse)   low oil temperature thermostat
thermostat *m* d'huile (sécurité haute)   high oil temperature thermostat
thermostat *m* de l'eau de refroidissement   cooling water thermostat
thermostat *m* de protection contre le bouchage congélation   freeze-up protection thermostat
thermostat *m* de radiateur   radiator thermostat
thermostat *m* de sécurité au refoulement   high discharge temperature cut-out
thermostat *m* de sûreté   high limit thermostat
thermostat *m* plongeur   immersion thermostat
thermostatique *adj*   thermostatic
thixotropique *adj*   thixotropic
thyristor *m* à té   thyristor
tiède *adj*   lukewarm, tepid
tierceron *m*   rib springing from corner of a vault, tierceron
tiers-central *m*   middle third
tiers-point *m*   point of intersection of two arc(he)s, tierce-point, three-cornered/triangular file
tige *f*   rod [metal], shaft, shank, stem [of a bolt, rivet, etc], spindle [of a valve, tap], stalk, stem
tige *f* d'ancrage   tie-bar
tige *f* de botte   Spanish or mission tile
tige *f* de nivellement   levelling rod/staff [surveying]
tige *f* de paratonnerre   lightning rod
tige *f* de sondage   sounding rod/probe
tige *f* de soupape   valve spindle
tige *f* filetée   stud bolt, threaded rod, threaded shank of a bolt
tilleul *m*   lime [tree]
timbre *m*   total measured quantity in a bill, test pressure [of a boiler], (postage) stamp
timbre *m* fiscal   excise/revenue stamp
timbre *m* officiel   official stamp
tir *m* à l'air/l'oxygène liquide   liquid oxygen blasting

tir *m* à la mèche   firing with a safety fuse [blasting]
tir *m* électrique   electric firing [blasting]
tirage *m*   draught [of a chimney, flue], print [of a drawing, photo, etc]
tirage *m* de cheminée   chimney/flue draught
tirage *m* équilibré   balanced draught
tirage *m* mécanique   forced draught
tirage *m* par aspiration   induced draught
tirant *m*   collar beam [roof], tension member, tie, tie member, tie-rod [across a truss or building]
tirant *m* d'air   clearance [overhead]
tirant *m* d'ancrage   anchoring rod, tie-back
tirant *m* d'eau   draught, depth of water
tirant *m* de coffrage   formwork tie
tire-clous *m*   nail puller
tire-fils *m*   draw wire [in a conduit, etc]
tire-fond *m*   coach bolt or screw
tire-joint *m*   jointer [tool for mortar joints]
tire-ligne *m*   ruling pen
tirer *vb* un chèque   draw or write a cheque
tiroir *m*   drawer [pull out]
tisonnier *m*   poker [fire iron]
tissé *adj*   woven
tissu *m* d'ameublement   upholstery fabric
tissu *m* filtrant   filter cloth
titre *m*   security [stock exchange], title [deeds]
titulaire *m*   bearer, holder [of a card, document, etc]
titulaire *m* du marché   appointed contractor, successful bidder/tenderer
toile *f*   cloth
toile *f* de fibre de verre   glass fibre cloth
toile *f* émeri   emery cloth
toile *f* filtrante   filter cloth
toile *f* goudronnée   bitumen/roofing felt
toile *f* métallique   wire gauze
toilette *f*   wash stand
toilettes *f,pl*   toilets (WCs), public convenience/lavatories (WCs)
toit *m, voir aussi* toiture   roof
toit *m*, profil *m* en forme de   cambered cross-section [of a road]
toit *m*, profil *m* en travers en   crossfall [profile]
toit *m* à chevrons   rafter roof
toit *m* à croupe faîtière   hipped-gable roof
toit *m* à deux égouts   ridge roof
toit *m* à deux pans   rafter roof
toit *m* à deux pentes   ridge roof
toit *m* à deux versants   rafter roof
toit *m* à la mansarde/mansart   French or mansard roof
toit *m* à pignon   gable roof
toit *m* à quatre arêtiers   hipped roof
toit *m* à redans/redents   sawtooth roof
toit *m* à un pan/versant   mono-pitch roof
toit *m* conique   conical/turret roof
toit *m* d'ardoises   slated roof
toit *m* de chaume   thatched roof
toit *m* de tuiles à talon   plain tiled roof
toit *m* en appentis   lean-to/mono-pitch roof
toit *m* en croupe   hipped roof

toit *m* **en dents de scie**   sawtooth roof
toit *m* **en dos d'âne**   gable roof
toit *m* **en pente**   gable roof
toit *m* **en poivrière**   conical (turret) roof
toit *m* **en selle**   gable roof
toit *m* **en shed**   sawtooth roof
toit *m* **en tente**   tent roof
toit *m* **en terrasse**   flat roof
toit *m* **plat**   flat roof
toiture *f*, *voir aussi* toit   roof structure and covering, roof(ing)
toiture *f* **à la mansarde/mansart**   French or mansard roof
toiture *f* **à redans/redents**   sawtooth roof
toiture *f* **à quatre pentes**   hipped roof
toiture *f* **à une seule pente**   mono-pitch roof
toiture *f* **en appentis**   lean-to roof
toiture *f* **flamande**   pantile(d) roof
toiture *f* **inclinée**   pitched roof
toiture *f* **romaine**   mission/Spanish tiled roof
toiture-terrasse *f*   sun roof
tôle *f*   metal sheet, plate [of metal], sheet metal
tôle *f* **d'acier**   steel plate/sheet
tôle *f* **d'acier galvanisée**   galvanized steel sheet
tôle *f* **émaillée**   enamelled sheet (metal)
tôle *f* **en zinc**   zinc sheet
tôle *f* **fine**   thin sheet (metal)
tôle *f* **forte**   heavy or thick plate
tôle *f* **galvanisée**   galvanized steel sheet
tôle *f* **gaufrée**   checquer plate, embossed sheet [metal]
tôle *f* **laminée**   rolled sheet (metal)
tôle *f* **mince**   thin sheet (metal)
tôle *f* **ondulée**   corrugated iron/steel sheet
tôle *f* **plate**   flat metal sheet
tôle *f* **striée**   checquer plate
tôle *f* **zinguée**   galvanized steel sheet
tolérance *f*   clearance, allowable permissible deviation or variation, tolerance
tolérance *f* **d'usinage**   machining allowance
tolérances, hors   off/out of gauge/tolerance
tomber *vb*   drop [fall, an object]
tomber *vb* **en panne**   fail [break down]
tombereau *m* **(automoteur)**   dumper/dump truck
tommette *f*   floor tile [hexagonal], quarry tile
ton *m*   shade, tone [of colour]
tondeuse *f*   grass cutter, lawnmower
tondin *m*   torus moulding
tonnage *m* **maximum**   maximum load [on a road or bridge]
tonne *f*   metric tonne
tonne *f* **à carburant**   fuel bowser, tank(er) lorry
tonne *f* **à eau**   water tanker/bowser
tonneau *m*   butt [containing liquid, e.g. rainwater]
tonnelle *f*   barrel/tunnel vault
topographe *m*   land or topographical surveyor
topographie *f*   surveying, topography
topographique *adj*   topographic(al)
topographiques, instruments *m,pl*   surveying instruments
torche *f* **à braser**   brazing torch

torche *f* **à souder**   blowtorch
torcher *vb*   build in cob
torchis *m*   cob
tordre *vb*   distort [by twisting, buckling], twist
tordu *adj*   buckled, distorted, out of true, twisted
tore *m*   torus moulding
toron *m*   strand [of a cable]
toron *m* **de câble**   cable strand
torrent *m*   mountain stream, torrent
torsadage *m*   twisting [action and result]
torsadé *adj*   twisted
torsader *vb*   twist
torsion *f*   torsion
total *m*   total
touche *f*   electrical contact key, key [typewriter or computer]
touche *f* **d'appel**   call button
touche *f* **et frottis** *m*   rubbing down and touching up [of paintwork]
toucher *vb* **un chèque**   cash a cheque
toupie *f*   plumber's tool for expanding/bellmouthing pipe ends, router, spindle moulder or moulding machine
toupie *f* **à béton**   truckmixer
toupie-pompe *f* **(à béton)**   truckmixer with concrete pump
tour *f*   tower
tour *f* **à courants croisés**   cross-flow (cooling) tower
tour *f* **d'angle**   corner tower
tour *f* **d'habitation**   high-rise apartment building, block of flats, tower block [flats/apartments]
tour *f* **de contrôle**   control tower
tour *f* **de flanquement**   flanking tower
tour *f* **de guet**   watch tower
tour *f* **de réfrigération/refroidissement**   cooling tower
tour *f* **de refroidissement à convection forcée**   forced draught water cooling tower
tour *f* **de refroidissement à pulvérisation**   spray cooling tower
tour *f* **de refroidissement à tirage forcé/naturel**   forced/natural draught water cooling tower
tour *f* **de refroidissement d'eau/par film**   water/film cooling tower
tour *f* **réfrigérante**   cooling tower
tour *m*   circumference, revolution, rotation, turn, lathe
tour *m* **à bois/à dégauchir**   wood (turning) lathe
tourbe *f*   peat
tourbeux, sol *m*   peat
tourbillon *m*   vortex
tourbillon *m* **d'eau**   whirlpool
tourbillonner *vb*   swirl
tourelle *f*   turret
touret *m*   cable drum, reel
tourillon *m*   bearing, link pin [of a chain], swivel pin [of a gate]
tournant *adj*   spiral/winding [of a staircase], swivel *adj*
tournant *m*   bend, twist [in a road], turning point
tourne-à-gauche *m*   saw setting tool, tap wrench
tourner *vb*   go round, revolve, rotate, turn, run [motor], swivel, turn round [an object]

tourner *vb* **sans balourd**   run true

tournevis *m*   screwdriver

tournevis *m* **à spirale et à rochet**   pump/spiral ratchet/yankee screwdriver

tournevis *m* **cruciforme**   cross-point screwdriver

tournisse *f*   stud [timber structure]

tournures *f,pl*   shavings [wood turning]

tours *m,pl* **par minute**   revolutions/rotations/turns per minute (rpm)

tous les x mètres   every x metres, at x metre centres

tout à l'égout *m*   main drainage

tout corps *m* **d'état**   all trades

tout corps *m* **de métier**   all trades

tout-venant *adj*   quarry run [ungraded material], ungraded, unsorted

toutes taxes comprises (TTC)   tax included

toxicité *f*   toxicity

toxique *adj*   toxic

toyère *f*   eye [of an axe head, etc]

traçage *m*   laying/setting out, marking off/out, tracing

traçage *m* **électrique**   trace heating

tracé *m*   alignment, layout, line, plan, route or setting out [of a road, railway, etc], diagram, drawing, outline [drawn], sketch, plot(ting) [of curves, data]

trace-circles/ellipses *m*   circle/ellipse template/stencil

trace-lettres *m*   lettering stencil

tracer *vb*   draw, plot [a curve, etc], sketch, lay out, mark off/out, set out [peg or mark out]

traceur *m*   plotter [computer], tracer [in a drawing office]

traceur *m* **à plat**   flat-bed plotter

traceur *m* **thermique**   thermal plotter

traceuse *f* **de lignes**   line painter or striper [road marking]

tracté *adj*   towed [equipment]

tracter *vb*   tow [equipment, a trailer]

tracteur *m*   tractor

tracteur *m* **à chenilles**   track-laying tractor

tracteur *m* **à lame droite**   bulldozer

tracteur *m* **à lame orientable**   angledozer

tracteur *m* **de chantier**   off-highway tractor

tracteur *m* **de semi-remorque**   semi-trailer tractor unit

tracteur *m* **débardeur**   skidder

tracteur *m* **forestier**   skidder

tracteur *m* **poseur de canalisations**   pipelayer/pipelaying tractor

tracteur *m* **sur pneu(matique)s**   wheeled tractor

traction *f*   pulling, tension, traction

traction, effort *m* **de**   pull

traction, résistance *f* **à la**   tensile strength

traction *f* **électrique**   electric traction

tracto-pelle *f*   backhoe-loader, digger-loader

trafic *m*   traffic

trafic, écoulement *m* **du**   traffic flow

trafic *m* **de transit**   through traffic

trafic *m* **ferroviaire**   rail traffic

trafic *m* **intérieur/local**   local traffic

trafic *m* **interurbain**   intercity traffic

trafic *m* **lourd**   heavy traffic

trafic *m* **moyen**   medium-heavy or moderate traffic

trafic *m* **très faible**   light traffic

trafic *m* **urbain**   urban traffic

train *m* **d'engrenages**   train of gears

train *m* **de marchandises**   freight train

train *m* **de reconditionnement/retraitement de chaussée**   road retreatment plant/train

train *m* **de voyageurs**   passenger train

traînant *adj*   dragging

traînard *m*   lining brush

traîner *vb*   drag [e.g. a mould board], line [painting, with a lining brush], lag [behind]

trait *m*   line, mark, streak

trait *m* **de cote**   dimension line [on a drawing]

trait *m* **de niveau**   horizontal line marked on walls 1m above finished floor level

trait *m* **de scie**   line of cut, saw-line, saw cut, saw mark [timber surface defect]

traité *adj*   coated [road material], treated

traitement *m*   processing, treatment

traitement *m* **après prise**   curing [[of concrete]

traitement *m* **autoclave**   pressure treatment [for timber]

traitement *m* **d'air/de l'air**   air handling, air treatment

traitement *m* **de données**   data processing

traitement *m* **de l'eau/d'eau**   water treatment

traitement *m* **de surface**   surface finish(ing)/treatment

traitement *m* **de texte**   word processing

traitement *m* **des données**   data processing

traitement *m* **des eaux d'égouts**   waste water treatment

traitement *m* **des eaux de chaudière**   boiler water treatment

traitement *m* **des eaux potables**   water purification/treatment

traitement *m* **des eaux usées**   sewage treatment

traitement *m* **des ordures ménagères**   refuse disposal/treatment

traitement *m* **des sols**   soil stabilization/treatment

traitement *m* **par flottation**   flotation

traitement *m* **thermique**   heat treatment

traitement *m* **ultérieur**   after-/post-treatment, secondary treatment

traiter *vb*   handle [a subject], do a deal, negotiate, treat, process

trajet *m*   course, route [itinerary], distance [of journey, travel], journey

trajet *m* **d'infiltration**   seepage path

trame *f*   grid system [of lines, structure, etc], network of grid lines

trame *f* **de structure**   structural grid

tramway *m*   tram(way)

tramway, voie *f* **de (circulation du)**   tramway lane

tranchant *adj*   cutting [of a knife], sharp

tranché *adj*   clear, distinct

tranché, bois *m*   wood with grain disturbed by knots

tranche *f*   mason's wedge [for splitting stone], slab [cut stone, etc], slice

tranche *f* **d'un plancher**   vertical end of a floor

tranche *f* **de travaux**   bid package, lot, work section

tranchée *f*   cut(ting), ditch, trench(ing) [result, not the process]

tranchée *f*   groove, notch [in masonry]
tranchée, remplissage *m* **de**   trench backfilling
tranchée, sondage *m* **par**   ground investigation by test/trial pits
tranchée *f* **à câbles**   cable trench
tranchée *f* **de fondation**   foundation trench
tranchée *f* **drainante**   drainage trench
tranchée *f* **ouverte**   open trench
tranchée *f* **pour conduites**   pipe trench
trancher *vb*   cut [excavate or slice], settle [a question, finally]
tranchet *m*   wallpaper cutting knife
trancheur *m*   ditcher, trencher/trench excavator [worker]
trancheur *m* **sur chenilles**   tracked ditcher
trancheur *m* **sur pneus**   wheeled ditcher
trancheuse *f*   ditcher, trencher/trench excavator [machine]
tranchis *m*   angled cut(ting) of slates/tiles to fit next to ridges/valleys
transducteur *m*   transducer
transférer *vb*   transfer
transfert *m*   transfer
transfert *m* **de chaleur**   heat transfer
transformateur *m*   transformer
transformateur *m* **élévateur**   step-up transformer
transformateur *m* **monophasé**   single-phase transformer
transformateur *m* **pour sonnerie**   bell transformer
transformation *f*   alteration, conversion
transformer *vb* **(en)**   convert (into), transform (into)
transistor *m*   transistor
transit, en   in transit
transit, trafic *m* **de**   through traffic
transitoire *adj*   provisional, transient, transitional, transitory
translation *f*   translation [e.g. of a crane]
transmetteur *m*   transmitter
transmettre *vb*   hand over or pass on [powers, property sold], transmit
transmission *f*   assignment [of shares, patents, rights, etc], conveyance, making over [of property, etc], transmission [of drive, heat, sound, signals]
transmission *f* **à friction**   drive, friction, friction drive
transmission *f* **analogique**   analogue transmission
transmission *f* **d'effort**   transmission of force
transmission *f* **de chaleur**   heat transmission
transmission *f* **de pompe**   pump drive
transmission *f* **de ventilateur**   fan drive
transmission *f* **funiculaire**   points cable
transmission *f* **latérale**   flanking transmission
transmission *f* **numérique**   digital transmission
transmission *f* **par courroie (trapézoïdale)**   belt drive
transmission *f* **par pignons**   bevel drive
transmission *f* **phonique**   crosstalk [sound transmission between spaces]
transmission *f* **quatre-roues**   four-wheel drive
transmissivité *f*   transmissivity
transparence *f*   transparency [of air, water, etc]
transparent *adj*   transparent [air, glass, etc]

transport *m*   carriage or conveyance [of goods], transport(ation)
transport, compagnie *f* **de**   forwarding/transport company
transport *m*, **frais** *m,pl* **de**   costs of carriage/transport
transport *m* **de banlieue**   suburban transport
transport *m* **des cendres**   ash transport
transports *m,pl*, **plan** *m* **général des**   general transport/traffic plan
transports (en commun)   public transport
transportable *adj*   portable
transporter *vb*   carry, convey [goods, etc], transport
transporteur *m*   conveyor, haulier
transporteur *m* **à air comprimé**   pneumatic conveyor
transporteur *m* **à bande/courroie/tapis**   belt conveyor
transporteur *m* **à godets**   bucket conveyor
transporteur *m* **aérien à câble**   aerial ropeway, blondin, cableway
transporteur *m* **des documents**   document transporter
transversal *adj*   cross-, transverse
trapillon *m*   cover [e.g. manhole], trapdoor [in a stage]
trappe *f*   hatch, trapdoor
trappe *f* **d'accès**   access hatch
trappe *f* **d'évacuation**   escape hatch
trappe *f* **de désenfumage**   fire damper, roof/smoke vent
trappe *f* **de fumée**   control damper throat restrictor [in a fireplace chimney]
trappe *f* **de nettoyage**   cleaning door [in the side of a duct]
trappe *f* **de visite**   access/entry/inspection cover/hatch
trappe *f* **en fonte**   cast iron cover
trappe *f* **en fonte série lourde**   heavy duty cast iron cover
travail *m*   machining, work, workmanship
travail *m* **à la journée**   daywork
travail *m* **d'étude**   design work
travaillé *adj*   worked, wrought
travailler *vb*   work
travailler *vb* **au tour**   turn [on a lathe]
travailleur *m*   labourer, worker
travailleur *m* **indépendant**   self-employed worker
travaux *m,pl*   works
travaux *m,pl*, **grands**   construction project
travaux *m,pl* **d'amélioration**   improvement (works)
travaux *m,pl* **dans l'air comprimé**   compressed air working
travaux *m,pl* **de bétonnage**   concrete works, concreting
travaux *m,pl* **de creusement**   excavation works
travaux *m,pl* **de coupage**   cutting (work) [metal cutting]
travaux *m,pl* **de décoration**   decoration (works)
travaux *m,pl* **de démantèlement**   dismantling (works)
travaux *m,pl* **de démolition**   demolition (works)
travaux *m,pl* **de finissage**   finishing operations/works
travaux *m,pl* **de fondation**   foundation work(s)
travaux *m,pl* **de fouilles en tranchées**   trench cutting, trenching
travaux *m,pl* **de peinture**   painting
travaux *m,pl* **de réhabilitation**   refurbishment, rehabilitation, restoration (works)
travaux *m,pl* **de soudage**   welding (work)

travaux *m,pl* **de soutènement**   support(ing) works [provision of support]

travaux *m,pl* **de terrassement**   earthworks, excavation works

travaux *m,pl* **de toiture**   roofing works

travaux *m,pl* **de voirie**   roadworks

travaux *m,pl* **en caisson**   compressed air working

travaux *m,pl* **en milieu hyperbare**   compressed air working

travaux *m,pl* **en plongée**   diving (work)

travaux *m,pl* **en régie**   cost-plus-percentage [type of contract], daywork

travaux *m,pl* **souterrains**   underground work(s)

travée *f*   bay [of a bridge or building], panel [bay or section], distance between supports [e.g. battens for metal sheet roofing], spacing [between joists, rafters, etc], span [part between supports]

travée *f* **centrale**   central/centre span

travée *f* **de rive**   end panel/span

travée *f* **latérale**   side span

travée *f* **levante/mobile**   lift span [bridge]

travers, en   across, crosswise, transversely

travers *m*   breadth

traverse *f*   cross-bar/-piece, horizontal member of railings/window bars, rail [of door or shutter], rung [of ladder], sleeper [of railway track], transom, short cut

traverse *f* **de liaison**   tie-plate

traverse *f* **de serrure**   lock rail [of a door]

traverse *f* **dormante**   window head (rail) [timber structure]

traverse *f* **inférieure**   bottom rail [of a door]

traverse *f* **supérieure**   top rail [of a door]

traverser *vb*   cross, span (over)

tréfilé *m*   extrusion

tréfilerie *f*   wire drawing

trèfle *m*   cloverleaf [road junction]

treillage *m*   lattice, trellis

treillageur *m*   steelfixer

treille *f* **à l'italienne**   pergola

treillis *m*   lattice, trellis, grid [maps, surveying]

treillis, poutre *f* **à**   lattice girder

treillis *m* **d'armature**   mesh reinforcement

treillis *m* **de fil d'acier**   steel wire mesh

treillis *m* **métallique**   wire netting

treillis *m* **soudé**   welded fabric/mesh reinforcement

treillis *m* **tridimensionnel**   space-frame

tremblement *m* **de terre**   earthquake

tremblement *m* **de terre, léger**   earth tremor

tremblement *m* **tectonique**   tectonic quake

tremblement *m* **volcanique**   volcanic quake

trémie *f*   bin/bunker/hopper [for solid fuel, aggregates, etc], chimney, funnel [for loading], light well, opening in a floor for vertical access/circulation

trémie *f* **(d'aération)**   ventilation shaft/well

trémie *f* **d'alimentation**   feed hopper

trémie *f* **d'escalier**   stair well

trémie *f* **de stockage**   storage hopper

trempe *f*   quenching, soaking

trempage *m*   dip treatment [for timber]

tremper *vb*   quench, soak, steep

trépan *m*   rock drill bit

trépan *m* **à molette**   roller bit

trépan *m* **au métal dur**   hard metal bit

trépied *m*   tripod

trésillon *m*   drying batten [between pieces of green timber]

tresse *f*   braid

tresse *f* **de filasse**   hemp fibre

tréteau *m*   horse, stand or trestle

treuil *m*   winch, windlass

triangle, couplage *m* **en**   delta connection

triangulation *f*   triangulation

triaxial *adj*   triaxial

tribenne *f*   three-way tipper

tribunal *m*   court

tribunal *m* **administratif**   administrative court

tribunal *m* **d'instance**   civil court [dealing with minor matters]

tribunal *m* **de commerce**   commercial court

tribunal *m* **de grande instance**   civil court [dealing with major matters]

tribune *f*   grandstand

tricoises *f,pl*   large pincers

tridimensionnel *adj*   three-dimensional

trier *vb*   pick out, separate [sort], sort

trimestre *m*   quarter [of year]

tringle *f*   bar [small cross-section iron/steel], rod [curtain, etc], square bead [wood]

tringle *f* **à ressort**   curtain wire [plastic covered spring]

tringle *f* **d'escalier**   stair rod

tringle *f* **de commande**   control rod

tringle *f* **(de rideau)**   curtain rod/rail

tringler *vb*   chalk [a line]

tringlerie *f*   linkage [mechanical]

triphasé *adj*   three-phase

triplex *m*   Triplex laminated/safety glass

triquet *m*   pair of steps

trommel *m*   revolving screen

trompe *f*   pendentive, squinch

trompe *f* **d'entrée**   bellmouth entry/intake

trompillon *m*   point of meeting of glazing bars of a fanlight, small pendentive/squinch

tronc *m*   trunk [of a tree]

tronçon *m*   length [i.e. a particular length, of pipe, road, etc], section [length, stretch], stretch [of road or route]

tronçonner *vb*   cut into lengths/pieces/sections, cross-cut

tronçonneuse *f*   abrasive (wheel) cut-off saw, chainsaw, crosscut saw

trop-plein *m*   overflow pipe

trottoir *m*   footway or walkway [e.g. on a bridge], pavement, or footpath [beside a road]

trottoir, bordure *f* **de**   kerb

trottoir *m* **roulant**   travolator

trou *m*   hole, pit

trou *m* **alésé**   drill(ed) hole

trou *m* **d'aération**   vent [air]

trou *m* **d'homme**   inspection hole, manhole

trou *m* **d'injection**   grout(ing) hole

trou *m* **de buée**   condensation drain/weephole [in joinery]
trou *m* **de fixation**   fixing hole
trou *m* **de forage/sondage/sonde**   bore [ground investigation], borehole
trou *m* **de poing**   handhole
trou *m* **de regard**   inspection hole
trou *m* **de serrure**   keyhole
trou *m* **de visite**   manhole
trou *m* **fileté**   tapped hole
trou *m* **taraudé**   tapped hole
troussequin *m*   carpenter's beam compass
truelle *f*   trowel
truelle *f* **à mortier**   brick(laying) trowel
truelle *f* **mécanique**   power float
trumeau *m*   pier [length of wall between two openings]
trusquin *m*   carpenter's marking gauge, mortise gauge, scriber
tubage *m*   casing, lining or tubing [of a bore-hole, the action and the result], lining [of a flue]
tubage *m* **de cheminée**   flue liner
tube *m* *voir aussi* tuyau   pipe, tube
tube *m* **à ailettes/ailetté**   finned tube
tube *m* **à ailettes hélicoïdales**   spiral fin tube
tube *m* **d'acier/de cuivre/etc**   steel/copper/etc pipe/tube
tube *m* **d'égout**   soil pipe
tube *m* **d'évent**   vent pipe
tube *m* **de bétonnage**   tremie pipe [concreting]
tube *m* **de Bourdon**   Bourdon gauge or tube
tube *m* **de compensation**   expansion pipe
tube *m* **de condenseur**   condenser tube
tube *m* **de dérivation**   bypass (pipe)
tube *m* **de descente**   downpipe
tube *m* **de descente (pluviale)**   rainwater pipe
tube *m* **de fumée**   smoke pipe
tube *m* **de niveau**   level gauge
tube *m* **de niveau d'eau**   water gauge
tube *m* **de Pitot**   pitot tube
tube *m* **de rallonge**   extension pipe, make-up piece [for extending pipe]
tube *m* **de revêtement**   casing pipe [in borehole, etc]
tube *m* **de sécurité/sûreté**   vent pipe [emergency, safety]
tube *m* **de vidange**   drain(ing) pipe/tube
tube *m* **en acier/cuivre/etc**   steel/copper/etc pipe/tube
tube *m* **en acier sans soudure**   seamless steel tube
tube *m* **en laiton**   brass tube
tube *m* **fileté**   screwed/threaded pipe/tube
tube *m* **flexible**   flexible pipe, hosepipe
tube *m* **fluorescent**   fluorescent tube
tube *m* **sans emboîtement**   socketless pipe
tube *m* **sans soudure**   seamless tube
tube *m* **soudé**   welded pipe or tube
tube *m* **taraudé**   screwed or threaded pipe or tube
tubes *m,pl*   tubing [pipes, etc]
tuber *vb*   case or line [a borehole]
tubulaire *adj*   tubular
tubulure *f* **de dérivation/raccordement**   take-off piece
tuf *m*   tufa
tuffeau *m*   limestone [very soft]
tufté, tapis *m*   tufted carpet

tuile *f*   tile [roofing]
tuile *f* **à emboîtement**   interlocking tile
tuile *f* **arêtière**   ridge tile
tuile *f* **canal**   Spanish or mission tile
tuile *f* **chaperonne**   wall-capping tile [special French]
tuile *f* **cornière**   corner/hip/valley tile
tuile *f* **creuse sans emboîtement**   pantile
tuile *f* **crochet**   nib(bed) tile
tuile *f* **de croupe**   ridge tile
tuile *f* **de dessous**   under tile
tuile *f* **de dessus**   over tile
tuile *f* **de rive**   edge tile
tuile *f* **en S**   pantile
tuile *f* **faîtière**   ridge tile
tuile *f* **faîtière d'about**   hip tile
tuile *f* **faîtière de dernier rang**   ridge course tile
tuile *f* **faîtière mécanique**   interlocking ridge tile
tuile *f* **femelle**   under tile
tuile *f* **flamande**   pantile
tuile *f* **mâle**   over tile
tuile *f* **mécanique**   interlocking tile
tuile *f* **plate écaille**   plain tile with rounded end
tuile *f* **plate (ordinaire)**   flat/plain tile
tuile *f* **romaine**   interlocking Spanish tile
tuile *f* **ronde**   Spanish or mission tile
tuile *f* **sarrasine**   Spanish tile [broad type used in Provence]
tuileau *m*   thin fired clay decoration
tuilerie *f*   tile kiln, tile works
tungstène *m*   tungsten
tunnel *m*   tunnel
tunnel *m* **aérodynamique**   wind tunnel
tunnel *m* **d'essai aéraulique**   wind tunnel
tunnel *m* **de congélation**   tunnel freezer
tunnel *m* **de réfrigération**   tunnel cooler
tunnel *m* **ferroviaire**   railway tunnel
tunnel *m* **routier**   road tunnel
tunnelier *m*   mole, tunnel boring machine, tunnelling machine
turbidité *f*   turbidity
turbine *f*   turbine
turbine *f* **à gaz**   gas turbine
turbine *f* **à vapeur**   steam turbine
turbo-compresseur *m*   turbo-compressor
turbo-compresseur *m* **radial**   centrifugal turbo-compressor
turbo-générateur *m*   turbo-generator
turbosoufflante   turbo-blower
turbulence *f*   turbulence
turbulent *adj*   turbulent
turcie *f*   dike, flood bank, river wall
tuyau *m* *voir aussi* tube   pipe
tuyau *m* **à brides**   flanged pipe
tuyau *m* **à trois voies**   three-way (pipe) fitting
tuyau *m* **à vide**   vacuum pipe/tube
tuyau *m* **coudé**   swan-neck
tuyau *m* **d'acier/de cuivre/etc**   steel/copper/etc pipe
tuyau *m* **d'aération**   vent(ilation) pipe
tuyau *m* **d'alimentation**   feed/supply pipe

tuyau *m* **d'alimentation d'eau**   water supply pipe
tuyau *m* **d'arrosage**   garden hose, hosepipe
tuyau *m* **d'échappement**   waste pipe, exhaust pipe [vehicle]
tuyau *m* **d'écoulement**   sewer pipe [small diameter], drain [pipe]
tuyau *m* **d'égout**   sewer pipe [small], soil pipe
tuyau *m* **d'entrée**   inlet pipe
tuyau *m* **d'équilibrage**   balance/balancing pipe
tuyau *m* **d'évacuation**   drainage/waste pipe
tuyau *m* **d'évent**   vent pipe
tuyau *m* **d'incendie**   fire hose
tuyau *m* **de branchement**   branch pipe
tuyau *m* **de caoutchouc**   rubber hose/pipe
tuyau *m* **de chauffage**   heating pipe
tuyau *m* **de cheminée**   chimney flue [pipe], flue pipe
tuyau *m* **de chute**   soil stack
tuyau *m* **de chute unique**   soil and waste stack
tuyau *m* **de circulation**   circulation pipe
tuyau *m* **de décharge**   outlet/waste pipe, discharge
tuyau *m* **de dérivation**   bypass pipe
tuyau *m* **de descente**   downpipe
tuyau *m* **de descente d'eaux ménagères**   soil and waste stack
tuyau *m* **de descente (pluviale)**   rainwater pipe
tuyau *m* **de distribution**   distributor pipe
tuyau *m* **de drainage**   drainage pipe
tuyau *m* **de drainage perforé**   perforated drain pipe
tuyau *m* **de gaz**   gas pipe
tuyau *m* **de prise d'eau**   standpipe
tuyau *m* **de purge**   blow-off pipe
tuyau *m* **de raccordement**   connecting pipe
tuyau *m* **de refoulement**   delivery pipe
tuyau *m* **de refroidissement**   cooling or refrigerating pipe
tuyau *m* **de remplissage**   filling pipe
tuyau *m* **de retour**   return pipe
tuyau *m* **de sortie du robinet chef**   gas pipe from mains valve to consumer meter
tuyau *m* **de trop-plein**   expansion/overflow pipe
tuyau *m* **de vapeur**   steam pipe
tuyau *m* **de ventilation**   ventilation pipe
tuyau *m* **de vidange**   drain down pipe
tuyau *m* **en acier/cuivre/etc**   steel/copper/etc pipe

tuyau *m* **en acier revêtu de bitume**   bitumen coated steel pipe
tuyau *m* **en béton précontraint**   prestressed concrete pipe
tuyau *m* **en caoutchouc**   rubber hose/pipe
tuyau *m* **en béton/céramique/grès**   concrete/clay/earthenware pipe
tuyau *m* **en croix**   four-way (pipe) fitting
tuyau *m* **en fer forgé/en fonte/en fonte malléable**   wrought/cast/malleable iron pipe
tuyau *m* **en fibrociment**   asbestos cement pipe
tuyau *m* **en plastique/polyéthylène**   plastic/polyethylene pipe
tuyau *m* **fendu**   slotted pipe [drainage]
tuyau *m* **filtrant**   filter pipe [drainage]
tuyau *m* **flexible**   flexible pipe
tuyau *m* **galvanisé**   galvanized pipe
tuyau *m* **métallique flexible**   flexible metal pipe, metal hose
tuyau *m* **perforé**   filter pipe [drainage]
tuyau *m* **préfabriqué**   prefabricated pipe
tuyau *m* **soudé en bout**   butt-welded pipe
tuyau *m* **souple**   flexible pipe
tuyau *m* **soutirage**   bleed pipe
tuyauterie *f*   pipework, piping
tuyauterie *f* **d'aération**   ventilating pipework
tuyauterie *f* **d'aspiration**   suction pipework
tuyauterie *f* **d'eau condensée**   condensate line or pipework
tuyauterie *f* **d'équilibr(ag)e**   balancing pipework
tuyauterie *f* **de chauffage**   heating pipework
tuyauterie *f* **de vapeur**   steam pipework
tuyauterie *f* **des eaux usées**   foul water drain pipework
tuyauterie *f* **enterrée**   buried pipework
tuyère *f*   nozzle
tuyère *f* **d'injecteur**   injector nozzle
tympan *m*   spandrel [of door of cathedral], tympanum
type *adj*   typical
type *m*   type
type, coupe *f*   typical section
type *m* **de route**   highway type
type *m* **spécial**   special design
typique *adj*   typical

ultérieur *adj*   later [subsequent]
ultime *adj*   ultimate [e.g. at failure]
ultraviolet *adj*   ultraviolet
ultraviolet, rayonnement *m*   ultraviolet radiation
uni *adj*   even, level, smooth [ground]
unidirectionnel *adj*   unidirectional
uniforme *adj*   uniform

uniformément réparti *adj*   uniformly distributed
uniformité *f*   uniformity
uniformité, facteur *m* **d'**   uniformity ratio [lighting]
uniformité *f* **de l'éclairage**   uniformity of illumination
union *f*   union [pipe joint]
unipolaire *adj*   single-pole
unique *adj*   single

unitaire *adj* common [as in common flue]

unitaire, système *m* **d'assainissement** combined drainage system

unité *f* unit [mathematical, physical], unity [oneness]

unité *f* **centrale** central processor, mainframe [computer]

unité *f* **d'habitation** dwelling unit

unité *f* **de condensation refroidie par l'air** air-cooled condensing unit

unité *f* **de disques** disk unit [computer]

unité *f* **de disquette** diskette unit [computer]

unité *f* **de mesure** unit of measurement

unité *f* **de puissance** unit of power

unité *f* **de temps** unit of time

unité *f* **de traitement d'air** air handling unit (AHU)

unité *f* **de ventilation** ventilating unit

unité *f* **esclave** slave unit

universel *adj* all-purpose [e.g. spanner]

urbain *adj* urban

urbaniser *vb* urbanize

urbanisme *m* town planning

urbanisme, administration *f*/**service** *m* **de l'** planning department/office [town planning]

urbaniste *m/f* town planner

urbanistique *adj* town planning

d'urgence *adv* immediately, urgently

urgent *adj* urgent

urinoir *m* urinal

usagé *adj* old, used, worn

usage *m* use

usage *m* **intensif** intensive use

usager *m* user

usé *adj* worn away/out [by use]

user *vb* wear away/out, exercise [permission, powers, etc]

usinage *m* machining, manufacture

usine *f* factory, plant [factory rather than equipment]

usine *f* **à béton** concrete mixing/batching plant

usine *f* **à gaz** gasworks

usine *f* **à purification des eaux d'égouts** sewage disposal/treatment plant

usine *f* **d'incinération** incinerator plant

usine *f* **de bétonnage** concrete mixing/batching plant

usine *f* **de pâte à papier** paper mill

usine *f* **de raffinage** refinery

usine *f* **de traitement des eaux** water treatment plant

usine *f* **marémotrice** tidal power plant/station

usine *f* **métallurgique** ironworks

usine *f* **sidérurgique** steel mill, steelworks

usiner *vb* machine, manufacture

usufruit *m* usufruct

usufruitier *m* usufructuary

usure *f* deterioration, wear (and tear), wearing away

usure *f* **normale** fair wear and tear

usure *f* **par frottement** abrasion [wear], frictional wear

utile *adj* effective, serviceable, useful, usable

utile, effet *m* usable or useful power

utile, plan *m* working plane

utilisable *adj* usable, working [of equipment, etc]

utilisation *f* use

utiliser *vb* employ [force, a piece of equipment, etc], make use of, use, utilize

utilité *f* usefulness

vaction *f* fees based on time to complete an item of work

vacherie *f* cowshed

va-et-vient *adj* two-way

vague *f* wave [water]

valet *m* carpenter's clamp

valeur *f* value

valeur *f* **approximative** approximate value

valeur *f* **calorifique** calorific value

valeur *f* **crête** peak value

valeur *f* **de base** basic value

valeur *f* **de consigne** set value

valeur *f* **de contrat** contract value

valeur *f* **de défaut** default value

valeur *f* **de pic/de pointe (maximale)** peak value

valeur *f* **de référence** reference value, set point

valeur *f* **de saturation** saturation value

valeur *f* **désirée** desired value

valeur *f* **du pH** pH value

valeur *f* **effective** root mean square (RMS) value

valeur *f* **empirique** empirical value

valeur *f* **étalonnée** calibrated value

valeur *f* **extrême** limit value

valeur *f* **limite** limiting value

valeur *f* **limite admissible** threshold limit value

valeur *f* **maximum** peak value

valeur *f* **mesurée** measured value

valeur *f* **moyenne** mean value

valeur *f* **nominale** nominal value

valeur *f* **numérique** numerical value

valeur *f* **réelle** actual value

valeur *f* **théorique** theoretical value

valeur *f* **type** typical value

valeurs *f,pl* securities [stock exchange]

vallée *f* valley

vallée *f* **en auge/en fond de bateau** U-shaped valley

vallée *f* **évasée/en berceau** synclinal valley

vallée *f* **fluviale** river valley

vallée *f* **morte/sèche** dry valley

vallée *f* **submergée**   drowned valley
vallée *f* **suspendue**   hanging valley
vallon *m*   vale, small valley
vallonné *adj*   undulating [of land, countryside]
valoriser *vb*   develop [the economy of an area], enhance the value of
valve *f, voir aussi* robinet, soupape, vanne   valve [of a machine, tyre]
valve *f* **de radiateur**   radiator valve
vanadium *m*   vanadium
vanne *f, voir aussi* robinet, soupape   valve [generally large, with a turning or sliding gate], butterfly valve, gate valve, lock-/sluice-gate
vanne *f* **à disque**   disc valve
vanne *f* **à frigorigène**   refrigeration valve
vanne *f* **à passage direct**   full-way valve
vanne *f* **(à) papillon**   butterfly valve
vanne *f* **à soufflet**   bellows valve
vanne *f* **à trois voies**   three-way valve
vanne *f* **d'arrêt**   gate or stop valve
vanne *f* **d'arrêt d'air**   air shut-off valve
vanne *f* **d'écluse**   sluice-valve [in a canal lock]
vanne *f* **d'extraction**   extract(ion) valve
vanne *f* **d'isolement**   isolating/isolation valve
vanne *f* **de barrage**   stop valve
vanne *f* **de communication**   sluice-valve
vanne *f* **de décharge**   flood gate
vanne *f* **de dérivation**   bypass/diverter valve
vanne *f* **de fermeture**   shut-off valve
vanne *f* **de guidage**   guide vane
vanne *f* **de mélange**   mixing valve
vanne *f* **de non-retour**   non-return flap valve
vanne *f* **de pression d'aspiration**   back-pressure valve
vanne *f* **de réglage**   control/regulating valve
vanne *f* **de réglage de circuit**   circuit control valve
vanne *f* **de régulation**   regulating valve
vanne *f* **de régulation de débit**   flow control valve
vanne *f* **de sectionnement**   isolating stop valve
vanne *f* **de secours**   emergency water valve
vanne *f* **de sécurité et de décharge**   emergency relief valve
vanne *f* **de soutirage**   bleed valve
vanne *f* **de sûreté**   relief valve
vanne *f* **de tête d'eau**   penstock [penning gate]
vanne *f* **de trop plein**   overflow valve
vanne *f* **de vidange**   drain valve, sluice-valve
vanne *f* **deux voies**   two-way valve
vanne *f* **mélangeuse**   mixing valve
vanne *f* **motorisée**   motorized valve
vanne *f* **principale de contrôle**   main control valve
vanne *f* **régulatrice**   regulating valve
vanne *f* **solénoïde**   solenoid valve
vanne *f* **thermostatique**   thermostatic valve
vannes, lever *vb* **les**   open the flood-gates/sluices
vannes, mettre *vb* **les**   close the flood-gates/sluices
vannelle *f*   sluice-valve
vantail *m*   leaf [of a door, window, etc]
vapeur *f*   steam, vapour
vapeur, lampe *f* **à**   vapour lamp [mercury, sodium, etc]

vapeur *f* **à basse/haute pression**   low/high pressure steam
vapeur *f* **d'eau**   water vapour
vapeur *f* **d'échappement**   exhaust/waste steam
vapeur de mercure à haute pression, lampe *f* **à**   high pressure mercury vapour lamp
vapeur *f* **de saturation**   saturated water vapour
vapeur métallique, lampe *f* **à**   metal vapour lamp
vapeur *f* **saturée**   saturated steam
vapeur *f* **sèche**   dry steam
vapeur *f* **surchauffée**   superheated steam
vapeur *f* **vive**   live steam
vaporisation *f*   vaporization
vaporisation *f* **instantanée**   flash vaporization
variable *adj*   changeable, variable
variable *f*   variable *n*
variable *f* **aléatoire**   independent variable
variable *f* **de contrôle**   control variable
variable *f* **dépendante**   dependent variable
variable *f* **régulée**   controlled variable
variant *adj*   alternat(iv)e
variante *f*   alternative, option
variateur *m*   variator
variateur *m* **de lumière**   dimmer
variateur *m* **de vitesse**   speed variator
variation *f*   fluctuation, range [between limits], variation
variation *f* **annuelle**   annual variation
variation *f* **de charge**   load variation
variation *f* **de débit**   flow/output/yield variation
variation *f* **de pression/température/vitesse/etc**   pressure/temperature/speed/etc variation
variation *f* **journalière/quotidienne**   daily range/variation
variation *f* **mensuelle**   monthly range/variation
varié *adj*   varied, varying
varier *vb*   change, vary
varlope *f*   trying plane
vase *f*   silt, mud [river bed]
vase *m*   receptacle, vessel, tank [of expansion unit]
vase *m* **d'expansion/de dilatation**   expansion tank/vessel
vase *m* **de dilatation à membrane**   membrane expansion vessel/tank
vasistas *m*   fanlight
vastringue *f*   spokeshave
vasque *f*   handbasin
vasque *f* **à poser**   handbasin for building-in
veilleuse *f*   pilot flame or light [of boiler, gas stove, etc]
veinage *m*   veining [in wood or stone]
veine *f*   vein [in wood or stone]
veine *f* **d'air**   air stream
veinette *f*   paintbrush used for imitating wood grain
vélum *m*   awning
Velux *m*   rooflight [Velux, patented]
vendeur *m*   seller, vendor
vénéneux *adj*   poisonous
vénitien *adj*   venetian [of a blind]
vent *m*   wind
vente *f*   sale
vente *f* **à l'essai**   sale on approval

vente *f* **aux enchères**   sale by auction
vente *f* **ferme**   firm sale
ventelle *f*   sluice-valve
ventilateur *m*   (air) blower, fan, ventilator
ventilateur *m* **à aubes profilées**   airfoil fan
ventilateur *m* **à deux/double ouïes**   double entry or inlet fan
ventilateur *m* **à fourche**   bifurcated fan
ventilateur *m* **à une ouïe**   single inlet fan
ventilateur *m* **axial**   axial (flow) fan
ventilateur *m* **basse pression**   low-pressure fan
ventilateur *m* **bifurqué**   bifurcated fan
ventilateur *m* **brasseur d'air**   circulating fan
ventilateur *m* **centrifuge**   centrifugal fan
ventilateur *m* **d'aspiration d'air frais**   ventilating fan [drawing in air]
ventilateur *m* **d'extraction (d'air)**   extract(or) fan
ventilateur *m* **d'extraction de cuisine**   kitchen extract(or) fan
ventilateur *m* **de désenfumage**   smoke extractor fan
ventilateur *m* **de plafond**   ceiling fan
ventilateur *m* **de soufflage**   supply fan
ventilateur *m* **de tirage par aspiration**   induced draught fan
ventilateur *m* **de toiture**   roof ventilator
ventilateur *m* **extracteur**   extractor fan
ventilateur *m* **hélico-centrifuge**   mixed flow fan
ventilateur *m* **mural**   wall fan
ventilateur *m* **radial**   radial fan
ventilateur *m* **tangentiel**   cross-flow/tangential fan
ventilation *f*   ventilation, vent(ilation) pipe, vent stack
ventilation, puits *m* **de**   ventilation shaft/well
ventilation *f* **à haute pression**   high pressure ventilation
ventilation *f* **à surpression**   pressurized ventilation
ventilation *f* **contrôlée**   controlled ventilation
ventilation *f* **mécanique**   mechanical ventilation
ventilation *f* **naturelle**   natural ventilation
ventilation *f* **par extraction**   exhaust ventilation
ventilation *f* **secondaire**   vent(ilation) pipe
ventiler *vb*   ventilate
ventilo-convecteur *m*   fan coil unit
ventouse *f*   air hole/valve/vent, force cup, suction disc/cup [plumbing], underfloor draught pipe to a fireplace, vent opening [built into a partition or wall]
ventrière *f*   purlin
véranda *f*   sun lounge, veranda
verdunisation *f*   water sterilization with sodium hypochlorite
verger *m*   orchard
vergette *f*   steel bar providing extra support to glazing
vérificateur *m*   checker [checks the measurements of a *métreur*]
vérificateur *m* **de circuit**   circuit tester [electrical]
vérificateur *m* **de continuité**   continuity tester [electrical]
vérificateur *m* **de haute tension**   high-voltage tester
vérificateur *m* **de tension**   voltage detector [electrical]
vérification *f*   audit, check(ing), control [check], examination, scrutiny, recalculation, testing
vérification *f* **des contraintes**   stress analysis

vérifier *vb*   check, recalculate, verify
vérin *m*   jack [lifting], ram
vérin *m* **à crémaillère**   rack-and-pinion, ratchet or tooth-and-pinion jack
vérin *m* **(à) double-effet**   double-acting jack
vérin *m* **à vis**   screw jack
vérin *m* **de mise en tension**   stressing jack
vérin *m* **Freyssinet**   flat jack
vérin *m* **hydraulique**   hydraulic jack/ram
vérin *m* **plat**   flat jack
vérin *m* **pneumatique**   pneumatic jack
vérinage *m*   jacking
vériner *vb*   jack
vermiculures *f,pl*   vermiculations [decoration on stonework]
vermiculite *f*   vermiculite
vermoulu *adj*   decrepit, worm-eaten
vermoulure *f*   bore-dust/hole [of wood-boring insects], worm-eaten state [of wood]
verni *adj*   varnished
vernier *m*   vernier
vernir *vb*   varnish
vernis *m*   varnish
vernissage *m*   varnishing
vernisseur *m*   painter specializing in varnish work
verre *m*   glass
verre, laine *f* **de**   fibreglass, glass wool
verre, mousse *f* **de**   foam glass
verre *m* **à glace**   plate glass
verre *m* **à vitres**   window glass
verre *m* **armé**   wire reinforced glass
verre *m* **brut**   unpolished glass
verre *m* **casilleux**   glass of low quality that cannot be diamond cut
verre *m* **catathermique**   glass opaque to infra-red light
verre *m* **cathédrale**   stained glass
verre *m* **cellulaire**   cellular glass, foam glass
verre *m* **coloré**   stained glass
verre *m* **dalle**   pavement light
verre *m* **de lampe**   lamp glass
verre *m* **de sécurité**   safety glass
verre *m* **dépoli**   frosted glass
verre *m* **dur**   hard glass [borosilicate, for lamps]
verre *m* **en cul de bouteille**   bull's eye glass
verre *m* **expansé**   foam glass
verre *m* **feuilleté**   laminated glass
verre *m* **grillagé**   wired glass
verre *m* **incassable**   unbreakable glass
verre *m* **mousse**   foam glass
verre *m* **mousseline**   patterned glass
verre *m* **multicellulaire**   cellular or foam glass
verre *m* **non poli**   unpolished glass
verre *m* **opale**   white/opal glass
verre *m* **plat**   sheet glass
verre *m* **protecteur**   protective glass [used by welders]
verre *m* **strié**   reeded glass
verre *m* **tendre**   soft glass [lime-soda, for lamps]
verre *m* **trempé**   case-hardened/toughened glass
verre *m* **triplex**   Triplex laminated/safety glass

verre-regard *m*   sight glass
verrerie *f*   glassware
verrière *f*   glass/glazed roof, roof glazing, glazed panel, stained glass window
verrou *m*   bar [door], bolt [cupboard, door, gate]
verrou *m* **à ressort**   spring bolt
verrou *m* **glissant**   sliding bolt
verrouillage *m* **de porte**   door fastening
verrouiller *vb*   bolt [a door, window]
verrous, mettre *vb* **les**   bolt [the door, window]
vers le bas   to below [on pipework drawing]
vers le haut   to above [on pipework drawing]
versant *m*   side [of a hill, roof, etc], slope [the ground or roof, not the angle]
versant *m* **de colline**   hillside
versant *m* **de vallée**   valley side
versement *m*   instalment, pouring out [of liquid, etc]
verser *vb*   overturn [vehicle], pour out/forth, pour [liquids]
vert *m* **anglais**   green pigment [Prussian blue + chrome yellow mixture]
verterelle/vertevelle *f*   bolt staple
vertical *adj*   vertical
vestiaire *m*   cloakroom
vestiaires *m,pl*   changing rooms, locker rooms
vestibule *m*   hall [as in entrance hall], lobby, vestibule
vestibule *m* **d'ascenseur**   lift lobby
vestige *m*   relic, remnant
vestiges *m,pl*   remains
vêtement *m* **protecteur**   protective clothing
viabilisation *f*   site servicing [provision of basic services]
viabilisé *adj*   serviced [of a site for building]
viabilité *f*   provision of roads and main services to a building site
viaduc *m*   viaduct
viager *m*   life interest [generally in land/property]
vibrage *m*   vibration [of concrete to compact it]
vibrant *adj*   vibrating
vibrateur *m*   vibrator
vibrateur *m* **à béton**   concrete vibrator
vibrateur *m* **de fonçage et d'arrachage**   piling vibrator
vibration *f*   structure-borne sound, vibration
vibration, compactage *m* **par**   vibration compaction
vibration *f* **forcée**   forced vibration
vibration *f* **libre**   free vibration
vibratoire *adj*   vibrating, vibratory
vibrer *vb*   vibrate
vibreur *m*   vibrator
vibreur *m* **à béton**   concrete vibrator
vibreur *m* **de coffrage**   formwork vibrator
vibreuse-finisseuse *f*   vibrator-finisher [road works]
vibreuse-surfaceuse *f* **à coffrage glissant**   slipform paver/paving machine
vibro-compacteur *m*   vibrating plate, vibro-compactor
vibro-flottation *f*   vibroflotation
vibro-fonçage *m*   pile driving by vibration
vibro-fonceur *m*   piling vibrator, vibratory driver, vibro-hammer

vibrolance *m*   vibrating probe
vibro-mouleuse *f*   vibro-mould equipment for precast concrete
vice *m*   defect [general]
vice *m* **apparent**   visible defect
vice *m* **caché/latent/occulte**   concealed/hidden/latent defect
vice *m* **de construction**   building defect
vidage *m*   emptying
vidange *f*   draining, emptying
vidange *f* **d'huile**   oil drain(age)
vidange *f* **de fond**   bottom outlet
vidanger *vb*   drain, empty (out) [drain off]
vide *adj*   vacant
vide *adj* **(de)**   empty (of)
vide, à   no-load, unloaded
vide *m*   empty space, gap, void, vacuum
vide, courant *m* **à**   no-load current
vide, frein *m* **à**   vacuum brake
vide, lampe *f* **à**   vacuum lamp
vide, marcher *vb* **à**   run on/at no-load
vide, pompe *f* **à**   vacuum pump
vide-ordures *m*   waste/rubbish chute [domestic rubbish]
vide *m* **poussé**   high vacuum
vide *m* **sanitaire**   crawl way, space under a suspended floor, underfloor space
vide *m* **sous plafond**   ceiling void
vides, teneur *f* **en**   voids content
vider *vb*   clear, vacate [empty], empty, drain, pour away, quit or vacate [premises], strip [empty, a house, etc]
se vider *vb*   drain or empty [itself]
vie, durée *f* **de**   life [e.g. of a lamp]
vieillir *vb*   age
vieillissage *m*   ageing [to make sth look old]
vieillissement *m*   age hardening, ageing
vieillissement, résistance *f* **au**   durability
vieillissement *m* **mécanique**   work hardening
vif *adj*   bright or light [of a colour]
vif, à   back to solid material [when dressing stone, plaster, etc]
vigne *f*   vine
vigne *f* **vierge**   virginia creeper
vignoble *m*   vineyard
vigueur, en   current, *adj* in force/use [of code of practice, etc]
vilebrequin *m*   brace-and-bit
vilebrequin *m* **à cliquet**   ratchet brace
villa *m*   detached/small house, villa
village *m*   village
ville *f*   city, town
violon *m*   painter's shave hook
virage *m*   bend, curve, twist or twisting [of a road], turn [action of altering direction]
virer *vb*   turn [change direction]
virgule *f*   decimal point [comma in French numbers]
virole *f*   ferrule, metal flue connection ring built into masonry, removable sheet metal cover/plug
vis *f*   vice screw, screw
vis *f* **à bois**   wood screw

vis *f* à ciment   cement screw conveyor
vis *f* à œillet   eye-bolt
vis *f* à quatre pans   square-headed screw
vis *f* à sable   sand screw [conveyor]
vis *f* à tête hexagonale   hexagon head screw
vis *f* (auto)taraudeuse   self-tapping screw
vis *f* calante   levelling screw
vis *f* de blocage   clamping/fixing/locking screw
vis *f* de fixation   fixing/securing screw
vis *f* de mise à la terre   earthing screw
vis *f* de rappel/réglage   adjusting screw
vis *f* de serrage   clamping screw
vis *f* sans tête   grub screw
visa *m*   stamp [e.g. on approved drawing]
viscomètre *m*   viscometer
viscosité *f*   viscosity
visée *f*   sighting [surveying]
viseur *m*   eyepiece [theodolite], sight glass
visibilité   visibility *f* [highways]
visibilité *f*, distance de   sight/visibility distance
visible *adj*   visible
visible, rayonnement *m*   visible radiation [light]
visière *f*   lamp hood [traffic lights, railway signals]
vision *f*   vision
visite, passerelle *f* de   inspection walkway
visite, trou *m* de   manhole
visser *vb*   screw [fix with a]
visser *vb* à fond   screw home
visuel *adj*   visual
visuel, examen *m*   visual examination
vitesse *f*   speed, velocity
vitesse *f* axiale   centre-line velocity
vitesse *f* d'alimentation   feed rate
vitesse *f* d'augmentation   rate of increase
vitesse *f* d'échappement de sortie   outlet velocity
vitesse *f* d'écoulement   flow rate
vitesse *f* d'entrée   inlet velocity
vitesse *f* de charge(ment)   loading rate
vitesse *f* de convection   convection velocity
vitesse *f* de décharge   discharge velocity
vitesse *f* de déformation   strain rate
vitesse *f* de détérioration   decay rate
vitesse *f* de filtrage   filtration rate
vitesse *f* de l'air   air velocity
vitesse *f* de refroidissement   cooling rate
vitesse *f* de sortie   exit velocity
vitesse *f* des gaz d'échappement   exhaust gas velocity
vitesse *f* du son   speed of sound
vitesse *f* du ventilateur   fan speed
vitesse *f* frontale   face velocity
vitesse *f* périphérique   tip speed
vitesse *f* terminale   terminal velocity
viticole *adj*   wine-growing/producing
vitrage *m*   glazing [the action and the result], glass partition, roof light, windows
vitrage *m*, double   double glazing
vitrage *m* extérieur collé   structural silicone glazing
vitrail *m*   leaded/stained glass window
vitré *adj*   glass, glazed

vitre *f*   glass/window pane
vitrer *vb*   glaze
vitrier *m*   glazier
vitrine *f*   shop window
vitrine *f* réfrigérante   refrigerated display cabinet
vivant, sol *m*   organic soil, topsoil
vive arête, à   square-edged
vive, arête *f*   sharp edge
voie *f*   lane [of road], road, track, way, set [of saw teeth]
voie *f* à circulation rapide   high-speed road
voie *f* d'accélération   acceleration lane
voie *f* d'accès   access/approach road
voie *f* d'air   air passage, airway
voie *f* d'écoulement d'air   air flow path
voie *f* d'évacuation   escape route
voie *f* d'insertion   slip road, weaving lane
voie *f* de banlieue   suburban line (railway)
voie *f* de circulation   taxiway [airfield], traffic lane
voie *f* de (circulation du) tramway   tramway lane
voie *f* de circulation normale   general-purpose lane
voie *f* de collecte   local distributor [collector] road
voie *f* de décélération   deceleration lane
voie *f* de desserte   access and service road
voie *f* de distribution   distributor (road)
voie *f* de garage   siding [railway]
voie *f* de grue   crane track
voie *f* de scie   kerf [when sawing wood]
voie *f* de stockage   storage/waiting lane
voie *f* de tourne-à-gauche/droite   (left/right) turning lane
voie *f* de triage   siding [railway]
voie *f* ferrée   permanent way [railway]
voie *f* navigable   navigable river
voie *f* piétonnière   footpath
voie *f* pour véhicules lents   crawler lane
voie *f* privée   private drive
voie *f* publique   public highway
voie *f* rapide   high-speed road
voie *f* secondaire   minor road
voie *f* supplémentaire   auxiliary traffic lane [i.e. hard shoulders, etc]
voies, route *f* à deux/trois/quatre/etc   two-/three-/four-/etc lane highway or road
voilé *adj*   buckled [of a wheel], warped [wood]
voile *m* armé   reinforced wall (thin)
voile *m* d'étanchéité   diaphragm [watertight], grout curtain [dam]
voile *m* d'injection   grout curtain
voile *m* (de béton)   thin (concrete) wall
voile *m* de sécurité   safety net
voile *m* en béton   concrete shell
voile *m* mince   thin concrete shell roof
voile *m* non armé   thin unreinforced wall
voilement *m*   warp(ing) [of wood]
voilement *m* longitudinal de face   bow [of timber]
voilement *m* longitudinal de rive   crook, spring [of timber]
voilement *m* transversal   cupping [of timber]

voiler *vb*   buckle [a wheel], warp [wood]

voir   see [i.e. cross reference in document]

voirie *f*   highway or road network or system, highway maintenance [work itself or department responsible], roadworks

voirie *f* **et réseaux divers (VRD)**   highways and miscellaneous external works

voirie *f* **principale/rurale**   primary/rural road network

voiture *f* **de pompiers**   fire engine

volant *m*   steering wheel, wheel [handwheel]

volant *m* **à main**   handwheel

volant *m* **de manœuvre**   operating handwheel

volant *m* **de réglage**   adjusting/adjustment (hand)wheel

volant *m* **de réglage en hauteur**   rise-and-fall/height adjusting (hand)wheel

volant *m* **de serrage**   clamping/work clamp (hand)wheel

volant *m* **mécanique/moteur**   flywheel

volatil *adj*   volatile

volée *f* **(d'escalier)**   flight (of stairs)

volet *m*   damper, shutter, cover flap [on a socket outlet]

volet *m* **anti-retour**   non-return damper

volet *m* **brisé**   folding shutter

volet *m* **coulissant**   roller shutter

volet *m* **coupe-feu**   fire damper

volet *m* **d'aération**   air damper, louvre

volet *m* **d'aération réglable**   adjustable louvre

volet *m* **d'équilibrage**   equalizing damper

volet *m* **de dérivation**   bypass damper

volet *m* **de réglage**   control damper

volet *m* **de réglage d'air**   air regulating damper

volet *m* **de répartition**   splitter damper

volet *m* **roulant**   roller shutter

volige *f*   fascia board, roof sheathing/sarking board

voligeage *m*   roof boarding/sheathing, sarking

volt *m*   volt

voltage *m*   voltage

voltage *m* **du réseau**   mains/supply voltage

voltampèremètre *m*   multimeter, universal test meter

voltmètre *m*   voltmeter, voltage meter

volume *m*   bulk [volume], volume

volume *m* **à transporter**   volume to be transported

volume *m* **brut**   gross volume

volume *m* **de retenue**   reservoir capacity

volume *m* **habitable**   habitable volume [floor area × ceiling height]

volume *m* **utile**   net or useful volume

volumétrique *adj*   volumetric

volute *f*   volute [spiral ornamentation]

voussoir *m*   arch stone, segment [of an arch], voussoir

voussoir *m* **conjugué**   match-cast segment

voussoir *m* **d'articulation**   hinge segment [of an arch]

voussure *f*   arching [of a ceiling, etc], bend or curve [of an arch], cove between wall and ceiling, upper part of the reveal above an opening in a wall

voûtain *m*   arched brickwork between steel beams, jack/small arch

voûté *adj*   vaulted, arched

voûte *f*   arch(way), vault

voûte *f* **d'arêtes**   cross or groin(ed) vault

voûte *f* **en arc de cloître**   cloister(ed) vault

voûte *f* **en berceau**   barrel or tunnel vault

voûte *f* **plate**   flat arch or vault

voûte *f* **sarrasine**   lightweight brick or tile vault

voûte *f* **sur croisée d'ogives**   rib(bed) vault

voyage *m* **d'affaires**   business trip

voyant *m*   inspection window, sight glass [e.g. of a boiler], sighting-board/target [surveying]

voyant *m* **(d'avertissement)**   warning light/signal

voyant *m* **de fusion**   blow-out indicator [electrical]

voyant *m* **du brûleur**   burner sight glass

vrac, en   in bulk

vraie grandeur, en   full-scale, full-sized

vrille *f*   gimlet

vriller *vb*   bore (with a gimlet)

vrillette *f*   death-watch beetle

vue *f*   outlook, scenery, view

vue *f* **de côté**   side view/elevation

vue *f* **de dessus**   view from above

vue *f* **de face**   front view/elevation

vue *f* **éclatée**   exploded view

vue *f* **en coupe**   (cross-)sectional view

vue *f* **en plan**   plan view

vue *f* **imprenable**   unobstructed/unrestricted outlook/view

vue *f* **latérale**   side elevation/view

vue *f* **transversale**   (cross-)sectional view

wabstringue *f*   spokeshave

wagon *m*   prefabricated conduit section for building into a wall

waters *m,pl*   toilet (WC)

watt *m*   watt

wattmètre *m*   wattmeter

WC *m, pl*   toilet (WC)

white spirit *m*   white spirit

X, Y, Z

xénon *m* xenon

yttrium *m* yttrium

zéolite *m* zeolite
zéro *m* **absolu** absolute zero
zinc *m* zinc
zingage *m* galvanizing, zinc coating
zingué *adj* galvanized, zinc coated
zinguer *vb* flash, galvanize
zinguerie *f* flashing, rain water goods [galvanized metal]
zonage *m* zoning
zone *f* area, zone
zone *f* **à urbaniser en priorité** housing priority area
zone *f* **bâtie** built-up area
zone *f* **butoir** buffer zone
zone *f* **climatique** climatic zone [defined in French regulations]
zone *f* **comprimée** compression zone
zone *f* **d'accès** access point [for entry to/exit from motorways]
zone *f* **d'activité commerciale (ZAC)** shopping centre
zone *f* **d'aménagement concerté (ZAC)** planned development area
zone *f* **d'aménagement différé (ZAD)** planned development area where the public authority has compulsory purchase powers
zone *f* **d'alluvionnement** alluvial plain
zone *f* **d'arrêt** bus (stopping) bay
zone *f* **d'emprunt** borrow area/zone
zone *f* **d'incendie** fire zone
zone *f* **d'inondation** flood plain
zone *f* **d'interventions multiples** multi-activity zone

zone *f* **d'occupée/d'occupation** occupied zone
zone *f* **dangereuse** danger(ous) area/zone
zone *f* **de chargement** loading area/bay
zone *f* **de confort** comfort zone
zone *f* **de contrôle local** local control zone
zone *f* **de décélération** deceleration section [at a road junction]
zone *f* **de drainage** drainage area
zone *f* **de livraison** delivery area
zone *f* **de manutention** handling area
zone *f* **de rupture** failure zone
zone *f* **de rupture par cisaillement** shear zone
zone *f* **de stationnement** parking lane
zone *f* **de stockage** storage/waiting section/space [roads]
zone *f* **de stockage et d'entreposage** storage area
zone *f* **de transition** transition zone
zone *f* **de travail blanche** clean working area
zone *f* **injectée** grout(ed) zone
zone *f* **injectée de la fondation** grout blanket [beneath dam]
zone *f* **interdite** prohibited area
zone *f* **morte** dead zone
Zone *f* **Naturelle d'Intérêt Écologique Faunistique et Floristique (ZNIEFF)** [en France] Site of Special Scientific Interest (SSSI) [near equivalent in UK]
zone *f* **neutre** neutral zone
zone *f* **non-aedificandi** area with no building development
zone *f* **sociale réservée aux travailleurs** worker recreation area
zone *f* **tendue** tension zone
zone *f* **trempée** hardened zone

APPENDICES

APPENDIX A
FALSE FRIENDS OR *FAUX AMIS*

A false friend is a word in one language that looks like the same word in another, but that actually has other, related or completely different meanings in the second language. These verbal traps are of two kinds: words that do mean what they look like, but also something different; and words that do not mean what they look like at all. Examples are the best way to show you what we mean.

Of the first type, take the French word **action** *f*. It does mean **act** or **action** in English, but in another context it can also mean **a share** in a company. Of the second kind, **damage** *m* does not mean **damage**, or anything like it, in English. Its correct translation is **ramming** or **tamping down**. These two examples alone can show what a mire (**mire** *f* = surveyor's rod or staff) one can land up in (**lande** *f* = sandy heath or moor).

The list that follows gives several hundred false friends that you may come across in the activities covered by this dictionary. For simplicity, French words are always given in **bold** or *italics*, English in plain type. Where the English meanings are very much everyday words, we have not given all of them, for space reasons. They can easily be found in standard dictionaries.

Some may argue that there is a third class of false friends: the common everyday words in a language that specialists pick up and use in technical senses, but which do not sound like a word in the other language. An example is **hérisson** *m*, which doesn't sound like an English word and normally means **hedgehog**. However, in French it has also acquired the meanings of a metal leafed chimney sweep's brush, a special brush for cleaning venetian blinds and a bed of rough, large blocks of hard material laid on the formation in foundation or road works. Such words as this are not given in the list below. You are invited to see how many you can find in the main body of the dictionary – and if you meet any elsewhere which we have missed out, please let us know. Feedback on more true false friends in the areas covered by this dictionary will always be welcome too.

APPENDICE A

FAUX AMIS OU *FALSE FRIENDS*

Un faux ami est un mot qui ressemble dans une langue à un mot d'une autre langue mais qui, en fait, a un autre sens, proche ou totalement différent dans la deuxième langue. Ces pièges auditifs ou visuels sont de deux types: des mots qui veulent dire ce dont ils ont l'air, mais qui veulent aussi dire quelque chose d'autre; et des mots qui ne veulent pas dire du tout ce à quoi ils ressemblent. La meilleure façon d'expliquer ceci est de donner des exemples.

Du premier type, prenons le mot français **action**. Ce mot signifie **act** ou **action** en anglais, mais aussi **a share** dans une compagnie dans un autre contexte. Du deuxième type, **damage** en français ne signifie pas **damage** en anglais. Sa traduction correcte est **ramming** ou **tamping down**. Ces deux exemples à eux seuls prouvent que les mots peuvent prêter à confusion.

La liste qui suit répertorie des centaines de faux amis que vous risquez de rencontrer dans les activités décrites dans ce dictionnaire. Pour raison de simplicité, les mots français sont toujours écrits en caractères gras ou italiques, les mots anglais en caractères normaux. Lorsque les sens anglais sont d'usage très courant, nous ne les avons pas tous listés pour raison de place. Ils peuvent se trouver facilement dans un dictionnaire conventionnel.

Certains pourraient avancer qu'il existe un troisième type de faux amis, quand des mots d'usage courant qui ne ressemblent pas à un mot de l'autre langue sont utilisés par des spécialistes à des fins techniques. Par exemple, **un hérisson** ne ressemble à aucun mot anglais et signifie **hedgehog**. Cependant, en français, hormis un petit animal dans le jardin, **hérisson** a acquis d'autres sens: une brosse métallique sphérique de ramoneur, une brosse spéciale pour nettoyer les stores vénitiens et un blocage de pierres concassées dans l'assise inférieure d'une fondation ou d'une chausée.

Ce type de mots n'est pas répertorié dans la liste ci-dessous. Vous êtes invités à voir combien de ces mots vous trouverez dans la partie principale de ce dictionnaire, et si vous en trouvez d'autres ailleurs, que nous avons omis, faites-le nous savoir. Nous serons aussi toujours heureux de recevoir plus d'information sur d'autres vrais faux amis concernant les domaines mentionnés dans ce dictionnaire.

French	English and Notes
about *m*	butt [end], end or extremity [of a piece of timber or steel, a wall]
actif *m*	assets [of a company]
action *f*	share [in a company]; also act, action
adjudication *f*	allocation, auction sale, award [of a tender]; adjudication = *arrêt m, décision f, jugement m*
admission *f*	induction, inlet, intake; also admission
agrément *m*	acceptance, approval, consent; agreement = *accord m, entente f*
aire *f*	area, flat space; air = *air m*
alignement *m*	boundary [between public and private domains], building line; also alignment
allure *f*	pace, speed
altération *f*	deterioration, weathering [of soil]; alteration = *changement m*
arc *m*	arch; also arc or curve
are *m*	100m^2
armature *f*	reinforcement (bar); also armature
articulation *f*	hinge [structural], pin joint
aspersion *f*	irrigation, sprinkling
aspiration *f*	induction [of air]; also sucking, suction
assignation *f*	writ
attache *f*	clip, fastener
attachement *m*	as-built record of work subsequently covered up
bail *m*	lease; bail = *liberté f provisoire*
ballon *m*	cylinder [water], tank [expansion]
blindé *adj*	armoured, reinforced
bombé *adj*	domed, raised [rounded]
bout *m*	butt, end, tip
bride *f*	flange or collar [on a pipe or duct]
but *m*	aim, purpose
butter *vb*	to ridge [ground]
cabinet *m*	closet, office [e.g. architect's or lawyer's], small room
cage *f*	carcass or shell [of a house], stair or lift well
calamine *f*	mill scale
came *f*	cam; came [of a leaded light] = *baguette f*

canal *m*	channel, conduit, culvert, flume, pipe; also canal
carotte *f*	borehole or core sample
carrier *m*	quarryman; see *carrière* below
carrière *f*	gravel pit, quarry; carrier [goods] = *transporteur m, camionneur m, entreprise f de transports*
carter *m*	casing
case *f*	hut, box [on a form]; case [instance] = *cas m*
caution *f*	guarantee, deposit; caution = *prudence f, précaution f*
cavalier *m*	spoil or waste bank
cave *f*	cellar; cave = *caverne f, grotte f*
cémentation *f*	case hardening; also cementation
centrale *f*	power station
cession *f*	transfer [legal]; session = *séance f;* ceasing = *cessation f*
chaînage *m*	ring beam, wall ties, clamps; also chainage
chandelier *m*	candlestick; chandelier = *lustre m*
chant *m*	edge [of a board]
charge *f*	load [physical]; also electrical charge; charge [fee] = *prix m;* charge [legal] = *inculpation f;* charge [a battery] = *mettre vb en charge*
chariot *m*	cart, trolley, truck, wagon
châssis *m*	casement or sash, frame [of a door,etc]; also chassis
cheminée *f*	fireplace; also chimney
chute *f*	drop or fall, head of water, offcut [waste], wastepipe below WC; rubbish chute = *vide-ordures m*
circulation *f*	traffic; also circulation
cité *m*	housing estate; also city, but can be applied to quite small towns; *cité m universitaire* = grouped student hostel(s)
classé *adj*	listed as of architectural and/or historic interest
cliché *m*	negative [photo]
coin *m*	corner, chock, wedge, quoin; coin = *pièce f de monnaie*
collier *m*	pipe clip or collar
commande *f*	control, drive, order [for goods, services, etc]
commode *f*	chest of drawers
commodité *f*	comfort, convenience; commodity = *marchandise f, produit m*

compas *m*	pair of compasses; compass = *boussole f*
condamnation *f*	locking, fixing
console *f*	cantilever, corbel, (wall) bracket; also console
consommation *f*	consumption [of fuel, etc]
construction *f*	structure; also building, construction, erection
contour *m*	bend or twist [of a road], circumference, line, outline; contour [on a map] = *courbe f de niveau*
contrôle *m*	check, checking, testing; also control; control = *autorité f, maîtrise f;* controls = *commandes f,pl*
cordon *m*	belt/string course [of stone or brickwork], line or row [of posts, trees], cord, strand [of rope], welded seam
coup *m*	blow, impact, knock, shock
courbe *f*	diagram, graph; also curve, bend
course *f*	stroke [of a jack, a piston, etc], travel [of a tool, etc]; race course = *champ m de courses, hippodrome m*
crédit *m*	loan, mortgage; also credit
cric *m*	jack [for lifting]
cumulus *m*	diaphragm or expansion tank [of a central heating system]
cylindre *m*	roller [compactor]; also cylinder
dais *m*	canopy; dais = *estrade f*
damage *m*	ramming down, tamping; damage [visible] = *dégats m,pl;* damages = *dommages m,pl*
dame *f* **vibrante**	vibrating rammer
débit *m*	flow, output, retail selling; also debit
délai *m*	extension of time, time allowed, time limit; also delay; delay also = *retard m*
délayer *vb*	to water/thin down; to delay = *retarder vb*
dent *f*	tooth; dent = *bosse f, bosselure f*
départ *m*	beginning, start; also departure
département *m*	county [nearest French equivalent]; also department
descente *f*	downpipe; also descent
désigner *vb*	to point out; to design = *dessiner vb, créer vb;* designer = *dessinateur m, créateur m*
déterrer *vb*	to dig up; to deter = *décourager vb*
détour *m*	curve, bend; also detour [roundabout way]
déviation *f*	diversion; also deviation

dévider *vb*	to unwind; to divide = *diviser vb*
différer *vb*	to defer, to postpone, to put off; also to differ [be different from]; to differ from = *se distinguer vb de*
diffusion *f*	broadcasting, spreading [of news]; also diffusion, e.g. of heat
diffuser *vb*	to spread [news]; diffuser = *diffuseur m*
dilapidation *f*	misappropriation; delapidation = *délabrement m, dégradations f,pl*
direction *f*	management, running [of a business], strike [geological]; also direction
donjon *m*	keep [of a fortress], turret; dungeon = *cachot m souterrain*
dormant *adj*	fixed, immovable [of a window, fanlight, etc]
dormant *m*	casing, frame [of a door or window], dead bolt, fixed fanlight, dead light
dresser *vb*	to draw up [plans, a schedule, etc], to make out [plans, a schedule, etc], to erect, put up, to align, true or line up [a machine], to square [a block of stone]; dresser [furniture] = *buffet m, vaisselier m*
effectif *m*	complement [number of people]
effort *m*	stress; also effort
embrasse *f*	curtain loop; also kiss!
émisssaire *m*	outlet [of a lake, reservoir, etc]
engagement *m*	agreement, contract, investing, liability; also engagement in the sense of commitment, undertaking or promise
engager *vb*	to mortgage; to engage [labour] = *embaucher vb*
engin *m*	machine plant, piece of equipment
ergot *m*	cam, lug, stop
essence *f*	petrol
essai *m*	test, trial
estimation *f*	appraisal, quotation, quote, valuation; also estimate
évacuation *f*	discharge, drain, drainage, eviction [of tenants], extraction [of dust, air, liquid, etc]; also evacuation
évent *m*	vent [tap or cock]
eventuel *adj*	possible
expédition *f*	despatch [e.g. of goods]
expérience *f*	experiment, test; also experience
expert *m*	appraiser, surveyor, valuer; also expert

expertise *f*	appraisal, survey [normally by independent expert], expert opinion
exploitation *f*	development [e.g. industrial or mining], enterprise [business], operation or running [of a business, a railway, etc]
exploiter *vb*	to operate, run [e.g. a railway, a quarry]
exposé *m*	account, report, statement [of facts, etc]
exposer *vb*	to exhibit; also to expose; exposed [as beams, stonework] = *apparent adj*
exposition *f*	exhibition; also exposition
expulsion *f*	extrusion; also expulsion, etc
fabrication *f*	manufacture, production; also fabrication
file *f*	grid, line, row; also file [row]; file [tool] = *lime f*, file [documents] = *dossier m*, file [computer] = *fichier m*
filet *m*	fillet, net, screw thread, trickle [i.e. small flow]
filtration *f*	percolation; also filtration
fixation *f*	fastening, fixing
flache *f*	depression or pothole [in a road, pathway, etc], dishing of a horizontal surface, wane [on the edge of sawn timber]
fontaine *f*	spring [of water]; also fountain; fountain also = *source f, jet m d'eau*
forage *m*	boring, drilling [e.g. tunnel, pile, hole], bore-hole; sinking [e.g. a well]
force *f*	strength; also force
foret *m*	twist drill bit; forest = *forêt f*
formation *f*	training; also formation, forming
forme *f*	formation [ground], mould, shape; also form [shape]
former *vb*	to train [i.e. to teach a skill to], to build up [e.g. a collection, a deposit]
formulaire *m*	form [to fill in]
formule *f*	form [to fill in]; also formula
fort *adj*	strong; also fort [fortress]
fosse *f*	hole, pit, sump; fosse [trench] = *fossé m*
fossé *m*	ditch, drain, moat [of a fortress]
four *m*	furnace, oven
fourniture *f*	goods, materials [as opposed to labour], providing, supplies, supplying; furniture = *meubles m, pl, mobilier m*
foyer *m*	firebox, fireplace, grate, hearth(stone), home [in the sense of household unit]; also foyer; foyer also = *vestibule m, hall m*
franchise *f*	excess [on insurance claim], exemption; also franchise
fret *m*	freight; to fret [wood] = *découper vb, chantourner vb*

fruit *m*	batter, lean inwards or inclination [opposite of overhang]
fuel *m*	heating oil; also fuel oil
funiculaire *m*	catenary curve; also cable railway or funicular
fuser *vb*	to spread or run [light, colour, etc], to sound forth; to fuse [electrical] = *sauter vb* or *faire vb sauter*
gaine *f*	conduit, duct, shaft, sheath(ing), sleeve; to gain = *gagner vb*
galerie *f*	balcony [theatre], culvert, tunnel, underground duct [for services, etc]; also gallery [mining, theatre, etc]
garage *m*	storage [vehicles]; also garage, but *not* a petrol station [= *station-service f*]
gel *m*	freezing, frost
gorge *f*	concave circular moulding, throat of a chimney; also gorge
gousset *m*	bracket [supporting a shelf], haunch [in a concrete structure]; also gusset [plate]
gouttière *f*	rainwater downpipe, groove, gulley; also gutter
gradation *f*	gradual process; also gradation
grandeur *f*	height, size [of a tree, etc]
grange *m*	barn
grave *f*	aggregate [for road construction]
graver *vb*	to cut or engrave [letters or a design on a material]
grillage *m*	grating, netting, wire fencing: also grillage
grille *f*	grating, railing; also grille
groupe *m*	portable generator [short for *groupe m électrogène*]; also group [of houses, piles, trees, etc]
hall *m*	bay [of a building], concourse [e.g. of station], lobby; also hall
halle *f*	covered market
hangar *m*	barn or shed [generally open-sided], shed, warehouse; also hangar
herbe *f*	grass, plant; also herbs
herse *f*	batten [for stage lights], metal division across a balcony between two flats
hôtel *m*	mansion, town house; also hotel; *hôtel m de ville* = town hall; *hôtel m particulier* = private mansion or town house
hotte *f*	cooker/fume hood
if *m*	yew
illumination *f*	floodlighting; also illumination; lighting = *éclairage m*
immobilisation *f*	capitalisation [of assets], fixed asset
immobiliser *vb*	to lock or tie up [capital]; also to immobilize
impasse *f*	cul-de-sac; also impasse, dilemma

implantation *f*	layout, setting or marking out
imposition *f*	taxation
imposte *f*	fanlight; also impost [of arch]
impropre *adj*	unsuitable
infiltration *f*	percolation, seepage; also infiltration
infirmerie *f*	health centre, sick bay or room [e.g. infirmary in a school]; infirmary = *hôpital m*
infrastructure *f*	bed [of a road or railway], substructure; also infrastructure
inhabité *adj*	uninhabited; inhabited = *habité adj*
injection *f*	grouting
insensible à *adj*	resistant or insensitive to
interrupteur *m*	breaker, switch [electrical]
inusable *adj*	hard wearing; unusable = *inutilisable adj*
invalide *adj*	invalid; also infirm
isolation *f*	insulation [thermal, acoustic]
issue *f*	exit, outlet; also issue in the sense of outcome
itinéraire *m*	route; also itinerary
jalousie *f*	adjustable louvres [of a window, blind, etc], sun blind
joint *m*	gasket, splice, pointing [in block-, brick- or stonework]; also joint
justesse *f*	accuracy, precision
justification *f*	supporting calculations
label *m*	seal or stamp [commercial]; label = *étiquette f*; circuit label = *label m de circuit*
labourer *vb*	to plough; labourer = *ouvrier m, travailleur m, manœuvre m*
lame *f*	blade [of a knife, a saw], strip of wood [e.g. for flooring]
lamelle *f*	small strip [of metal or plastic]; also lamella
laminage *m*	rolling [as of steel]
lande *f*	sandy heath or moor(land); land = *terre f, terrain m*
large *adj*	wide; large = *fort adj, grand adj, gros adj, important adj*, etc
latte *f*	batten or slat; also lath
lattis *m*	roof battens, laths [of lath and plaster work]
lavabo *m*	hand/wash basin; also, in plural, toilets

lecture *f*	reading [of a meter, instrument, etc]
lest *m*	ballast [load, e.g. for pile test], counterweight
levé *m*	survey; levee [embankment] = *levée f*
lever *vb*	to lift or raise; lever = *levier m*
levier *m*	crow-bar; also lever
librairie *f*	bookshop; library = *bibliothèque f*
lice *f*	hand-rail, rail [of post-and-rail fencing]
licencié *adj*	graduate *adj*; licenced [product] = *sous licence*
licencier *vb*	to make redundant [an employee]
lien *m*	bond, strap, tie [generally structural]; lien = *privilège m*
ligne *f*	contour, outline, electrical wires; also line
lime *f*	file [a rasp, not one containing data]; lime = *chaux f*, lime tree = *tilleul m*
lingerie *f*	linen room
local *m*	building, premises; *local m technique* = plant room
location *f*	rental, renting; location = *emplacement m, situation f*
lot *m*	contract or work package; also building plot
magasin *m*	shop, warehouse
mail *m*	mall [avenue], sledgehammer, square [on a grid, of a mesh]; mail [post] = *courrier m*
maille *f*	mesh [individual opening in a mesh]
manœuvre *m*	labourer [bricklayer's or builder's], operation; also manoeuvre
manoir *m*	country house; also manor
marche *f*	running [of equipment], step, stair, tread
mare *f*	pond
marge *f*	margin
masse *f*	earth [electrical], overall layout, sledgehammer; also mass, massing
massif *adj*	bulky, solid [e.g. solid timber, etc]; also massive
massif *m*	body or mass [of concrete, masonry, a structure, etc]
mât *m*	mast or pole; (door)mat = *essuie-pieds m, paillasson m, tapis-brosse m*
mat *adj*	dull [finish], unpolished; also matt
matériel *m*	equipment, plant; material = *matériau m, matière f*

membrane *f*	diaphragm; also membrane
mesure *f*	extent, size; also measure(ment)
métré *m*	estimate of costs, measurement [of quantities and work], quantity surveying; *not* metre or meter
mètre *m*	rule [metre]; also metre; meter = *compteur m*
mezzanine *f*	small window at mezzanine level; also mezzanine
mince *adj*	thin
minute *f*	original [of a plan, document]; also minute [of time, or a note of something]
mire *f*	surveyor's rod or staff; mire = *boue f, bourbe f*; swampy ground = *bourbier m, molets m,pl*
mitre *f*	large chimney pot, chimney cowl
mobilier *m*	furniture
module *m*	modulus; also module
moment *m*	momentum; also moment [statics]
montage *m*	assembling or assembly, erecting, erection, hoisting, wiring up
monter *vb*	to assemble, to erect, to rise
net *adj*	clean; also net *adj*
noise à, chercher *vb*	to pick a quarrel with; noise = *bruit m*; sound = *son m*
norme *f*	standard or code of practice; also norm
note *f*	account [financial]; also note
notice *f*	account [report on or about something]; notice is translated into French in many ways depending on context
obligation *f*	bond, debenture
obscur *adj*	dark [room, etc, rather than colour]
obscurité *f*	darkness
occasion *f*	bargain; also occasion, opportunity; *d'occasion* = second-hand
ordinateur *m*	computer
ordonnance *f*	edict, layout or general arrangement
ordures *f,pl*	refuse [rubbish]; ordure = *ordure f* [singular]
organe *m*	instrument, medium; also organ
pair *adj*	even [of a number]; pair = *paire f*
palace *f*	large luxury hotel; palace = *palais m*

pale *f*	blade; also pale [as in a fence]; pale *adj* = *pâle*
pan *m*	face or side [of angular object, building, etc], section or length [of a wall, a roof plane]; pan = *casserole f*
parabole *f*	satellite dish [aerial]; also parabola
parachute *m*	emergency or safety arrestor [of lift or hoist]
parc *m*	large garden or grounds, depot or store, stock [of equipment, vehicles, etc], storage yard; also park; carpark = *parking m, parc m de stationnement*
parcelle *f*	compartment [of a forest], plot of land; also parcel of land; parcel = *colis m, paquet m*
parquet *m*	timber or wood(en) floor(ing); also parquet flooring
particulier *adj*	private, special; also particular
particulier *m*	private individual or person; particulars = *détails m,pl, renseignements m,pl*
partie *f*	part [fraction, portion]; *faire vb partie de* = to be part of
passage *m*	crossing, shopping arcade, route of journey, way through; also passage(way)
passer *vb*	to conclude [an agreement], to cross [a bridge], to enter into/sign [a contract], to apply, spread [a coat of paint]; also to pass; to pass [overtake] = *dépasser vb*
passif *m*	liabilities [of a company]
patente *f*	trading licence or dues; patent = *brevet m*; patented = *breveté adj*
pavillon *m*	villa or detached house, gate lodge, ward [of a hospital], wing [of a building], and several more meanings; also pavilion
perception *f*	collection [of fines, taxes, tolls, etc], tax office
perfectionner *vb*	to improve; also to perfect
performant *adj*	high performing or highly productive [of equipment, plant, etc], high return [of an investment]
pétrifier *vb*	to calcify
pétrole *m*	petroleum [crude oil, *not* petrol, which = *essence f*]
photographe *m*	photographer; photograph = *photo(graphie) f*
pièce *f*	paper [document], part, room; also piece; piece has many French translations depending on context
pigeon *m*	lump [in lime], tongue [in a mitred joint]
pile *f*	electric battery, bridge pier; also pile or stack

pilon *m*	rammer or tamper; pylon = *pylône m*
pilot *m*	(wooden) pile; pilot [driver] = *pilote m*
pin *m*	pine [tree or the timber]
pince *f*	clamp, crowbar, pinch bar, pair of pliers
pistolet *m*	drawing or French curve, spray gun
place *f*	space, room [amount of space], square [as in a town]
placement *m*	investment; also placing of
plan *adj*	even, level or plane [of a surface]
plan *m*	level, layout/outline, drawing, programme; also plan [map]
plane *f*	draw(ing) knife, drawshave; plane = *rabot m*
planer *vb*	to smooth [wood]; also to plane
planning *m*	programme; planning [programming] = *planification f*; town planning = *urbanisme m*
plaque *f*	fireback, hot-plate, plate, sheet of metal, slab, street name sign; also plaque [on a wall]
plate-forme *f*	level area, wall plate [supporting rafters]; also platform; station platform = *quai m*; platform [rostrum] = *estrade f*
plateau *m*	cradle [of a flying scaffolding], flat (bed) trailer; also plateau
plier *vb*	to bend (over), to bend or sag, to fold (up, etc); pliers = *pinces f,pl*, *tenailles f,pl*
plinthe *f*	skirting (board); also plinth
plot *m*	electrical contact; plot of ground = *lot m de terrain*, *lotissement m*
plot *m* **réflectorisé**	reflective stud
plus-value *f*	capital gain, cost increase
point *m*	place or spot, stage [in a process, etc]; also point [in position, space, a programme]
pointe *f*	small headed nail, panel pin, tack, peak [as in peak hours, etc]; also point or tip
pointer *vb*	to tick/check (off) [on or from a list], to clock in or out [of an employee], to mark, to level, point or train [a surveying instrument]; French has many words for to point, depending on context
police *f*	policy [of insurance]
port, franc de	carriage or postage paid
pose *f*	fitting, fixing, laying
poser *vb*	to fit, fix, lay, place

position *f*	position [location, job], site [of a house, town, etc], situation [of a house, town, etc, or job]
poste *m*	element, item [e.g. in bill of quantities], telephone extension, installation; post [mail] = *courrier m*; post [stake, etc] = *poteau m*
pourchasser *vb*	to chase after; purchase = *acheter vb*
prairie *f*	grassland, meadow; also prairie; prairie also = *plaine f herbeuse*
pratiquer *vb*	to exercise [a profession], to construct [a component, e.g. a door, in a partition, i.e. to contrive or fit something in]
pré *m*	meadow; also prefix pre-
préciser *vb*	to make clear, to specify; precise = *précis, exact adj*
précisions *f,pl*	further or exact details; also precision; precision also = *exactitude f*
premier *adj*	first, initial
prescription *f*	order, instruction
pression *f*	compression; also pressure
prévision *f*	estimate, forecast, prediction
prime *f*	premium [of insurance], grant, subsidy; prime number = *nombre m premier*; prime rate = *taux m de base*; to prime [pump] = *amorcer vb*
primeur *m*	something new, early or fresh; primer [paint] = *couche f de fond, couche f d'impression*
principe *m*	principle; principal *adj* = *principal adj*; principal [capital] = *principal m, capital m*
prise *f*	plug socket, setting [of concrete, glue, etc], tapping [for a sensor, a current take-off]; to prise off/up = *enlever vb en faisant levier*; to prise open = *forcer vb*
prix *m*	value; also price
procès *m*	action or proceedings at law; process = *procédure f, processus m, procédé m* or *méthode f*, depending on context; to process has many translations, again depending on context
processus *m*	procedure; also process, see above
produit *m*	income, profit, yield; also product
professionnel *adj*	trade *adj*; also professional
profil *m*	contour, section; also profile [section]
programme *m*	project content; also programme; (computer) program = *logiciel m*
projecteur *m*	flood/spotlight; also (ciné) projector
projet *m*	design, scheme, plan [design, or intention]; also project

projeteur *m*	designer, planner [of projects]; see also *projecteur*
proportion *f*	ratio; also proportion
propreté *f*	clean(li)ness; property = *propriété f*
propriétaire *m*	landlord, owner; also proprietor
prospect *m*	open space in front of a building in French highway regulations; also prospect [view]
provision *f*	reserve or stock; also provision
pulvériser *vb*	to atomize, spray [liquids]; also to pulverize or grind to powder
purge *f*	bleed(ing) [e.g. of brakes], draining [e.g. pipework], flushing out; also purge, purging
quartier *m*	area, district [of a town, e.g. office, commercial]
quitter *vb*	to discharge [someone of a debt]; also to quit or vacate [premises, property, etc]
rampant *adj*	sloping, or, as a noun, sloping part of a building
rampe *f*	balustrade, lighting battery, footlights, slipway [of a shipyard]; also ramp, access ramp
range *f*	row [of paving slabs of equal size]; range [mountains] = *chaîne f*; range [scope] = *portée f*; range [row] = *rangée f, rang m*
rangée *f*	array, line, row; also see *range* above
rangement *m*	cupboard or storage space, putting away
ranger *vb*	to arrange, to put away/back in place/in order, to stow away, to tidy up, etc
rapide *m*	fast train; also, in plural, rapids
rapport *m*	income, profit, return, revenue, yield [from an investment]; also report
rayon *m*	radius, shelf
réalisation *f*	achievement, carrying out, implementation; also realization
rebut *m*	reject, throw-out [the noun, not the verb]
recharger *vb*	to reload; also to recharge
rechercher *vb*	to inquire into [causes, etc], to search for; researcher = *chercheur m*
récipient *m*	container, vessel; also receiver [for gas, fluids, etc]
réclamation *f*	claim [contractual], complaint; reclamation [of waste, etc] = *récupération f*; reclamation [of wetlands] = *dessèchement m*; reclamation [of land generally] = *mise f en valeur*
recommandation *f*	registration [of post]; also recommendation
rectification *f*	(re-)adjustment [of an instrument], straightening [e.g. of a road]; also rectification

rectifier *vb*	to adjust, amend, correct, put right; also to rectify; rectifier [electrical] = *rectificateur m*
regard *m*	manhole [for sewer, etc]
régime *m*	rate of flow, speed [of a motor], administrative system
régime *m* **permanent**	steady state [of conditions, a process, loading, etc]
registre *m*	damper; also register, record
relais *m*	sandbank or flats, temporary site for dumping earth; also relay
relief *m*	higher ground [e.g. hilly area, mountains]; also relief, in general sense of projection above a surface
rentabilité *f*	profitability [*not* lettability]
rente *f*	annuity, pension; rent [of premises] = *loyer m*; to rent [equipment, premises] = *louer vb*; renting [equipment, premises] = *location f*; rent [in cloth, fabric] = *déchirure f*; rent [in ground, rock] = *fissure f*
repasser *vb*	to recoat [of renderings, finishes]; see also *passer*
repos *m*	bearing [of a beam or column]; also rest or repose
repose *f*	refixing or relaying; see also *repos*
résistance *f*	resistor, strength; also resistance
ressort *m*	competence, scope, mechanical spring; (holiday) resort = *lieu m de séjour*
reste *m*	remainder, remains
résurgence *f*	spring [of water]
retenue *f*	deduction [from salary, wages], impounding dam
retiré *adj*	secluded; retired = *en retraite*
retirer *vb*	to remove, to take off or out; to retire to/from = *se retirer vb à/de*
réunion *f*	meeting [of several people, to discuss, etc]
revenu *m*	tempering of steel; also revenue
reviser *vb*	to overhaul, to renew, to review, to service [e.g. a car]; also to revise
ripe *f*	scraper
riper *vb*	to polish, scrape [stone]
river *vb*	to clinch [a nail], to rivet; river = *rivière f*; river [flowing into the sea] = *fleuve m*
rosée *f*	condensation, dew
route *f*	road(way); also route
ruine *f*	decay, downfall [of a building]; also ruin
ruiner *vb*	to destroy; also to ruin
rupture *f*	breach(ing) or bursting [of a dam, etc], failure, fracture, parting [of a cable, rope]; also rupture

schéma *m*	diagram, outline, schematic drawing, sketch
scène *f*	stage [of a theatre]
secrétaire *m*	writing desk; also secretary [modern usage]
sécurité *f*	safety or safety device; also security [safety]; security [for loan] = *caution f, garantie f*; securities [stock exchange] = *valeurs f,pl, titres m,pl*
self *f*	choke [electrical]
sens *m*	direction, meaning; also sense
sensibilité *f* à	sensitivity or sensitiveness to, susceptibility to
sensible *adj*	sensitive
séparateur *m*	central reserve [on a highway]; also separator
serpentin *m*	coil [of pipe, tubing]
service *m*	administrative department; also service
servitudes *f,pl*	charges [attached to a property], constraints [on the design or use of a property], easements affecting a property
shunt *m*	smoke extract or ventilation duct [for a whole building]; also electrical shunt
siège *m*	head office, headquarters, seat or seating [of a valve or tap]
signaler *vb*	to indicate, to point out; also to signal
signe *m*	indication, symbol, symptom; also sign
simple *adj*	single; also simple
situation *f*	location, position, record of work and payments; also situation
société *f*	company, firm
solidaire *adj*	jointly liable [e.g. of joint venture partners]
solidarisé *adj*	interlocked
solide *adj*	robust, sound, strong, sturdy, tough; also solid
solidité *f*	reliability, robustness, etc [see *solide* above]; also solidity
sollicitation *f*	externally applied effect [e.g. load or stress]
sommaire *m*	list of contents; also summary
source *f*	spring [of water]; also source
spatule *f*	stripping knife
spire *f*	hoop [reinforcement]; spire = *flèche f, aiguille f*
stabilisation *f*	annealing, balancing; also stabilization
stabiliser *vb*	to anneal; stabilizer = *stabilisateur m*
staff *m*	fibrous plaster; staff [employees] = *personnel m*; staff [rood] = *bâton m, mire f*

standard *m*	telephone switchboard; see also *table*, *tableau*
station *f*	plant, position, resort, spa; also station
station-service *f*	petrol station [UK]; also service station
stockage *m*	stocking, storage
store *m*	awning, blind, shade; store [stock] = *réserve f, stock m, provision f*; store [depot] = *entrepôt m*
stratifié *m*	laminate; also stratified *adj*
succession *f*	sequence [of operations, etc]
superficie *f*	area, surface or surface area; superficial = *superficiel adj*
support *m*	bearer or bearing, bracket, prop; also support
supporter *vb*	to bear, to carry, to hold up; also to support
sûreté *f*	safety, security; surety = *caution f, garantie f*
surveillance *f*	control, inspection, supervision; also surveillance
suspension *f*	hanging or suspended light fitting or fixture; also suspension
syndicat *m*	association or trade union; also syndicate; syndicate also = *coopérative f*
table *f*	face [of a hammer], flange [of a girder], plane area of wall, or panel, slab or tablet [of stone], telephone switchboard [see also *standard*, *tableau*]; also table
tableau *m*	board [for notices, etc], chart, diagram, panel, reveal [of a door or window], switchboard [see also *standard*, *table*], table [e.g. of figures]
tablette *f*	coping stone, mantlepiece or shelf, stone handrail
talon *m*	cheque stub or counterfoil, seam end closure [in metallic roofing], ogee moulding, nib [of a beam, a plain tile], sole [of a plane]
tambour *m*	revolving door; also tambour [architectural]
tampon *m*	buffer, stopper, cover, end plate [of a pipe, etc], pad [e.g. for spreading varnish], plug [for a screw into a wall], rubber stamp
taper *vb*	to dab or lay on [paint, etc], to plug or stop up [a hole, etc], to prime [with paint], to tap; to taper = *effiler vb, terminer/finir vb en pointe*; to tape (record) = *enregistrer vb sur bande*
targette *f*	sash bolt, or small flat bolt [for a door, gate, etc]
tartre *m*	boiler fur or scale
teint *m*	colour or dye
teinte *f*	hue, shade, stain, tint
tempérament *m*, **achat** *m* **à**	hire/instalment purchase
tempérament *m*, **vente** *f* **à**	sale on deferred payment terms

temporisé *adj*	slow acting; temporary = *temporaire adj*
tenacité *f*	toughness [of metals]
tension *f*	tensile stress, voltage; also tension
terminer *vb*	to complete/finish; also to terminate
terrasse *f*	nearly flat roof; also terrace
terrasser *vb*	to dig out, to excavate
terrassier *m*	earthworks labourer, navvy
tiède *adj*	lukewarm, tepid
toilette *f*	wash stand; toilets = *toilettes f,pl*
ton *m*	shade or tone [of colour]; metric tonne = *tonne f*
touche *f*	electrical contact, key [of a typewriter or computer]
tour *f*	tower
tour *m*	circumference, lathe, turn; also tour
tracé *m*	alignment [of a road, etc], diagram, drawing, layout or line [of a road, railway, etc], outline, route, setting out [of a road, railway, etc], sketch, plot(ting) [of curves, data]
tracer *vb*	to draw, lay out, mark off/out, plot, set out, sketch; to trace with tracing paper = *décalquer vb*; to trace [draw] = *dessiner vb, tracer vb*
traceur *m*	plotter [computer]; also drawing office tracer
trait *m*	line, mark, streak
traite *f*	stretch [of a journey]
traitement *m*	processing; also treatment
traiter *vb*	to handle [a subject], to do a deal, to negotiate, to process; also to treat
tranche *f*	slab or slice [of cut stone, etc]
tranchée *f*	cut(ting) [a long excavation, not the process], ditch, groove or notch [in masonry]; also trench
transmission *f*	assignment [of shares, patents, rights, etc], conveyance [of property, etc]; also transmission of drive, heat, etc
traverse *f*	crossbar, cross piece, rail [of a door], rung [of ladder], short cut, sleeper [railway], transom; also traverse [horizontal member]
treillis *m*	lattice; also trellis (work)
trémie *f*	bin or bunker [for solid fuel, etc], chimney or funnel [for loading], hopper, light well, vent shaft or well; also [concrete] trémie
trépan *m*	rock drill bit

trier *vb*	to pick out, to separate or sort
tuber *vb*	to case [a borehole, well, etc]
type *adj*	typical; typical also = *typique adj*; type = *type m*
ultérieur *adj*	later, subsequent
unité *f*	unit; also unity
urbanisme *m*	town planning; urban = *urbain adj*; to urbanize = *urbaniser vb*
user *vb*	to wear away/out
usure *f*	deterioration, wear and tear, wearing away
vacation *f*	fees based on time to complete an item of work
vague *f*	wave [water]
valeurs *f,pl*	securities [stock exchange]; also values; see also *titres*
vase *m*	receptacle, vessel, tank [of expansion unit]
vent *m*	wind; vent = *évent m, orifice m*
vente *f*	sale
ventilateur *m*	air blower or fan; also ventilator
verger *m*	orchard
vernir *vb*	to varnish; vernier = *vernier m*
vibration *f*	structure-borne sound; also vibration
vice *m*	defect [general term]; [see also *vice m apparent/caché/de construction*]; vice [on a workbench] = *étau m*
vue *f*	outlook [visual]; also view

APPENDIX B

REFERENCES

General French–English dictionaries

Atkins, B.T., Duval, A., Milne, R.C. *et al.* (1993) *Collins Robert* French–English English–French Dictionary, 3rd edn, Harper Collins, London, and Harper & Row, New York.

The Oxford-Hachette French Dictionary (1994) Oxford University Press, Oxford.

Kettridge, J.O. and Arden, Y.R. (1980) *Dictionary of Technical Terms and Phrases, French/English and English/French*, revised edn, Routledge, London.

Moskowitz, D. and Pheby, J. (eds) (1983) *The Oxford-Duden Pictorial French and English Dictionary*, Oxford University Press, Oxford and New York.

Multilingual dictionaries

Publishing Committee of REHVA (1994) *The International Dictionary of Heating, Ventilating and Air Conditioning* (12 languages), E & FN Spon, London.

English specialist dictionaries

Curl, J. (1992) *Encyclopedia of Architectural Terms*, Donhead.

BSI (1993) *Glossary of Building and Civil Engineering Terms*, British Standards Institution, London.

Marsh, P. (1982) *Illustrated Dictionary of Building*, Longman Scientific and Technical, Harlow.

McMullan, R. (1988) *Macmillan Dictionary of Building*, Macmillan, London.

Fleming, J., Honour, H. and Pevsner, N. (1972) *Penguin Dictionary of Architecture*, 2nd edn, Penguin, London.

Maclean, J.H. and Scott, J.S. (1993) *Penguin Dictionary of Building*, 4th edn, Penguin, London.

Scott, J.S. (1991) *Penguin Dictionary of Civil Engineering*, 4th edn, Penguin, London.

French specialist dictionaries

Vigan, J. de (1994) *Dictionnaire Général du Bâtiment: Le Petit Dicobat*, Éditions Arcature, Ris-Orangis.

Barbier, M., Cadiergues, R., Delefosse, J. *et al.* (1992) *Dictionnaire Technique du Bâtiment et des Travaux Publics*, 11th edn, Éditions Eyrolles, Paris.

General guides

Meikle, J.L. and Hillebrandt, P.M. (1989) *The French Construction Industry: A Guide for UK Professionals*, Construction Industry Research and Information Association, London.

The Franco-British Construction Industry Group (1993) *The Business of Building in France*, Chambre de Commerce Française de Grande-Bretagne and The Royal Institution of Chartered Surveyors, London.

Hampshire, D. (1993) *Living and Working in France: A Survival Handbook*, Survival Books, Fleet.